Astronomy: A Physical Perspective

This fully revised and updated text is a comprehensive introduction to astronomical objects and phenomena. By applying some basic physical principles to a variety of situations, students will learn how to relate everyday physics to the astronomical world. Starting with the simplest objects, the text contains thorough explanations of how and why astronomical phenomena occur, and how astronomers collect and interpret information about stars, galaxies and the Solar System. The text looks at the properties of stars, star formation and evolution; neutron stars and black holes; the nature of galaxies; and the structure of the universe. It examines the past, present and future states of the universe; and final chapters use the concepts that have been developed to study the Solar System and its formation; the possibility of finding other planetary systems; and the search for extraterrestrial life. This comprehensive text contains useful equations, chapter summaries, worked examples and end-of-chapter problem sets. It is suitable for undergraduate students taking a first course in astronomy, and assumes a basic knowledge of physics with calculus.

Marc L. Kutner obtained his doctorate in physics from Columbia University in 1972. He has been a Visiting Scientist in the Department of Astronomy at the University of Texas at Austin since 1998, prior to which he was Professor in the Department of Physics and Astronomy at the Rensselaer Polytechnic Institute, New York, and Visiting Scientist at the National Radio Observatory, Tucson, Arizona. His main area of research involves the use of radio astronomy to study of star formation in the Milky Way and other galaxies. He has also done some research in cosmology. Professor Kutner has published three successful textbooks and over one hundred research papers.

Astronomy: A Physical Perspective

Marc L. Kutner

CAMBRIDGE UNIVERSITY PRESS
Cambridge, New York, Melbourne, Madrid, Cape Town, Singapore, São Paulo
Delhi, Tokyo, Mexico City

Cambridge University Press
The Edinburgh Building, Cambridge CB2 8RU, UK

Published in the United States of America by Cambridge University Press, New York

http://www.cambridge.org
Information on this title: www.cambridge.org/9780521529273

First edition published by John Wiley & Sons, 1987
Second edition published 2003
4th printing 2011

Printed in the United Kingdom at the University Press, Cambridge

A catalog record for this publication is available from the British Library

Library of Congress Cataloging in Publication data

Kutner, Marc Leslie.
Astronomy: a physical perspective / Marc Kutner.
    p.   cm.
Includes bibliographical references and index.
ISBN 0 521 82196 7–ISBN 0 521 52927 1 (pb.)
1. Astronomy. I. Title.

QB45.2 .K87 2003
520–dc21   2002034946

ISBN 978-0-521-82196-4 Hardback
ISBN 978-0-521-52927-3 Paperback

Contents

Abbreviations used in the figure credits

Figure credits are given in the captions. Abbreviations used are as follows.

2MASS	Two Micron All Sky Survey
AUI	Associated Universities Inc.
AURA	Association of Universities for Research in Astronomy
Caltech	California Institute of Technology
CFA	Center for Astrophysics
ESA	European Space Agency
ESO	European Southern Observatory
GSFC ADF	Goddard Space Flight Center Astrophysics Data Facility
HST	Hubble Space Telescope
IFA	Institute for Astronomy
IRAM	Institut de Radioastronomie Millimétrique
ISO	Infrared Space Observatory
JCMT	James Clerk Maxwell Telescope
MIT	Massachusetts Institute of Technology
MPIFR	Max Planck Institut für Radioastronomie
NASA	National Aeronautics and Space Administration
NMTech	New Mexico Institute of Mining and Technology
NOAA	National Oceanographic and Atmospheric Observatory
NOAO	National Optical Astronomy Observatory (operated by AURA under contract with the NSF, all rights reserved)
NRAO	National Radio Astronomy Observatory (operated by AUI, under contract with the NSF)
NSF	National Science Foundation
ONR	Office of Naval Research
SCUBA	Submillimeter Common User Bolometer Array
STScI	Space Telescope Science Institute (operated by AURA under contract with NASA)
UCLA	University of California at Los Angeles
USGS	US Geological Survey

Preface

The study of astronomy has blossomed in a variety of ways in the last decade of the 20th century. Every part of the electromagnetic spectrum has seen a revolution in observing techniques. While much of this has been on the ground, space-based observing has come into its own, as we are seeing the results of second and third generation space-based telescopes. These have provided sensitivity and clarity that have revolutionized all subfields in astronomy and created some new ones. These observational developments have been supplemented by massive improvements in computing power, allowing for the processing of large amounts of astronomical data, and the theoretical modeling of the results.

The most amazing aspect of all of this progress is that we can still provide reasonable answers to the naive question, 'How does it all work?' As our astronomical horizon expands, we can still use familiar physics to explain the wealth of phenomena. Even when the explanation at the research level requires a complex application of certain physical laws, there is usually still a way of understanding the phenomena based on introductory level physics. Perhaps this is just the realization that the laws of physics are small in number but apply universally. There are a few exceptions, where the astronomical problems help drive back the frontiers of physics, but these can be explained in more familiar terms.

This book is dedicated to the student who would like more out of even a brief study of astronomy than a list of what there is. It is for the student who wants to understand why certain phenomena occur, and how astronomical objects work. In addition, it addresses the question of how we collect and interpret information about remote objects.

The primary audience of this book will be science majors who have taken a year of college physics (classical) with calculus. We therefore presume that the student has seen the classical physics needed for the astronomy course, but do not presume a knowledge of 'modern' physics.

This book is the successor to *Astronomy: a Physical Perspective*, published by Wiley in 1986. I am grateful to the loyal audience that book developed, and for their encouragement to work on this new version.

I am grateful to Simon Mitton at Cambridge University Press, who shared my view that a 'higher level' book could still be visually attractive. I am also grateful to Jacqueline Garget, who believed in this project, seeing it through a few rough early reviews to its completion. At every stage, she always knew exactly how to answer my email questions to keep me going. I would also like to thank an extraordinary copy editor, Irene Pizzie, for always knowing what I meant to say, and production manager, Carol Miller, for keeping the project moving along, and always keeping me in the loop.

Three professors, Stephen Boughn (Haverford), James Houck (Cornell) and Judith Pipher (Rochester) class-tested various versions of this manuscript. I appreciate their patience and their feedback. I also appreciate their students taking the time to use a 'book' in a non-standard form, and to give comments.

Special thanks go to Nadine Dinshaw, a friend/ colleague, who read the whole manuscipt in an early form. Her comments and support were very helpful at that early stage.

At every stage the manuscript benefited greatly from the feedback from reviewers who read all or various parts of the manuscript. Some were anonymous, and others were: Imke DePater (University of California at Berkeley), Debra Elmegreen (Vassar), Andrea Ghez (UCLA), Steven Gottesman (University of Florida), Richard Griffiths (Carnegie Mellon), David Helfand (Columbia), Lee Mundy (University of Maryland), James Napolitano (Rensselaer), and Heidi Newberg (Rensselaer).

Many astronomers and physicists have contributed data and illustrations which I have used directly. They are too numerous to mention here, but are credited in the figure captions. My special thanks go to those who were anxious for me to have the most recent data or best pictures.

Gathering these figures proved to be frustrating sometimes. However, the contact that I had with the vast majority was very rewarding.

This project started during my three-year stay at the National Radio Astronomy Observatory, in Tucson. I am thankful to Paul Vanden Bout (NRAO director) for helping me settle into that position, and to all the people in Tucson who provided a stimulating atmosphere and a view of the Santa Catalina Mountains. The project has finished during my stay at the University of Texas, Austin. I am grateful to Frank Bash (McDonald Observatory director) for arranging that position and always having an open door. I thank my colleagues here in Austin for providing a stimulating environment also.

On the personal level, I got my start in astronomy when my mother encouraged me to take courses at the Hayden Planetarium, in New York. I am also grateful to my two sons, Eric and Jeff, who never stop asking questions.

Most important, at many levels, this book would not be here without my best colleague and best friend, Kathryn Mead. She encouraged me to tackle hard tasks, from running marathons, to biking centuries, to refereeing soccer, to writing books. Her drive and curiosity led to our most important discovery (molecular clouds in the outer Milky Way). More immediately, she also helped dress up those figures for this book that needed it the most.

Chapter 1

Introduction

1.1 | An understandable universe

Our curiosity about the world around us is most naturally manifested when we look up at the night sky. We don't need any special instruments to tell us something interesting is going on. However, only with the scrutiny afforded by a variety of instruments can these patches of light, and the dark regions between them, offer clues about their nature. We have to be clever to collect those clues, and just as clever to interpret them. It is the total of these studies that we call *astronomy*.

We are fortunate to live in an era of extraordinary astronomical discovery. Some have even called this the 'Golden Era of Astronomy'. For centuries astronomers were restricted to making visual observations from the surface of the Earth. We can now detect virtually any type of radiation given off by an astronomical object, from radio waves to gamma rays. Where necessary, we can put observatories in space. For the Solar System, we can even visit the objects we are studying.

For all of these capabilities, there is a major drawback. We cannot do traditional experiments on remote astronomical objects. We cannot change their environment and see how they respond. We must passively study the radiation that they give off. For this reason, we refer to astronomy as an *observational* science rather than an *experimental* one. It is because of this difference that we must be clever in using the information that we do receive. In this book, we will see what information we can obtain and how the clues are processed. We will see that, in exchange for the remoteness of astronomical objects, we get to study a large number of objects under a variety of conditions.

One of the most fascinating aspects of astronomy is that many phenomena can be understood in terms of relatively simple physics. This does not mean that we can explain every detail. However, we can explain the basic phenomena. In this book, we emphasize the application of a few physical principles to a variety of situations. For this purpose, some background in physics is needed. We assume that the reader has had an introductory course in classical physics (mechanics, electricity and magnetism, thermodynamics). We also use quite a bit of modern physics (relativity, atomic and nuclear physics). The modern physics will be developed as we need it. In addition, a familiarity with the concepts of calculus is assumed. While most of the material can be mastered without actually taking derivatives and working out integrals, the concepts of derivatives as representing changes and integrals as representing sums are used. The reader may also note a variation in the mathematical level from subject to subject. This is because the goal in writing this book is to present each astronomical subject at the simplest level that still provides for a reasonable understanding.

In organizing an astronomy text, one important question is where to put the material on the Solar System. The traditional approach has been to place the Solar System first. This allows the student to start with familiar, nearby objects first and work out from there. The disadvantage is that we use techniques to study the Solar System that we cannot use on more distant objects. In this book we place the Solar System last. This allows the student to form a better idea of how astronomy is done on remote objects. We can also use the physics that we develop in studying stars and other astronomical objects to give us a better appreciation for how the Solar System works. Finally, putting the Solar System at the end allows for a discussion of the formation of the Solar System, utilizing things that we learn about star formation.

We start with stars, those points of light in the night sky. This allows us to develop physical ideas (radiation, gravity, etc.) that we will use throughout the book. We will see how we obtain information about the basic properties of stars: temperatures, sizes, masses, compositions. The Sun will then be looked at as an example of a typical star. We will then put these stellar properties together, and describe a theoretical picture of how stars work. In Part II we will develop the special and general theories of relativity, to allow us to understand better the unusual states that are reached when stars die. We will discuss the normal lifetime of stars and stellar old age and death in Part III. In stellar death, we will encounter a variety of exotic objects, including neutron stars and black holes.

In Part IV, we will look at the contents of our own galaxy, the Milky Way. We will start by looking at the interstellar medium and then at how stars are formed. Finally, we will look at how stars, gas and dust are organized into a galaxy.

In Part V, we will look at the overall structure of the universe, including the arrangement of galaxies and their motions. We will start by looking at other galaxies. We will also study active galaxies, which give off much more energy than our own. We will follow the trail of active galaxies from starburst galaxies to quasars. The early history of the universe (the big bang) will be described, and we will see can how we look for clues about the past and its ultimate fate. In talking about the early universe, we will encounter one of the most fascinating recent developments, the merging of physics on the largest and smallest scales. This involves blending theories on the ultimate structure of matter with theories of the overall structure of the universe.

In the final part, Part VI, we will study the Solar System. We will see how the formation of the Solar System can be fit into ideas already developed about star formation. We will encounter a variety of surfaces, atmospheres and rings that can be explained by using the physical ideas already developed. We will also look at the origin of life on the Earth and the search for life elsewhere in the Solar System and in our galaxy.

Although the organization of the book is around astronomical objects, the presentation of the topics emphasizes the application of the underlying physics. Almost all of the physical tools will apply to several topics. A great strength of physical theories is the great range of their applicability. For example, orbital mechanics can tell us about the masses of binary stars or help us plan a probe to Mars. Radiative transfer helps us understand the appearance of the Sun, the physical conditions in interstellar clouds or the temperatures of planetary atmospheres. Tidal effects help us explain the appearance of certain galaxies, rings around some planets and the internal heating of Io, one of Jupiter's moons.

Though understanding how astronomical objects work is our goal, astronomy's foundation is observation. We will see how observations often define a problem – the discovery of new phenomena. Observations usually provide a check on theories that are developed. In this book, we will therefore emphasize the interplay between observation and understanding the physics. We will see how some observations yield numbers with great precision, while others only give order of magnitude estimates, but both types can be equally important for deciding between theories.

With the current pace of astronomical discovery, there is an important caution to keep in mind. When you read an introductory text on classical physics, you are reading about theories that were worked out and tested over a century ago. No question is raised about the correctness of these theories. In astronomy, new ideas or new observations are constantly changing the thinking about various problems. Many of the topics discussed in this book are far from being settled. Sometimes, more than one explanation is presented for a given phenomenon. This is done either because we don't know which is correct, or to show how one theory was eliminated in favor of another. Just because this is a "text" it doesn't mean that it has the final word. If you understand where the problems lie, and the reasoning behind the explanations, then you will be able to follow future developments as they appear in scientific magazines or journals.

This, then, is the plan. As you study the material that follows, see how far you can go with a little bit of physics and a lot of curiosity and ingenuity.

1.2 | The scale of the universe

The objects that we encounter in astronomy are, for the most part, so large or distant that it is hard to comprehend their size or distance. We will take a brief look at the distances involved when we study different astronomical objects. We will talk about these sizes in more detail when we encounter the objects in the rest of the book. In Fig. 1.1, we show a selection of objects on the various scales.

We start by looking at the Earth and Moon (Fig. 1.1a). Earth has a radius of about 6000 km. Its mass is about 6×10^{27} grams. The Moon is about 4×10^5 km from the Earth. It takes about one second for light to travel from the Moon to the Earth.

We next look at the Sun (Fig. 1.1b). It is 1.5×10^8 km from the Earth, meaning it takes light over eight minutes to get here from the Sun. We call this distance the Astronomical Unit. Its mass is 2×10^{33} g. This turns out to be average for a star, and we even use it as a convenient measure. The Sun's radius is 6×10^5 km.

We see how far out the planets are by looking at Pluto (Fig. 1.1c). It is almost 40 astronomical units from the Sun, meaning it takes light almost six hours to reach us from Pluto.

By the time we reach the nearest stars, they are so far away that it takes light years to reach us. So we measure their distance in light years

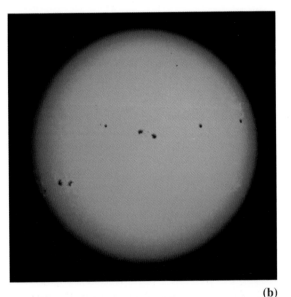

(b)

(a)

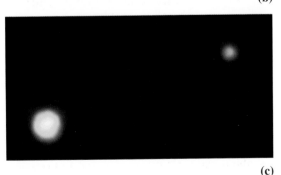

(c)

Fig 1.1. Photographs to show different astronomical scales. (a) The Earth and Moon from space. [NASA] (b) The Sun. [NOAO/AURA/NSF] (c) Pluto and its moon, Charon. [STScI/NASA]

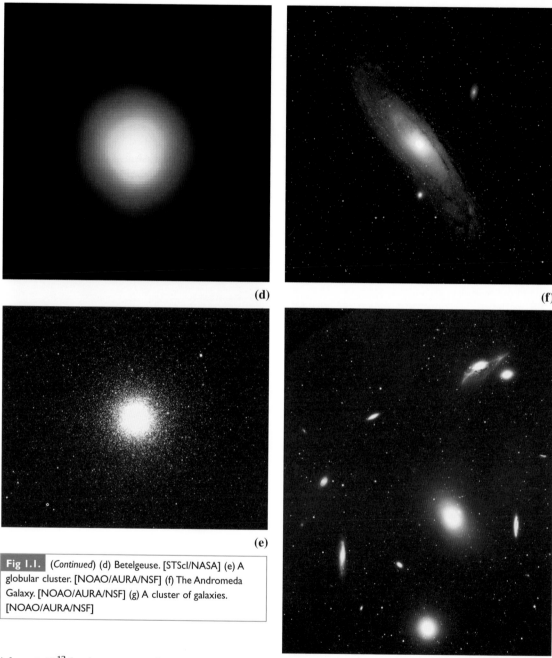

Fig 1.1. (Continued) (d) Betelgeuse. [STScI/NASA] (e) A globular cluster. [NOAO/AURA/NSF] (f) The Andromeda Galaxy. [NOAO/AURA/NSF] (g) A cluster of galaxies. [NOAO/AURA/NSF]

(almost 10^{13} km) or parsecs (one parsec is about three light years). Fig. 1.1(d) shows a star about as far away as we can take direct picture of its disk. It is the giant star Betelgeuse in the constellation of Orion, some 500 parsecs away, meaning it took the light for that image about 1500 years to reach us.

The next largest scale are groupings of stars called clusters, such as the globular cluster in Fig. 1.1(e). These objects may contain 10^5 stars, and have extents of tens of parsecs. Because of their collective brightness, we can see them far away, even on the other side of our galaxy. In fact, they tell us that we are 8500 parsecs from our galactic center. That means it takes light from the galactic center 25 000 years to reach us.

In Fig. 1.1(f), we leave the Milky Way Galaxy and look at one of our neighbors, the Andromeda Galaxy, which we think looks a lot like our galaxy would look if we could view it from outside. It is so far away that we measure its distance in thousands of parsecs, kiloparsecs. It is 700 kiloparsecs away, meaning it takes light about 2.1 million years to reach us. It is about 20 kiloparsecs across. It has a mass equal to more than 10^{11} Suns. When we look at larger scales, we will see that galaxies are like the molecules of the universe.

Our final step is to a cluster of galaxies, such as the Virgo Cluster, which is shown in Fig. 1.1(g). These clusters are groupings of thousands of galaxies, and are typically millions of parsecs across. We detect some clusters so far away that their light has taken a significant fraction of the age of the universe (which we think is about 14×10^{9} yr) to reach us.

As we have said, this description is just to give you a flavor of the sizes involved. The individual objects will be discussed in detail throughout this book.

Part I

Properties of ordinary stars

Chapter 2

Continuous radiation from stars

2.1 | Brightness of starlight

When we look at the sky, we note that some stars appear brighter than others. At this point we are not concerned with what causes these brightness differences. (They may result from stars actually having different power outputs, or from stars being at different distances.) All we know at first glance is that stars *appear* to have different brightnesses.

We would like to have some way of quantifying the observed brightnesses of stars. When we speak loosely of brightness, we are really talking about the *energy flux, f*, which is the energy per unit area per unit time received from the star. This can be measured with current instruments (as we will discuss in Chapter 4). However, the study of stellar brightness started long before such instruments, or even telescopes, were available. Ancient astronomers made naked eye estimates of brightness. Hipparchus, the Greek astronomer, and later Ptolemy, a Greek living in Alexandria, Egypt, around 150 AD, divided stars into six classes of brightness. These classes were called *magnitudes*. This was an ordinal arrangement, with first-magnitude stars being the brightest and sixth-magnitude stars being the faintest.

When quantitative measurements were made, it was found that each jump of one magnitude corresponded to a fixed *flux ratio*, not a flux difference. Because of this, the magnitude scale is essentially a logarithmic one. This is not too surprising, since the eye is approximately logarithmic in its response to light. This type of response allows us to see in very low and very high light levels. (We say

that the eye has a large dynamic range; this range is achieved at a sacrifice in our ability to discriminate small brightness differences.)

The next step was to make the scale continuous, so that, for example, we could accurately describe the brightness of a star that is between second and third magnitude. In addition we would like to extend the scale, so that the brightnesses of stars that we can see only through telescopes can be included. It was found that a difference of five magnitudes corresponds to a factor of 100 in brightness. In setting up the magnitude scale, this relation is defined to be exact.

Let b_1 and b_2 be the observed brightnesses of two stars, and let m_1 and m_2 be the corresponding magnitudes. The statement that a five-magnitude difference gives a flux ratio of 100 corresponds to

$$b_1/b_2 = 100^{(m_2 - m_1)/5} \tag{2.1}$$

We can see that this equation guarantees that each time $m_2 - m_1$ increases by five, b_1/b_2 decreases by a factor of 100. Remember, *increasing* the brightness *decreases* the magnitude. This point sometimes confuses even professional astronomers. That is why you will often hear astronomers talking about being so many magnitudes "brighter" or "fainter" than something else, without worrying about whether that makes m larger or smaller.

Equation (2.1) gives brightness ratios in powers of 100, but we usually work in powers of ten. To convert this we write 100 as 10^2, so equation (2.1) becomes

$$b_1/b_2 = 10^{(m_2 - m_1)/2.5} \tag{2.2}$$

This equation can be used to calculate the brightness ratio for a given magnitude difference. If we want to calculate a magnitude difference for a given brightness ratio, we take the logarithm (base 10) of both sides, giving

$$m_2 - m_1 = 2.5 \log_{10}(b_1/b_2) \qquad (2.3)$$

To see how this works, let's look at a few simple examples. On the original scale, the magnitude range for stars visible to the naked eye is 1 to 6 mag. This corresponds to a brightness ratio

$$b_1/b_2 = 10^{(6-1)/2.5} = 10^2$$

The largest ground-based telescopes extend our range from 6 to 26 mag. This corresponds to an additional brightness ratio

$$b_1/b_2 = 10^{(26-6)/2.5} = 10^8$$

We can also find the magnitude difference, Δm, corresponding to a factor of 10^6 in brightness:

$$\Delta m = 2.5 \log_{10}(10^6) = (2.6)(6) \text{ mag} = 15 \text{ mag}$$

So, we have taken the original six magnitude groups and come up with a continuous scale that can be extended to fainter or brighter objects. Objects brighter than magnitude 1 can have magnitude 0 or even negative magnitudes.

2.2 | The electromagnetic spectrum

Thomas Young first demonstrated interference effects in light, showing that light is a wave phenomenon. If we pass light through a prism (Fig. 2.1), we can see that the light is spread out into

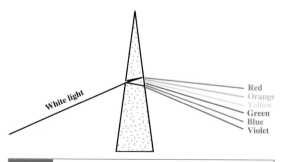

Fig 2.1. The colors of visible light. When light passes through a prism, the rays of different colors are deflected by different amounts. The colors are listed, from top to bottom, in order of decreasing wavelength.

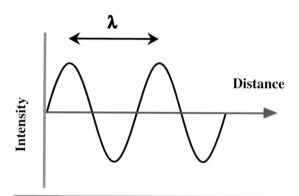

Fig 2.2. The wavelength λ is the distance between the corresponding points of a wave in successive cycles. For example, it can be from peak to peak.

different colors. We call this range of colors the *visible spectrum*. These colors have different wavelengths (Fig. 2.2). For example, the red light has a wavelength around 650 nm ($= 650 \times 10^{-9}$ m $= 6.5 \times 10^{-7}$ m $= 6.5 \times 10^{-5}$ cm). (We used to express this in terms of angstrom units, after the Swedish physicist A. J. Ångstrom, but this is not part of the official metric system. The angstrom was a convenient unit, since it is about the size of a typical atom.) At the opposite end of the visible spectrum from red is violet, with a wavelength of about 400 nm.

In a vacuum, all wavelengths of light travel at the same speed $c = 3.0 \times 10^{10}$ cm/s (3.0×10^8 m/s, 3.0×10^5 km/s). At this speed light can travel a distance equal to the Earth's circumference 7.5 times per second. A light pulse take 1.3 s to reach the Moon. The speed of light is so large that measuring it requires the accurate measurement of time over short intervals, or the passage of light over long distances. Until late in the 19th century, the large distances between astronomical objects were used to provide reasonably long travel times. More recently, accurate timing devices have made laboratory measurements feasible.

All waves have a frequency associated with them. The frequency tells us the number of oscillations per second, or the number of crests that pass per second. The product of the wavelength λ and the frequency ν gives the speed of the wave. That is,

$$\lambda \nu = c \qquad (2.4)$$

Table 2.1. | The electromagnetic spectrum.

Region	Wavelength	Frequency (Hz)
Radio	> 1 mm	$< 3 \times 10^{11}$
Infrared	700 nm–1 mm	3×10^{11}–4.3×10^{14}
Visible	400–700 nm	4.3×10^{14}–7.5×10^{14}
Ultraviolet	10–400 nm	7.5×10^{14}–3×10^{16}
X-ray	0.1–10 nm	3×10^{16}–3×10^{18}
Gamma-ray	< 0.1 nm	$> 3 \times 10^{18}$

The higher the frequency, the shorter the wavelength. For example, we can find the frequency for light at a wavelength of 600 nm:

$$\nu = c/\lambda$$

$$= \frac{3.0 \times 10^8 \text{ m/s}}{600 \times 10^{-9} \text{ m}}$$

$$= 5.0 \times 10^{14} \text{ cps}$$

For 1 cycle per second (cps), we use the unit 1 *hertz (Hz)*.

When we talk about light waves with the above frequency, what is actually varying at 5×10^{14} cycles per second? This question was answered more than 100 years ago by *James Clerk Maxwell*, who pointed out the unity between electric and magnetic fields. The behavior of these fields, and their relationship to charged particles is described by four equations known as *Maxwell's equations*.

In these equations, Maxwell was mostly summarizing the work of others, but it was he who put the whole picture together. For example, one of Maxwell's equations is Faraday's law of induction, which describes how a changing magnetic field can produce an electric field. (This is the basis for the production of electricity in a generator.) Maxwell realized that if there is a symmetry between electric and magnetic fields, then a varying electric field should be able to produce a magnetic field.

This realization serves as the basis for our understanding of *electromagnetic waves*. An electric field that varies sinusoidally (as a sine wave) produces a sinusoidally varying magnetic field, which in turn produces a varying electric field, and so on. These varying fields can propagate through space, even empty space. All wavelengths are possible. The speed of these waves can be predicted from Maxwell's equations. The speed of these waves in a vacuum is the same at all wavelengths, and turns out to be numerically equal to *c*, the speed of light. Light is just one form of electromagnetic wave. Other forms have wavelengths that fall in different ranges.

The full set of electromagnetic waves is called the *electromagnetic spectrum* (see Table 2.1). The visible spectrum is just a small part of the electromagnetic spectrum. At longer wavelengths are infrared and radio waves. At shorter wavelengths are the ultraviolet, X-ray and gamma-ray parts of the spectrum. Even though there is no difference between the waves in various parts of the spectrum, we use the divisions because different techniques are used to detect electromagnetic waves in various wavelength ranges. For example, our eyes are sensitive to wavelengths between 400 nm and 700 nm. This is not too surprising, since this is where the Sun gives off most of its energy. It makes sense that we have evolved with our eyes able to make the best use of the illuminating light.

We now know that astronomical objects give off radiation in all parts of the spectrum. However, the Earth's atmosphere limits what we can actually detect (Fig. 2.3). Ultraviolet and shorter wavelengths are blocked by the atmosphere. Visible light passes through the clear atmosphere (but is blocked by clouds). Most infrared wavelengths are blocked by the atmosphere, but some wavelengths get through. For the most part, radio waves pass through the atmosphere with little absorption. We speak of *visible* and *radio windows* in the atmosphere, as well as

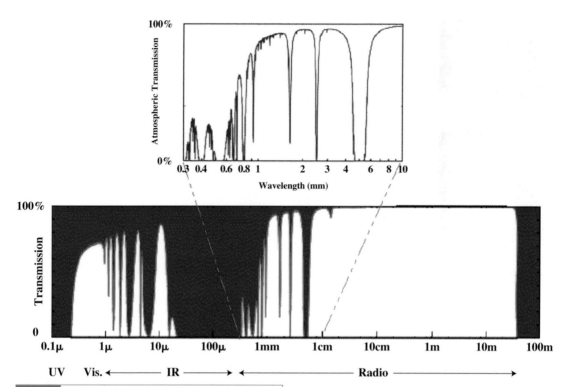

Fig 2.3. Atmospheric transmission as a function of wavelength. The curve shows the fraction of transmission at each wavelength. Note the good transmission in the radio and visible parts of the spectrum. Also note a few narrow ranges, or "windows" of relatively good transmission in the infrared. [IRAM]

some narrow windows in the infrared. A window is simply a wavelength range in which the atmosphere is at least partially transparent.

Until relatively recently, astronomers could only gather information in the visible part of the spectrum, because of the lack of equipment. Much of the development of astronomy was biased by this handicap. In the middle of the 20th century, the radio part of the spectrum was opened for astronomical observations (taking advantage of equipment developed for radar in WW II). Even more recently, other parts of the spectrum have become available to us, due in part to observatories orbiting the Earth. Observing in various parts of the spectrum will be discussed throughout this book.

2.3 | Colors of stars

2.3.1 Quantifying color

When we look at a star, we would like to know how much energy it gives off at various wavelengths. We sometimes refer to a graph, or some equivalent representation, showing intensity as a function of wavelength (or frequency) as a *spectrum*. It is not really proper to talk about the energy given off at a particular wavelength. If we can specify a wavelength to an arbitrary number of decimal places then even a small wavelength range has an infinite number of wavelengths. If there was even a little energy "at" each wavelength, then there would be an infinite amount of energy.

Instead, we talk about the energy given off over some wavelength (or frequency) range. For example, we define the intensity function $I(\lambda)$ such that $I(\lambda)\,d\lambda$ is the energy/unit time/unit surface area given off by an object in the wavelength range λ to $\lambda + d\lambda$. Similarly, $I(\nu)\,d\nu$ is the energy/unit time/unit surface area given off by an object in the frequency range ν to $\nu + d\nu$.

Fig 2.4. Star cluster H and Chi Persei. (We will talk more about clusters of stars in Chapter 13.) Notice the wide range of star colors. [NOAO/AURA/NSF]

When we make a plot of $I(\lambda)$ vs. λ for a star we find that the graph varies smoothly over most wavelengths. There are some wavelength ranges at which there is a sharp increase or decrease in $I(\lambda)$ over a very narrow wavelength range. These sharp increases and decreases are called *spectral lines* and will be discussed in the next chapter. In this chapter, we will be concerned with the smooth or *continuous* part of the spectrum. This is also called the *continuum*.

When we look at stars we see that they have different colors. Stars with different colors have different continuous spectra. In Fig. 2.4, we look at a cluster of stars, and note a wide range of colors. If we took a continuous spectrum of various colored stars, we would find that stars that appear blue have continuous spectra that peak in the (shorter wavelength) blue. The color of a star depends on its temperature. We know that as we heat an object, first it glows in the red, then turns yellow/green, and then it turns blue as it becomes even hotter.

We can therefore measure the temperature of a star by measuring its continuum. In fact, it is not necessary to measure the whole spectrum in detail. We can measure the amounts of radiation received in certain wavelength ranges. These ranges are defined by *filters* that let a given wavelength range pass through. By comparing the intensity of radiation received in various filters, we can come up with a quantitative way of determining the color of a star and therefore its temperature.

2.3.2 Blackbodies

We can understand the relationship between color and temperature by considering objects called *blackbodies*. A blackbody is a theoretical idea that closely approximates many real objects in *thermodynamic equilibrium*. (We say that an object is in thermodynamic equilibrium with its surroundings when energy is freely interchanged and a steady state is reached in which there is no *net* energy flow. That is, energy flows in and out at the same rate.) A blackbody is an object that absorbs all of the radiation that strikes it.

A blackbody can also emit radiation. In fact, if a blackbody is to maintain a constant temperature, it must radiate energy at the same rate that it absorbs energy. If it radiates less energy than it absorbs, it will heat up. If it radiates more energy than it absorbs, then it will cool. However, this does not mean that the spectrum of emitted radiation must match the spectrum of absorbed radiation. Only the total energies must balance. The spectrum of emitted radiation is determined by the temperature of the blackbody. As the temperature changes, the spectrum changes. The blackbody will adjust its temperature so that its emitted spectrum contains just enough energy to balance the absorbed energy. When the temperature

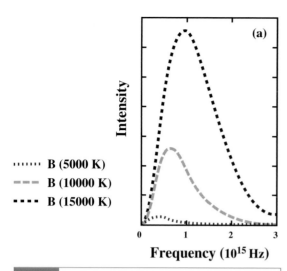

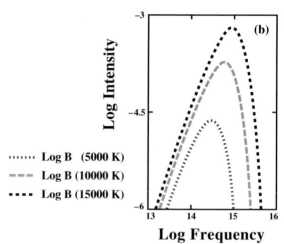

Fig 2.5. Blackbody spectra. Note the shift of the peak wavelength to higher frequency (shorter wavelength) at higher temperature. Note also that, at any frequency, a hotter blackbody gives off more radiation than a cooler one. (a) Intensity as a function of frequency. Notice the big change in intensity with only a factor of three change in temperature. For this reason, we often find it useful to make a plot such as the set of curves in (b), which show the log of the intensity as a function of the log of the frequency.

which allows this balance is reached, the blackbody is in equilibrium.

Figure 2.5 shows some sample blackbody spectra. If we compare these spectra to those of actual stars, we see that the actual spectra are very much like blackbody spectra. Notice that in any wavelength range, a hotter blackbody gives off more energy than a cooler blackbody of the same size. We also see that as the temperature increases the peak of the spectrum shifts to shorter wavelengths.

The relationship between the wavelength at which the peak occurs, λ_{max}, and temperature, T, is very simple. It is given by *Wien's displacement law*:

$$\lambda_{max}T = 2.90 \times 10^{-1}\,\text{cm K} = 2.90 \times 10^6\,\text{nm K} \quad (2.5)$$

In this law, we must use temperature on an absolute (Kelvin) scale. (The temperature on the Kelvin scale is the temperature on the Celsius scale plus 273.1.)

Example 2.1 Using Wien's displacement law
(a) Find the temperature of an object whose blackbody spectrum peaks in the middle of the visible

part of the spectrum, $\lambda = 550$ nm. (b) The Earth has an average temperature of about 300 K. At what wavelength does the Earth's blackbody spectrum peak?

SOLUTION
(a) Given the wavelength, we solve equation (2.5) for the temperature:

$$T = \frac{2.9 \times 10^6\,\text{nm K}}{550\,\text{nm}} = 5270\,\text{K}$$

This is close to the temperature of the Sun.
(b) Given the temperature, we solve equation (2.5) for the wavelength:

$$\lambda = \frac{2.9 \times 10^6\,\text{nm K}}{300\,\text{K}}$$

$$= 1 \times 10^4\,\text{nm}$$

$$= 10 \times 10^{-6}\,\text{m}$$

$$= 10\,\mu\text{m}$$

This is in the infrared part of the spectrum. Even though the Earth is giving off radiation, we don't see it glowing in the visible part of the spectrum. Similarly, objects around us that are at essentially the same temperature as the Earth give off most of their radiation in the infrared part of the spectrum, with very little visible light. The visible light that we see from surrounding objects is partially reflected sunlight or artificial light.

We could have solved (b) by scaling a known result, such as the answer in (a):

$$\frac{\lambda_{\text{Earth}}}{\lambda_A} = \frac{T_A}{T_{\text{Earth}}}$$

$$\lambda_{\text{Earth}} = \left(\frac{T_A}{T_{\text{Earth}}}\right)\lambda_A$$

$$= \left(\frac{5270 \text{ K}}{300 \text{ K}}\right)(550 \text{ nm})$$

$$= 10 \times 10^3 \text{ nm}$$

$$= 10 \text{ } \mu\text{m}$$

Scaling results can be useful because they show how different physical parameters are related to each other. It also provides us with a way of using an equation even if we don't remember the constants.

Suppose we are interested in the total energy given off by a blackbody (per unit time per unit surface area) over the whole electromagnetic spectrum. We must add the contributions at all wavelengths. This amounts to taking an integral over blackbody curves, such as those in Fig. 2.5. Since a hotter blackbody gives off more energy at all wavelengths than a cooler one, and is particularly dominant at shorter wavelengths, we would expect a hotter blackbody to give off much more energy than a cooler one. Indeed, this is the case. The total energy per unit time, per unit surface area, E, given off by a blackbody is proportional to the *fourth* power of the temperature. That is

$$E = \sigma T^4 \qquad (2.6)$$

This relationship is called the *Stefan–Boltzmann law*. The constant of proportionality, σ, is called the *Stefan–Boltzmann constant*. It has a value of 5.7×10^{-5} erg/(cm^2 K^4 s). This law was first determined experimentally, but it can also be derived theoretically. The T^4 dependence means that E depends strongly on T. If we double the temperature of an object, the rate at which it gives off energy goes up by a factor of 16. If we change the temperature by a factor of ten (say from 300 K to 3000 K), the energy radiated goes up by a factor of 10^4.

For a star, we are interested in the total *luminosity*. The luminosity is the total energy per second (i.e. the power) given off by the star. The

quantity σT^4 is only the energy per second per unit surface area. Therefore, to obtain the luminosity, we must multiply it by the surface area. If the star is a sphere with radius R, the surface area is $(4\pi R^2)$, so the luminosity is

$$L = (4\pi R^2)(\sigma T^4) \qquad (2.7)$$

Example 2.2 Luminosity of the Sun
The surface temperature of the Sun is about 5800 K and its radius is 7×10^5 km (7×10^{10} cm). What is the luminosity of the Sun?

SOLUTION
We use equation (2.7) to find the luminosity:

$$L = 4\pi(7 \times 10^{10} \text{ cm})^2 \text{ } [5.7 \times 10^{-5} \text{ erg/(cm}^2 \text{ K}^4 \text{ s)]}$$
$$\times (5.8 \times 10^3 \text{ K})^4$$

$$= 4 \times 10^{33} \text{ erg/s}.$$

This quantity is called the *solar luminosity*, $L_\odot$, and serves as a convenient unit for expressing the luminosities of other stars.

2.4 | Planck's law and photons

2.4.1 Planck's law
The study of blackbody radiation plays an important role in the development of what we refer to as "modern" physics (even though these developments took place early in the 20th century). When physicists tried to apply classical ideas of radiation, they could not derive blackbody spectra that agreed with the experimental results. The classical calculations yielded an intensity $I(\nu, T)$ given by

$$I(\nu, T) = 2kT\nu^2/c^2 \qquad (2.8)$$

This is known as the *Rayleigh–Jeans law*. The constant k that appears in this law is the *Boltzmann constant* (not to be confused with the Stefan–Boltzmann constant). Its value is 1.38×10^{-16} erg/K. (The quantity kT is proportional to the kinetic energy per particle in the gas.) The Rayleigh–Jeans law agrees with experimental results at low frequencies (long wavelengths), but disagrees at high frequencies. In fact, you can see from equation (2.8) that, as we go to higher and higher frequencies, the energy given off becomes arbitrarily large.

However, in looking at the blackbody curves (as in Fig. 2.5) you see that there is a peak and then there is less energy at higher frequencies. (The classical prediction of arbitrarily large energies at high frequencies was sometimes referred to as the "ultraviolet catastrophe".)

The first step in solving the problem of theoretically predicting blackbody curves was to deduce an *empirical* formula for the observed spectra. By "empirical" we mean a formula that is arbitrarily put together to describe the observations, but which is not derived from any theory. Once that is done, then we try to find a theory that can be used to derive the formula.

In 1900, *Max Planck*, a German physicist, produced an empirical formula that accurately describes the experimental blackbody spectra:

$$I(\nu, T) = \frac{2h\nu^3/c^2}{e^{h\nu/kT} - 1} \qquad (2.9)$$

In this equation, h is called *Planck's constant*, and has the numerical value 6.63×10^{-27} erg s. This value was determined to provide the best agreement with observed blackbody spectra.

Since the Rayleigh–Jeans law adequately describes blackbody spectra at low frequencies, the Planck law must reduce to the Rayleigh–Jeans law in the limit of low frequencies. We can see this if we take low frequencies to mean that $h\nu \ll kT$ (or equivalently $h\nu/kT \ll 1$). In that case, we can take advantage of the fact that for $x \ll 1$, $e^x \cong 1 + x$. The Planck function then becomes

$$I(\nu, T) \cong \frac{2h\nu^3}{c^2} \frac{kT}{h\nu}$$

$$= \frac{2kT\nu^2}{c^2}$$

which is the Rayleigh–Jeans law.

Equation (2.9) gives the Planck function in terms of frequency. How do we find it as a function of wavelength? Your first guess might be simply to substitute c/λ for each occurrence of ν in equation (2.9). However, we must remember that $I(\nu, T)$ gives the energy per second per frequency interval, whereas $I(\lambda, T)$ gives the energy per second per wavelength interval. The functions must reflect that difference (especially since they will need different units). We therefore require that

$$I(\lambda, T) \, d\lambda = I(\nu, T) \, d\nu$$

Solving for $I(\lambda, T)$ gives

$$I(\lambda, T) = I(\nu, T) \, (d\nu/d\lambda)$$

To find $I(\lambda, T)$ we must be able to evaluate $d\nu/d\lambda$. We do this by remembering that $\nu = c/\lambda$, so that

$$d\nu/d\lambda = -c/\lambda^2 \qquad (2.10a)$$

We don't care about the minus sign, which just tells us that frequency increases when wavelength decreases. Using this result gives

$$I(\lambda, T) = I(\nu, T) \, (c/\lambda^2) \qquad (2.10b)$$

Now we can substitute c/ν for ν to obtain the final result:

$$I(\lambda, T) = \frac{2hc^2/\lambda^5}{e^{hc/\lambda kT} - 1} \qquad (2.10c)$$

Remember, the Planck function accurately describes blackbody spectra, but it was originally presented as an empirical formula. There was still no theoretical understanding of the origin of the formula. Planck continued his work in an effort to derive the formula from some theory. Planck found that he could derive the formula from classical physics if he inserted a mathematical trick. The trick amounted to taking a sum rather than an integral. The trick corresponds to the physical statement that a blackbody can only emit radiation at a frequency ν in multiples of the quantity $h\nu$. That is, the energy could only be emitted in small bundles or *quanta* (singular *quantum*). The quanta have energy $h\nu$. Even though Planck was able to derive the blackbody formula correctly, he was still not satisfied. There was no justification for the restriction that energy must be quantized.

2.4.2 Photons

An explanation for why energy must be quantized was proposed by *Albert Einstein*, in 1905. (It was for this explanation that Einstein was later awarded the Nobel Prize in physics.) Einstein was trying to explain a phenomenon known as the *photoelectric effect*, in which electrons can be ejected from a metal surface if light falls on the surface. (This is the basis for photocells, which are used in many applications.) Laboratory studies had shown that increasing the intensity of the light falling on the surface increased the number of electrons ejected from the surface, but not their energy. Einstein said that all radiation (whether from a blackbody or otherwise) must come in small bundles, called *photons*. The energy, E, of a photon with a frequency ν is given by

$$E = h\nu \tag{2.11}$$

This explains the observed properties of the photoelectric effect by stating that each electron is ejected by a single photon striking the surface. Increasing the intensity of light increases the number of photons striking the surface per second, and therefore increases the *rate* at which electrons are ejected. Increasing the frequency of the light increases the *energy* at which the electrons are ejected. (This latter prediction was finally tested by *Robert Millikan* in 1916.) A further test of the photon hypothesis came in an analysis of collisions between light (photons) and electrons, by *A. H. Compton*. The fact that light is composed of photons explains why Planck had to assume that energy is quantized in deriving the formula for blackbody spectra.

The assertion that light is essentially a particle went against the then accepted ideas about light. The question of whether light is a particle or a wave had been going on for centuries. For example, Newton believed that it is a particle, and he worked out a theory of refraction – the bending of light when it passes, for example, from air to glass – on the basis of light speeding up when it enters the glass. (We now know, however, that as a wave it slows down.) The wave theory became dominant with the demonstration of interference effects by Young and the explanation of electromagnetic waves by Maxwell. In explaining the photoelectric effect, Einstein was saying that the particle picture must be revived. The explanation that, somehow, light can exhibit both wave and particle properties is referred to as the wave–particle duality. This concept is the foundation of what we refer to as the *quantum revolution*, since it was such a radical departure from previous theories. We will discuss this point farther in the next chapter.

2.5 | Stellar colors

We have seen that the color of a star can tell us about the star's temperature. However, we now need a way of quantifying a color, rather than just saying something is red, green or blue. For example, if we compare two blue stars, how do we decide which one is bluer?

We define two standard wavelength ranges, centered at λ_1 and λ_2 and take the ratio of the observed brightnesses, $b(\lambda_1)/b(\lambda_2)$. We then convert that brightness ratio into a magnitude difference (using equation (2.3)), giving

$$m_2 - m_1 = 2.5 \log_{10}[b(\lambda_1)/b(\lambda_2)] \tag{2.12}$$

We define the quantity $m_2 - m_1$ as the color, measured in magnitudes, corresponding to the wavelength pair, λ_1, λ_2. For definiteness, let's assume that $\lambda_2 > \lambda_1$. As we increase the temperature, $b(\lambda_1)/b(\lambda_2)$ increases. This means that the quantity $m_2 - m_1$ increases since the magnitude scale runs backwards. If we know that an object is radiating exactly like a blackbody, we need only take the ratio of brightnesses at any two wavelengths to determine the temperature.

As we have said, we don't really measure the intensity of radiation at a wavelength. Instead, we measure the amount of energy received in some wavelength interval. We can control that wavelength interval by using a *filter* that only passes light in that wavelength range. When we use a filter, we are actually measuring the integral of $I(\lambda, T)$ over some wavelength range. Actually, the situation is more complicated. The transmission of any real filter is not 100% over the selected range, and this must be factored in (see Problem 2.23).

Another complication is that continuous spectra of stars do not exactly follow blackbody curves. Therefore, observations through two filters are not generally sufficient to tell us the temperature of the star. Over the years, a system of standard filters has been developed, so that astronomers at various observatories can compare their results. The wavelength ranges of the various filters are shown in Table 2.2. The most commonly discussed filters are U (for *u*ltraviolet) B (for *b*lue) and V (for *v*isible, meaning the center of the visible part of the spectrum). More recently,

Table 2.2.	Filter systems.	
Filter	Peak wavelength (nm)	Width (nm)[a]
U	350	70
B	435	100
V	555	80
R	680	150
I	800	150

[a]Full width at half maximum.

R (for red), and I (for infrared) have been added. (There are actually a couple of filters in different parts of the infrared.)

For example, the $B - V$ color is defined by

$$B - V = 2.5 \log_{10} [I(\lambda_V) / I(\lambda_B)] + \text{constant}$$

where $I(\lambda_V)$ and $I(\lambda_B)$ are the intensities averaged over the filter ranges. (The constant is adjusted so that $B - V$ is zero for a particular temperature star, designated A0. These designations will be discussed in the next chapter.) As the temperature of an object increases, the ratio of blue to visible increases. This means that the $B - V$ color decreases (again because the magnitude scale runs backwards.)

2.6 | Stellar distances

So far we have discussed how bright stars appear as seen from Earth. However, the apparent brightness depends on two quantities: the intrinsic luminosity of the star and its distance from us. (As we will see in Chapter 14, starlight is also dimmed when it passes through clouds of interstellar dust.) Two identical stars at different distances will have different apparent brightnesses. If we want to understand how stars work, we must know their total luminosities. This requires correcting the apparent brightness for the distance to the star.

If we have a star of luminosity L, we can calculate the observed energy flux at a distance d. If no radiation is absorbed along the way, all the energy per second leaving the surface of the star will cross a sphere at a distance d in the same time. It will just be spread over a larger area. Therefore, the energy per second reaching d is

still L, but it is spread over an area of $4\pi d^2$ so the energy flux, f, is

$$f = L/4\pi d^2 \tag{2.13}$$

The received flux falls off inversely as the square of the distance.

Unfortunately, distances to astronomical objects are generally hard to determine. There is a direct method for determining distances to nearby stars. It is called *trigonometric parallax*, and amounts to triangulation from two different observing points. You can demonstrate parallax for yourself by holding out a finger at arm's length and viewing it against a distant background. Look at the finger alternately using your left and right eye. The finger appears to shift against the distant background. Bring the finger closer and repeat the experiment. The shift now appears larger. If you could move your eyes farther apart, the effect would be even greater.

Even the closest stars are too far away to demonstrate parallax when we just use our eyes. However, we can take advantage of the fact that the Earth orbits the Sun at a distance defined to be one *astronomical unit (AU)*. Therefore, if we observe a star and then observe it again six months later, we have viewing points separated by 2 AU. The situation is illustrated in Fig. 2.6. We note the position of the star against the background of distant stars, and then six months later we note the angle by which the position has shifted. If we take half of the value of this angle, we have the *parallax angle, p*.

Once we know the value of p, we can construct a right triangle with a base of 1 AU and the other leg being the length, d, the unknown distance to the star. From the right triangle, we can see that

$$\tan p = 1 \text{ AU}/d \tag{2.14}$$

Since p is small, $\tan(p) \cong p$ (rad), which is the value of p, measured in radians. Equation (2.14) then gives us

$$p(\text{rad}) = 1 \text{ AU}/d \tag{2.15}$$

It is not very convenient measuring such small angles in radians, so we convert to arc seconds (see Box 2.1):

$$p('') = 2.06 \times 10^5 p(\text{rad})$$

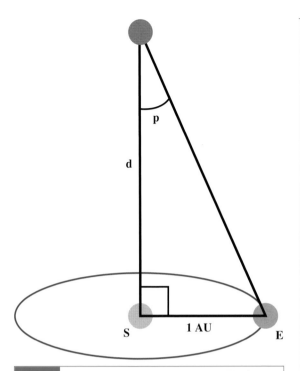

Fig 2.6. Geometry for parallax measurements. The figure is not to scale. In reality the distance to the star, d, is much greater than 1 AU, so the parallax angle, p, would normally be very small.

where $p('')$ is the parallax angle measured in arc seconds. Substituting this into equation (2.15) gives

$$d/1 \text{ AU} = 2.06 \times 10^5/p('') \tag{2.16}$$

This gives the distance to the star in AU (1 AU = 1.50×10^8 km).

This method suggests a convenient unit for measuring distances. We define the *parsec* (abbreviated pc) as the distance of a star that produces a parallax angle p of 1 arc sec. From equation (2.16), we can see that 1 pc = 2.06×10^5 AU (or 3.09×10^{13} km, or 3.26 light years). We rewrite equation (2.16) as

$$d(\text{pc}) = 1/p('') \tag{2.17}$$

Remember, as an object moves farther away, the parallax angle decreases. Therefore, a star at a distance of 2 pc will have a parallax angle of 0.5 arc sec.

Box 2.1. | Angular measure

The natural unit for measuring angles is the *radian*. If we have a circle of radius R, and two lines from the center making an angle θ with each other, then the length of the arc bounded by the two lines is

$$L = \theta(\text{rad})R$$

where $\theta(\text{rad})$ is the value of θ measured in radians. Since the full circumference of a circle is $2\pi R$, the angle corresponding to a full circle must be 2π radians. This tells us that a full circle, that is 360°, is equal to 2π radians, or 180° is equal to π radians. In astronomy, we often deal with very small angles, and measurements in arc seconds ('') are convenient. We can convert measurements by saying

$$\theta('') = \theta(\text{rad}) \frac{180°}{\pi \text{ rad}} \frac{60'}{1°} \frac{60''}{1'}$$

$$= 2.06 \times 10^5 \theta(\text{rad})$$

When we take the derivatives of trigonometric functions (for example, $d(\sin\theta)/d\theta = \cos\theta$), it is assumed that the angles are in radians. If not, a conversion factor must be carried through the differentiation.

When angles expressed in radians have a value that is much less than unity, we can use a Taylor series to approximate them:

$$\sin(\theta) \cong \sin(0) + \theta \left.\frac{d(\sin\theta')}{d\theta'}\right|_{\theta=0}$$

$$= \theta \cos(0)$$

$$= \theta$$

$$\tan(\theta) \cong \tan(0) + \theta \left.\frac{d(\tan\theta')}{d\theta'}\right|_{\theta=0}$$

$$= \theta \sec^2(0)$$

$$= \theta$$

$$\cos(\theta) \cong \cos(0) + \theta \left.\frac{d(\cos\theta')}{d\theta'}\right|_{\theta=0}$$

$$= 1 - \theta \sin(0)$$

$$= 1$$

Note that for small θ, $\sin\theta$ and $\tan\theta$ are both approximately equal to θ, so they must be equal to each other. (Remember, in each of the above expressions, θ must be expressed in radians.)

Example 2.3 Distance to the nearest star
The nearest star (Proxima Centauri) has a parallax $p = 0.76$ arc sec. Find its distance from Earth in parsecs.

SOLUTION
We use equation (2.17) to give

$$d(\text{pc}) = 1/(0.76)$$

$$d = 1.32 \text{ pc}$$

With current ground-based equipment, we can measure parallax to within a few hundredths of an arc second. Parallax measurements are therefore useful for the few thousand nearest stars. They are a starting point for a very complex system of determining distances to astronomical objects. We will encounter a variety of distance determination methods throughout this book. The trigonometric parallax method is the only one that is direct and free of any assumptions. For this reason, astronomers would like to extend their capability for measuring parallax. The Hipparcos satellite measures parallaxes to 10^{-3} arc sec.

2.7 | Absolute magnitudes

The magnitudes discussed in Section 2.1, based on observed energy fluxes, are called *apparent magnitudes*. In order to compare intrinsic luminosities of stars, we define a system of *absolute magnitudes*. The absolute magnitude of a star is that magnitude that it would appear to have as viewed from a standard distance, d_0. This standard distance is chosen to be 10 pc. From this definition, you can see that if a star is actually at a distance of 10 pc, the absolute and apparent magnitudes will be the same.

To see how this system works, consider two identical stars, one at a distance d and the other at the standard distance d_0. We let m be the apparent magnitude of the star at distance d, and M that of the star at distance d_0. (Of course, M will be the absolute magnitude for both stars.) The energy flux falls off inversely as the square of the distance, therefore the ratio of the flux of the star at d to that from the star at d_0 is $(d_0/d)^2$. Equation (2.3) then gives us

$$m - M = 2.5 \log_{10}(d/d_0)^2$$

Using the fact that $\log(x^2) = 2 \log(x)$ gives

$$m = M + 5 \log_{10}(d/10 \text{ pc}) \tag{2.18}$$

The quantity $5 \log_{10}(d/10 \text{ pc})$, which is equal to $(m - M)$, is called the *distance modulus* of the star. It indicates the amount (in magnitudes) by which distance has dimmed the starlight. If you know any two of the quantities (m, M or d) you can use equation (2.18) to find the third. For any star that we can observe, we can always measure m, its apparent magnitude. Therefore, we are generally faced with knowing M and finding d or knowing d and finding M. In the next chapter we will look at some ways of determining M.

Example 2.4 Absolute magnitude
A star is at a distance of 100 pc, and its apparent magnitude is $+5$. What is its absolute magnitude?

SOLUTION
We use equation (2.18) to find

$$M = m - 5 \log (d/10 \text{ pc})$$

$$= 5 - 5 \log (100 \text{ pc}/10 \text{ pc})$$

$$= 5 - 5 \log (10)$$

$$= 0$$

We should note that changing the distance of a star changes its apparent magnitude, but it does not change any of its colors. Because colors are defined to be differences in magnitudes, each is changed by the distance modulus. For example, using equation (2.18)

$$m_B = M_B + 5 \log (d/10 \text{ pc})$$

$$m_V = M_V + 5 \log (d/10 \text{ pc})$$

Taking the difference gives

$$m_B - m_V = M_B - M_V$$

Therefore, the distance modulus never appears in the colors.

When we talk about determining an absolute magnitude, we are really only determining it over some wavelength range, corresponding to the wavelength range of the observations. We would like to have an absolute magnitude that corresponds to the total luminosity of the star. This magnitude is called the *bolometric magnitude* of the star. (As we will see in Chapter 4, a bolometer is a device for measuring the total energy received from an object.) For any type of star, we can define a number, called the *bolometric correction* (abbreviated BC), which relates the bolometric magnitude to the absolute visual magnitude M_V. Therefore

$$M_{\mathrm{BOL}} = M_V + \mathrm{BC} \tag{2.19}$$

Chapter summary

We saw in this chapter what can be learned from the brightness and spectrum of the continuous radiation from stars.

We introduced a logarithmic scale, the magnitude scale, for keeping track of brightness. Apparent magnitude is related to the observed energy flux from the star, and the absolute magnitude is related to the intrinsic luminosity of the star.

We saw how, even though stars are obvious to us in the visible part of the spectrum, they, and other astronomical objects, give off radiation in other parts of the spectrum. The richness of information in other parts of the spectrum is a theme that we will come back to throughout the book.

We introduced the concept of a blackbody, which is useful because the continuous spectrum of a star closely resembles that of a blackbody. Hotter bodies give off more power per unit surface area than cooler ones (as described by the Stefan–Boltzmann law), and also have their spectra peaking at shorter wavelengths (as described by Wien's displacement law). We saw how attempts to understand the details of blackbody spectra (Planck's law) contributed to the idea of light coming in bundles, called photons, with specific energies. With a knowledge of blackbody spectra, we saw how stellar colors can be used to deduce stellar temperatures.

We saw how finding the distances to astronomical objects is very important, but can be quite difficult. If we don't know the distance to an object, we cannot convert its apparent brightness into a luminosity. We introduced one method of measuring distances – trigonometric parallax. It is the most direct method, but only works for nearby stars. The problem of distance determination will come up throughout the book.

Questions

2.1. Why is the magnitude scale logarithmic?

2.2. Are there any other types of measurements that we encounter in the everyday world that are logarithmic? (Hint: Think of sound.)

2.3. Why are astronomical observations potentially useful in measuring the speed of light?

2.4. What are the factors that have resulted in early astronomical observations being in the "visible" part of the spectrum?

2.5. What do we mean by "atmospheric window"?

2.6. Why was Maxwell's realization that a varying electric field can create a magnetic field important in understanding electromagnetic waves?

2.7. (a) Estimate the number of people on Earth who are *exactly* 2 m tall. (By "exactly" we mean to an arbitrary number of decimal places.) (b) How does this relate to the way we define the intensity function I(λ)?

2.8. What are the different ways in which the word "spectrum" is used in this chapter?

2.9. Give some examples of objects whose spectra are close to that of blackbodies.

2.10. How can we determine the temperature of a blackbody?

2.11. If the peak of a blackbody spectrum shifts to shorter wavelengths as we reach higher temperatures, how can it be that a hotter blackbody gives off more energy at all wavelengths than a cooler one?

2.12. What is the evidence for the existence of photons?

2.13. Explain how we quantify the concept of color.

2.14. What is the value of using standard filters in looking at stellar spectra?

2.15. Suppose you could communicate with an astronomer on a planet orbiting a nearby star. (The astronomer is native to that planet, rather than having traveled from Earth.) You determine the distance to the star (by trigonometric parallax) to be 2 pc. The distant astronomer says that you are wrong; the distance is only 1 pc. What is the problem?

2.16. How would parallax measurement improve if we could do our observations from Mars?

2.17. As we determine the astronomical unit more accurately, how does the relationship between the AU and the parsec change?

Problems

2.1. What magnitude difference corresponds to a factor of ten change in energy flux?

2.2. One star is observed to have $m = -1$ and another has $m = +1$. What is the ratio of energy fluxes from the two stars?

2.3. The apparent magnitude of the Sun is -26.8. How much brighter does the Sun appear than the brightest star, which has $m = -1$?

2.4. (a) What is the distance modulus of the Sun? (b) What is the Sun's absolute magnitude?

2.5. Suppose two objects have energy fluxes, f and $f + \Delta f$, where $\Delta f \ll f$. Derive an approximate expression for the magnitude difference Δm between these objects. Your expression should have Δm proportional to Δf. (Hint: Use the fact that $\ln (1 + x) \cong x$ when $x \ll 1$.)

2.6. Show that our definition of magnitudes has the following property: If we have three stars with energy fluxes, f_1, f_2 and f_3, and we define

$$m_2 - m_1 = 2.5 \log_{10}(f_1/f_2)$$

$$m_3 - m_2 = 2.5 \log_{10}(f_2/f_3)$$

then

$$m_3 - m_1 = 2.5 \log_{10}(f_1/f_3)$$

2.7. Suppose we measure the speed of light in a laboratory, with the light traveling a path of 10 m. How accurately do you have to time the light travel time to measure c to eight significant figures?

2.8. Let λ_1 and $\lambda_2(\nu_1, \nu_2)$ be the wavelength (frequency) limits of the visible part of the spectrum. Compare $(\lambda_1 - \lambda_2)/(\lambda_1 + \lambda_2)$ with $(\nu_1 - \nu_2)/(\nu_1 + \nu_2)$. Comment on the significance.

2.9. (a) Calculate the frequencies corresponding to the wavelengths 500.00 nm and 500.10 nm. Use these to check the accuracy of equation (2.10a). (b) Repeat the process for the second

wavelength being 501.00 nm and 510.00 nm. What do you conclude?

*2.10. (a) Use equation (2.9) to derive ν_{max}, the frequency at which $I(\nu, T)$ peaks. Convert this ν_{max} into a wavelength λ_{max}. (b) Use equation (2.10c) to find the wavelength at which it peaks. (c) How do the results in (a) and (b) compare? It is necessary to solve this numerically.

2.11. For a 300 K blackbody, over what wavelength range would you expect the Rayleigh–Jeans law to be a good approximation?

2.12. Derive an approximation for the Planck function valid for high frequencies ($h\nu \gg kT$).

2.13. As we will see in Chapter 21, the universe is filled with blackbody radiation at a temperature of 2.7 K. (a) At what wavelength does the spectrum of that radiation peak? (b) What part of the electromagnetic spectrum is this?

2.14. (a) We observe the blackbody spectrum from a star to peak at 400 nm. What is the temperature of the star? (b) What about one that peaks at 450 nm?

2.15. Derive an expression for the shift $\Delta\lambda$ in the peak wavelength of the Planck function for a blackbody of temperature T, corresponding to a small shift in temperature, ΔT.

2.16. Calculate the energy per square centimeter per second reaching the Earth from the Sun.

2.17. How does the absolute magnitude of a star vary with the size of the star (assuming the temperature stays constant)?

2.18. (a) What is the energy of a photon in the middle of the visible spectrum ($\lambda = 550$ nm)? (b) Approximately how many photons per second are emitted by (i) a 100 W light bulb, (ii) the Sun?

2.19. If we double the temperature of a blackbody, by how much must we decrease the surface area to keep the luminosity constant?

---

*An asterisk denotes a harder Problem or Question. The convention continues throughout the book.

2.20. (a) How does the absolute bolometric magnitude vary with the temperature of a star (assuming the radius stays constant)? (b) Does the absolute visual magnitude vary in the same way?

*2.21. For a star of radius R, whose radiation follows a blackbody spectrum at temperature T, derive an expression for the bolometric correction.

2.22. Suppose we observe the intensity of a blackbody, I_0, in a narrow frequency range centered at ν_0. Find an expression for T, the temperature of the blackbody in terms of I_0 and ν_0. (a) First do it in the Rayleigh–Jeans limit and (b) in the general case.

*2.23. Suppose we receive light from a star for which the received energy flux is given by the function $f(\lambda)$. Suppose we observe the star through a filter for which the fraction of light transmitted is $t(\lambda)$. Derive an expression for the total energy detected from the star. (Hint: Start by thinking of the energy detected in a small wavelength range.)

2.24. What is the distance to a star whose parallax is 0.1 arc sec?

2.25. Derive an expression for the distance of an object as a function of the parallax angle seen by your eyes?

2.26. (a) If we can measure parallaxes as small as 0.1 arc sec, what is the greatest distance that can be measured using the method of trigonometric parallaxes? (b) By what factor will the volume of space over which we can measure parallax change if we can measure to 0.001 arc sec? (c) Why is the volume of space important?

2.27. If we lived on Mars instead of the Earth, how large would the parsec be?

2.28. Suppose we discover a planet orbiting a nearby star. The distance to the star is 3 pc. We observe the angular radius of the planet's orbit to be 0.1 arc sec. How many AU from the star is the planet? (Hint: You can solve this problem by "brute force", converting all the units. For an easier solution, think about what the answer would be if the star were 1 pc from us and the angular radius of the orbit were 1 arc sec, and then scale the result accordingly.)

2.29. Derive an expression for the distance to a star in terms of its distance modulus.

2.30. If we make a 0.05 magnitude error in measuring the apparent magnitude of a star, what error does that introduce in our distance determination (assuming its absolute magnitude is known exactly)?

Computer problems

2.1. Make a fourth column for Table 2.1, showing the range of photon frequencies for each part of the spectrum. Make a fifth column showing the range of photon energies for each part of the spectrum. Make a sixth column showing the temperatures that blackbodies would have to peak at the wavelengths corresponding to the boundaries between the parts of the spectrum

2.2. Make a graph of the magnitude difference $M_B - M_V$ as a function of temperature for a temperature range of 3000 K to 30 000 K. To simplify the calculation you may assume that magnitudes are determined in a narrow range of wavelengths around the peak of each filter.

2.3. For the Sun, plot the difference between the Rayleigh–Jeans approximation and the Planck formula, as a function of wavelength, for wavelengths in the visible part of the spectrum.

2.4. For the Sun, calculate the energy given off over the wavelength bands that correspond to the U, B and V filters. Use this to estimate the colors $U - B$ and $B - V$.

Chapter 3

Spectral lines in stars

In Chapter 2 we discussed the continuous spectra of stars and saw that they could be closely described by blackbody spectra. In this chapter, we will discuss the situations in which the spectrum shows an increase or decrease in intensity over a very narrow wavelength range.

3.1 | Spectral lines

We know that if we pass white light through a prism, light of different colors (wavelengths) will emerge at different angles with respect to the initial beam of light. If we pass white light through a slit before it strikes the prism (Fig. 3.1), and then let the spread-out light fall on the screen, at each position on the screen we get the image of the slit at a particular wavelength.

Both *William Hyde Wollaston* (1804) and *Josef von Fraunhofer* (1811) used this method to examine sunlight. They found that the normal spectrum was crossed by dark lines. These lines represent wavelengths where there is less radiation than at nearby wavelengths. (The lines are only dark in comparison with the nearby bright regions.) The linelike appearance comes from the fact that, at each wavelength, we are seeing the image of the slit. It is this linelike appearance that leads us to call these features *spectral lines*. If we were to make a graph of intensity vs. wavelength, we would find narrow dips superimposed on the continuum. The solar spectrum with dark lines is sometimes referred to as the *Fraunhofer spectrum*. Fraunhofer

gave the strongest lines letter designations that we still use today.

The origin of these lines was a mystery for some time. In 1859, the German chemist *Gustav Robert Kirchhoff* noticed a similar phenomenon in the laboratory. He found that when a beam of white light was passed through a tube containing some gas, the spectrum showed dark lines. The gas was absorbing energy in a few specific narrow wavelength bands. In this situation, we refer to the lines as *absorption lines*. When the white light was removed, the spectrum showed bright lines, or *emission lines*, the wavelengths where absorption lines had previously appeared. The gas could emit or absorb energy only in certain wavelength bands.

Kirchhoff found that the wavelengths of the emission or absorption lines depend only on the type of gas that is used. Each element or compound has it own set of special wavelengths. If two elements which don't react chemically are mixed, the spectrum shows the lines of both elements. Thus, the emission or absorption spectrum of an element identifies that element as uniquely as fingerprints identify a person. This identification can be carried out without understanding why it works.

Whether we see absorption or emission depends in part on whether or not there is a strong enough background source providing energy to be absorbed (Fig. 3.2). The strength of the spectral lines also depends on how much gas is present and on the temperature of the gas. Sample emission and absorption spectra of stars are shown in Fig. 3.3.

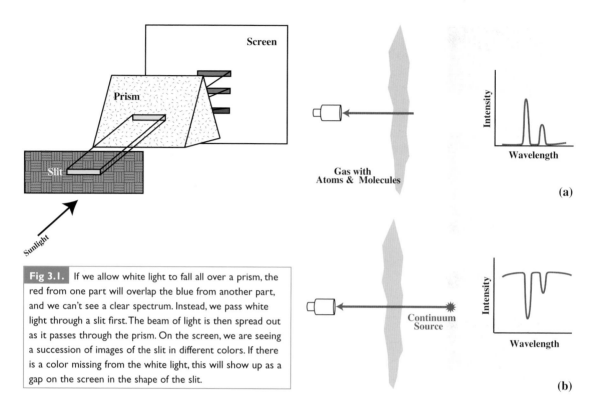

Fig 3.1. If we allow white light to fall all over a prism, the red from one part will overlap the blue from another part, and we can't see a clear spectrum. Instead, we pass white light through a slit first. The beam of light is then spread out as it passes through the prism. On the screen, we are seeing a succession of images of the slit in different colors. If there is a color missing from the white light, this will show up as a gap on the screen in the shape of the slit.

Fig 3.2. Conditions for the formation of emission and absorption lines. (a) We look at a cloud of gas with the atoms or molecules capable of producing spectral lines. Since there is no continuum radiation to absorb, we can only have emission. (b) We now look through the gas at a background continuum source. This can produce absorption lines.

3.2 | Spectral types

When spectra were taken of stars other than the Sun, they also showed absorption spectra. Presumably, the continuous radiation produced in a star passes through an atmosphere in which the absorption lines are produced. Not all stars have absorption lines at the same wavelength.

Astronomers began to classify and catalog the spectra, even though they still did not understand the mechanism for producing the lines. This points out an important general technique in astronomy – studying large numbers of objects to look for general trends. In one very important study, over 200 000 stars were classified by *Annie Jump Cannon* at the Harvard College Observatory. The benefactor of that study was *Henry Draper*, and the catalog of stellar spectra was named after him. The stars in this catalog are still known by their HD numbers.

One set of spectral lines common to many stars was recognized as belonging to the element hydrogen. The stars were classified according to the strongest hydrogen absorption lines. In this system, A stars have the strongest hydrogen lines,

B stars the next strongest, and so on. These letter designations were called *spectral classes* or *spectral types*. We now know that the different spectral types correspond to different surface temperatures. However, the sequence A, B, ... is not a temperature-ordered sequence. For reasons we will discuss below, hydrogen lines are strongest in intermediate temperature stars.

The spectral classes we use, in order of decreasing temperature, are O, B, A, F, G, K, M. We break each of these classes into ten subclasses, identified by a number from zero to nine; for example, the sequence O7, O8, O9, B0, B1, B2, . . . , B9, A0, A1, ... (For O stars the few hottest subclasses are not used.) For some of the hotter spectral types, we even use half subclasses, for example, B1.5. It was originally thought that stars became cooler as they evolved, so that the temperature sequence

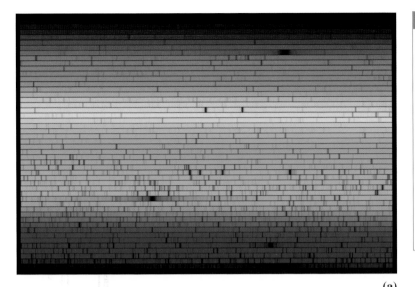

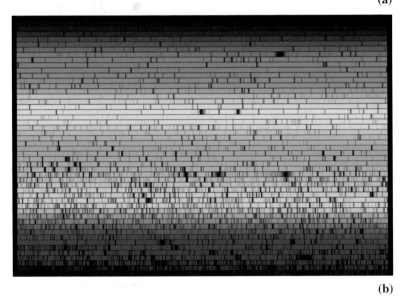

Fig 3.3. Samples of stellar spectra. These are high resolution spectra, with the visible part of the spectrum (400 to 700 nm) broken into 50 slices. Wavelength increases from left to right along each strip and from bottom to top. (a) Procyon, also known as Alpha Canis Majoris (the brightest star in Canis Major). It has spectral type F5 (see Section 3.2), making it a little warmer than the Sun. (b) Arcturus, also known as Alpha Bootes. It is spectral type K1, being cooler than the Sun. [NOAO/AURA/NSF]

(a)

(b)

was really an evolutionary sequence. Therefore, the hotter spectral types were called *early* and the cooler spectral types were called *late*. We now know that these evolutionary ideas are not correct. However, the nomenclature still remains. We even talk about a B0 or B1 star being 'early B' and a B8 or B9 as being a 'late B'.

3.3 | The origin of spectral lines

The processes that result in atoms being able to emit or absorb radiation at certain wavelengths are tied to the nature of matter and light. In Chapter 2, we saw the beginnings of the quantum revolution with the realization that light exhibits both particle and wave properties. We now see how the ideas of quantization apply to the structure of the atom.

The modern picture of the atom begins with the experiments of *Ernest Rutherford*, who studied the scattering of alpha particles (helium nuclei) off gold atoms. Most of the alpha particles passed through the gold atoms without being deflected, suggesting that most of the atom is empty space! Some alpha particles were deflected through large angles, suggesting a concentration of positive charge at the center of each atom. This concentration is called the *nucleus*. A sufficient number of electrons orbit the nucleus to keep the atom electrically neutral.

There were still some problems with this picture. It did not explain why electron orbits were stable. Classical electricity and magnetism tells us that an accelerating charge gives off radiation. An electron going in a circular orbit is accelerating, since its direction of motion is always changing. Therefore, as the electrons orbit, they should give off radiation, lose energy and spiral into the nucleus. This is obviously not happening. The second problem concerns the origin of spectral lines. There is nothing in the Rutherford model of the atom that allows for spectral lines.

The arrangement of spectral lines in a particular element is not random. For example, in 1885, *Johann Jakob Balmer*, a Swiss teacher, realized that there was a regularity in the wavelengths of the spectral lines of hydrogen. They obeyed a simple relationship which became known as the *Balmer formula*:

$$1/\lambda = R(1/2^2 - 1/n^2)$$

The constant R is called the *Rydberg constant*, and its value is given by $1/R = 91.17636$ nm. The quantity n is any integer greater than two. By setting n to 3, 4, . . . , we obtain the wavelengths for the visible hydrogen lines (also known as the *Balmer series*). Of course, this was just an empirical formula, with no theoretical justification.

Example 3.1 First Balmer line
Calculate the wavelength of the longest wavelength Balmer line. This line is known as the Balmer-alpha, or simply Hα.

SOLUTION
We let $n = 3$ in equation (3.1) to obtain

$$1/\lambda = R(1/2^2 - 1/3^2)$$

Substituting for R and inverting gives

$$\lambda = 656.47 \text{ nm}$$

This is the wavelength as measured in a vacuum. We generally refer to the wavelength in air, since that is how we measure it at a telescope. The wavelength in air is that in vacuum divided by the

index of refraction of air, 1.000 29, giving 656.28 nm. (It is interesting to note that when spectroscopists tabulate wavelengths, those longer than 200 nm are given as they would be in air, since that is how they usually will be measured. Radiation with wavelengths less than 200 nm doesn't penetrate through air, and its wavelengths are usually measured in a vacuum, so the vacuum values are tabulated.)

3.3.1 The Bohr atom
The next advancement was by the Danish physicist *Neils Bohr* who tried to understand hydrogen (the simplest atom), illustrated in Fig. 3.4. He postulated the existence of certain *stationary states*. If the electron is orbiting in one of these states, the atom is stable. Each of these states has a particular energy. We can let the energy of the nth state be E_n and the energy of the mth state be E_m. For definiteness, let $E_n > E_m$.

Under the right conditions, transitions between states can take place. If the electron is in the higher energy state, it can drop down to the lower energy state, as long as a photon is emitted with an energy equal to the energy difference between

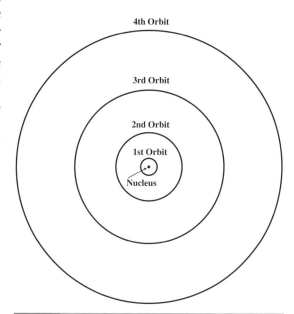

Fig 3.4. The Bohr atom. Electrons orbit the nucleus in allowed orbits. The relative sizes of the orbits are correct, but on this scale the nucleus should be much smaller.

the two states. If the frequency of the photon is ν, this means that

$$h\nu = E_n - E_m$$

If the electron is in the lower energy state, it can make a transition to the upper state if the atom absorbs a photon with exactly the right energy. This explanation incorporated Einstein's idea of photons.

Bohr pointed out that one could calculate the energies of the allowed states by assuming that the angular momentum J of the orbiting electrons is quantized in integer multiples of $h/2\pi$. The combination $h/2\pi$ appears so often we give it its own symbol, $\hbar$ (spoken as 'h-bar'). We apply this to a hydrogen atom, with an electron, with charge $-e$, orbiting a distance r from a nucleus with charge $+e$. We assume that the nucleus is much more massive than the electron, so we can ignore the small motion of the nucleus, since both the nucleus and electron orbit their common center of mass.

We first look at the kinetic energy

$$\text{KE} = (1/2)mv^2$$

The potential energy, relative to the potential energy being zero when the electron is infinitely far from the nucleus, is given by

$$\text{PE} = -e^2/r$$

(By writing the potential energy in this form, rather than with a factor of $1/4\pi\epsilon_0$, we are using cgs units. This means that charges are expressed in electrostatic units, esu, with the charge on the electron being 4.8×10^{-10} esu.) The total energy is the sum of kinetic and potential:

$$E = (1/2)mv^2 - e^2/r \qquad (3.2)$$

We can relate v and r by noting that the electrical force between the electron and the nucleus, e^2/r^2, must provide the acceleration to keep the electron in the circular orbit, v^2/r. This tells us that

$$mv^2/r = e^2/r^2$$

Multiplying both sides of the equation by r gives

$$mv^2 = e^2/r \qquad (3.3)$$

We put this into equation (3.2) to find the total energy in terms of r:

$$E = -(1/2)e^2/r \qquad (3.4)$$

The minus sign indicates that the total energy is negative. To see what this means, remember that we have defined the potential energy such that it is zero when the electron and proton are infinitely far apart. The electron and proton being far apart with no motion is the minimal condition for the electron being free of the proton. So, if the electron is barely free of the proton, the total energy would be zero. So, if the total energy is negative, as in equation (3.4), then we must add energy $[(1/2)e^2/r]$ if we want to bring it up to zero, which would free the electron. So the negative energy means that the electron is not free. In this case we say that the system is *bound*.

We still have to find the allowed values of r. The angular momentum is $J = mvr$. The quantization condition becomes

$$mvr = nh/2\pi$$

Solving for v gives

$$v = nh/2\pi\,mr$$

Squaring and multiplying by m gives

$$mv^2 = n^2h^2/4\pi^2\,mr^2$$

By equation (3.3), we have

$$e^2/r = n^2h^2/4\pi^2\,mr^2$$

We now solve for r, giving the radius of the nth orbit:

$$r_n = n^2h^2/4\pi^2me^2 \qquad (3.5)$$

Substituting into equation (3.4) gives the energy of the nth state:

$$E_n = -(1/2)e^4m(4\pi^2)/n^2h^2 \qquad (3.6)$$

Note that this has the $1/n^2$ dependence that we would expect from the Balmer formula.

One modification that we should make is to account for the motion of the nucleus (since it is not infinitely massive). We should replace the mass of the electron, m, in equations (3.5) and (3.6) by the *reduced mass* of the electron and proton. The reduced mass, m_r, is defined such that the motion of the electron, as viewed from the (moving) proton, is as if the proton were fixed and the electron's mass is reduced to m_r. An expression for m_r is (see Problem 3.2)

$$m_r = \frac{m_e m_p}{m_e + m_p}$$

$$= 0.9995\, m_e$$

where m_e and m_p are the masses of the electron and proton, respectively.

Example 3.2 Hydrogen atom energy

Compute the energy of the lowest (ground) energy level in a hydrogen atom. Also, find the radius of the orbit of the electron in that state.

SOLUTION

We use equation (3.6) with $n = 1$ to give

$$E_1 = \frac{-(1/2)(4.8 \times 10^{-10}\,\text{esu})^4(9.11 \times 10^{-28}\,\text{g})(4\pi^2)}{(1)^2(6.63 \times 10^{-27}\,\text{erg s})^2}$$

$$= -2.2 \times 10^{-11}\,\text{erg}$$

An erg is not a convenient unit to use to keep track of such small energies, so we convert to electron volts, eV (1 eV $= 1.6 \times 10^{-12}$ erg, is the energy acquired by an electron in being accelerated through a potential difference of 1 volt), giving

$$E_1 = -13.6\,\text{eV}$$

The radius is given by equation (3.5):

$$r_1 = \frac{(6.23 \times 10^{-27}\,\text{erg s})^2}{(9.11 \times 10^{-28}\,\text{g})(4.8 \times 10^{-10}\,\text{esu})^2(4\pi^2)}$$

$$= 5.25 \times 10^{-9}\,\text{cm}$$

$$= 0.0525\,\text{nm}$$

Note that if we take $n = \infty$ in equation (3.6) we get $E = 0$. However, $n = \infty$ corresponds to a free electron. Therefore, to move the electron far from the nucleus, we must add 13.6 eV. The energy that we must add to an atom to break it apart is called the *binding energy*. The energy goes to do work against the electrical attraction between the electron and the nucleus as you try to pull the electron away.

Now that we have evaluated E_1, we can rewrite equation (3.6) as

$$E_n = -13.6\,\text{eV}/n^2 \qquad (3.7a)$$

It is then very easy to calculate the energies of the other levels. For example, $E_2 = -3.4$ eV. Therefore the energy difference, $E_2 - E_1$ is equal to 10.2 eV. The energy levels are shown in Fig. 3.5. This diagram is a convenient graphical representation of energy levels, called an *energy level diagram*. In this diagram the levels are plotted as horizontal lines with vertical locations proportional to the energy. We can draw vertical arrows indicating possible transitions between the levels. The length of the arrow would then indicate the energy change associated with that transition. Note that the levels are closer together as one goes to higher values of n.

Since the zero of potential energy is arbitrarily defined, we sometimes choose to shift the energy scale by the binding energy (13.6 eV for hydrogen). This would make the energy of the ground state ($n = 1$) zero, and for the $n = 2$ state, +10.2 eV. To become a free electron, it

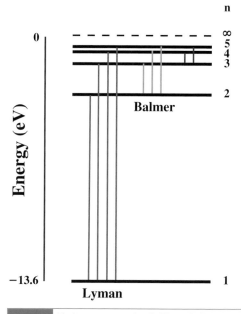

Fig 3.5. Hydrogen energy levels. The right hand column gives the principal quantum number, n. The energies are relative to the state in which the electron and proton are infinitely far apart, so the ground state energy is -13.6 eV. Transitions (which can be either emission or absorption) are grouped according to the lower level of the transition. For example, the Balmer series consists of emissions with the electrons ending in state $n = 2$, and absorptions starting in the $n = 2$ state.

must acquire an energy of $+13.6$ eV, or greater. The values of the *energy differences* between these states are unaffected by this shift in the zero point of the energy.

We can use equation (3.6) to derive the Balmer formula. First, we rewrite the equation as

$$E_n = -hcR/n^2 \qquad (3.7b)$$

where

$$R = (1/2)e^4 m(4\pi^2)/ch^3 \qquad (3.7c)$$

The energy of an emitted or absorbed photon must equal the energy difference between the two states:

$$E_n - E_m = h\nu$$
$$= hc/\lambda$$

Taking the energies from equation (3.7b) gives

$$1/\lambda = R(1/m^2 - 1/n^2) \qquad (3.8)$$

which looks very similar to the Balmer formula, except that the Balmer formula has a 2 instead of the m. This means that the Balmer series all have the second energy level as their lower level.

We can use equation (3.8) to divide the hydrogen spectrum into different series. A given series is characterized by having the same lower energy state. For example, the Balmer series consists of absorptions accompanying transitions from level 2 to any higher levels, and emissions accompanying transitions from higher levels down to level 2. The first Balmer transition (involving levels 2 and 3) has the smallest energy difference of the series. (Clearly the energy difference between levels 2 and 3 is less than the energy difference between levels 2 and 4, or between levels 2 and 5, and so on.) The Balmer series is important because the first few transitions fall in the visible part of the spectrum. The series with the lower energy level being level 1 is called the *Lyman series*. Even the lowest transition in the Lyman series is in the ultraviolet.

We have developed a labeling system for various transitions. First we give the chemical symbol for the element (e.g. H for hydrogen). Then we give the m for the lowest level that characterizes the series (1 for Lyman, 2 for Balmer, etc.). Finally, we give a Greek letter denoting the number of levels jumped. For example, if $n = m + 1$, we have

an alpha (α) transition; if $n = m + 2$, we have a beta (β) transition. The first Balmer line is then designated H2α. (Note that for the Balmer series of hydrogen only, we sometimes drop the 2 and just say Hα, Hβ, etc.)

3.3.2 Quantum mechanics

The Bohr model of the atom allowed physicists to understand the organization of energy levels. However, it was far from a complete theory. One shortcoming was that it did not explain why some spectral lines are stronger than others. More fundamentally, it was an *ad hoc* theory. Bohr had no explanation of why stationary states exist, or why angular momentum must be quantized in some particular way. These were just postulates. A much deeper understanding was needed.

An important step was made by *Louis de Broglie*, who proposed the revolutionary idea that if light could exhibit a wave–particle duality, then maybe all matter could. That is, an electron orbiting a nucleus has certain wavelike properties, and it is those properties that determine the states that are stable. One could think of the electron as having a certain wavelength. Stationary states could be those whose circumference contained an integral number of wavelengths, producing a pattern that reinforced during each orbit (like a standing wave). It was necessary to have expressions for the wavelength and frequency of a particle, and de Broglie noted that if the wavelength was taken as h/p (where p is the momentum of the particle) and the frequency as E/h, then the orbits allowed by the standing wave idea were the same as the orbits that Bohr found from his postulates (see Problem 3.8).

This is clearly a departure from our normal way of looking at matter around us, and we cannot go through all of the ramifications here. To this point, we have gone far enough to understand stellar spectra. The picture as presented by Bohr and de Broglie is quantum theory in its most naive form. It was realized that if particles behave, in some fashion, like waves then the description of particle motions (mechanics) must be changed from Newton's laws of motion to laws of motion involving waves. (Of course, in the limit of large objects, such as apples falling to Earth, these new laws of motion must reduce to

Newton's laws, because we know that Newton's laws work quite well for apples and planets.) Theories that describe the mechanics of waves are called *wave mechanics* or *quantum mechanics*. One such theory was presented in 1925 by the German physicist *Erwin Schrödinger*. In his theory the information about the motion of a particle is contained in a function, called a *wave function*. Schrodinger's interpretation of the wave function was that it is related to the *probability* of finding a particle in a particular place with a particular momentum. This replaced the absolute determinism of classical physics, with the statement that we can only predict where a particle is *likely* to be, but not *exactly* where it will be. However, we can predict the average positions and momenta of a large group of particles, and it is these average properties that we see (and measure) in our everyday world. Many physicists (including Einstein) were not comfortable with this probabilistic interpretation, but quantum theory has been very successful in predicting the outcome of a wide variety of experiments. We will pick up on some of the threads of the quantum revolution later in this book.

3.4 | Formation of spectral lines

Now that we have some idea of how atoms can emit or absorb radiation, we can return to stellar spectra. The first point to realize is that in a star we are not talking about the radiation from a single hydrogen atom, but from a large number of them. We see a strong Hα absorption line in stars because many photons are removed from the continuum by this process. It is clear, however, that having a lot of hydrogen does not assure us of a strong Hα absorption. In order for such absorption to take place, a significant number of atoms must be in level 2, ready to absorb a photon. If all the hydrogen is in level 1, you will not see the Balmer series, no matter how much hydrogen is present.

3.4.1 Excitation
In general, the strength of a particular transition (emission or absorption) will depend on the number of atoms in the initial state for that transi-

tion. The number of atoms per unit volume in a given state is called the *population* of that state. In this section we look at the factors that determine the populations of the various states. We refer to processes that can alter the populations as *excitation* processes. We have already seen one type of excitation process – the emission and absorption of photons. Electrons can jump to a higher level when a photon is absorbed or they can jump to a lower level when a photon is emitted.

Populations can also be changed by collisions with other atoms, as illustrated in Fig. 3.6. For example, atom 1 can be in state i. It then undergoes a collision with atom 2, and makes a transition to a higher state, j. In the process the kinetic energy of atom 2 is decreased by the difference between the energies of the two states in atom 1, $E_j - E_i$. The reverse process is also possible, with atom 2 gaining kinetic energy and atom 1 dropping from state j to state i.

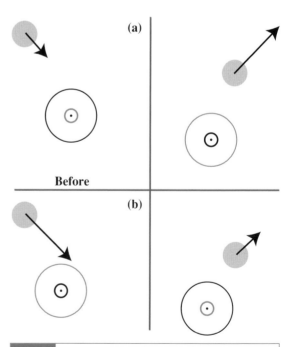

Fig 3.6. Collisional excitation. In each case, the left frame shows the atoms before the collision and the right frame shows them after. In each frame, the occupied level is indicated by a heavier line. (a) To a lower state. After the collision, atom 1 is in a lower state and atom 2 is moving faster. (b) To a higher state. After the collision atom, 1 is in a higher state and atom 2 is moving slower.

The collisional excitation rates will depend on the kinetic temperature of the gas. The higher the temperature the faster the atoms are moving. For atoms of kinetic temperature T_k the average kinetic energy per atom is $(3/2)kT_k$. As the temperature increases more energy is available for collisions. This makes higher energy states easier to reach. Also, since the particles are moving faster, they spend less time between collisions. There are more collisions per second.

When a gas is in thermodynamic equilibrium (which we discussed in the previous chapter), with a kinetic temperature T_k, the ratios of the level populations are given by a *Boltzmann distribution*. If we let n_i and n_j be the populations of levels i and j, respectively, their ratio is given by

$$\frac{n_j}{n_i} = \frac{g_j}{g_i} e^{-[(E_j - E_i)/kT_k]} \tag{3.9}$$

In this equation g_i and g_j are called *statistical weights*. They are needed because certain energy levels are actually groupings of sublevels that have the same energy. The statistical weight of a level is just a count of the number of sublevels in that level. Typically, g are small integers.

To help us understand the Boltzmann distribution, Fig. 3.7 shows how the ratio of populations for an atom with just two levels depends on temperature. When the temperature is zero, all the atoms are in the ground state, so the ratio is zero. As the temperature increases, the quantity in square brackets gets smaller, so the exponent becomes less negative, and the ratio increases. If we let T_k go to infinity the ratio of populations approaches the ratio of statistical weights. For a given temperature, increasing the energy separation between the two levels makes the exponent more negative, lowering the ratio. This makes sense, since the greater the energy separation, the harder it is to excite the atom to the higher level.

The Boltzmann distribution provides us with a convenient reference point, even for a system that is not in thermodynamic equilibrium. For any given population ratio n_j/n_i, we can always find some value of T to plug into equation (3.9) to make the equation correct. We call such a temperature the *excitation temperature*. When they are not in equilibrium, each pair of levels can have a different excitation temperature. In thermodynamic equilibrium all excitation temperatures are equal to each other and to the kinetic temperature.

3.4.2 Ionization

If we know the temperature in the atmosphere of a star, we can use the Boltzmann equation to predict how many atoms will be in each state, i, and predict the strengths of various spectral lines. However, there is still an additional effect that we have not taken into account – *ionization*. If the temperature is very high, some of the colliding particles will have kinetic energies greater than the ionization energy of the atom, so the electron will be torn away in the collision. Once a hydrogen atom is ionized, it can no longer participate in line emission or absorption.

When the gas is ionized, electrons and positive ions will sometimes collide and recombine. When the total rate of ionizations is equal to the total rate of recombinations, we say that the gas is in *ionization equilibrium*. If the gas is in thermal equilibrium and ionization equilibrium, then the *Saha equation* tells us the relative abundances of various ions. We let $n(X_r)$ and $n(X_{r+1})$ be the densities of the r and $r + 1$ ionization states, respectively, of element X. (For example, if $r = 0$, then we are comparing the neutral species and the first ionized state.) The ionization energy to go from r to $r + 1$ is E_{ion}. The electron density is n_e,

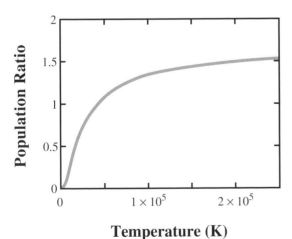

Fig 3.7. Level populations as a function of temperature for a two-level system. In this case we have put in energies and statistical weights (3, 5) for the $n = 2$ and $n = 3$ states of hydrogen (first Balmer transition).

and the kinetic temperature is T_k. Finally, g_r and g_{r+1} are the statistical weights of the ground electronic states of X_r and X_{r+1} (assuming that most of each species is in the ground electronic state). The Saha equation tells us that

$$\frac{n_e\, n(X_{r+1})}{n(X_r)} = \frac{2\, g_{r+1}}{g_r}\left(\frac{2\pi m_r\, kT_k}{h^2}\right)^{3/2} e^{-[E_i/kT_k]} \qquad (3.10)$$

The Saha equation has the same exponential energy dependence as the Boltzmann distribution. However, there is an additional factor of $T_k^{3/2}$. This comes from the fact that a free electron has more states available to it at higher T_k than at lower T_k. In addition there is a factor of n_e on the left. This is because a higher abundance of electrons leads to a higher rate of recombinations, driving down the fraction of atoms that are ionized. Just as we did with the excitation temperature in the Boltzmann equation, we can define an *ionization temperature* T_i, which makes the Saha equation correct, even if the gas is not in thermodynamic equilibrium.

In this equation n_e is the number of electrons from all sources, since any electron can combine with a hydrogen ion (for example) no matter where that electron came from (hydrogen, helium, etc.) In many situations, virtually all of the ions are hydrogen. That is because hydrogen is by far the most abundant element, and because the next most abundant element, helium, is very hard to ionize. In that case, the number of electrons is equal to the number of positive ions, n_+,

Atom	Singly ionized	Doubly ionized
H	13.6	–
He	24.6	54.4
C	11.3	24.4
N	14.5	29.6
O	13.6	35.1
Na	5.1	47.3
K	4.3	31.8
Ca	6.1	11.9
Fe	7.9	16.2

Table 3.1. | Ionization energies (eV).

so the left side of equation (3.10) simplifies to n_e^2/n_0, where n_0 is the number of neutrals. This extra factor of n_e makes even this simpler form of the Saha equation harder to solve for (n_e/n_0) than the Boltzmann equation is to solve for the ratio of level populations. In Fig. 3.8, we show the ratio n_e/n_0 as a function of temperature, for a value of n_e reasonable for stars like the Sun.

The ionization energies of some common atoms are given in Table 3.1. This table is useful in deciding which ions you are likely to encounter at various temperatures. In designating ionized atoms, there is a shorthand that has been adopted. The roman numeral I is used to designate the neutral species, II the singly ionized species, III the doubly ionized species, and so on. For example, neutral hydrogen is H(I), ionized hydrogen (H^+) is H(II), doubly ionized carbon is C(III).

3.4.3 Intensities of spectral lines

We are now in a position to discuss the intensities of various absorption lines in stars. We will take Hα as an example to see the combined effects of excitation and ionization. At low temperatures, essentially all of hydrogen is neutral, and most of it is in the ground state. Since little H will be in the second state, there will be few chances for Hα absorption. The Hα line will be weak.

As we go to moderate temperatures, most of the hydrogen is still neutral. However, more of the hydrogen is in excited states, meaning that a reasonable amount will be in level 2. Hα absorption is possible. As the temperature increases, the Hα absorption becomes stronger.

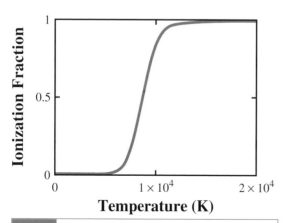

Fig 3.8. The ratio of electrons to the total number of hydrogen atoms (neutral plus ion), for an electron density appropriate to stars like the Sun.

At very high temperatures, the hydrogen becomes ionized. Since there is less neutral hydrogen, the Hα line becomes weaker. This explains why the Hα line is strongest in middle-temperature stars, and why the original scheme of classifying by hydrogen line strengths did not produce a sequence ordered in temperature.

We can apply a similar analysis to other elements. The details will differ because of different energy level structures and different ionization energies. It should be noted that, after hydrogen and helium, the abundances of the elements fall off drastically (see Appendix F for the abundances of the elements). In fact, astronomers often refer to hydrogen, helium and 'everything else'. The 'everything else' are collectively called *metals*, even though many of the elements don't fit our common definition of a metal.

We now look at the properties of different spectral types, in order of increasing temperature. Sample spectra are shown in Fig. 3.9, and the behaviors of a few spectral lines are shown in Fig. 3.10.

M Temperatures in M stars are below 3500 K, explaining their red color. The temperature is not high enough to produce strong Hα absorption, but some lines from neutral metals are seen. The stars are cool enough for simple molecules to form, and many lines are seen from molecules such as CN (cyanogen) and TiO (titanium oxide). If cool stars show strong CH lines, we designate them as C-type or 'carbon stars'. If any M

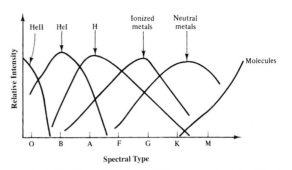

Fig 3.10. The relative strengths of spectral lines from important species as a function of spectral type. Each species shows the effects of excitation and ionization. For example, the increase in H line strengths from K to A stars occurs because the increasing temperature results in more hydrogen in the $n = 2$ (and higher) levels. However, the higher temperatures of the B and O stars ionize much of the hydrogen and the lines get much weaker.

star has strong ZrO (zirconium oxide) lines as opposed to TiO lines, we call it an S-type.

K Temperatures range from 3500 to 5000 K. There are many lines from neutral metals. The H lines are stronger than in M stars but most of the H is still in the ground state.

G Temperatures in the range 5000–6000 K. The Sun is a G2 star. The H lines are stronger than in K stars, as more atoms are in excited states. The temperature is high enough for metals with low ionization energies to be partially ionized. Two prominent lines are from Ca(II). When Fraunhofer studied the solar spectrum, he gave the strongest lines letter designations. These Ca(II) lines are the H and K lines in his sequence.

F Temperatures range from 6000 to 7500 K. The H lines are a little stronger than in G stars. The ionized metal lines are also stronger.

A Temperatures range from 7500 to 10 000 K. These stars are white–blue in color. They have the strongest H lines. Lines of ionized metals are still present.

B Temperatures are in the range 10 000–30 000 K, and the stars appear blue. The H lines are beginning to weaken because the temperatures are high enough to ionize a significant fraction of the hydrogen. The lines of neutral and singly ionized helium begin to appear. Otherwise there are relatively few lines in the spectrum.

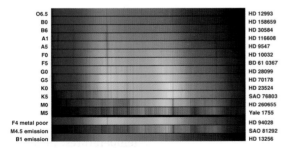

Fig 3.9. Samples of spectra from stars of different spectral types. The name of the star appears on the right of each spectrum, and the spectral type appears on the left. In each spectrum, the wavelength increases from left to right. Hotter stars are at the top. [NOAO/AURA/NSF]

O Temperatures range from 30 000 to over 60 000 K, and the stars appear blue. The earliest spectral types that have been seen are O3 stars and there are very few O3 and O4 stars. The hydrogen lines fall off very sharply because of the high rate of ionization. The lines of singly ionized helium are still present, but there are very few lines overall in the visible part of the spectrum. There are several lines in the ultraviolet.

Some stars have emission as well as absorption lines in their spectra. These stars are designated with an 'e' after the spectral class, for example, Oe, Be, Ae, etc. O stars with very broad emission lines are called *Wolf–Rayet stars*. These stars probably have circumstellar material that has been ejected from the star. (Wolf–Rayet stars are not the only stars with such outflowing material.)

3.5 | The Hertzsprung–Russell diagram

Even though we cannot study any one star (except for the Sun) in great detail, we can compensate somewhat by having a large number of stars to study. From statistical studies we learn about general trends. For example, if we find that brighter stars tend to be both hotter and larger, then any theory of stellar structure would have to explain that trend. Also, we think that any property that is common to many stars must be telling us about the laws of physics that are important in understanding the structure of stars.

One of the earliest statistical studies was carried out in 1910 independently by the Danish astronomer *Enjar Hertzsprung*, and the American astronomer *Henry Norris Russell*. They plotted the properties of stars on a diagram in which the horizontal axis is some measure of temperature (e.g. color or spectral type) and the vertical axis is some measure of luminosity. We call such a diagram a *Hertzsprung–Russell diagram,* or simply an *HR diagram*.

If a random group of stars is chosen, all at different distances, a comparison of apparent magnitudes is not very meaningful. The apparent magnitude must be corrected to give the absolute magnitude. However, if we find a group of stars all at the same distance, we can plot their apparent magnitudes, since the distance modulus would be the same for all the stars. For this purpose, we use clusters of stars.

An HR diagram for over 40 000 nearby stars is shown in Fig. 3.11(a). These stars were studied by the Hipparcos satellite, which was designed to measure trigonometric parallaxes, so distances to these stars are well known. So, apparent magnitudes can be converted into absolute magnitudes. This allows us to compare, on the same basis, the properties of stars that are not all in a cluster. The first thing we notice is that stars appear only in certain parts of the diagram. Arbitrary combinations of temperature and luminosity are not allowed. Remember, for a given temperature, the luminosity depends on the radius of the star, so the HR diagram is telling us that *arbitrary combinations of radius and temperature are not allowed*.

Most of the stars are found in a narrow band, called the *main sequence*. The significance of the main sequence is that most stars of the same temperature have essentially the same luminosity, and hence essentially the same size. This close relationship between size and temperature must be a result of the laws of physics as applied to stars. It gives us hope that we can understand stellar structure by applying the known laws. It also gives us a crucial test: any theory of stellar structure must predict the existence of the main sequence.

Not all stars appear on the main sequence. Some appear above the main sequence. This means that they are more luminous than main sequence stars of the same temperature. If two stars have the same temperature but one is more luminous, it must be larger than the other. Stars appearing above the main sequence are therefore larger than main sequence stars. We call these stars *giants*. By contrast, we call the main sequence stars *dwarfs*. We subdivide the giants into three groups: *subgiants, giants, supergiants*.

To keep track of the size of a star of a given spectral type, we append a *luminosity class* to the spectral type. The luminosity class is denoted by a roman numeral. Main sequence stars are luminosity class V. The Sun, for example, is a G2 V star. Subgiants are luminosity class IV, giants are luminosity class III. Luminosity class II stars are

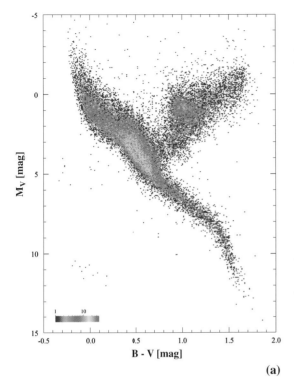

(a)

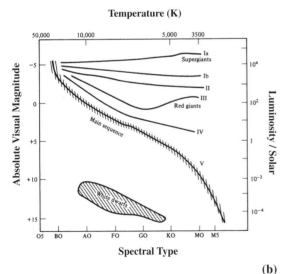

(b)

somewhere between giants and supergiants. Supergiants are luminosity class I. We further divide supergiants into Ia and Ib, with Ia being larger. When we look at the spectral lines from a star we can actually tell something about the size. Stars of different sizes will have different accelerations of gravity near their surface. The surface gravity affects the detailed appearance of certain spectral lines.

There are also stars that appear below the main sequence. These stars are typically 10 mag fainter than main sequence stars of the same temperature. They are clearly much smaller than main sequence stars. Since most of these are in the middle spectral types, and therefore appear white, we refer to them as *white dwarfs*. (Do not confuse dwarfs, which are main sequence stars, with white dwarfs, which are much smaller than ordinary dwarfs.)

Example 3.3 Size of white dwarfs

Suppose that some white dwarf has the same spectral type as the Sun, but has an absolute magnitude that is 10 mag fainter than the Sun. What is the ratio of the radius of the white dwarf, R_{wd}, to that of the Sun, $R_\odot$?

SOLUTION

The luminosity is proportional to the square of the radius, so

$$L_{wd}/L_\odot = (R_{wd}/R_\odot)^2$$

We use equation (2.2) to find the luminosity ratio for a 10 mag difference:

$$L_{wd}/L_\odot = 10^{(M_\odot - M_{wd})/2.5}$$

$$= 10^{-4}$$

Combining these two results to find the ratio of the radii yields

$$R_{wd}/R_\odot = (L_{wd}/L_\odot)^{1/2}$$

$$= (10^{-4})^{1/2}$$

$$= 10^{-2}$$

The radius of a white dwarf is 1% of the radius of the Sun!

For any cluster for which we plot an HR diagram, we only know the apparent magnitudes, not the absolute magnitudes. If we know the absolute magnitude for one spectral type, then we can find the distance modulus for stars of that spectral type in the cluster. The distance modulus is the same for all the stars in the cluster, so we can calibrate the whole HR diagram in terms of absolute magnitudes. To obtain a reliable calibration, we would like to carry it out for many stars. We have already seen that there is a growing group of nearby stars for which trigonometric parallax can give us a good distance measurement. In Chapter 13, we will see how we can improve on this sample by looking at the motions of clusters.

Once we know the absolute magnitude for a given spectral type, we have a very useful way of determining distances. For any given star, we measure m, the apparent magnitude. We take a spectrum of the star to determine its spectral type. From the spectral type we know the absolute magnitude, M. Since we know m and M, we know the distance modulus, $m - M$, and therefore the distance. This procedure is called spectro-

scopic parallax. The word 'spectroscopic' refers to the fact that we use the star's spectrum to determine its absolute magnitude. The word 'parallax' refers to the fact that this is a distance measurement (just as trigonometric parallax was a distance measurement using triangulation).

Example 3.4 Spectroscopic parallax
For a B0 star ($M = -3$), we observe an apparent magnitude $m = 10$. What is the distance to the star, d?

SOLUTION
The distance modulus is

$$m - M = 10 - (-3) = 13 \text{ mag}$$

We use equation (2.17) to find the distance:

$$5 \log_{10}(d/10 \text{ pc}) = 13 \text{ mag}$$

$$\log_{10}(d/10 \text{ pc}) = 2.6$$

Solving for d gives

$$d/10 \text{ pc} = 400$$

$$d = 4000 \text{ pc}$$

Chapter summary

In this chapter we looked at how spectral lines are formed, and how spectral lines can tell us about the physical conditions in the atmosphere of a star.

We saw that stars were originally classified into spectral types before the nature of the temperature sequence was understood.

We saw how an explanation of spectral lines, in general, requires an atomic theory in which the electrons can occupy only certain energy states. An atom can go from one state to another by emitting or absorbing a photon with the appropriate energy. We saw that a relatively simple theory could explain the spectrum of the hydrogen atom.

In a star, the strength of a spectral line depends on the abundance of the particular atom, and on the relative number in the appropriate ionization and orbital states. The populations of orbital states is described by the Boltzmann equation.

The distribution among ionization states is described by the Saha equation. In general, the higher the temperature, the higher the level of ionization and the more we find electrons in higher orbital states.

Finally, we saw what could be learned from a Hertzsprung–Russell diagram, in which the horizontal axis is some measure of temperature and the vertical axis is some measure of luminosity. Most of the points representing stars on an HR diagram fall along a narrow band, called the main sequence. This tells us that, for most stars, there is a simple relationship between size and temperature. Stars that do not lie along the main sequence are identified as being various classes of giants, for the brighter ones, and white dwarfs, for the fainter ones.

We saw how we can determine distances to a star using its apparent magnitude and a spectral type to deduce its absolute magnitude.

Questions

3.1. What is responsible for the linelike appearance of spectral lines?

3.2. Arrange the standard spectral sequence – O, B, A, F, G, K, M – in order of decreasing Hα strength.

3.3. What are (a) the strong points (b) the weak points of (i) Rutherford's atom and (ii) Bohr's atom?

3.4. What evidence supports the idea of photons?

3.5. What evidence supports the idea that electrons behave as waves?

*3.6. Consider a neutral carbon atom that has six electrons orbiting the nucleus. Suppose that five of the electrons are in their lowest states, but the sixth is in a very high state. Why might the energy levels for the outermost electron be similar to those for the single electron in hydrogen. (Hint: Think of what is exerting an electrical force on the outermost electron.)

3.7. (a) What do we mean when we say that a system is bound? (b) If you looked at a electron moving near a nucleus, how would you decide if the system is bound?

3.8. When we looked at the hydrogen atom, we said that if it is in the ground state, the binding energy is 13.6 eV. What happens to the binding energy if the electron is not in the ground state?

3.9. Explain how raising the temperature of the gas increases the rate of collisional excitations.

3.10. (a) Explain how the Hα absorption strength changes as we raise the temperature of a star. (b) Explain how the Lyman α absorption strength changes as we raise the temperature of a star.

3.11. Why do we not see helium absorption lines in stars like the Sun?

3.12. Explain the advantage of studying the HR diagram for a cluster, as opposed to a random group of stars.

3.13. What is the significance of the main sequence?

3.14. (a) How do we know that giants are larger than main sequence stars of the same temperature? (b) How do we know that white dwarfs are smaller than main sequence stars of the same temperature?

3.15. (a) Explain how the method of spectroscopic parallax works. (b) What are its advantages and disadvantages relative to trigonometric parallax?

Problems

3.1. Find the wavelengths of the H1α (Lyman-alpha), H1β and H2β transitions.

3.2. What is the wavelength of a photon that will barely ionize hydrogen in the ground state?

3.3. (a) How much energy is required to ionize hydrogen already in the $n = 2$ state? (b) At what temperature would the average kinetic energy of the particles in the gas equal that energy?

3.4. Show that if we add a constant to all of the energies in hydrogen, the energies of the various transitions are unaffected.

3.5. An electron in a hydrogen atom is in a high n state. It drops down one state at a time. What is the first transition to give a visible photon?

*3.6. What is the wavelength of the 2α transition in singly ionized helium? (Hint: The difference between this case and hydrogen is that the charge on the helium nucleus is twice that for hydrogen. Ignore the difference in reduced masses.)

3.7. Using the de Broglie wavelength, h/p, show that orbits whose angular momentum is quantized according to the Bohr quantization condition ($J = nh/2\pi$) correspond to orbits whose circumference is an integer number of wavelengths.

*3.8. Rederive equation (3.6) without making the assumption of an infinitely massive nucleus, and show that one obtains the same expression except for the reduced mass replacing the electron mass. (Hint: This problem is the electrical analog of the gravitational problem in binary stars, discussed in Chapter 5.)

3.9. What are the radii of the $n = 2$ and $n = 3$ orbits in hydrogen?

3.10. Consider only the two lowest levels in hydrogen with $g_1 = 2$, $g_2 = 6$. (a) Find the ratio of their populations at a temperature of 5000 K; (b) at a temperature of 10 000 K.

3.11. Consider only the three lowest levels in hydrogen with $g_1 = 2$, $g_2 = 6$, $g_3 = 10$. Find the three population ratios at a temperature of 5000 K.

3.12. If the populations of two levels of energies E_i and E_j and statistical weights g_i and $g_j (E_i < E_j)$ are found to be n_i and n_j, respectively, find an expression for the excitation temperature of this transition.

3.13. Assuming that all the level populations are given by equation (3.9), derive an expression for f_i, the fractional population in the ith level, defined as

$$f_i = \frac{n_i}{\sum\limits_{j=1}^{N} n_j}$$

where N is the highest populated level.

3.14. At what temperature will the average kinetic energy of the gas be equal to the hydrogen ionization energy?

3.15. For an atom whose populations are given by equation (3.9), (a) in what temperature limit is the ratio of the populations equal to the ratio of the statistical weights, and (b) in what temperature range would you expect the ratio of the populations to be greater than the ratio of the statistical weights?

*3.16. Assume that we are considering only the ionization of hydrogen, so that the electron density is equal to the positive ion density, and the Saha equation simplifies to $n_e^2/n_0 = F(T)$, where $F(T)$ is the right side of equation (3.10). Assuming that the total amount of hydrogen, $n_{TOT} = n_e + n_0$, is known and constant, find an expression for n_e^2/n_{TOT}, the fraction of hydrogen ionized, in terms of n_{TOT} and $F(T)$.

3.17. How much larger is an M0Ia star than an M0V star? (See Appendix E for stellar properties.)

3.18. For an A3 star, we measure an apparent magnitude $m = 12$. How far away is the star (assuming it is a main sequence star)? (See Appendix E for stellar properties.)

3.19. We observe a cluster, in the constellation Orion, whose distance is 500 pc. We find a star whose spectrum is that of an A0, but we cannot tell the luminosity class from the spectrum. (a) If the apparent magnitude of the star is +9, what is its luminosity class? (b) What if the apparent magnitude is +4?

Computer problems

3.1. Tabulate the electron–nucleus reduced mass for the nucleus being H, He, C, Fe.

3.2. Find all of the H$n\alpha$ transitions that fall in the visible part of the spectrum.

3.3. Consider only the three lowest levels in hydrogen with $g_1 = 2$, $g_2 = 6$, $g_3 = 10$. Plot the fraction of hydrogen in level 2, $[n_2/(n_1 + n_2 + n_3)]$ vs. T, for T covering the range from the coolest M to the hottest O stars discussed in this chapter.

3.4. Make additional columns to Table 3.1 showing, for each element in the table, the wavelength of a photon that would just (singly) ionize the atom, and the temperature of the gas for which the average kinetic energy is equal to the ionization energy.

3.5. Make a table showing, for the mid-range temperature of each spectral type, the wavelength at which the blackbody spectrum peaks.

3.6. If we are limited to $m = 6$ or brighter for making naked eye observations, make a table of the maximum distance we can see a star for the mid-range of each spectral type (O5, B5, etc) for main sequence stars.

3.7. For the mid-range temperature for each spectral type, draw a graph of log $B(\lambda, T)$ vs. λ for wavelengths ranging from IR to UV.

3.8. For the mid-range temperature for each spectral type, find the number of H ionizing photons emitted per second.

Chapter 4

Telescopes

The past decades have seen dramatic improvements in our observing capabilities. There have been improvements in our ability to detect visible radiation, and there have also been exciting extensions to other parts of the spectrum. These improved observing capabilities have had a major impact on astronomy and astrophysics. In this chapter we will first discuss the basic concepts behind optical observations. We will then discuss observations in other parts of the spectrum.

4.1 | What a telescope does

An optical telescope provides two important capabilities:

(1) It provides us with *light-gathering* power. This means that we can see fainter objects with a telescope than we can see with our naked eye.
(2) It provides us with *angular resolution*. This means that we can see greater detail with a telescope than without.

For ground-based optical telescopes, light-gathering power is usually the most important feature.

4.1.1 Light gathering
We can think of light from a star as a steady stream of photons striking the ground with a certain number of photons per unit area per second. If we look straight at a star, we will see only the photons that directly strike our eyes. If we can somehow collect photons over an area much larger than our eye, and concentrate them on the eye, then the eye will receive more photons per

second than the unaided eye. A telescope provides us with a large collecting area to intercept as much of the beam of incoming photons as possible, and then has the optics to focus those photons on the eye, or a camera, or onto some detector.

Example 4.1 Light-gathering power
Compare the light-gathering power of the naked eye, with a pupil diameter of 5 mm, to that of a 1 m diameter optical telescope.

SOLUTION
Let d_1 be the diameter of the pupil and d_2 be the diameter of the telescope. The collecting area is proportional to the square of the diameter. The ratio of areas is

$$\left(\frac{d_2}{d_1}\right)^2 = \left(\frac{1.0 \text{ m}}{5.0 \times 10^{-3} \text{ m}}\right)^2$$

$$= 4.0 \times 10^4$$

This is the ratio of luminosities that we can see with the naked eye and with the telescope. We can express this ratio as a magnitude difference

$$m_1 - m_2 = -2.5 \log_{10}(4.0 \times 10^4)$$

$$= -11.5 \text{ mag}$$

This means that the faintest objects we can see with the telescope are 11.5 mag fainter than the faintest objects we can see with the naked eye. If the naked eye can see down to 6 mag, the telescope-aided eye can see down to 17.5 mag. This illustrates the great improvement in light-gathering power with the telescope.

A major advantage of film or a photoelectric detector over the eye is its ability to collect light for a long time. In the eye, 'exposures' are fixed at about 1/20 s. With modern detectors, exposures of several hours are possible. Therefore, the limiting magnitude for direct visual observing is not as faint as for photography or photoelectric detectors.

4.1.2 Angular resolution

We now look at resolving power. Resolution is the ability to separate the images of stars that are close together. It also allows us to discern the details in an extended object.

One phenomenon that affects resolution is *diffraction*. Diffraction is the bending or spreading of waves when they strike a barrier or pass through an aperture. As they spread out, waves from different parts of the aperture or barrier interfere with one another, producing maxima and minima, as shown in Fig. 4.1. As the aperture size, relative to the wavelength, increases, there are more waves to interfere, so the pattern is less spread out. Most of the power is in the *central maximum*, whose angular width $\Delta\theta$ (in radians) is related to the wavelength of the wave λ and the diameter of the aperture, D, by

$$\Delta\theta \text{ (rad)} \cong \lambda/D \qquad (4.1a)$$

Diffraction results in the images of stars being smeared out by this angle. That means that if two stars are closer than $\Delta\theta$, their images will blend together. We consider the images of two stars to just be resolved when the maximum of one diffraction pattern falls on the first minimum of the other. This condition is called the Rayleigh criterion. While equation (4.1a) is an approximation good for all shapes of aperture, the actual size of the diffraction pattern depends on the shape of the aperture. You may remember that, for circular apertures, the resolution is given by

$$\Delta\theta \text{ (rad)} = (1.22)\,\lambda/D \qquad (4.1b)$$

Example 4.2 Angular resolution

Estimate the angular resolution of the eye for light of wavelength 550 nm.

SOLUTION

We use a diameter $D = 5$ mm for the pupil. We use equation (4.1a) to find the angular resolution in radians. We convert from radians to arc seconds to

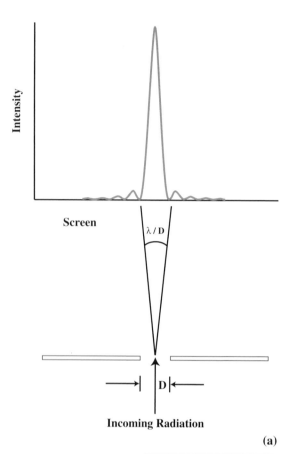

(a)

(b)

Fig 4.1. Diffraction. (a) A light ray enters from the bottom, and passes through a slit of length D. Diffraction spreads the beam out and it falls on a screen. The intensity as a function of position on the screen is shown at the top. Most of the energy is in the main peak, whose angular width is approximately λ/D (in radians). Smaller peaks occur at larger angles. (b) The effect in a real image. [ESO]

convert the result to a convenient unit (1 rad = 2.06×10^5 arc sec; see Example 2.3).

$$\Delta\theta(") \cong \frac{(2.06 \times 10^5)(5.5 \times 10^{-7}\,\mathrm{m})}{(5.0 \times 10^{-3}\,\mathrm{m})}$$

$$= 23 \text{ arc sec}$$

The eye's resolution is not quite this good, since the full diameter of the pupil is not generally used.

From equation (4.1a) we can see that we can improve the resolution if we use a larger aperture. A larger telescope will give us better resolution. A 10 cm diameter telescope (20 times the diameter of the pupil of the eye) will give an angular resolution of 1 arc sec. However, diffraction is not the only phenomenon that limits resolution. The Earth's atmosphere also distorts images.

When light passes through the atmosphere from above, it is passing through increasingly dense air. As the density of air increases, its index of refraction increases. Therefore, the light encounters an increasing index of refraction as it passes through the atmosphere. We can think of the atmosphere as having a large number of thin layers (as shown in Fig. 4.2) each with a slightly different index of refraction. As the light passes from one layer to the next it is bent slightly towards the vertical. The star appears to be higher above the horizon than it actually is.

This would not be a problem if the atmosphere were stable. However, variations on time scales shorter than a second cause changes in the index of refraction in some places. The image moves around. If we take a picture, we just see a blurred image. This effect is called *seeing* and usually limits resolution to a few arc seconds. We refer to the numerical value of the blurring as 'the seeing'. At a good observatory site, on a good night, the seeing might be as good as 1/3 arc sec or better. This corresponds to the diffraction limit of a 30 cm diameter telescope. Building a larger telescope does not help us past the seeing limitation on resolution, but it improves the light-gathering power. Hence our earlier statement that light gathering is the main purpose of large ground-based optical telescopes. We will also see later in this chapter that there are techniques for overcoming the effects of seeing to produce dif-

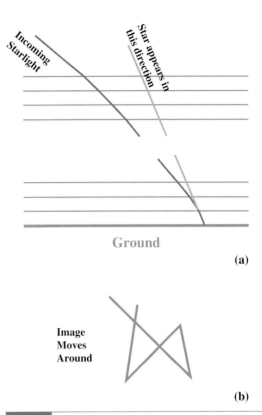

Fig 4.2. Seeing. (a) Bending of a light ray as it passes through the atmosphere. We can think of the atmosphere as being made up many thin layers, each with a slightly larger index of refraction as you get closer to the ground. The amount of bending is actually much less than in this picture. (b) Effect of changes in the amount of bending on the image of a star.

fraction limited images for telescopes with diameters from 1 to 2 m.

4.1.3 Image formation in a camera

To illustrate some basic points about the formation of images in optical systems, we look at the operation of a simple camera (Fig. 4.3). For astronomical situations, we are dealing with objects that are 'at infinity', so the light rays from a point on the sky are traveling parallel to each other. In the figure, we show bundles of rays coming from two different stars. The rays within each bundle arrive at an angle with each other equal to the angular separation of the stars on the sky.

For a camera with a lens of focal length f, the rays in each bundle are brought together at a

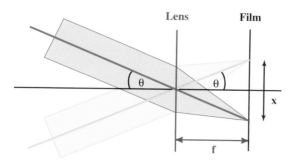

Fig 4.3. The optics of a camera. Bundles of rays from two distant points enter, making an angle θ with each other. The focal length of the lens is f.

distance f behind the lens. (The image is one focal length behind the lens when the object is at infinity. That is the definition of the focal length.) The images of all the stars in a field lie in a plane, called the *focal plane*. The images of two stars are at different points in the focal plane. We can locate the image of each star by following the *chief ray* of each bundle (the ray that passes through the center of the lens, undeflected) until it intersects the focal plane.

If stars have an angular separation θ on the sky, then, as viewed from the lens, the two images have an angular separation θ on the focal plane. This is simply the angle between the two chief rays. The camera provides no *angular magnification*. As viewed from the lens, the angular separation between the stars is the same as the angular separation of the images.

We can also find the linear separation x between the two images. From the right triangle in the figure, we see that

$$\tan (\theta/2) = x/2f$$

If θ is small, then $\tan (\theta/2)$ is approximately $\theta/2$, in radians. This gives us

$$\theta/2 = x/2f$$

Solving for x gives

$$x = f\theta \qquad (4.2)$$

This tells us that the linear size of the image is proportional to the focal length. To obtain a larger image, we use a longer focal length lens. (This is what we are doing when we put a telephoto lens in a camera.)

Apart from image size, we are also concerned with the brightness of the image. We can see that the amount of light entering the camera is proportional to the area of the lens. If D is the diameter of the lens, then its area is $\pi D^2/4$. This means that the image brightness is proportional to D^2. The brightness of the image also depends on the image size. The more the image is spread out, the less light reaches any small area of the film or detector. The linear image size is proportional to f, so the image area is proportional to f^2. This means that the image brightness is proportional to $1/f^2$.

Combining these two results, we find that the image brightness is proportional to $(D/f)^2$. The quantity f/D is called the *focal ratio*, so the brightness is proportional to $(1/\text{focal ratio})^2$. We adjust the focal ratio in a camera by changing f-stops. Since the focal length of the lens is fixed, we change the focal ratio by changing the diameter of a diaphragm that controls the fraction of the total lens diameter that is actually used. Each f-stop corresponds to a factor of $\sqrt{2}$ in the focal ratio, meaning that the image brightness changes by a factor of 2.

The discussions so far on image formation are really only appropriate for thin lenses, as well as optical systems where all of the angles are small. In real optical systems, rays that enter parallel do not all leave parallel to each other. Imperfections in the images formed by optical systems are called *aberrations*. Some of the aberrations are reduced by using the central part. The less we use the edges of the lens the better the images. That is why we might choose to use a diaphragm in a camera to block out the outer part of the lens. In a real optical system there is a tradeoff between image brightness and image quality.

One type of aberration is called *spherical aberration*. It arises from the fact that spherical curves are the easiest to grind on glass surfaces. These spherical shapes are close to the shapes required for proper image formation, but differ slightly, so the images are imperfect. Another type of aberration is called *astigmatism*. It occurs when the focal length depends on where around the lens the light strikes.

One aberration that occurs in lenses but not in mirrors is called *chromatic aberration* (Fig. 4.4).

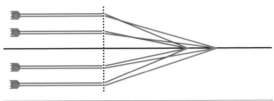

Fig 4.4. Chromatic aberration. The focal length is different for different wavelengths.

This happens because a material's index of refraction depends on the wavelength. The focal length of a lens is therefore different at different wavelengths. The images at different wavelengths are formed in different places. We can correct, somewhat, for chromatic aberration with a two-lens system, called an *achromat*. The two lenses are made of different materials, with different indices of refraction, and different variations in the indices of refraction with wavelength. An achromat only brings the images at two wavelengths together, but images for intermediate wavelengths are not far off.

Now that we have seen some of the basics of optical systems, we can look at astronomical telescopes. Most current astronomical research is done on *reflecting telescopes*. However, the basic ideas of image formation in reflecting and *refracting telescopes* are the same. It is easier to visualize refracting telescopes so we consider them first.

4.2 | Refracting telescopes

In a refracting telescope, the light first passes through a large lens, called the *objective lens*. The objective is the part that intercepts the incoming light, so it determines the light-gathering power of the telescope. The larger the objective is, the greater the light-gathering power. The light passing through the objective is concentrated on a second lens, called the *eyepiece*. The eyepiece is used to inspect the image formed by the objective. The image formed by the eyepiece is viewed either by the eye or by a camera. In practice, either the objective or the eyepiece may be a multiple lens, to correct for aberrations, but we will treat each as a single optical element. It is also possible to just have a film holder with no camera lens.

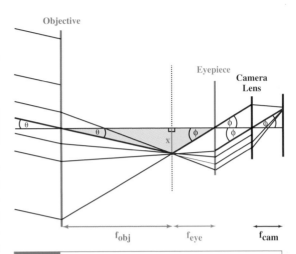

Fig 4.5. Image formation in a refracting telescope. Light from a star enters from the left, making an angle θ with the axis, and leaves the eyepiece making a larger angle φ with the axis. The focal lengths of the objective, eyepiece and camera lens are indicated. For each lens, the ray that goes through the center undeflected (the chief ray) is indicated as a heavier line. In a real telescope, the angles would be much smaller.

The basic arrangement of the refracting telescope is shown in Fig. 4.5. We follow the formation of the images of two stars, just as we did with the camera. Let's assume that the focal length of the objective is f_{obj}. Since the stars are at infinity, the objective forms their images this distance behind the objective. The eyepiece has a focal length f_{eye}. We place the eyepiece this distance behind the images formed by the objective. (This means that the objective and eyepiece are separated by a distance equal to the sum of their focal lengths.) Since the initial images of the stars are f_{eye} from the eyepiece, the eyepiece will focus the light at infinity. This means that all of the rays in a given bundle emerge from the eyepiece parallel to one another.

If you now look through the eyepiece, and focus your eyes at infinity (by relaxing the muscles around your eye), the rays in each bundle will be brought back together on your retina. Similarly, if you use a camera, you focus the camera at infinity, and the images of the stars will fall on the film. The need to focus your eyes at infinity means that the best way to look through the eyepiece is to relax both eyes and cover the unused eye, rather than squinting to close the unused eye.

Lets's go back to the two bundles of rays emerging from the eyepiece. Even though the rays within a given bundle are parallel to one another, the bundles make some angle with each other. If the two stars are an angle θ apart on the sky, then the two bundles will enter the objective, making this angle with each other. The bundles leave the eyepiece, making a larger angle φ with each other. We can find the angle φ by following the chief ray through the eyepiece. Note that the chief ray at the eyepiece is not the same ray that was the chief ray at the objective. However, all rays in a given bundle will emerge from the eyepiece parallel to the new chief ray.

From the two right triangles in the diagram with the common side x, we see that

$$\tan \theta = x/f_{obj}$$

$$\tan \varphi = x/f_{eye}$$

If the angles are small, we can replace the tangent of the angle with the value of the angle in radians. If we also eliminate x in the equations, we find

$$\varphi/\theta = f_{obj}/f_{eye} \qquad (4.3)$$

This means that we have an angular magnification equal to the ratio of the value of the focal lengths of the two optical elements.

In general, when we want to do work with good detail in the image, we use a telescope with a long focal length objective. Of course, we can change the angular magnification of a telescope by changing the eyepiece. There is a practical limit. You don't want to magnify the image so much that you blow up the blurring caused by atmospheric seeing.

There are some limitations in the use of a refracting telescope. One problem is the chromatic aberration of the objective. Also, the objective must be made from a piece of glass that is perfect throughout its volume, since the light must pass through it. This is harder as you try to make larger objectives. Larger objectives are also harder to support. The objective can only be supported at its edges, since light must pass through. Also, in many modern applications, we want to place instruments near the eyepiece. However, the telescope must be supported closer to the center of mass, which means far from the eyepiece. Any instrument hung at the eyepiece will exert a large torque about the mount, limiting the weight of the instrument. As a practical matter, the largest refractors, such as that shown in Fig. 4.6, have objectives with diameters of, at most, 1 m.

4.3 | Reflecting telescopes

Many of the difficulties with refracting telescopes are avoided with reflecting telescopes. In reflectors, the objective lens is replaced by an objective mirror. With a mirror, there is no problem of chromatic aberration, since light of all wavelengths is

Fig 4.6. The 1 m refracting telescope at the Yerkes Observatory. Note the long distance over which the observer must move to keep up with the eyepiece. [Yerkes Observatory photograph]

(a)

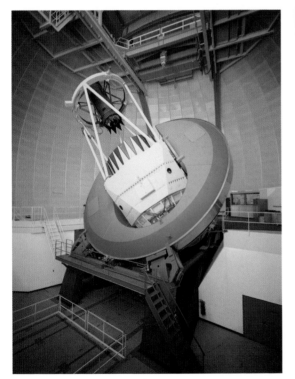

(b)

Fig 4.7. (a) The 5 m diameter Hale telescope on Mt Palomar (California). For almost four decades it was the largest useful telescope in the world. The caged part is the telescope. It has an equatorial mount. The solid piece in the foreground is part of the fork shaped support for the telescope. To track an object, as the Earth rotates, the whole fork rotates in the opposite direction. The prime focus cage is near the top of the telescope. (b) The 4 m diameter Mayall telescope of the National Optical Astronomy Observatory, on Kitt Peak, Arizona. There is an identical telescope located on Cerro Tololo, Chile. The Cassegrain focus is in a cage below the telescope. The observer does not stay in that cage for observing; that is done from a control room, where a television is used to keep track of where the telescope is pointing. [(a) Palomar Observatory/California Institute of Technology; (b) NOAO/AURA/NSF]

reflected at the same angle. The mirrors are made by shaping and then polishing a large piece of glass. While the polished surface has some reflective ability, it is not enough for a good mirror. Therefore a thin layer of reflecting material (usually aluminum) is deposited on the surface. The process of applying the reflective coating is called *aluminization*. This is best done under very clean conditions and under close to vacuum conditions, to avoid impurities on the surface. The chamber in which this is done is called an *aluminization chamber*. Typically the effects of dust and oxidation result in telescopes needing a new coating every few years. So, large telescopes generally have aluminizing chambers near the telescope.

Since the light doesn't pass through the glass, the requirements are for a good surface, not a good volume. Moreover, the glass can be supported from behind. It is therefore possible to make reflectors larger than refractors. For many years the largest reflector was the 5 m (200 inch) diameter Hale telescope on Palomar Mountain (Fig. 4.7a).

One advantage of the wave nature of electromagnetic radiation is that the radiation is essentially unaffected by objects much smaller than the wavelength. When electromagnetic waves reflect off a metal surface, they do it by inducing an oscillating current in the surface. This oscillating current then produces the reflected wave. If the surface is much smaller than the wavelength, there will not be enough room to produce a reflected wave at this wavelength. This means that to have good image formation, the surface of the mirror must be perfect to within approximately $\lambda/20$, where λ is the wavelength of the light being observed. For example, if you are observing with a wavelength of 500 nm, the surface must be accurate to within 25 nm. (This is about 250 atoms.)

Various shapes are possible for the mirror. It turns out that spherical ones are the easiest to grind. You may remember that a parabola focuses to a single point all rays coming in parallel to the axis. This means that a *paraboloid*, where any cross section of the mirror will be a parabola, is a useful shape. Paraboloids are generally easy to grind, if you start with a spherical shape and then make a slight adjustment (taking a little glass off the center). Current grinding technologies (discussed below) allow customized shaping of the mirror to optimize for various applications (e.g. better imaging over a wide field).

We now look at what happens to the image formed by the objective. Replacing the lens with a mirror doesn't change any of the basic ideas of image formation. There is, however, a problem caused by the reflection of the light back along the direction from which it came. To examine the image, the eyepiece (and observer) must be placed between the stars and the mirror, blocking some of the incoming light. If an eyepiece is put at this location, we call the arrangement a *prime focus*. The advantage of the prime focus is that no more mirrors are required, so light is not lost (or images distorted) in additional reflections. It provides for a 'fast' system (small focal ratio) with a large field of view. However, there is some blockage of the objective. If the telescope is very large, this blockage is a small fraction of the total collecting area of the objective.

Example 4.3 Blockage in prime focus
Consider a 5.0 m diameter telescope, with a 1.0 m diameter prime focus cage. What fraction of the incoming light is blocked by the cage?

SOLUTION
The ratio of the areas will be the square of the ratio of the diameters. The fraction of the mirror blocked is therefore

$$\text{fraction blocked} = (1.0 \text{ m}/5.0 \text{ m})^2$$

$$= 0.04$$

This means that only 4% of the incoming light is blocked. If we make the telescope smaller, but keep the cage the same size, the blockage worsens. Clearly, prime focus arrangements are only suitable in larger telescopes.

This problem was recognized by Newton, who devised a mirror arrangement, called the *Newtonian focus*, in which a flat diagonal mirror is used to direct the image formed by the objective to the side. This is shown in Fig. 4.8(a). The eyepiece is then mounted on the side of the tube. There is still some blockage but it can be kept small even for small telescopes. For a larger telescope, the Newtonian arrangement is difficult to use, since the eyepiece is at the top end of the telescope. Also, the eyepiece is farther from the mount's point of support, and equipment placed at the focus exerts a large torque about the support.

An alternative solution is called the *Cassegrain focus*, shown in Fig. 4.8(b). The prime focus cage is replaced with a mirror that directs the rays back through a hole in the center of the primary mirror. Little light is lost by removing the center of the mirror, since it would be blocked by the prime focus cage or the secondary mirror. The secondary mirror in a Cassegrain arrangement is diverging (convex), so the telescope seems to have a longer focal length than the objective. Since the eyepiece is just behind the primary mirror, it is a convenient arrangement. Also, if you want to place a lot of equipment at the eyepiece position, this is not too far from the point of support of the telescope.

Sometimes an astronomer will want equipment that cannot conveniently be mounted on a telescope. It might be too large or it might require

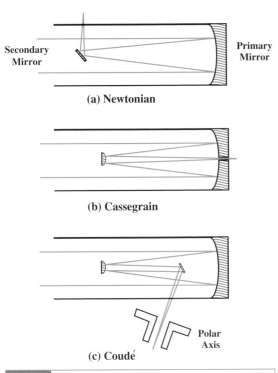

(a) Newtonian

(b) Cassegrain

(c) Coudé

Fig 4.8. Focal arrangements in (a) Newtonian, (b) Cassegrain and (c) coudé telescopes. In each case the light enters the telescope from the left.

a room in which the temperature can be kept constant. It may also be necessary to have no mechanical flexure of the instrument that moving it would cause. For this purpose, some telescopes have *coudé* focal arrangements (Fig. 4.8c). (The term, pronounced *coo-DAY*, comes from the French word for elbow, since the light beam is bent many times.) A series of mirrors is used to direct the image into a laboratory under the telescope mount. One disadvantage of this arrangement is the large number of mirrors that must be used. No mirror is perfectly reflective, and a little light is lost at each reflection (see Problem 4.10).

A general problem with any of these arrangements is that they all involve some blockage of the objective. In addition to reducing the light striking the objective, the blocking element must also be supported. Starlight passing by the element and its supports is diffracted, creating unusual stellar images (as shown in Fig. 4.9).

Some telescopes follow the basic layout of the Cassegrain system, but have some differences in their optics to optimize them for a certain type of

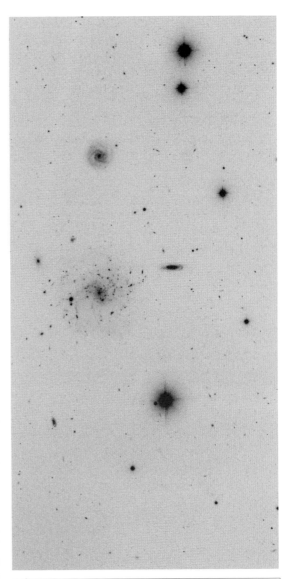

Fig 4.9. Stars act as true point sources, and their images have a diffraction pattern resulting from the supports for the secondary mirror. The pattern is evident as a cross on the brightest stars. [NOAO/AURA/NSF/Co.WIYN Consortium].

observation. Often the goal is to provide a large field of view that is relatively free of aberrations. For example, a *Schmidt camera* incorporates a glass plate shaped to provide corrections for some aberrations. This plate is placed at the front end of the telescope and the light passes through it before striking the primary mirror. Schmidt cameras are very good for wide field photography. Many newer telescopes are of the *Ritchey–Cretien* design (named after the two telescope designers who came up with the idea in 1910), which incorporates hyperbolic mirrors as an alternative way for correcting for aberrations.

While the 5 m telescope was the largest for many years, there have been a number of breakthroughs in telescope design and fabrication in the past decade, and we have seen a progression of larger and more sensitive telescopes. For example, the old large telescopes are all equatorially mounted. This means that they keep up with the Earth's rotation by rotating at a constant rate about an axis parallel to the Earth's rotation axis, the polar axis. This is convenient, but requires a large counterweight (Fig 4.6) or a fork to support the telescope on both sides of the polar axis (Fig 4.7a). The alignment of the polar axis is not perfect, and the motion of the telescope is not smooth. It is therefore necessary to make small corrections to the position of the telescope. This process, called

guiding, is done with the aid of a small auxiliary telescope and a control to adjust the position of the telescope about two axes to keep the object of interest in the center of your field. With the advent of computers to control telescopes in real time (and television systems to fine tune the guiding), it is now easier to use alt-azimuth mounts, which move in azimuth and elevation, and are light and symmetric about the local vertical.

There has been a growing realization that thermal currents in and just above the dome can create bad seeing. Some telescopes built in the 1970s had designs that tried to reduce these effects by using massive mirrors and large domes, to assure that they change temperature slowly. Newer designs have mirrors that are very light with good airflow, and minimal domes, so that the systems quickly equilbrate with the outside conditions when the dome is opened. The New Technology Telescope (Fig. 4.10) of the European

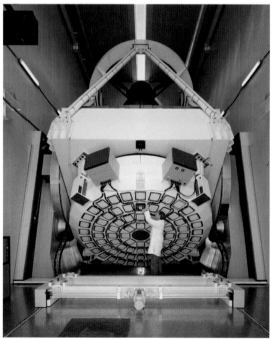 

(a) (b)

Fig 4.10. The 3.5 m New Technology Telescope of the European Southern Observatory, located at Cerro Paranal. It is mounted in azimuth and elevation, and must move in two axes to track a source. As the telescope tilts to different elevations, the shape of the mirror is adjusted with a grid of motors mounted on the back, visible in their casings in (a). The telescope is placed in a small dome (b) that allows for quick equilibration with the outside air, reducing currents within the dome that produce bad seeing. [ESO]

Southern Observatory (Chile) was one of the first to utilize that design. It is now becoming standard for new large telescopes.

One potential problem with large telescopes is that, as they tilt at different angles, gravity acts at different angles relative to the surface of the telescope. This causes the surface to deform as the tilt is changed. To get around this problem, some newer telescopes have a grid of remotely controlled motors on the back of the telescope. These motors turn screws that adjust the shape of the surface, from the back, in a preprogrammed way. The NTT was also one of the first to utilize this concept. An even more ambitious idea is to overcome some of the effects of bad seeing by using real time signals from a bright star to distort the third mirror in a coudé arrangement. This reshapes the wavefronts, compensating for the distortions induced by seeing. These processes are called active optics and adaptive optics. Together they are producing diffraction limited images in telescopes up to 2 meters in diameter.

Recently a group at the University of Arizona has developed a technique for making high quality mirrors with diameters as large as 8 meters (Fig. 4.11). It involves heating the glass and then spinning the glass while it cools. The surface of the spinning molten glass takes on the shape of a paraboloid. The mirror is cast so that most of the glass on the back side is missing, leaving a honeycomb pattern. This means that the mirror can be lighter than ones made using conventional designs. Also, the honeycomb allows air to flow through the back of the mirror meaning that the mirror can quickly reach the temperature of the outside. As we just saw, this cuts down on air currents in the telescope, a major source of bad seeing.

Even with the technology to build 8 m mirrors, astronomers need even larger telescopes. A different approach was pioneered by the Multiple Mirror Telescope (MMT), in Arizona. Instead of one large mirror the telescope had six moderate sized mirrors. The images from all six mirrors are brought together to produce an image that is six times as bright as the image from one mirror. While the multiple mirror approach sounds like an obvious idea, a number of technical obstacles had to be overcome before it could work. Among these are issues of aligning the mirrors, and combining the images properly. In the last few years a number of multiple mirror telescopes have been developed with larger and larger collecting areas.

The first of the newer generation multiple mirror telescopes is the Very Large Telescope (VLT) operated by ESO on Cerro Paranal (2635 m) in

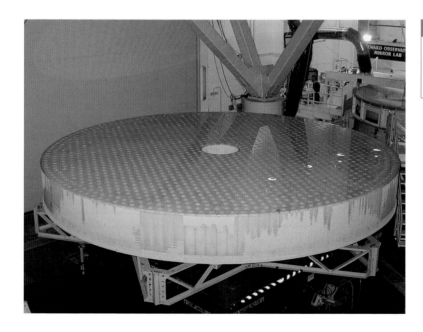

Fig 4.11. Polishing an 8 m mirror in the University of Arizona Mirror Laboratory. [Steward Observatory Mirror Lab.]

Chile (Fig. 4.12a, b). It has four telescopes, each with an 8.2 m reflector. Eventually the light from all four telescopes will be combined into a single beam. Still under construction is the Large Binocular Telescope, on Mt Graham in Arizona. It uses two 8.4 m mirrors made in the Arizona Mirror Laboratory (Fig. 4.11). It will have the resolution of a single telescope with 23 m diameter,

in one direction. That direction changes as the Earth rotates, just as for radio interferometers, discussed in Section 4.8.

Another approach to large collecting areas is a variation on the multiple mirror approach. The mirror is broken into a number of smaller segments, all on the same mount. This segmented mirror approach allows all of the mirrors to be pointed collectively, with fine tuning of their positions as the telescope tilts. Located on Mauna Kea (Fig. 4.12c, d) are the two Keck telescopes, each 9.8 m, which utilize this design. The Hobby-Eberly Telescope at McDonald Observatory, also uses this design (Fig. 4.12e).

(a)

(b)

(c)

Fig 4.12. (a) One element of the Very Large Telescope (VLT) on Cerro Paranal in Chile, built by ESO. (b) Exterior view of the four domes. The telesopes are named after the Mapuche (a pre-Columbian tribe in northern Chile) words for Sun, Moon, Southern Cross and Sirius. (c) The Keck telescope, located on Mauna Kea in Hawaii. Its mirror consists of a number of individually controlled segments. It is operated by the California Association for Research in Astronomy, which is a partnership among the University of California, the California Institute of Technology and NASA.

(d)

(e)

Fig 4.12. *(Continued)* (d) An outside view of Keck. (e) The Hobby-Eberly telescope at McDonald Observatory, also a multisegment telescope. It is operated by the University of Texas, Austin, and Pennsylvania State University. [(a), (b), ESO; (c), (d) W. M. Keck Observatory; (e) McDonald Observatory]

4.4 | Observatories

4.4.1 Ground-based observing

In the past, the convenient location of observatories was considered important. Observatories were built near universities that had astronomers, and those astronomers used whatever clear nights were available. Today, the considerable investment in large telescopes and sophisticated equipment requires more regular utilization of the facilities. Moreover, high quality telescopes are now built at the sites that best allow them to take advantage of their capabilities. Observatories are now built only after there has been an extensive investigation of the quality of the site.

Instruments have become more expensive; in the 1960s and 1970s there was a trend away from privately financed observatories to publically financed *national observatories*. National observatories are available to any qualified astronomer. An astronomer who has a project will be required to write a proposal, explaining the scientific justification and the details of the observations. Generally, there is not enough observing time for all of the submitted proposals, and a panel of astronomers decides which projects are to be done. More recently, with developments to cut the cost of telescopes, there has been a trend back to private observatories. Many of these are cooperative efforts by, typically, two to four universities with some public support. Keck is an example of such an effort.

The selection of an observatory site depends on a number of considerations. Obviously, good weather is important. However, clear weather is not enough. The air should be dry, since water vapor can attenuate signals. This suggests a desert. Also, the higher you go in altitude, the less air you have to look through. An altitude of 3 km (10 000 ft) puts you above a significant amount of atmospheric water vapor. This suggests a mountain in the desert. Even with a mountain in the desert, good seeing is not guaranteed. Seeing often varies with local conditions, depending on air flow and terrain. Before an optical observa-tory is built, seeing tests are done, with test observations being done over the course of a number of years.

An additional consideration is *light pollution*. Light from nearby cities is reflected up into the sky, making the sky appear to glow. The brighter this glow, the harder it is to see faint astronomical objects. Astronomers have found that certain lights are better than others. For example, low pressure sodium vapor lights, which have a yellow appearance, give off most of their light in a narrow wavelength range, and this range can be filtered out at the telescope. For any light, a hood

(a)

(b)

(c)

(d)

Fig 4.13. Observatories. (a) Kitt Peak National Observatory (operated by NOAO), southwest of Tucson, Arizona. Notice the large number of telescopes. The 4 m telescope is in the background. (b) Mauna Kea, on the island of Hawaii. At 4.3 km (14 000 ft), its summit is one of the best ground-based astronomical sites. (c) Cerro Tololo Interamerican Observatory (operated by NOAO) in Chile. The largest dome is a twin to the 4 m telescope on Kitt Peak. (d) The European Southern Observatory, located on La Silla in Chile, about 100 km from Cerro Tololo. [(a), (c) NOAO/AURA/NSF; (b) Richard Wainscoat, Institute of Astronomy, University of Hawaii; (d) ESO]

that reflects light back to the ground rather than letting it into the sky is very helpful. Such hoods also essentially double the brightness of the light on the ground.

Once a good site is found, it is likely that many telescopes will be built there. A good example is *Kitt Peak in Arizona*, operated by the *National Optical Astronomy Observatory (NOAO)*, which is shown in Fig. 4.13(a). This observatory has a number of different-sized telescopes, the largest being the 4 m Mayall telescope (shown in Fig. 4.7a). To make maximum use of the site, there are even telescopes on Kitt Peak operated by individual universities or groups of universities, not directly affiliated with NOAO.

Surprisingly, one of the best observing sites is in the middle of the Pacific Ocean. It is on the island of Hawaii, at an elevation of 4.3 km (14 000 ft) on a dormant volcano, Mauna Kea (Fig. 4.13b). The island often has clouds, but they are generally below the altitude of the observatory, and the air above the clouds is very dry. However, the lack of oxygen at this altitude makes work very difficult. Many astronomers report headaches and other discomforts. Clear thinking is also difficult, and there are many stories about simple mistakes made by experienced observers. For that reason, observing is conducted remotely, typically with the observer at sea level.

The development of observatory sites in the Chilean Andes has had a major impact on astronomy since the 1990s. First, it is important to have telescopes in the southern hemisphere, since there are large parts of the sky that cannot be seen from the northern hemisphere. The northern part of the Andes runs next to the Atacama Desert, which is dry even as deserts go. (There are places in the Atacama Desert, some not too far from the Pacific, where there has been no recorded rain in over a century.) There is precipitation in the mountains, as is evidenced by the snowy peaks, but a typical site in the Andes has half the amount of water vapor overhead of a comparable (in latitude and elevation) site in the US. Three major observatories have been developed in the Andes. One is the *Cerro Tololo Interamerican Observatory (CTIO)* (Fig. 4.13c), which is operated by NOAO (in cooperation with the University of Chile). Another is the *European Southern Observatory (ESO)*. ESO operates under a treaty among member European countries. Its primary location in Chile is on La Silla (Fig. 4.13d), which is about 100 km from Cerro Tololo. La Silla is the site of the NTT (Fig. 4.7a). ESO has recently gained another site, Cerro Paranal, further into the desert. It is the site of the VLT (Fig. 4.12a). The third is *Las Campanis*, which is near La Silla. All of these Chilean sites are quite far from major population centers so that light pollution is virtually non-existent.

The availability of spectacular sites in Chile has driven astronomers to make the best use of those sites, by getting the best possible seeing. As we have said, an important part of this is in the site selection. However, astronomers have long known that, at good sites, where the seeing is about 1 arc sec on a good night, about half of that comes from air in and directly above the telescope. Turbulence, caused by the ground, dome and telescope are important contributions to seeing. As we mentioned above, new telescope and dome designs are improving seeing. For example, seeing at the NTT is frequently better than 1 arc sec and, on really good nights, is better than 0.5 arc sec.

The lastest NOAO push to take advantage of excellent sites in the northern and southern hemispheres is *Project Gemini* (Fig. 4.14). Both telescopes are 8.1 m in diameter. The northern telescope is on Mauna Kea, and started operation in 1999. The southern telescope is on Cerro Pachon (2715 m) in Chile.

There is one other place that has recently been developed for astronomy. This is Antarctica. It is more than 2000 m above sea level, so it is at a good altitude. The air is so cold that it is very dry. In fact, once you are more than 150 km from the coast, you lose the ocean as a source of water in the air, and there is very little precipitation. The snow that you see far from the coast is blown there. This brings up one of the major problems, wind. Telescopes would have to be put in protective domes. This is reasonable for the infrared and millimeter parts of the spectrum. There is also an international science station, which is supported during the summer, so there is logistical support. Astronomers are investigating various sites near the South Pole.

(a)

(b)

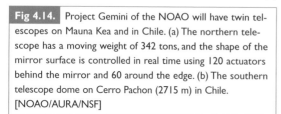

Fig 4.14. Project Gemini of the NOAO will have twin telescopes on Mauna Kea and in Chile. (a) The northern telescope has a moving weight of 342 tons, and the shape of the mirror surface is controlled in real time using 120 actuators behind the mirror and 60 around the edge. (b) The southern telescope dome on Cerro Pachon (2715 m) in Chile. [NOAO/AURA/NSF]

4.4.2 Observations from space

One of the major advances in observational astronomy has been the ability to place telescopes in space. This is particularly important for observing in parts of the spectrum that don't penetrate the Earth's atmosphere. However, a telescope in space can even be important in the visible part of the spectrum. It allows us to make observations free of the blurring caused by atmospheric seeing conditions.

Example 4.4 Diffraction-limited optical telescope
What is the resolution of a 1 m diameter telescope in space for observations at a wavelength of 550 nm?

SOLUTION
We find the diffraction limit from equation (4.1a):

$$\Delta\theta = \frac{(2.06 \times 10^5)(5.5 \times 10^{-7}\,\text{m})}{(1\,\text{m})}$$

$$= 1.1 \times 10^{-1}\,\text{arc sec}$$

A 1 m diameter telescope on the ground will never realize this resolution because of the seeing limitations (typically worse than 1 arc sec and sometimes as good as 0.5 arc sec). By putting a 1 m telescope in space we can realize a factor of 5 improvement over the best ground-based conditions. With a 2 m telescope, we would have a factor of 10 improvement.

This is the reason for the development, by the *National Aeronautics and Space Administration (NASA)* and the *European Space Agency (ESA)* of the *Hubble Space Telescope (HST)*, launched in 1990. HST, shown in Fig. 4.15, has a 2.4 m diameter mirror providing an angular resolution of about 0.05 arc sec. The telescope is equipped with a full complement of instruments so that it can carry out a full range of astronomical observations: imaging, photometry,

(a)

(b)

Fig 4.15. Views of the Hubble Space Telescope (HST). (a) HST being deployed. (b) After deployment on service mission. [STScI/NASA]

tific support comes from the *Space Telescope Science Institute (STSCI)*, in Baltimore, MD. Observers can view data at computer work stations at their home institutions.

Shortly after HST was launched, astronomers discovered a serious flaw in the optics, which degraded the images. An error in fabrication had produced a severe spherical aberration. This resulted in a server degradation in the image

(a)

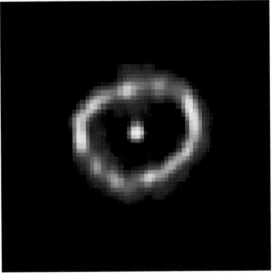

(b)

Fig 4.16. Images of a shell around a star, taken by HST (a) before and (b) after servicing. [STScI/NASA]

spectroscopy. It is an international facility, with time available on the basis of proposals, just as with ground-based national observatories. The telescope is controlled from NASA's Goddard Spaceflight Center in Greenbelt, MD. The scien-

quality, and in the sensitivity. Astronomers were excited about the successful completion of a servicing mission that compensated for the error. We will come back to HST results throughout this book, but in Figs. 4.16 and 4.17, we show images taken with HST before and after the servicing. In 2002 there was a scheduled servicing mission in which many of the instruments were upgraded.

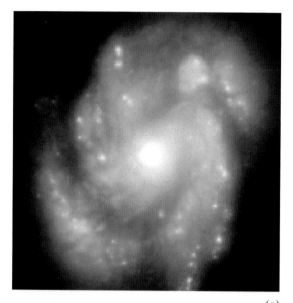

(a)

(b)

Fig 4.17. Images of the spiral galaxy M100 from HST (a) before and (b) after servicing. [STScI/NASA]

Since the servicing, HST has been so successful that astronomers are now planning a successor, the Next Generation Space Telescope (NGST).

4.5 | Data handling

In the previous sections we concentrated on bringing as many photons to the eyepiece as possible. Now we will look at what we do with these photons once they reach the eyepiece. We consider three different types of observations:

(1) **Imaging**. This is probably the most familiar type of observation. The goal of these observations is to obtain a picture of some part of the sky.
(2) **Photometry**. The name implies the measurement of light. The goal of the observations is to measure the brightness of some object. This may include measuring the brightnesses through certain filters to measure colors. It may also include measuring time variations in brightness.
(3) **Spectroscopy**. The goal of these observations is to obtain a spectrum of some object, generally with sufficient detail to allow the study of spectral lines.

4.5.1 Detection

Whatever the type of observation, the data must be recorded in some way. In the past, the most common way was to use a photographic plate. These plates contain an emulsion with light sensitive grains. Each grain serves as a little detector of radiation, or picture element (*pixel*). One advantage of photographic plates is that there are many pixels. We say that the plate has a *panoramic* quality. This means that we can simultaneously record many parts of the image. There are some disadvantages to photographs. One is that a very small fraction of the photons that strike the plate are actually detected. We call the fraction of photons that are detected the *quantum efficiency* of the detector. For most emulsions, this efficiency is only a few percent.

A much higher efficiency can be obtained with *photoelectric* devices. A photon strikes a surface, causing an electron to be ejected. This is the photoelectric effect that we discussed in Chapter 3. The electron is accelerated towards another surface, where more electrons are ejected. The process is repeated many times, and eventually a sufficient number of electrons are moving for a current to be detected. In the past, these *photomultiplier* devices allowed only a single detection element in the focal plane. That is, there is one pixel. So, they provided higher efficiency, but at the cost of the panoramic quality.

One problem with any detector is that it produces some background level. This is sometimes called the *dark current*, because it is present even when no light is shining on the detector. This generally results from thermal emission from the detector. For most astronomical observations, this background is much stronger than the signal you are trying to detect. The background has two deleterious effects. The first is that it must be subtracted from any measurement, to just give the value of the astronomical signal. The second is that it produces a random fluctuation in the measurement. The stronger the background, the higher the fluctuation. This fluctuation produces an uncertainty (statistical error) in any result.

To see how this effect works, it is easiest to think in terms of numbers of counts in a given measurement. For a photomultiplier, the number of counts in the signal is the number of photons striking the detector, multiplied by the quantum efficiency. For the background, we can think of the number of photons that would be equivalent to the emission from the background. If the background is thermal emission, that number would be $kT/h\nu$, where T is the temperature of your background emission and ν is the frequency at which you are measuring. From this, you see that we can reduce the background by cooling the detector.

The effect of this fluctuation is to cause a scatter in the results of counting experiments. If you want to determine the average rate at which you are counting photons from a given source, you might measure for some time, say ten seconds. If you repeat this experiment many times, you will find that the number of counts is not always the same. If you plot a histogram of the number of times each result comes out, you will find a gaussian centered on some value. That

value is the best estimate of the number of photons detected in ten seconds, but the spread in the gaussian – the standard deviation – tells you the uncertainty. For a counting experiment, in which you measure N events, that uncertainty is $\sqrt{N}$. So the ratio of the signal to the uncertainty, sometimes called the *signal-to-noise ratio*, is $\sqrt{N}$. So, as you increase N, the signal-to-noise ratio increases but only as $\sqrt{N}$. For example, if you want to improve the signal-to-noise ratio by a factor of two, you would have to increase the number of counts by a factor of four.

More recently, astronomy has been revolutionized by the development of *charge-coupled-devices (CCDs)*. They provide a grid of detectors all with high quantum efficiency (greater than 50% and sometimes as high as 80% to 90%). Each element of the grid is one pixel, and keeps an electronic record of the intensity of light striking its position. The array is on a single silicon chip. There are typically over 1 million pixels (1000 × 1000) on astronomical CCDs. (Commercial digital cameras use CCDs.) Each pixel is a potential well that traps electrons. For the most part, the electrons are the result of photons striking the pixel. In order to use the information on the chip, there has to be some way of reading it into a computer. This is illustrated in Fig. 4.18.

An advantage to a CCD is that it is nearly linear in its response. This means that the number of electrons is proportional to the number of striking photons. This is true over a wide range of intensities. As with photomultipliers, there is some dark current. This can be reduced by cooling the detector. The dark current is generally measured by taking exposures with no light entering. There is also a variation in sensitivity from pixel to pixel. This can be measured by making an exposure of a uniform field, such as the twilight sky or the inside of the dome. This process is called *flat fielding*. CCDs are so stable that dark current measurements and flat field measurements only need to be done a few times a night.

If a cosmic ray (cosmic rays are charged particles that permeate interstellar space) strikes a CCD during an exposure, it will make a few pixels look very bright. These can easily be removed in the computer processing of the image. To minimize their effects, it is better to take a few short

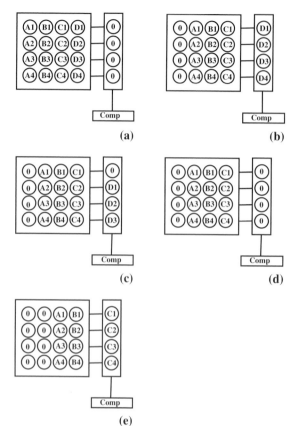

Fig 4.18. How a CCD readout works. In this example, we just look at 16 pixels, with rows numbered and columns lettered. The readout device is on the right. (a) An exposure is finished. Each pixel has a value indicated by the label for that pixel. The readout device has levels zero. (b) The numbers are all shifted one pixel to the right. The first row is now zeroed, and the readout device has the contents of the last row. (c) The contents of the readout row are shifted down one, as the first number (bottom right value) is read. (d) The process continues until that whole row is read. (e) Everything shifts one more to the right, and the readout process of the last row repeats. This then continues until all values have been read and pixels have been set to zero.

exposures and add them together (using a computer) than one long exposure. There is also an error introduced in the readout process, called readout noise. This can put a limit on the faintest signals CCDs can see.

Having the image in computer readable form is actually very convenient, because many new techniques are being used to computer enhance very faint images. This provides a large *dynamic*

range, meaning that we can see faint objects in the presence of bright ones.

When photometric observations are being made, we generally compare the brightness of the star under study with the brightnesses of stars whose properties have already been studied. By changing filters we can measure, for example, the U, B and V magnitudes of a star, one after another. Some method of recording the data is still needed. One option is photographic. The brighter the star is, the larger its image on a photographic plate. (This is an artifact of the photographic process and atmospheric seeing.) We can measure the brightness of a star by measuring the size of its image. (Remember the actual extent of the star is too small to detect in our images.) Photoelectric devices are well suited for photometry. Almost all photometry is now done using photomultipliers or CCDs. Some of the standard colors even account for the wavelength responses of various commercially available photomultipliers.

4.5.2 Spectroscopy

In spectroscopy we need a means of bringing the image in different wavelengths to different physical locations on our detector. We have already seen that this can be done with a prism. Since a prism does not spread the light out very much, we say that the prism is a *low dispersion* instrument. Dispersion is a measure of the degree to which the spectrum is spread out. Low dispersion spectra are sometimes adequate for determining the spectral type of a star. Sometimes a thin prism is placed over the objective of the telescope and a photograph is taken of the whole field. Instead of seeing the individual stars, the spectrum of each star appears in its place. These *objective prism spectra* are quite useful for classifying large numbers of stars very quickly.

When better resolution is needed, we generally use a *diffraction grating*, illustrated in Fig. 4.19. For any wavelength λ, the grating produces a maximum at an angle given by

$$d \sin \theta = m\lambda \qquad (4.4)$$

where d is the separation between the slits and m is an integer, called the *order* of the maximum. The higher the order is, the more spread out the

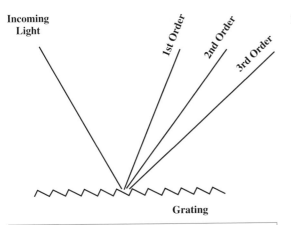

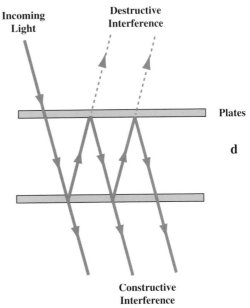

Fig 4.19. Diffraction grating. Light comes in from the upper left. The beam reflected off each step spreads out due to diffraction. However, interference effects result in maxima in the indicated directions. The angles of the steps can be adjusted (blazed) to throw most of the light into the desired order.

Fig 4.20. Operation of an interference filter.

spectrum. Suppose our grating just lets us separate (resolve) two spectral lines that are $\Delta\lambda$ apart in wavelength. The *resolving power* of the grating is then defined as

$$R \equiv \lambda/\Delta\lambda \qquad (4.5)$$

If the grating has N lines in it, then in the order m the resolving power is given by

$$R = Nm \qquad (4.6)$$

Some gratings have over 10 000 lines per centimeter over a length of several centimeters. This means that resolving powers of 10^5 can be achieved. In general, light will go out into several orders. It is possible to cut the lines of a grating so that most of the light goes into a particular order. This process is called blazing.

It is possible to use *interference filters* such as that shown in Fig. 4.20. There are two flat parallel reflecting surfaces placed close to each other. There is a maximum in the transmitted radiation when twice the spacing between the surfaces, d, is equal to an integral number of wavelengths. That is

$$2d = m\lambda \qquad (4.7)$$

One problem with this approach is that we can only measure a small wavelength range at a time, and must keep changing the spacing, d, to obtain a complete spectrum. Another problem is that different orders (m) of different wavelengths can get through at the same time. You can solve this problem by adding a second filter with a different spacing, set to pass the desired wavelength and remove the unwanted orders. A device with multiple interference filters is called a *Fabry–Perot interferometer*.

A major recent improvement has been the development of devices that produce a Fourier transform of the spectrum. These devices provide astronomers with a great deal of flexibility and sensitivity. Fig. 4.21 shows the operation of one such device, called a *Michelson interferometer*. The incoming radiation is split into two beams, which are reflected off mirrors so that they come back to the same location and interfere with each other. The path length of one of the beams can be altered by moving a mirror. This changes the phase of the incoming beams. By seeing how the intensity changes as we move the mirror, we form an idea of the relative importance of longer and shorter wavelength radiation.

According to Fig. 4.21, the total path length difference is x. We look at the electric field for each wave. In this case it is convenient to write

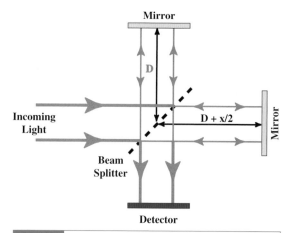

Fig 4.21. Michelson interferometer. Light enters from the left, and strikes a beam splitter. The split beams bounce off mirrors and are brought back together to interfere with each other. One of the mirrors has its position fixed, and the other is movable.

the waves as $E_0 \exp[i(kx - \omega t)]$, where $k = 2\pi/\lambda$ and $\omega = 2\pi\nu$, and E_0 is the electric field amplitude of the wave. So, the two waves that will be recombined can be written as

$$E_1 = E_0 \exp[-i\omega t] \tag{4.8a}$$

$$E_2 = E_0 \exp[i(kx - \omega t)] \tag{4.8b}$$

Taking the total electric field, $E = E_1 + E_2$, and the intensity, $I = EE^*$, we can write (see Problem 4.34)

$$I(k, x) = 2E_0^2 \left[1 + \frac{\exp(ikx) + \exp(-ikx)}{2} \right] \tag{4.9}$$

For any position of the mirror, corresponding to a path length x, we will receive the contributions from all wavelengths (k). So, to find the total intensity as a function of x, we integrate over all k:

$$I(x) = \int_0^\infty I(k, x) dk$$

$$= 2 \int_0^\infty I(k) \left[1 + \frac{\exp(ikx) + \exp(-ikx)}{2} \right] dk$$

$$= 2 \int_0^\infty I(k) dk + \int_0^\infty I(k) \exp(ikx) dk + \int_0^\infty I(k) \exp(-ikx) dk$$

If we let

$$I_0 = \int_0^\infty I(k) dk \tag{4.10}$$

be the total power, and we define $I(-k) = I(k)$, this simplifies to

$$I(x) = 2I_0 + \int_{-\infty}^\infty I(k) \exp[-ikx] dk \tag{4.11}$$

The integral in this expression is the Fourier transform of $I(x)$. So by measuring $I(x)$, we are also measuring the Fourier transform . This means that we can find $I(k)$, power as a function of wavelength, from the inverse Fourier transfrom of $I(x)$, which is

$$\int_{-\infty}^\infty I(x) \exp[ikx] dk$$

In a real measurement, we don't measure $I(x)$ for all values of x. There are two limitations. One is the total range over which we move the mirror. This limits our ability to do the integral from minus to plus infinity. The other is that we can only move the mirror in finite steps. This means that we only measure $I(x)$ at those positions, so this limits our ability to approximate the inverse transform integral as a sum. The closer together we measure $I(x)$, the shorter wavelengths (higher frequencies) we are sensitive to. The greater the largest value of x at which we measure $I(x)$, the more information we have on the longer wavelengths (lower frequencies), so this sets the limit in the frequency resolution in the computed spectrum.

4.6 | Observing in the ultraviolet

The visible part of the spectrum only gives us access to a small fraction of the radiation given off by astronomical objects. For centuries, however, this was the only information available to astronomers. We will see throughout this book that observations in other parts of the spectrum have revealed entirely new types of objects or provided us with information crucial to understanding objects that are already observed in the visible. In discussing other parts of the spectrum, we start with ultraviolet observations, because the techniques are very similar to those in optical observations.

In many ways, we can think of ultraviolet observations as being short wavelength visible observations. The basic imaging ideas are the same. Of course, since the wavelength is shorter, mirror surfaces must be more accurate in the ultraviolet than in the visible. The normal coatings that we use to make mirrors reflective in the visible do not work as well in the ultraviolet, and different coatings are needed. Since ultraviolet photons have more energy than visible photons, the uv photons can easily be detected with photographic plates, photomultipliers or CCDs.

The major problem is that ultraviolet radiation does not penetrate the Earth's atmosphere. If you don't go too far into the ultraviolet, some observations are possible at high altitudes. However, we have become increasingly dependent on ultraviolet satellites. Some pioneering satellites were Copernicus (1972–1981) and International Ultraviolet Explorer (IUE, 1978–1996). IUE had a 0.45 m mirror, 3 arc sec angular resolutions and an $R = 12\,000$ spectrograph. Currently, we can use HST, whose mirror was designed to work in the ultraviolet as well as the visible. Far Ultraviolet Spectroscopic Explorer (FUSE) was launched in 1998, with a 0.64 m mirror and a high resolution spectrograph.

4.7 | Observing in the infrared

In this section we briefly look at some of the techniques for observing in the infrared part of the spectrum. For some purposes we can simply think of infrared radiation as being long wavelength visible radiation. In fact, much of infrared astronomy is done on normal optical telescopes. The long wavelength means that surface accuracy of mirrors is not a problem. A surface accurate enough for optical observations is certainly accurate enough for infrared observations. However, the longer wavelength makes diffraction more of a problem. For example, for a 1 m diameter telescope working at a wavelength of 10 μm, the diffraction limit is 2 arc sec, slightly worse than the seeing limit at a good site.

One problem with infrared observations not common with optical observations is radiation from the telescope itself. Parts of the telescope that are not perfectly reflective radiate like black-bodies at temperatures close to 300 K, with a peak

at 10 μm. This is not a problem for optical detectors, but it is a problem for infrared detectors. (See Problem 4.23.) In an infrared telescope, the radiation paths must be carefully designed so that the detector cannot 'see' any hot surface. Some reduction in the problem can be obtained by cooling surfaces that can radiate into the detector. Such infrared optimization techniques are being incorporated into the northern hemisphere (Mauna Kea) part of Project Gemini (Fig. 4.14).

Detectors used in the infrared are generally different from those used in the visible. Infrared photons are not energetic enough to expose normal photographic emulsion. Recently, infrared sensitive emulsions have been developed. Infrared photons also have a hard time causing electrons to be ejected from metals. One of the great advances of the past few years has been the development of efficient infrared arrays of detectors. They read voltage, rather than the current in a CCD, but have readout schemes similar to CCDs. Infrared arrays also have a smaller number of pixels, typically 32 $\times$ 32. The detectors are cooled to reduce background noise. (As an aside, these arrays were first developed by the military to put into satellites looking down at the Earth. With the end of the cold war, this technology became declassified.)

Originally, the most common type of infrared detector was called a bolometer. A bolometer is a device that heats up in a known way when radiation falls on it. We generally use a material whose electrical properties change with temperature. For example, if the resistance of a bolometer changes with temperature, we can measure temperature changes by measuring resistance changes. By measuring the temperature increase, we can determine the total amount of energy striking the bolometer. (Remember, in Chapter 2, we defined a bolometric magnitude based on the total amount of energy given off by a star.)

Spectroscopy in the infrared is different than in the visible. One problem is that the longer wavelength means that objects must be physically larger to provide the same spectral resolution. Another problem is the thermal emission from the material used to construct the devices. If the surfaces can be cooled, then this problem is reduced. Prisms are of some value for low resolution. Cooled gratings are used. It is possible to use tunable interference filters (Fabry–Perot interferometers) or Michelson interferometers, as discussed earlier.

The major problem in the infrared is the Earth's atmosphere. The atmosphere is totally opaque at some infrared wavelengths, and is, at best, only partially transparent at all other infrared wavelengths. The opacity of the atmosphere causes two problems: (1) the atmosphere blocks the infrared radiation from the sources we are studying; (2) the atmosphere emits its own infrared radiation, which can be much stronger than that received from the astronomical objects. To observe with this atmospheric emission it is generally necessary to compare the astronomical source you are looking at with some empty sky nearby, thereby canceling the effects of the atmosphere. However, this limits you to studying relatively small sources.

We call the wavelengths at which some observations are possible from the ground *infrared windows* in the atmosphere. Fig. 4.22 shows some of the major infrared windows. At the very least, the 2 km altitude of many optical observatories is required. In general, 2 km is only sufficient for working in the near infrared, at wavelengths of a few micrometers. If we want to work farther into the infrared, higher altitudes are necessary. Some observatories have been placed at altitudes as high as 4.3 km (even though higher elevations result in difficult working conditions). For example, there are a number of infrared telescopes on Mauna Kea (Fig. 4.13b).

For many studies, even higher altitudes are needed. For 20 years NASA operated the *Kuiper Airborne Observatory (KAO)*. The KAO was a converted military transport (a C141), that carries a 0.9 m infrared optimized telescope to altitudes up to 45 000 ft, for 7 hour observing sessions. The KAO was operated as a national facility, with qualified astronomers submitting proposals for observing time. It made approximately 80 flights per year. It operated out of the NASA Ames Research Center (Moffet Field, CA), but could change its base when the astronomical need dictated. For example, there were regular observing sessions in the southern hemisphere from Christchurch, NZ. There were also customized flights to look at transient astrophysical phenomena such as solar eclipses. The KAO was taken out of service in 1997.

To replace KAO, NASA is building, in cooperation with astronomers in Germany, the *Stratospheric Observatory for Infrared Astronomy (SOFIA)*, shown in Fig. 4.23. To allow for a larger (2.6 m) telescope it will be made from a converted Boeing 747 SP. This aircraft is also capable of cruising at higher altitudes and providing for longer (up to 16 hour) flights. For even higher altitude work, balloons are used up to 100 000 ft.

For some observations, even a minimal atmosphere causes problems, and we carry out observations from space. Fig. 4.24 shows two of the primary infrared space missions. One of the important early missions was the *Infrared Astronomy Satellite (IRAS)*, a joint American–Dutch–British project launched in January 1983. The 0.6 m diameter cooled telescope was primarily

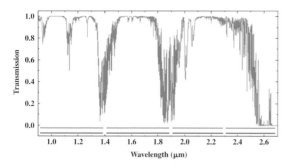

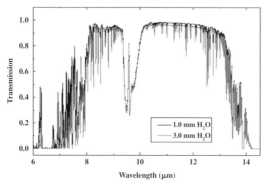

Fig 4.22. Infrared windows. Transmission as a function of wavelength is shown for an observatory at sea level. At low altitudes good transmission (close to unity) is in only a few narrow wavelength ranges, or windows. As one goes to higher observing altitudes, the windows cover a wider range of wavelengths. [NOAO/AURA/NSF]

Fig 4.23. Artist's impression of the Stratospheric Observatory for Infrared Astronomy (SOFIA), which will be made from a converted 747 SP. [NASA]

designed for imaging observations. It contained arrays of detectors operating in four wavelength ranges, centered roughly at 12, 25, 60 and 100 μm. From its high (560 mile) polar orbit, much of IRAS's lifetime was devoted to a systematic survey of the sky. A certain fraction of the time was devoted to specific objects. The large scale survey revealed over 100 000 point sources and a network of extended infrared emission. The whole set of data is available as a resource to the general astronomical community. An astronomer interested in

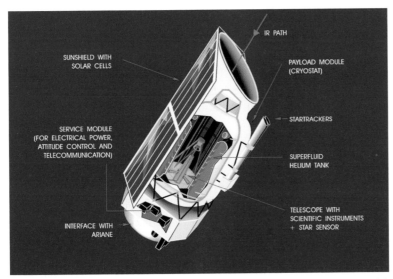

Fig 4.24. Infrared satellites. (a) Artist's conception of Infrared Space Observatory (ISO)

IR PATH

SUNSHIELD WITH SOLAR CELLS

PAYLOAD MODULE (CRYOSTAT)

STARTRACKERS

SERVICE MODULE (FOR ELECTRICAL POWER, ATTITUDE CONTROL AND TELECOMMUNICATION)

SUPERFLUID HELIUM TANK

TELESCOPE WITH SCIENTIFIC INSTRUMENTS + STAR SENSOR

INTERFACE WITH ARIANE

(a)

(b)

Fig 4.24. (Continued) (b) Space Infrared Telescope Facility (SIRTF) in the laboratory for testing. [(a) ESA/ISO; (b) NASA]

a particular object can check the data at the Infrared Processing and Analysis Center (IPAC), at Caltech, which is not too far from the Jet Propulsion Laboratory (JPL). IPAC has also become the curator of other infrared data.

More recently, astronomers have been able to utilize the *Infrared Space Observatory (ISO)*, Fig. 4.24(a), a project of the European Space Agency. ISO provides a wavelength coverage extending farther into the infrared, a larger telescope, arrays of more sensitive detectors for good imaging, and for the first time the ability to make high quality spectra. During the HST servicing mission, NASA added an infrared camera and spectrometer *(NICMOS)*. NASA is now making plans for the *Space Infrared Telescope Facility (SIRTF)*, Fig. 4.24(b).

Box 4.1. | Methods of displaying images.

When we look at a normal optical photograph of some astronomical object, we have a sense of how our brains should interpret that image. In a sense it is how the object would look if we could view it through a large telescope, or if we could somehow be transported close enough to the object so we could see it with this detail with the unaided eye. However, what does it mean when we display a radio image like that in Fig. 4.25?

There is even a terrestrial analogy to this question. You have seen 'night vision glasses', which allow you to 'see' even with no illuminating light. Remember, we normally see earthbound objects as they reflect sunlight or roomlight, our eyes being sensitive to the range of wavelengths at which the Sun's emission is strongest. The night vision glasses work differently: they actually detect the infrared radiation given off by objects (most of which are usually close to 300 K). So, the night vision glasses have infrared detectors, but our eyes are not sensitive to that infrared radiation. Therefore the glasses also convert that infrared image into an optical image, usually with the brightest part of the optical image corresponding to the strongest infrared emission. So, the image you see is an optical representation of the infrared image.

We can do the same thing with astronomical infrared (or ultraviolet, radio, etc.) images. We can make a *false gray-scale image*, by creating an optical image where brighter regions correspond to stronger infrared emission. It is important to remember that, while such images are often constructed to have a true-looking appearance, they are just a particular representation for that image. Sometimes our eyes are better at picking information out of a color image than a black and white (or gray) image. It is therefore sometimes useful to make *false color images*. In this case, we arbitrarily assign a color to each level of infrared emission. Often the colors will run through the spectrum from red to blue (or the other way). Often, a sample bar will be placed next to the image, showing what intensity level each color represents.

So far, we have been talking about what we do when we have one piece of data at each location, say the average intensity in some particular wavelength band. Suppose we have observed in more than one band (for example, IRAS observed in four infrared bands). We could certainly make a separate false gray or color image of each band (and we often do this), but what if we want to compare the bands, or simply display all the information together?

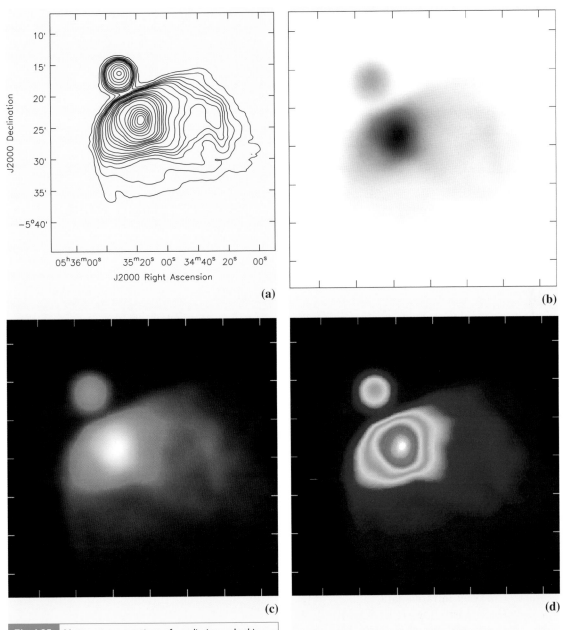

Fig 4.25. Various representations of a radio image. In this case it is a 6 cm wavelength image of the Orion Nebula (which we will discuss in Chapter 15), made with the Byrd GBT (which we will discuss in the Section 4.8). (a) Contour map. (b) Gray-scale map. (c) False color image. (d) Color contours, in which colors change where there would be a contour line. [D. Shephard, R. Maddalena, J. McMullin, NRAO/AUI/NSF]

We then make a false gray-scale image of each band. We then tint each band a different color, generally making the longest wavelength band red and the shortest wavelength band blue (mimicking what happens in the visible part of the spectrum). We then have a false color image in which the color has some intrinsic meaning (in that hotter objects will appear bluer).

We should point out that this technique can also be used to make a true color visible image. You might say that if you want a color image you simply use color film and take a picture. However, no two types of color film are the same. Some are meant to enhance skin tones and are set to emphasize reds, for example. Therefore the way to make a color photograph that looks like what you would see with your eyes, we take a series of black and white images, through red, green and blue filters. We then combine the images, utilizing the various wavelength ranges in the same proportion as the eye uses them. This technique is well suited to making 'true color' images with CCDs.

There is another method of displaying two-dimensional images, *contour maps*. You should be familiar with topological maps on Earth, which are normally displayed as contour maps. All points within a given contour level have a value (e.g. average intensity in a particular wavelength range) greater than the value assigned to that level. Contour maps give a good feel for how the quantity you are displaying changes over some region. The closer together the contours, the more rapid the variation in the plotted quantity.

4.8 | Radio astronomy

Radio observations provide us with very different information from optical observations and use very different techniques. The long wavelength means that the wave nature of the radiation is very apparent in the observations. The long wavelength also corresponds to low energy photons. This means that radio regions can tell us about cool regions. For example, we will see how radio observations tell us about star formation in Part IV. We will also see that there are high energy sources that give off much of their energy at longer wavelengths. Thus, radio observations also give us a way of studying high energy phenomena.

Radio astronomy owes its origins to an accidental discovery by *Karl Jansky*, an engineer at the Bell Telephone Laboratories in New Jersey. In 1931, Jansky detected a mysterious source of radio interference. He noticed that this interference reached its peak four minutes earlier each day. This timing suggests an object that is fixed with respect to the stars. (This four minute per day shift is caused by the Earth's motion around the Sun. This and other aspects of astronomical timekeeping are discussed in Appendix G.) The time of maximum interference coincided with the galactic center crossing the local meridian. Jansky concluded that he was receiving radio waves from the galactic center. It was realized that astronomical objects can be strong radio sources.

The discovery was not followed up immediately. In fact, for a long time there was only one active radio astronomer. *Grote Reber* was an amateur radio astronomer in Illinois, who carried out observations on his back yard radio telescope in the 1930s and early 1940s. (When Reber submitted his first paper for publication in *The Astrophysical Journal*, it was sent to a referee, a normal procedure. To make sure that the data were to be believed, the referee, *Bart Bok*, a Dutch astronomer, then living in the US, took the abnormal step of visiting Reber and his telescope, and taking the editor along. Bok recommended publication of the paper, and was the first traditional optical astronomer to understand the importance of radio astronomy. Following WW II, radio astronomers benefitted from the development of radar equipment during the war. Radio observations were pursued by the British, Dutch, Australians, and a small group of Americans at Harvard. A major advancement was the ability to observe spectral lines in the radio part of the spectrum. We will discuss these lines in Chapter 14.

By the mid-1950s, it was clear that a major radio observatory had to be a cooperative effort, and the *National Radio Astronomy Observatory (NRAO)* was founded. (This was the first US national observatory, being formed a little before the optical observatory on Kitt Peak.) Bart Bok played a major role in the founding of the NRAO. The first telescopes of the NRAO were in Green Bank, West Virginia, far away from sources of man made interference (in the National Radio Quiet Zone). Since the Earth's atmosphere is virtually transparent through much of the radio part of the spectrum, it is not necessary to place radio observatories at high altitudes or clear sites. We can even observe through clouds. We can also observe day or night, since the sky does not scatter radio waves from the Sun the way it scatters light from the Sun, making the sky appear bright (blue).

We now take a look at how a radio telescope works. A radio telescope consists of some element that collects the radiation and a receiver to detect the radiation. Most modern radio telescopes have a large dish to collect the radiation and send it to

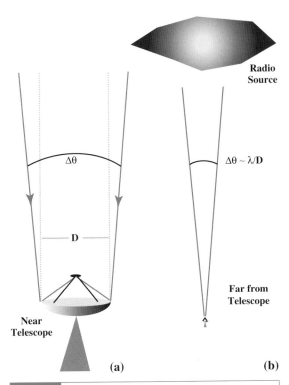

Fig 4.26. Resolution for a radio telescope. (a) The short dashed lines show what would happen if there were no diffraction. Only radiation traveling parallel to the telescope axis would reach the focus. The solid lines show the effects of diffraction. Radiation coming in at a slight angle with the telescope axis can still be reflected on to the focus. This means that when the telescope is pointed in one direction, it is sensitive to radiation from neighboring directions. This is shown in (b), as the telescope is sensitive to radiation coming from within a cone of angle approximately λ/D (in radians).

of smaller panels that are easier to machine accurately. The panels are then aligned to produce the best surface. The alignment is at least adjusted for the effects of gravity as the telescope tilts at different angles, and techniques are being developed to control the surface actively by monitoring the panels at all times during observations. The best resolution for single radio telescopes is about 30 arc sec, slightly better than the naked eye for visible viewing.

Example 4.5 Strength of radio sources
We measure the strength of radio sources in a unit called a *Jansky* (Jy). It is defined as 10^{-26} W/m²/Hz reaching our telescope. For a 1 Jy source, calculate the power received by a perfect antenna with an area of 10^2 m², using a frequency range (bandwidth) of 10^6 Hz.

SOLUTION
The total power received is the power/area/Hz, multiplied by the frequency range (in Hz), and the surface area of the telescope:

$$P = (10^{-26} \text{ W/m}^2/\text{Hz})(10^6 \text{ Hz})(10^2 \text{ m}^2)$$

$$= 10^{-18} \text{ W}.$$

This is 10^{-20} of the power of a 100 W light bulb. Note that the larger the dish, the larger the total power detected.

We have already seen that for making maps of extended sources, larger dishes are important because they provide us with better angular resolution. The weakness of radio sources gives us another reason for building large telescopes. A larger telescope intercepts more of the radiation, and allows us to detect weaker sources. A few large telescopes are shown in Fig. 4.27. The largest single dish is at the *National Atmospheric and Ionospheric Center* in Aricebo, Puerto Rico (Fig. 4.27a). The dish is made of a mesh surface that is set in a natural bowl. The holes in the surface are large enough that we can only use this dish for long wavelength observations. Also, the dish cannot be steered in any direction; it looks straight up. However, by moving the detectors around you can actually view a reasonable amount of sky. For many years, the largest fully steerable antenna was at the *Max Planck Institüt für Radioastromie* in

a focal point. (They are like optical reflectors.) The long wavelength becomes important in this process. We have already seen that the resolution of a telescope depends on the size of the telescope, relative to the wavelength (Fig. 4.26). (In the radio part of the spectrum, atmospheric seeing is not a problem.) Since the wavelengths are large, to achieve good resolution you need a large collector. However, that surface doesn't have to be perfect. It can have imperfections as long as they are smaller than approximately $\lambda/20$. For example, at a wavelength of 20 cm, 1 cm diameter holes have no effect on the performance of the telescope. We are hindered by the fact that it is hard to make large telescopes with very accurate surfaces. Most large telescopes are made up

(a)

(b)

(c)

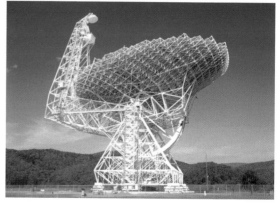

(d)

Fig 4.27. Large radio telescopes. (a) The 300 m (1000 ft) dish in Aricebo, Puerto Rico. The dish always points straight up, but moving the receiver to different off axis positions allows looking away from overhead. (b) The 100 m diameter telescope of the Max Planck Institüt für Radioastronomie, Effelsberg, Germany. It operates in azimuth and elevation. Azimuth is controlled by moving the whole structure on the circular track. As the telescope changes its elevation angle it deforms under gravity. However, it is designed to deform from one paraboloid into another, so only its focal length changes. (c), (d) The 100 m Byrd Telescope at the NRAO in Green Bank, WV. The offset arm to support receivers results in no blockage of the dish. This optimizes sensitivity and imaging quality. The telescope surface (c) and back-up structure (d) are shown. [(a) The Aricebo Observatory is part of the National Astronomy and Ionosphere Center operated by Cornell University under a cooperative agreement with the National Science Foundation; (b) MPIFR; (c),(d) NRAO/AUI/NSF]

Effelsberg, Germany (Fig. 4.27b). That telescope is 100 m in diameter. The inner 80 m has a surface made of solid panels so it can be used at wavelengths of a few centimeters. The outer 20 m is a mesh and is used to make the dish larger at longer wavelengths where diffraction is worse. The newest large telescope is the Byrd telescope, just being completed at the NRAO in Green Bank. It is a fully steerable 100 m telescope (Fig. 4.27c). It is designed so that more of the collecting area is used than in the German telescope. This is done in part by reducing the blockage by the support structure. All of the surface is accurate enough for observations at wavelengths as short as 7 mm.

The detection of the radio waves takes place in a *radio receiver*. In general, the size of the receiver limits us to only one receiver operating on a telescope at a given time. This is equivalent to doing optical observations with only one detector in your CCD. At any given time, the telescope is receiving radiation from a piece of sky determined by the diffraction pattern of the antenna. If we want to build up a radio image of part of the sky, we must point the telescope at each position and take a separate observation. Recently, improvements in receiver technology have allowed limited multireceiver systems.

The receivers in radio astronomy are similar in concept to home radios. Like your home radio, the incoming signal is first mixed with a signal from a reference oscillator, and the resulting lower frequency beat note is then amplified. We change the frequency we are observing (like changing radio stations) by changing the frequency of the reference oscillator. However, the signals from astronomical sources are so weak by the time they reach us that receivers for radio astronomy must be much more sensitive than your home radio. Sometimes the receivers are cooled to a few degrees above absolute zero to minimize sources of background instrumental noise. Unlike bolometers, they do not simply detect all of the energy that hits them; they are also capable of preserving spectral information.

Just as with optical observations, in radio astronomy we can make continuum and spectral line observations. Continuum studies are like optical photometry. We tune our receivers to receive radiation over a wide range of frequencies, and we measure the total amount of power received. From this information, we obtain the general shape of the continuous spectrum (intensity vs. frequency).

In spectral line observations the radiation is detected in small frequency intervals, so the shapes of spectral lines can be determined. The spectrometers for radio observations have traditionally been large numbers of electronic filters, tuned to pass narrow frequency ranges. More recently, the ability of very fast computer chips has allowed for very flexible digital spectrometers. They measure the auto-correlation function of the incoming signal, which is the result comparing the signal with a slightly delayed version of the signal, and doing this for different delay times. We can think of this as the digital analog of the Michelson interferometer spectrometer discussed above. It therefore produces a Fourier transform of the spectrum. As with the Michelson interferometer, it is limited by its ability to measure this Fourier transform at only a finite range of time delays, with a step size limited by how fast we can run the computer. The computer speed determines the total bandwidth of the spectrometer, and the largest delay time determines the frequency resolution. By changing one or the other, we can adjust the bandwidth or the frequency resolution. That is why we say this is a very flexible system.

With either technology it is relatively easy to make high resolution spectral observations, with up to a few thousand frequency channels observed simultaneously. So, compared with optical observations, in radio observations we have to work harder to build up an image, but it is easier to make a spectrum at each position in our image.

One of the most important advances in radio astronomy in the last three decades of the 20th century has been the development of the millimeter (or shortest radio wave) part of the spectrum. As we will see throughout this book there are certain observations which are only possible at millimeter wavelengths. There also are some inherent benefits in working at millimeter wavelengths. If we can observe at 1 mm, for example, we can achieve the same angular resolution as at 10 cm, with a 100 times smaller dish! Of course, the dish must have a surface that is 100 times more accurate, meaning that it is hard to make a very large dish. This has restricted the size of millimeter telescopes to a few tens of meters in diameter, providing resolutions of ~10 to 20 arc sec at best.

At millimeter wavelengths the atmosphere blocks some of the incoming radiation (being somewhere between the totally clear radio and totally blocked infrared). This means that it is useful to put millimeter telescopes at high altitudes and dry sites (just as with optical or infrared telescopes). One of the first (and until its closure, in 2000, one of the most heavily used) millimeter telescopes was the *12 m telescope* of the NRAO, located on Kitt Peak, Arizona, just below the site of the optical observatory (Fig. 4.28a). ESO

(a)

(c)

(b)

(d)

Fig 4.28. Millimeter telescopes. (a) The NRAO 12 m telescope on Kitt Peak, Arizona. (b) The Swedish–ESO Submillimetre Telescope (SEST) on La Silla, Chile. (c) The 30 m telescope, Spain. (d) The Nobeyama 45 m telescope. [(a) Jeffrey Mangum, NRAO/AUI/NSF; (b) ESO; (c) IRAM; (d) Nobeyama Radio Observatory/National Astronomical Observatory of Japan]

has operated the 15 m *Swedish–ESO Submillimetre Telescope (SEST)* at their optical site on La Silla, Chile (Fig. 4.28b). The largest millimeter telescopes are the *30 m telescope* operated by French and German institutes, and located in Spain (Fig. 4.28c) and the *Nobeyama 45 m telescope* in Japan (Fig. 4.28d).

The problem of poor angular resolution for radio observations has been solved, in part, by using combinations of telescopes, called *interferometers* (Fig. 4.29). Interferometers utilize the information contained in the phase difference between the signals arriving at different telescopes from the same radio source. Any pair of telescopes provides information on an angular scale approximately equal (in radians) to the wavelength, divided by the separation between the two telescopes in a direction parallel to a line

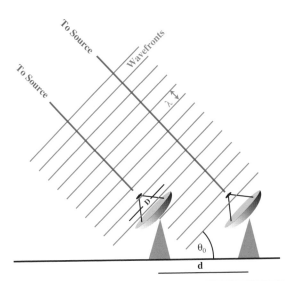

Fig 4.29. Radio interferometer. Here only two telescopes are shown, but an interferometer with any number of telescopes can be treated as a number of pairs of telescopes. The separation between the telescopes produces a phase delay, which depends on the separation, d, and the position of the source. The phase difference can be detected, providing information about source structures whose angular size is approximately λ/d (in radians). By using different telescope spacings and the Earth's rotation, information about structures on different angular sizes can be accumulated and eventually reconstructed into a map of the source.

ence between signals from objects at the center of each field of view. But objects off the field center will have varying phase differences as we track the source across the sky. To extract a map of our field of view, we first multiply the two signals together (actually the delayed signal from the first telescope times the complex conjugate of the signal from the second telescope). This product is called the *visibility*. When you work out the details, the visibility turns out to be the Fourier transform of the two-dimensional intensity distribution on the sky, $I(x, y)$. The visibility is a function of two variables (u, v), where $u = (d/\lambda)\cos\theta_0$ and v is defined for the corresponding angle in the perpendicular direction.

So, we measure the (2D) visibility at as many (u, v) points as possible, and then calculate $I(x, y)$ by taking the 2D Fourier transform. Obviously, the more (u, v) points we can measure, the better we can estimate the visibility and the better we can estimate its Fourier transform. This is similar to the Michelson interferometer, discussed earlier in this chapter, where the more mirror positions at which we could measure the interference pattern, the more accurately we could compute the spectrum, which is the Fourier transform.

How do we measure many (u, v) points? For any pair of telescopes, we let the Earth's rotation change the elevation angle of the source, and also the orientation on the sky, changing how much of u and how much of v we are changing. So, if we do a series of observations, of say, 5 minutes each, and track a source for 8 hours, we can take many measurements. It also helps to have many telescopes. For N telescopes there are $N(N-1)/2$, independent pairs of telescopes, so, for large N, the number of pairs goes up roughly as N^2. To make optimum use of these pairs, we don't simply have equally spaced telscopes, since every pair of spacing d will give redundant information. It is also useful to not have all the telescopes in a line, which would just give a lot of values of u or v, but not both. In general the shortest spacings give information on the large angular scales on the sky, and the longest spacing provide information on the smallest angular scales.

The most useful interferometer over the past several years has been the *Very Large Array* or *VLA*, near Socorro, New Mexico, operated by the NRAO

connecting the two telescopes. To obtain information on different angular scales, it is necessary to have pairs of telescopes with different spacings. In addition, different orientations are needed. For this reason, interferometers generally have a number of telescopes. The Earth's rotation also helps change the orientation of any pair of telescopes, as viewed from the direction of the source. Unlike single dish observations, you don't have to point the telescope at different parts of the source to make a map.

To see some of the limitations of using interferometers to make images, we look a little more at how they work. We again look at any pair of telescopes, as indicated in Fig. 4.29. Before combining the signals from the two telescopes, we delay the signal from the nearer telescope by the extra time it takes the waves to reach the second telescope. That delay will change as we point the telescope pair at different angles above the horizon. This allows us to zero out the phase differ-

(a)

(b)   **(c)**

Fig 4.30. Views of the Very Large Array (VLA), on the Plains of St Agustine (at an altitude of about 2.3 km) southwest of Socorro, New Mexico. There are 27 telescopes, each 25 m in diameter. At any instant there are 351 pairs of telescopes. Depending on the project, the spacings can be adjusted by moving the telescopes along railroad tracks. Moving all of the telescopes takes a few days and is done four times a year. The VLA is operated by the National Radio Astronomy Observatory. (a) The whole array. (b) The central section, showing a better view of the telescopes and the railroad tracks. (c) The transporter used to move telescopes on the tracks. [NRAO/AUI/NSF]

(Fig. 4.30). The VLA has 27 telescopes, 351 pairs (each 25 m in diameter), arranged in a 'Y' configuration, to allow a wide range of both (u, v) values. Each arm of the Y is 21 km long. The telescopes are placed alongside railroad tracks, so that the telescope spacings can be changed, depending on the resolution needed for a particular project. These changes can take up to two weeks and are only done a few times a year. The shortest wavelength at which the VLA operates is 7 mm. It can be used for both continuum and spectral line observations. It has proved to be a powerful tool, providing images of radio sources, with many observing sessions ranging from a few minutes to a few hours. (The amount of data taken is so large that it takes the computers much longer to process the data than it does to observe.)

For observations requiring the best possible resolution, telescopes on opposite sides of the Earth are used. This is called *very long baseline interferometry (VLBI)*. VLBI observations have provided angular resolutions of 10^{-4} arc seconds! In regular interferometry, the signals from the various telescopes are combined in real time, as the data comes in. In VLBI, signals at each telescope are recorded along with a time signal from a very accurate atomic clock. Later, the records are brought together, and the time signals are used

Fig 4.31. Arrangement of telescopes for the Very Long Baseline Array (VLBA), operated by the NRAO. There are ten telescopes, each 25 m in diameter. It operates down to a wavelength of 1 cm. [NRAO/AUI/NSF]

to coordinate the records from different telescopes, and the interferometry is then done by computer. To provide a dedicated group of telescopes for VLBI, the NRAO has recently built the *Very Long Baseline Array (VLBA)*, which extends over much of North America (Fig. 4.31).

The success of interferometry and the importance of millimeter observations have led astronomers to begin working with millimeter interferometers. Millimeter waves provide many technical challenges for interferometry. For example, the effects of the Earth's atmosphere on the millimeter signals are important. Following the demonstration of successful millimeter interferometers operated by Caltech and Berkeley, the NRAO has started development of the *Atacama Large Millimeter Array (ALMA)*, shown in Fig. 4.32. The final details are still being worked out, but the array, which is being built in collaboration with ESO and the Chilean government, will have approximately 75 telescopes, each approximately 12 m in diameter. They will work down to a wavelength of 0.8 mm. In order to get around atmospheric problems, a very high (5000 m) dry site in the Atacama Desert in Chile

has been chosen. It is expected that operation will start in 2010.

4.9 | High energy astronomy

X-ray astronomy is one of the youngest fields in observational astronomy. Since X-rays do not penetrate the Earth's atmosphere, the history of X-ray astronomy is the history of high altitude (balloon) and space astronomy. Early X-ray observations were done with sounding rockets (which provided very brief flights with only a few minutes of data taking) and high altitude balloons. Of course, the balloons still do not rise above all the atmosphere, and, in the X-ray part of the spectrum, even the little bit that is left matters.

One problem in observing X-rays is that it is very hard to make a mirror that works for these short wavelengths, less than 1 nm. That is because the typical spacing between atoms in a solid is about 0.1 nm. So the incoming radiation sees a rough surface, with reflections off each atom producing a scattering in some essentially random direction. There is one possibility. If we arrange for the X-rays to come in at a very shallow angle, only a degree or two, with the surface, the atoms appear to be closer together, and we achieve normal reflection. This is called *grazing incidence*. Of course, being constrained to only grazing angles makes it difficult to design a telescope that will collect and focus a reasonable amount of radiation. A diagram of the imaging system in one X-ray satellite is shown in Fig. 4.33(b).

X-ray satellites are able to provide both continuous and spectral line observations. Originally, the spectral information came from detectors similar to those used by high energy physicists, called *proportional counters*, which register the energy of the photons as they hit. Better spectral resolution was obtained by using a type of grating called a *Bragg crystal*, in which the 'slits' are the individual atoms in a solid. More recently, X-ray astronomers have been able to use solid state detectors that give good spectral resolution.

More recently a number of satellites have opened our eyes to the X-ray sky. One of the satellites is shown in Fig. 4.33(a). A number of the

(a)

(b)

(c)

Fig 4.32. The Atacama Large Millimeter Array (ALMA). (a) Views of the site. (b) Artist's conception of the arrangement of telescopes. (c) Artist's conception of the antenna appearance. [NRAO/AUI/NSF and ESO]

tions was *Uhuru* (launched in 1970). It was also the first to survey the whole sky, and, compared to previous missions, had very sensitive detectors. It found 339 objects, showing astronomers that many different types of objects give off strong X-ray emission. Following Uhuru, which was a relatively small satellite, NASA launched a number of larger satellites in the *High Energy Astronomy Observatory (HEAO)* program. The second in the HEAO series was the *Einstein Observatory*, which was the first to utilize the grazing incidence imaging, and so produced the first real X-ray images. The Einstein images had a profound impact on our thinking about many types of astronomical objects. We went from being able to probe small sections of objects to forming whole images. In many ways it was like having a blindfold removed. The next major jump in sensitivity and angular resolution was the *Roentgen Satellite (ROSAT)*, launched in 1990 (Fig. 4.33a). *Chandra* was launched in 1999, and provides sub-arcsecond imaging (Fig. 4.33c) and grating spectrometry, so it does high quality imaging spectroscopy (Fig. 4.33d).

All of the telescopes that we have discussed so far have been for electromagnetic radiation. High energy phenomena also make their presence known in other ways. One way is by the emitting beams of cosmic rays, charged particles, often with very high energies. They also give off neutrinos, subatomic particles that are very difficult to detect. They also give off gravitational radiation,

earliest satellites incorporated X-ray detectors, which shared space with other experiments. The first satellite devoted entirely to X-ray observa-

distortions in the fabric of spacetime, which again are very difficult to detect. We will talk about detecting each of these in the chapters that discuss their astrophysical origin.

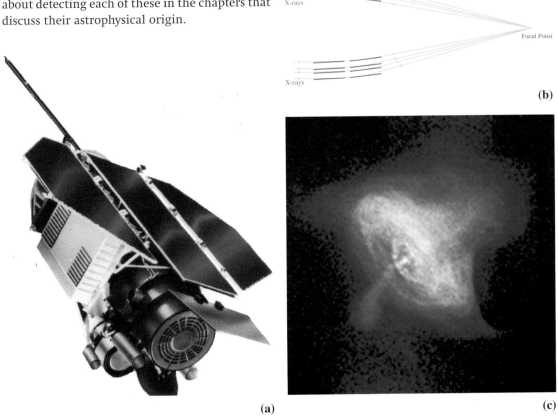

(a)

(b)

(c)

Fig 4.33. X-ray satellites. (a) ROSAT. (b) Imaging system in the Chandra X-ray satellite, utilizing grazing incidence. (c) Chandra image of the Crab Nebula showing the great detail achievable in current X-ray systems. (d) Chandra test spectrum, showing the good spectral resolution. [NASA]

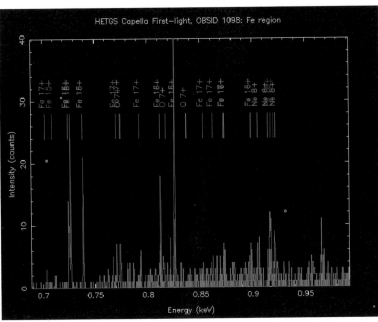

(d)

Chapter summary

In this chapter we saw how different types of telescopes are used to collect data across the electromagnetic spectrum. We saw the differences and similarities among the techniques used in different parts of the spectrum. Much of the progress in astronomy in the past few decades has come from our ability to make high quality observations in parts of the spectrum other than the visible.

We looked at the important features of any telescope, the collecting area and the angular resolution. The collecting area determines how sensitive the telescope is to faint objects. The angular resolution is limited by diffraction (especially in

the radio), and in other cases is limited by atmospheric seeing (especially in the visible).

We saw the various techniques for extracting information from the radiation collected by our telescopes. Improving detector efficiency and panoramic ability has been important in all parts of the spectrum.

We saw the importance of site selection for an observatory. As an ultimate site, we saw the advantages of telescopes in space. In space, we can observe at wavelengths where the radiation does not penetrate the Earth's atmosphere. Even in the visible, we can achieve improved sensitivity and angular resolution.

Questions

4.1. Describe the factors that limit the angular resolution of an optical telescope. Include estimates of the size of each effect.

4.2. What do we mean when we say that the main reason for building large ground-based optical telescopes is light-gathering power?

*4.3. Explain how improving the seeing at a site might allow you to detect fainter stars. (Hint: Think of what happens to the photons on your detector (film or CCD) when the image is smeared.)

*4.4. Suppose you are observing two stars that are 2 arc sec apart. Draw a diagram illustrating what you observe under conditions of 4, 2 and 1 arc sec seeing. Your diagram should be a series of graphs showing intensity as a function of position on a detector.

4.5. Suppose two stars are 5 arc sec apart on the sky. We clearly cannot resolve them with our eyes, but the angular resolution of even a modest sized telescope is sufficient to resolve them. However, the light from the telescope must still pass through the narrow pupil of the eye. Why doesn't the diffraction of the light entering the eye smear the images too much for us to resolve the two stars?

4.6. 'Faster' photographic emulsions can be made by making the grains larger. Why do you

think this works? What are the possible drawbacks to this?

4.7. What are the advantages of CCDs over photographic emulsions and photomultipliers?

4.8. Compare image formation (similarities and differences) in the eye and in a camera.

4.9. Why is chromatic aberration a problem even for black and white photographs?

4.10. If there is no angular magnification in a simple camera, how can using a longer focal length lens give a larger image?

4.11. A higher quality (more expensive) camera lens generally has smaller f-stops than an inferior lens. Why is this?

4.12. Why do some people need to wear eyeglasses while driving at night but not during the day? (What is it about the lower light level that degrades images?)

4.13. Compare the advantages and disadvantages of reflecting and refracting telescopes.

4.14. Compare the advantages and disadvantages of various focal arrangements in reflecting telescopes.

4.15. What does it mean to focus the eye or a camera 'at infinity'?

4.16. If you want to photograph a planet you use your long focal length telescope; if you want to do photometry on a faint star, you use a short focal length telescope. Explain.

4.17. For many observations (both imaging and spectroscopy) it is becoming important to read the data into a computer. Briefly discuss the techniques for doing this that we have mentioned in this chapter.

4.18. What are the important considerations in choosing an observatory site?

4.19. What are currently the best methods of reducing seeing effects at a given site?

4.20. What advantages does HST have over ground-based telescopes for optical observations?

4.21. What are the similarities and differences between ultraviolet and optical observations?

4.22. What are the similarities and differences between infrared and optical observations?

4.23. For infrared observations, we must still live with the fact that parts of a telescope radiate like blackbodies at about 300 K. Why isn't this a problem for optical observations?

4.24. How does a bolometer work? How would you use a bolometer to measure the power received in a small wavelength range, for example between 10 and 11 μm?

4.25. Why can't balloons get above all of the atmosphere?

4.26. Explain the similarities between trying to display astronomical images and displaying things such as topographical maps.

4.27. Explain why simply using color film in your camera does not give you a true color astronomical photograph.

4.28. In what ways are radio observations similar to and different from (a) optical and (b) infrared observations?

4.29. Suppose we want to use a radio telescope with a transmitter rather than a receiver of radio waves. Draw a diagram (similar to Fig. 4.26) showing how the transmitted radiation would be spread out on the sky.

4.30. Why is possible to observe with the Aricebo dish even though the surface has holes in it?

4.31. Why is it possible to do radio observations during the day, but not optical observations?

4.32. What are the advantages of observing in the millimeter part of the spectrum? What are the additional difficulties?

4.33. How would an image made by the VLA compare with one made with a single telescope as large as the VLA (assuming you could build one)?

4.34. How does VLBI differ from normal interferometry?

4.35. What are the difficulties in making a mirror to work for X-rays?

Problems

4.1. What is the limiting magnitude for naked eye viewing with a 5 m diameter telescope?

4.2. Estimate the angular resolution of a 5 m diameter telescope in space.

4.3. Compare the collecting areas of 5 and 8 m diameter reflectors. Comment on the significance of this comparison.

4.4. The full Moon subtends an angle of approximately 30 arc min. How large would the image of the Moon be on your film if you used a 500 mm focal length lens for your camera?

4.5. If we have two objects $\theta('')$ apart on the sky, how far apart, x, are their images on the film of a camera with a focal length f. (Assume that we wish to express x and f in the same units.)

4.6. The focal length of the objective on your telescope is 0.8 m. You are using a 25 cm focal length eyepiece. In the image you find that the angular separation between two stars is 10 arc sec. What is the actual angular separation on the sky between the two stars?

4.7. The focal length of the objective on your telescope is 0.5 m. (a) What focal length eyepiece would you have to use to have the image of the full Moon (whose actual size is 30 arc min) subtend an angle of 2°? (b) If you then took a photograph with a 500 mm focal length camera lens, how large would the image be on the film?

4.8. Scale the results in Example 4.1 to write an expression for the limiting magnitude of a telescope of diameter, assuming that you will be viewing directly with your eye.

4.9. Suppose some star is at the limit of naked eye visibility ($m = 6$). How much farther away can we see the same object with a telescope of diameter D? Evaluate your answer for $D = 5$ m.

4.10. Suppose that we use a reflector in the coudé focus, and each of the three mirrors reflects 95% of the light. What fraction of the light is lost as a result of these three reflections?

*4.11. (a) Using the fact that the limiting magnitude of the eye is 6, derive an expression for the limiting magnitude for direct viewing with a telescope of diameter D. (Ignore the effects of sky brightness.) (b) Use this result to derive an expression for the farthest distance at which a telescope of diameter D can be used to see an object of absolute magnitude M.

4.12. (a) What is the diameter of a single telescope with the same collecting area as the Multiple Mirror Telescope? (b) Astronomers have proposed a new telescope with a total collecting area equal to that of a single 25 m diameter telescope. How many 4 m diameter telescopes would be needed to make up this new telescope? (c) The Very Large Telescope, being built by ESO in Chile, has four telescopes, each with an 8 m diameter area. What would be the diameter of a single telescope with the same collecting area?

4.13. Suppose you have a Cassegrain telescope at home, with a 0.25 m diameter primary mirror and a secondary mirror with a diameter of 5 cm. What fraction of the primary is blocked by the secondary?

4.14. If we want to double the image size in a particular observation, by what amount would we have to change the exposure length to have a properly exposed photo?

4.15. Scale the results in Example 4.2 to write an expression for the angular resolution, in seconds of arc ("), of a telescope of diameter D for viewing the middle of the visible part of the spectrum.

4.16. What is the angular resolution of the HST at 200 nm wavelength?

4.17. Generally CCDs have fewer picture elements (pixels) than do photographic plates, so if you want to image a large field, a single photograph might suffice, but you would need a number of CCD images. Suppose you had to make a 3 × 3 square of CCD images to cover your single photographic field, and that your photograph has a quantum efficiency of 5%

while your CCD has an efficiency of 80%. How long will the needed CCD images take relative to the time for the photograph?

4.18. The sodium D lines in the Sun's spectrum are at wavelengths of 589.594 and 588.997 nm. (a) If a grating has 10^4 lines/cm, how wide must the grating be to resolve the two lines in first order? (b) Under these conditions what is the angular separation between the two lines? (c) How would the results in (a) and (b) change for second order?

4.19. A diffraction grating has N lines, a separation d apart. The spectrum is projected on a screen a distance D ($\gg d$) from the grating. Two lines are λ and $\lambda + \Delta\lambda$ apart. How far apart are they on the screen?

4.20. If we want to observe at a wavelength of 10 μm, what are the largest fluctuations that the mirror surface can have?

4.21. What are the angular resolutions of the KAO, SOFIA, IRAS, ISO and SIRTF at wavelengths of 100 μm? The diameter of ISO is 0.60 m, and the diameter of SIRTF is 0.85 m.

4.22. Two infrared sources in the Orion Nebula are 500 pc from us and are separated by 0.1 pc. How large a telescope would you need to distinguish the sources at a wavelength of 100 μm?

*4.23. Suppose we are observing an infrared source that is 500 pc away. It radiates like a 50 K blackbody and is 1 pc in extent. (a) What is the total energy per second per square meter reaching the Earth from this source? How does that compare with the total amount of solar radiation reaching the Earth per second per square meter. (b) Suppose we observe this source using a satellite with a 1 m diameter mirror, and we observe at a wavelength of 100 μm. What is the energy/s/Hz striking the telescope? (c) Suppose the telescope radiates like a blackbody at 300 K, but with an efficiency of 1%. (That is, the spectrum looks like that of a blackbody but with an intensity reduced by a factor of 100.) What is the energy/Hz/s given off by the telescope at this wavelength? How does your answer compare with that in (b). (d) Redo part (c), assuming that we can cool the mirror to 30 K (still with a 1% emission efficiency).

4.24. Suppose we are using an interference filter at a wavelength of 10 μm. (a) How far do you have to move the plates to go from one order maximum to the next? (b) For any given order, how far do you have to move the plates for the peak wavelength to shift from 10.00 μm to 10.01 μm?

4.25. What is the angular resolution (in arc minutes) of (a) a 100 m diameter telescope operating at 1 cm wavelength, and (b) a 30 m telescope operating at 1 mm wavelength?

4.26. Two radio sources in the Orion Nebula are 500 pc from us and are separated by 0.1 pc. How large a telescope would you need to distinguish the sources at a wavelength of (a) 21 cm? (b) 1 mm?

4.27. We sometimes use as a measure of the quality of a radio telescope the diameter d, divided by the limiting wavelength λ_{min}. (a) Why is this quantity important? (b) If a telescope has surface errors of size Δx, give an expression for this quantity in terms of d and Δx.

4.28. How large a collecting area would you need to collect 1 W from a 1 Jy source over a bandwidth of 10^9 Hz (1 GHz)?

4.29. If a radio source emits a solar luminosity $(4 \times 10^{33}$ erg/s) in radio waves, (a) what would be the power per surface area reaching us if we were (i) 1 AU away, (ii) 1 pc away? (b) If that power is uniformly spread out over a frequency range of 10^{11} Hz, what is the flux density (power/surface area/Hz) in each case? (c) How does this compare with the power you receive from a 50 kW radio station 10 km from you?

4.30. What are the angular resolutions of (a) the VLA in its largest configuration (baselines up to 13 km) at a wavelength of 21 cm; (b) the VLA in its most compact configuration (baselines up to 1 km) at a wavelength of 1 cm; (c) the VLBA (with baselines up to 3000 km) at a wavelength of 21 cm; (d) the Millimeter Array (with baselines up to 1 km) at 1 mm wavelength?

4.31. How many pairs of telescopes are there in (a) the VLA with 27 telescopes, (b) the VLBI, with ten telescopes, (c) the proposed Millimeter Array, which may have 40 or 75 telescopes depending on the final design?

4.32. What is the total collecting area of (a) the VLA (b) the ALMA with 75 10 m diameter telescopes? (c) For ALMA, how many 12 m diameter telescopes would you need to have the same collecting area as 75 10 m diameter telescopes.

4.33. (a) Show that for an interferometer with N telescopes, the number of independent pairs of telescopes, at any instant, is $N(N - 1)/2$. (b) Evaluate this for the VLA, the VLBA and ALMA.

4.34. Derive equation (4.9).

Computer problems

4.1. Draw a graph of the angular resolution vs. wavelength, over the infrared part of the spectrum, for IRAS, KAO, SOFIA, ISO.

4.2. Suppose we have a radio telescope whose (one-dimensional) beam pattern is a gaussian with a full width at half maximum (FWHM) of 1 arc min. Calculate the observed source intensity as a function of (one-dimensional) position on the sky, assuming that the source has a uniform intensity over a finite size of (a) 1 arc sec (b) 1 arc min (c) 1 degree, and the source has zero intensity outside that region.

Chapter 5

Binary stars and stellar masses

If we are to understand the workings of stars, it is important to know their masses. The best way to measure the mass of an object is to measure its gravitational influence on another object. (When you stand on a bathroom scale, you are measuring the Earth's gravitational effect on your mass.) For stars, we are fortunate to be able to measure the gravitational effects from pairs of stars, called *binary stars*.

5.1 | Binary stars

Many stars we can observe appear to have companions, the two stars orbiting their common center of mass. It appears that approximately half of all stars in our galaxy are in binary systems. By studying the orbits of binary stars, we can measure the gravitational forces that the two stars exert on each other. This allows us to determine the masses of the stars.

We classify binaries according to how the companion star manifests its presence:

Optical double. This is not really a binary star. Two stars just happen to appear along almost the same line of sight. The two stars can be at very different distances.

Visual binary. These stars are in orbit about each other and we can see both stars directly.

Composite spectrum binary. When we take a spectrum of the star, we see the lines of two different spectral type stars. From this we infer the presence of two stars.

Eclipsing binary. As we observe the light from such a system, it periodically becomes brighter and fainter. We interpret the dimming as occurring when the companion passes behind or in front of the main star. For us to see these eclipses, we must be aligned in the plane of the orbit (Fig. 5.1). A famous example of such a star is *Algol* (also known as β Persei, indicating that it is the second brightest star in the constellation Perseus).

Astrometric binary. Astrometry is the branch of astronomy in which the positions of objects are measured very accurately. In an astrometric binary, we can only see the brighter star. However, when we follow its path on the sky, we see that, instead of following a straight line, it 'wobbles' back and forth across the straight line path. This means that the star is moving in an orbit, so we can infer the presence of a companion (Fig. 5.2).

Spectroscopic binary. When we study the spectrum of a star we may see that the wavelengths of spectral lines oscillate periodically about the average wavelength. We interpret these variations as being caused by a Doppler shift (discussed in the next section). When the star is coming towards us in its orbit, we see the lines at shorter wavelengths, and when the star is moving away from us in its orbit, we see the lines at longer wavelengths.

It is possible for a given binary system to fit into more than one of these categories, depending on what we can observe.

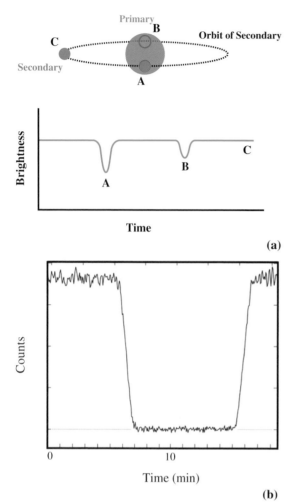

(a)

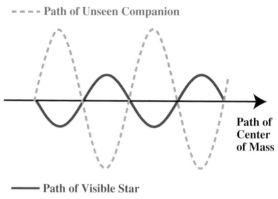

Fig 5.2. Astrometric binary. Two stars orbit about a common center of mass, which in turn moves across the sky. The fainter star is too faint to see, so we only see the brighter star, moving back and forth across the path of the center of mass.

(b)

Fig 5.1. Eclipsing binary. (a) The binary system is shown above; the light curve below. The fainter secondary passes alternately in front of and behind the primary. Most of the time, as at position C, we see light from both stars. When the secondary eclipses the primary, part of the primary light is blocked and there is a dip in the intensity, as at point A. When the secondary passes behind the primary, as at B, its light is lost. Since the secondary is not as hot as the primary, the loss of brightness is not as great as at A. (b) Part of the light curve for the eclipsing binary, NN Ser, around the eclipse part of the orbit. [(b) ESO]

5.2 | Doppler shift

A *Doppler shift* is a change in the wavelength (and frequency) of a wave, resulting from the motion of the source and/or the observer. It is most easily visualized for a sound wave or a water wave, where the waves are moving through a particular elastic medium (Fig 5.3).

5.2.1 Moving sources and observers

We first look at the case of the moving observer. If the observer is moving toward the source, then the waves will be encountered more frequently than if the observer were standing still. This means that the observed frequency of the wave increases. If the frequency increases, then the wavelength decreases. If the observer is moving away from the source then the situation is reversed. Waves will be encountered less frequently; the frequency decreases; the wavelength therefore increases. It should be noted that if the observer moves perpendicular to the line joining the source and observer, no shift will be observed.

We now look at the case of the moving source. Each wavefront is now emitted in a different place. If the source is moving toward the observer, the waves will be emitted closer together than if the source were standing still. This means the wavelength decreases. The decreased wavelength results in an increased frequency. If the source is moving away from the observer, the waves will be emitted farther apart than if the source were standing still. The wavelength increases and the frequency decreases. Again, if the source is moving perpendicular to the line joining the source and observer, no shift results.

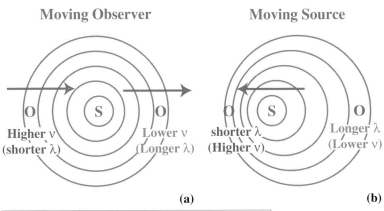

Moving Observer   **Moving Source**

Higher v
(shorter λ)

Lower v
(Longer λ)

shorter λ
(Higher v)

Longer λ
(Lower v)

(a)   **(b)**

Fig 5.3. Doppler shift for waves in an elastic medium, such as sound waves. (a) Moving observer. On the left the observer is moving toward the source, encountering wave crests more frequently than for a stationary observer. The frequency appears to increase (and the wavelength to decrease). On the right, the observer is moving away from the source, encountering waves at a lower frequency, corresponding to a lower frequency, corresponding to a longer wavelength. (b) Moving source. The motion of the source distorts the wave pattern, so the circles are no longer concentric. The observer on the left has the source approaching, producing a shorter wavelength (and a higher frequency). The observer on the right has the source receding, producing a longer wavelength (and a lower frequency).

It is possible for both the source and the observer to be moving. If their combined motion brings them closer together, the wavelength will decrease and the frequency will increase. If their combined motion makes them move farther apart, the wavelength will increase, and the frequency will decrease. If there is no instantaneous change in their separation, there is no shift in wavelength or frequency.

The shift only depends on the component of the relative velocity along the line joining the source and observer, since this is the only component that can change the distance, r, between them. We call this component the *radial velocity* (Fig. 5.4). We refer to the line joining the source and observer as the *line of sight*. From our definition of radial velocity, v_r, we can see that it is given by

$$v_r = dr/dt \qquad (5.1)$$

Note that if the source and observer are moving apart, r is increasing, and $v_r > 0$. If the source and observer are moving together, r is decreasing, and $v_r < 0$.

Suppose the source is moving with a speed v_s in a direction that makes an angle θ with the line of sight, and the observer is moving with a speed v_o in a direction making an angle ϕ with the line of sight. Taking the components of the two velocities along the line of sight as $v_s \cos \theta$ and $v_o \cos \phi$, and subtracting the get the relative radial velocity, gives

$$v_r = v_s \cos \theta - v_o \cos \phi \qquad (5.2)$$

In astronomy we are interested in the Doppler shift for electromagnetic waves. The underlying physics is a little different, because there is no mechanical medium for these waves to move through. They can travel even in a vacuum. (We will discuss this point further in Chapter 7.) For sound waves, the actual amount of Doppler shift depends on whether the source or observer (or

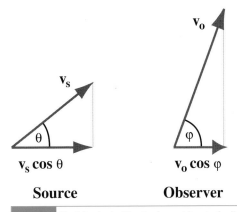

v_o

v_s

θ

$v_s \cos \theta$

φ

$v_o \cos \varphi$

Source   **Observer**

Fig 5.4. Radial velocity. The horizontal line is the line of sight between the source and observer. The radial velocity is the difference between the line of sight components of the observer and source velocities.

both) is moving. For electromagnetic waves, only the relative motion counts.

As long as the relative speed of the source and the observer is much less than the speed of light, the results for electromagnetic radiation are relatively simple. If λ is the wavelength at which a signal is received, and λ_0 is the wavelength at which it was emitted, called the *rest wavelength*, the wavelength shift $\Delta\lambda$ is defined by

$$\Delta\lambda = \lambda - \lambda_0 \qquad (5.3)$$

The simple result is that the wavelength shift, expressed as a fraction of the original wavelength, is equal to the radial velocity, expressed as a fraction of the speed of light. That is

$$\Delta\lambda/\lambda_0 = v_r/c \qquad (v_r \ll c) \qquad (5.4)$$

If $v_r > 0$, then $\Delta\lambda > 0$. For a spectral line in the middle of the visible part of the spectrum, a shift to longer wavelength is a shift to the red, so this is called a *redshift*. The name applies even if we are in other parts of the spectrum. A positive radial velocity always produces a redshift. If $v_r < 0$, then $\Delta\lambda < 0$, and we have a *blueshift*.

We now look at what happens to the frequency. We remember that $\lambda = c/v$, so

$$d\lambda/dv = -c/v^2$$
$$= -\lambda/v$$

This means that $\Delta v = (-v/\lambda)\,\Delta\lambda$. Substituting into equation (5.4) gives

$$\Delta v/v_0 = -v_r/c \qquad (v_r \ll c) \qquad (5.5)$$

The shift in frequency, expressed as a fraction of the rest frequency, is the negative of the radial velocity, expressed as a fraction of the speed of light. (So $\Delta v/v_0 = -\Delta\lambda/\lambda_0$.)

Example 5.1 Doppler shift
The Hα line has a rest wavelength, $\lambda_0 = 656.28$ nm. What is the observed wavelength for a radial velocity (a) $v_r = 10$ km/s, and (b) $v_r = -10$ km/s?

SOLUTION
(a) We use equation (5.4) to find the wavelength shift:

$$\Delta\lambda = \lambda_0(v_r/c)$$
$$= \frac{(656.28\ \text{nm})(10.0\ \text{km/s})}{(3.0 \times 10^5\ \text{km/s})} = 0.022\ \text{nm}$$

We now add this to the original wavelength to find the observed wavelength:

$$\lambda = \lambda_0 + \Delta\lambda$$
$$= 656.28\ \text{nm} + 0.022\ \text{nm}$$
$$= 656.30\ \text{nm}$$

(b) If we take the negative of v_r, we just get the negative of $\Delta\lambda$, so

$$\Delta\lambda = -0.022\ \text{nm}$$

This gives a wavelength of

$$\lambda = 656.26\ \text{nm}$$

If we observe two spectral lines, their wavelength shifts will be different, since each is shifted by an amount proportional to its own rest wavelength. Thus, the spacing between spectral lines will be shifted.

5.2.2 Circular orbits
We now look at Doppler shifts produced by a star in a circular orbit. The orbital speed is v, and the radius is r. The angular speed of the star in its orbit (in radians per second) is given by $\omega = v/r$. The situation is shown in Fig. 5.5. Suppose the star is moving directly away from the observer at time $t = 0$. At that instant the radial velocity $v_r = v$. As the stars moves, the component of its

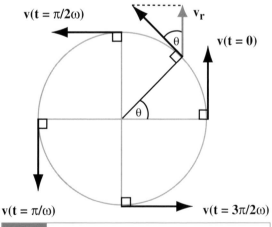

Fig 5.5. Doppler shift for a circular orbit. The speed v of the source remains constant, but the direction changes, so the radial velocity v_r changes. The angle θ keeps track of how far around the circle the source has gone. The source velocity is shown for five different values of θ.

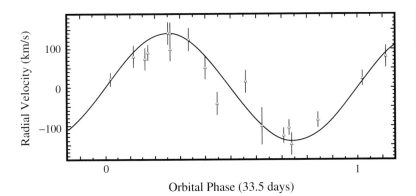

Fig 5.6. Radial velocity curve for the spectroscopic binary GRS1915+105, which has an orbital period of 33.5 days. The horizontal axis is the fraction through that orbit. Data points and error bars are indicated. The smooth curve is the best fit to the data.[ESO]

velocity along the line of sight is $v \cos \theta$. However, $\theta = \omega t$, so

$$v_r = v \cos(\omega t) \qquad (5.6)$$

The radial velocity changes sign every half-cycle, and repeats periodically. This is shown in Fig. 5.6. The period of the motion is the circumference, $2\pi r$, divided by the speed v, so $P = 2\pi/\omega$. If we substitute equation (5.6) into equation (5.4), we find that the spectral lines shift back and forth, with a shift given by

$$\Delta\lambda/\lambda_0 = (v/c) \cos (\omega t)$$

So far we have been considering the situation in which the observer is in the plane of the orbit. If the observer is not in the plane of the orbit, the Doppler shift will be reduced (Fig. 5.7). If i is the angle between the plane of the orbit and the plane of the sky, then the projection of any velocity in the plane of the orbit into the line of sight is $v \sin i$. The angle i is known as the *inclination* of

the orbit. This gives us a radial velocity for an orbiting star

$$v_r = v \sin i \cos(\omega t) \qquad (5.7)$$

5.3 | Binary stars and circular orbits

In this section we will see how Newton's laws of motion and gravitation can be applied to binary stars in circular orbits. Circular orbits are not the most general case of orbital motion, but the analysis is most straightforward, and most of the basic points are clearly illustrated. In the next section we will go to the general case of elliptical orbits.

We consider two stars, of masses m_1 and m_2, orbiting their common center of mass at distances r_1 and r_2, respectively (Fig. 5.8). From the

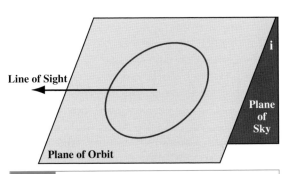

Fig 5.7. Inclination of an orbit. The orbit is an ellipse, which lies in a plane. The plane makes some angle i with a plane of the sky. The plane of the sky is defined to be perpendicular to the line of sight.

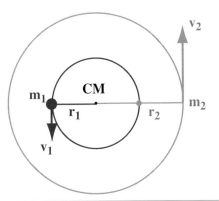

Fig 5.8. Binary system with circular orbits. Both stars orbit the center of mass (CM). The more massive star is closer to the center of mass. The center of mass must always be between the two stars, so the stars lie on opposite sides of it.

definition of center of mass, these quantities are related by

$$m_1 r_1 = m_2 r_2 \qquad (5.8)$$

The center of mass moves through space subject only to the external forces on the binary star system. The forces between the two stars do not affect the motion of the center of mass. We will therefore ignore the actual motion of the center of mass, and view the situation as it would be viewed by an observer sitting at the center of mass.

Since the center of mass must always be along the line joining the two stars, the stars must always be on opposite sides of the center of mass. This means that the stars orbit with the same orbital period P. In general, the period of the orbit is related to the radius, r, and the speed, v, by

$$P = 2\pi r/v \qquad (5.9)$$

Solving for v gives

$$v = 2\pi r/P \qquad (5.10)$$

Since the periods of the two stars must be the same, equation (5.9) tells us that

$$r_1/v_1 = r_2/v_2 \qquad (5.11)$$

Combining these with equation (5.8) gives

$$v_1/v_2 = r_1/r_2 = m_2/m_1 \qquad (5.12)$$

(Note: We could have also obtained $m_1 v_1 = m_2 v_2$ directly from conservation of momentum. This is not surprising since the properties of the center of mass come from conservation of momentum.)

We now look at the gravitational forces. The distance between the two stars is $r_1 + r_2$, so the force on either star is given by Newton's law of gravitation as

$$F = G\frac{m_1 m_2}{(r_1 + r_2)^2} \qquad (5.13)$$

This force must provide the acceleration associated with the change of the direction of motion in circular motion, v^2/r. For definiteness, we look at the force on star 1, so

$$F = m_1 v_1^2/r_1 \qquad (5.14)$$

Combining equations (5.13) and (5.14) gives

$$\frac{m_1 v_1^2}{r_1} = G\frac{m_1 m_2}{(r_1 + r_2)^2} \qquad (5.15)$$

Note that we can divide both sides by m_1. If we also use equation (5.10) to relate v_1 to P, we find that

$$\frac{4\pi^2 r_1}{P^2} = \frac{G m_2}{(r_1 + r_2)^2} \qquad (5.16)$$

This can be simplified if we introduce the total distance R between the two stars:

$$R = r_1 + r_2$$
$$= r_1(1 + r_2/r_1) \qquad (5.17)$$

Using equation (5.12), this becomes

$$R = r_1(1 + m_1/m_2) \qquad (5.18)$$
$$= (r_1/m_2)(m_1 + m_2) \qquad (5.19)$$

Substituting this into equation (5.16) gives

$$4\pi^2 R^3/G = (m_1 + m_2)P^2 \qquad (5.20)$$

Let's look at how equation (5.20) can be used to give us stellar masses. For any binary system, we can determine the period directly if we watch the system for long enough. If the star is a spectroscopic binary, we can see how long it takes for the Doppler shifts to go through a full cycle. If it is an astrometric binary we can see how long it takes for the 'wobble' to go through a full cycle. If it is an eclipsing binary, we can see how long it takes the light curve to go through a full cycle. If we can see both stars, we can determine R. Once we know R and P, we can use equation (5.20) to determine the sum of the masses, $(m_1 + m_2)$. We can also obtain the ratio of the masses m_1/m_2, either from r_1/r_2, if both stars can be seen, or v_1/v_2, if both Doppler shifts are observed. Once we know the sum of the masses and the ratio of the masses, the individual masses can be determined. The situation we have outlined here is the ideal one, however. Usually, we don't have all of these pieces of information (as we will see below).

Example 5.2 Mass of the Sun
We can consider the Sun and Earth as a binary system, so we should be able to apply equation (5.20) to

find the mass of the Sun. It turns out that this is the most accurate measure we have of the Sun's mass.

SOLUTION

Since the mass of the Sun is so much greater than that of the Earth, we can approximate the sum of the masses as being the mass of the Sun, $M_\odot$. Equation (5.20) then becomes

$$M_\odot = \frac{4\pi^2 R^3}{GP^2}$$

$$= \frac{4\pi^2 (1.5 \times 10^{13} \text{ cm})^3}{(6.67 \times 10^{-8} \text{ dyn cm}^2/\text{g}^2)(3.16 \times 10^7 \text{ s})^2}$$

$$= 2 \times 10^{33} \text{g}$$

We call this quantity a *solar mass*. It becomes a convenient quantity for expressing the masses of other stars.

From the Earth and Sun we know that for a pair of objects orbiting with a period of 1 yr, at a distance of 1 AU (defined in Section 2.6), the sum of the masses must be one solar mass. This suggests a convenient system of units for equation (5.20). If we express masses in solar masses, distances in astronomical units, and periods in years, the constants must equal one to give the above result. We can therefore rewrite equation (5.20) as

$$\left[\frac{R}{1 \text{ AU}}\right]^3 = \left[\frac{m_1 + m_2}{1\, M_\odot}\right]\left[\frac{P}{1 \text{ yr}}\right]^2 \tag{5.21}$$

For the Solar System, we write the sum of the masses as one in these units, so the equation simply says that the cube of the radius (in AU) is equal to the square of the period (in yr). This is also known as *Kepler's third law of planetary motion*. The law was originally found by Kepler observationally, and Newton used it to show that gravity must be an inverse square law force. (See Problem 5.10.) We will discuss planetary motions in more detail in Chapter 22.

For a visual binary, we don't directly measure R, the linear separation. We actually measure the angular separation on the sky, θ. If d is the distance to the binary, then R is equal to $\theta(\text{rad})d$, where $\theta(\text{rad})$ is the value of θ measured in radians. When we use this relation, R and d will come out in the same units. Therefore

$$R(\text{AU}) = d(\text{AU})\theta(\text{rad})$$

The values of θ are so small that radians are an inconvenient quantity. We can convert to arc seconds (equation (2.16)) to give

$$R(\text{AU}) = d(\text{AU})\theta('')(2.06 \times 10^5)$$

The factor of 2.06×10^5 was to convert radians to arc seconds, but it is also the factor to convert astronomical units to parsecs, so we have

$$R(\text{AU}) = d(\text{pc})\theta('')$$

If we use equation (2.17) to relate the distance in parsecs to the parallax in arc seconds, this becomes

$$R(\text{AU}) = \theta('')/p('') \tag{5.22}$$

This can then be substituted directly into equation (5.21).

We will now look at the behavior of the Doppler shifts. Applying equation (5.10) to both speeds, v_1 and v_2, and remembering that the period of the orbit is the same for both stars, we have

$$r_1 + r_2 = (P/2\pi)(v_1 + v_2)$$

Using this to eliminate R in equation (5.20) gives

$$(P/2\pi G)(v_1 + v_2)^3 = m_1 + m_2 \tag{5.23}$$

If the orbit is inclined at an angle i, then the Doppler shifts only measure the components $v_r = v\sin(i)$. In terms of the radial velocities v_{1r} and v_{2r}, equation (5.23) becomes

$$(P/2\pi G)(v_{1r} + v_{2r})^3/\sin^3 i = m_1 + m_2 \tag{5.24}$$

If a binary happens to be an eclipsing binary, then we know that we are close to the plane of the orbit, and i is close to 90°. Otherwise we don't know i. If a circular orbit is projected at some angle on the plane of the sky it will appear elliptical. We will see in the next section that there are ways to determine i if we can trace that projected orbit on the sky. If i is unknown all we can do is solve equation (5.24) with $i = 90°$. This will give us a value of $m_1 + m_2$ that is a *lower limit* to the true value. The true value would be this lower limit divided by $\sin^3 i$, and since $\sin^3 i$ is less than or equal to unity, the value assuming $i = 90°$ is less than the true value. Finding lower limits is not as useful as finding actual values. However, if we study enough binary systems, we will encounter a full range of inclination angles.

These statistical studies can be used to relate mass to spectral type.

Example 5.3 Binary star Doppler shifts

A binary system is observed to have a period of 10 yr. The radial velocities of the two stars are determined to be $v_{1r} = 10$ km/s and $v_{2r} = 20$ km/s, respectively. Find the masses of the two stars (a) if the inclination of the orbit is 90°, and (b) if it is 45°.

SOLUTION

From equation (5.24), we have

$$\frac{m_1 + m_2}{M_\odot} =$$

$$\frac{(10\ \text{yr})(3.16 \times 10^7\ \text{s/yr})[(10 + 20)(1 \times 10^5\ \text{cm/s})]^3}{2\pi(6.67 \times 10^{-8}\ \text{dyn cm}^2/\text{g}^2)(2 \times 10^{33}\ \text{g})(\sin^3 i)}$$

$$= 10.2\ M_\odot/\sin^3 i$$

This means that

$$m_1 + m_2 = 10.2\ M_\odot/\sin^3 i$$

If $i = 90°$, $\sin^3 i = 1$, so

$$m_1 + m_2 = 10.2\ M_\odot$$

We find the ratio of the masses from the ratio of the radial velocities

$$m_1/m_2 = v_2/v_1$$

$$= 2.0$$

This means that $m_1 = 2m_2$, so

$$2m_2 + m_2 = 10.2\ M_\odot$$

giving $m_1 = 6.8\ M_\odot$ and $m_2 = 3.4\ M_\odot$. If $i = 45°$, $1/\sin^3 i = 2.8$. The ratio of the masses does not change, since the sin i drops out of the ratio of the radial velocities. This means that we can just multiply each mass by 2.8 to give 19.2 and 9.5 $M_\odot$, respectively.

It is often the case that only one Doppler shift can be observed. Let's assume that we measure v_1 but not v_2. We must therefore eliminate v_2 from our equations. We can write v_2 as

$$v_2 = v_1(m_1/m_2)$$

The sum of the velocities then becomes

$$v_1 + v_2 = v_1(1 + v_2/v_1)$$

$$= v_1(1 + m_1/m_2)$$

$$= (v_1/m_2)(m_1 + m_2)$$

If we now substitute into equation (5.24), we find

$$\left(\frac{P}{2\pi G}\right)(v_{ir})^3 = \frac{m_2^3 \sin^3 i}{(m_1 + m_2)^2} \tag{5.25}$$

The quantity on the right-hand side of equation (5.25) is called the *mass function*. If we can measure only one Doppler shift, we cannot determine either of the masses. We can only measure the value of the mass function. We can, however, obtain information on the masses of various spectral types through extensive statistical studies.

We can also look at energy of a binary system with circular orbits. It is the sum of the kinetic energies of the two stars plus the gravitational potential energy, which we take to be zero when the two objects are infinitely far apart. The energy is

$$E = (1/2)m_1 v_1^2 + (1/2)m_2 v_2^2 - Gm_1 m_2/R \tag{5.26}$$

From equation (5.15), we have

$$m_1 v_1^2 = Gr_1 \frac{m_1 m_2}{(R)^2}$$

and, using equation (5.12), we obtain

$$m_2 v_2^2 = Gr_2 \frac{m_1 m_2}{(R)^2}$$

Substituting these into equation (5.26) and simplifying gives

$$E = -(1/2)\,Gm_1 m_2/R \tag{5.27}$$

Compare this with equation (3.4), which has the energy for circular orbits with electrical (rather than gravitational) forces.

The negative energy means that the system is bound. We would have to add at least $(1/2)Gm_1 m_2/R$ to break up the binary system. (We can think of this as being analogous to the binding energy of a hydrogen atom.) For any pair of masses, as you make the orbits of the binary system smaller, the energy becomes more negative.

Example 5.4 Binding energy of a binary system.

What is the binding energy of a binary system with two 1 $M_\odot$ stars orbiting with each 100 AU from the center of mass?

SOLUTION

Using equation (5.27)

$$E = -\left(\frac{1}{2}\right)\frac{Gm_1 m_2}{R}$$

$$= -\left(\frac{1}{2}\right)\frac{(6.67 \times 10^{-8}\mathrm{dyn\ cm^2/g^2})(2 \times 10^{33}\,\mathrm{g})^2}{(200)(1.5 \times 10^{13}\mathrm{cm})}$$

$$= -4.5 \times 10^{43}\ \mathrm{erg}$$

5.4 | Elliptical orbits

5.4.1 Geometry of ellipses

In general, orbiting bodies follow elliptical paths. A circle is just a special case of an ellipse. In this section, we generalize the results from the previous section from circular orbits to elliptical orbits. The basic underlying physical ideas are the same.

We first review the geometry of an ellipse, as shown in Fig. 5.9. We describe the ellipse by its semi-major axis a and semi-minor axis b. Each point on the ellipse satisfies the condition that the sum of the distances from any point to two fixed points, called *foci* (singular *focus*), is constant. If r and r' are these two distances then

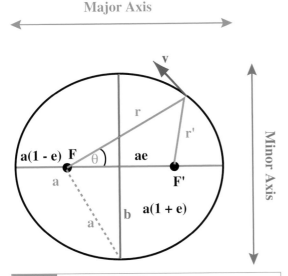

Fig 5.9. Geometry of an ellipse. The length of the semi-major axis is a; the length of the semi-minor axis is b. The two foci are at F and F'. The eccentricity is e, and the distances from the two foci to points on the ellipse are r and r'.

$r + r'$ is a constant. We can see that for a point on the semi-major axis (and the ellipse), this sum is $2a$, so it must be $2a$ everywhere. That is

$$r + r' = 2a \tag{5.28}$$

The *eccentricity* of an ellipse is the distance between the foci, divided by $2a$. A circle is an ellipse of eccentricity zero (both foci are at the same point, the center of the circle). The eccentricity of any ellipse must be less than unity. From the point where the curve crosses the minor axis, $r = r' = a$, so

$$b^2 = a^2 - (ae)^2$$

$$= a^2(1 - e^2) \tag{5.29}$$

In a binary system, the center of mass of the two stars will be at one focus on the ellipse. The farthest point from that focus is called the *apastron*. From the figure, we see that the distance from the focus to this point is

$$r(\mathrm{apastron}) = a(1 + e) \tag{5.30a}$$

The closest point to the focus is called the *periastron*. Its distance from the focus is

$$r(\mathrm{periastron}) = a(1 - e) \tag{5.30b}$$

The average of these two values is a, the semi-major axis. This is the quantity that replaces the radius of a circular orbit in our study of binary stars.

It is useful to have an expression for the ellipse, relating the variables r and θ. From the law of cosines, we see that

$$r'^2 = r^2 + (2ae)^2 + 2r(2ae)\cos\theta \tag{5.31}$$

Using equation (5.28) gives

$$r = \frac{a(1 - e^2)}{1 + e \cos\theta} \tag{5.32}$$

5.4.2 Angular momentum in elliptical orbits

The gravitational force between two objects always acts along the line joining the two objects. The center of mass also lies along this line. This means that the force on either object points directly from that object towards the center of mass. Therefore, these forces can exert no torques about the center of mass. If there are no torques about the center of

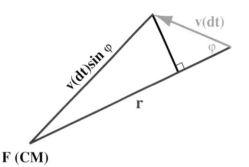

F (CM)

Fig 5.10. Angular momentum in an elliptical orbit. In this case the object is a distance r from the focus F (which is also the center of mass). Its velocity makes an angle ϕ with the line from F to the object. (If this were a circle, ϕ would be 90°.) The time interval over which we mark the motion is dt.

mass, then the angular momentum about the center of mass is conserved.

To consider the angular momentum of an object in an elliptical orbit, we look at Fig. 5.10. For an object of mass m, the angular momentum about the center of mass (which corresponds to one of the foci) is r times the component of its linear momentum perpendicular to the direction of r. That is

$$L = mvr \sin \phi \qquad (5.33)$$

where ϕ is between the line from the center of mass to the object (of length r) and its direction of motion (the direction of the velocity vector, $\mathbf{v}$).

We look at the area swept out by the line from the center of mass to r, in the time interval dt. The area is the thin triangle shown in the figure. The long side of the triangle is r and the short side is $v\,dt$. The small right triangle shows that the height of the larger triangle is $v\,(dt)\sin\phi$. The area is then half the base × height, so

$$dA = (1/2)\,r(v\,dt)\sin\phi$$

The rate at which the area is swept out is

$$dA/dt = (1/2)r\,v\sin\phi$$

However, we can use equation (5.33) to eliminate $r\,v\sin\varphi$, giving

$$dA/dt = (1/2)\,(L/m)$$

Since L is constant, the rate at which area is swept out is constant. Equal areas are swept out in equal times. (When applied to planetary motions, this is known as *Kepler's second law*, as we will discuss in Chapter 22.)

The major consequence of angular momentum conservation is therefore that the objects move slower when they are farther apart and faster when they are closer together. (As we will see, the Earth's orbit is only slightly eccentric but it moves faster when it is closer to the Sun, in January, and slower when it is farther away. This results in winter and spring in the northern hemisphere being shorter than summer and winter. Check a calendar to see this.)

5.4.3 Energy in elliptical orbits

We next look at the total energy of the binary system. Adding the kinetic energies of the two stars and their gravitational potential energy (defined as zero when the stars are infinitely far apart) gives

$$E = (1/2)m_1 v_1^2 + (1/2)\,m_2 v_2^2 - Gm_1 m_2/R \qquad (5.34)$$

In this expression v_1 and v_2 are the speeds relative to the center of mass. Remember, from the definition of the center of mass, we found that $m_1 v_1 = m_2 v_2$. We introduce the relative speed of the two stars

$$v = v_1 + v_2 \qquad (5.35a)$$

The two speeds are added because the two stars are always moving in opposite directions, so their relative speed will be the sum of the magnitudes of their individual speeds. In terms of v, the two speeds are

$$v_1 = \frac{m_2 v}{m_1 + m_2} \qquad (5.35b)$$

$$v_2 = \frac{m_1 v}{m_1 + m_2} \qquad (5.35c)$$

Substituting into equation (5.32) gives

$$E = m_1 m_2 \left[\frac{v^2}{2(m_1 + m_2)} - \frac{G}{R} \right] \qquad (5.36)$$

Since energy is conserved, we can evaluate it at any point we want. For simplicity, we choose apastron (speed v_a) and periastron (speed v_p). We can relate the speeds v_a and v_p by conservation of angular momentum. Angular momentum conservation is easy to apply at the apastron and periastron because the velocities are perpendicular to the line from the focus to the star. Since $\phi = 90°$ at these points, the angular momentum (equation (5.33)) is just mvr. Using equations (5.30a) and

(5.30b) we find that

$$a(1 + e)v_a = a(1 - e)v_p \qquad (5.37)$$

Solving for the ratio v_p/v_a gives

$$\frac{v_p}{v_a} = \frac{1 + e}{1 - e} \qquad (5.38)$$

Now that we have the ratio v_p/v_a, we need another relation between them to be able to solve for v_a and v_p individually. We can use conservation of energy to equate the energies at the apastron and periastron. Using equation (5.36) gives

$$\frac{v_a^2}{2(m_1 + m_2)} - \frac{G}{a(1 + e)} = \frac{v_b^2}{2(m_1 + m_2)} - \frac{G}{a(1 - e)} \qquad (5.39)$$

Rearranging gives

$$\left[\frac{G(m_1 + m_2)}{a} \right] \left(\frac{1}{1 - e} - \frac{1}{1 + e} \right) = \frac{1}{2}(v_p^2 - v_a^2)$$

$$= \frac{1}{2} v_a^2 \left[\left(\frac{v_p^2}{v_a^2} \right) - 1 \right]$$

We can use equation (5.38) to eliminate the ratio v_p/v_a. Solving for v_a gives

$$v_a^2 = \left[\frac{G(m_1 + m_2)}{a} \right] \left(\frac{1 - e}{1 + e} \right) \qquad (5.40a)$$

$$v_p^2 = \left[\frac{G(m_1 + m_2)}{a} \right] \left(\frac{1 + e}{1 - e} \right) \qquad (5.40b)$$

If we put these into equation (5.36), the total energy simplifies to

$$E = -Gm_1m_2/2a \qquad (5.41)$$

We can now use this in the left-hand side of equation (5.36). We can then solve for v at any point r:

$$v^2 = G(m_1 + m_2)(2/r - 1/a) \qquad (5.42)$$

5.4.4 Observing elliptical orbits

In studying the Doppler shifts of elliptical orbits as compared with circular orbits, there are three important differences:

(1) The speed along an elliptical orbit is not constant.
(2) In an elliptical orbit the velocity is not perpendicular to the line from the center of mass to the orbiting object.

(3) Even if you are in the plane of the orbit, the radial velocity curve depends on where you are relative to the major axis of the ellipse.

These points are illustrated in Fig. 5.11.

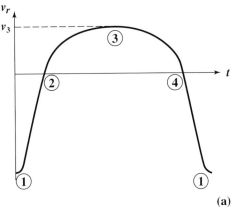

(a)

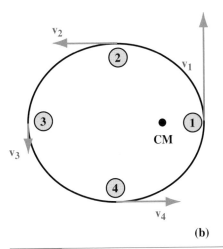

(b)

Fig 5.11. (a) Radial velocity vs. time, t, for an elliptical orbit. (b) In contrast to the circular orbit both the magnitude and direction of **v** change throughout the orbit. (We assume for this figure that the observer is in the plane of the orbit, or that $i = 90°$.) Four points are shown in the orbit and in the radial velocity curve. At points 2 and 4 the motion is perpendicular to the line of sight, so $v_r = 0$. For point 1 the motion is directly toward the observer, producing the maximum negative v_r, and, for point 3, the motion is away from the observer, producing the maximum positive v_r. The motion is also faster at 1 than at 3. In addition, going from 4 to 1 to 2 takes less time than going from 2 to 3 to 4. This accounts for the distorted shape of the radial velocity curve.

As we said above, we must correct any Doppler shift for the inclination of the orbit, i. If we take a tilted ellipse and project it onto the sky, we still have an ellipse. However, that ellipse will have a different eccentricity than the true ellipse. When we look at an elliptical orbit, how can we tell if it is tilted or not? For a tilted orbit the foci will not appear in the right place for the projected ellipse. Therefore, if we see two stars orbiting a point different from the center of mass, we will know that the orbit is inclined. We can determine the inclination from the degree to which the foci appear to be displaced. In this technique, we don't actually know where the center of mass is, so a process must be used in which we try one position for the center of mass and try to match the projected orbit, repeating the process until a good fit is achieved.

5.5 | Stellar masses

As a result of studying many binary systems, astronomers have a good idea of the masses of main sequence stars. These results are summarized in Table 5.1. Just as the Sun's temperature places it in the middle of the main sequence, its mass is in the middle of the range of stellar masses. The lowest mass main sequence stars have about 0.07 of a solar mass, and the most massive stars commonly encountered have about 60 solar masses. When we think of how large or small stars might have turned out to be, the observed range of stellar masses is not very large. This range is an important constraint on theories of stellar structure.

Table 5.1. | Mass and spectral type (MS).

Spectral type	$M/M_\odot$
O5	40.0
B5	7.1
A5	2.2
F5	1.4
G5	0.9
K5	0.7
M5	0.2

An even more stringent constraint is the relationship between mass and temperature on the main sequence. The cooler stars are less massive and the hotter stars are more massive. We have already said that the existence of the main sequence implies a certain relationship between size and temperature. This means that if a star is on the main sequence, once its mass is specified, its radius and temperature are determined. Another way of looking at this to say that a star's mass determines where on the main sequence it will fall.

Since the mass determines the radius and temperature of a main sequence star, it should not be surprising that it also determines the luminosity. The exact dependence of the luminosity on mass is called the *mass–luminosity relationship*. This relationship is also explainable from theories of stellar structure. This relationship is shown in Fig. 5.12. We can summarize it by saying that the luminosity varies approximately as some power, α, of the mass. If we express luminosities in terms of solar luminosities, and masses in terms of solar masses, this means that

$$L/L_\odot = (M/M_\odot)^\alpha \qquad (5.43a)$$

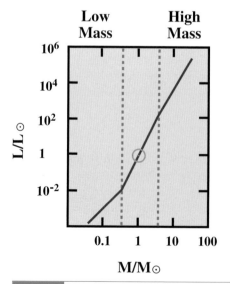

Fig 5.12. Mass–luminosity relationship.

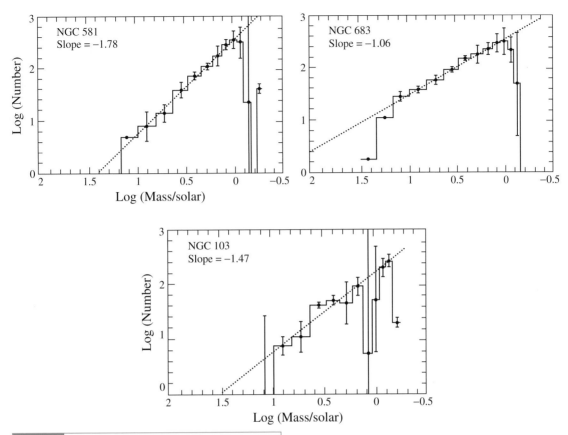

Fig 5.13. Initial mass function for three young clusters. Note that masses are in solar masses, so log(mass) = 0 corresponds to one solar mass. [John Scalo, University of Texas, Austin]

A single value of α does not work for the full range of masses along the main sequence. The approximate values are:

$$\alpha = 1.8 \text{ for } M < 0.3\,M_\odot \qquad \text{(low mass)}$$

$$\alpha = 4.0 \text{ for } 0.3\,M_\odot < M < 3\,M_\odot \text{ (intermediate mass)} \tag{5.43b}$$

$$\alpha = 2.8 \text{ for } 3\,M_\odot < M \qquad \text{(high mass)}$$

In understanding how stars are formed (Chapter 14) we would like to know the distribution of stellar masses. That is, we would like to know the proportions of stars of various masses. Since we now know how to relate mass and spectral type, we can carry out such studies by looking at the relative numbers of different spectral types or luminosities. These studies are difficult,

because we can see brighter stars to greater distances than faint stars. This is called a *selection effect*, because it makes it difficult to select an unbiased sample.

It is of interest to study the number of stars in different mass ranges. This is called the mass function. This is a good check on various theories of star formation, which we will discuss further in Chapter 15. When we discuss stellar evolution (Chapters 10 and 11), we will see that low mass stars live longer than high mass stars. So, if we study the mass function in a group of stars (usually clusters, as we will discuss in Chapter 13), it will change with time. As the higher mass stars die, they leave behind a cluster with a depleted number of high mass stars. So, if we want to really see what the mix of masses are when the cluster forms, we need to look at young clusters. This allows us to study what we call the *initial mass function (IMF)*. Data for some sample clusters are shown in Fig. 5.13. Note: the number of low mass stars is so much greater than the number of high mass stars, that we show the results on a log–log plot. In general, we find that we can fit the data with power laws of the form $N(m) \sim m^{-\gamma}$. The best fit values of γ are shown as slopes in the figure. These figures don't have much data for stars much less massive than the Sun, but other studies indicate that the function doesn't rise as fast as it does for the mass range shown in Fig. 5.13.

We can also look at the number of stars in different luminosity ranges, or the *luminosity function*. We find that there are also many more low luminosity stars than high luminosity stars. However, the luminosity of each high mass star is so much greater than that for each low mass star, that most of the luminosity of our galaxy comes from a relatively small number of high mass stars.

5.6 | Stellar sizes

We have alluded so far to stellar sizes, but we have not discussed how they are determined. In this section, we will look at various methods for measuring stellar radii.

The star whose size is easiest to measure is the Sun. This is actually quite useful. We have seen that the Sun is intermediate in its mass and temperature, so its radius is probably a fairly representative stellar radius. The angular radius of the Sun, $\Delta\theta$, is 16 arc min. The Sun is at a distance $d = 1.50 \times 10^8$ km, so its radius, $R_\odot$, is given by

$$R_\odot = d \tan \Delta\theta$$

$$= 6.96 \times 10^5 \text{ km}$$

The Sun is the only star whose disk subtends an angle larger than the seeing limitations of ground-based telescopes. We therefore need other techniques for determining radii. If we know the luminosity (from its absolute magnitude) and the surface temperature (from the spectral type) of a star, we can calculate its radius using equation (2.7). Solving for the radius gives

$$R = (L/4\pi \, \sigma T^4)^{1/2} \tag{5.44}$$

Example 5.5 Luminosity radius
Estimate the radius of an A0 star. (Use Appendix E for the stellar properties.)

SOLUTION
We can express the various quantities in solar units. Taking ratios, we can use equation (5.44) to give

$$R/R_\odot = (L/L_\odot)^{1/2}(T/T_\odot)^{-2}$$

$$= (80)^{1/2}(1.69)^{-2}$$

$$= 3.1$$

Eclipsing binaries (such as in Fig. 5.1) provide us with another means of determining stellar radii. This method involves analysis of the shape of the light curve and a knowledge of the orbital velocities from Doppler shift measurements. (In an eclipsing binary, we don't have to worry about the inclination of the orbit.) Particularly important is the rate at which the light level decreases and increases at the beginning and end of eclipses.

We can also estimate the radii of rotating stars. If there are surface irregularities, such as hot spots or cool spots, the brightness of the star will depend on whether these spots are facing us or are turned away from us. The brightness variations give us the rotation period P. From the broadening of spectral lines, due to the Doppler shift, we can determine the rotation speed v. This speed is equal to the circumference $2\pi R$, divided by the period. Solving for the radius gives

$$R = P\,v/2\pi \tag{5.45}$$

Sometimes the Moon passes in front of a star bright enough and close enough for detailed study. An analysis of these *lunar occultations* tells us about the radius of the star. The larger the star is, the longer it takes the light to go from maximum value to zero as the lunar edge passes in front of the star. Actually, since light is a wave, there are diffraction effects as the starlight passes the lunar limb. The light level oscillates as the star disappears. The nature of these oscillations tells us about the radius of the star.

There is another observational technique, called *speckle interferometry*, that has been quite successful recently. If it were not for the seeing fluctuations in the Earth's atmosphere, we would be able to obtain images of stellar disks down to the diffraction limits of large telescopes. However, the atmosphere is stable for short periods of

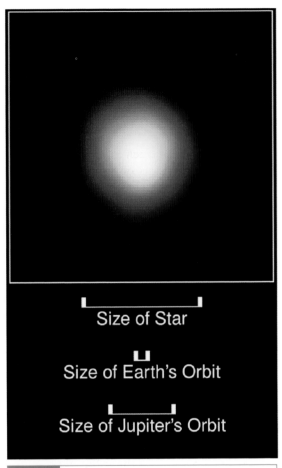

Size of Star

Size of Earth's Orbit

Size of Jupiter's Orbit

Fig 5.14. Hubble Space Telescope image of the red giant star Betelgeuse. You can see that it is barely resolved, but by understanding how the telescope response smears the image, we can achieve a very accurate estimate for the size of the star. [STScI/NASA]

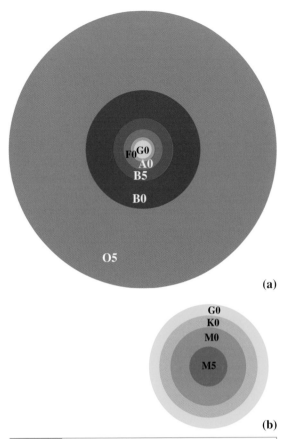

(a)

(b)

Fig 5.15. Stellar sizes for different spectral types. (a) The largest, down to G0. (b) The smaller ones, from G0 down, but blown up. The G0 circle is repeated to give the relative scale for the two parts of the figure.

time, of the order of 0.01 s. The blurring of images comes from trying to observe for longer than this. If an image were bright enough to see in this short time, we could take a picture with diffraction-limited resolution. Unfortunately, 0.01 s is not long enough to collect enough photons from a star. However, we can collect a series of 0.01 s images, observing interference between light coming through slightly different paths in the atmosphere. The final images must be reconstructed mathematically.

Even more recently, work on stellar sizes has been done on the Hubble Space Telescope, as shown in Fig. 5.14. The results of these various techniques are shown in Fig. 5.15. We can see that, on the main sequence, stars become larger with increasing surface temperature. This is why the luminosity of stars increases with increasing surface temperature at a rate greater than T^4.

Chapter summary

We saw in this chapter how the gravitational interactions in binary systems can be studied to determine the masses of stars.

We saw how the Doppler shift can be used to determine the radial velocities of objects. In binary systems, we can observe the wavelengths of spectral lines varying periodically as the stars orbit the center of mass. We saw what information could be obtained even if we don't know the orbital inclination.

We saw how the masses of orbiting objects are related to the orbital radii and speeds, and the period of the system for circular orbits. We extended these ideas to elliptical orbits. For elliptical orbits, we saw how conservation of angular momentum means a change in the speed with distance from the center of mass, and how this affects the Doppler shift we see in different parts of the cycle. We also found the total energy for elliptical orbits, and saw how the kinetic energy varied as the speed varies.

We saw that the range of masses for stars along the main sequence is much less than the range of luminosities. There is also a close relationship between mass and luminosity for main sequence stars.

We saw how eclipsing binaries can be used to tell us something about stellar sizes. We also looked at other techniques for determining stellar sizes, including knowing the luminosity and temperature, using lunar occultations and speckle interferometry.

Questions

5.1. Under what conditions can we determine the masses of both stars in a binary system? Think of as many combinations of situations as you can.

5.2. For a binary that is *only* detected as an astrometric binary, what are the conditions under which we can determine the masses of both stars?

5.3. If you observe two stars close together on the sky, how would you decide if they were an optical double or a true binary?

5.4. If we see an eclipsing binary, how do we know the inclination of the orbit?

5.5. Describe situations in which a source and/or observer are moving and no Doppler shift for electromagnetic waves is observed.

5.6. When you hear the Doppler shift in a train whistle, the effect is most obvious as the train goes by you. Why is that?

5.7. For the following situations, indicate whether the radial velocity is positive, negative or zero:
(a) Observer moving towards (stationary) source.
(b) Source moving towards (stationary) observer.
(c) Source moving away from (stationary) observer.
(d) Source and observer moving towards each other.
(e) Source and observer moving away from each other.
(f) Source moving away from observer at 10 km/s; observer moving towards source at 5 km/s.

(g) Source moving perpendicular to line of sight. Observer moving towards source.
(h) Source moving and observer moving perpendicular to line of sight but in opposite directions.

5.8. Why must the center of mass of two stars be along the line joining the two stars?

5.9. Why is it not possible for the orbits of the two stars in a binary system to have different periods?

5.10. Is the Earth–Sun center of mass closer to the center of the Earth or the center of the Sun?

*5.11. Suppose that a binary system is moving under the external influence of other stars (in a cluster for example) and we observe the center of mass to be accelerating as a result. Can we still apply the analyses in this chapter to such a system?

5.12. Why do we say that for the Earth–Sun system, the sum of the masses is approximately equal to the mass of the Sun?

5.13. In a binary system, the gravitational force that star 1 exerts on star 2 must produce an acceleration. How is that acceleration manifested?

5.14. Discuss how the Sun is a 'typical' main sequence star.

5.15. How would you measure the mass of the Earth?

5.16. Use a calendar to find out how much longer it takes to go from the first day of Spring to the first day of Autumn than from the first day of Autumn to the first day of Spring (in the northern hemisphere).

5.17. What is the relationship between Kepler's second law and angular momentum?

*5.18. As an object approaches periastron in an elliptical orbit, it is moving faster and faster. This means that there must be a force in the direction of motion (in addition to the force that is perpendicular to the direction of motion, which is responsible for the curving of the path). What is the source of the force that makes the object go faster? Use a diagram, showing the forces and components, to illustrate your answer.

5.19. At apastron the objects in elliptical orbits are moving slower than at periastron, so their kinetic energy must be lower. What happens to the 'missing' energy?

5.20. How does the range of masses for main sequence stars compare with the range of temperatures and the range of luminosities?

5.21. What do we mean when we talk about a selection effect?

5.22. If high mass stars are more massive than low mass stars, how can most of the mass in our Galaxy be in the form of low mass stars? If most of the mass is in low mass stars, how can most of the luminosity come from high mass stars?

5.23. When we study stellar structure, in Chapter 9, we will see that once we know the mass of a main sequence star all of its other properties (size, temperature, luminosity) are determined. How does that show up in the observational results of this chapter?

5.24. Why is it difficult to measure stellar sizes?

Problems

5.1. A police radar, operating at a frequency of 5 GHz, detects a Doppler shift of 2.5 MHz as you approach. How fast are you going?

5.2. (a) A source is moving away from an observer with a speed of 10 km/s, along a path that makes a 30° angle with the line of sight. At what wavelength will the Hα line be observed? (b) If the observer is also moving directly away from the source at a speed of 20 km/s, at what wavelength will the Hβ line be observed?

5.3. Some source is moving away from an observer along a line making a 30° angle with the line of sight at a speed of 80 km/s. The observer is moving towards the source along a line making a 20° angle with the line of sight, at a speed of 10 km/s. (a) What is the radial velocity? (b) By how much is the Hα line shifted? (c) How would the answer be changed if the observer were moving in the opposite direction?

5.4. We observe a source with a radial velocity of -25 km/s. By how much is the separation between the Hα and Hβ lines changed?

5.5. If the Hα line in a source is shifted by $+0.01$ nm, by how much is the Hβ line shifted?

5.6. If the Hα line in a source is shifted by -0.10 nm, what is the radial velocity of the source?

5.7. You are standing a distance d from a railroad track. A train comes past you at a constant speed v, passing you at time $t = 0$. (a) What is the radial velocity of the train as a function of time? (b) Draw a graph of your result for both positive and negative times; take $v = 20$ m/s and $d = 1000$ m.

5.8. By how much can the Hα line in some object be shifted as a result of the Earth's motion around the Sun?

5.9. Show that, if we had taken the force on the second star in equation (5.13), we would still have obtained the result in equation (5.20).

5.10. In deriving the law of gravitation, Newton actually started with Kepler's third law as being observationally given and worked backwards (from the derivation in this chapter). Show how that derivation would be done.

5.11. For a binary system with stars of equal masses m, in circular orbits, with a total separation r, what is the orbital speed v?

5.12. For a binary system with stars of masses 5 and 10 $M_{\odot}$, in circular orbits with a period of 3 yr, what is the total energy of the system?

5.13. For a binary system with stars of masses m_1 and m_2, in circular orbits, with a total separation r, find an expression for the ratios of the *kinetic* energies of the two stars.

5.14. Let $M = m_1 + m_2$, and $x = m_1/m_2$. (a) Find expressions for m_1 and m_2 in terms of M and x. (b) What is the significance of your result?

5.15. We observe a binary system in which the two stars are 1 and 2 arc sec, respectively, from the center of mass. The system is 10 pc from us. The period is 33 yr. What are the masses of the two stars, assuming that $i = 90°$?

5.16. Suppose we can measure the positions of stars to 0.01 arc sec. How far away could we detect an astrometric binary where the separation is 100 AU?

5.17. A star in a circular orbit has a speed of 30 km/s. The period is 10 yr. The star is 2 arc sec from the center of mass. How far away is this star from us?

5.18. Derive a form of equation (5.23) that relates mass, in solar masses, period, in years, and velocity, in kilometers per second.

5.19. The Hα lines from two stars in a binary system are observed to have Doppler shifts of 0.022 and 0.044 nm, respectively. The period of the system is 20 yr. What are the masses of the two stars (a) if $i = 90°$ (b) if $i = 30°$?

5.20. An astrometric binary is 10 pc from the Sun. The visible star orbits 2 arc sec from the center of mass with a period of 30 yr. What is the mass of the unseen companion?

5.21. An eclipsing binary system has a period of 1.7 days. One star has a Doppler shift of 100 km/s. What can you conclude about the mass of the companion?

5.22. Derive equations (5.40a) and (5.40b).

*5.23. A star moves in an elliptical orbit of eccentricity e. The plane of the orbit makes an angle i with the plane of the sky. The orbit is oriented so that the line joining the foci is perpendicular to the line formed by the intersection of the plane of the orbit and the plane of the sky. Show that the projected orbit is an ellipse.

5.24. What is the luminosity (in solar luminosities) of (a) a 0.5 $M_\odot$ and (b) a 5.0 $M_\odot$ star?

5.25. For an elliptical orbit, calculate the angular momentum L in terms of G, m_1, m_2, a and e.

Computer problems

5.1. Plot ellipses with the same major axes, but with eccentricities from 0.1 to 1.0 in steps of 0.1.

5.2. Consider a binary system with stars of masses 5 and 10 $M_\odot$ in elliptical orbits, with eccentricity $e = 0.8$. The period is 8 yr. Assume that the more massive star is at periastron at angle $\theta = 0$. (a) Plot the speed of the more massive star as a function of position in the orbit, θ. (b) If this system is viewed from along the major axis, such that the more massive star is closer to the observer when it is at periastron, plot the radial velocity v_r as a function of θ. *(c) Plot v_r as a function of time t, taking $t = 0$ when $\theta = 0$.

5.3. For the system in the previous problem, draw the orbit and draw an arrow showing the force on the more massive star at positions every 45° around the orbit.

5.4. Suppose that many binary star systems, with randomly distributed inclinations, are observed.

(a) Draw a graph showing the relative probabilities of finding values of $\sin^3 i$ in various small ranges from zero to one. (b) The average value of some function $f(x)$ over the interval 0 to L is

$$\langle f(x) \rangle = \left(\frac{1}{L}\right) \int_0^L f(x)\, dx$$

What is the average value of $\sin^3 i$ over the angle range 0 to $\pi/2$ radians? (c) What does this tell you about the lower limits on masses and the actual masses of binary systems?

5.5. For stars in the middle range of each spectral type (05, B5, etc.), calculate the average density, and express it as a ratio to that for the Sun.

5.6. For stars in the middle range of each spectral type (05, B5, etc.), calculate the luminosity, and express it as a ratio to that for the Sun.

Chapter 6

The Sun: a typical star

The Sun (Fig. 6.1) is the only star we can study in any detail. It therefore serves as a guide to our pictures of other stars. Any theory of stellar structure must first be able to explain the Sun before explaining other stars. As we have seen, the Sun's spectral type places it in the mid-range of main sequence stars. If we understand the Sun, we have the hope of being able to understand a significant number of other types of stars.

6.1 | Basic structure

We have already seen that the mass of the Sun is 2.0×10^{33} g and that its radius is 7×10^{10} cm (7×10^5 km). Its average density is

$$\rho = \frac{M_{\odot}}{(4\pi/3)R_{\odot}^3}$$

$$= \frac{2 \times 10^{33}\,\text{g}}{(4\pi/3)(7 \times 10^{10}\,\text{cm})^3}$$

$$= 1.4\,\text{g/cm}^3$$

For comparison, the density of water is 1 g/cm^3. The Sun is composed mostly of hydrogen (94% by the number of atoms), with some helium (6% by number of atoms), and only 0.1% other elements. The abundances of the elements are given in Appendix G. Our best measurement of the effective temperature of the Sun is 5762 K. The solar luminosity is 3.8×10^{33} erg/s. (The effective temperature is the temperature that we use in the Stefan–Boltzmann law to give the solar luminosity.)

When we look at the Sun, we see only the outermost layers. We have to deduce the internal structure from theories of stellar structure (which we will discuss in Chapter 9). The basic structure of the Sun is shown in Fig. 6.2. The center is the *core*. It is the source of the Sun's energy. Its radius is about 10% of the full solar radius. The outermost layers form the atmosphere. We divide the atmosphere into three parts. Most of the light we see comes from the *photosphere*, the bottom layer of the atmosphere. Above the photosphere is the *chromosphere*, named because it is the source of red light seen briefly during total eclipses of the Sun. The chromosphere is about 10^4 km thick. The outermost layer is the *corona*, which extends far into space. It is very faint, and, for most of us, can only be seen during total solar eclipses. Beyond the corona, we have the *solar wind*, not strictly part of the Sun, but a stream of particles from the Sun into interplanetary space.

6.2 | Elements of radiation transport theory

Radiation is being emitted and absorbed in the Sun in all layers. However, we see radiation mostly from the surface. Most radiation from below is absorbed before it reaches the surface. To understand what we are seeing when we look at the Sun, we need to understand about the interaction between radiation and matter. For example, much of what we know about the solar atmosphere comes from studying spectral lines as well as the continuum. In studying how radiation interacts with matter, known as *radiation transport theory*, we see how to use spectral lines

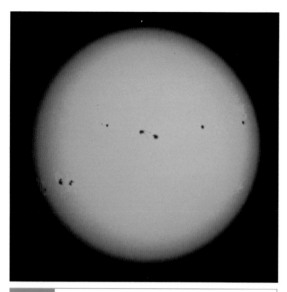

Fig 6.1. The Sun. [NOAO/AURA/NSF]

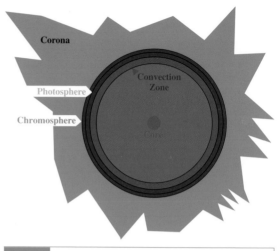

Fig 6.2. Basic structure of the Sun.

to extract detailed information about the solar atmosphere.

We first look at the absorption of radiation by atoms in matter. We can think of the atoms as acting like small spheres, each of radius r (Fig. 6.3). Each sphere absorbs any radiation that strikes it. To any beam of radiation, a sphere looks like a circle of projected area πr^2. If the beam is within that circle, it will strike the sphere and be absorbed. We say that the *cross section* for striking a sphere is $\sigma = \pi r^2$. The concept of a cross section carries over into quantum mechanics. Instead of the actual size of an atom, we use the effective area over which some process (such as absorption) takes place. So then r would be how close the photon would have to be to the atom in order to be absorbed.

We consider a cylinder of these spheres, with the radiation entering the cylinder from one end. We would like to know how much radiation is absorbed, and how much passes through to the far end. We let n be the number of spheres per unit volume. The cylinder has length l and area A, so the volume is Al. The number of spheres in the cylinder is

$$N = n\,Al \tag{6.1}$$

We define the total cross section of all the spheres as the number of spheres multiplied by the cross section per sphere:

$$\sigma_{\text{tot}} = N\,\sigma \tag{6.2}$$

$$\sigma_{\text{tot}} = n\,A\,l\,\sigma \tag{6.3}$$

In making this definition, we have assumed that the incoming beam can "see" all the spheres. No sphere blocks or shadows another. We are assured of little shadowing if the spheres occupy a small fraction of the area, as viewed from the end. That is

$$\sigma_{(\text{tot})} \ll A \tag{6.4}$$

Under these conditions, the fraction of the incoming radiation that will be absorbed, f, is just that fraction of the total area A that is covered by the spheres. That is

$$f = \sigma_{(\text{tot})}/A \tag{6.5}$$

Using equation (6.3), this becomes

$$f = n\sigma\,l \tag{6.6}$$

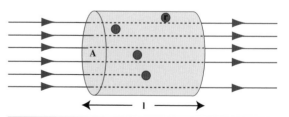

Fig 6.3. Absorption of radiation. Radiation enters from the left. Any beam striking a sphere is absorbed.

We define the *optical depth* to be this quantity:

$$\tau = n\sigma l \tag{6.7}$$

Our requirement in equation (6.4) reduces to $\tau \ll 1$. Under this restriction, the optical depth of any section of material is simply the fraction of incoming radiation that is absorbed when the radiation passes through that material. (For example, if the optical depth is 0.01, then 1% of the incoming radiation is absorbed.)

In general, σ will be a function of wavelength. For example, we know that at a wavelength corresponding to a spectral line, a particular atom will have a very large cross section for absorption. At a wavelength not corresponding to a spectral line, the cross section will be very small. To remind us that σ is a function of λ (or ν), we write it as σ_λ (or σ_ν). This means that the optical depth is also a function of λ (or ν), so we rewrite equation (6.7) as

$$\tau_\lambda = n l \sigma_\lambda \tag{6.8}$$

In our discussions, the quantity nl occurs often. It is the product of a number density and a length, so its units are measured in number per unit area. It is the number of particles along the full length, l, of the cylinder per unit surface area. For example, if we are measuring lengths in centimeters, it is the number of particles in a column whose face surface area is 1 cm^2, and whose length is l, the full length of the cylinder. We call this quantity the *column density*.

We can see that the optical depth depends on the properties of the material – e.g. cross section and density of particles – and on the overall size of the absorbing region. It is sometimes convenient to separate these two dependencies by defining the *absorption coefficient*, which is the optical depth per unit length through the material,

$$\kappa_\lambda = \frac{\tau_\lambda}{l} \tag{6.9a}$$

$$= n\sigma_\lambda \tag{6.9b}$$

If κ_λ gives the number of absorptions per unit length, then its inverse gives the mean distance between absorptions. This quantity is called the *mean free path*, and is given by

$$L_\lambda = 1/\kappa_\lambda \tag{6.10}$$

$$= 1/n\sigma_\lambda \tag{6.11}$$

In terms of these quantities, the optical depth is given by

$$\tau_\lambda = \frac{l}{L_\lambda} \tag{6.12a}$$

$$= \kappa_\lambda l \tag{6.12b}$$

In the above discussion, we required that the optical depth be much less than unity. Our interpretation of τ as the fraction of radiation absorbed only holds for $\tau \ll 1$. What if that is not the case? We then have to divide our cylinder into several layers. If we make the layers thin enough, we can be assured that the optical depth for each layer will be very small. We then follow the radiation through, layer by layer, looking at the fraction absorbed in each layer (Fig. 6.4).

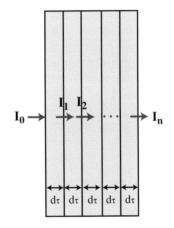

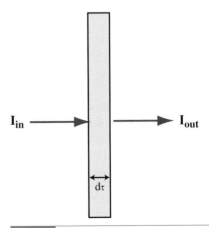

Fig 6.4. Radiation passing through several layers. Each layer has an optical depth dτ. The bottom of the figure is for calculating the effect of each layer.

Let's look at the radiation passing through some layer with optical depth $d\tau$. Since $d\tau \ll 1$, it is the fraction of this radiation that is absorbed. The amount of radiation absorbed in this layer is $I\,d\tau$. The amount of radiation passing through to the next layer is $I(1 - d\tau)$. The change in intensity, dI, while passing through the layer, is

$$dI = I_{\text{out}} - I_{\text{in}} \qquad (6.13)$$

$$= -I\,d\tau \qquad (6.14)$$

Notice that dI is negative, since the intensity is decreased in passing through the layer.

Now that we know how to treat each layer, we must add up the effect of all the layers to find the effect on the whole sample of material. We can see that we have formulated the problem so that we are following I as a function of τ. We let τ' be the optical depth through which the radiation has passed by the time it reaches a particular layer, and I' be the intensity reaching that layer. Then τ' ranges from zero, at the point where the radiation enters the material, to τ, the full optical depth where the radiation leaves the material. Over that range, I' varies from I_0, the incident intensity, to I, the final intensity. Using equation (6.14),

$$dI'/I' = -d\tau' \qquad (6.15)$$

In this form, all of the I' dependence is on the left, and the τ' dependence is on the right. To add up the effect of the layers, we integrate equation (6.15) between the limits given above:

$$\int_{I_0}^{I} \frac{dI'}{I'} = -\int_0^\tau d\tau'$$

$$\ln(I) - \ln(I_0) = -\tau \qquad (6.16)$$

Using the fact that $\ln(a/b) = \ln(a) - \ln(b)$, this becomes

$$\ln(I/I_0) = -\tau \qquad (6.17)$$

Raising e to the value on each side, remembering that $e^{\ln x} = x$, and multiplying both sides by I_0 gives

$$I = I_0\,e^{-\tau} \qquad (6.18)$$

We can check this result in the limit $\tau \ll 1$, called the *optically thin limit*, using the fact that

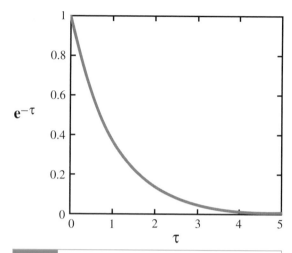

Fig 6.5. $e^{-\tau}$ vs. τ, showing the fall-off in transmitted radiation as the optical depth increases. Note that the curve looks almost linear for small τ. For large τ, it approaches zero asymptotically.

$e^x \cong 1 + x$, for $x \ll 1$. In this case, equation (6.18) becomes

$$I = I_0\,(1 - \tau) \qquad (6.19)$$

This is the expected result for small optical depths, where τ again becomes the fraction of radiation absorbed.

As shown in Fig. 6.5, $e^{-\tau}$ falls off very quickly with τ. This means that to escape from the Sun, radiation must come from within approximately one optical depth of the surface. This explains why we only see the outermost layers. Since the absorption coefficient κ_λ is a function of wavelength, we can see to different depths at different wavelengths. At a wavelength where κ_λ is large, we don't see very far into the material. At wavelengths where κ_λ is small, it takes a lot of material to make $\tau_\lambda = 1$. We take advantage of this to study conditions at different depths below the surface.

So far we have only looked at the absorption of radiation passing through each layer. However, radiation can also be emitted in each layer, and the amount of emission also depends on the optical depth. In general, we must carry out complicated radiative transfer calculations to take all effects into account. To solve these problems, we use powerful computers to make mathematical models of stellar atmospheres. In these calculations, we input the distribution of temperature, density and composition and predict the spectrum that we will see, including emission and absorption lines. We vary the input parameters until we find models that produce predictions that agree with the observations. The results of these calculations are not unique, but they do give us a feel for what processes are important in stellar atmospheres. The more observational data we can predict with the models, the more confident we can be that the temperatures, densities and compositions we derive are close to the actual ones.

6.3 | The photosphere

Most of the visible photons we receive from the Sun originate in the photosphere. One question you might ask is why we see continuum radiation at all. We have already seen how atoms can emit or absorb energy at particular wavelengths, producing spectral lines. However, we have not discussed the source of the emission and absorption of the continuum. It turns out that the continuum opacity in the Sun at optical wavelengths comes from the presence of H^- ions. An H^- ion is an H atom to which an extra electron has been added. As you might guess, this extra electron is held only very weakly to the atom. Very little energy is required to remove it again. H^- ions are present because there is so much hydrogen, and

there are a large number of free electrons to collide with those atoms.

If we have an H^- ion and a photon (γ) strikes it, the photon can be absorbed, and the electron set free:

$$H^- + \gamma \rightarrow H + e^-$$

The H^- ion is a bound system. The final state has an H atom and a free electron. We call this process a *bound–free* process. In such a process, the wavelength of the incoming photon is not restricted, as long as the photon has enough energy to remove the electron. The electron in the final state can have any kinetic energy, so a continuous range of photon energies is possible. This process then provides most of the continuum opacity of the photosphere. The continuum emission results from the inverse process.

6.3.1 Appearance of the photosphere

We have said that the Sun is the one star that we can study in great detail. To do this, we try to observe the photosphere with the best resolution possible. When we observe the Sun, the light-gathering power of our telescope is not usually a problem. Therefore, we can try to spread the image out over as large an area as possible, making it easier to see detailed structure. We therefore want a telescope with a long focal length to give us a large image scale. The solar telescope shown in Fig. 6.6 provides this type of detailed picture.

Fig 6.6. Solar telescope on Kitt Peak (operated by NOAO). The telescope has a very long focal length, so that two can produce a large image and study the detailed appearance. Since the tube is so long, it is not reasonable to move it. Instead, the large mirror at the top (the objective) is moved to keep the sunlight directed down the tube. [NOAO/AURA/NSF]

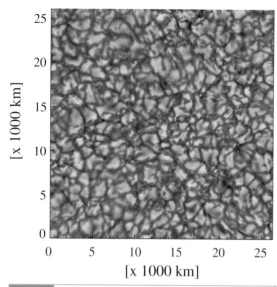

Fig 6.7. Granulation in the Sun. Remember, the darker areas are not really dark. They are only a little cooler than the bright areas. [NOAO/AURA/NSF]

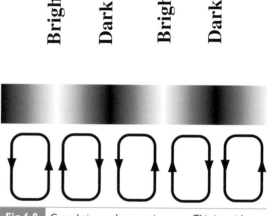

Fig 6.8. Granulation and convection zone. This is a side view to show what is happening below the surface. Hotter gas is being brought up from below, producing the bright regions. The cooler gas, which produces the darker regions, is carried down to replace the gas that was brought up.

When we look on a scale of a few arc seconds, we see that the surface of the Sun does not have a smooth appearance (Fig. 6.7). We see a structure, called *granulation*, in which lighter areas are surrounded by darker areas. The darker areas are not really dark. They are just a little cooler than the lighter areas, and only appear dark in comparison to the light areas. The granules are typically about 1000 km across. The pattern of granulation also changes with time, with a new pattern appearing every 5 to 10 min.

We interpret this granulation as telling us about the underlying structure we cannot see directly. The granulation can be explained by circulating cells of material, called *convection zones* (Fig. 6.8). (Convection is the form of energy transport in which matter actually moves from one place to another. Strong convection on the Earth is responsible for the updrafts that produce thunderstorms. A pot of boiling water also has energy transport by convection.) The brighter regions are warmer gas rising up from below. The dark regions are cooler gas falling back down.

In addition to the granulation variations, there is also a variation called the *five minute oscillation*, in which parts of the photosphere are moving up and down. We think this convection results

from sound waves in the upper layers of the convection zone. This type of oscillation is one of many that are studied for clues to the Sun's interior structure. This area of research is called *solar seismology*. (On the Earth, seismologists study motions near the surface to learn about the interior.)

One interesting question about the photosphere concerns the sharpness of the solar limb. The Sun is a ball of gas whose density falls off continuously as one moves farther from the center. There is no sharp boundary (like the surface of the Earth), yet we see a definite edge on the Sun. In Fig. 6.9, we see some lines of sight through the

Fig 6.9. Lines of sight through the solar limb. For clarity, we think of the Sun as being composed of a series of spherical shells. The density in each shell decreases as we move farther from the center. This decreasing density is indicated by the shading; two lines of sight are indicated. Note that most of each line of sight is in the densest layer through which that particular line passes. Even though line 2 is not shifted very far from line 1, line 1 passes through much more material.

1 2

photosphere. As the line of sight passes farther from the center of the Sun, the opacity decreases because (1) the path length through the Sun is less, and (2) it passes through less dense regions. Since the amount of light getting through is proportional to $e^{-\tau}$, the effect of τ changing from line of sight to line of sight is enhanced by the exponential behavior. Therefore the transition from the Sun being mostly opaque to being mostly transparent takes place over a region that is small compared with our resolution, and the edge looks sharp.

6.3.2 Temperature distribution

Another interesting phenomenon near the solar limb can be seen in the photograph in Fig. 6.1. The Sun does not appear as bright near the limb as near the center. This *limb darkening* is also an optical depth effect, as shown in Fig. 6.10. We compare two lines of sight: (1) toward the center of the Sun from the observer, and (2) offset from the center of the Sun. In each line of sight, we can only see down to $\tau \cong 1$. Line of sight 1 is looking straight down into the atmosphere, so it gets closer to the center of the Sun before an optical depth of unity is reached than does line of sight 2, which has a longer path through any layer. We see deeper into the Sun on line of sight 1 than we do on line of sight 2. If the temperature decreases with increasing height in the photosphere, we are seeing hotter material on line of sight 1 than on 2, so line of sight 1 appears brighter than line of sight 2. Since line of sight 1 takes us the deepest into the Sun, we define the point at which τ reaches unity on this line as being the *base of the photosphere*. When we talk about the temperature of the Sun, we are talking about the temperature at the base of the photosphere.

When we look at the Sun, it appears brighter along line of sight 1 than it does along line of sight 2. This means that the Sun is hotter at the end of 1 than at the end of 2. From this, we conclude that the photosphere cools as one moves farther from the center of the Sun. If the photosphere became hotter as one moves farther from the center, we would see limb brightening.

We obtain more useful information about the photosphere by studying its spectral lines (Fig. 6.11). The spectrum shows a few strong absorption lines and a myriad of weaker ones. The stronger lines were labeled A through K by Fraunhofer in 1814. These lines have since been identified. For example, the C line is the first Balmer line (Hα); the D line is a pair of lines belonging to neutral sodium (NaI); and the H and K lines belong to singly ionized calcium (CaII). Sodium and calcium are much

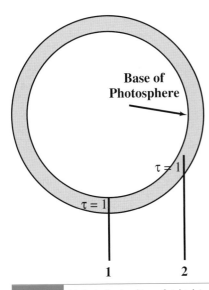

Fig 6.10. Limb darkening. Line of sight 1 is directed from the observer toward the center of the Sun, so it takes the shortest path through the atmosphere. This line allows us to see the deepest into the photosphere, and the base of the photosphere is defined to be where the optical depth τ along this line reaches unity. Line of sight 2 is closer to the edge, so it doesn't allow us to see as deep into the photosphere. If the temperature decreases with increasing height in the photosphere, then line of sight 1 allows us to see hotter material than does line of sight 2, and the edge of the Sun appears darker than the center.

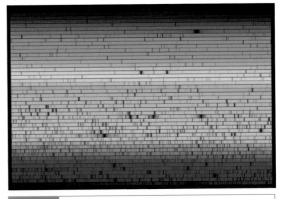

Fig 6.11. The solar spectrum. [NOAO/AURA/NSF]

less abundant than hydrogen but their absorption lines are as strong as Hα. We have already seen in Chapter 3 that this can result from the combined effects of excitation and ionization.

6.3.3 Doppler broadening of spectral lines

When we study the lines with good spectral resolution, we can look at the details of the line profile. The lines are broadened by Doppler shifts due to the random motions of the atoms and ions in the gas (Figs. 6.12 and 6.13). If all the atoms were at rest, all the photons from a given transition would emerge with a very small spread in wavelength. However, the atoms are moving with random speeds in random directions. We therefore see a spread in the Doppler shifts, and the line is broadened. This process is called *Doppler broadening*. If the gas is hotter, the spread in speeds is greater, and thus the Doppler broadening is also greater. If in addition to these random motions all the objects

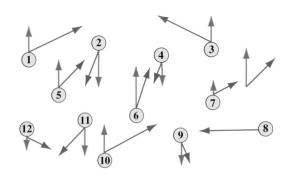

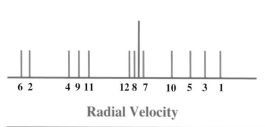

Radial Velocity

Fig 6.12. Doppler broadening. The top shows the (random) motions of a group of particles. The purple vectors are the actual velocities; the green vectors are the radial components, which produce the Doppler shift. For each particle, identified by a number, the radial velocities are plotted below. The line profile is the sum of all the individual Doppler shifted signals, and with many more particles it would have a smooth appearance.

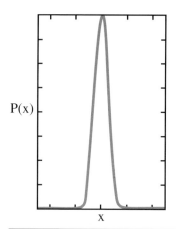

P(x)

X

Fig 6.13. Line profile. We plot intensity as a function of wavelength.

containing the particles have some overall motion that just shifts the center wavelength of the line, the broadening would still be the same.

We can estimate the broadening as a function of temperature. If $\langle v^2 \rangle$ is the average of the square of the random velocities in a gas, and m is the mass per particle, the average kinetic energy per particle is $(1/2)m\langle v^2 \rangle$. If we have an ideal monatomic gas, this should equal $(3/2)kT$, giving

$$(1/2)m\langle v^2 \rangle = (3/2)kT \tag{6.20}$$

Solving for $\langle v^2 \rangle$ gives

$$\langle v^2 \rangle = 3kT/m \tag{6.21}$$

Taking the square root gives the *root mean square (rms) speed*

$$v_{\rm rms} = \left(\frac{3kT}{m} \right)^{1/2} \tag{6.22}$$

This gives us an estimate of the range of speeds we will encounter. (The range will be larger, since we can have atoms coming toward us or away from us, and since some atoms will be moving faster than this average, but the radial velocity is reduced since only the component of motion along the line of sight contributes to the Doppler broadening.) To find the actual wavelength range over which the line is spread out, we would use the Doppler shift expression in equation (5.4).

Example 6.1 Doppler broadening
Estimate the wavelength broadening in the Hα line in a gas composed of hydrogen atoms at $T = 5500$ K.

SOLUTION

Using equation (6.22) gives

$$v_{rms} = \left[\frac{(3)(1.38 \times 10^{-16}\ erg/K)(5.5 \times 10^3\ K)}{1.7 \times 10^{-24}\ g} \right]^{1/2}$$

$$= 1.2 \times 10^6\ cm/s$$

From equation (5.4), we estimate the linewidth as

$$\Delta\lambda = \lambda\left(\frac{v_{rms}}{c}\right)$$

$$= \left(\frac{(656.28\ nm)(1.2 \times 10^6\ cm/s)}{(3 \times 10^{10}\ cm/s)}\right)^{1/2}$$

$$= 0.03\ nm$$

In any spectral line, smaller Doppler shifts relative to the line center are more likely than larger ones. Therefore, the optical depth is greatest in the line center, and falls off to either side. At different Doppler shifts away from the line center, we will see different distances into the Sun. The farther we are from the line center, the deeper we see. By studying line profiles in detail we learn about physical conditions at different depths. Also, each spectral line has a different optical depth in the line center, so different lines allow us to see down to different depths in the photosphere. When we perform model stellar atmosphere calculations, we try to predict as many of the features of the spectrum as possible, including the relative strengths of various lines and the details of certain line profiles.

From observations of various spectral lines, the temperature profile for the photosphere has been derived. It is shown in Fig. 6.14. Note that the temperature falls as one goes up from the base of the photosphere. This is what we might expect, since we are moving farther from the heating source. However, an interesting phenomenon is observed. The temperature reaches a minimum at 500 km above the base of the photosphere, and then begins to rise with altitude. We will see below that this temperature rise continues into the higher layers.

6.4 | The chromosphere

At most wavelengths the chromosphere is optically thin, so we can see right through it to the photosphere. Under normal conditions the continuum radiation from the photosphere overwhelms that from the chromosphere. However, during a total eclipse of the Sun, just before and after totality, the Moon blocks the light from the photosphere, but not from the chromosphere. For that brief moment, we see the red glow of the chromosphere. The red glow comes from Hα emission. The optical depth of the Hα line is sufficiently large that we can study the chromosphere by studying that line. At the center of the Hα line we see down only to 1500 km above the base of the photosphere.

We can also study the large scale structure of the chromosphere by taking photographs through filters that only pass light from one line. One such photograph is shown in Fig. 6.15. In this picture we see granulation on an even larger scale than in the photosphere. This is called *supergranulation*. The supergranules are some 30 000 km across. As with the granules in the photosphere, the matter in the center of the supergranules is moving up and the matter at the edges is moving down. These motions can be determined from Doppler shifts of spectral lines. We also see smaller scale irregularities in the chromosphere, called *spicules*. These are protrusions from the surface some 700 km across and 7000 km high. An Hα image of the chromosphere is shown in Fig. 6.16.

When the chromosphere is visible just before and after totality, we see only a thin sliver of light.

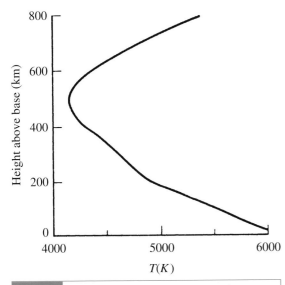

Fig 6.14. Temperature vs. height in the photosphere.

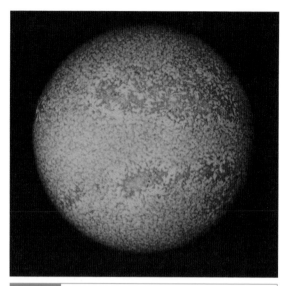

Fig 6.15. This spectroheliogram shows large scale motions at the solar surface based on the Hα emission. [NOAO/AURA/NSF]

The effect is the same as if the light had been passed through a curved slit. If we then use a prism or grating to spread out the different wavelengths, we will obtain a line spectrum, though each line will appear curved. This spectrum is called a *flash spectrum* because it is only visible for the brief instant that the Moon is covering all of the photosphere but not the chromosphere. Note that the

spectrum shows emission lines. This is because there is no strong continuum to be absorbed.

When we study the spectra of the chromosphere we find that it is hotter than the photosphere. The chromospheric temperature is about 15 000 K. (The Sun doesn't appear this hot because the chromosphere is optically thin and doesn't contribute much to the total radiation we see.) We are faced with trying to explain how the temperature rises as we move farther from the center of the Sun. We will discuss this point in the next section, when we discuss the corona.

6.5 | The corona

6.5.1 Parts of the corona

The corona is most apparent during total solar eclipses (Fig. 6.17), when the much brighter light

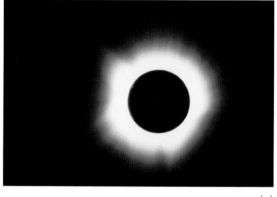

(a)

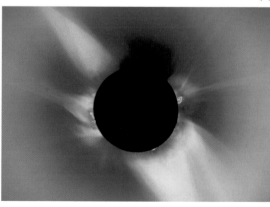

(b)

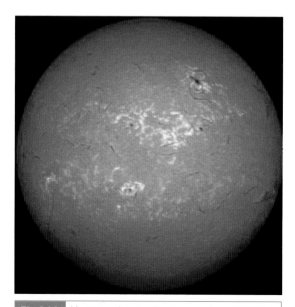

Fig 6.16. Hα image showing the chromosphere. [NASA]

Fig 6.17. Two views of total solar eclipses, showing the corona. [(a) NOAO/AURA/NSF; (b) NASA]

from the photosphere and chromosphere is blocked out. The corona is simply too faint to be seen when any photospheric light is present. You might think that we can simulate the effect of an eclipse by holding a disk over the Sun. If you try this, light that would come directly from the photosphere to your eye will be blocked out. However, some photospheric light that is originally headed in a direction other than directly at you will scatter off the atoms and molecules in the Earth's atmosphere, and reach you anyway (Fig. 6.18). This scattered light is only a small fraction of the total photospheric light, but it is still enough to overwhelm the faint corona. This is not a problem during solar eclipses, because the Moon is outside the atmosphere, so there is nothing to scatter light around it.

Therefore, solar eclipses still provide us with unique opportunities to study the corona. For this purpose, we are fortunate that the Moon subtends almost the same angle as the Sun, as viewed from the Earth. The Moon can exactly block the photosphere and chromosphere, but not the corona. Unfortunately, we do not have an eclipse of the Sun every month. We would if the Moon's orbit

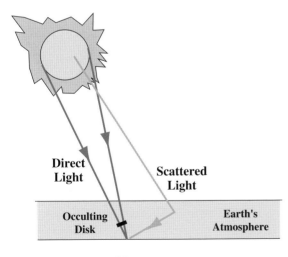

Direct Light

Scattered Light

Occulting Disk

Earth's Atmosphere

Observer

Fig 6.18. Effects of scattered light on corona studies. We can use an occulting disk, but sunlight can scatter off particles in the Earth's atmosphere and reach our telescope. Since the corona is very faint, and the Earth's atmosphere is very efficient at scattering, especially for blue light, this scattered light can overwhelm the direct light from the corona.

were in the same plane as the Earth's orbit around the Sun. However, the Moon's orbit is inclined by about 5°, so total eclipses of the Sun are rare events. The average time between total solar eclipses is about one and a half years. Even when one occurs, the total eclipse is observable from a band not more than 300 km wide, and totality lasts only a few minutes.

Solar astronomers take advantage of eclipses whenever possible, but also look for other ways to study the corona. It turns out that some ground-based non-eclipse observations are possible. Telescopes, called *coronagraphs*, have disks to block out the photospheric light, and reduce as much scattered light as possible. They are placed at high altitude sites, with very clear skies. For example, the Haleakala Crater on the island of Maui (operated by the University of Hawaii) is at an altitude of 3000 m (10 000 ft).

One way to get around the scattering in the Earth's atmosphere is to put a coronagraph in space. This is not quite as good as a solar eclipse, since space probes are not totally free of escaping gases. Some photospheric light is scattered by these gases. However, the results are much better than for a ground-based telescope. They also have an advantage over eclipse studies in that they allow for continuous study of the corona. For example, the Orbiting Solar Observatory 7 (OSO 7) provided observations of the corona for the period 1971–4.

There is an additional technique for studying the corona. It involves studying radio waves. Radio waves pass through the Earth's atmosphere and are not appreciably scattered. We therefore don't have to worry about the radio waves behaving like the visible light in Fig. 6.18.

In discussing what we have learned so far about the corona, we divide it into three parts:

(1) *The E-corona* is a source of emission lines directly from material in the corona. These lines come from highly ionized species, such as Fe XIV (13 times ionize iron). If we look at the Saha equation, discussed in Chapter 3, we see that highly ionized states are favored under conditions of high temperature and low density. The high temperature provides the energy necessary for the ionization.

The low density means that collisions leading to recombinations are rare.

(2) *The K-corona* (from the German *Kontinuierlich*, for continuous) is the result of photospheric light scattered from electrons in the corona.

(3) *The F-corona* (for Fraunhofer) is not really part of the Sun. It comes from photospheric light scattered by interplanetary dust. Since we are just seeing reflected photospheric light, the light still has the Fraunhofer spectrum. Both the F- and K-coronas appear at approximately the same angular distance from the center of the Sun, but there are experimental ways of separating their contributions to the light we see.

6.5.2 Temperature of the corona

When we analyze the abundance of highly ionized states, the Doppler broadening of lines and the strength of the radio emission, we find that the corona is very hot, about 2×10^6 K. As we have stated, the density is very low, approximately 10^{-9} times the density of the Earth's atmosphere.

Again, we must explain why a part of the atmosphere farther from the center of the Sun is hotter than a part closer to the center. We should note that it is not necessarily hard to keep something hot if it can't lose heat efficiently. For example, in a well insulated oven, once the required temperature has been reached, the heat source can be turned off and the temperature of the oven still stays high. An ionized gas (plasma) like the corona can lose energy only through collisions between the particles. For example, an electron and an ion could collide, with some of the kinetic energy going to excite the ion to a higher state. The ion can then emit a photon and return to its lower energy state. The emitted photon escapes and its energy is lost to the gas. If the lost energy is not replaced, the gas will cool.

Since collisions play an important role in the above process, the rate at which the gas can lose energy will depend on the rate at which collisions occur. The rate of collisions should be proportional to the product of the densities of the two colliding species. However, the density of each species is roughly proportional to the total gas density ρ (each being some fraction of the total). This means that the collision rate is pro-

portional to ρ^2. Of course the amount of gas to be cooled is proportional to ρ. We can define a *cooling time* which is the ratio of the gas volume to be cooled (which is proportional to ρ) to the gas cooling rate (which is proportional to ρ^2). The cooling time is then proportional to $1/\rho$. Therefore, low density gases take longer to get rid of their stored heat.

Further, if the density is very low, a very high temperature doesn't necessarily mean a lot of stored energy. The energy is $(3/2)kT$ per particle. Even if the quantity kT is very large, if the total number of particles is very small, the energy stored is not as large as if the density were much higher.

Example 6.2 Energy density in the corona and the Earth's atmosphere

Compare the energy density (energy per unit volume) in the corona with that in the Earth's atmosphere.

SOLUTION

For each, the energy density is proportional to the density of particles n and the temperature T. (All other constants will drop out when we take the ratio.) Therefore

$$\frac{\text{energy density (corona)}}{\text{energy density (Earth)}} = \left[\frac{n(\text{corona})}{n(\text{Earth})} \right] \left[\frac{T(\text{corona})}{T(\text{Earth})} \right]$$

$$= [1 \times 10^{-9}] \left[\frac{2 \times 10^6}{3 \times 10^2} \right]$$

$$= 7 \times 10^{-6}$$

Even though the corona is so hot, its very low density gives it a lower energy density.

We must still come up with some explanation for getting energy into the corona. Some have suggested mechanisms in which oscillations near the surface of the Sun send supersonic sound waves (shock waves) into the Sun's upper atmosphere. In addition, there are mechanisms for heating that involve the Sun's magnetic field. These theories are still under study, and we still do not have a definitive picture of the energy balance in the corona.

When we study photographs of the corona, we find its structure is very irregular. We often see long streamers whose appearance varies with time. We think that these phenomena are related to the Sun's magnetic field, as are other aspects of solar activity (to be discussed in the following section).

6.6 | Solar activity

6.6.1 Sunspots

When we look at photographs of the Sun (Figs. 6.1 and 6.19) we note a pattern of darker areas. As with the granulation, the dark areas are not really

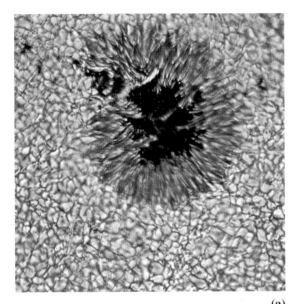

(a)

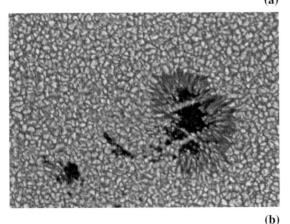

(b)

Fig 6.19. Sunspot images. The darker inner area of each spot is the umbra and the lighter outer area of each spot is the penumbra. The spots appear dark because the surrounding areas are brighter. [(a) NOAO/AURA/NSF; (b) NASA]

dark. They are just not as bright as the surrounding areas. The gas in these darker areas is probably at a temperature of about 3800 K. A closeup of these *sunspots* shows that they have a darker inner region, the *umbra*, surrounded by a lighter region, the *penumbra*.

The number of sunspots on the Sun is not even approximately constant. It varies in a regular way, as shown in Fig. 6.20. It was realized in the mid-19th century that sunspot numbers follow an 11-year cycle. The number of sunspots in a peak year is not the same as in another peak year. However, the peaks are easily noticeable.

We see more regularity when we plot the height of each sunspot group above or below the Sun's equator as a function of when in the sunspot cycle they appear. This was done in 1904 by *E. Walter Maunder*. An example of such a diagram is shown in Fig. 6.21. Early in an 11-year cycle, sunspots appear far from the Sun's equator. Later in the cycle, they appear closer to the equator. This results in a butterfly-like pattern, and in fact these diagrams are sometimes called "butterfly diagrams".

When Maunder investigated records of past sunspot activity, he found that there was an extended period when no sunspots were observed. This period, from 1645 to 1715 is known as the *Maunder minimum*. This minimum has recently been reinvestigated by John A. Eddy, who found records to indicate that no aurorae were observed for many years during this period. Also during this time, weak coronae were reported during total solar eclipses. An unusual correlation was also found in the growth rings in trees, suggesting that altered solar activity had some effect on growth on Earth. The mechanism by which this takes place is poorly understood, but we can use the growth rings as an indicator of solar activity farther into the past than other records. A study of growth rings in old trees indicates that the Maunder minimum is not unique. There may have been several periods in the past with extended reduced solar activity.

Sunspots appear to be regions where the magnetic field is higher than on the rest of the Sun. In discussing sunspots, we should briefly review properties of magnetic fields and their interactions with matter. Magnetic fields arise from moving

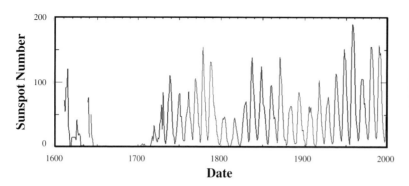

(including spinning) charges. Magnetic field lines form closed loops (Fig. 6.22). This is equivalent to saying that there are no point sources of magnetic charge, *magnetic monopoles*. (We will discuss the implications of the possible existence of magnetic monopoles in Chapter 21.) We call the closed loop pattern in Fig. 6.22 a *magnetic dipole*.

The magnetic field strength **B** is defined such that the magnetic force on a charge q, moving with velocity **v**, is (in cgs units)

$$\mathbf{F} = q\,\mathbf{v} \times \mathbf{B}/c \tag{6.23}$$

where the **x** indicates a vector cross product. There is no magnetic force on a charge at rest, or

on a charge moving parallel to the magnetic field lines. The force is maximum when the velocity is perpendicular to the magnetic field. The force is perpendicular to both the direction of motion and to the magnetic field, and its direction is given by the so-called "right hand rule". This means that the component of motion parallel to the field lines is not altered, but the component perpendicular to the field lines is. So, charged particles will move so that they spiral around magnetic field lines.

Through the forces that it exerts on moving charges, a magnetic field will exert a torque on a current loop. We can think of a current loop as having a magnetic dipole moment (as in Fig. 6.22), and the torque will cause the dipole to line up with the field lines. In this way, the dipole moment of a compass needle lines up with the Earth's magnetic field.

How do we measure the Sun's magnetic field? For certain atoms placed in a magnetic field, energy

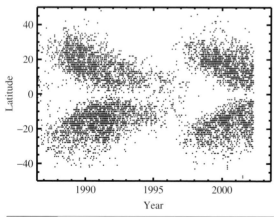

Fig 6.21. Butterfly diagram. This shows the distribution of sunspots in solar latitude as a function of time in the sunspot cycle. Early in a cycle, spots appear at higher latitudes; late in a cycle, they appear close to the equator. Note that a new cycle starts before the old cycle completely ends. [Roger Ulrich, UCLA, Mt Wilson. This study includes data from the synoptic program at the 150 ft Solar Tower of the Mt Wilson Observatory. The Mt Wilson 150 ft Solar Tower is operated by UCLA, with funding from NASA, ONR and NSF, under agreement with the Mt Wilson Institute]

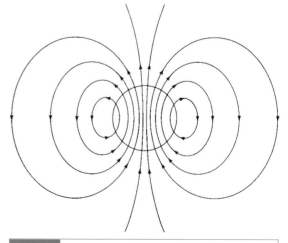

Fig 6.22. Field lines for a dipole magnetic field.

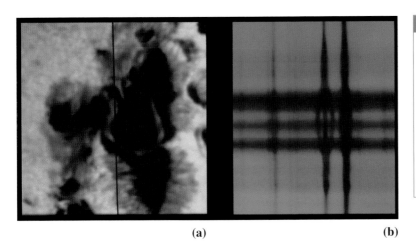

(a)   (b)

Fig 6.23. Zeeman effect in sunspots. (a) The placement of the spectrometer slit across a sunspot. (b) The spectrum at various positions along the slit. In the spectrum, away from the spot, the spectral lines are unsplit. In the center, near the spot, some of the spectral lines are split into three. The stronger the magnetic field the greater the splitting. [NOAO/AURA/NSF]

levels will shift. This is known as the *Zeeman effect* (Fig. 6.23). Different energy levels shift by different amounts. Some transitions that normally appear as one line split into a group of lines. The amount of splitting is proportional to the strength of the magnetic field. (We can check the amount of splitting in different atoms in various magnetic fields in the laboratory.) An image of the magnetic field strength over the whole Sun is shown in Fig. 6.24.

Measurements by *George Ellery Hale* in 1908 first showed that the magnetic fields are stronger in sunspots. He also found that sunspots occur in pairs, with one corresponding to the north pole of a magnet and the other to the south pole. In each sunspot pair, we can identify the one that "leads" as the sun rotates. In a given solar hemisphere, the polarity (i.e. N or S) of the leading spot of all pairs is the same. The polarity is different in the two hemispheres.

This polarity reverses in successive 11-year cycles. During one cycle all of the leading sunspots in the northern hemisphere will be magnetic north, while those in the south will be magnetic south. In the next cycle all the leading sunspots in the northern hemisphere will be magnetic south. The Sun's magnetic field reverses every 11 years! This means that the sunspot cycle is really a 22-year cycle in the Sun's magnetic field.

If the Sun's magnetic field arose in the core, as the Earth's does, we would expect it to be quite stable. (There is geological evidence that the Earth's magnetic field reverses periodically, but on geological time scales, not every 11 years.) We now

think that the Sun's magnetic field arises below the surface, rather than in the core, as the Earth's does. We also know that the Sun does not rotate as a rigid body. By following sunspot groups (Figs. 6.25 and 6.26), we see that material at the equator takes less time to go around than material at higher latitudes. For example, it takes 25 days for

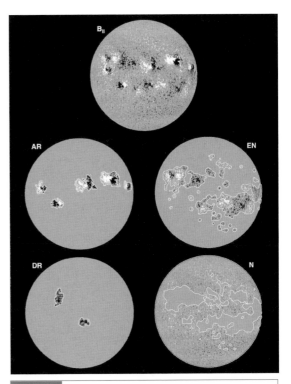

Fig 6.24. Solar magnetic field. This shows the result of observations of the Zeeman effect. Brighter areas correspond to stronger fields. [NOAO/AURA/NSF]

North Pole

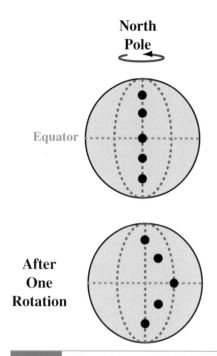

Equator

After One Rotation

Fig 6.25. Differential rotation of the Sun, as traced out by sunspots. Since the Sun rotates faster at the equator than at higher latitudes, sunspots at the equator take less time to make one rotation than do those at higher latitudes. In this schematic diagram, a selection of sunspots starts at the same meridian, but, after one rotation of the Sun, they are on different meridians.

material at the equator to make one circuit, while it takes 28 days at 40° latitude.

As the Sun rotates differentially, the magnetic field lines become distorted. This is because the charged particles in the matter cannot move across field lines, so the field lines are carried along with the material. We say that the magnetic field is frozen into the material. The development of the magnetic field is shown in Fig. 6.27. As the field lines wind up, the field becomes very strong in places. Kinks in the field lines break through the surface. The sunspots apparently arise through some, as yet poorly understood, dynamo motion, involving convective motion and the magnetic fields.

6.6.2 Other activity

Sunspots are just one manifestation of solar activity related to the Sun's magnetic field. Another form of activity is the *solar flare*, shown in Fig. 6.28. A flare involves a large ejection of particles. Flares develop very quickly and last tens of minutes to a few hours. Temperatures in flares are high, up to 5×10^6 K. They also give off strong Hα emission, and flares are seen when the Sun is photographed through an Hα filter. Flares have been detected to give off energy in all parts of the

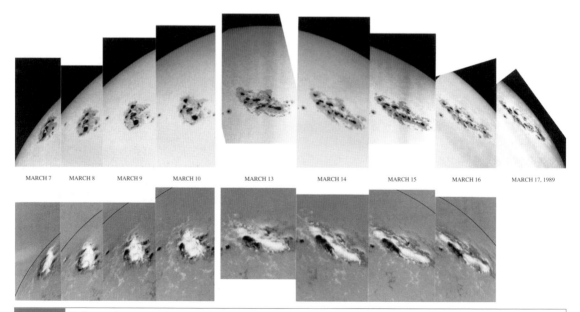

MARCH 7 MARCH 8 MARCH 9 MARCH 10 MARCH 13 MARCH 14 MARCH 15 MARCH 16 MARCH 17, 1989

Fig 6.26. Differential rotation of the Sun, as traced out by sunspots. This sequence of images shows this effect on the Sun. [NOAO/AURA/NSF]

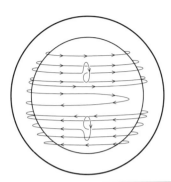

Fig 6.27. This shows how the solar magnetic field is twisted and kinked by the differential rotation of the Sun. It takes about 8 months for the field to wrap around once. The kinked parts of the field lines represent places where these loops come outside the Sun, allowing charged particles to follow those looped paths. [NASA]

electromagnetic spectrum. The cause of flares is not understood, but they appear to be related to strong magnetic fields and the flow of particles along field lines.

Solar activity is also manifested in *plages* (from the French word for "beach"). Plages are bright regions around sunspots. They show up in Hα images of the Sun. They remain after sunspots disappear. Plages are apparently chromospheric brightening caused by strong magnetic fields.

Filaments are dark bands near sunspot regions. They can be up to 10^5 km long. Filaments appear to be boundaries between regions of opposite magnetic polarity. When filaments are projected into space at the limb of the Sun, they appear as *prominences* (Fig. 6.29). Some prominences vary on short times scales while others evolve more slowly.

The *solar wind* is a stream of particles that are emitted from the Sun into interplanetary space. We can see the effects of the solar wind when we look at a comet that is passing near the Sun (Chapter 26). The tail of the comet always points away from the Sun. This is because the material in the tail is driven out of the head of the comet by the solar wind. The rate at which the Sun is losing mass is 10^{-14} $M_\odot$/yr. The wind is still accelerating in its five- to ten-day trip from the Sun to Earth. At the Earth's orbit, the speed is about 400–450 km/s, and the density of particles is 5 to 10 cm^{-3}. The particles in the solar wind are positive ions and electrons. It is thought that the

solar wind originates in lower density areas of the corona, called *coronal holes*.

The solar wind can have an effect on the Earth. Most of the solar wind particles directed at Earth never reach the surface of the Earth. The Earth's magnetic field serves as an effective shield, since the charged particles cannot travel across the field lines. Some of the particles, however, travel along the field lines and come closest to the Earth near the magnetic poles. These charged particles are responsible for the aurora displays. This explains why aurorae are seen primarily near the magnetic poles. When solar activity is increased, the aurorae become more widespread. The increased abundance of charged particles in our atmosphere also creates radio interference.

(a)

(b)

Fig 6.28. Images of solar flares. (a) Looking obliquely as the flare stands out from the surface. (b) In this photograph, taken through an Hα filter, we are looking down on a large solar flare from the side. This flare appears brighter than its surroundings in the Hα line. [NOAO/AURA/NSF]

(a)

Fig 6.29. Images of large prominences. (a) This shows the large loop structure. (b) This Hα image shows an eruptive prominence, where some of the material may actually escape from the Sun. (c) The X-ray image of the Sun gives another view of regions of activity. [(a), (b) NOAO/AURA/NSF; (c) NASA]

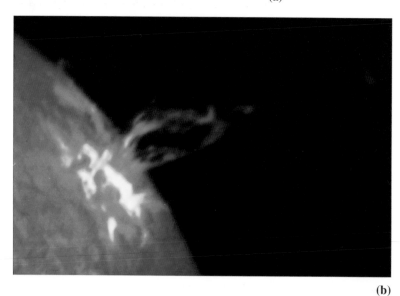

(b)

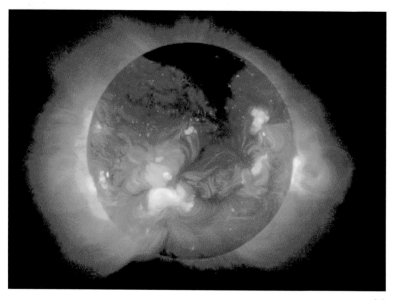

(c)

Chapter summary

We looked in this chapter at the one star we can study in detail, the Sun.

Most of what we see in the Sun is a relatively thin layer, the photosphere. In studying radiative transfer, we saw how we can see to different depths in the photosphere by looking at different wavelengths. The distance we can see corresponds to about one optical depth.

The photosphere doesn't have a smooth appearance. Instead, it has a granular appearance, with the granular pattern changing on a time scale of several minutes. This suggests convection currents below the surface. Supergranulation suggests even deeper convection currents.

The chromosphere is difficult to study. In the chromosphere the temperature begins to increase as one moves farther from the center of the Sun. This trend continues dramatically into the corona. The best studies of the corona have come from total solar eclipses or from space.

We saw how sunspots appear in place of intensified magnetic fields, as evidenced by the Zeeman effect. The number of spots goes through a 22-year cycle, which includes a complete reversal in the Sun's magnetic field. The structure of the magnetic field is also related to other manifestations of solar activity, such as prominences.

Questions

6.1. In Fig. 6.3, what would be the effect on the total cross section of (a) doubling the density of particles, (b) doubling the length of the tube, (c) doubling the radius of each particle?

*6.2. (a) What does it mean to say that some sphere has a geometric cross section of 10^{-16} cm^2? (b) Suppose we do a quantum mechanical calculation and find that at some wavelength some atom has a cross section of 10^{-16} cm^2. What does that mean?

6.3. In studying radiative transfer effects, we let τ be a measure of where we are in a given sample, rather than position x. Explain why we can do this.

6.4. What does it mean when we say that something is optically thin?

6.5. Suppose we have a gas that has a large optical depth. We now double the amount of gas. How does that affect (a) the optical depth, (b) the amount of absorption?

6.6. Why do we need the "mean" in "mean free path"?

6.7. (a) Explain how we can use two different optical depth spectral lines to see different distances into the Sun. (b) Why is Hα particularly useful in studying the chromosphere?

6.8. What other situations have you encountered that have exponential fall-offs?

6.9. Explain how absorption and emission by the H$^-$ ion can produce a continuum, rather than spectral lines?

6.10. Explain why we see a range of Doppler shifts over a spectral line.

6.11. How does Doppler broadening affect the separation between the centers of the Hα line and the Hβ line in a star?

6.12. What do granulation and supergranulation tell us about the Sun?

6.13. If the corona has $T = 2 \times 10^6$ K, why don't we see the Sun as a blackbody at this temperature?

6.14. Why can't you see the corona when you cover the Sun with your hand?

6.15. What advantages would a coronagraph on the Moon have over one on the Earth?

6.16. (a) Why does the F-corona still show the Fraunhofer spectrum? (b) Would you expect light from the Moon to show the Fraunhofer spectrum?

6.17. Why is the low density in the corona favorable for high levels of ionization?

6.18. (a) Why are collisions important in cooling a gas? (b) Why does the cooling rate depend on the square of the density? (c) How would you expect the heating rate to depend on the density?

6.19. Explain why charged particles drift parallel to magnetic field lines.

6.20. How are the various forms of solar activity related to the Sun's magnetic field?

Problems

6.1. Appendix G gives the composition of the Sun, measured by the fraction of the number of nuclei in the form of each element. Express the entries in this table as the fraction of the mass that is in each element. (Do this for the ten most abundant elements.)

6.2. Calculate the effective temperature of the Sun from the given solar luminosity, and radius, and compare your answer with the value given in the chapter.

6.3. Assume that for some process the cross section for absorption of a certain wavelength photon is 10^{-16} cm^2, and the density of H is 1 g/cm^3. (a) Suppose we have a cylinder that is 1 m long and has an end area of 1 cm^2. What is the total absorption cross section? How does this compare with the area of the end? (b) What is the absorption coefficient (per unit length)? (c) What is the mean free path? (d) How long a sample of material is needed to produce an optical depth of unity.

6.4. Suppose we have a uniform sphere (radius $R_\odot$) of 1 $M_\odot$ of hydrogen. What is the column density through the center of the sphere?

6.5. How large must the optical depth through a material be for the material to absorb: (a) 1% of the incident photons; (b) 10% of the incident photons; (c) 50% of the incident photons; (d) 99% of the incident photons?

6.6. If we have material that emits uniformly over its volume, what fraction of the photons that we see come from within one optical depth of the surface?

*6.7. Suppose we divide a material into N layers, each with optical depth $d\tau = \tau/N$, where τ is the total optical depth through the material and $d\tau \ll 1$. (a) Show that if radiation I_0 is incident on the material, the emergent radiation is

$$I = I_0 (1 - d\tau)^N$$

(b) Show that this reduces to $I = I_0 e^{-\tau}$ (equation 6.19) in the limit of large N. (Hint: You may want to look at various representations of the function e^x.)

6.8. For what value of x does the error in the approximation $e^x \cong 1 + x$ reach 1%?

*6.9. Suppose we have a uniform sphere of radius R and absorption coefficient κ. We look along various paths, passing different distances p from the center of the sphere at their points of closest approach to the center. (a) Find an expression for the optical depth τ as a function of p. (b) Calculate $d\tau/dp$, the rate of change of τ with p. (c) Use your results to discuss the sharpness of the solar limb.

*6.10. Consider a charge Q near a neutral object. If the object is a conductor, charge can flow within it. The presence of the charge Q induces a dipole moment in the conductor, and there is a net force between the dipole and the charge. (a) Show that this force is attractive. (b) How does this apply to the possible existence of the H$^-$ ion?

6.11. What is the thermal Doppler broadening of the Hα line in a star whose temperature is 20 000 K?

6.12. We observe the Hα line in a star to be broadened by 0.05 nm. What is the temperature of the star?

6.13. Compare the total thermal energy stored in the corona and photosphere.

6.14. (a) At what wavelength does the continuous spectrum from sunspots peak? (b) What is the ratio of intensities at 550 nm in a sunspot and in the normal photosphere? (c) What is the ratio of energy per second per surface area given off in a sunspot and in the normal photosphere?

6.15. How long does it take before material at the solar equator makes one more revolution than that at 40° latitude?

6.16. Calculate the energy per second given off in the solar wind?

*6.17. (a) What is the pressure exerted on the Earth by the solar wind? (Hint: Calculate the momentum per second on an object whose cross sectional area is that of the Earth.) (b) How large a sail would you need to give an object with the mass of the space shuttle an acceleration of 0.1 g at the distance of the Earth from the Sun?

*6.18. To completely describe the radiative transfer problem, we must take emission into account as well as absorption. The *source function* S is defined so that $S \, \mathrm{d}\tau$ is the increase in intensity due to emission in passing through a region of optical depth $\mathrm{d}\tau$. This means that the radiative transfer equation should be written

$$\mathrm{d}I/\mathrm{d}\tau = -I + S$$

(a) If S is a constant, solve for I vs. τ, assuming an intensity I_0 enters the material. (b) Discuss your result in the limits $\tau \ll 1$ and $\tau \gg 1$.

Computer problems

6.1. Consider the situation in Fig. 6.4 with 1000 layers. Draw a graph of the fraction of the initial beam emerging from each layer, for total optical depths (a) 0.1, (b) 1.0, (c) 10.0. Show that the fraction emerging from the final layer agrees with equation (6.18).

6.2. Estimate the Doppler broadening for the Hα lines from the atmospheres at the mid-range of each spectral type (e.g. O5, B5, etc.). (Hint: Scale from the result in example (6.1).)

Part II

Relativity

Einstein's theory of relativity caused us to rethink the meaning of both space and time, concepts that had been taken for granted for centuries. The foundation of this revolution is the *special theory of relativity*, which Einstein published in 1905. The *general theory of relativity*, published in 1916, is really a theory of gravitation set in the foundations of the special theory; it also allows us to analyze the properties of frames of reference that are accelerating.

Chapter 7

Special relativity

7.1 | Foundations of special relativity

7.1.1 Problems with electromagnetic radiation

The problems that lead to special relativity start with Maxwell's theory of electromagnetic radiation. Maxwell's equations, presented in 1873, allow for the existence of waves of oscillating electric and magnetic fields. All waves known before electromagnetic waves required a medium in which to travel. For example, sound waves can travel through air, but not through a vacuum. There is no obvious medium necessary for the propagation of electromagnetic waves. Physicists postulated a medium that is difficult to detect, called the *luminiferous ether*, or simply the *ether*. The ether supposedly fills all of space. Once we have a medium, then we have a reference frame for the motion of the waves. For example, the speed of sound is measured with respect to the air through which it is moving. An observer moving through the medium will detect a different speed for the waves than an observer at rest in the medium.

Einstein's questions about Maxwell's equations involve the appearance of electromagnetic waves to different observers, who are moving at different speeds. Einstein started with the *postulate of special relativity*, that, *the laws of physics, properly stated, should be independent of the velocity of the observer*. It may be that the values of certain quantities change with the motion of the observer, but the relationships among the physical quantities do not change.

Einstein examined Maxwell's equations to see if they obeyed this simple rule. His reasoning is illustrated in Fig. 7.1. The solution of Maxwell's equations gives us waves that vary sinusoidally in both space and time. That is, the waves vary with position, repeating each wavelength and, with time, repeating each cycle. How would an electromagnetic wave appear to an observer moving along with the wave at the speed of light? The wave would appear sinusoidal in space but constant in time, since the observer is moving along with the wave (crest for example). However, there is no mathematical solution to Maxwell's equations that is constant in time, but which varies sinusoidally. (Remember, it is precisely the time variation of electric and magnetic fields that allows the propagation of the waves.) This seems to be a contradiction.

Two possibilities were left: (1) Maxwell's equations were correct in only one reference frame, that of the ether, or (2) there was something wrong with the basic concepts of space and time. The first possibility violates Einstein's postulate of special relativity, so he chose the second, saying that, for some reason, it must be impossible to move at an arbitrary speed relative to an electromagnetic wave. He concluded that *the speed of light in a vacuum is the same for all observers, independent of their motion*. This suggests that electromagnetic waves must be different from the familiar mechanical waves. There must also be something wrong with the concept of the ether.

When Einstein was working on this problem, experiments had already been done which cast the existence of the ether into doubt. An experiment,

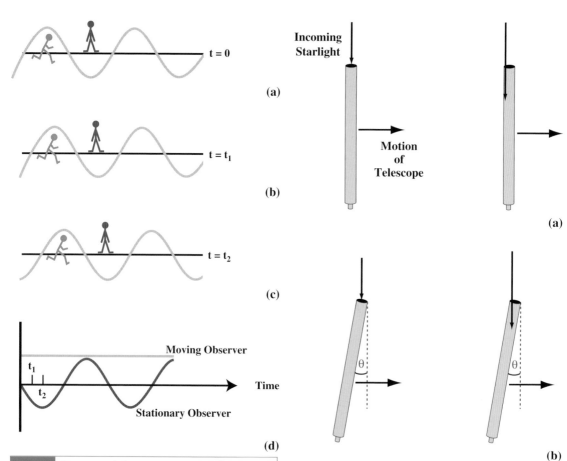

Fig 7.1. Observers of electromagnetic waves. One observer is stationary and the other observer is moving with the wave. (a) The moving observer is at a crest, and the stationary observer is at a null. (b) The moving observer is still at a crest, and the stationary observer sees a negative value. (c) The moving observer is still at the crest and the stationary observer is at a dip. (d) We plot what each sees as a function of time. The moving observer sees a constant value while the stationary observer sees a sinusoidally varying value.

Fig 7.2. Aberration of starlight. (a) Assume that the telescope is moving to the right as the beam of light enters, with the telescope tube lined up with the beam of light. Since the speed of light c is finite, the telescope moves as light passes through, and the light strikes the side. (b) To observe the light, we must tilt the telescope slightly. Thus, as the telescope moves over, the beam is always centered in the tube. We must tip the telescope in the direction in which it is moving.

originally done by A. A. Michelson in 1881, and in an improved fashion by Michelson and Morley in 1887, was designed to measure the motion of the Earth through the ether by measuring the speed of light in two directions perpendicular to each other. No change was found. This meant that the Earth could not be moving through the ether. If the ether existed, it must be dragged along with the Earth. However, there is another observation which rules out the dragging of the ether by the Earth. It is illustrated in Fig. 7.2, and is called *aber-*

ration of starlight. (This has nothing to do with aberrations in optical systems.) It is a slight change in the angle at which light from stars appears to be arriving due to the motion of the observer, in this case the motion of the Earth about the Sun. (It is analogous to the change in the apparent angle at which rain is falling when you start to move.) The shift is always in the direction of the motion of the Earth, so it changes throughout the year. The positions of some stars are shifted by as much as 20 arc seconds from their true positions. (This effect has been used in the past to measure the speed of light.) In the ether theory there is no way for aberration to be observed if the Earth is dragging the ether.

The fact that the speed of light is independent of the velocity of the observer contradicts our everyday experience, in which relative velocities are additive. Einstein began to look at the underlying cause for the speed appearing to be constant. In measuring a speed, we measure a distance and a time interval. Einstein suspected that the problem lay in our traditional concepts of space and time. Physicists such as Newton simply assumed that space and time were given. Einstein suggested that they might not be absolute but might depend on the motion of the observer. Einstein examined the idea of an absolute time and looked at whether time might actually be a quantity that depends on the motion of the observer.

7.1.2 Problems with simultaneity

Einstein realized that an absolute time was tied to the concept of *absolute simultaneity*. By absolute simultaneity we mean that if two events appear simultaneous to one observer, they appear simultaneous to all other observers. This is important because telling time is actually noting the simultaneity of two events. For example, if we say the train left the station at 7:00, we are saying that two events are simultaneous. The first event is the train leaving the station, and the second event is the clock showing 7:00. If those events are simultaneous for one observer, but not for all observers, then the concept of absolute time has no meaning.

An experiment depicted in Fig. 7.3 shows that two events can be simultaneous for one observer, but not another. The two observers are at the centers of identical railroad cars. One car is at rest with respect to the station. The other is moving past at some speed. When the two observers are opposite each other, two flashes go off at the ends of one car. The flashes are judged to be simultaneous by the observer at rest. How are they seen by the other observer? The figure shows that the flashes that the observer is moving toward is seen first. The flashes are not simultaneous for the moving observer. *Simultaneity is not absolute. Therefore, time is not absolute.*

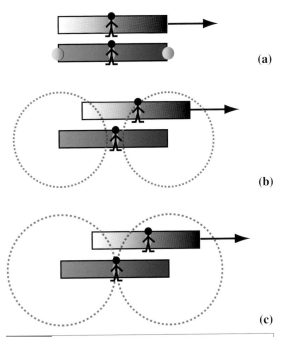

Fig 7.3. Flashes in railroad cars and simultaneity. The top car is moving past the bottom car. (a) When the observers at the centers of each car are closest to each other, flashes go off at opposite ends of the lower car. (b) The motion of the top car means that the observer in that car sees the right flash first. (c) The observer in the bottom car sees both flashes at the same time.

With this as a starting point, we now go on to investigate how different types of situations appear to observers with different velocities. In special relativity, we deal only with reference frames that are not accelerating with respect to each other or in which there are no external gravitational forces. Such a reference frame is called an *inertial reference frame*. An inertial frame might be provided by a space station far from any mass and with the engines off so there is no acceleration. Einstein's postulate about the laws of physics being the same in different reference frames only applies to inertial frames. (We know that accelerating frames must be different, because they have pseudoforces, such as centrifugal force.) Another way of stating Einstein's postulate is that *There is no experiment we can perform to tell us which inertial frame is moving and which is at rest. There is no 'preferred' inertial frame.* All we can talk about is the relative motion of two inertial frames.

7.2 | Time dilation

Now that we know that time is relative, we can see how a clock appears to two different observers. One observer is at rest with respect to the clock, and the other observer sees the clock moving. The time viewed in the frame in which the clock is at rest is called the *proper time* for that clock. The word 'proper' does not denote anything superior about this frame; it just happens to be the frame in which the clock is at rest. We can think of proper time as being the time interval between two events at the same place.

A simple clock is shown in Fig. 7.4. In this clock a light beam bounces back and forth between two mirrors, separated by a distance L. We would keep time by counting the light bounces. The time for the light pulse to make a round trip is

$$t_0 = 2L/c \tag{7.1}$$

In the frame in which the clock is moving, the light beam takes a longer path. Since the speed of light is the same in both frames, the beam must take longer to make the round trip. From the figure, we see that the distance traveled is $2[L^2 + (vt/2)^2]^{1/2}$, so the time is

$$t = (2/c) [L^2 + (vt/2)^2]^{1/2}$$

Squaring this gives

$$t^2 = (4L^2/c^2) [1 + (vt/2L)^2]$$

We use equation (7.1) to eliminate L, giving

$$t^2 = t_0{}^2 + (v^2/c^2) t^2$$

We want to solve for t in terms of t_0:

$$t^2 [1 - (v/c)^2] = t_0{}^2$$

Taking the square root of both sides and solving for t, we have

$$t = \frac{t_0}{[1 - (v/c)^2]^{1/2}} \tag{7.2}$$

The significance of this result is that the time interval measured in the frame in which the clock is moving is greater than that measured in the frame in which the clock is at rest. Suppose we have two identical clocks. If we keep one at rest (with respect to us) and let the other one move, the moving clock appears to run slow. It is important to realize that the situation is perfectly symmetric. If there is an observer traveling with each clock, *each observer sees the other clock running slow*. This effect is called *time dilation*.

From equation (7.2), we can see that the amount of time dilation depends on the quantity $1/[1 - (v/c)^2]^{1/2}$. This quantity is generally designated γ and is plotted as a function of (v/c) in Fig. 7.5. Note that this quantity is close to unity for small velocities, and only becomes large when v is very close to c. This confirms our intuition that the results of

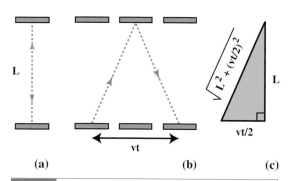

(a)   **(b)**   **(c)**

Fig 7.4.   Light clock. (a) In the rest frame of the clock, the light bounces back and forth. (b) In the laboratory frame, with the clock moving, the light beam travels a longer path. (c) Calculating the extra distance traveled.

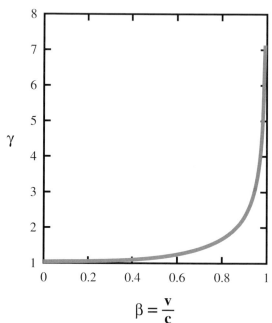

$$\beta = \frac{v}{c}$$

Fig 7.5.   The quantity γ vs. v/c. For v/c small, γ is close to unity. As v/c approaches unity, γ approaches infinity.

special relativity should reduce to familiar everyday results when speeds are much less than c.

Time dilation is not an artifact of the light clock that we have depicted in Fig. 7.4. It applies to all clocks. For example, it applies to the decay of unstable elementary particles. Particles moving close to the speed of light should appear to live longer than the same particles at rest. This is tested almost daily in particle accelerators around the world. A dramatic example is in cosmic rays, which contain unstable particles which can decay as they pass through the Earth's atmosphere. If we measure the flux of cosmic rays at high altitude and near the ground, we find that many more survive this trip than we would expect, unless we account for the effect of time dilation.

Example 7.1 How fast must a particle be traveling to live ten times as long as the same particle at rest?

SOLUTION
We simply set

$$10 = \frac{1}{[1 - (v/c)^2]^{1/2}}$$

Squaring gives

$$100 = \frac{1}{[1 - (v/c)^2]}$$

Solving for $(v/c)^2$ gives

$$(v/c)^2 = 0.99$$

Taking the square root gives $v/c = 0.995$. The particle must be within one-half of one percent of the speed of light!

Time dilation applies to biological clocks. A person traveling at a high speed will not age as fast as a person at rest. Of course, the situation must be symmetric. Each person sees the other age slower. This leads to a puzzle known as the *twin paradox*. Two twins are on Earth. One is an astronaut who goes on a trip at a speed close to c. The other stays on the Earth. From the point of view of the one that stayed on Earth, the astronaut is moving and will not age as fast as the one on Earth. The astronaut will appear younger upon returning. However, the astronaut sees the one on Earth moving away at high speed. Therefore the

one on Earth should appear younger. It is alright for two moving observers to see each other age slower. However, we have a problem if we try to bring the twins together – both at rest. We can see which one is really younger and decide which was really moving. This would seem to violate Einstein's postulate. However, if the twins start and end together at rest, then one twin must accelerate to get to very high speeds. That acceleration produces pseudoforces which can be felt by only one twin. This breaks the symmetry of the problem without any logical contradiction. (Remember, a pseudoforce is really an inertial response to an acceleration of the reference frame.)

7.3 | Length contraction

Once the concept of time becomes suspect, the concept of length must also be reinvestigated. Think of how we measure the length of an object. We measure the positions of the two ends and take the difference between the two positions. For this procedure to have any meaning, the measurements must be carried out simultaneously. (If I measure the position of the front of an airplane, when it is in NY, and the position of the tail 6 hours later when it is in LA, I should not conclude that the airplane is 5000 km long.) Unfortunately, we have seen that observers in different inertial frames cannot agree on the simultaneity of events separated in space.

It is therefore not surprising that lengths will appear different to observers in different inertial frames. In fact, physicists had been playing with this idea before Einstein's 1905 paper. *H. Lorentz* had proposed it as a way around the results of the Michelson–Morley experiment. He said that the ether could be saved if the lengths of objects depended in a particular way on their state of motion.

In considering changes in length, we look separately at lengths perpendicular to and parallel to the direction of motion. We can first show that there can be no length changes perpendicular to the direction of motion. Let's assume that there were such a change and that moving objects shrink. We now consider an experiment. Two people of identical height are standing, as in Fig. 7.6.

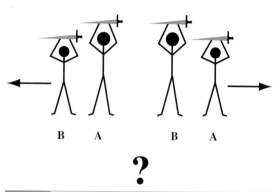

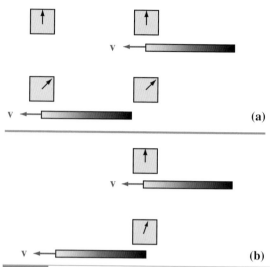

Fig 7.6. The effect of possible length contraction perpendicular to the direction of motion. Assume that objects shrink perpendicular to the direction of motion. A and B are the same height when both are at rest. Both hold swords parallel to the ground, and B moves past. If B shrinks, A's sword will miss B, but B's sword will cut A. However, from B's point of view, as shown in the right figure, A is moving, and it is A who shrinks. That would result in an injury to B not A.

Fig 7.7. The effect of length changes parallel to the direction of motion. (a) To measure the length of a stick, we must first measure its speed. We do this by measuring the time for one point on the stick to go a known distance between two stationary clocks. In the upper frame, the front of the stick starts at the right clock. In the lower frame it reaches the left clock. The time difference is noted, and the speed is calculated. (b) Knowing the speed of the stick, we measure its length by seeing how long it takes the stick to pass a single stationary clock. In the top half of the frame the front of the stick is at the clock, and the measurement starts. In the bottom half of the frame, the back of the stick reaches the clock and the measurement ends.

Each has a sword held out at the level of the top of the head. Now person B is carried past person A at a high speed. According to A, B is moving, and B gets shorter. B's sword cuts A's head while A's sword passes safely over B. According to B, A is moving and the situation is reversed. We have a true contradiction. Each person is wounded in their own rest frame but not in the other. The only way out of this is to say that there can be no length change perpendicular to the direction of motion, and no one gets hurt. (A similar argument would rule out expansion perpendicular to the direction of motion as well as contraction.)

We can think of no such examples to rule out changes parallel to the direction of motion. Here, there is actually a change of length. Moving objects appear to shrink. We call this effect *Lorentz contraction*. To see this we use Fig. 7.7 to show how we might measure the length of a moving object. The length of an object, measured in the frame in which it is at rest, is called the *proper length*, L_0 . This can be measured in the usual way, since its ends are not going anywhere. We now measure its length in a frame in which it is moving. We can tell its speed v by having two markers at rest in our frame, and measuring the time for the front of the object to travel from one marker to the other. We can then measure its length by

timing the passage of the object past one marker. The time interval between the two measurements at the one marker Δt, as measured in the frame of the object, is

$$\Delta t = L_0/v$$

In our frame the time interval is different because of time dilation, so the interval is

$$\Delta t' = \Delta t/\gamma$$

We now say that the length of the object is

$$L = v\,\Delta t'$$

$$= v\,\Delta t/\gamma$$

Finally, substituting $\Delta t = L_0/v$ gives

$$L = L_0/\gamma \qquad (7.3)$$

Not surprisingly, the length contraction has the same dependence on v/c as does the time dilation.

As with time dilation, length contraction is symmetric. If A and B are carrying meter sticks parallel to the direction of their relative motion, A will see B's stick shrink, and B will see A's stick shrink. There is no contradiction here since we cannot compare the ends of the sticks simultaneously for both observers. This symmetry has provided some interesting puzzles that start with seeming contradictions, but end up with logical resolutions. (See Problem 7.5.)

7.4 | The Doppler shift

With lengths and times appearing different to different observers, it is also necessary to take a closer look at the Doppler shift, since wavelengths obviously involve length, and frequencies obviously involve time. Since there is no ether, the Doppler shift for electromagnetic waves can only depend on the relative motions of the source and observer. This is different from the case of sound waves, for which the shift depends on which is moving. We can show that the result for electromagnetic waves doesn't depend on which is moving by considering separately the case of the moving source and the moving observer. In both cases, we denote quantities measured in the rest frame of the receiver as primed (').

7.4.1 Moving source
Let's assume the source is moving away from the receiver at a speed v. The source emits N waves in time $\Delta t'$, as measured by the receiver. In this time, the first wave travels a distance $c\,\Delta t'$ and the source travels a distance $v\,\Delta t'$. The wavelength will then be the distance between the source and the first wave, divided by the number of waves. That is

$$\lambda' = \frac{c\Delta t' - v\Delta t'}{N}$$

The frequency is then given by

$$\nu' = \frac{c}{\lambda'}$$

$$= \frac{c}{c - v}\frac{N}{\Delta t'} \qquad (7.4)$$

We would like to relate this to the frequency in the source frame, ν. It is given as the number

of waves, N, divided by the time interval, Δt, as measured in the source frame:

$$\nu = N/\Delta t$$

We can use the time dilation formula, $\Delta t' = \gamma\,\Delta t$, to make this

$$\nu = N\,\gamma/\Delta t'$$

This can now be used to eliminate $\Delta t'$ in equation (7.4), giving

$$\nu' = \frac{1}{1 - v/c}\frac{v}{\gamma}$$

$$= \nu\frac{[1 - (v/c)^2]^{1/2}}{[1 - (v/c)]}$$

$$= \nu\frac{[(1 - v/c)^{1/2}(1 + v/c)^{1/2}]}{[1 - (v/c)]}$$

If we multiply the numerator and denominator by $(1 + v/c)^{1/2}$, this simplifies to

$$\nu' = \frac{v(1 + v/c)}{[1 - (v/c)^2]^{1/2}}$$

$$= \nu\left(1 + \frac{v}{c}\right)\gamma \qquad (7.5)$$

This is like the classical Doppler shift formula, except for the extra factor of γ which comes from time dilation.

7.4.2 Moving observer
We now consider the receiver moving away from the source. In the source's frame, in time Δt, the receiver will receive all waves in a length $(c + v)\Delta t$. This number is the length divided by the wavelength:

$$\nu' = \frac{N}{\Delta t'}$$

$$= \frac{(1 + v/c)\Delta t}{\Delta t'}$$

$$= \frac{(1 + v/c)v}{[1 - (v/c)^2]^{1/2}}$$

$$= \nu\left(1 + \frac{v}{c}\right)\gamma$$

This is identical to equation (7.5), proving that the Doppler shift is independent of whether the source or the receiver is moving.

7.4.3 General result

We will now generalize the result. If we had considered the source and observer moving apart, we would have $1 - v/c$ in the numerator of equation (7.5). The v in the $1 \pm v/c$ is just the radial velocity, the component of the velocity along the line of sight. We should therefore replace *that* v in equation (7.5) with v_r. However, the other v in equation (7.5), in the γ, comes from time dilation, which is independent of the direction of motion. It must remain as the total relative speed of the source and receiver. This means that it is possible to have a Doppler shift, even when the motion is perpendicular to the line of sight. This is simply a result of time dilation, and is not important until v is close to c. With these generalizations, equation (7.5) becomes

$$\nu' = \nu/(1 + v_r/c)\,\gamma \tag{7.6}$$

We can derive a corresponding expression for wavelength, remembering that $\lambda' = c/\nu'$,

$$\lambda' = \lambda(1 + v_r/c)\,\gamma \tag{7.7}$$

Later in this book, we will encounter objects moving away from us at speeds close to c. For these, the radial velocity is very close to the total speed v. This allows us to make the simplification

$$\lambda' = \frac{\lambda[1 + \beta]}{[1 - \beta^2]^{1/2}}$$

$$= \frac{\lambda[1 + \beta]}{[(1 - \beta)(1 + \beta)]^{1/2}}$$

which simplifies to

$$\lambda' = \lambda \left(\frac{1 + v/c}{1 - v/c} \right)^{1/2} \tag{7.8}$$

Example 7.2 Relativistic Doppler shift
Find the wavelength at which we will observe the Hα line if it is emitted by an object moving away with $v/c = 0.3$.

SOLUTION
From equation (7.8), we find that

$$\frac{\lambda'}{\lambda} = \left(\frac{1 + 0.3}{1 - 0.3} \right)^{1/2}$$

$$= 1.36$$

This means that

$$\lambda' = (1.36)\,(656.28 \text{ nm})$$

$$= 892.54 \text{ nm}$$

The line is shifted from the visible into the near infrared!

7.5 | Space-time

Phenomena such as time dilation and length contraction are not simply illusions. They are real effects. Our failure to appreciate this previously comes from a failure to appreciate the true nature of space and time. Classical physicists assumed that space and time were simply there, just like a blank piece of graph paper, and that the laws of physics were laid down on top of them. Einstein realized that the laws of physics were intimately entwined with space and time. We can understand the nature of this relationship by abandoning our normal three-dimensional world and replacing it with the four-dimensional world of *space-time*.

7.5.1 Four-vectors and Lorentz transformation

In space-time we simply treat time as another coordinate. To remind us that time is just another way of measuring distance, we sometimes write the time coordinate as ct, so that it has the same dimensions as the other coordinates. In this way, we could measure time in meters. What is a time of one meter? It is the time that light takes to travel one meter. (Note that we have previously used time as a measure of distance when we introduced the light-year.)

An interesting aside to this has come from the organizations that set international standards such as the meter and the second. It used to be that such units were defined independently, and c was just a measured quantity. The speed of light is now taken to have a defined value, where all decimal places beyond the most accurate measured value are taken to be zero. It now gives the conversion from meters to seconds. This means that we only need a standard for the second or the meter, but not both.

In space-time we speak of *four-vectors* to distinguish them from ordinary three-dimensional vectors. Any event is characterized by the four coordinates (ct, x, y, z). Observers in different inertial frames will note different coordinates for events, but the coordinates are related. If one inertial system is moving with respect to another at a speed v, in the x-direction, the coordinates in the transformations between the two coordinate systems are found by assuming they are linear in the coordinates, and must give the correct results for length contraction and time dilation. The result is (letting $\beta = v/c$)

$$ct = \gamma (ct' + \beta x')$$

$$x = \gamma (x' + \beta ct')$$

$$y = y'$$

$$z = z' \qquad (7.9)$$

The reverse transformation is given by

$$ct' = \gamma (ct - \beta x)$$

$$x' = \gamma (x - \beta ct)$$

$$y' = y$$

$$z' = z \qquad (7.10)$$

These relationships together are called the *Lorentz transformation*.

We interpret the Lorentz transformation as telling us that the rules of geometry are different in space-time than they are for ordinary space. To illustrate this point, we use a *space-time diagram*, like that shown in Fig. 7.8. For simplicity, we plot only one space coordinate, x, as well as the time coordinate. We can keep track of events in such a diagram by plotting the coordinates of the event. By convention, we have time running vertically. The effect of the Lorentz transformation is to rotate the axes through an angle whose tangent is v/c. The unusual feature is that the x-axis and t-axis rotate in opposite directions, so that the axes are no longer perpendicular to each other. Note that $v = c$ puts both axes in the same place. It should not surprise us that something funny happens when $v = c$, because this is where the quantity γ becomes infinite.

We know that in ordinary three-dimensional space, a rotation changes the coordinates of an object, but the lengths of things are unchanged. That is, if we have two objects, as shown in Fig. 7.9, whose separations are given in one coordinate system by $(\Delta x, \Delta y, \Delta z)$ and $(\Delta x', \Delta y', \Delta z')$ in another, then the distance between the two, which is the square root of the sum of the squares of the coordinate differences, doesn't change. That is

$$(\Delta x)^2 + (\Delta y)^2 + (\Delta z)^2 = (\Delta x')^2 + (\Delta y')^2 + (\Delta z')^2 \quad (7.11)$$

We say that the *length is invariant under rotation*.

Since the Lorentz transformation has properties of a rotation, is there a corresponding concept

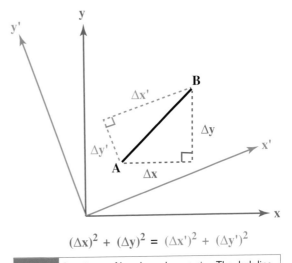

$$(\Delta x)^2 + (\Delta y)^2 = (\Delta x')^2 + (\Delta y')^2$$

Fig 7.9. Invariance of lengths under rotation. The dark line represents the distance between points A and B. The components of this length with respect to the x- and y-axes are Δx and Δy, respectively, and with respect to the x'- and y'-axes are $\Delta x'$ and $\Delta y'$, respectively. Independent of the components used, the length of the dark line is the same.

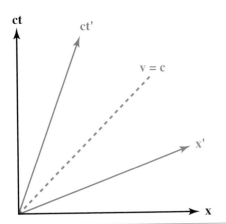

Fig 7.8. Lorentz transformation on a space-time diagram. The transformation looks like a rotation of the axes except that the time and space axes rotate in opposite directions.

in space-time? The answer is yes, but the invariant quantity is slightly different, because the time axis rotates in the opposite direction to the space axes. We define the *space-time interval* as

$$(\Delta s)^2 = (c\,\Delta t)^2 - (\Delta x)^2 - (\Delta y)^2 - (\Delta z)^2 \qquad (7.12)$$

This is the quantity that is invariant under a Lorentz transformation. Note that the Lorentz transformation can be derived by assuming that this quantity is invariant, and that the transformations be linear in the coordinates. When this is done, time dilation and length contraction can be derived from the Lorentz transformation rather than the other way around. This reinforces the idea that length contraction and time dilation are not artifacts of some particular measurement, but are an integral part of the nature of space-time.

To get a feeling for the physical meaning of Δs, consider an observer moving from one place to another in time Δt, as measured in the observer's rest frame. This means that Δt is the proper time interval. In the observer's rest frame, there is no change in position, so $\Delta x = \Delta y = \Delta z = 0$. This means that $\Delta s = c\Delta t$. Therefore, Δs is just the proper time interval (in units of length). Moreover, since it is an invariant, for any two events and any inertial reference frames Δs will always equal the proper time interval between the two events.

We can define three types of space-time intervals (Fig. 7.10a), depending on whether $(\Delta s)^2$ is zero, positive or negative. Suppose our two events are the emission and absorption of a photon. A photon will move on the sphere whose surface is given by, $(\Delta x)^2 + (\Delta y)^2 + (\Delta z)^2 = (c\,\Delta t)^2$. This means that $(\Delta s)^2$ is zero for a photon traveling any distance in any direction. We call such intervals *lightlike*. Intervals for which $(\Delta s)^2$ is positive are called *timelike*. The positions are close enough in space that a photon would have had more than enough time to travel from the first event to the second. This means that the first event could have caused the second. In the opposite case, when $(\Delta s)^2$ is negative, we call the interval *spacelike*. A photon cannot traverse the distance in the time given. Unless a signal can be sent faster than the speed of light, there is no way the one event could have caused the other.

If we extend our space-time diagram to more dimensions, we call the surface defined by $(\Delta s)^2 = 0$

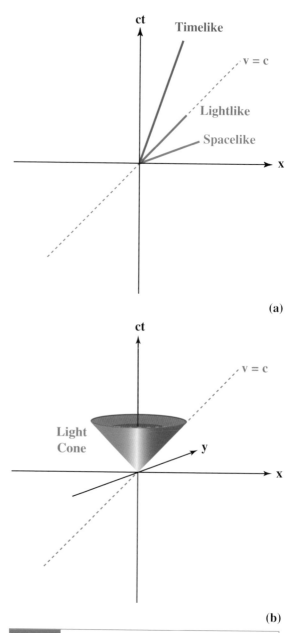

Fig 7.10. (a) Space-time intervals. (b) Light cone.

the *light cone* (Fig. 7.10b). Events that could have caused the event at the origin of the cone are inside the past light cone. Events that could be caused by events at the origin of the cone are inside the future light cone. Events that are outside the light cone can have no causal connection with the event at the origin.

It should be noted that in our discussion of a space-time interval we could have defined it to be the negative of what it was in equation (7.12), and not changed any of the interpretation (apart from carrying through the minus sign). It is just a matter of convention to do it one way or the other, and you will find some authors who do it one way and some who do it the other. As long as they are internally consistent there is no problem. People who use one convention or the other then also differ in how they 'count' the time coordinate. That is, we can write (x, y, z) as (x_1, x_2, x_3). If we use the space-time interval as given in equation (7.12), then we write ct as x_0 and think of time as the 'zeroth coordinate'; if we use the space-time interval as the negative of that given in equation (7.12), then we write ct as x_4 and think of time as the 'fourth coordinate'.

7.5.2 Energy and momentum

The space-time coordinates of an event are not the only quantities that transform according to the Lorentz transformation. For example, another important four-vector involves energy and momentum. To see the analogy with (ct, x, y, z), remember that for a photon moving in the x-direction, $x = ct$. The energy and momentum of a photon are related by $E = cp$, so, for a photon moving in the x-direction, $E = cp_x$. This suggests that the *energy-momentum four vector* should be (E, cp_x, cp_y, cp_z). These should then obey the Lorentz transformations:

$$E = \gamma (E' + \beta cp_x')$$

$$cp_x = \gamma (cp_x' + \beta E')$$

$$cp_y = cp_y'$$

$$cp_z = cp_z' \qquad (7.13)$$

The reverse transformation is given by

$$E' = \gamma (E - \beta cp_x)$$

$$cp_x' = \gamma (cp_x - \beta E)$$

$$cp_y' = cp_y$$

$$cp_z' = cp_z \qquad (7.14)$$

If we let the (′) reference frame be one in which the particle is at rest, so that $p_x' = 0$, then the first thing we note is that $E = \gamma E'$, so, if the energy of the particle at rest, E', were equal to zero, then the energy, E', would always be zero. This is obviously not the case, since we know that moving particles must have some energy. This means that the *rest energy*, E_0, cannot be zero. So, this gives us an expression for the relativistic energy:

$$E = \gamma E_0 \qquad (7.15)$$

We can then find the relativistic momentum as

$$cp_x = \gamma \beta E_0 \qquad (7.16)$$

In the non-relativistic (γ close to one) limit, the momentum must give the classical expression, $p_x = mv_x$. From equation (7.16) this can only occur if

$$E_0 = m_0 c^2 \qquad (7.17)$$

where m_0 is the *rest mass* of the particle.

We can rewrite the expressions for relativistic energy and momentum:

$$E = \gamma m_0 c^2 \qquad (7.18)$$

$$\mathbf{p} = \gamma m_0 \mathbf{v} \qquad (7.19)$$

We can also define a kinetic energy as the difference between the total energy and the rest energy:

$$E_k = (\gamma - 1) m_0 c^2 \qquad (7.20)$$

In the limit $v \ll c$, we can write

$$\gamma = [1 - (v/c)^2]^{1/2}$$

$$\cong 1 + v^2/2\,c^2$$

where we have used the fact that, for $x \ll 1$, $(1 + x)^n \cong 1 + nx$. The kinetic energy for $v \ll c$ then becomes

$$E_k = m_0 c^2 (1 + v^2/2c^2 - 1)$$

$$= (1/2)m_0 v^2$$

which is the classical expression.

Since the energy-momentum four-vector obeys the Lorentz transformations, it must have an invariant length associated with it. It is

$$E^2 - (cp_x)^2 - (cp_y)^2 - (cp_z)^2 \qquad (7.21)$$

To give this quantity a physical meaning, we evaluate it in the rest frame of some particle. In that case, the momentum is zero and the energy is the rest energy. So the invariant length is simply $m_0 c^2$. Since this quantity is invariant, its value must be $m_0 c^2$ for any observer. (We just choose to work it out in an easy frame.)

Example 7.3 Rest energy of a proton

What is the rest energy of a proton?

SOLUTION
By equation (7.20)

$$E_0 = (1.67 \times 10^{-24}\,\text{g})(3.0 \times 10^{10}\,\text{cm/s})^2$$

$$= 1.5 \times 10^{-3}\,\text{erg}$$

To form an idea of how large this is, we express the answer in eV, to get 939 MeV (as compared, for example, with the 13.6 eV needed to ionize a hydrogen atom).

Note that, as v/c approaches unity, γ approaches infinity. This means that it takes an infinite amount of energy to accelerate an object with non-zero rest mass to the speed of light. This means that the speed of light is a limiting speed.

Some physicists have speculated on particles that can travel faster than light. These particles have been given the name *tachyons*. The trick is that these particles, if they exist, can never go slower than the speed of light. The speed of light would seem to be a barrier for them as well, only from above. If tachyons do exist they can interact with photons, and make their presence known. All experiments to look for tachyons have indicated that they do not exist.

Chapter summary

In this chapter we saw how the special theory of relativity has changed our thinking about the nature of space and time.

We saw how the requirement that the laws of physics be the same in all inertial frames leads to the idea that the speed of light is the same for all observers. This, in turn, leads us to the phenomena of time dilation and length contraction. The phenomena are only large when the speeds involved are close to c.

We saw that we can no longer think of space and time as being separate entities, but must consider a four-dimensional coordinate system, called space-time. We defined a space-time interval which is invariant under the Lorentz transformation (and is equal to the proper time).

We saw that energy and momentum must be treated like space and time. This leads to a relativistic energy $E = \gamma m_0 c^2$ and the idea of a rest energy, $m_0 c^2$.

Questions

7.1. What are the differences between sound waves and electromagnetic waves?

7.2. How does the speed of light being independent of the velocity of the observer eliminate the problem that Einstein found with Maxwell's equations?

7.3. What is the relationship between simultaneity and absolute time?

7.4. What do we mean by the terms 'proper length' and 'proper time'?

*7.5. A painter's assistant is carrying a 10 m ladder parallel to the ground. The assistant is moving at 0.99c. The painter is up on a high ladder and drops a cloth with a 5 m diameter hole in it, parallel to the ground. From the point of view of an observer on the cloth, the ladder shrinks to less than 5 m in length and fits through the hole. From the point of view of the assistant, the hole shrinks, so it is even smaller than the ladder. Yet we know that the ladder must get through in all reference frames if it gets through in one. How does it get through as viewed by the assistant? (Hint: Think of how the cloth appears to the assistant.)

7.6. How is the geometry of space-time in special relativity different from the geometry of three-dimensional space?

Problems

7.1. The angular displacement of an image (in radians) due to aberration is approximately v/c, as long as $v \ll c$. Use the fact that the Earth orbits the Sun once per year at a distance of 1.50×10^8 km to find the maximum displacement of a star's image due to the motion of the Earth. Express your answer in arc seconds.

7.2. You and your friend carry identical clocks. Your friend passes by in a rapidly moving train. As your clock ticks off 1.00 s, you see your friend's clock tick off 0.50 s. How much time would your friend see your clock tick off in the time it takes your friends clock to tick off 1.00 s?

7.3. How fast must a clock be moving to appear to run at half the rate of an identical clock at rest?

7.4. Some radioactive particles are traveling at $0.999c$. If their lifetime is 10^{-20} s when they are at rest, what is their lifetime at this speed? How far do they travel in that time (as viewed in the frame at which they are moving at $0.999c$?

7.5. You and your friend carry identical meter sticks and identical clocks. Your friend goes by on a fast moving train, holding the meter stick parallel to the direction of motion of the train. If in the time it takes your clock to tick off 1.00 s you see 0.5 s tick off on your friend's clock, how long does your friend's

stick appear to you? How long does your stick appear to your friend, assuming the sticks are parallel to each other.

7.6. How fast does an object have to be going so that it is found to be 10% of its original length?

7.7. A source of radiation is moving away from you at 10% of the speed of light. At what wavelength is the Hα line seen?

7.8. We define the redshift, z, as the shift in wavelength, $\Delta\lambda$ divided by the rest wavelength λ_0. On the assumption that only radial motions are involved, find an expression for z as a function of v/c.

7.9. Show that equation (7.8) reduces to the classical result when $v \ll c$.

7.10. Think about how the length of an object is determined and show that the Lorentz transformations give Lorentz contraction.

7.11. Suppose we have two events that take place at (ct_1, x_1, y_1, z_1) and (ct_2, x_2, y_2, z_2) in one reference frame and at $(ct'_1, x'_1, y'_1, z'_1)$ and $(ct'_2, x'_2, y'_2, z'_2)$ in the other reference frame. The coordinates in the two frames are related by the Lorentz transformations. Show that the space-time interval between the two events is the same in both reference frames.

7.12. Show that if tachyons exist, their rest mass must be an imaginary number if the energy is to be real for $v > c$. (An imaginary number is the square root of a negative number.)

Computer problems

7.1. Make a table showing the speeds (v/c) for which time dilation is a 1%, 10%, 50%, 90%, 99%, 99.9% effect.

7.2. Make a graph of $1/\gamma$ vs. (v/c) for (v/c) ranging from 0 to 1.

7.3. For the relativistic Doppler shift, make a graph of (λ'/λ) vs. (v/c) for (v/c) ranging from 0 to 1.

7.4. For the relativistic Doppler shift, make a graph of (v/c) vs. (λ'/λ) for (λ'/λ) ranging from 1 to 10.

Chapter 8

General relativity

General relativity is Einstein's theory of gravitation that builds on the geometric concepts of space-time introduced by special relativity. Einstein was looking for a more fundamental explanation of gravity than the empirical laws of Newton. Besides coming up with a different way of thinking about gravity (in terms of geometry), general relativity makes a series of specific predictions of observable deviations from Newtonian gravitation, especially under strong gravitational fields. These predictions provide a stringent test of Einstein's theory (e.g. Fig. 8.1).

8.1 Curved space-time

A central tenet of general relativity is that the presence of a gravitational field alters the rules of geometry in space-time. The effect is to make it seem as if space-time is "curved". To see what we mean by geometry in a curved space, we look at geometry on the surface of a sphere, as illustrated in Fig. 8.2. The surface is two-dimensional. We need only two coordinates (say latitude and longitude) to locate any point on the surface. However, it is curved into a three-dimensional world, and that curvature can be detected.

To discuss the geometry of a sphere, we must first extend our concept of a straight line. In a plane, the shortest distance between two points is a straight line. On the surface of the sphere it is a great circle. Examples of great circles on the Earth are the equator and the meridians. (A great circle is the intersection of the surface of the sphere with a plane passing through the center of

the sphere.) In general, on any surface, the shortest distance between two points is called a *geodesic*.

People on the surface of the Earth can tell that it is curved, and can even measure the radius, without leaving the surface. For example, two observers can measure the different position of the Sun as viewed from two different places at the same time. (Thus, even the ancient Greeks knew the Earth was round. When Columbus sailed the only issue being seriously debated was how big the Earth is, since there was some confusion in interpreting the Greek results, which had been given in "stadia". Columbus believed the "small Earth" camp, explaining why he thought he had reached India.)

Surveying the surface will also tell you that the rules of geometry are different. For example, consider the triangle in Fig. 8.2. In a plane, a triangle has three sides, each made up of a straight line. The sum of the angles is 180°. On the surface of a sphere, we replace straight lines by great circles. A triangle should therefore be made up of parts of three great circles. In the figure we use sections of two meridians and a section of the equator. Each meridian intersects the equator at a right angle, so the sum of those two angles is 180°. When we add the third angle, between the two meridians, that makes the sum of the angles greater than 180°. The results of Euclidean (flat space) geometry no longer apply. The greater the curvature of the sphere, the more non-Euclidean the geometry appears. On the other hand, if we stick to regions on the surface that are much smaller than the radius of the sphere, the geometry will be very close to Euclidean.

We now look at what we mean when we say that gravity curves the geometry of space-time. This is illustrated in the space-time diagram in

Fig. 8.3. In the absence of gravity, objects move in straight lines at constant speeds. If we throw a ball straight up with no gravity, the world line for the ball is a straight line. If we turn on gravity, the world line looks like a parabola. We can

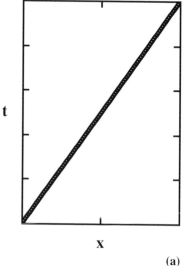

(a)

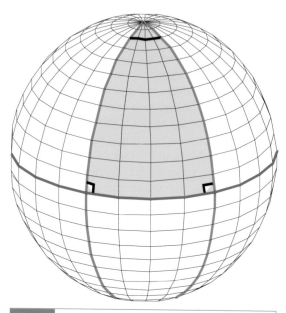

Fig 8.2. Geometry on the surface of a sphere. The shortest distance between two points is along a great circle. We look at the triangle bounded by the equator and two meridians. The meridians cross the equator at right angles, so the sum of the angles in the triangle is greater than 180°.

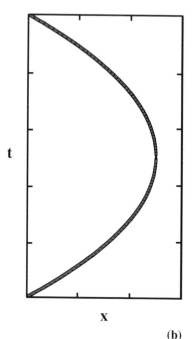

(b)

Fig 8.3. Space-time diagram for a ball thrown up from the ground. (a) With no gravity, the space-time trajectory is a straight line. (b) With constant gravity, the trajectory is a parabola.

say that it follows this path because the space-time surface on which it must stay is curved. Ultimately, to represent fully the trajectory of the ball we would have to consider all of the four space-time dimensions. The effect of gravity is then to curve that four-dimensional world into a fifth dimension. It is hard to represent that dimension in pictures, but we can still measure the curvature by doing careful geometric measurements.

In this geometric interpretation of gravitation, we need two parts to a theory. The first is to calculate the curvature of space-time caused by the presence of a particular arrangement of masses. The second is to calculate the trajectories of particles through a given curved space-time. Einstein's theory of general relativity provides both. However, the mathematical complexity goes well beyond the level of this book. (Supposedly, even Einstein was upset when he realized the area of formal mathematics into which the theory had taken him.) However, we can still appreciate the underlying physical ideas, and we can even carry out some simple calculations that bring us close to the right answers.

8.2 | Principle of equivalence

The starting point for general relativity is a statement called the *principle of equivalence*, which states that *a uniform gravitational field in some direction is indistinguishable from a uniform acceleration in the opposite direction.* Remember, an accelerating reference frame introduces pseudo-forces in the direction opposite to the true acceleration of the reference frame. For example, if you are driving in a car and step on the brakes, the car has a backward acceleration. Inside the car, you have a forward acceleration relative to the car.

We can illustrate the principle of equivalence by looking at the forces on a person standing on a scale in a elevator, as illustrated in Fig. 8.4. In the first case, we have the elevator being supported so there is no acceleration, but there is gravity. We take the acceleration of gravity to be $-g$. (Upward forces and accelerations are positive; downward forces and accelerations are negative, and we have

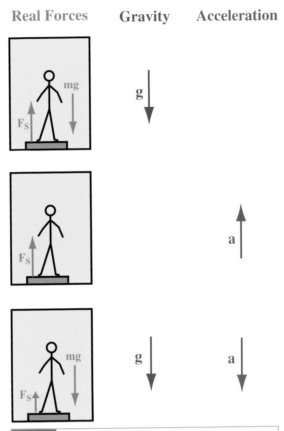

Real Forces   **Gravity**   **Acceleration**

Fig 8.4. Person in an accelerating elevator. When gravity is present it is indicated by a downward arrow, marked g. When the elevator is accelerating it is indicated by an arrow marked a.

take g as a positive number.) We now want to add up all of the forces on the person, and equate them to ma, where m is the person's mass and a is the person's acceleration. The forces are the person's weight, $-mg$, and the upward force of the scale on the person's feet, F_S. The acceleration is zero, so

$$mg + F_S = 0$$

Solving for F_S gives us

$$F_S = mg$$

By Newton's third law, the force the scale exerts on the person has the same magnitude as the force the person exerts on the scale. Therefore, F_S also gives the reading of the scale. In this case it is simply the weight of the person – the expected result.

We now look at the case of no gravity, but with an upward acceleration a. The only force on the person is F_S. Applying $F = ma$ gives

$$F_S = ma$$

If we arrange for the acceleration so its value is equal to g, we have

$$F_S = mg$$

This is the same result we had in the first case. As far as the person in the elevator is concerned, there is no way to tell the difference between a gravitational field with an acceleration g downward and an upward acceleration g of the reference frame.

To illustrate the point farther, we look at a third case, in which there is gravity, but the elevator is in free-fall. The forces on the person are F_S upward and mg downward, and the acceleration is mg downward. This gives us

$$F_S - mg = -mg$$

This tells us that F_S is zero. The person is "weightless". The acceleration of the elevator has exactly canceled the gravitational field. For the person inside the elevator, there is no way to distinguish this situation from that of a non-accelerating elevator and no gravitational field. This is the same weightlessness felt by astronauts in orbiting space vehicles (Fig. 8.5). Orbiting objects are also in free-fall, but the horizontal component of their velocity is so great that they never get closer to the ground; they just follow the curvature of the Earth.

If you look carefully at the above discussion, you will see that we have really used the concept of mass in two different ways. In one case we said that a body of mass m, subjected to a force F, will have an acceleration F/m. In this sense, mass is the ability of an object to resist the effects of an applied force. We call this resistance inertia. When we use mass in this sense, we refer to it as *inertial mass*. The second use of mass is as a measure of the ability of an object to exert and feel a gravitational force. In this context, we speak of *gravitational mass*. In the same sense, we use electric charge as a measure of an object to exert and feel electrical forces. (So, we should think of gravitational mass as being like a gravitational charge.)

Fig 8.5. Pseudo-force in an accelerating space station. In this case the station is accelerating towards Earth (like the free-falling elevator) so the astronaut appears weightless. [NASA]

The principle of equivalence is really a statement that *inertial and gravitational masses are the same for any object*. If the two masses are equal then they do cancel in the above examples, as we have done. This also explains why *all objects have the same acceleration in a gravitational field*, a point first realized by Galileo. It is not obvious on the surface of the Earth, since air resistance affects how objects fall. However, a hammer and a feather fall with the same acceleration on the surface of the Moon, where there is no air resistance.

It is important to remember that just because we call both quantities "mass" there is no obvious reason for gravitational and inertial mass to have the same numerical value. In the same way, we expect no equality between the electric charge of an object and its inertial mass. If inertial and gravitational mass are the same, this tells us that gravity must somehow be special. As we will see in the next section, considerable effort has gone into verifying the principle of equivalence.

8.3 | Tests of general relativity

Over the years since Einstein's publication of general relativity, a number of exacting tests have been carried out to test observational predictions of the theory. Some of the tests are really only tests of the principle of equivalence, while others are true tests of the full theory.

A direct test of the principle of equivalence involves the measurement of the attraction of two different objects by some third body. A class of such experiments are called *Eotvos experiments*, after the person who devised the original experiment around the turn of the 20th century. The most accurate recent versions of the experiment were carried out by a group at Princeton University in the 1960s and a group at Moscow University in the 1970s. Their findings indicate that the principle of equivalence is accurate to one part in 10^{11}.

The equivalence principle we have discussed applies strictly to objects that are so small that we can ignore the differences from one side to the other in the gravitational field they feel. We can treat them as point objects. However, there is a stronger form of the principle of equivalence that says that it also applies to objects with substantial gravitational binding energy, such as planets or stars. This has been tested by closely measuring the motion of the Moon (Fig. 8.6). A series of mirrors have been left on the Moon by the Apollo astronauts. Laser signals can be sent from Earth, bounced off these small mirrors, and then detected as very weak return signals. By timing the round trip we can measure the distance to the Moon very accurately, to within a few centimeters. These studies have indicated that the Earth and Moon fall towards the Sun with the same acceleration to within seven parts in 10^{12}.

8.3.1 Orbiting bodies
One series of tests of general relativity involves the behavior of orbiting bodies. The paths are slightly different than predicted by Newtonian

Fig 8.6. The 2.7 m telescope of the McDonald Observatory, Texas, has been used to fire a laser beam at a reflector on the Moon, then they detect the weak return. By timing the round trip, the distance to the Moon is very accurately determined. [McDonald Observatory]

gravitation. An important feature involves elliptical orbits. In an elliptical orbit, the distance of the orbiting body from the body exerting the force is changing. The orbiting body is therefore passing through regions of different space-time curvature. (See Fig. 8.7, which may help in visualizing this.) The effect of the changing curvature is to cause the orbit not to close. After each orbit, the position of perihelion (closest approach) has moved around slightly.

The effect will be greatest for orbits of highest eccentricity, since the widest range of curvatures will be covered. Also, the smaller the semi-major axis, the greater the effect. This is because the gravitational field changes faster with distance when you are closer to the object exerting the force. In the Solar System, both of these points make the effect most pronounced for Mercury.

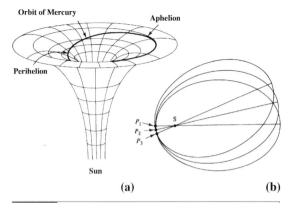

Fig 8.7. (a) Curved space-time for Mercury's orbit around the Sun. The closer to the Sun you get, the greater the curvature of space-time. Since Mercury's orbit is elliptical, its distance from the Sun changes. It therefore passes through regions of different curvature. (b) This causes the orbit to precess. We can keep track of the precession by noting the movement in the perihelion, designated P_1, P_2 and P_3 for three successive orbits. (The amount of the shift is greatly exaggerated.)

It is closest to the Sun, and, except for Pluto, has the most eccentric orbit.

The perihelion of Mercury's orbit advances by some 5600 arc seconds per century. However, of this, all but 43 arc seconds per century can be accounted for by Newtonian effects and the perturbations due to motions of other planets. The Newtonian effects could be calculated accurately and subtracted off. Einstein was able to explain the 43 arc seconds per century exactly in his general relativity calculations. This was considered to be an interesting result for general relativity, but not a crucial test, since Einstein explained something that had been observed. A crucial test involves predicting things that haven't been observed yet.

In recent years a controversy has grown out of this test of general relativity. A group at Princeton in the 1960s measured the shape of the Sun and found a slight flattening. A flattened Sun would also have an effect on the orbit of Mercury, reducing the general relativistic effects by enough to say that Einstein's calculation is wrong. Further measurements have indicated that the original experiment on the Sun's shape was in error, but some experiments suggest that there is some flattening. While some of this research is continuing,

at this point it appears that there is not enough solar flattening to challenge Einstein's results.

8.3.2 Bending electromagnetic radiation

Einstein's chance to predict an effect that had not been seen came in the bending of light passing by the edge of the Sun. He said that the warping of space-time alters the path of light as it passes near the source of a strong gravitational field. According to general relativity, photons follow geodesics. The light will then appear to be coming from a slightly different direction. If the light is coming from a star, the position of the star will appear to be slightly different than if the bending had not taken place, as indicated in Fig. 8.8.

According to Einstein, the angle θ (in radians) through which the light passing a distance b from an object of mass M is given by

$$\theta = 4GM/bc^2 \qquad (8.1)$$

If we set b equal to the radius of the Sun (6.96×10^{10} cm) we get an angle of 8.47×10^{-6} rad, which is equal to 1.74 arc seconds. This is a very small angle and is hard to measure.

The measurement is made even more difficult by the fact that we cannot see stars close to the

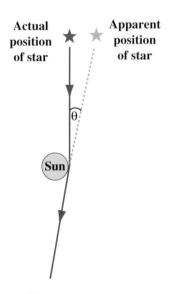

Fig 8.8. Bending of starlight passing by the Sun. The observer thinks that the star is straight back along the received ray.

Sun on the sky. Therefore, the test must be made during a total eclipse of the Sun, when the sky is photographed, and then the same part of sky is photographed approximately six months later. The positions of the stars on the two photographs are then compared. The first attempt to carry this out was by a German team trying to get to a Russian viewing site for a 1914 eclipse. They were thwarted by the state of war between the two countries. The next try was in 1919, in an effort headed by *Sir Arthur Eddington*. In the intervening years, Einstein had found an error in his calculations, so it is probably just as well that the observations weren't done until the theoretical prediction was finalized. The result was a confirmation of Einstein's prediction. The recognition of the magnitude of Einstein's contribution was immediate, both among physicists and the general public.

The solar eclipse experiment is a hard one, and the original one had a 10% uncertainty associated with it. More recent tries have reduced the uncertainty to about 5%. Different types of experiments are needed for greater accuracy. A major improvement can be made by using radio waves. The bending applies equally to electromagnetic radiation of all wavelengths. The advantage of radio waves is that the Earth's atmosphere does not scatter them. We can observe any radio source as the Sun passes in front of it and watch the position of the source change. These tests have confirmed Einstein's predictions to greater accuracy than the eclipse experiments.

There is another effect related to the bending of light. The longer path that results from the curvature of space-time around the Sun causes a delay in the time for a signal to pass by the Sun. Two types of observations have been done to test this. One involves the reflection of radio waves from Mercury and Venus as they pass behind the Sun. We know the positions of the planets very accurately, so we know how long it should take for the signal to make a round trip. The other type of experiment involves spacecraft that have been sent to various parts of the Solar System, especially Mariners 6, 7 and 9, and Viking orbiters and landers on Mars. We simply follow the signals from the spacecraft. Since we know where the spacecraft should be, we can determine the time

delay as the spacecraft pass behind the Sun. Using this technique, Einstein's predictions have been confirmed to an accuracy of 0.1%.

There is another interesting result related to the bending of the paths of electromagnetic waves. A massive object can bend rays so well that it can act as a *gravitational lens*. Physicists have speculated on this possibility for some time. Recent observations of quasars, to be discussed in Chapter 19, have revealed a number of sources in which double images are seen as a result of this gravitational lens effect (e.g. Fig. 8.1).

8.3.3 Gravitational redshift

The wavelengths of photons change as they pass through a gravitational field. This effect is called the *gravitational redshift* (Fig. 8.9). It is really a consequence of the principle of equivalence.

We can make a plausibility argument to estimate the magnitude of the effect. We have already seen in the previous section that the gravitational effect of some mass is to alter the trajectories of photons (i.e. they follow geodesics that are not straight lines). This makes it plausible that the gravitational field can do work on the photon, changing its energy. In order to estimate the gravitational potential energy of a photon $(-GMm/r)$ we assign an "effective mass", E/c^2, and since $E = hc/\lambda$, this effective mass is $h/c\lambda$. So if a photon moves

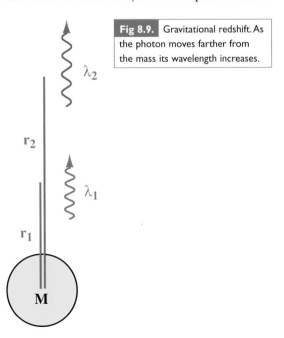

Fig 8.9. Gravitational redshift. As the photon moves farther from the mass its wavelength increases.

from r_1 out to r_2, conservation of energy would give us

$$\frac{hc}{\lambda_1} - \frac{GMh}{r_1 c \lambda_1} = \frac{hc}{\lambda_2} - \frac{GMh}{r_2 c \lambda_2} \qquad (8.2)$$

Solving for the ratio of the wavelengths gives

$$\frac{\lambda_2}{\lambda_1} = \frac{\left[1 - \dfrac{GM}{r_2 c^2}\right]}{\left[1 - \dfrac{GM}{r_1 c^2}\right]} \qquad (8.3)$$

As we have said, this derivation should not be considered rigorous – it is more of a dimensional analysis. However, it gives a result that agrees with the full general relativistic calculation for shifts that are not too large. The actual result is

$$\frac{\lambda_2}{\lambda_1} = \left[\frac{1 - \dfrac{2GM}{r_2 c^2}}{1 - \dfrac{2GM}{r_1 c^2}}\right]^{1/2} \qquad (8.4)$$

If we use the fact that $(1 - x)^{1/2} \cong (1 - x/2)$ for $x \ll 1$, equation (8.4) gives the same result as equation (8.3).

If we use equation (8.4) and take $r_2 = \infty$, and use the approximation for small shifts, we obtain

$$\frac{\lambda_2}{\lambda_1} = 1 + \frac{GM}{r_1 c^2} \qquad (8.5)$$

If we compute the wavelength shift, $\Delta\lambda$, we find

$$\frac{\Delta\lambda}{\lambda} = \frac{GM}{r c^2} \qquad (8.6)$$

Example 8.1 Gravitational redshift
Find the gravitational redshift for radiation emitted from the surface of the Sun and for radiation emitted from a $1\,M_\odot$ white dwarf, whose radius is 1% that of the Sun.

SOLUTION
For $M = 2.0 \times 10^{33}$ g we use equation (8.5) to get

$$\frac{\lambda_2}{\lambda_1} = 1 + \frac{(6.67 \times 10^{-8}\ \mathrm{dyn\ cm^2/g^2})(2 \times 10^{33}\ \mathrm{g})}{r\,(3 \times 10^{10}\ \mathrm{cm/s})^2}$$

$$= 1 + \frac{1.48 \times 10^5\ \mathrm{cm}}{r}$$

For $r = 7 \times 10^{10}$ cm,

$$\frac{\lambda_2}{\lambda_1} = 1 + 2.12 \times 10^{-6}$$

For $r = 7 \times 10^8$ cm,

$$\frac{\lambda_2}{\lambda_1} = 1 + 2.12 \times 10^{-4}$$

The shift for spectral lines in the Sun is very small. The shift for white dwarfs is measurable. The two best cases so far are Sirius B (3×10^{-4}) and 40 Eridani (6×10^{-5}).

There is an interesting way to measure the gravitational redshift on Earth. It utilizes a phenomenon known as the *Mössbauer effect* (Fig. 8.10). This involves the emission of a gamma-ray by a nucleus held firmly in place by a solid crystal. In a free nucleus, the gamma-ray would lose a little

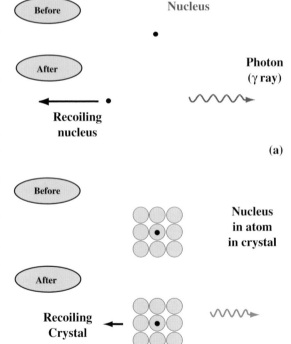

Fig 8.10. The Mössbauer effect. (a) Emission of a gamma-ray by a free nucleus. To conserve momentum, the nucleus recoils. The recoiling nucleus carries away some energy. Therefore, the energy of the gamma-ray is less than the energy difference between the two levels involved in the particular transition. (b) If the nucleus is part of an atom, which is, in turn, part of a crystal, the whole crystal must recoil. Since the crystal is much more massive than the nucleus, its recoil is negligible. This means that the energy of the gamma-ray is always equal to the difference between the two levels involved in the transition.

energy due to the recoil of the nucleus. (The recoil is to conserve momentum.) When the nucleus is in a crystal, the whole crystal takes up the recoil. It moves very little because of its large mass, and the energy loss by the gamma-ray is small. This means that the gamma-ray energy is well defined.

If the gamma-ray is emitted by a nucleus in one crystal, it can be absorbed by a nucleus in an identical crystal, as long as there is no wavelength shift while the photon is in motion. A group of physicists tried an arrangement in which the gamma-rays were emitted in the basement and absorbed on the roof. The small gravitational redshift was enough for the gamma-rays to arrive at the roof with the wrong wavelength to be absorbed. The gamma-rays could be blueshifted back to the right wavelength by moving the crystal on the roof towards that in the basement. By seeing what Doppler shift is necessary to offset the gravitational redshift, the size of the gravitational redshift can be measured. The result agrees with the theoretical prediction.

A phenomenon related to the gravitational redshift is *gravitational time dilation*. All oscillators or clocks run slower in a strong gravitational field than they do in a weaker field. If we have two clocks at r_1 and r_2, the times they keep will be related by the same expression as the gravitational redshift. That is

$$\frac{t_2}{t_1} = \left[\frac{1 - \dfrac{2GM}{r_2 c^2}}{1 - \dfrac{2GM}{r_1 c^2}} \right]^{1/2} \tag{8.7}$$

From this we see that $t_2 > t_1$. This effect has been tested by taking identical clocks, leaving one on the ground and placing the other in an airplane. (Of course, you must first correct for the special relativistic effect due to the motion of the airplane.) The airplane experiments have yielded results that agree with theory. Even more recently, tests on rockets have yielded even more accurate results.

8.3.4 Gravitational radiation

Just as the classical theory of electricity and magnetism predicts that accelerating charges will give off electromagnetic radiation, general relativity predicts that certain types of systems should give off gravitational radiation. Gravitational radiation is more complicated than electromagnetic radiation. When a gravity wave passes by the geometry of space-time briefly distorts. The types of systems that might produce gravitational radiation are orbiting systems with objects close together or collapsing objects.

Some groups have attempted to detect gravitational radiation directly from astronomical sources, hoping to detect small changes in very large detectors. None of these has been successful yet. More recently, an ambitious program involving laser interferometers to detect small changes is being developed (Fig. 8.11).

The best evidence for gravitational radiation at the time of writing has been indirect. Astronomers have been studying a binary pulsar system (to be discussed in Chapter 12), in which the system appears to be losing orbital energy at exactly the rate that would be predicted for gravitational radiation.

Fig 8.11. To detect gravity waves, physicists are hoping to measure very small changes in very large objects. This shows a view of one station of the Laser Interferometer Gravity Observatory (LIGO), in Hanford, WA. A passing gravity wave would cause a small shift in the interference patterns in the laser. [Photo courtesy of Caltech/LIGO]

8.3.5 Competing theories

One of the reasons that there has been so much interest in testing general relativity as accurately as possible is that there are some competing theories to Einstein's. These theories generally have the same starting point, but differ in their details. The result of the tests of these competing theories is that the experimental foundations of Einstein's theory are now much stronger than they were when the theory was initially studied.

8.4 | Black holes

One of the exciting aspects of astronomy is the possibility of studying a variety of fascinating objects. By our earthly standards, even a normal star contains extreme conditions. However, there are other objects that make the conditions on stars seem commonplace. Among these objects are *black holes*, objects from which no light can escape.

8.4.1 The Schwarzschild radius

Shortly after Einstein published his general theory of relativity, *Karl Schwarzschild* worked out the solution for the curvature of space-time around a point mass. He found that there is a critical radius at which a *singularity* occurs. A singularity is a place where some quantity becomes infinite. This critical radius is called the *Schwarzschild radius*. For a mass M this radius, R_S, is given by

$$R_S = 2GM/c^2 \qquad (8.8)$$

Real objects are not pointlike, but have some finite extent. An interpretation of Schwarzschild's result is that *if an object is completely contained within its Schwarzschild radius, the singularity will occur.*

We can understand the significance of this critical radius by recalling the discussion of gravitational redshift in Section 8.2. We saw that if a photon is emitted at a wavelength λ_1 at a distance r_1 from a mass M, and is detected at r_2, its wavelength λ_2 is given by

$$\frac{\lambda_2}{\lambda_1} = \left[\frac{1 - \dfrac{2GM}{r_2 c^2}}{1 - \dfrac{2GM}{r_1 c^2}} \right]^{1/2} \qquad (8.9)$$

If we set $r_1 = 2GM/c^2$ (the Schwarzschild radius), we find that λ_2 is infinite, even if r_2 is only slightly greater than r_1. This means that no electromagnetic energy can escape from within the Schwarzschild radius. We call an object that is contained within its Schwarzschild radius a *black hole*.

Example 8.2 Schwarzschild radius

Find the Schwarzschild radius for an object of one solar mass.

SOLUTION

From equation (8.8) we have

$$R_s = \frac{(2)(6.67 \times 10^{-8} \text{ dyn cm}^2/\text{g}^2)(2 \times 10^{33} \text{ g})}{(3 \times 10^{10} \text{ cm/s})^2}$$

$$= 3.0 \times 10^5 \text{ cm}$$

$$= 3.0 \text{ km}$$

Since the Schwarzschild radius varies linearly with mass and has a value 3 km for a 1 $M_\odot$ object, we can write an expression for R_S for an object of any mass. It is

$$R_S = (3.0 \text{ km})(M/M_\odot) \qquad (8.10)$$

Remember, every object has its Schwarzschild radius. However, it can only be a black hole if it is contained within this radius. For example, the Sun is much larger than 3 km, so it is not a black hole.

The density of a 1 $M_\odot$ black hole would be quite high, almost 10^{17} g/cm^3. It is higher than the density of the nucleus of an atom. However, as we consider more massive black holes, the density goes down. This is because the radius is proportional to the mass, but the volume is proportional to the radius cubed (and therefore to the mass cubed). This means that the density will be proportional to $1/M^2$. Since we know the density for a 1 $M_\odot$ black hole, we can write the density for any other mass black hole as

$$\rho = (1 \times 10^{17} \text{ g/cm}^3)(M/M_\odot)^{-2} \qquad (8.11)$$

By the time the mass reaches $10^8 M_\odot$, the density is only a few grams per centimeter cubed, just a few times the density of water.

We would expect the region just outside a black hole to be characterized by a large change in gravitational force over a small distance.

On the Earth the tides result from the changing of the gravitational forces exerted by the Sun and Moon from one side of the Earth to the other. By extension, we refer to any effect of the variation of a gravitational force as a *tidal effect*. Near a black hole, the gravitational force should fall off very quickly with small changes in distance from the surface. We write the acceleration of gravity as a function of radius

$$g(r) = GM/r^2 \qquad (8.12)$$

Differentiating with respect to r gives

$$dg(r)/dr = -2GM/r^3 \qquad (8.13)$$

Though the gravitational force falls off as $1/r^2$, the tidal effects fall off as $1/r^3$, meaning that they are most important for small values of r.

Example 8.3 Black hole tidal forces
Find the difference between the acceleration of gravity at the feet and head of an astronaut just outside a $1M_\odot$ black hole.

SOLUTION
The change in g, Δg, is given by

$$\Delta g = (dg/dr)\, \Delta r$$

where Δr is the distance over which the change is to be found. In this case, Δr is the height of the astronaut, which we will take to be 2m. We find dg/dr from equation (8.13) to be

$$dg/dr = -(2)(6.67 \times 10^{-8}\ \text{dyn cm}^2/\text{g}^2)(2.0 \times 10^{33}\ \text{g})/ \\ (3 \times 10^5\ \text{cm})^3$$

$$= -1 \times 10^{10}\ \text{cm/s}^2/\text{cm}$$

(Note that the minus sign means that gravity is stronger at the feet than at the head.) For $\Delta r = 2 \times 10^2$ cm, we have

$$\Delta g = -2 \times 10^{12}\ \text{cm/s}^2$$

This is two billion times the acceleration of gravity at the surface of the Earth. The astronaut would be pulled apart with a force of over a billion times the astronaut's weight!

The tidal force, dg/dr, is proportional to M/r^3, just as is the density. Therefore, the tidal force will be less for more massive black holes, falling to more tolerable values for very massive black holes (see Problem 8.9.)

8.4.2 Approaching a black hole

What is it like to fall into a black hole? We consider two astronauts. One approaches the black hole, and the other stays a safe distance away. The various steps are indicated in Fig. 8.12.

We assume that the astronaut approaching the black hole can send out signals in various directions, including back to the other astronaut. As the first astronaut approaches the black hole, the first thing the distant astronaut would notice is the redshift in the signals received. The magnitude of the redshift increases as the first astronaut becomes closer to the Schwarzschild radius.

Before the Schwarzschild radius is reached, another effect becomes noticeable. The paths of photons sent out by the first astronaut are not straight lines. They bend. The only direction in which the astronaut can aim a beam and not have

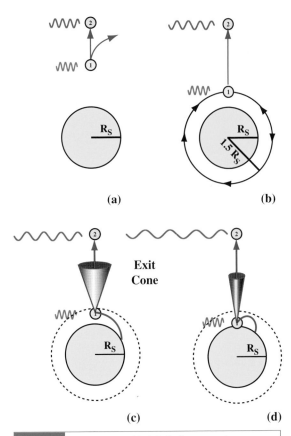

Fig 8.12. Approaching a black hole. Astronaut 1 approaches the black hole while astronaut 2 stays behind. In each frame, the emitted and received wave correspond to a beam sent from 1 to 2.

it bend is straight up. If the beam is not aimed sufficiently close to the vertical, the bending will be so great that the light will not escape. Only light aimed into a cone about the vertical, called the *exit cone*, will escape. As the first astronaut moves closer to the Schwarzschild radius, the exit cone becomes smaller. At a distance equal to $(3/2)R_S$, photons aimed horizontally go into orbit around the black hole. The sphere of orbiting photons is called the *photon sphere*. If you were to look straight out, along the horizon, you would see the back of your head.

The second astronaut never actually sees the first astronaut reach the Schwarzschild radius. The gravitational time dilation is so great that, as R_S is approached, the second astronaut thinks that it takes the first astronaut an infinite amount of time to reach R_S. The time dilation makes the first astronaut appear to slow down as R_S is approached.

From the point of view of the first astronaut, there is no such respite. The Schwarzschild radius is reached very quickly. If the black hole is of sufficiently small mass, the tidal forces would tear the first astronaut apart. However, if the black hole is massive enough, the tidal forces might be survived and the astronaut crosses R_S. When this happens, we say that the astronaut has crossed the *event horizon*. If the black hole is massive enough, the astronaut might not notice anything unusual, except that escape is impossible!

Once inside the black hole, the inevitable journey to the center continues. The gravitational time dilation is so great that time passes slowly. However, the headlong rush through space continues. Outside the black hole, it is time that rushes on while distance is covered slowly. It is as if crossing the event horizon has interchanged to roles of space and time.

The second astronaut can tell nothing about what is going on inside the black hole. In fact, the only properties of a black hole that can be deduced are its mass, radius, electric charge and angular momentum. (So far, we have assumed zero angular momentum. We will discuss rotating black holes below.) The external simplicity of black holes is summarized in a theorem that states that *black holes have no hair*.

So far we have been discussing non-rotating black holes. The structure of a rotating black hole

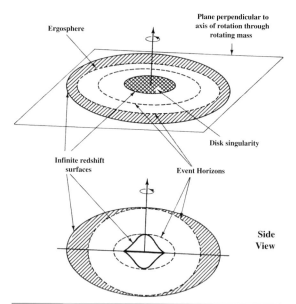

Fig 8.13. Rotating black hole. The structure of the surface is complicated, so we show two cuts. In the upper figure, we see the intersection of various surfaces with the plane perpendicular to the axis of rotation. At the center is a disk singularity. (This is just a disk and doesn't extend above or below the plane.) There are two infinite redshift surfaces and two event horizons. Between the event horizons, the roles of space and time are reversed. The region between the outer infinite redshift surface and the outer event horizon is called the ergosphere. The lower figure shows a side view.

is somewhat more complicated than that of a non-rotating black hole, and is depicted schematically in Fig. 8.13. The situation shown is for the case in which the angular momentum per unit mass, J/M, is less than GM/c. For the case shown, there are two infinite redshift surfaces instead of a single event horizon. Between the two surfaces, the roles of space and time are reversed, just as inside the event horizon in the non-rotating case. The region between the outer infinite redshift surface and the event horizon is called the *ergosphere*. The name results from the fact that there is a way to extract energy from the black hole by moving particles through the ergosphere in the correct trajectory.

8.4.3 Stellar black holes

In Chapter 11 we will see that some types of stars evolve to a point were nothing can support them. Such a star will collapse right through the

Schwarzschild radius for its mass, and will become a black hole. Black holes would be a normal state for the evolution of some stars. How would we detect a stellar black hole? We obviously could not see it directly. We could not even see it in silhouette against a bright source, since the area blocked would be only a few kilometers across. We have to detect stellar black holes indirectly. We hope to see their gravitational effects on their surrounding environment. This is not a hopeless task, since we might expect to find a reasonable number in binary systems. We will discuss the probable detection of black holes in binary systems in Chapter 12.

8.4.4 Non-stellar black holes

Black holes that have masses much less than a solar mass are called *mini black holes*. We think that mini black holes might have formed when the density of the universe was much higher than it is now. (The conditions in the early universe will be discussed in Chapter 21.) These may still exist. The British physicist *Stephen Hawking* has found that there is a mechanism by which mini black holes could actually evaporate. Hawking is studying the relationship between gravity and quantum mechanics, and the process he has proposed is a quantum mechanical one.

This mechanism involves a different concept of a vacuum than we are accustomed to seeing. In classical physics, a vacuum is simply nothing. In quantum mechanics it is possible to make something out of nothing, if you don't do it for long. It amounts to borrowing energy for a brief time interval. The more energy you borrow, the less time you can borrow it for. It is related to the uncertainty principle (which we discussed in Chapter 3). We have talked about the uncertainty principle as it relates to momentum and position. However, it also relates to energy and the lifetime of a state. It says that if the state has a lifetime Δt, then the energy of the state is uncertain by an amount ΔE, given by

$$\Delta t \, \Delta E \geq h/2\pi \qquad (8.14)$$

The longer lived a state, the more accurately its energy can be determined. Since the energy of a state is uncertain by ΔE, it is possible for us to have this extra amount of energy and not detect it.

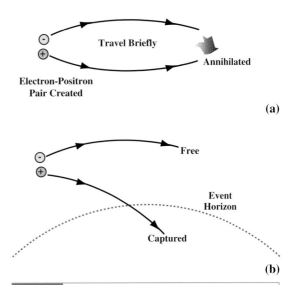

Fig 8.14. Pair production. (a) The process in free space. An electron and positron are created out of nothing, but quickly come back together to annihilate. (b) Near a black hole, one of the particles can be captured before they can annihilate, and the other escapes, carrying energy away from the black hole.

As a result of the uncertainty principle, a quantum mechanical vacuum is a very busy place. At any place it is possible to create a particle–antiparticle pair (Fig. 8.14). (We will discuss antiparitcles in Chapter 21.) It requires an energy equal to $2mc^2$, where m is the mass of the particle (and the antiparticle). The pair can exist for at most a time $h/[(2\pi)mc^2]$. Before the time is up, they must find each other and annihilate. Since electrons have masses that are much less than protons, an electron–positron (antielectron) pair will live longer than a proton–antiproton pair. We can therefore think of a vacuum as being made up of continuously appearing and disappearing electron–positron pairs (with a small contribution from heavier particle–antiparticle pairs). The phenomenon is called *vacuum polarization*.

When an electron–positron pair is created just outside a black hole, it is possible for one of the particles to be pulled into the black hole before the two recombine. The other particle will continue moving away from the black hole. The two cannot recombine. The particles then exist for much longer than the time limit for violating conservation of energy. We must therefore make up the energy from somewhere. This process actually reduces the mass of the black hole. The black hole shrinks slightly. For mini black holes this energy loss can be a significant fraction of the mass of the black hole. Eventually, the black hole shrinks to the point where it disappears in a small burst of gamma radiation. The more massive a black hole is when it starts out, the longer it will live. An estimate for the lifetime of a black hole of mass M (in grams) is $(10^{-26}\,\text{s})(M^3)$. So, a black hole of about 10^{14} g would have a liftime of about 10^{10} yr, a little less than the age of the universe. In the lifetime of the universe, black holes smaller than some given mass should have disappeared. Those at that mass should just be dying now. Some physicists have suggested that when this happens we should be able to see the burst of gamma radiation.

At the other end of the mass scale, much larger than stellar black holes, are *maxi black holes*. They probably result from large amounts of material gathering together in a small region. In Chapter 19, we will see evidence for 10^8 to 10^9 $M_\odot$ black holes being present in the centers of many galaxies.

Chapter summary

In this chapter we saw how the general theory of relativity has changed our thinking about the nature of space and time.

We then saw how the ideas of space-time carry over to a theory of gravitation – general relativity. The interpretation of gravitational fields is that they alter the geometry of space-time, causing it to behave like that on a curved surface. The starting point for general relativity is the principle of equivalence, which tells us that inertial and gravitational masses are the same.

We saw that there are several effects of general relativity that can be tested. These include the advancement of the periastron of orbiting bodies, the bending of electromagnetic radiation, gravitational redshift and time dilation, and gravitational radiation.

We also saw how the gravitational redshift leads to a concept of black holes, objects from which nothing can escape.

Questions

8.1. What do we mean when we say that gravity alters the geometry of space-time?
8.2. Re-do the analysis of the person on a scale in the elevator (all three cases) explicitly noting the uses of inertial mass and gravitational mass.
8.3. Why don't you need a solar eclipse to measure the bending of radio waves past the edge of the Sun?
8.4. Briefly describe the tests of general relativity discussed in this chapter.
8.5. Since we cannot run a rule from the center of a black hole to the Schwarzschild radius, how would you "measure" the radius of a black hole? (Hint: Think in terms of a measurement that doesn't involve crossing the event horizon.)
8.6. (a) What is the exit cone? (b) When the exit cone closes, what happens to photons aimed straight up?
8.7. Is there a place near a black hole where you could look straight ahead and see the back of your head? Explain.
8.8. If the Schwarzschild radius of the Sun is 3 km, does that mean that the inner 3 km of the Sun is a black hole?

Problems

8.1. Consider an object with the same density as the Sun. Find an expression for the bending of starlight past the edge of this object as a function of the size of the object.

8.2. For a neutron star (to be discussed in Chapter 11) with 1 $M_\odot$ and a radius of 10 km, through what angle is light bent as it passes the edge?

8.3. For a white dwarf with 1 $M_\odot$ and a radius of 5×10^3 km, find the wavelength to which the Hα line will be shifted by the time it is seen by a distant observer.

8.4. (a) For the test of the gravitational redshift involving the Mössbauer effect, calculate the shift in going from the Earth's surface to 50 m above the Earth's surface. (b) How fast would the receiver have to move toward the source to compensate for the redshift? (c) Compare your answer in (b) with the speed that an object falling from the roof would acquire just before striking the ground.

8.5. Show that the Schwarzschild radius can also be found by taking the escape velocity from an object of mass M and radius R, and setting it equal to c.

8.6. (a) Compute your Schwarzschild radius. (b) What would the density be for a black hole of your mass?

8.7. For what mass black hole does the density equal 1 g/cm^3 ?

8.8. For what mass black hole does the difference between the acceleration of gravity at an astronaut's feet and head equal the acceleration of gravity on the Earth (1000 cm/s^2)?

8.9. Find an expression for dg/dr at the surface of a black hole as a function of the mass of a black hole. Your expression should be a scaling relationship as in equation (8.10).

8.10. How does the rate of a clock 1.5 R_S from a $3M_\odot$ black hole compare with the rate of a clock far from the black hole?

8.11. How close must you be (in terms of R_S) to a $3M_\odot$ black hole to find that a clock runs at 10% the rate it runs when it is far away?

8.12. If an electron–positron pair forms from the vacuum, how long can they live before they must annihilate?

Computer problems

8.1. For stars in the mid-range of each spectral type (O5, B5, . . .), make a table showing the angle of bending of starlight just passing the limb of that star. Also include in your table a white dwarf which has 1 solar mass in an object the size of the Earth.

8.2. For gravitational redshift, make a graph of $\Delta\lambda/\lambda$ vs. r, for an interesting range of r (assuming 1 $M_\odot$).

8.3. Make a table showing the Schwarzschild radii and average density for black holes of 10^n $M_\odot$ where n goes from 0 to 10. Also include a column showing the acceleration of gravity at the surface.

Part III

Stellar evolution

Now that we know the basic properties of stars, we look at how the laws of physics determine those properties, and then how stars change with time – how they evolve. Stars go through a recurring full life cycle. They are born, they live through middle age, and they die. In their death, they distribute material into interstellar space to be incorporated into the next generation of stars.

In describing the life cycle, we can start anywhere in the process. In Chapter 9, we discuss the most stable part of their life cycle, life on the main sequence – stellar middle age. In Chapters 10, 11 and 12 we will look at the deaths of different types of stars. After discussing the interstellar medium in Chapter 14, we will look at star formation in Chapter 15.

Chapter 9

The main sequence

In this chapter we look at the inner workings of a star once it has settled into a main sequence existence. We start by looking at the sources of stellar energy, and then we look at the physical processes that govern stellar structure.

9.1 | Stellar energy sources

When material collapses to form a star, there is gravitational potential energy stored from the (negative) work done by gravity in bringing the material together. We might wonder how long this energy supply will last. We start by calculating the gravitational potential energy in a uniform sphere.

9.1.1 Gravitational potential energy of a sphere

We are dealing with systems of large numbers of particles. Therefore, rather than thinking in terms of individual particles of mass m, we can think of a fluid of average density ρ. (The density is simply the mass per particle, multiplied by the number of particles per unit volume.) In this section, we will evaluate the potential energy for a uniform (constant density) sphere. Even though real objects might not be exactly uniform or spherical, the results will generally only change by numerical factors of order unity.

We begin by calculating the gravitational potential energy of a uniform sphere of mass M, radius R, and density ρ. These quantities are related by

$$M = (4\pi/3)\, R^3 \rho \tag{9.1}$$

The gravitational potential energy is the work required to bring all of the material from far away (infinity) to the final configuration. The final result does not depend on the order in which the various parts of the sphere are assembled, so we do the calculation in the easiest way that we can envision. We can think of the sphere as being made up of shells (Fig. 9.1). We can assemble the sphere one shell at a time, starting with the smallest.

Let's assume that we have already assembled shells through radius r. We now want to calculate the work done to bring in the next shell. The thickness of the shell is dr. The volume of the (thin) shell is its surface area multiplied by its thickness:

$$dV = 4\pi r^2\, dr \tag{9.2}$$

The mass contained in the shell is the volume multiplied by the density:

$$dM = 4\pi r^2 \rho\, dr \tag{9.3}$$

The total mass of material already assembled is

$$M(r) = (4\pi/3)R^3\rho \tag{9.4}$$

The quantity $M(r)$ is important since the shell that ends up at radius r will only feel a net force from material inside it. Even after we bring in more material outside this shell, the net force exerted on any particle in the shell by any matter outside radius r is zero. Also, for mass in the shell, the mass $M(r)$ exerts a force equal to that which would be exerted by the same mass all located at the center of the sphere.

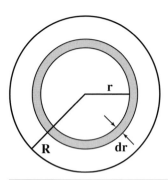

Fig 9.1. We model stars by studying spherical shells.

For any two point masses, remember the gravitational potential energy (relative to infinity) is given by

$$U = -Gm_1m_2/r \qquad (9.5)$$

We let $m_1 = M(r)$ and m_2 equal the mass of the shell, dM. The work to bring this shell is

$$dU(r) = -GM(r)\, dM/r$$

$$= -G(4\pi/3)(4\pi)\rho^2 r^4\, dr \qquad (9.6)$$

To find the effect of all the shells, we integrate the quantity dU from $r = 0$ to $r = R$:

$$U = \int_0^R dU(r)$$

$$= -G(4\pi/3)(4\pi)\,\rho^2 \int_0^R r^4\, dr$$

$$= -G(4\pi/3)(4\pi)\,\rho^2\,(1/5)\,R^5 \qquad (9.7)$$

$$= -(3/5)G[(4\pi/3)\,\rho\, R^3]^2/R$$

$$= -\left(\frac{3}{5}\right)\frac{GM^2}{R}$$

For other shaped objects, the gravitational potential energy is generally proportional to $-GM^2/R$, where R is some average length. The constant of proportionality is generally close to unity.

The thermal energy is $(3/2)kT$ per particle. The total thermal energy is then

$$K = (3/2)NkT \qquad (9.8)$$

where N is the total number of particles in the cloud. If the mass per particle is m, and the total mass of the cloud is M, then

$$N = M/m \qquad (9.9)$$

or

$$K = (3/2)(M/m)kT \qquad (9.10)$$

9.1.2 Gravitational lifetime for a star

It is possible for stars to use their stored gravitational potential energy to power the star. In this section, we calculate how long this process can go on. For example, could the Sun be powered in this way even now? To see, we estimate the gravitational lifetime, t_g. This lifetime is the stored energy divided by the rate at which the energy is being lost. The rate at which the energy is being lost is the luminosity. That is

$$t_g = \frac{E}{dE/dt}$$

$$= \frac{E}{L} \qquad (9.11)$$

where L is the luminosity.

We estimate E as the negative of the Sun's current gravitational potential energy:

$$E \cong \frac{(3/5)GM^2}{R}$$

$$= \frac{(0.6)(6.67 \times 10^{-8}\ \text{dyn cm}^2/\text{g}^2)(2 \times 10^{33}\text{g})^2}{(7 \times 10^{10}\ \text{cm})}$$

$$= 2 \times 10^{48}\ \text{erg} \qquad (9.12)$$

This gives a lifetime, called the *Kelvin time*,

$$t_g = \frac{(2 \times 10^{48}\ \text{erg})}{(4 \times 10^{33}\ \text{erg/s})}$$

$$= 5 \times 10^{14}\ \text{s}$$

$$= 2 \times 10^7\ \text{yr}$$

This is a lifetime of 20 million years. It may sound like a long time. However, we know from geological evidence that the Earth has been around for over four billion years. This means that the Sun must be at least that old. Therefore, the Sun (and presumably other stars) cannot exist in a stable configuration on stored gravitational energy.

9.1.3 Other energy sources

An alternative source of energy could be chemical reactions. After all, we use chemical reactions to make automobiles go on Earth. We can estimate the amount of energy stored in the Sun, capable of being released in chemical reactions. Typical energies of these chemical reactions should be equivalent to some fraction of the binding energy of molecules that might be formed and destroyed in these reactions. This means that we might expect something like 1 eV per atom in the Sun. The total chemical energy is then 1 eV multiplied by the total number of atoms in the Sun, M/m. The energy available is then

$$E = \frac{(1 \text{ eV})(1.6 \times 10^{-12} \text{ erg/eV})(2 \times 10^{33} \text{ g})}{(1.67 \times 10^{-24} \text{ g})}$$

$$= 2 \times 10^{45} \text{ erg}$$

This is much less than is stored in gravitational potential energy, so chemical reactions clearly cannot provide a longer term energy source for the Sun.

You might wonder if we have not been too casual in our dismissal of chemical reactions as a possible energy source for the Sun. After all, we didn't even say what chemical reactions might be involved. The important point is that *any* chemical reaction involves moving electrons from atom to atom, and the energies associated with this are a few eV, independent of what the reaction is. Suppose there were some very energetic reaction that produced 10 eV per atom, that would increase the chemical energy by a factor of ten but it would still be many orders of magnitude short of the required amount. If our estimate had shown that chemical reactions might work, then we would have to worry about the details, figure out what reactions were important and then do a more accurate calculation of the stored energy. However, our estimate tells us that it is not worth wasting our time on the details. These types of calculations, called *order of magnitude calculations*, are very important in astronomy. They help us eliminate processes that obviously don't work and allow us to focus our attention on possibilities that might.

The answer to the problem of stellar energy sources is nuclear reactions. The typical energies available in nuclear reactions are about 1 MeV per atom, instead of 1 eV. This is an improvement of a factor of 10^6. To see how nuclear reactions provide energy for the Sun, we look at some of the elements of nuclear physics.

9.2 | Nuclear physics

9.2.1 Nuclear building blocks

We have already seen that the positive charge in atoms is confined to the small nucleus (10^{-13} cm across). The nucleus is composed of protons and neutrons. Because they are the building blocks of the nucleus, we call them *nucleons*. The proton has charge $+e$ (where $-e$ is the charge on the electron). Since atoms are neutral, the number of electrons orbiting the nucleus equals the number of protons in the nucleus. The chemical properties of an atom depend on the number of orbiting electrons. Therefore, the number of protons in the nucleus ultimately determines the identity of the element. We designate the number of protons by the symbol Z, called the *atomic number*. The charge on the nucleus is $+Ze$. The highest naturally occurring value of Z is 92 (uranium), with approximately a dozen "man-made" elements with higher values of Z.

Neutrons are electrically neutral. The mass of the neutron is slightly greater than the mass of the proton. Nuclei with the same numbers of protons can have different numbers of neutrons N. The total number of nucleons in the nucleus is called the *mass number*, $A = Z + N$. The mass of a given nucleus is approximately Am_p. Two nuclei with the same number of protons but different numbers of neutrons are called *isotopes* of the same element. We generally designate an element by a letter symbol (e.g. H for hydrogen), preceded by a superscript giving A, followed by a subscript giving Z. (This subscript is redundant since the symbol tells us Z, and is sometimes left out.) For example, the three known isotopes of hydrogen are $^1\text{H}_1$, $^2\text{H}_1$, $^3\text{H}_1$ or simply ^1H, ^2H, ^3H.

For any given number of protons, we do not find an arbitrary number of neutrons. In general, we find that the stable elements have approximately equal numbers of protons and neutrons. For larger values of Z the stable nuclei have a slightly larger number of neutrons than protons.

We now look at the force that holds nuclei together. We generally refer to four forces in nature. In order of decreasing strength, they are *strong nuclear, electromagnetic, weak nuclear* and *gravity*. The nuclear forces are shorter ranged, as

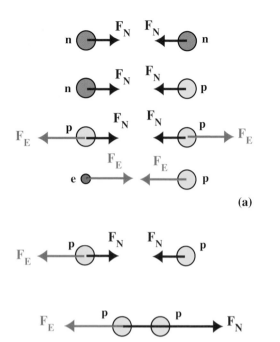

(a)

(b)

Fig 9.2. Properties of the nuclear and electrical forces between neutrons and protons. (a) The nuclear force, F_N, is the same for a proton and a proton, a neutron and a neutron, or a proton and a neutron. The electrical force only acts between the proton and the proton. The magnitude of the electrical force is the same as that between a proton and an electron. (b) How the forces vary with distance. As the two protons are brought closer together, the electric repulsion becomes stronger, but the nuclear attraction becomes stronger faster.

opposed to the electromagnetic and gravitational forces that can be felt at very large distances. For example, we look at the force between two protons. When they are far apart, the electric force dominates, and the protons repel each other. When they are close together, the attractive nuclear force dominates. The strong nuclear force between two protons is the same as between two neutrons and is the same as between a neutron and a proton (Fig. 9.2).

This gives us an idea of the role of a neutron in a nucleus. If we just have protons, the electric repulsion will be appreciable, and the nucleus will not be stable. If we add neutrons, we have additional binding from the nuclear force but no additional electrical repulsion. (The electric repulsion is actually reduced, since the neutrons

keep the protons farther apart.) This explains why a nucleus needs approximately as many neutrons as protons.

9.2.2 Binding energy

Since the nuclear force is attractive, we must do work to move two nucleons apart. The work required to disassemble a nucleus is the *binding energy* of the nucleus (Fig. 9.3). The binding energy is analogous to the binding energy of an atom – the energy required to separate an electron from the rest of the atom, doing work against the electrical attraction. Since the nuclear force is so strong, nuclear binding energies are greater than atomic binding energies by a factor of about 10^6. We measure nuclear binding energies in MeV, rather than eV. The greater the binding energy of a nucleus, the more work we must do to get the nucleus apart. Therefore, a larger binding energy means a more stable nucleus.

The binding energies of nuclei are so large that we can measure them directly by comparing the mass of a nucleus with the masses of its components. This follows from Einstein's relationship between mass and energy ($E = mc^2$) and conservation of energy

$$M_{\text{nucleus}}c^2 + \text{BE} = Zm_pc^2 + Nm_nc^2 \qquad (9.13)$$

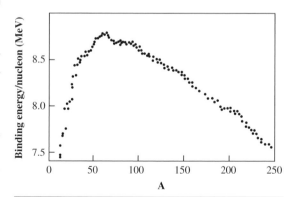

Fig 9.3. We must do work to break a nucleus apart. This work is the binding energy. Shown here are nuclear binding energies for nuclei that are found in nature. The horizontal axis is the mass number A, and the vertical axis the binding energy, divided by the number of nucleons in the nucleus (BE/A). Nuclei with higher binding energies per nucleon are more stable.

where BE is the binding energy. This means that the mass of a nucleus is less than the masses of its components by BE/c^2. This is also true for atoms. However, the binding energy for atoms is so small that the mass difference is negligible (about 10^{-6} of the nuclear binding energy).

Example 9.1 Proton rest energy
Compute the rest energy of a proton and express the result in MeV.

SOLUTION

$E = (1.67 \times 10^{-24}\,\text{g})(3.0 \times 10^{10}\,\text{cm/s})^2$

$\quad = 1.5 \times 10^{-3}\,\text{erg}$

$\quad = \dfrac{1.5 \times 10^{-3}\,\text{erg}}{1.6 \times 10^{-12}\,\text{erg/eV}}$

$\quad = 9.4 \times 10^{8}\,\text{eV}$

$\quad = 940\,\text{MeV}$

Example 9.2 Deuteron binding energy
Calculate the binding energy of the deuteron, an isotope of hydrogen containing one proton and one neutron, given the following data:

$m_p = 1.6726 \times 10^{-24}\,\text{g}$

$m_n = 1.6749 \times 10^{-24}\,\text{g}$

$m_d = 3.3436 \times 10^{-24}\,\text{g}$

SOLUTION
The binding energy is given by

$\text{BE} = (m_p + m_n - m_d)c^2$

$\quad = (3.9 \times 10^{-27}\,\text{g})(2.9979 \times 10^{10}\,\text{cm/s})^2$

$\quad = 3.6 \times 10^{-6}\,\text{erg}$

$\quad = 2.2\,\text{MeV}$

By comparing the result of this example with the previous example, we see that the nuclear binding energy is a few tenths of a percent of the total rest energy. This mass difference is relatively easy to measure.

9.2.3 Nuclear reactions

We have seen that there are large amounts of energy involved in holding a nucleus together. However, we must still have a way of liberating that energy. If we hope to power stars by liberating nuclear energy, a nuclear reaction in which the products have a greater binding energy than the reactants must be found. Then, the difference in energy will be available to heat the surroundings. Before we discuss the particular nuclear reactions that work in stars, we look at the types of nuclear reactions that can take place.

The first type of reaction is a *decay*. In a decay, a nucleus emits a particle. Depending on the type of particle that is emitted, the result is either a different nucleus, or the same nucleus in a lower energy state. We identify three types of decay, depending on the type of particle that is emitted. The three types are called alpha (α), beta (β), and gamma (γ).

An *alpha particle* is the nucleus of the most common form of helium, ^{4}He. This means that an alpha particle has two protons and two neutrons. This particular combination is very stable and can be emitted as a single group. The final nucleus has two fewer protons and two fewer neutrons than the original nucleus. This means that the element changes as the atomic number Z decreases by two. The mass number A decreases by four. This process can only take place if the resulting nucleus is more stable than the original nucleus. This means that the binding energy of the final nucleus must be greater than that of the original nucleus. Once the alpha particle is far enough from the nucleus, it no longer feels the nuclear attraction. However, since it has a charge of $+2e$, it feels the Coulomb repulsion of the remaining nucleus. The alpha particle accelerates away from the nucleus. The energy liberated in the reaction is carried away in the form of the kinetic energy of the alpha particle.

A *gamma-ray* is simply a high energy photon. Since it carries away neither mass nor charge, neither A nor Z change. The identity of the nucleus does not change in gamma decay. The photon does carry away energy. The protons and neutrons in a nucleus must also be treated quantum mechanically as being waves. This means that there are allowed states for the motions of the nucleons within the nucleus. A gamma-ray is emitted when a transition is made from a higher energy state to a lower energy state. The energy of the gamma-ray is equal to the energy difference between the states. We can learn a lot about the internal structure of nuclei by studying the energies of gamma-rays that are emitted, just as optical spectroscopy tells us about the structure of atoms.

A *beta particle* is an electron or a *positron*. A positron is the *antiparticle* to the electron. Every particle has a corresponding antiparticle. A particle and its antiparticle have identical masses, but all other properties, such as charge, are the negative of each other. When a particle and an antiparticle come together, they can convert all of their mass into energy – in other words they annihilate each other – without violating any conservation laws. In a beta decay, an electron or a positron is emitted. The charge of the nucleus increases by $+e$ (or $-e$ if a positron is emitted), but the mass number A remains unchanged. The net result is to change a neutron into a proton. (The reverse, changing a proton into a neutron is also possible, but requires a source of energy, because the neutron is more massive than the proton.) When a neutron changes to a proton, Z increases by one, but N deceases by one, again leaving A unchanged.

In beta decays, an additional particle, called a *neutrino*, is emitted. Neutrinos were originally postulated because an analysis of beta decays indicated that some energy and angular momentum were being lost in the process. It was therefore assumed that a massless neutral particle was carrying away this energy and angular momentum. If neutrinos are truly massless (and we will have more to say on this point in Part VI of this book), then they travel at the speed of light. The existence of the neutrino was verified experimentally in the 1950s. Neutrinos do not interact with matter via the strong nuclear or electromagnetic forces. This means that a reaction in which a neutrino is involved must proceed by the weak nuclear force. The weak force is so weak that neutrinos rarely interact with matter. Weak decays also proceed at much slower rates than do strong decays.

The basic beta decay reaction is

$$n \rightarrow p + e^- + \bar{\nu} \tag{9.14}$$

The bar over the ν indicates an antineutrino. In free space, the average time for this reaction to

take place is 11 min. In nuclei, the reaction can take place only if the resulting proton ends up in a lower energy state than the original neutron. This places an upper limit on the number of neutrons that a nucleus with a given Z can have and still be stable. If we have too many neutrons, some will beta decay and become protons in lower energy states.

Another type of nuclear reaction is called *fission*, in which a nucleus breaks into smaller parts. The controlled fission of uranium provides the energy in current nuclear power plants. Generally, in fission, a very heavy nucleus breaks into a form that is of more stable, middle mass nuclei. From Fig. 9.3, we see that the most stable nuclei are the ones with intermediate masses, with iron being the most stable. The fission of uranium or plutonium has been used in "atomic bombs".

The final type of nuclear reaction we will consider is *fusion*, in which lighter nuclei can come together to build heavy nuclei. Fusion is important in stellar energy generation, because there is an ample supply of lighter elements. One problem with fusion is that it is hard to start. When two nuclei are far apart the nuclear force between them vanishes. They feel only the electrical repulsion. To bring nuclei close enough together for the nuclear force to take over, we must do work against the electrical repulsion. This requires accelerating the particles to high energies and letting them collide. Since they have high energies, they will be slowed by the electrical repulsion, but not stopped. On the Earth, we can accomplish this acceleration for small quantities of matter in particle accelerators, but this is impractical for large quantities of material. This is the major problem that must be overcome before we can realistically think about using fusion as an economical source of energy on the Earth.

9.2.4 Overcoming the fusion barrier

One way to increase the energies of particles is to raise their temperature. If the temperature is high enough, particles will be moving fast enough to overcome the electrical repulsion, and fusion reactions can take place. On the Earth this poses problems, since we have trouble containing a gas at the required temperature, tens of millions of kelvin. However, in stars the material is confined by gravity. If sufficiently high temperatures are reached, the nuclear reactions can take place. If the temperatures are not high enough to support nuclear reactions, then we don't have a star.

We can estimate the temperature required for nuclear fusion to take place. Let's consider the case of two protons. Their electric potential energy when they are a distance r apart is

$$U = e^2/r \tag{9.15}$$

Suppose we start with two protons far apart, so we can take their potential energy to be zero. We let K be the kinetic energy of the particles at this point. We would like the particles to come to rest (zero kinetic energy) a distance r_p, the radius of the proton, apart. From conservation of energy, the kinetic energy when the protons are far apart must be

$$K = e^2/r_p$$

$$= \frac{(4.8 \times 10^{-10}\ \text{esu})^2}{(1 \times 10^{-13}\ \text{cm})}$$

$$= 2 \times 10^{-6}\ \text{erg}$$

If we divide this by the Boltzmann constant, k, we have an estimate of the temperature at which the average kinetic energy of the particles in the gas is equal to this energy. This gives a temperature of 2×10^{10} K, a very high temperature. If we had considered the case of nuclei with charges $+Z_1e$ and $+Z_2e$, the potential energy becomes $Z_1Z_2e^2/r$. This means that even higher temperatures are needed for the fusion of higher charged nuclei.

Actually, the temperature doesn't have to be as high as our calculation suggests. If the temperature is as high as we calculate, then the energy of the *average* particle will be high enough for it to participate in fusion. We have already seen that not all particles in a gas have the average energy. Even at lower temperatures there will be some particles with a high enough energy to undergo fusion. For particles with a Maxwell–Boltzmann velocity distribution, the probability of finding a particle with an energy between E and $E + dE$ is proportional to a term like the exp($-E/kT$) in the Boltzmann equation. That is

$$P(E)dE \propto E^{1/2}\ e^{-E/kT}\ dE \tag{9.16}$$

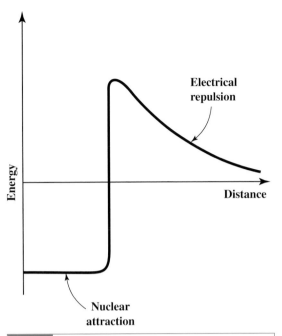

Fig 9.4. The potential energy for two protons as a function of distance. This includes the electrical force and an estimate of the nuclear force.

There is another effect which allows nuclear reactions to take place at a lower temperature than we have calculated. Fig. 9.4 shows the potential energy of two protons as a function of distance between them. For the most part, the force is repulsive (electrical), but if the protons are brought close enough together the force would be attractive. Suppose E_0 is the minimum energy for a particle to be able to overcome the Coulomb repulsion. We can see that this energy E_0 is the "height" of the potential energy barrier. Particles with energies greater than E_0 will pass through the barrier. However, according to classical physics, particles with energies less than E_0 will reach a point at which their energy equals the height of the barrier at some distance r_0. At this point all of the energy is in the form of potential, so the kinetic energy is zero. The particle has stopped and is about to head back in the other direction. The point at which the particle turns back is called the *turning point*.

Quantum mechanically, we should talk about the probability of finding a particle in various places. Let's look again at the particle with energy less than E_0. Initially, the particle has a high probability of being found farther from the proton than the classical turning point. As the classical turning point is approached, the probability cannot suddenly go from some finite value to zero, because that probability is related to a wave phenomenon. So, as we go into the barrier, the probability of finding the particle falls off gradually. This means that there is some probability of finding the particles closer together than the classical turning point. This phenomenon is called *barrier penetration* or *tunneling*.

In general, a particle can penetrate a distance approximately equal to its wavelength, h/mv. More precisely, the probability of penetrating a distance x is related to the wavelength by

$$P(x)dx \propto e^{-ax/\lambda} dx$$

$$= e^{-axmv/h} dx \qquad (9.17)$$

where a is some constant. Suppose we have to tunnel a distance x, equal to the classical turning point r_0, defined by the initial kinetic energy being equal to the electrostatic potential energy at r_0, or

$$\frac{1}{2}mv^2 = \frac{Z_1 Z_2 e^2}{r_0}$$

Substituting into equation (9.17), we have the probability of penetrating r_0:

$$P(r_0) \propto e^{-aZ_1 Z_2 e^2/2hv}$$

$$\propto e^{-b/E^{1/2}}$$

where b is a constant. The effective area of the nucleus for a reaction is approximately λ^2 and is proportional to $1/E$. When we combine this with the Maxwell–Boltzmann velocity distribution, we find the probability of a reaction by nuclei of energy E is proportional to

$$e^{(-E/kT - b/E^{1/2})}$$

This means that at a given T there will be a most likely value of E for reactions to take place. This is known as the *Gamow peak* shown in Fig. 9.5. It turns out that most reactions involve particles that are on the low energy side of the velocity distribution. These lower energy particles have longer wavelengths and can penetrate farther into the barrier. The exponential behavior of the reaction rates also makes them very sensitive to

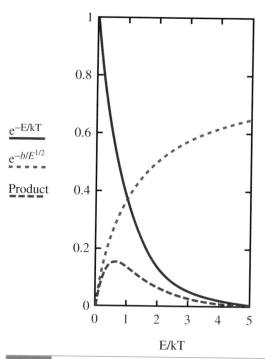

$e^{-E/kT}$

$e^{-b/E^{1/2}}$ - - - - -

Product - - - -

Fig 9.5. The probability of a nuclear fusion, as a function of the particle energy E, at a given gas temperature T. This shows the combined effects of the location of the classical turning point and the quantum-mechanical tunneling.

temperature. That is, small changes in T will produce large changes in the reaction rates.

9.3 | Nuclear energy for stars

When a star is on the main sequence, its basic source of energy is the conversion of hydrogen into helium. We start with four protons and end up with one ^{4}He nucleus. However, it is unlikely that four protons will get close enough to directly form a ^{4}He nucleus in a single reaction. There are different series of reactions that achieve this net result, and they will be discussed below.

We can calculate the energy released by converting four protons to one ^{4}He by comparing their masses. We find that

$$4m_p - m(^4\text{He}) = 0.007(4m_p) \qquad (9.18)$$

This means that 0.007 of the mass of each proton is converted into energy. (The 007 will be familiar to James Bond fans.)

Example 9.3 Lifetime of the Sun
Estimate the lifetime of the Sun for producing energy at its current rate from nuclear fusion.

SOLUTION
If 0.007 of the mass of each proton in the Sun is converted into energy, and if we assume that most of the mass of the Sun was originally in the form of protons, then 0.007 of the Sun's total mass is available for conversion into energy. The total energy available is therefore

$$E = 0.007 \, M_\odot c^2$$

$$= (0.007)(2.0 \times 10^{33} \, \text{g})(3.0 \times 10^{10} \, \text{cm/s})^2$$

$$= 1.3 \times 10^{52} \, \text{erg}$$

The lifetime is this energy divided by the luminosity:

$$t_n = E/L$$

$$= \frac{(1.3 \times 10^{52} \, \text{erg})}{(4 \times 10^{33} \, \text{erg/s})}$$

$$= 3.2 \times 10^{18} \, \text{s}$$

$$= 1 \times 10^{11} \, \text{yr}$$

However, we think that only 10% of the mass of the Sun is in a region hot enough for nuclear reactions – the core, so we must lower our estimate by a factor of ten. This leaves us with a lifetime of ten billion years. We think that the Sun has already lived half of this time.

We now look at actual nuclear reactions, the net result of which is to convert four protons in a ^{4}He nucleus, plus energy. The basic series of reactions in stars like the Sun is called the *proton–proton chain*, because it starts with the direct combination of two protons:

$$p + p \rightarrow d + e^+ + \nu \qquad (9.19)$$

which we can also write as

$$^1\text{H} + {}^1\text{H} \rightarrow {}^2\text{H} + e^+ + \nu$$

This process requires a temperature of 10^7 K. In this reaction the e^+ is a positron, the antiparticle to the electron. We can see by the presence of the neutrino that this process is a weak interaction, and therefore goes very slowly.

Once the deuteron has been created, it can quickly react with another proton:

$$d + p \rightarrow {}^3\text{He} + \gamma \qquad (9.20)$$

which we can also write as

$$^2\text{H} + {}^1\text{H} \rightarrow {}^3\text{He} + \gamma$$

If each of these reactions takes place twice, then we have started with six protons, and we now have two ^{3}He nuclei. (^{3}He is an isotope of helium with two protons and one neutron. It is one of the few stable nuclei with fewer neutrons than protons.) The ^{3}He nuclei combine to give

$$^3\text{He} + {}^3\text{He} \rightarrow {}^4\text{He} + {}^1\text{H} + {}^1\text{H} \qquad (9.21)$$

Note that we have two of the protons back. The net result is that we have converted four protons into a ^{4}He nucleus, along with two positrons, two gamma-rays and two neutrinos. The energy given off in this chain is carried away by the positrons, gamma-rays and neutrinos. The positrons will immediately scatter off (or annihilate) particles in the gas, resulting in heating of the gas. The gamma-ray will travel a small distance before being absorbed, also heating the gas. The hot gas will emit new photons, which are absorbed. This process of absorption and re-emission of the photons takes place until photons are close enough to the surface to escape. The neutrino interacts so weakly that it will escape from the star completely.

Stars more massive than the Sun have higher central temperatures than the Sun. This means reactions involving higher charged nuclei are possible. One such process is responsible for the buildup of elements heavier than helium. It is called the *triple-alpha process*, because the net result is to convert three alpha particles into a ^{12}C nucleus. The repulsion between two alpha particles is four times that between two protons, so higher temperatures, about 10^8 K, are needed. The first step in the chain is

$$^4\text{He} + {}^4\text{He} \rightarrow {}^8\text{Be} + \gamma \qquad (9.22a)$$

We should note that the binding energy of two ^{4}He nuclei exceeds that of the ^{8}Be because the ^{4}He is such a stable nucleus. This tells us that the ^{8}Be should be unstable and break up. However, there will be so many alpha particles around that some of the ^{8}Be nuclei will capture an alpha

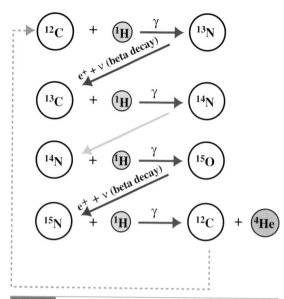

Fig 9.6. The CNO cycle. Solid arrows represent reactions. Symbols over these arrows indicate emitted particles. Dashed lines indicate when a created particle participates in another reaction.

particle before breaking up. If this did not happen, the buildup of heavier elements would be blocked. The combination of the ^{4}He and ^{8}Be gives

$$^4\text{He} + {}^8\text{Be} \rightarrow {}^{12}\text{C} + \gamma \qquad (9.22b)$$

The triple-alpha process is also important as solar mass stars age and leave the main sequence (discussed in Chapter 10).

In massive stars, there is another scheme that is important in converting four protons into one ^{4}He nucleus. It is called the *CNO cycle*. The cycle is indicated graphically in Fig. 9.6, and the steps are:

$^{12}\text{C} + {}^1\text{H} \rightarrow {}^{13}\text{N} + \gamma$	(strong force)	(9.23)
$^{13}\text{N} \qquad \rightarrow {}^{13}\text{C} + e^+ + \nu$	(beta decay)	(9.24)
$^{13}\text{C} + {}^1\text{H} \rightarrow {}^{14}\text{N} + \gamma$	(strong force)	(9.25)
$^{14}\text{N} + {}^1\text{H} \rightarrow {}^{15}\text{O} + \gamma$	(strong force)	(9.26)
$^{15}\text{O} \qquad \rightarrow {}^{15}\text{N} + e^+ + \nu$	(beta decay)	(9.27)
$^{15}\text{N} + {}^1\text{H} \rightarrow {}^{12}\text{C} + {}^4\text{He}$	(strong force)	(9.28)

We see that the net result is the conversion of four protons into one ^{4}He nucleus plus two positrons, two neutrinos, and three photons. All nuclei created as intermediate products are used in the next step. In addition, the last step returns the ^{12}C we need to start the cycle, so the cycle can go again. In a sense, we can think of the ^{12}C as a

catalyst for this cycle. (A catalyst helps something happen, but is not itself changed in the process.) There are more complicated versions of this cycle involving even heavier elements, but the basic ideas are the same.

As we try to build up heavier and heavier elements through fusion, the electrical repulsion becomes stronger. This becomes an effective barrier to the formation of heavier elements. However, no matter how high the Z a nucleus has, we can always get a neutron near it with no electrical repulsion. If the neutron is moving slowly, it can be captured by the nucleus. This can be important in stars, because some reactions provide free neutrons, so there are generally some free neutrons available. We can schematically represent what happens when the nucleus (Z, A) captures a neutron:

$$(Z, A) + \text{n} \rightarrow (Z, A + 1) \qquad (9.29)$$

What happens next depends on whether the rate of neutron capture is slow or rapid, compared to the rate of beta decay. If neutron capture is slow, we call the sequence of reactions an *s-process* (Fig. 9.7). In this situation the new nucleus

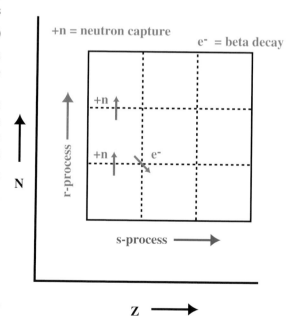

Fig 9.7 The r- and s-processes. The horizontal axis indicates increasing proton number Z, and the vertical axis indicates increasing neutron number N. In an r-process, a neutron is captured, and in an s-process the capture is followed by a beta decay. The +n next to an arrow indicates neutron capture. The e$^-$ next to an arrow indicates a beta decay.

(Z, A + 1) will beta decay before it can capture another neutron:

$$(Z, A + 1) \rightarrow (Z + 1, A) + e^- + \bar{\nu} \qquad (9.30)$$

If the neutron capture is rapid we call the sequence of reactions an *r-process* (Fig. 9.7). The nucleus (Z, A+1) will capture another neutron before it beta decays:

$$(Z, A + 1) + n \rightarrow (Z, A + 2) \qquad (9.31)$$

In either case, the resulting nucleus can either beta decay or capture a neutron, depending on the relative rates. When we have a string of nuclei for which neutron capture is favorable, the r-process allows the buildup of neutron rich nuclei. This will go on until so many neutrons are added that a beta decay breaks the chain. The r- and s- processes can explain the abundances of many of the heavier nuclei. (It should be noted that these are not equilibrium processes.)

The various nuclear processes that we have discussed are responsible for the presence of the heavy elements around us. We will see in later chapters how this material is spread into interstellar space. The net result is to produce the abundances shown in Fig. 9.8. Nuclear physics determines which elements are the most abundant.

When we discuss stellar structure in the next section, we will treat the nuclear physics as something that is known. We assume that once we know the composition of some region and the temperature, we can specify which nuclear reactions are important. Moreover, we assume that we know how the reaction rates depend on temperature. This is a very important point. We have already seen in this section that tunneling by those nuclei with higher than average energy leads to a strong temperature dependence on the reaction rates. For example, the rates of some important reactions depend on temperature as T^7 or even higher powers.

We say that a collapsing object makes a transition from protostar to star when its primary source of energy generation is fusion via the p–p chain, rather than gravitational collapse. The changeover is not a sudden one, as the material closer to the center heats faster. Eventually, enough energy is generated internally for the collapse to halt. When this happens, the star reaches a stable condition. It is on the main sequence.

As nuclear reactions take place in the star, the composition of the star is actually changing. This change could affect the spectral type and luminosity of the star, while the basic structure doesn't change very much. These changes result in a main sequence that is a band on the HR diagram rather than a thin line. However, there is a line that we can identify as connecting the points on the HR diagram where stars of each spectral type first appear on the main sequence. We call this line the *zero-age main sequence*, or ZAMS. This line is shown relative to the main sequence band in Fig. 9.9.

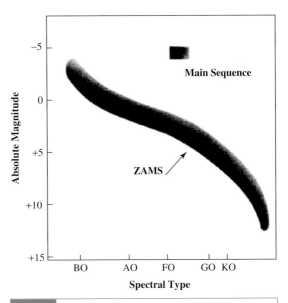

Fig 9.9 Zero-age main sequence. The main sequence on this HR diagram appears as a band, since stars on the main sequence become slightly brighter as their composition changes. The lower edge of this band represents the points where stars first appear on the main sequence, the ZAMS.

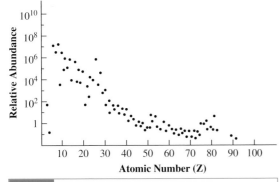

Fig 9.8 Cosmic abundances of the elements as a function of atomic number Z.

9.4 | Stellar structure

The basic philosophy of stellar structure studies is that stars obey the laws of physics, so we should be able to predict and explain their structure by applying those laws. To do this, we must identify the basic physical processes that are important in stars, such as nuclear physics for energy generation. We must also be able to perform large numbers of intricate calculations. This latter facility is provided by modern computers.

Once we carry out stellar structure calculations, we can compare the predictions of the theories with observations. For example, if we put 1 $M_\odot$ of material into model calculations, we should come out with a star whose radius, temperature and luminosity match those of the Sun. If we put in different amounts of material, our calculations should reproduce the main sequence. We should also find stars in the same mass range as those on the main sequence. Stellar models should allow us to predict stellar evolution. They should also be able to tell us how changes in composition lead to changes in structure.

Stars are easier to analyze than some other astronomical objects because they have simple shapes. They are spheres. We also assume that their structure is spherically symmetric. That is, the conditions – temperature, density and composition – only depend on distance from the center, not the direction in which you are going away from the center. We can rotate the star through any angle about any axis through the center and not change the result. (This is not strictly true if the star is rotating or has a strong magnetic field.) To study a star we divide it into spherical shells, each of thickness dr, as shown in Fig. 9.1. If the density a distance r from the center is $\rho(r)$, then the mass contained in a shell of radius r and thickness dr is

$$dM = \rho(r)\,dV \qquad (9.32a)$$

where dV ($= 4\pi r^2\,dr$) is the volume of the shell. This gives

$$dM = 4\pi r^2\,\rho(r)\,dr \qquad (9.32b)$$

Dividing both sides by dr gives us the rate at which we add mass as we go farther out from the center of the star:

$$dM/dr = 4\pi r^2\,\rho(r) \qquad (9.33)$$

This condition is called *mass continuity*, and simply tells us how the rate of change of M(r), mass interior to r, is related to the density at r. M(r) is important because, for a spherical object, the gravitational force on an object a distance r from the center only depends on M(r).

9.4.1 Hydrostatic equilibrium

The material in a shell of radius r is pulled toward the center of the star by the gravitational attraction of all the mass interior to that shell. Something must support the matter in the shell or else the star will collapse. That something is the pressure difference between the bottom of the shell and the top of the shell. This condition is called *hydrostatic equilibrium* (Fig. 9.10). Hydrostatic equilibrium applies to the Earth's atmosphere as well as in the oceans and in a glass of water. The weight of each layer of the fluid is supported by the pressure difference between the bottom and the top.

We can see how much of a pressure difference is needed by considering a small cylinder, of height dr and area dA, as shown in Fig. 9.10. The mass of the cylindrical element is

$$dm = \rho(r)\,dr\,dA \qquad (9.34)$$

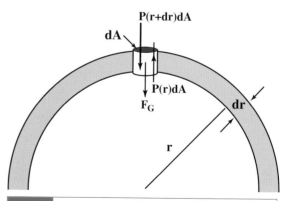

Fig 9.10 Hydrostatic equilibrium. The distance from the center of the star is r, and the thickness of the shell is dr. The density in the shell is $\rho(r)$. We consider the forces on a cylinder of height dr and end area dA. The pressure at r is P(r).

The gravitational force depends on $M(r)$. In any particular model, we can find $M(r)$ by integrating equation (9.33):

$$M(r) = 4\pi \int_0^r \rho(r')r'^2\, dr' \tag{9.35}$$

The gravitational force is given by

$$F_G = -GM(r)\, dm/r^2 \tag{9.36}$$

We use the minus sign $(-)$ to indicate that the force is directed downward. Taking dm from equation (9.34) gives

$$F_G = -[GM(r)/r^2]\rho(r)\, dr\, dA \tag{9.37}$$

We now look at the force exerted on the top and bottom of the cylinder by the pressure of the fluid. The difference between the upward force on the bottom and the downward force on the top is called the *buoyant force*, F_B. If $P(r)$ is the pressure at the bottom of the cylinder and $P(r + dr)$ is the pressure at the top, then

$$F_B = P(r)\, dA - P(r + dr)\, dA$$

$$= [P(r) - P(r + dr)]\, dA \tag{9.38}$$

We have used the fact that force is pressure × area. We define the pressure difference between the top and the bottom, dP, as

$$dP = P(r + dr) - P(r) \tag{9.39}$$

We see that the pressure will decrease as r increases (since the pressure at the bottom of the shell must be greater than the pressure at the top), so dP is a negative number. Equation (9.38) then becomes

$$F_B = \{P(r) - [P(r) + dP]\}\, dA$$

$$= -dP\, dA \tag{9.40}$$

The condition for hydrostatic equilibrium is

$$F_G + F_B = 0 \tag{9.41}$$

Substituting from equations (9.37) and (9.40) gives

$$-[GM(r)/r^2]\, \rho(r)\, dr\, dA - dP\, dA = 0 \tag{9.42}$$

We are interested in the rate at which the pressure changes with radius dP/dr, so we divide

both sides by dA, add dP to both sides, and then divide both sides by dr to give

$$dP/dr = -[GM(r)/r^2]\, \rho(r) \tag{9.43}$$

This is sometimes called the *equation of hydrostatic equilibrium*. We can rewrite it more simply by noting that the quantity $GM(r)/r^2$ is equal to the local acceleration of gravity $g(r)$, so

$$dP/dr = -\rho(r)\, g(r) \tag{9.44}$$

The equation of hydrostatic equilibrium tells us that the denser the fluid is, the more rapidly P changes with r. This is because a denser fluid means a higher mass shell, and a stronger gravitational force pulling it in. This requires a larger pressure difference between the top and the bottom to support it. Also, the greater $g(r)$ is, the greater the gravitational force is pulling the shell in. This means that a larger $g(r)$ also requires a faster rate of change in the pressure.

Example 9.4 Central pressure of the Sun
Use the equation of hydrostatic equilibrium to estimate the central pressure of the Sun by considering the whole Sun as one shell.

SOLUTION
If we consider a whole star to be one shell, then $dR = R$, the radius of the star, and $dP = P_C$, the central pressure (taking the pressure at the surface to be zero). The equation of hydrostatic equilibrium then gives

$$P_C = (GM/R^2)\, \rho\, R$$

where ρ is the average density, and is approximately M/R^3, so

$$P_C \cong (GM/R)(M/R^3)$$

$$= GM^2/R^4$$

Substituting for the Sun gives

$$P_C = \frac{(6.67 \times 10^{-8}\ \text{dyn cm}^2/\text{g}^2)(2 \times 10^{33}\ \text{g})^2}{(7 \times 10^{10}\ \text{cm})^4}$$

$$= 1 \times 10^{16}\ \text{dyn/cm}^2$$

The actual value, obtained from stellar models, is about 20 times this value

---

Another equation that we use in modeling stars is the *equation of state*. The state of a gas is described by the pressure, density and temperature. The equation of state relates those three quantities. We can write it in the general form

$$P = f(\rho, T) \tag{9.45}$$

The actual form of the function depends on the nature of the gas. For an ideal gas, the equation of state has the simple form

$$P = (\rho/m)\, kT \tag{9.46}$$

where m is the mass per particle. For a gas of relativistic particles, the equation of state is different, and we will discuss it in Chapter 10.

Example 9.5 Central temperature of the Sun
Use the result of the previous example and the equation of state for an ideal gas to estimate the central temperature of the Sun.

SOLUTION
To use the ideal gas law, we need to estimate the density. We simply use the average density, which is the mass divided by the volume. Since the hydrogen is completely ionized, there are an equal number of electrons and protons, so the average mass per particle is $(1/2)m_p$.

$$T = \frac{mP}{\rho k}$$

$$= \frac{mP(4\pi/3)\, R_\odot^3}{M_\odot\, k}$$

$$= \frac{(0.5)(1.67 \times 10^{-24}\,\mathrm{g})(1 \times 10^{16}\,\mathrm{dyn/cm^2})(4\pi/3)(7 \times 10^{10}\,\mathrm{cm})^3}{(2 \times 10^{33}\,\mathrm{g})(1.38 \times 10^{-16}\,\mathrm{erg/K})}$$

$$= 4.4 \times 10^7\,\mathrm{K}$$

This is a little larger than the model results, with $T \sim 1.5 \times 10^7\,\mathrm{K}$.

9.4.2 Energy transport
In making a stellar model we must also consider how energy gets from the inside of the star to the outside. In general, energy can be transported by conduction, convection and radiation. In stellar interiors conduction is not generally important. The energy transport must be such that the temperature $T(r)$ does not change with time. If it did, the star would not be stable for the multibillion year lifetime for stars like the Sun. For the temperature distribution to be constant with time,

the energy entering any shell per second must equal the energy leaving that shell per second.

If radiation transport dominates, we can calculate the required temperature distribution, $T(r)$. We let $f(r)$ be the flux of radiation through a surface at radius r. If the surface emits like a blackbody, then

$$f(r) = \sigma\, T(r)^4 \tag{9.47}$$

We can find the rate at which $f(r)$ changes with $T(r)$,

$$df/dT = 4\sigma\, T(r)^3 \tag{9.48}$$

If we interpret df as the small change in f due to a small change in T, dT, then

$$df = 4\sigma\, T(r)^3\, dT \tag{9.49}$$

However, the change in flux passing through a given layer of the star must depend on the ability of the layer to absorb radiation. In Section 6.2 we saw that this is given by the absorption coefficient κ. In stellar structure it is more convenient to deal with the absorption coefficient per density of material. We therefore let $\kappa'(r)$ be the opacity per unit mass at r. This means that $\kappa'(r)\, \rho(r)$ gives the fraction of radiation absorbed per centimeter. Using these definitions

$$df = -\kappa'(r)\rho(r)f(r)dr \tag{9.50}$$

The minus sign $(-)$ tells us that the flux is decreased by the absorption.

We define the luminosity of a given layer as the flux f multiplied by the surface area of the layer,

$$L(r) = 4\pi r^2\, f(r) \tag{9.51}$$

If we use equations (9.49) and (9.50) to eliminate df and equation (9.51) to eliminate $f(r)$, we have

$$4\sigma T(r)^3\, dT = -\frac{\kappa'(r)\, \rho(r)\, L(r)\, dr}{4\pi r^2} \tag{9.52}$$

Solving for the luminosity gives

$$L(r) = -\left[\frac{16\pi r^2 \sigma T(r)^3}{\kappa'(r)\, \rho(r)} \right] \frac{dT}{dr} \tag{9.53}$$

This tells us how the rate of energy flow depends on the rate at which the temperature changes with r. In general, $\mathrm{d}T/\mathrm{d}r$ is negative (the temperature drops with distance from the center), so the luminosity is positive. A more exact calculation gives essentially the same result as equation (9.53), but with the 16 replaced by 64/3.

From equation (9.53) we can see that the opacity per unit mass $\kappa'(r)$ is very important in determining the energy transfer, and therefore the structure of the star. The opacity depends on the composition of the star. Accurate stellar structure calculations require good knowledge of the opacity as a function of composition and temperature.

In addition to energy transport, we must also consider energy generation. If energy is generated in a particular shell, then the energy leaving that shell will exceed the energy entering the shell by the amount of energy generated. We let $\varepsilon(r)$ be the energy generated per unit mass within the shell at radius r. The increase in luminosity in that layer, due to energy generation, is then

$$\mathrm{d}L = 4\pi r^2 \, \rho(r)\varepsilon(r) \, \mathrm{d}r \qquad (9.54)$$

The rate at which the luminosity changes with radius is then

$$\mathrm{d}L/\mathrm{d}r = 4\pi r^2 \, \rho(r)\varepsilon(r) \qquad (9.55)$$

To carry out model calculations we must be able to specify $\varepsilon(r)$ as a function of composition and temperature. This is where the input from nuclear physics is important.

If, at any r, the radiative temperature gradient becomes too large, then convection will set in. The quantitative effect of convection is discussed in Section 23.3.2. Convection is an adiabatic process, no energy is lost from parcels of material as they move outward. At that point the temperature gradient becomes the one appropriate for adiabatic processes. This is called the adiabatic lapse rate.

We might wonder why stars don't explode with all the energy they produce. The answer is that their stability comes from their negative heat capacity. (Heat capacity is the energy required to raise the temperature by a given amount.) To see this, we look at the total energy $E = U + K$ (where K is now the kinetic energy of the thermal motions of particles in the gas). In Chapter 13, we will derive a relationship between total, kinetic and potential energies for bound systems, called the virial theorem, which tells us that $E = -K$. Suppose we now add energy. This increases E, making it less negative. This makes K less positive, so the total thermal kinetic energy decreases. The gas cools. Therefore, if the star produces too much energy for its equilibrium configuration, it can expand and cool to adjust. The above argument doesn't apply to degenerate gases, since their thermal energy is essentially independent of the temperature. Therefore, as we encounter degenerate stars, or parts of stars, we will see that explosions are possible.

9.5 | Stellar models

In the previous section we saw a group of equations that describe the physics that governs stellar structure: (1) mass continuity, (2) hydrostatic equilibrium, (3) equation of state, (4) energy transport, and (5) energy generation.

The inputs to stellar models are the mass of the star and the composition. We must also specify the nuclear physics, which gives the energy generation as a function of these conditions. We must also put in the information about the opacities. For the purposes of calculation, we break the star into spherical shells. We must solve for the distribution of density $\rho(r)$ and temperature $T(r)$ which satisfy the conditions imposed by the equations of stellar structure. This is such a complicated process that realistic stellar models are all calculated on computers.

Stellar model calculations can also be used to predict stellar evolution. As nuclear reactions take place in the hot central core of the star, the composition changes. This changes the energy generation rate and the opacity, meaning the structure will change. We do a model calculation for the initial composition. We then determine the rate at which various nuclei are produced or destroyed. We then know what the composition will be some time, say 50 000 years, later. We now calculate a model with the altered composition, and then repeat the process. We follow the evolution of the star in these time steps. We choose the time steps so that the composition changes somewhat, but not too much, during each time step.

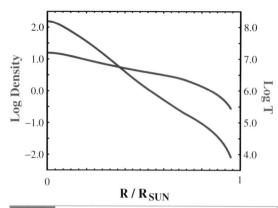

Fig 9.11 Temperature and density as a function of distance from the center of the Sun, as calculated from the solar model which best agrees with the global properties (radius, surface temperature) of the Sun.

To see the results of a model calculation, we look at a model for the Sun. The distribution of temperature and density is shown in Fig. 9.11.

9.6 | Solar neutrinos

Though our model for the Sun gives the correct radius and temperature, there are certain aspects we cannot check directly. Almost all of the direct information we receive from the Sun comes from photons emitted in the solar atmosphere. We cannot directly observe photons that are emitted in the nuclear reactions in the center. Those photons are quickly absorbed and their energy takes about 10^7 years to reach the surface. We have no direct observations of the solar core now.

There is one opportunity to make a direct observation of the solar core. Neutrinos created in nuclear reactions in the core escape at the speed of light, virtually unattenuated. They reach us 8.5 min after they are created. We could use our stellar models to predict the rate at which neutrinos are emitted, and then try to measure the flux of neutrinos at the Earth. This would be a direct test of the solar model. The problem is that neutrinos are very hard to detect. If the whole Sun cannot absorb many neutrinos, a detector on Earth will absorb even fewer. The neutrino produced in the first reaction of the proton–proton chain ($p + p \rightarrow d + e^+ + \nu$) has a

relatively low energy, and this makes it particularly hard to detect.

There is, however, a source of higher energy neutrinos. Once ^{3}He is formed, most of it reacts to form ^{4}He, as discussed in Section 9.2. However, in a small fraction of cases, the following reaction can take place:

$$^3\text{He} + {}^4\text{He} \rightarrow {}^7\text{Be} + \gamma \tag{9.56}$$

The ^{7}Be then captures a proton:

$$^7\text{Be} + \text{p} \rightarrow {}^8\text{B} + \gamma \tag{9.57}$$

The boron then beta decays, emitting a neutrino in the process:

$$^8\text{B} \rightarrow {}^8\text{Be} + \text{e}^+ + \nu \tag{9.58}$$

The ^{8}Be then breaks apart into two alpha particles:

$$^8\text{Be} \rightarrow {}^4\text{He} + {}^4\text{He} \tag{9.59}$$

The neutrino emitted in the beta decay of the ^{8}B has enough energy to provide us with some chance of detecting it.

We can detect this neutrino using an isotope of chlorine, ^{37}Cl. About 25% of all naturally occurring chlorine is in the form of this isotope. When struck with a sufficiently high energy neutrino, it can absorb the neutrino:

$$^{37}\text{Cl} + \nu \rightarrow {}^{37}\text{Ar} + \text{e}^- \tag{9.60}$$

This particular isotope of argon is radioactive, and its decay can be detected in normal particle detectors in the laboratory. If we start with a tank of chlorine (and no radioactive argon), and we end up with a small amount of ^{37}Ar, we can determine the rate at which neutrinos were hitting the tank.

This is the basic idea behind an experiment conducted by R. Davis of the Brookhaven National Laboratory. The source of liquid chlorine is perchloroethylene (cleaning fluid). Since neutrinos interact infrequently, a large quantity of "detector" is used – about 10^5 gallons. The argon is produced as an inert gas, so it will form bubbles in the fluid, allowing it to be removed. The experiment is run for some period of time, typically a month. The gas is then collected, and is measured to gauge ^{37}Ar activity.

Even if some radioactive argon is found, we don't know that it results from solar neutrinos.

There are also high energy particles in the Earth's atmosphere, called cosmic rays, which can produce a similar result. The Earth can shield the tank from cosmic rays but not from neutrinos. The tank is therefore placed 1.5 km underground in the Homestake gold mine, near Lead, South Dakota. Another source of possible contamination is natural radioactivity from the rocks in the mine. For this reason, the tank is surrounded by a larger tank containing water. The water blocks the high energy particles from radioactive decays in the mine, but doesn't block neutrinos. You can see from this brief description that this experiment is a very difficult one.

The results of the experiment from some 20 years observing have been astounding. It is convenient to express the rate of detections in *solar neutrino units* or *SNU*. The standard model of the Sun predicts that we should be measuring 8 SNU. The experiment, which detects about one event every two days, yields 2.6 SNU, a number that is only about one-third of this value. This leads to what we call *the solar neutrino problem*. Astrophysicists have so much faith in our understanding of stellar interiors that such a large discrepancy is an indication of a severe problem.

A number of solutions have been suggested. One possibility is that there is something wrong with the experiment itself. Possibly, the detector is not as sensitive to neutrinos as originally thought. However, various aspects of the experiment have been checked and refined over the years, and there is a general feeling that the experiment is correct.

If the experiment is correct, then the solar model must be examined. It is possible that some of the inputs are not correct. For example, the nuclear reaction rates have a very strong temperature dependence. If we have that dependence slightly wrong, then significant changes can occur in the solar model. It is also possible that the opacities (as a function of composition and temperature) are slightly off. The solar neutrino problem has stimulated work in these areas. We now have improved nuclear physics data and opacity data. A better solar model has been calculated, with a much smaller uncertainty, but the predicted neutrino flux remains essentially the same. There is an additional check on the solar model. It comes from helioseismology (discussed in Chapter 6), in which the oscillations of the Sun are studied. The solar model (temperature, density and composition vs. radius) can be used to predict the details of these oscillations. The agreement between theory and experiment is very good, meaning that the solar model is probably not the cause of the solar neutrino problem.

It has been suggested that the Sun may go through a cycle in its energy generation, and that right now it is generating less than the average amount of energy. In this cycle, at some point the core cools, reducing the rate of nuclear reactions. The pressure decreases and the core contracts. As the core contracts, it converts gravitational potential energy into kinetic energy, and begins to heat. As the core becomes hotter the rate of nuclear reactions increases. The pressure increases and the core expands. The cycle then starts again. If this is the answer, then the neutrino observations are giving us a good view of what is happening in the Sun now. In this picture, it is possible for the cycle to take place with very little variation in the solar luminosity. Since photons are scattered many times before they can go from the center to the photosphere, the light from the photosphere reflects the average energy production over the Kelvin time. It should be pointed out that no mechanism for such oscillations has been found.

It has been suggested that we don't know as much as we thought we did about neutrino physics. We now know that there are three different neutrino types, each with its own antineutrino. (We'll discuss particle physics more in Chapter 21.) The type of neutrino that we have been discussing is called the electron neutrino, because it always appears in reactions with electrons (or positrons). This is the type of neutrino produced in the Sun, and the type that will interact with ^{37}Cl. However, some theories have suggested that neutrinos can change their identities. If this is the case, then a neutrino can be created as an electron neutrino in the Sun, but change its identity by the time it reaches the Earth. According to this idea, as few as one-third of the neutrinos produced in the Sun might be capable of being absorbed by the ^{37}Cl. These identity changes, or *neutrino oscillations*, are also related to the suggestions that neutrinos have a very small (but not zero) mass. The experimental evidence for this is being studied, and new experiments are underway.

In the meantime two new types of solar neutrino experiments are being done. The first involves gallium as a detector. Since the ^{8}B neutrino comes from a relatively unimportant branch in the Sun's nuclear reaction chain, it may be that there is a small error in our calculations that just happens to be magnified for this minor branch. There is another reaction that also produces a detectable neutrino. The reaction is not as important as the proton–proton chain, but it is more significant than the branch involving ^{8}B. This reaction is called the *p–e–p reaction*:

$$p + e^- + p \rightarrow d + \nu \tag{9.61}$$

It occurs once for every 400 direct p–p reactions. It is rare because it is much harder to bring three particles together at the same time than it is to bring two particles together.

The neutrino produced by the p–e–p reaction is not as high in energy as the ^{8}B decay neutrino, and is not absorbed by ^{37}Cl. However, it has a higher energy than the p–p neutrino. The p–e–p neutrino can be absorbed by gallium. As with chlorine, large quantities of gallium are needed for the experiments. Two such experiments are being carried out, a European–Israeli–USA collaboration known as GALLEX (30 tons of gallium), and a Russian–USA collaboration known as SAFE (60 tons of gallium). These experiments yield about one event per day. The measured rates are about half of the predicted values.

The second set of new experiments involves larger versions of traditional detectors used in elementary particle physics experiments. These involve large amounts of water as detectors, with each reaction producing a small flash of light, and that light is detected by an array of photomultipliers. The first set of experiments was carried out by a Japanese group at Kamiokande. The experiment was originally set up as a particle physics experiment, but it turned out that it was also capable of detecting solar neutrinos. An important feature of this detector is that it can measure the direction from which the neutrinos are arriving, and the group were able to verify that the neutrinos are, indeed, coming from the Sun (and not, for example, from some background contamination). Following their original success, they built a second version, known as SuperKamiokande, more closely designed for observing solar neutrinos. The results of these observations is that observed flux is about half the theoretical predictions. For their work on this problem, Davis and M. Koshiba, who headed the Japanese project, shared the 2002 Nobel Prize in Physics.

At this point, it appears that the best solution to the solar neutrino problem involves the neutrinos. If they have a very small mass, and can change identities, then the theory and experiment can be brought into agreement. In 2002, a group in Sudbury, Canada (Sudbury Neutrino Observatory, SNO, shown in Fig. 9.12), provided strong evidence that neutrino oscillations are the answer. An analysis of their results

Fig 9.12 The large detector of the Sudbury (Canada) Neutrino Observatory (SNO). [Photo courtesy of Sudbury Neutrino Observatory]

suggests that the neutrino mass is in the 10^{-2} eV range. (Remember, for comparison, the mass of the proton is almost 10^9 eV.) The analysis also shows that there is a significant chance (of the order of 50% with a large uncertainty) for neutri-nos created in the Sun to change their identity by the time they reach our detectors. The next major test of these ideas would be an experiment to detect directly the p–p neutrinos (by far the most abundant, and lowest energy).

Chapter summary

In this chapter we looked at the processes responsible for the structure of main sequence stars.

We started by looking at energy sources. Nuclear reactions are the only source capable of giving stars their inferred lifetimes. We saw that the temperatures required for nuclear reactions to take place are in excess of 10^7 K, even with tunneling to help bring the nuclei together.

The basic source of energy on the main sequence is the conversion of hydrogen to helium. In low mass stars this takes place primarily via the proton–proton chain. In more massive stars, with higher central temperatures, other cycles, such as the CNO cycle, are important.

We also looked at the basic processes that govern stellar structure. We saw that normal stars are in hydrostatic equilibrium, with each layer supported by the pressure difference between its bottom and top. We also saw that the temperature distribution is determined by the requirement that the temperature of each layer be constant.

Once the basic laws of stellar structure are outlined, stellar models can be computed, generally using computers. In a model, we start with a certain mass and composition, and calculate the equilibrium configuration.

We saw how much of stellar structure seems to be understood, but encountered the puzzle of the solar neutrino experiment. The neutrinos allow us to see what the core of the Sun is doing now, and it does not appear to be doing as much as models predict. The answer to this problem seems to lie in neutrons changing their identity in their trip from the Sun.

Questions

9.1. (a) Why is 1 eV/atom a reasonable estimate for the energy available in chemical reactions? (b) Is the estimate likely to be wrong by more than a factor of ten in either direction? Explain. (c) If the estimate is wrong by a factor of ten in either direction, will it change the conclusion that the Sun cannot exist on chemical reactions?

9.2. It has been said that if we did not know that $E = mc^2$, then we would not know about nuclear energy. Discuss this.

9.3. Explain the factors that place upper and lower limits on the number of neutrons that go into a nucleus with some specific number of protons.

9.4. Why can't a free proton beta decay into a neutron?

9.5. What are the similarities between gamma emission by nuclei and visible light emission by atoms?

9.6. Why is there no Coulomb barrier to fission if there is one to fusion?

9.7. Why are the rates of certain fusion reactions very sensitive to temperature?

9.8. Why are the r- and s-processes important?

9.9. (a) What are the parameters that we put into a stellar model? (b) What calculations do we perform? (c) How do we test the results?

9.10. (a) What do we mean by spherical symmetry? (b) Why will a rotating star not be spherically symmetric?

*9.11 In equation (9.26) we have used the form of the gravitational force between two point particles. These particles have masses $M(r)$ and dm and are a distance r apart. However, $M(r)$ represents an extended mass and dm represents a shell, so neither represents a point. How can we use the simple formula? (Hint: Treat the justification for $M(r)$ and dm separately.)

*9.12 What effect would a slight increase in opacity at all layers have on the structure of a star?

9.13 Explain how we simulate stellar evolution.

9.14 Why is the solar neutrino problem so important?

9.15 When we considered explanations of the solar neutrino experiment, we said that the Sun may be generating some energy now through gravitational collapse. However, earlier in the chapter, we ruled out gravitational collapse as a stellar energy source. Why isn't this a contradiction?

*9.16 The equilibrium structure of a star is ultimately determined by its mass and composition. Show that the structure of a star determines the rate of energy generation and not the other way around.

Problems

9.1. What is the gravitational potential energy of (a) the Sun, (b) a 1 $M_\odot$ black hole?

9.2. (a) What is the gravitational potential energy of an interstellar cloud, with a density of 1000 H atoms/cm^3 and radius of 10 pc? (b) What is its kinetic energy if its temperature is 10 K?

*9.3. Find the gravitational potential energy of a sphere with a $1/r^2$ density distribution. Take the total mass of the sphere to be M, and let the density $\rho(r) = \rho_0/r^2$ out to a radius, R. Express your final answer in terms of M and R.

9.4. Estimate the lifetime of a 10 $M_\odot$ star on the main sequence to give off energy stored from gravitational collapse.

9.5. Calculate the mass corresponding to the binding energy of an H atom. What fraction of the mass of the atom is this?

9.6. What is the rate at which the Sun is converting mass to energy?

9.7. (a) What is the difference in mass between the neutron and the proton, expressed in MeV? (b) How does this relate to the energy available in the beta decay of a neutron?

9.8. Calculate the binding energy of a ^{4}He nucleus.

9.9. How much energy per proton is given off in the p–p chain? (Express your answer in MeV.)

*9.10. Suppose we have Z protons and have to distribute them into two nuclei, one with Z_1 protons and the other with Z_2 protons ($Z = Z_1 + Z_2$). (a) What arrangements give the maximum and minimum Coulomb repulsion between the two nuclei? (b) What does this tell you about the types of fusion that are most likely to take place in stars?

9.11. (a) How close can two protons get if one is at rest and the other has a kinetic energy equal to the average energy at $T = 10^7$ K? (b) What is the wavelength of the moving proton, and how does it compare with the minimum separation between the two protons? (c) Repeat the calculations for a proton with ten times the average energy at this temperature.

*9.12. Suppose the density of a star is given by

$$\rho(r) = \begin{cases} \rho_0 & r < r_0 \\ \rho_0(r_0/r)^2 & r_0 < r < R \\ 0 & R < r \end{cases}$$

(a) Find an expression for $M(r)$. (b) If the mass of the star is 1 $M_\odot$ and $R = R_\odot$, and $r_0 = 0.1\,R_\odot$, what is the value of ρ_0?

9.13. For the density distribution in the previous problem, find $P(r)$.

Computer problems

9.1. Calculate the gravitational potential energies for stars in the middle of each spectral class on the main sequence (O5, B5, . . .).

9.2. Write a routine that will calculate the gravitational potential energy for uniform density elliptical objects which have two axes the same. Evaluate the potential energy for the two following cases, both clouds with 100$M_\odot$ of material: (a) an oblate (flattened) ellipsoid with semi-axes 10, 10 and 5 pc; (b) a prolate (elongated) ellipsoid with semi-axes 5, 5, and 10 pc.

9.3. Estimate the nuclear reaction lifetime of a star as a function of its mass. Assume that 10% of the mass is in the core and available for nuclear reactions. (Hint: Use the mass–luminosity relationship.) Plot your results for the range of masses encountered on the main sequence.

9.4. Make a table showing the Kelvin time for stars in the mid-range of each spectral type (O5, B5, . . .).

9.5. Make a graph of the electrical potential energy between two protons as a function of their separation, r. Let r range from the size of a nucleus to ten times that.

Chapter 10

Stellar old age

We have already seen that the mass of a star is the most important property in determining a star's structure. For a main sequence star the mass determines the size and temperature. The lifetime of a star on the main sequence depends on the available fuel and the rate at which that fuel is being consumed – the luminosity. Both of these quantities depend on the star's mass, so the lifetime on the main sequence also depends on the mass. When the star uses up its basic supply of fuel, its ultimate fate also depends on its mass. In fact, the mass and the initial composition of a star completely determine its structure and evolution. This can be proven mathematically on the basis of the physical equations involved. This result is known as Vogt's theorem.

10.1 | Evolution off the main sequence

10.1.1 Low mass stars
We first look at stars whose mass is less than about 5 $M_\odot$. Eventually a star will reach the point where all the hydrogen in the core has been converted to helium. For a low mass star, the central temperature will not be high enough for the helium to fuse into heavier elements. There is still a lot of hydrogen outside the core, but the temperature is not high enough for nuclear reactions to take place. The core begins to contract, converting gravitational potential energy into kinetic energy, resulting in a heating of the core. The hydrogen just outside the core is heated to the point where it can fuse to form helium, and

this takes place in a shell at the outer edge of the core (Fig. 10.1). We refer to this as a *hydrogen-burning shell*, where the word "burning" refers to nuclear reactions, rather then chemical burning. As the core contracts, the rate of energy generation in the shell increases. The shell can easily give off energy at a greater rate than the core did during the star's normal lifetime.

While all of this is happening in the interior, the outer layers of the star are changing. Energy transport from the core is radiative, and is limited by the rate at which photons can diffuse through the star. The outer layers of the star become hotter and expand. As the gas expands, it cools. The star's radius has increased, but its temperature has decreased, so the luminosity increases slightly. The behavior of the star's track on the HR diagram is shown in Fig. 10.2. The track moves to the right (cooler), and the star appears as a subgiant.

There is a mechanism that keeps the surface temperature from becoming too low. The rate of photon diffusion increases as the absolute value of dT/dr increases. Remember, dT/dr is negative, so we are saying that the greater the temperature difference between some point on the inside and the surface, the greater the energy flow between those two points. (In winter, the larger temperature difference between the inside and outside of your house results in a faster heat loss, and higher fuel bills.) If the surface temperature of the star falls too much, the photon diffusion is faster, delivering more energy to the surface, raising the surface temperature. Therefore, as the radius continues to increase, the surface temperature remains approximately constant. The luminosity

H Envelope   **H Envelope**

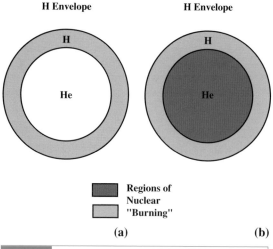

Regions of
Nuclear
"Burning"

(a)   **(b)**

Fig 10.1. Star with an H-burning shell. (a) The temperature in the star is not hot enough to fuse the helium in the center, but is hot enough to keep the H in the shell burning. (b) In this star, the temperature is hot enough to keep both burning. (Remember, by "burning" we are talking about nuclear reactions.)

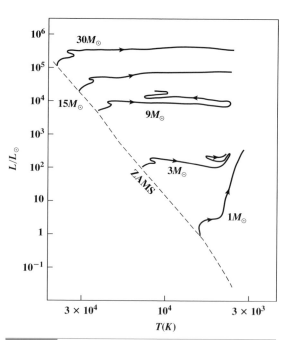

Fig 10.2. Evolutionary tracks away from the main sequence on an HR diagram. Each track is marked by the mass for the model. The dashed line is the zero-age main sequence (ZAMS).

therefore increases, and the evolutionary track moves vertically. The star is then a *red giant*.

By the time the star becomes a red giant, the energy transport in the envelope is convective. This is because of the large value of $-dT/dr$. The analogous situation on Earth involves the heating of the atmosphere. Sunlight heats the ground, and then infrared radiation from the ground heats the air. (This explains why the air is cooler at high altitudes; it is farther from the direct heat source, the ground.) In this situation, we say that the energy transport is radiative. However, if $-dT/dr$ becomes larger, then $-dP/dr$, the rate at which the pressure falls, also becomes large. The air that is heated near the ground expands slightly, and becomes very buoyant, being driven upward by the pressure difference between the bottom and top of any parcel of air. The hot air rising, being replaced by cool air falling, known as convection, becomes the dominant mode of energy transport.

We now look at the evolution of the core while the star is becoming a red giant. The temperature of the core climbs to 10^8 K. This is hot enough for the triple-alpha process to take place (equations 9.11 and 9.12), fusing the helium into carbon. The density is so high that the material no longer behaves like an ideal gas. This is called a *degenerate*

gas. We will discuss degenerate gases in Section 10.4, but for now we note that the equation of state is very different for a degenerate gas. In an ideal gas, when the triple-alpha process starts, the extra energy generated causes an increase in pressure, which causes the gas to expand, slowing the reaction rate. This keeps the reactions going slowly. In a degenerate gas the pressure doesn't depend on temperature and no such safety valve exists. The conversion of helium to carbon takes place very quickly. We call this sudden release of energy the *helium flash*. The energy released causes a brief increase in stellar luminosity.

Following the helium flash the energy production decreases. The core is no longer degenerate, and steady fusion of helium to carbon takes place. This region is surrounded by a shell in which hydrogen is still being converted into helium. At this point the star reaches the horizontal branch on the HR diagram. The outer layers of the star are weakly held to the star, since they are so far from the center. The star begins to undergo mass loss. The subsequent evolution depends on the amount of mass that is lost.

Eventually all the helium in the core is converted into carbon and oxygen. The temperature is not high enough for further fusion, and the core again begins to contract. A helium-burning shell develops, and the rate of energy production again increases. The envelope of the star again expands. On the HR diagram the evolutionary track ascends the giant branch again, reaching what is called the *asymptotic giant branch*. Stars on the asymptotic giant branch are more luminous than red giants. The star can briefly become large enough to become a red supergiant at this stage. The star can also undergo oscillations in the rate of nuclear energy generation.

10.1.2 High mass stars

More massive stars live a shorter lifetime on the main sequence than do lower mass stars. As with the lower mass stars, the main sequence lifetime for higher mass stars ends when the hydrogen in the core is used up. The core then begins to contract, and the temperature for helium fusion to heavier elements is quickly reached. The helium fusion takes place before the core can become degenerate. Therefore, in contrast with the helium flash in lower mass stars, the helium burning in more massive stars takes place steadily. At this point, the star has a helium-burning core with a hydrogen-burning shell around it (Fig. 10.3).

When the helium in the core is exhausted, the temperature is high enough for the carbon and oxygen to fuse into even heavier elements. At this time, we have a carbon- and oxygen-burning core, surrounded by a helium-burning shell, which in turn is surrounded by a hydrogen-burning shell. As heavier elements are built up, the core develops more layers.

As the luminosity of the core increases, the outer layers of the star expand. The atmosphere cools with the expansion, but the size increases sufficiently for the luminosity to increase. At this point the envelope is convective, and the temperature gradient is limited by the adiabatic lapse rate. So the envelope must grow to a large size to accommodate the large temperature difference between the core and the surface. Eventually, the radius of the star reaches about $10^3\,R_\odot$. At this point the star is called a *red supergiant*.

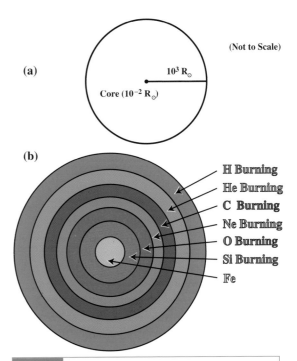

Fig 10.3. Shells in the core of a high mass star as it evolves away from the main sequence. (a) The core is only a small fraction of the total radius. (b) In the core, there is a succession of shells of different composition. Each shell has exhausted the fuels that are still burning in shells farther out.

10.2 | Cepheid variables

10.2.1 Variable stars

If we monitor the brightnesses of certain stars, we find that many oscillate with time. These are known as *variable stars*. The periods of variability range from seconds to years. We have already seen that eclipsing binaries appear as variables. However, many stars have luminosity variations associated with physical changes in the stars themselves (rather than simply by eclipsing one another).

Since we will be using specific stars as examples, we will briefly mention systems for naming normal and variable stars. The bright stars are named, in order of brightness within their constellation, by a Greek letter, followed by the Latin genitive form of the constellation name. An example is α Orionis (abbreviated as α Ori). Some of the brightest stars are also known by their ancient names. For example, α Ori is Betelgeuse. Variable stars are listed in order of discovery within a given constellation. The first is designated R (e.g. R Ori), the next S, and so on to Z. After that, two letters are used, starting with RR, RS to RZ, then SS to SZ, and so on, until ZZ is reached. Then comes AA through AZ, BB through BZ, and so on to QZ. (The letter J is never used because of possible confusion with I.) This gives a total of 334 variable stars per constellation. Beyond that, numbers starting with 335, preceded by a V (for variable), are used (e.g. V335 Ori, V336 Ori, etc.)

For any particular star, we are interested in producing a light curve, a graph of its magnitude as a function of time. Studies of variable stars often require very long term monitoring. In some cases, it is possible to recover information on a star's variability from plate archives. When photographic plates are taken at an observatory, the astronomer who took them is often required to return the plates when that astronomer's work has been completed. The astronomer may be interested in only one star on the plate, but it contains a record of many stars. With the advent of CCD observations, archives are no longer being kept in the same manner. Observations of many variable stars can be so time consuming that it has become an area of astronomy where amateur observers have been able to make major contributions, generally coordinated by the American Association of Variable Star Observers (AAVSO). (In measuring light curves, we often measure time in Julian days, the number of days since noon on January 1 4713 BC, or modified Julian days, the number of days since the beginning of the Besselian year 1950 (see Appendix F for a further discussion of timekeeping).

We distinguish different types of variable stars by such things as their period and the magnitude range. A particular class of variable is generally named after the prototype of the class, either the first or most prominent star with the distinguishing properties of the class. In this section, we look at a few examples of the most important types of variables. Different types of variables appear in different parts of the HR diagram, as shown in Fig. 10.4. These bright stars were named before their variable nature was known, so they do not follow the naming convention discussed above.

Mira variables are named after the prototype (a star also known as O Ceti). These stars have periods of about three months to two years, or even

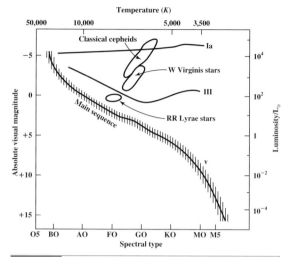

Fig 10.4. HR diagram, showing the locations of various types of variable stars.

longer, and are called *long period variables*. Any individual Mira variable may show fluctuations in its period. The stars change their brightness by about 6 mag, or over a factor of 250 in brightness. For example, the apparent magnitude of Mira ranges from 9 mag to 3 mag. These changes in brightness are accompanied by changes in spectral type. Mira changes from M9 to M5. This means that a temperature change is accompanying the luminosity change.

Cepheid variables are named after the prototype δ Cephei. Its period is 5.4 days and its apparent magnitude varies from 3.6 to 4.3 mag. In general, Cepheids have periods from 1 to 100 days. We know of more than 1000 in our galaxy. Another familiar Cepheid is Polaris (the North Star), which changes by only 0.1 mag, from 2.5 to 2.6. (For a long time, astronomers did not know that Polaris is a variable, and it was used as a reference for measuring the magnitudes of other stars.) Cepheids have masses of approximately 6 $M_\odot$, and radii of about 25 $R_\odot$.

10.2.2 Cepheid mechanism

When we study the spectral lines in Cepheids, we can detect Doppler shifts that vary throughout the light cycle (Fig. 10.5). The Doppler shifts go through a cycle in the same period as the light. This means that the surface of the star is moving. The size of the star changes as the luminosity changes. The spectral type also changes throughout the cycle.

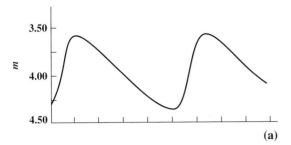

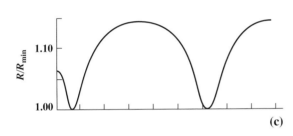

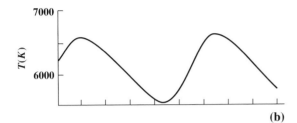

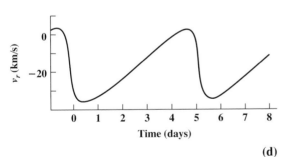

Fig 10.5. Radial velocity and light curves for δ Cep, the prototypical Cepheid. (a) Apparent magnitude as a function of time within the period. (b) Temperature as a function of time. (c) Radius, relative to the minimum radius, as a function of time. (d) Radial velocity of the surface as a function of time. Note that the radial velocity is one quarter cycle (90°) out of phase with the radius.

The luminosity change is then associated with changes in the surface temperature and in the radius.

A star may become a Cepheid variable when it reaches the stage described at the end of the pre-

ceding section. To see how a Cepheid oscillates, lets consider the oscillations of a normal star. These oscillations are radial. They involve inward and outward motions of the outer layers of the star. Suppose we are able to perturb a star by decreasing its radius R. The density then increases, and the pressure increases. The excess pressure will make the outer layers expand back. However, just as a swing overshoots its lowest point as it returns from its maximum height, the star can overshoot its equilibrium radius R_0. Now the star is larger than its equilibrium radius, and the pressure decreases, allowing material to fall back. This process then repeats itself.

In the above analysis, we have ignored the effects of opacity. In a normal star, the opacity decreases as the temperature increases. Now, we again start the perturbation by reducing R below R_0. This causes P and T to increase. The increase in T decreases the opacity. The reduction of the opacity allows some of the excess pressure to be relieved by allowing heat to flow out of the denser regions as radiation. This reduces the tendency of the star to overshoot. If we had started with a perturbation in which $R > R_0$, P and T would have decreased. The opacity would have increased, and the tendency to fall back too fast would be reduced. The result of the opacity is to quench the oscillation.

For a narrow range of conditions, the opacity increases as the temperature increases. The source of the opacity is the ionization of He^+ to form He^{++}. If we now start with a perturbation in which $R < R_0$, the pressure and temperature increase. Now the opacity also increases, so the excess pressure is not relieved, except by driving the star back. The tendency to overshoot is enhanced. Similarly, with $R > R_0$, the pressure and temperature decrease. The opacity also decreases, reducing the pressure even further. The material falls back quickly and overshoots. This oscillation can continue indefinitely, rather than being quenched. These are the conditions that produce a Cepheid variable.

10.2.3 Period–luminosity relation

An important feature of Cepheids is that they provide us with a method of measuring distances. The method involves a *period–luminosity relation* (Fig. 10.6). This relation was discovered by *Henrietta*

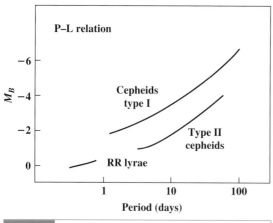

Fig 10.6. Period–luminosity relation for Cepheids.

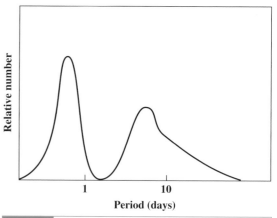

Fig 10.7. Distribution of periods for Cepheids. Note that there are two distinct groupings.

Leavitt, who was studying Cepheids in the Large and Small Magellanic Clouds, two small galaxies near the Milky Way. The advantage of studying Cepheids in either of these galaxies is that the Cepheids are all at the same distance. For the Small Magellanic Cloud it was found that there is a relationship between the period of the Cepheid and its mean apparent magnitude. Since all of the stars are at essentially the same distance, this means that there is a relationship between the period and the mean absolute magnitude.

If we know the exact relationship between period and absolute magnitude, then, when we observe a Cepheid, we can measure its period and convert that into an absolute magnitude. We can always measure the apparent magnitude. The difference $m - M$ is the distance modulus, and gives us the distance. This technique is important because Cepheids are bright enough to be seen in other galaxies, providing us with distances to those galaxies.

However, before we can use the period–luminosity relationship, it must be calibrated. We need independent methods of measuring distances to some Cepheids. This is difficult, since there are none nearby. Statistical studies have been used to achieve this calibration. More recently, the Hiparcos satellite, which was designed to provide more accurate trigonometric parallaxes than had previously been available, made great strides on this problem.

If we plot a histogram indicating how many Cepheids have various periods (Fig. 10.7), we find

an interesting result. The distribution has two peaks in it. This suggests that there are actually two different types of Cepheids. The group with the shorter periods are typical of those studied in the Magellanic Clouds, and are called *classical Cepheids*. The group with longer periods are called *type II Cepheids* or *W Virginis stars* (named after their prototype). Type II Cepheids are found in globular clusters in our galaxy (see Chapter 13 for a discussion of clusters). In general, a type II Cepheid is 1.5 mag fainter than a classical Cepheid of the same period. Also, the period–luminosity relation is slightly different for the two types.

The original calibration of the Cepheid distance scale was carried out for type II Cepheids, since we can study them in our galaxy. However, when we look at a distant galaxy, we can more easily study the brighter classical Cepheids. Therefore, the Cepheids studied in other galaxies were 1.5 mag brighter than assumed. This means that the galaxies are farther away than originally assumed.

Example 10.1 Cepheid distance scale
By how much does the calculated distance to a galaxy change when we realize that we are looking at classical, rather than type II, Cepheids?

SOLUTION
We have already seen that the Cepheids originally studied are 1.5 mag brighter than assumed. This increases the distance modulus, $m - M$, by 1.5 mag. By equation (2.18), this increases the distance by a

factor of $10^{(1.5/5)} = 2$. Thus, these galaxies are twice as far away as originally thought. The difference between the two types of Cepheids was realized in the 1950s, and people talked about the size of the universe doubling.

Another type of variable star that is useful in distance determinations is the *RR Lyrae* variable. These are found in globular clusters and are sometimes called *cluster variables*. They have short periods, generally less than one day. The absolute magnitudes of all RR Lyrae stars are very close to zero. Actually, they fall between zero and unity, and obey a weak period–luminosity relation of their own. The absolute magnitudes were established by using clusters whose distances were known from other techniques. Once the absolute magnitudes are calibrated, we can use RR Lyrae stars as distance indicators.

It should not be surprising that stars with pulsations have period–luminosity relations. For radial oscillations, we expect the period to be roughly equal to $(G\rho)^{-1/2}$, where ρ is the average density of the star. We can understand this qualitatively by noting that a star pulsating under its own gravity is like a large pendulum. The period of a pendulum is $2\pi\,(L/g)^{1/2}$. For a star, $L = R$ and $g = GM/R^2$, so the period is approximately $(GM/R^3)^{-1/2}$, and M/R^3 is approximately the density. Therefore, since the period is related to ρ (which is approximately M/R^3), and the luminosity is related to the radius, the period should be related to the luminosity (see Problem 10.4). $(G\rho)^{-1/2}$ is also approximately the period of a satellite orbiting near the surface of a mass M, or the period of a small mass dropped through a hole in a larger mass. In short, if gravity dominates, $(G\rho)^{-1/2}$ is the time scale.

10.3 | Planetary nebulae

We have already said that the outer layers of a red giant are held together very weakly. Remember, the gravitational force on a mass m in the outer layer is GmM/R^2, where M is the mass of the star and R is its radius. As the star expands, M stays constant, so the pull on the outer layer falls off as $1/R^2$. Since the outer layer is weakly held, it is subject to

being driven away. The actual mechanism for driving material away is still not fully understood. It may involve pressure waves moving radially outward. It may also involve radiation pressure. Photons carry

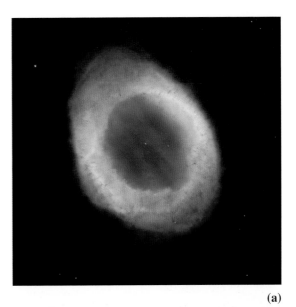

(a)

(b)

Fig 10.8. Images of planetary nebulae. (a) HST image of the Ring Nebula (M57), in the constellation Lyra. It is at a distance of 1 kpc, and is about 0.3 pc across. This image reveals elongated dark clumps of material at the edge of the nebula. (b) The Dumbbell Nebula (M27), in a ground-based image. This is 300 pc away and is 0.5 pc across, in the constellation of Vulpecula.

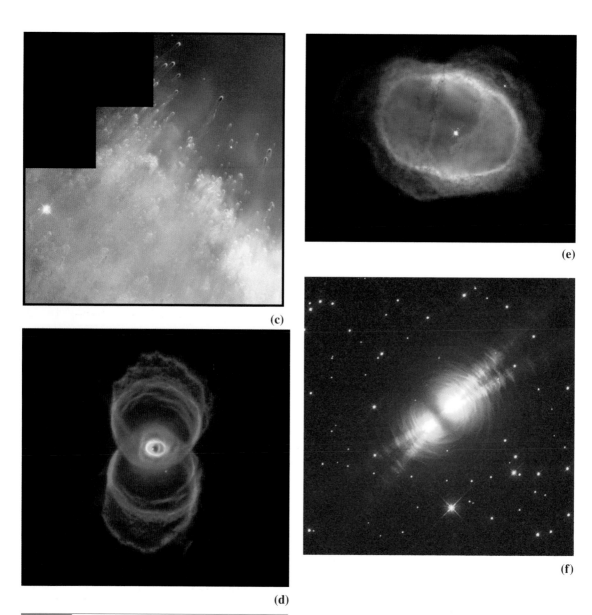

(c)

(e)

(d)

(f)

Fig 10.8. (*Continued*) (c) HST image of the Helix Nebula, which is a little more than 100 pc away. Notice the dark knots with glowing edges. These knots may be the result of faster clumps overtaking the main part of the nebula. (d) The Hourglass Nebula, in an HST image. This is approximately 2.5 kpc away. This picture was from three separate images, taken in the light of ionized nitrogen (represented in red), hydrogen (green) and doubly ionized oxygen (blue). (e) HST image of NGC 3132, at a distance of 0.8 kpc with a diameter of 200 pc. The gases are expanding from the central star at a speed of 15 km/s. (f) HST image of the Egg Nebula (CRL 2688) which is about 0.1 pc across, and is believed to be a star making the transition to a planetary nebula (a proto-planetary nebula). [(a), (c)–(f) STScI/NASA; (b) ESO]

energy and momentum. (Remember, the momentum of a photon of energy E is E/c.) When photons from inside the star strike the gas in the outer layers, and are absorbed, their momentum is also absorbed. By conservation of momentum, the shell must move slightly outward.

We do observe shells that are ejected. They are fuzzy in appearance in small telescopes, just like planets; when originally observed, they were called *planetary nebulae* (Fig. 10.8). (Their name has nothing to do with their properties, but with their appearance as viewed with small telescopes.) From the photograph in Fig. 10.8, we see that

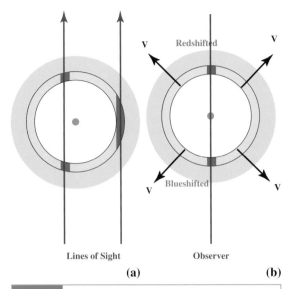

Lines of Sight   Observer

(a)   **(b)**

Fig 10.9. Lines of sight through a planetary nebula. (a) Appearance. The shaded regions represent the places where the lines of sight pass though the shell. The line of sight near the edge passes through more material than that through the center. This is responsible for the ringlike appearance. (b) Doppler shifts. Material on the near side is moving toward the observer, producing a blueshift, and material on the far side is moving away, producing a redshift.

some planetary nebulae have a ringlike appearance. However, they are spherical shells. We see them as rings because our line of sight through the edge of the shell passes through more mate-

rial than the line of sight through the center (Fig. 10.9). Thus, the center appears to be quite faint. When we look at spectral lines in planetary nebulae, we see two Doppler shifts. One line is redshifted and the other is blueshifted. The blueshifted one comes from the part of the shell that is moving towards us, and the redshifted line comes from the part of the shell that is moving away from us. From the Doppler shifts we find that the shells are expanding at velocities of a few tens of kilometers per second.

The physical conditions in planetary nebulae are determined from observations of various spectral lines. Different lines are sensitive to different temperature and density ranges. For example, Fig. 10.10 shows photographs of a planetary nebula, taken at wavelengths of certain emission lines. Different lines reveal different aspects of the nebula structure. Information is also obtained from studies of radio waves emitted by the nebulae.

From these studies, we find that masses of planetary nebulae are of the order of 0.1 $M_\odot$. The temperatures are about 10^4 K. The mass tells us that up to 10% of the stellar mass is returned to the interstellar medium in the ejection of the nebula. This material will be included in the next generation of stars to form out of the interstellar medium. (This is in addition to mass lost through stellar winds.)

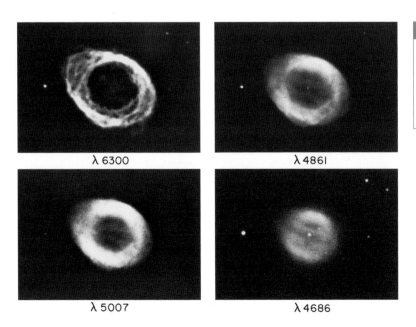

λ 6300

λ 4861

λ 5007

λ 4686

Fig 10.10. The Ring Nebula, shown in Fig. 10.8(a), is shown here in four different colors of light, highlighting gas with different physical conditions. The wavelengths are indicated below each frame. [NOAO/AURA/NSF]

10.4 | White dwarfs

10.4.1 Electron degeneracy

The material left behind after the planetary nebula is ejected is the remnant of the core of the star. It is mostly carbon or oxygen, and its temperature is not high enough for further nuclear fusion to take place. The gas pressure is not high enough to support the star against gravitational collapse. This collapse would continue forever if not for an additional source of pressure when a high enough density is reached. This pressure arises from *electron degeneracy*.

Electron degeneracy arises from the *Pauli exclusion principle*, which states that no two electrons can be in the same state. For two electrons to be in the same state, all of the quantum numbers describing that state must be the same. For example, in an atom, there is a quantum number describing which orbit the electron is in, and another describing how that orbit is oriented (by giving the component of the angular momentum along some axis). In addition, we must take into account the fact that the electron has intrinsic angular momentum, called "spin". The spin can have two opposite orientations. For convenience, we call them "up" and "down" (depending on the direction of the angular momentum vector). An up electron and a down electron in the same energy level are considered to be in different states. However, two is the limit. We can only put two electrons into each energy level.

We can see how this affects the properties of atoms with many electrons. Suppose we build the atom by adding electrons one at a time. The first electron goes into the lowest energy level. The second also goes into the lowest level, but with the opposite spin orientation. The first level is now full. The third electron must go into the next level. After we have added all of the electrons, we can add up the excitation energies of all of the electrons. We will find that the average excitation energy of the electrons in the atom is much greater than kT. This means that electrons are in higher levels than we would guess by just considering the thermal energy available. The problem is that the electron cannot jump into the filled lower states.

We can apply the same idea to a solid, which we can think of as a structure with many energy levels. The electrons fill the lowest energy levels first, but as they fill the electrons end up in higher and higher levels, as shown in Fig. 10.11. The average energy of the electrons is, again, much greater than kT. In fact, the distribution of energies of electrons in a solid at room temperature is negligibly different from that in a solid at absolute zero. We call an electron gas in which all of the electrons are in the lowest energy states allowed by the exclusion principle a *degenerate gas*.

In a degenerate gas, most of the electrons will have energies much greater than they would in an ordinary gas. These high energy electrons also have high momenta. They can therefore exert a pressure considerably in excess of the pressure exerted by an ideal gas at the same temperature. The higher pressure is called *degeneracy pressure*. We have everyday examples of this pressure. For example, it is responsible for the hardness of metals. (Metals consist of a regular arrangement of positive ions, held together by many shared electrons. The exclusion principle results in those shared electrons

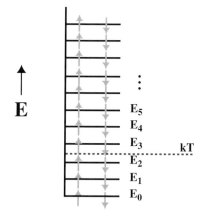

$$E = 2\{E_0 + E_1 + E_2 + E_3 + E_4 + \ldots\} \gg NkT$$

Fig 10.11. Energy levels in a degenerate gas. The energies of the levels are indicated on the right. In each level an upward arrow represents an electron with its spin in one sense, and a downward arrow represents an electron with its spin in the opposite sense. The dashed line indicates the average thermal energy per particle. The total energy is the sum of the energies of the individual electrons.

being in higher states than one would expect just based on temperature.)

We can also describe this pressure in terms of the uncertainty principle. In Chapter 3, we saw that we must think of electrons as having wave properties. We can only talk about the probability of finding an electron in a given place, or moving with a given speed. As a result of this wave property, we cannot simultaneously describe the position and momentum of the electron. If we can determine the momentum with an uncertainty Δp, and the position with an uncertainty Δx, the uncertainty principle tells us that

$$\Delta p \, \Delta x \geq h/2\pi \tag{10.1}$$

For a given Δx, the uncertainty in the momentum is

$$\Delta p \geq h/2\pi \, \Delta x \tag{10.2}$$

When the density becomes very high, we are trying to force the electrons close together. This means that we are trying to confine them to a small Δx. Therefore, the uncertainty principle tells us that the uncertainty in the momentum is large. This means that large momenta are possible. These high momentum electrons are responsible for the increased pressure. Fig. 10.12 shows a container with density n and particles moving with speed v_x in the x-direction. The number of particles hitting a wall per second per unit surface area is $n \, v_x$. The momentum per second per unit surface area delivered to the wall is then $n \, v_x \, p_x$, where p_x is the x-component of the momentum. The momentum per second per unit surface area is just the pressure exerted by the gas on the wall:

$$P = n \, v_x \, p_x \tag{10.3}$$

If we have n_e electrons per unit volume, then there is one electron per box with volume $1/n_e$. The side of such a box is $(1/n_e)^{1/3}$, so the average spacing between electrons is

$$\Delta x = (1/n_e)^{1/3} \tag{10.4}$$

If we say that the average momentum is of the order of the momentum uncertainty, then

$$p_x \cong h/2\pi \, \Delta x$$
$$= (h/2\pi)n_e^{\,1/3} \tag{10.5}$$

The speed of each electron is its momentum divided by its mass, so

$$v_x = p_x/m_e \tag{10.6}$$

This gives a pressure, using equation (10.3), of

$$P \cong (h/2\pi)^2 \, n_e^{5/3}/m_e \tag{10.7}$$

This is just an estimate of the pressure. A more detailed calculation yields a pressure that is a factor of about two higher than that given in equation (10.7).

Equation (10.7) gives the pressure in terms of electron density. We would like to express it in terms of the total mass density ρ. If the density of positive ions with charge Ze is n_Z, then in a neutral gas the density of electrons must be

$$n_e = Zn_Z \tag{10.8}$$

Each positive ion has a mass of Am_p, if we ignore the difference between the proton and neutron masses. The total density of the gas is then

$$\rho = Am_p \, n_Z + m_e n_e$$
$$\cong Am_p \, n_Z \tag{10.9}$$

In going to the second line we have ignored the mass of the electrons relative to the mass of the nucleons. Using equations (10.8) and (10.9), the electron density is related to the total density by

$$n_e = (Z/A) \, (\rho/m_p) \tag{10.10}$$

Substituting this into equation (10.7) and adding the factor of two to account for the difference

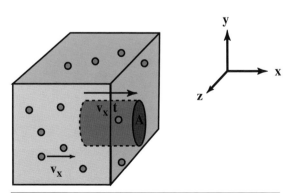

Fig 10.12. Pressure in a degenerate gas. We consider the force on the section of area A of the right-hand wall of the box, due to the x-component of the motions of the particles.

between our estimate and the detailed calculation, we have

$$P = 2(h/2\pi)^2 (Z/A)^{5/3} (\rho/m_p)^{5/3}/m_e \quad (10.11)$$

Note that, in a degenerate gas, the pressure depends on the density, but not on the temperature. We have already seen that this point is important in deciding whether the triple-alpha process will take place in a controlled way (normal gas) or in a flash (degenerate gas).

10.4.2 Properties of white dwarfs

A star supported by electron degeneracy pressure will be quite small, since it must collapse to a high density before the degeneracy pressure is high enough to stop the collapse. These objects are quite hot, being the remnant of the core of a star. These objects are the stars that appear on the HR diagram as *white dwarfs*.

Example 10.2 White dwarf density
Estimate the density of a white dwarf if it has a solar mass packed into a sphere with approximately $10^{-2}\,R_\odot$ (approximately the size of the Earth) as found in Section 3.5.

SOLUTION
We find the density by dividing the mass by the volume:

$$\rho = \frac{(2 \times 10^{33}\,\text{g})}{(4\pi/3)(7 \times 10^8\,\text{cm})^3}$$

$$= 1 \times 10^6\,\text{g/cm}^3$$

(Remember, the density of water is only 1 g/cm³, so a white dwarf is very dense.)

Example 10.3 White dwarf degeneracy pressure
For a white dwarf of density 1.0×10^6 g/cm³, and $Z/A = 0.5$, estimate the degeneracy pressure and compare it with the thermal pressure of a gas at a temperature of 1.0×10^7 K.

SOLUTION
We find the pressure from equation (10.11):

$$P = 2\left[\frac{(1.05 \times 10^{-27}\,\text{erg s})^2}{9.11 \times 10^{-28}\,\text{g}}\right]$$
$$\times \left[\frac{(0.5)(1.0 \times 10^6\,\text{g/cm}^3)}{1.67 \times 10^{-24}\,\text{g}}\right]^{5/3}$$

$$= 3.2 \times 10^{22}\,\text{dyn/cm}^2$$

For an ideal gas, the pressure is given by

$$P = (n_e + n_Z)\,kT$$

where $n_e + n_Z$ represents the total density. However, each atom of atomic number Z contributes Z electrons, so

$$n_e = Z n_Z$$

We therefore have

$$P = (Z + 1)n_Z\,kT$$

$$\cong Z\,n_Z\,kT$$

We can now relate this to the density ρ. If A is the mass number of the nuclei, then (ignoring the difference between proton and neutron masses)

$$\rho = A\,n_Z\,m_p$$

This gives

$$P = (Z/A)(\rho/m_p)kT$$

$$= (0.5)\left(\frac{1.0 \times 10^6\,\text{g/cm}}{1.67 \times 10^{-24}\,\text{g}}\right)(1.38 \times 10^{-16}\,\text{erg/K})$$
$$\times (1.0 \times 10^7\,\text{K})$$

$$= 4.1 \times 10^{20}\,\text{dyn/cm}^2$$

The degeneracy pressure is a factor of about 100 higher than the normal thermal pressure, even at this very high temperature.

We can also use the expression for degeneracy pressure (equation 10.11) to relate the mass and radius of a white dwarf. We saw in Example 9.4 that we could use hydrostatic equilibrium to approximate the central pressure by

$$P = GM^2/R^4 \quad (10.12)$$

If we put this into equation (10.11), and substitute $M/4R^3$ for the density, we find

$$G\frac{M^2}{R^4} = 2\left(\frac{h^2}{4\pi m_e}\right)\left(\frac{Z}{Am_p}\right)^{5/3}\left(\frac{M}{4R^3}\right)^{5/3} \quad (10.13)$$

Rearranging gives

$$GM^{1/3}R = 2\left(\frac{h^2}{4\pi m_e}\right)\left(\frac{Z}{4Am_p}\right)^{5/3} \quad (10.14)$$

The right-hand side is all constants, so for a given R a mass can be calculated (Problem 10.12).

A degenerate gas has a very low opacity to radiation. For a photon to be absorbed, an electron would have to jump to a higher energy state.

Path of White Dwarf

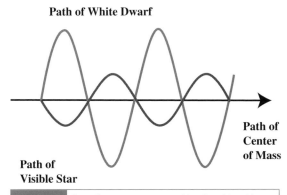

Path of Center of Mass

Path of Visible Star

Fig 10.13. Even if we cannot see a white dwarf in a binary system, we can detect its presence. The visible star will appear to wobble as it and the white dwarf orbit their common center of mass.

However, such a transition would have to be an already empty state, and may require more energy than is carried by an optical photon. In addition, degenerate gases are good heat conductors, which explains why metals are good heat conductors. The low opacity and high thermal conductivity mean that a degenerate gas cannot support a large temperature difference. The internal temperature of a white dwarf is approximately constant across that star, at about 10^7 K. The outermost 1% is not degenerate, and it is in that thin layer that the temperature falls from 10^7 K to roughly the 10^4 K indicated by its spectral type. These conditions make a white dwarf very different from a normal star. In addition, Zeeman splitting measurements suggest very strong magnetic fields, about 10^7 gauss in some cases.

What happens to a star after it becomes a white dwarf? As it radiates it must get cooler. This is because it is giving off energy, but has no source of new energy. The degeneracy pressure does not depend on the temperature, so the star will maintain its size as it cools. Eventually, it will be too cool to see. It will take tens of billions of years for a star that is now visible as a white dwarf to become that cool. We can understand the long lifetime when we realize that a white dwarf radiates like a 10^4 K blackbody, but has more energy than a thermal reservoir at 10^7 K (see Problem 10.13).

Even at their current temperatures, their small sizes make white dwarfs difficult to see. We some-

times detect their presence in binary systems by measuring their gravitational influence on a star that we can see. We deduce the mass of the white dwarf from its influence on the visible companion's orbit (Fig. 10.13).

10.4.3 Relativistic effects

The treatment of degeneracy pressure must be modified by considerations introduced in the special theory of relativity (Chapter 7). This was first realized by *S. Chandrasekhar* (who shared the 1983 Nobel Prize in physics for his work on stellar structure). Chandrasekhar found that these corrections reduce the degeneracy pressure. This provides an upper limit to the mass that can be supported by electron degeneracy pressure.

The modification arises from the fact that electrons cannot travel faster than the speed of light. In using equation (10.5), we can still say that p_x is $(h/2\pi)(n_e)^{1/3}$. However, we can no longer say that $v_x = p_x/m_e$. We have to use the correct relativistic expression, as discussed in Chapter 7. To find the maximum degeneracy pressure, we take $v_x = c$. This gives

$$P_{max} = (h/2\pi)\, c\, (n_e)^{4/3} \tag{10.15}$$

A more detailed calculation gives approximately 0.8 times this. Using this, the expression analogous to equation (10.11) is

$$P = (0.8)(h/2\pi)\, c\, (Z/A)^{4/3}\, (\rho/m_p)^{4/3} \tag{10.16}$$

If we use the same assumptions to find the mass–radius relation (equation 10.14), we find that the radius drops out, and we simply have an expression for the mass (Problem 10.14):

$$M = (0.05)(hc/2\pi G)^{3/2}(Z/A\, m_p)^2 \tag{10.17}$$

This mass corresponds to the maximum pressure, so it is the maximum mass that can be supported by electron degeneracy pressure. A more accurate calculation, which takes into account variations in pressure and density with distance from the center of the star, gives a mass that is a factor of about 60 higher. The resulting mass, called the *Chandrasekhar limit*, is 1.44 $M_\odot$. A star whose nuclear processes have stopped, and whose mass is greater than 1.44 $M_\odot$, will continue to collapse beyond the white dwarf phase. The fate of such stars will be discussed in Chapter 11.

Chapter summary

In this chapter we saw what happens to stars that use up their supply of hydrogen fuel in their cores. Low mass stars evolve into red giants. Higher mass stars evolve into red supergiants.

We saw how some stars at this stage are unstable to pulsations and become Cepheid variables. The Cepheids are particularly important because of a period–luminosity relation. This relation allows us to use Cepheids as important distance indicators.

In the red giant or supergiant phase, the outer layers are loosely bound to the star. Mass loss takes place. One form of mass loss is ejection of a planetary nebula.

We saw how the remnant of a low mass star is a white dwarf, a star supported by electron degeneracy pressure. There is a limit, the Chandrasekhar limit, on the mass that can be supported by electron degeneracy pressure. Best estimates of this limit are 1.44 $M_\odot$.

Questions

10.1. In which aspects of post-main sequence evolution does the mass enter into consideration? How are low mass stars really different from high mass stars?

10.2. Why is there hydrogen left in a shell around the core after it has been used up in the core?

10.3. For the surface oscillations of a Cepheid, show how the energy goes from kinetic to potential throughout the stellar oscillation cycle, as in the analogy of a swing.

10.4. How does electron degeneracy pressure support a white dwarf?

10.5. (a) If we try to confine an electron to within an atom, $\Delta x \cong 1$ Å, what is Δp, the uncertainty in its momentum? (b) What velocity would the electron need to have this momentum?

10.6. List the distance measurement techniques that we have encountered so far in this book, and estimate the distance range over which each is useful.

Problems

10.1. Suppose we observe a classical Cepheid with a period of 10 days and an apparent magnitude of +6. How far away is it?

10.2. Suppose we observe a classical Cepheid with a period of 8 days and $m = +7$. What is the period of a type II Cepheid at the same distance with the same apparent magnitude?

*10.3. Suppose a classical Cepheid has a period $P = 10$ days, and assume that the surface oscillates sinusoidally with a maximum amplitude $\Delta R_{max} = 10^3$ km. (a) Write an expression for ΔR vs. t, assuming that $\Delta R = 0$ at $t = 0$, and that the star is expanding at $t = 0$. (b) Find an expression for the speed of the surface, $v(t)$. (c) Find the wavelength observed for the Hα line vs. t.

*10.4. Use the fact that $P \propto \rho^{-1/2}$, $\rho \propto M/R^3$ and $L \propto R^2 T^4$ to derive a relationship between period and luminosity.

*10.5. The momentum of a photon is E/c. (a) Calculate the momentum per second delivered to the outer layers of a $10^2\,L_\odot$ star if all the photons are absorbed in that layer. (b) How does the force on the layer compare with the gravitational force on the layer if the layer has a radius $R = 100\,R_\odot$ and a mass $M = 0.1\,M_\odot$ and the rest of the star has a mass of 1 $M_\odot$.

10.6. (a) What is the kinetic energy in a 0.1 $M_\odot$ planetary nebula, expanding at 10^3 km/s? (b) Compare that with the gravitational potential energy of the shell when it was at $R = 100\,R_\odot$, assuming that there is 1 $M_\odot$ of material left behind.

10.7. Suppose a planetary nebula is a spherical shell whose thickness is 10% of its radius. Compare the length of the longest line of sight with that through the center, and relate your answer to the appearance of the planetary nebula.

10.8. Suppose a planetary nebula expands at a speed v, determined from Doppler shift observations. The angular size θ is observed to be increasing at the rate $d\theta/dt$. In terms of these quantities, find an expression for the distance to the planetary nebula.

10.9. Suppose we observe the Hα line in a planetary nebula. The maximum wavelength is 656.32 nm, and the minimum wavelength is 656.24 nm. How fast is the nebula expanding?

10.10. Calculate the escape speed from a 1 $M_\odot$ white dwarf. Compare it with that of a main sequence star of the same mass.

10.11. (a) At white dwarf densities, what is the average separation between electrons? (b) What is their momentum uncertainty Δp? (c) What is the velocity corresponding to that momentum?

10.12. (a) Use the mass–radius relationship for white dwarfs to calculate the radius of a 1 $M_\odot$ white dwarf. (b) Use this result to rewrite equation (10.14) in a form that gives R in kilometers when M is expressed in solar masses.

10.13. (a) What is the thermal energy stored in a 10^7 K white dwarf of 1 $M_\odot$? (b) Assuming that it radiates like a 10^4 K blackbody, estimate its lifetime as a luminous object.

10.14. Derive equation (10.17).

Computer problems

10.1. Suppose a planetary nebula is a spherical shell whose thickness is 10% of its radius. Find the length of a line of sight through the material as a function of position across the nebula.

10.2. Use the mass–radius relationship for white dwarfs to calculate the radii of white dwarfs with masses ranging from 0.1 $M_\odot$ to the Chandrasekhar limit, and make a graph of size vs. mass over that range.

10.3. Tabulate the radii of white dwarfs with masses corresponding to the mid-range of each spectral type (O5, B5, . . .).

Chapter 11

The death of high mass stars

In Chapter 10 we saw how stars evolve to the red giant or red supergiant stages, and how low mass stars (less than 5 $M_\odot$) lose enough mass to leave behind a white dwarf as the final stellar remnant. We also saw that electron degeneracy pressure can only support a 1.44 $M_\odot$ remnant. In this chapter we will see what happens to higher mass stars.

It is important to remember that stars lose mass as they evolve. This mass loss can be through winds, or the ejection of planetary nebulae. (In the next chapter, we will see that stars in close binary systems can transfer mass to a companion.) Though we only have estimates for the total amount of mass loss, it seems likely that massive stars can lose more than half of their mass by the time they pass through the red supergiant phase. A star's evolution will depend on how much mass it starts with, and how much mass it loses along the way.

11.1 | Supernovae

11.1.1 Core evolution of high mass stars

In the core of a high mass star the buildup of heavier elements continues. If we look at nuclear binding energies (Fig. 9.3) we see that the isotope of iron ^{56}Fe has the highest binding energy per nucleon. This makes it the most stable nucleus. This means that any reaction involving ^{56}Fe, be it fission or fusion, requires an input of energy. When all of the mass of the core of the star is converted to ^{56}Fe (and other stable elements, such as nickel), nuclear reactions in the core will stop.

At this stage, the core will start to cool and the thermal pressure will not be sufficient to support the core. As long as the mass of iron in the core is less than the Chandrasekhar limit, the core can be supported by electron degeneracy pressure. However, once the core goes beyond that limit, there is nothing to support it, and it collapses. In the collapse, some energy, previously in the form of gravitational potential energy, is liberated. Since that energy is available, the ^{56}Fe can react by using up the energy. This means that the core does not get any hotter. It continues to collapse. A runaway situation develops in which the iron and nickel consume liberated energy. As the iron is destroyed, protons are liberated from nuclei. The electrons in the star can combine with these protons to form neutrons and neutrinos. This reaction can be written

$$e^- + p \rightarrow n + \nu \tag{11.1}$$

The core is driven to a very dense state in a short time, about one second. What happens next is not completely understood, but the collapse results in an explosion in which most of the mass of the star is blown away. The neutrons created in reaction (11.1) probably play a role in this. They also obey the exclusion principle, and exert a degeneracy pressure (the details of which we will discuss in the next section). This pressure can stop the collapse and cause the material to bounce back. In addition, so many neutrinos are created, and the material is so dense, that a sufficient number of neutrinos interact with the matter forcing the material outward.

Such an exploding star is called a *supernova*. This type of supernova is actually called a *type II supernova*. Another type of supernova, type I, seems to be associated with older objects in our galaxy. (The mechanism for type I supernovae probably involves white dwarfs in close binary systems,

Fig 11.1. A supernova in another galaxy can be almost as bright as the whole galaxy. This shows a supernova in the spiral galaxy NGC4603. [STScI/NASA]

and is discussed in Chapter 12.) During the explosion, nuclear reactions take place very rapidly, and elements much heavier than iron are created. This material is then spread out into interstellar space, along with the results of the normal nucleosynthesis during the main sequence life of the star. This enriched material is then incorporated into the next generation of stars.

The light from a supernova explosion can exceed that of an entire galaxy (Fig. 11.1). The energy output in a type II supernova is about 10^{53} erg. About 1% of this shows up as kinetic energy of the shell, and 0.1% as light. (Most of the energy is in the escaping neutrinos.) After a rapid increase in brightness, the supernova fades gradually, on a time scale of several months (Fig. 11.2).

11.1.2 Supernova remnants

The material thrown out in a supernova explosion is called a *supernova remnant*. It contains most of the material that was once the star. In young supernova remnants we can actually see the expansion of the ejected material. These remnants are important because they spread the products of nucleosynthesis in stars throughout the interstellar medium. There, this material enriched in "metals" will be incorporated into the next generation of stars. This explains why stars that formed relatively recently in the history of our galaxy have a higher metal abundance than the older stars. In the later stages of a supernova remnant's expansion, we still see a glowing shell, like those in Fig. 11.3. These shells also serve to stir up

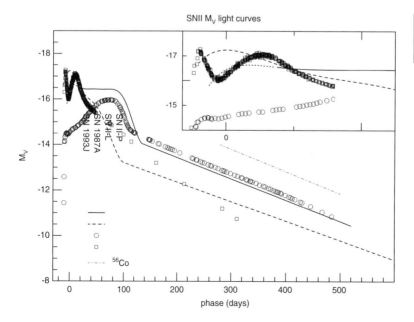

Fig 11.2. Light curves for type II supernovae. [Craig Wheeler, University of Texas, Austin]

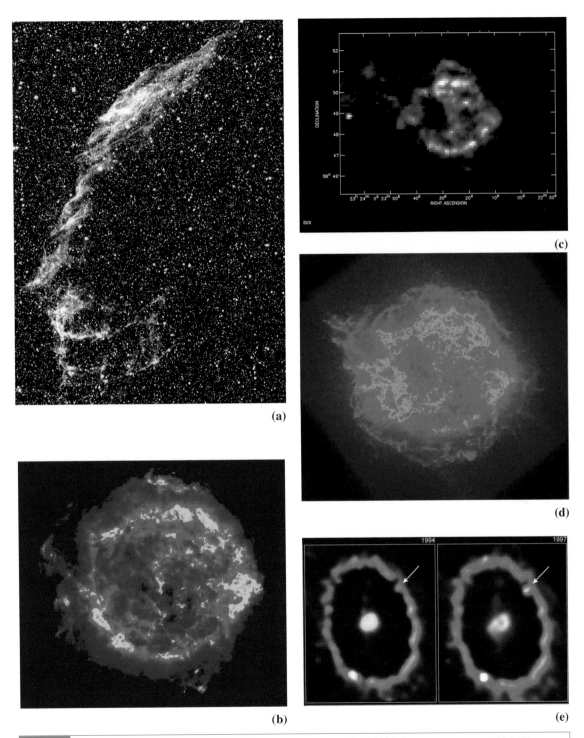

Fig 11.3. Views of supernova remnants. (a) An optical image of a portion of the Veil Nebula supernova remnant. (b) A (false color) radio image (6 cm wavelength) of the source known as Cas A (the brightest radio source in Cassiopeia), taken with the VLA, providing an angular resolution of 0.1 arc sec. (c) A far infrared image of Cas A, taken with the ISO satellite. (d) An X-ray image of Cas A, made with the Chandra Observatory. (e) HST image of the region of SN1987A in the LMC. The small bright ring shows the interaction of the expanding supernova remnant with the surrounding medium. [(a) NOAO/AURA/NSF; (b) NRAO/AUI/NSF; (c) ESA/ISO,ISOCAM/CEA and P. Lagage *et al.* (d) NASA; (e) STScI/NASA]

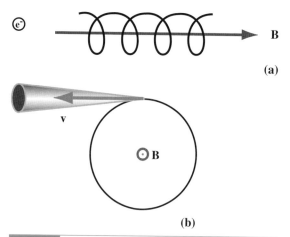

(a)

Fig 11.4. Synchrotron radiation. (a) Electrons spiral around magnetic field lines. (b) In their circular motion, they are constantly accelerating, and therefore give off radiation. At any instant, most of the radiation is emitted in a small cone centered on the velocity of the electron. The beaming of the radiation is a relativistic effect.

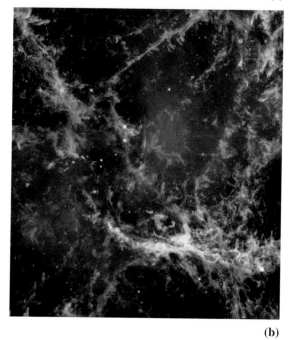

(b)

Fig 11.5. The Crab Nebula, a young supernova remnant. (a) In normal light from the ground. (b) From HST showing great detail. For comparison, an X-ray image is shown in Fig. 4.33(c). [(a) NOAO/AURA/NSF; (b) STScI/NASA]

the interstellar medium. This point will be discussed further in Chapter 15.

Supernova remnants generally have magnetic fields that are strong by interstellar standards, and a supply of high energy electrons. In Chapter 6, we discussed the motion of charged particles in a magnetic field. The component of the electron's velocity along the field lines is not changed, since the force is perpendicular to the field direction and to the electron's velocity. The electron traces out a helix as it circles around the field lines, and drifts along the direction of the field, as shown in Fig. 11.4. Since the velocity of the electron is changing (in direction), the electron must be accelerating. Classical electromagnetic theory tells us that accelerating charges must give off radiation. This radiation is called *synchrotron radiation*.

One of the best studied supernova remnants is the *Crab Nebula* in the constellation of Taurus. (This supernova was observed as a bright object in 1054 AD.) Some optical images are shown in Fig. 11.5. An X-ray image is shown in Fig. 4.33(c). All of the radiation is synchrotron radiation.

Synchrotron radiation has a number of distinguishing characteristics. One is that the radiation is polarized. If we have a radio telescope receiver that detects only one direction of linear polariza-tion, and we change the orientation of the detector, we see an intensity of radiation that varies with the angle of the detector. This means that the electric field vector of the incoming radiation is

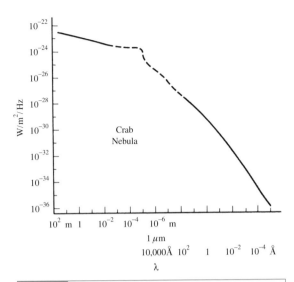

Fig 11.6. Spectrum of the Crab Nebula, from radio to gamma-rays.

preferentially aligned along some direction. Even the visible light is polarized.

The spectrum of synchrotron radiation is also quite distinctive. Most of the radiation is given off at long wavelengths. The intensity of radiation falls off as the wavelength raised to some power. This is called a *power law spectrum*. Fig. 11.6 shows the spectrum of the Crab Nebula. The intensity of radiation is much greater in the radio part of the spectrum than in other parts of the spectrum, but the source is so strong that we are able to detect synchrotron radiation in all parts of the spectrum, even at gamma-ray energies. So much energy is given off in synchrotron radiation that the high energy electrons should be losing their energy over a very short time. However, they are still radiating strongly some 1000 years after the supernova explosion. Until recently, it was a problem for astronomers to explain where the energy comes from to maintain the high energy electrons in the Crab Nebula.

11.2 | Neutron stars

In this section we look at what happens to the core that is left behind in the supernova explosion. The core is compressed so that normal gas pressure cannot support it. We have already seen

that if the mass is more than 1.44 $M_\odot$ electron degeneracy pressure cannot support it. The collapse of the core continues beyond even the high densities associated with a white dwarf. As the density increases, electrons and protons are forced together to make neutrons. The resulting object is called a *neutron star*.

11.2.1 Neutron degeneracy pressure

Neutrons have spin properties similar to those of electrons. They therefore also obey the Pauli exclusion principle. (We have already seen how this affects the energy levels of neutrons in a nucleus, in Chapter 9.) Neutrons are therefore capable of exerting a degeneracy pressure if the density is high enough. We can estimate the neutron degeneracy pressure as we did the electron degeneracy pressure in the preceding chapter. The result corresponding to equation (10.7), including the factor of two, is

$$P \cong 2(h/2\pi)^2 n_n^{5/3}/m_n \qquad (11.2a)$$

Since the star is all neutrons, $\rho = n_n m_n$, so

$$P \cong 2(h/2\pi)^2 \rho^{5/3}/m_n^{8/3} \qquad (11.2b)$$

In comparing this with equation (10.7), we see that at a given density electron degeneracy pressure will be greater by a factor of m_n/m_e, or about 2000. However, in a neutron star there are no free electrons, so neutron degeneracy pressure is all we have to support the star. Because the density of a neutron star is so much higher than that of a white dwarf, the neutron degeneracy pressure in a neutron star is greater than the electron degeneracy pressure in a white dwarf. The neutron degeneracy pressure will halt the collapse of the core.

Let's consider some properties of neutron stars. We can estimate the radius from a mass–radius relationship, like that for white dwarfs (Problem 11.3). Neutron stars will be smaller than white dwarfs by a factor of about m_e/m_n. We find the radius to be about 15 km for a mass of 1 $M_\odot$. This means that a neutron star concentrates more than a solar mass in a sphere smaller than the island of Manhattan.

Example 11.1 Density of a neutron star
Estimate the density of a neutron star and compare it with that of a neutron. Take the mass of the star to be 1.4 $M_\odot$.

SOLUTION

The density is the mass divided by the volume:

$$\rho = \frac{(1.4)(2 \times 10^{33}\,\text{g})}{(4\pi/3)(1.5 \times 10^6\,\text{cm})^3}$$

$$= 2 \times 10^{14}\,\text{g/cm}^3$$

The density of a neutron is

$$\rho_n = \frac{(1.7 \times 10^{-24}\,\text{g})}{(4\pi/3)(1.0 \times 10^{-13}\,\text{cm})^3}$$

$$= 4 \times 10^{14}\,\text{g/cm}^3$$

So we see that the density of a neutron star is very close to that of a neutron. This means that the neutrons in a neutron star must be packed very close together, with very little empty space.

Example 11.2 Neutron degeneracy pressure
Compare the neutron degeneracy pressure in a neutron star with the electron degeneracy pressure in a white dwarf.

SOLUTION

Using equations (11.2b) and (10.11),

$$\frac{P_{ns}}{P_{wd}} = \left(\frac{\rho_{ns}}{\rho_{wd}}\right)^{5/3}\left(\frac{m_e}{m_n}\right)$$

$$= \left(\frac{2 \times 10^{14}}{2 \times 10^6}\right)^{5/3}\left(\frac{1}{2000}\right)$$

$$= 1 \times 10^{10}$$

Example 11.3 Acceleration of gravity on a neutron star
For the star in the above example, find the acceleration of gravity at the surface

SOLUTION

$$g = \frac{GM}{R^2}$$

$$= \frac{(6.67 \times 10^{-8}\,\text{dyn cm}^2/\text{g}^2)(1.4)(2 \times 10^{33}\text{g})}{(1.5 \times 10^6\,\text{cm})^2}$$

$$= 8.3 \times 10^{13}\,\text{cm/s}^2$$

This is almost 10^{11} times the acceleration of gravity at the surface of the Earth! (With such strong gravitational fields, we should really use general relativity to calculate particle motions.) You can calculate your weight on the surface of a neutron star.

Another interesting effect comes from the fact that g changes very quickly with radius R. Differentiating the expression for g gives

$$dg/dr = -2GM/R^3$$

If we use the numbers in the above example, we find $dg/dR = -1.1 \times 10^8$ (cm/s^2)/cm. This is equal to a change of 10^5 times the acceleration of gravity on the Earth per centimeter. If you were floating near the surface, your feet would be pulled in with a much greater acceleration than your head. Your body would be pulled apart by these tidal forces. By tidal forces, we mean effects that depend on the difference between forces on opposite sides of an object. Some astrophysicists have jokingly noted that if an astronaut visits a neutron star, it should be in a prone position to minimize the tidal effects.

The large acceleration of gravity also has another interesting effect. The equation of hydrostatic equilibrium (equation 9.44) tells us that the rate at which the pressure in the atmosphere of a neutron star falls off, dP/dR, is proportional to g. The atmospheric pressure on a neutron star thus falls off very quickly. This leads to an atmosphere that is only about 1 cm thick. (The thin atmosphere is another reason for an astronaut to stay in a prone position.)

11.2.2 Rotation of neutron stars

In the process of the collapse of a core to become a neutron star, any original rotation of the core will be amplified. If the angular momentum of the core is conserved, the core must rotate faster as it becomes smaller. Since the core shrinks by a large amount, the rotation speed is increased by a large amount.

The angular momentum is given by

$$J = I\omega \tag{11.3}$$

where I is the rotational inertia and ω is the angular speed. If we put in the rotational inertia for a uniform sphere, we find

$$J = (2/5)MR^2\omega \tag{11.4}$$

To get a feel for how fast a neutron star can rotate, let's assume that the angular momentum of a neutron star is equal to that of the Sun. (This is probably a conservative estimate since the Sun is not rotating very rapidly compared with other stars we see.) Using equation (11.4), we have

$$\omega_\odot R_\odot{}^2 = \omega R^2 \qquad (11.5)$$

Solving for $(\omega/\omega_\odot)$, we have

$$(\omega/\omega_\odot) = (R_\odot/R)^2 \qquad (11.6)$$

$$= [(7 \times 10^{10})/(1.5 \times 10^6)]^2$$

$$= 2 \times 10^9$$

The rotation period of the Sun is about 30 days. The period of a neutron star is therefore

$$P_{ns} = (2.6 \times 10^6\ \text{s})/(2 \times 10^9)$$

$$= 1.3 \times 10^{-3}\ \text{s}$$

In addition to its other extreme properties, a neutron star might be rotating 1000 times per second!

11.2.3 Magnetic fields of neutron stars

We might also expect a neutron star to have a strong magnetic field. This is a consequence of Faraday's law, which can be written as

$$\oint \mathbf{E} \cdot \mathbf{dl} = - \, d\Phi_B/dt$$

In this expression Φ_B is the magnetic flux through a surface. (For a small surface element, the flux is the dot product of the magnetic field and a vector whose magnitude is the surface area and whose direction is perpendicular to the surface.) The integral on the left is performed around a closed path that forms the boundary of the surface. E is the electric field induced by the change in flux. The integral on the left is the induced electromotive force (emf) around the closed path. If there is an induced emf around the path, and if the material has some conductivity, currents will flow to oppose the change in flux. When the flux through the surface stays constant, we say that the flux is *frozen* into the material.

The flux is proportional to the strength of the magnetic field and the surface area. This means that the quantity

$$BR^2 = \text{constant} \qquad (11.7)$$

The magnetic field should therefore be proportional to $1/R^2$, just as the rotation rate is. If the neutron star started off with a solar magnetic field, it will end up with a magnetic field of about

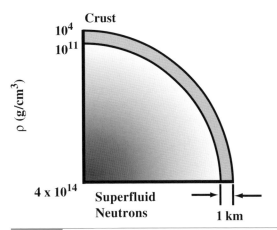

Fig 11.7. Structure of a neutron star.

2×10^9 times that of the Sun. It should be noted that, even though flux conservation arguments give us what might be an order of magnitude for B, there is some disagreement over the actual mechanism for the buildup of large fields in neutron stars. Some theories involve dynamo mechanisms, in which rotation plays an important role.

Most of the interior of a neutron star is thought to be a fluid. Because of certain quantum mechanical properties, this fluid can flow with no frictional resistance or viscosity. A fluid with no viscosity is called a *superfluid*. (This is analogous to a superconductor which allows the flow of electricity with no resistance.) Actually, the existence of vortices (such as whirlpools) in the rapidly rotating fluid leads to some viscosity. This is what couples the fluid to the outside layers. Outside the fluid is a solid crust made up of heavy elements, such as iron. The crust is probably less than 1 km thick (Fig. 11.7). We have already seen that the atmosphere is even thinner.

The possibility of the existence of neutron stars was first realized in the 1930s. However, they were objects that existed in theory only. It was felt that their small size would make them hard to see. In 1967 an accidental discovery in radio astronomy revived interest in neutron stars.

11.3 | Pulsars

11.3.1 Discovery

In 1967, *Antony Hewish* and his graduate student, *Jocelyn Bell Burnell*, were looking for the twinkling

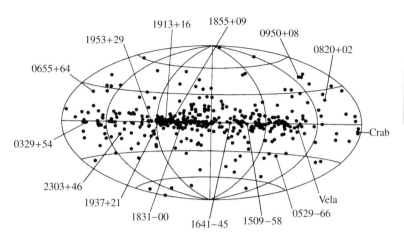

Fig 11.8. Distribution of pulsars on a galactic coordinate system. Their tendency to concentrate near the galactic plane indicates that they are within our own galaxy. We will talk more about galactic structure in Chapter 16.

of radio sources. This twinkling, called *scintillation*, is analogous to the twinkling of stars. However, it is not caused by the Earth's atmosphere. It is caused by charged particles in interplanetary space. The observations required very good time resolution. For most radio observations, the sources are so weak that we have to observe for a long time to detect a signal. There is generally no call to see how fast things are changing. However, Hewish and Burnell were looking for changes on time scales that were much shorter than for typical radio observations. This good time resolution is what allowed them to make an unusual accidental discovery. Hewish received the Nobel Prize in physics for this work.

Hewish and Burnell found a rapidly varying source. When they looked at it in more detail,

they noted that the signal was a string of pulses, with a very regular separation. By counting a large number of pulses, they determined the period to be 1.337 301 1 s. A few other *pulsars* were found, suggesting a general phenomenon. Since then, hundreds of pulsars have been discovered. Most of them are close to the plane of our galaxy, as shown in Fig. 11.8. This suggests that they are in the galaxy. If they were extragalactic, their distribution would not correlate with anything within our galaxy.

We see some properties of pulsar signals in Fig. 11.9. Note that each pulse is not exactly the same as the previous one. The period may even vary slightly from one pulse to the next. However, if we take the average of 100 pulses and compare it with the average of the next 100 pulses we find

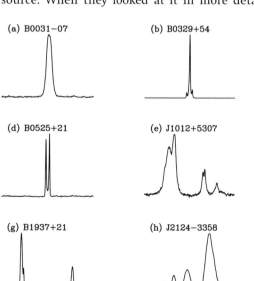

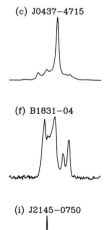

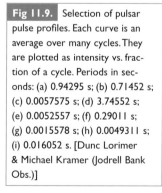

(a) B0031−07
(b) B0329+54
(c) J0437−4715
(d) B0525+21
(e) J1012+5307
(f) B1831−04
(g) B1937+21
(h) J2124−3358
(i) J2145−0750

Fig 11.9. Selection of pulsar pulse profiles. Each curve is an average over many cycles. They are plotted as intensity vs. fraction of a cycle. Periods in seconds: (a) 0.94295 s; (b) 0.71452 s; (c) 0.0057575 s; (d) 3.74552 s; (e) 0.0052557 s; (f) 0.29011 s; (g) 0.0015578 s; (h) 0.0049311 s; (i) 0.016052 s. [Dunc Lorimer & Michael Kramer (Jodrell Bank Obs.)]

that the two averages agree quite well. We can see that the pulsar is actually "off" most of the time. The pulse is on for only a small fraction of the period. We define the *duty cycle* as the fraction of the pulse period in which the pulse is actually on. For most pulsars the duty cycle is about 5%. This means that the peak brightness is about 20 times the average brightness (Problem 11.9). Notice that there may also be a small pulse somewhere in the middle of the cycle. This is called an *interpulse*, and fewer than 1% of pulsars have one.

11.3.2 What are pulsars?

When the discovery of pulsars was announced, astronomers immediately attacked the problem of what they are. Initial guesses even included the possibility that they were signals sent by intelligent civilizations (called LGM theories, for little green men). However, as more pulsars were discovered in different parts of the galaxy, it seemed that a more natural explanation was needed. Efforts were concentrated on trying to find the clock mechanism. Three basic mechanisms were tried: (1) pulsation, (2) orbital motion, and (3) rotation. Three different types of objects were considered: (1) normal stars, (2) white dwarfs, and (3) neutron stars.

Pulsation of stars is reasonably well understood. In Chapter 10, we saw that, for radial oscillations, the period is roughly proportional to $(G\rho)^{-1/2}$, where ρ is the average density of the star. Pulsations of normal stars have periods of hours to days, and would not explain pulsars. For a period of 0.1s, a density of about 10^9 g/cm^3 is required. This is about 10^3 times the density of a white dwarf, but only 10^{-5} of the density of a neutron star. No stellar object has a density even close to the right range. Therefore, the radial pulsations in stars are ruled out.

We next consider orbital motion. The period and radius of an orbit are related by (equation 5.20)

$$4\pi^2 R^3/G = (m_1 + m_2)P^2 \qquad (11.8)$$

Solving for the radius gives

$$R = [(G/4\pi^2)(m_1 + m_2)P^2]^{1/3} \qquad (11.9)$$

For a period of 1 s, the radius is 2000 km; for a period of 0.1 s, the radius is 400 km; for a period of 0.01 s, the radius is 100 km. For the range of pulsar periods, the orbital radii would have to be smaller than a normal star, or even a white dwarf. If we have orbiting objects, we would need two neutron stars.

There is a problem with orbiting systems. The energy of the orbit is (by equation 5.41)

$$E = -Gm_1m_2/2R$$

As the pulsar gives off radiation, to conserve energy, the orbital energy must decrease. This means that E becomes more negative, or the absolute value of E becomes larger. For this to happen, R must become smaller. As R becomes smaller the period must decrease, since $P \sim R^{3/2}$ (equation 11.8). However, we observe pulsar periods to increase, rather than decrease (see below). There is an additional problem with orbiting neutron stars. We saw in Chapter 8 that the general theory of relativity predicts that such a system would lose energy very quickly by giving off gravitational radiation. We would be able to detect this as a significant decrease in the orbital period. (We will discuss a binary pulsar system, in which this effect has been studied, in the next chapter.)

This leaves rotation as the mechanism for producing the regularity in the pulses. For any rotating star the rotation must not be so fast that objects on the surface lose contact with the surface. The gravitational force must be greater than the force required to keep a point moving in a circular path. If not, an object on the surface will go into orbit just above the surface. This gives the same constraint on the size of an object as equation (11.9), with the sum of the masses replaced by the mass of the single rotating star. This again rules out a rotating normal star and a rotating white dwarf. (White dwarfs might actually work for some of the slower pulsars, but not for the fastest.) This leaves rotating neutron stars as the best candidates for pulsars.

In describing the emission from pulsars, an analogy has been drawn with a *lighthouse*. The light in the lighthouse is always on, but you can only see it when the beam points in your direction. If you stay in one place, you will see the light appear to flash on briefly once per cycle. In this model of neutron stars, there is some emission point or "hot spot" on the surface of the star, producing a beam of radiation, like a lighthouse beam.

We can only see the radiation when the beam is pointed in our direction.

The emission mechanism may be related to the strong magnetic field that we think many neutron stars must have (Fig. 11.10). The magnetic axis of the star is probably not aligned with the rotation axis. (This is not unusual; it also happens on the Earth.) If the beam of radiation is somehow collimated along the magnetic axis, we only see it when the beam points in our direction. We may even see two pulses as opposite magnetic poles pass by.

The details of the mechanism are not clear. One possibility is that the rotation and strong magnetic field result in an observer near the surface seeing a rapidly changing magnetic field. Faraday's law tells us that a rapidly changing magnetic field can accelerate charged particles to large energies. These particles then spiral around the magnetic field lines, producing synchrotron radiation. This is only a very general outline of what might be taking place. A considerable effort is still going on to understand the emission of radiation under the extreme conditions that exist near the surface of a neutron star.

The lighthouse nature of the pulsar mechanism has another consequence. The beam only traces out a cone on the sky. If the observer is not

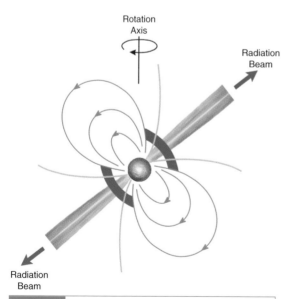

Fig 11.10. Role of a magnetic field in a pulsar emission mechanism. If the magnetic axis is not aligned with the rotation axis, then the magnetic axis can act like a searchlight beam. [NRAO/AUI/NSF]

on that cone, the pulsar will never be visible. This is also true for a lighthouse. If you are directly above the lighthouse, you will never see the light. Any given pulsar will only be seen by about 20% of the potential observers. This means that our galaxy contains many more pulsars than the few hundred that we actually observe. This is especially true since we cannot survey the whole galaxy for pulsars. Current estimates place the number of pulsars in the galaxy at about 200 000.

Probably the most extensively studied pulsar is one in the center of the Crab Nebula, the remnant of the supernova explosion observed in 1054. We have already seen that the emission from the Crab Nebula is polarized, suggesting synchrotron radiation, and that it gives off most of its energy in the radio part of the spectrum. We have also seen that the energy loss via synchrotron radiation is so great that the nebula should have faded considerably. Something is replacing the energy that is radiated away. The total energy loss rate of the nebula is about 3×10^{38} erg/s, or about $10^5 \, L_\odot$. It is known that there is a star with an unusual spectrum at the center of the nebula. This star is known as *Baade's star* after *Walter Baade* who first noted its peculiarity.

A very rapid pulsar was discovered at the position of this star. Its period is 0.033 s. This pulsar was very important in ruling out rotating white dwarfs as the source of pulsars. After the radio pulsar was discovered, it was suggested that it might be possible to observe optical pulses from the star. However, to catch the star in different parts of the cycle would require exposures of about 10^{-3} s, much too short to see anything. An interesting technique was used to get around this problem. The star was observed for many cycles, but the image was recorded only during a small part of each cycle. The part of the cycle was the same cycle after cycle. For example, we expose the same 10^{-3} s every period until we can see a good image. We then shift our exposure by 10^{-3} s and repeat the process. We use the radio signals to synchronize our observing with the pulsing star.

Optical pulses from the Crab pulsar are shown in Fig. 11.11. The optical pulsations are clearly visible. When the star is "on" it is the brightest star in the field. When it is "off" we cannot see it at all. If we just take a normal photograph of the

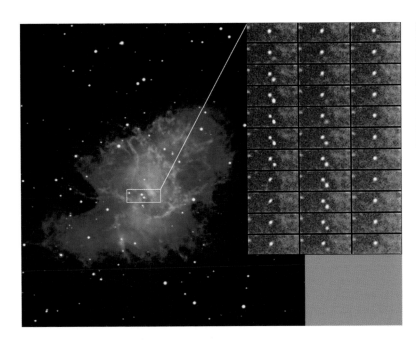

Fig 11.11. The Crab Pulsar (NP 0532) in visible light. Each frame shows an image of the field at equally spaced points in the cycle. When the pulsar is on, it is the brightest object in the field. When it is off, we cannot see it. [NOAO/AURA/NSF]

nebula, we see just the average brightness of the star. If the duty cycle is 5%, then the average brightness is about 5% of the peak brightness.

11.3.3 Period changes

Just as astronomers were becoming used to the idea of pulsars as dependable clocks in the sky, it was discovered that all pulsars are slowing down. Some sample data are shown in Fig. 11.12. It was also found that the fastest pulsars were slowing down with the greatest rate of change of the period. This suggests

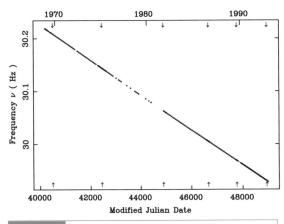

Fig 11.12. Period changes for the Crab pulsar. The general slowdown is clear. Glitches, brief period increases, are indicated by the locations of the arrows. [Lyne & Smith, Pulsar Astronomy, 2nd edn, CUP (1998) Fig. 6.5]

that the fastest ones must be younger. The increase in pulse period means that the star is not rotating as fast as it once was. This means that the kinetic energy of the star is decreasing. This is not surprising, since the star is giving off energy in the pulses.

The energy of a rotating sphere is given by

$$E = (1/2)I\omega^2 \tag{11.10}$$

The rate of change of the energy is found by differentiating both sides with respect to time:

$$dE/dt = (I\omega)(d\omega/dt) \tag{11.11}$$

We can relate the fractional change in energy, using equation (11.10):

$$(1/E)(dE/dt) = (2/I\omega^2)(I\omega)(d\omega/dt)$$
$$= 2(1/\omega)(d\omega/dt) \tag{11.12}$$

The fractional change in energy is therefore equal to twice the fractional change in the frequency. We can relate the change in frequency to the change in period:

$$\omega = 2\pi/P \tag{11.13}$$

Differentiating, we obtain

$$d\omega/dt = (-2\pi/P^2)(dP/dt)$$

Using equation (11.13) to eliminate one of the factors of P on the right-hand side gives

$$(1/\omega)(d\omega/dt) = -(1/P)(dP/dt) \tag{11.14}$$

The fractional rate of energy change then becomes

$$(1/E)(dE/dt) = -2(1/P)(dP/dt) \qquad (11.15)$$

Thus, when we observe the rate at which the pulsar is slowing down, we can directly relate it to the energy loss.

Example 11.4 Pulsar energy loss

Consider a rotating neutron star with a mass $M = 2$ $M_\odot$, and a radius $R = 15$ km, a period $P = 0.1$ s, and a rate of change of the period $dP/dt = 3 \times 10^{-6}$ s/yr. Find (a) the kinetic energy, (b) the rate at which the kinetic energy is decreasing, and (c) the lifetime of the pulsar if it loses energy at this rate.

SOLUTION

We find the energy from equation (11.10), remembering that the rotational inertia for a sphere is $(2/5)MR^2$. We also use the fact that the frequency is $\omega = 2\pi/P$:

$$E = (1/2)(2/5)MR^2\omega^2$$

$$= (1/5)(4 \times 10^{33} \text{ g})[(1.5 \times 10^6 \text{ cm})(2\pi/0.1 \text{ s})]^2$$

$$= 7 \times 10^{48} \text{ erg}$$

To find the rate at which is energy is changing, we find the fractional rate at which the period is changing. We first convert dP/dt into more useful units:

$$\frac{dP}{dt} = \frac{3 \times 10^{-6} \text{ s/yr}}{3 \times 10^7 \text{ s/yr}}$$

$$= 1 \times 10^{-13}$$

The fractional rate of change is

$$(1/P)(dP/dt) = (1/0.1 \text{ s})(1 \times 10^{-13})$$

$$= 1 \times 10^{-12}/\text{s}$$

This means that each second the period increases by 10^{-12} of a period. We use equation (11.15) to find the rate of change of the energy:

$$(1/E)(dE/dt) = -2(1/P)(dP/dt)$$

$$= -2 \times 10^{-12}/\text{s}$$

Multiplying both sides by E gives

$$dE/dt = (-2 \times 10^{-12}/\text{s})E$$

$$= -1.4 \times 10^{37} \text{ erg/s}$$

The lifetime of the pulsar, Δt, is just the energy, divided by the rate at which it is being lost:

$$\Delta t = E/(dE/dt)$$

$$= 1/(2 \times 10^{-12}/\text{s})$$

$$= 5 \times 10^{11} \text{ s}$$

$$= 1.7 \times 10^4 \text{ yr}$$

This means that pulsars have very short lifetimes by astronomical standards. (Actually, the loss rate is not constant, so the above calculation actually gives an upper limit to the lifetime.)

We should note that the rate at which the pulsar in the Crab Nebula is losing energy is equal to the rate at which the nebula itself is losing energy via synchrotron radiation. This solves the long-standing problem of the source of energy for the nebula. As the pulsar loses energy, somehow that energy ends up in the nebula. The connection between the pulsar and the nebula is probably the strong magnetic field. When we see radiation from the Crab Nebula, it is being indirectly fueled by the slowing down of the Crab pulsar.

Pulsars show another variation in their periods. These are sudden decreases in the period, called *glitches* (also shown in Fig. 11.12). After each glitch, the normal slowdown resumes. If the angular momentum is conserved, the rotation of the star can only speed up if the size of the star decreases. We think that the fast rotation and extreme conditions at the surface put a tremendous strain on the solid crust. Periodically, this strain is relieved by a quake, and the crust then settles into a more stable configuration. (Actually, this explanation of glitches may work for the Crab pulsar, which has small glitches, but it cannot explain the larger glitches seen in some other pulsars. It is thought that these larger glitches may involve a transfer of angular momentum from the superfluid interior to the crust.)

We can relate the change in period to the change in radius. The angular momentum is given by

$$J = I\omega$$

$$= I(2\pi/P)$$

$$= (2/5)MR^2(2\pi/P)$$

$$= (4\pi/5)MR^2/P \qquad (11.16)$$

Solving for P gives

$P = (4\pi/5)MR^2/J$

Differentiating with respect to R gives

$dP/dR = (4\pi/5)2MR/J$

$\quad\quad = (4\pi/5)2MR/(5/4\pi)P$

$\quad\quad = 2(P/R)$ (11.17)

This means that a small change in period, dP, will be caused by a small change in radius,

$dR/R = (1/2)(dP/P)$ (11.18)

The fractional change in radius is one-half the fractional change in period.

Example 11.5 Pulsar glitch
For the pulsar in the above example find the change in radius for a period change of 10^{-7} s.

SOLUTION
We find dR by multiplying both sides of equation (11.18) by R to give

$dR = (1/2)(dP/P)R$

$\quad = (1/2)(1 \times 10^{-7}/0.1)(1.5 \times 10^6 \text{ cm})$

$\quad = 0.75 \text{ cm}$

This is not a very large change.

On top of slowdown and glitches, there are additional irregular variations, such as those shown in Fig. 11.13. Understanding these transients is a current problem in pulsar studies.

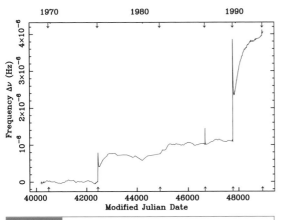

11.4 | Pulsars as probes of interstellar space

Let's look at what happens to pulsar signals as they propagate through interstellar space. The speed of the radiation depends on the index of refraction n, which is a function of wavelength λ. The speed of the radiation is given by

$c(n) = c/n$ (11.19)

where c is the speed of light in a vacuum (where $n = 1$). We can write the speed as a function of wavelength

$c(\lambda) = c/n(\lambda)$ (11.20)

Suppose we have two signals at wavelengths λ_1 and λ_2. Their speeds will be different. Their times t_1 and t_2 to travel a distance d are

$t_1 = dn(\lambda_1)/c$

$t_2 = dn(\lambda_2)/c$ (11.21)

The difference between the two times is

$\Delta t = (d/c)[n(\lambda_1) - n(\lambda_2)]$ (11.22)

This time delay for signals at different wavelengths is called *dispersion* (Fig. 11.14). If we know $n(\lambda)$, then we can measure Δt and find out the distance d that the signals have traveled. If d is known, measuring t tells us about the index of refraction.

For radio waves passing through interstellar space, the index of refraction results from the interaction of the radio waves with electrons, and

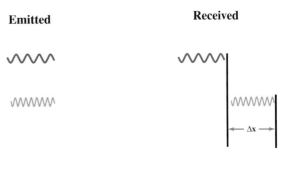

Emitted   **Received**

$\Delta x = c\,\Delta t$

Fig 11.14. Dispersion. Pulses are emitted at the same time at two different wavelengths on the left. They are received at different times on the right.

is proportional to the interstellar electron density n_e. That is,

$$n(\lambda) - 1 \propto n_e \qquad (11.23)$$

Then the time delay must be given by

$$\Delta t \propto d\, n_e \qquad (11.24)$$

When we observe a pulsar at two wavelengths, the signals will not arrive at the same time. However, we know they left the pulsar at the same time. We can attribute the time difference to interstellar dispersion. By measuring the time delay Δt, we can derive the quantity dn_e. If we observe objects for which we have another estimate of distance, we can derive the average interstellar electron density. This turns out to be 0.03 cm^{-3} on average. Once we know this average n_e, we can measure Δt for other pulsars, which gives us $d\, n_e$. Since we know n_e we can estimate d. This gives us a way of estimating the distances to pulsars that we cannot otherwise measure distances to.

11.5 | Stellar black holes

Just as special relativistic effects put a limit on the mass that can be supported by electron degeneracy, they put a limit on the mass that can be supported by neutron degeneracy pressure. There is some uncertainty about this limit. Neutron degeneracy alone can support about 2 $M_\odot$. Neutron–neutron interactions raise this to about 3 $M_\odot$, and this is the number that has been regularly used. However, some calculations show that this mass might be as large as 8 $M_\odot$. Whatever the limit, if a neutron star is more massive than this limit, it cannot be stable. We know of no other source of pressure that will halt the collapse. The star will eventually be small enough to be contained within its Schwarzschild radius. It will become a black hole, as described in Chapter 8. Therefore, unless all neutron stars are formed with masses less than about 3 $M_\odot$, black holes appear to be a normal stellar final state. Though this can happen for individual stars, or stars in binary systems, it is not likely that we can detect isolated stellar black holes. We have to detect its presence by its gravitational effects on another star. We will therefore discuss attempts to detect stellar black holes in the next chapter, where we discuss close binary systems.

Chapter summary

In this chapter we saw what happens to stars of higher mass after they leave the main sequence. The higher mass of the core means a higher temperature. This means that nuclear reactions can proceed farther (formation of heavier elements) in high mass stars than in low mass stars. A shell structure develops, with each layer closer to the center fusing heavier nuclei than the layer outside it.

When an iron core is built up, there is no longer a source of energy. The core begins to collapse. When a high enough density is reached, the star explodes in a supernova. We see the blown off material as a supernova remnant.

The left-over core of the star becomes a neutron star, supported by neutron degeneracy pressure. Neutron stars are observed as pulsars. We discussed the accidental discovery of pulsars, and the chain of reasoning to make the connection between pulsars and neutron stars. The conditions on neutron stars are quite extreme, often involving enormous magnetic fields. These strong fields are involved in the pulse emission mechanism. As pulsars give off energy, they slow down.

If the mass of a neutron star is greater than roughly 3 $M_\odot$, then neutron degeneracy pressure will not support it and it will become a black hole. Such objects must be detected in binary systems.

Questions

11.1. Why is the stability of ^{56}Fe important in the steps that lead to a supernova?

11.2. What are the distinguishing characteristics of synchrotron radiation?

11.3. At a given density, for equal numbers of electrons and neutrons, electrons will exert a greater degeneracy pressure. Why, then, are neutron stars supported by neutron degeneracy pressure?

11.4. (a) Suppose you have a drumhead vibrating at 10^3 Hz. You want to "freeze" its motion at different parts of the cycle. Explain how you would use a strobe light to obtain a sequence of photographs, each showing the drumhead at a slightly different point in the cycle. (b) How is this like the technique for studying optical pulses from pulsars?

Problems

11.1. Suppose a supernova explosion throws off 5 $M_\odot$ of material at an initial speed of 10^3 km/s. (a) Calculate the initial kinetic energy of the shell and the sum of the magnitudes of the momenta of all the pieces in the shell. (b) Suppose the shell slows down by conservation of momentum in sweeping up interstellar material. How much mass will be swept up before the shell slows to 10 km/s? (c) If the average density of interstellar material is 1 H atom/cm^3, what is the radius of the shell when it reaches 10 km/s?

11.2. Estimate the pressure in a 1.5 $M_\odot$ neutron star with a radius of 15 km.

11.3. We have already seen that hydrostatic equilibrium allows us to estimate the central pressure of a star as $P_C \cong GM^2/R^4$. (a) Show that this provides the following mass–radius relation for neutron stars:

$$RM^{1/3} = (2)(1/4)^{5/3}(h^2/4\pi^2 G)(1/m_p)^{8/3}$$

(b) Use this expression to find the radius of a 1.5 $M_\odot$ neutron star.

11.4. (a) If $F(r)$ is the force on an object as a function of radius r, show that the "tidal" force on an object of length Δr is

$$\Delta F = (dF/dr)\, \Delta r$$

(b) Calculate the tidal force on your body near the surface of a neutron star. (Do the calculations for standing and prone positions.)

11.5. For a 1.5 $M_\odot$ neutron star with $R = 15$ km, rotating 100 times per second, compare the gravitational force on an object at the surface with the force required to produce the circular motion for that object (at the star's equator). In other words, compare the weight of an object with the centripetal force on it.

11.6. A uniform density sphere of mass M has initial radius r and an angular speed ω_0. It collapses under its own gravity to a radius r, conserving angular momentum. (a) How do the initial and final kinetic energies compare? (b) Account for any difference.

11.7. (a) What is the escape speed from a 1.5 $M_\odot$ neutron star of radius 10 km? (b) How does it compare with the speed of light?

*11.8. Suppose we can measure the arrival time of pulses to within 10^{-8} s. (a) Explain how we can measure a pulsar period more accurately than this by timing a large number of pulses. (b) How many pulses do you have to measure to measure a period of 0.1 s with an accuracy of 10^{-10} s?

*11.9. Explain why the average brightness of an object is approximately equal to its duty cycle multiplied by its peak brightness. Use the fact that if $b(t)$ is the brightness as a function of time, then the average of $b(t)$ over some time interval T is given by

$$\langle b \rangle = \frac{1}{T} \int_0^T b(t)\, dt$$

11.10. (a) Examine the stability of a rotating object against centrifugal disruption. Show that, for a rotating object, the requirement that the gravitational force must at least balance the centrifugal force, produces an expression for the minimum radius like equation (11.9). (b) Using this result, what is the minimum rotation period for a white dwarf?

*11.11. Suppose we approximate the rate of change of a magnetic field for a stationary observer near a neutron star as the field strength divided by the rotation period. (a) Taking the magnetic field to be 10^{12} gauss, calculate the magnitude of the induced electric field. (b) Over what distance will an electron have to travel in this field to reach a speed of 0.1c?

*11.12. Suppose that, as a pulsar slows down, the quantity $(1/P)(dP/dt)$ stays constant (say at a value of $-b$, where b is a positive number). (a) If the initial period ($t = 0$) of the pulsar is P_0, find an expression for $P(t)$, the period as a function of time. (b) If the initial rotation energy is E_0, find an expression for $E(t)$, the energy as a function of time. (c) If the pulsar is formed with $P_0 = 10^{-3}$ s, how long will it take to reach $P = 3$ s?

*11.13. (a) For a pulsar of mass M, radius R, and period P, slowing at a rate dP/dt, at what rate is its angular momentum changing? (b) What provides the torque for this change in angular momentum?

*11.14. Find an expression, analogous to that for the Chandrasekhar limit, discussed in Chapter 10, for the maximum mass star that can be supported by neutron degeneracy pressure.

11.15. Compare the radius of a 1.4 $M_\odot$ neutron star with its Schwarzschild radius.

Computer problem

1. Construct a table showing neutron star sizes for masses 1, 2, 3 and 10 $M_\odot$. Also include columns for average density, the acceleration of gravity at the surface, and the rate of change of that acceleration, dg/dr.

Chapter 12

Evolution in close binaries

12.1 | Close binaries

If the two stars in a binary system are very close to each other, each has the effect of altering the structure of the other star. When this occurs we call the system a *close binary* system. The surface of a star can be distorted by the stronger gravitational force that the companion exerts on the near side than on the far side. Remember, we said that any effect that depends on variations in the gravitational force from one position to another is called a tidal effect. (A similar situation applies as the Sun and Moon distort the Earth's ocean surface, raising the tides.)

The distortion of stars results in internal dissipation of energy. As a star rotates, different material is incorporated in the bulge. Different layers of material rub against each other, in a fluid friction. This lost energy has to come from somewhere. It comes from both the orbital energy and the rotational energy of the star. As a result, eventually the orbits circularize and the two stars always keep the same sides towards each other. This is the lowest energy arrangement for the system (see Problem 12.1). We say that the spins are *synchronized*. (The Moon's spin and orbital motion around the Earth are synchronized, and the Moon keeps the same side towards the Earth.)

In certain situations, it is possible for material from one star to be pulled off the surface onto the other star. To see how this can happen, we look at a binary system from a coordinate system rotating with the same period as that of the orbits. If we look at the energy of a particle in this system, the rotation of the coordinate system introduces a term in addition to the gravitational potential. This term is equal to $J^2/2mr^2$, where J, m and r are the angular momentum, mass and distance, respectively, from the origin of some particle. (We can think of it as the term in the potential energy which gives rise to the pseudo "centrifugal force" in the rotating system.) When we add this term to the gravitational potential, we have an *effective potential* that can be used to describe the motions of particles. We can draw surfaces of constant effective potential, as in Fig. 12.1. The effective force (gravity plus "centrifugal") at any point on one of these surface is perpendicular to the surface. (This is analogous to contour maps of gravitational potential – elevation – on Earth. The gravitational force is perpendicular to the contour lines, and you don't have to do any work to move along an equipotential line.)

There are five points, called *Lagrangian points*, where the effective gravitational force is zero. These points are designated L_1, L_2, L_3, L_4, L_5. (Note that the L_5 Society wants to place a space station at the L_5 point for the Earth/Moon system.) The point L_1 lies between the two stars, at the intersection point of the "figure eight" shaped surface. L_1 is the dividing point between material being attracted to one star or the other. The two sides of the figure eight are called *Roche lobes* (Fig. 12.2).

The equipotential surfaces in a fluid must be surfaces of constant pressure. If they were not, there would be pressure differences forcing fluid along the surface, and these forces could not be balanced by any gravitational forces,

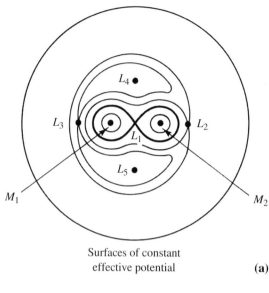

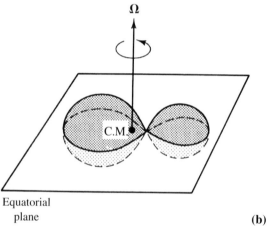

Surfaces of constant
effective potential
(a)

Ω

Equatorial
plane
(b)

Fig 12.1. Surfaces of constant effective potential. The effective potential is the true gravitational potential plus that resulting from the centrifugal force in the coordinate system rotating with the orbiting system. (a) We look from above and see the intersection of surfaces of constant potential and the plane of the orbit. The heavy 'figure eight' is the intersection of the Roche lobe and the plane of the orbit. (b) A three-dimensional view of the Roche lobe. (In this case, $M_1 \neq M_2$.)

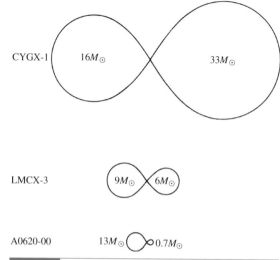

Fig 12.2. The sizes of the two Roche lobes depends on the mass and separations of the orbiting systems. This figure shows the Roche lobes for three mass combinations. The masses are the best estimates for the mass in three observed close binary systems.

which must be perpendicular to the surfaces. The equipotential surfaces must therefore also be surfaces of constant density. (Again, this is analogous to the situation on Earth. By the equation of hydrostatic equilibrium, we can only have a pressure gradient in the direction of the gravitation force. That would be perpendicular to the potential lines.)

In discussing the evolution of close binary systems, we divide them into three classes (Fig. 12.3): (1) *detached*, in which each star is totally contained within its own Roche lobe; (2) *semidetached*, in which the photosphere of one star exactly fills its side of the Roche lobe; and (3) *contact binaries*, in which both stars are at or over the Roche lobe.

So far in this book, we have discussed the evolution of isolated stars. However, we have already seen that approximately half of all stars are in binary systems. If the binaries are completely detached, and there is no mass transfer, then the evolution will not be altered by the presence of the companion. However, mass transfer in semidetached or contact systems can influence stellar evolution.

In general, the more massive star in a binary system will evolve off the main sequence first. When that star becomes a red giant, it may become large enough to fill its Roche lobe. In that case mass transfer to the companion will take place. This can alter the evolution of the companion. The degree to which it alters the evolution depends on the nature of the companion. As the more massive star continues to lose mass, its Roche lobe shrinks, but the Roche lobe for the companion grows. This means that mass transfer

Close Binaries

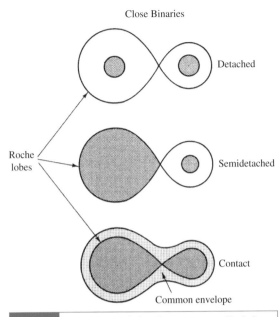

Detached

Roche lobes

Semidetached

Contact

Common envelope

Fig 12.3. Classification of close binary systems. Shaded areas are filled with material.

will take place until the masses of the stars are about equal. Some slow mass transfer may continue after that point.

At some point the star that was losing mass will become a white dwarf or some other collapsed object. In the following sections we look at examples of each type of collapsed object that we have discussed – white dwarf, neutron star and black hole.

12.2 | Systems with white dwarfs

We first consider systems in which the first star to evolve off the main sequence becomes a white dwarf. Eventually, the white dwarf's companion goes through its evolution off the main sequence. The companion becomes a red giant and fills its Roche lobe. At this point mass transfer starts back in the other direction from the original mass transfer. Now mass is falling in on the white dwarf. Not all the infalling matter strikes the white dwarf. Because of its angular momentum, some of the material goes into orbit around the white dwarf. This orbiting material forms a disk, called an *accretion disk* (Fig. 12.4). The disk forms because material can fall parallel to the axis of rotation but not perpendicular to that axis. (We will discuss disk formation in more detail in Chapter 15.)

As material falls in, its potential energy decreases, so its kinetic energy increases. The increase in kinetic energy will not equal the potential energy decrease, because some energy will be radiated away. We can expect roughly half of the change in potential energy to show up as kinetic energy. But this increase in kinetic energy will not produce much radiation on its own. This is where the accretion disk helps. As the faster moving gas strikes the accretion disk, it slows down, but its temperature increases. The now heated gas can then radiate.

Fig 12.4. Artist's conception of mass transfer leading to an accretion disk. [STScI/NASA]

We can estimate the energy available from the change in potential energy as mass falls in. If the mass starts at distance r_1 from the white dwarf, and ends up a distance r_2 from it, the luminosity is given by (see Problem 12.5)

$$L \cong \left(\frac{dM}{dt}\right)\left(\frac{GM}{r_2} - \frac{GM}{r_1}\right)$$

$$\cong GM\left(\frac{dM}{dt}\right)\left(\frac{1}{r_2} - \frac{1}{r_1}\right) \qquad (12.1)$$

In this equation dM/dt is the rate at which mass is falling in, and M is the mass of the white dwarf. We should think of equation (12.1) as the upper limit to the actual luminosity. That is because all of the energy gained in the infall is not converted into outgoing radiation. In fact, the formation of the accretion disk is crucial in this process. The accretion disk provides a place where the energy from the infalling material can be converted into heat. The heated disk then radiates.

Example 12.1 Mass accretion luminosity
Calculate the luminosity for a mass infall rate of $10^{-8}\ M_\odot/yr$, onto a $1\ M_\odot$ white dwarf. Assume that the material starts 1.0×10^{11} cm away from the white dwarf, and ends up 1.0×10^9 cm away.

SOLUTION
We first convert the mass loss rate into g/s:

$$\frac{dM}{dt} = \frac{(1.0 \times 10^{-8}\,M_\odot/yr)(2.0 \times 10^{33}\,g/M_\odot)}{(3.1 \times 10^7\,s/yr)}$$

$$= 6.5 \times 10^{17}\ g/s$$

The luminosity is then

$$L = (6.67 \times 10^{-8}\ dyn\ cm^2/g^2\,)(2.0 \times 10^{33}\ g)$$
$$\times (6.5 \times 10^{17}\ g/s)$$
$$\times \{[1/(1.0 \times 10^9\ cm)] - [1/(1.0 \times 10^{11}\ cm)]\}$$

$$= 8.6 \times 10^{34}\ erg/s$$

This is approximately 20 times the luminosity of the Sun.

Occasionally we observe a star that suddenly brightens by 5 to 15 magnitudes. These objects are called *novae* (Fig. 12.5). The name suggests the appearance of a new star where one was not previously seen. Some of these novae appear to be recurrent, on time scales of up to hundreds of

Fig 12.5. HST image of the shell around recurrent nova T Pyxidis. This is at a distance of 2000 pc. The shell is a little less than 1 parsec across. This image shows that it is made up of a large number of small objects. This is the material that has collected from the various nova outbursts. (Note that Fig. 4.16 shows an HST image of the shell expanding around Nova Cygni 1992, taken 467 days after that outburst.) [STScI/NASA]

years. There is evidence that mass is ejected in the process. In some cases this material can be seen expanding away from the star. The amount of mass ejected is about $10^{-5}\ M_\odot$.

We think that novae are the result of thermonuclear explosions on the surfaces of white dwarfs with mass falling in from a companion. The mass falling in is from the envelope of a red giant, and therefore contains hydrogen. (Remember, the white dwarf has used up all of the hydrogen in the core and has expelled the rest in its planetary nebula.) The surface of the white dwarf is hot enough for fusion of the hydrogen to take place. It takes place rapidly in a small explosion, which probably stops the mass transfer for a while. When the transfer resumes, another explosion can take place. Shells of material left around novae are shown in Figs. 12.5 and 4.16.

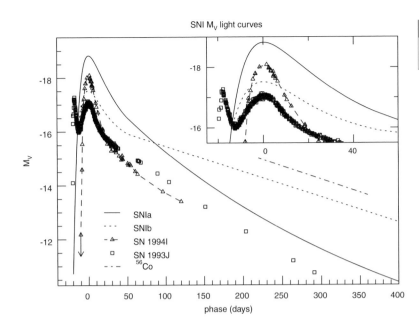

SNI M_V light curves

— SNIa
· · · SNIb
– ▵ – SN 1994I
□ SN 1993J
— · — ^{56}Co

M_V

phase (days)

Fig 12.6. Light curves for type I supernovae. [Craig Wheeler, University of Texas, Austin]

We think that mass transfer onto a white dwarf can also account for type I supernovae. Sometimes enough mass falls onto the white dwarf to make its mass greater than the Chandrasekhar limit, M_{CH}. In this case, electron degeneracy no longer supports the star, and it collapses. The energy from the collapse drives nuclear reactions, which eventually build up ^{56}Ni (with an even number of ^{4}He nuclei). The ^{56}Ni beta decays to form ^{56}Co, which, in turn, beta decays to ^{56}Fe. Type I supernovae have light curves with double exponential behavior. The time scales of the two exponentials turn out to be characteristic of these two beta decays. We think that this process accounts for most of the iron in the universe, since the iron created in more massive stars is destroyed in the type II supernova, as discussed in Chapter 11. Light curves for type I supernove are shown in Fig. 12.6. There is little variation in the peak brightness of type I supernovae. We will see in Chapter 18 that this makes them very good "standard candles" in distant galaxies.

The nuclear energy released in these reactions is greater than the binding energy of the white dwarf, and the star is destroyed, leaving no remnant. This explanation accounts for the light curves of type I supernovae, their spectra and luminosities, and their occurrence in what are thought to be old systems (as we will discuss in Chapter 13).

12.3 | Neutron stars in close binary systems

In the preceding section we saw how mass transfer in semidetached systems can alter the evolution of a star. In this section we consider a neutron star and a normal star in orbit around their common center of mass. As the normal star evolves towards a red giant, it fills its Roche lobe and matter starts to fall onto the neutron star.

At first it was thought that this situation could not develop. It was not clear how such a system could form. The problem is that, for a neutron star to be present, there must have been a supernova explosion in the past. The first star to go supernova in a binary is the more massive star. The supernova explosion drives away most of the mass of the more massive star, meaning that more than 50% of the original mass of the system was blown away. If this happens far too quickly for the system to adjust the orbit radius/period, it becomes unbound. To see this for circular orbits, we assume that the system starts with stars of mass m_1 and m_2 and a separation R, with speeds appropriate for a circular orbit, given by equation (5.15). Assume that star 1 explodes and is left with a mass m, but that R and the orbital speeds don't change. The new energy is then given by equation (5.26), with m substituted for m_1. For the new system to be on the boundary of being bound, its energy would be zero, so we can find the value of m that makes $E = 0$. It is (see Problem 12.6)

$$m = m_1 \left[\frac{1}{2 + (m_2/m_1)} \right] \qquad (12.2)$$

Note that for m_1 being much greater than m_2 this approaches $m_1/2$, meaning the massive star would have to lose half its mass. Note that, by the discussion in Section 5.4, this result doesn't change for elliptical orbits, since the equation for the energy is the same, with R replaced by the semi-major axis.

Therefore, the explosion should break up the binary system. This scenario explains the existence of "runaway stars". These are individual stars moving through space with much higher than average speeds.

X-ray observations have been made of systems in which it appears that a neutron star is still in orbit with a normal star. This means that there must be some way of forming such a system, and eventually theoreticians have come up with a number of plausible scenarios. Different scenarios might work in different types of systems that are found. (1) Theoretical simulations have shown that, while most systems in which the more massive star goes supernova first will become unbound, there are some combinations of initial conditions that will lead to bound systems after the supernova explosion. (2) Before the supernova explosion, the more massive star might have filled its Roche lobe and transferred mass to the less massive star. If enough mass is transferred before an explosion, the system can stay bound. (3) An alternative explanation is that the compact star may have originally been a white dwarf, not a neutron star. However, mass transfer from the companion may have pushed the mass of the white dwarf beyond the 1.44 $M_\odot$ limit. The electron degeneracy pressure could no longer support the star and it would collapse until it became a neutron star. So, the neutron star would have formed without a supernova explosion. (It is interesting to note that a 1.44 $M_\odot$ white dwarf that suddenly collapsed to form a neutron star would, according to general relativity, appear to exert the gravitational force of a 1.3 $M_\odot$ star. Thus, there may be stars we think are white dwarfs, but which are really neutron stars.)

We now suppose that we have a binary system with a neutron star and a normal star, with mass being transferred from the normal star to the neutron star. The mass falling in is heated and gives off irregular bursts of X-ray emission. To see how this works, we look at the case of a well studied X-ray source, *Her X-1* (Fig. 12.7a). (The name implies the brightest X-ray source in the constellation Hercules.) It is also coincident with a variable star HZ Her. The star is a binary with a period of 1.7 days. The mass of the unseen companion is estimated to be in the range 0.4 to 2.2 $M_\odot$. The X-rays are observed to pulse with a period of 1.24 s.

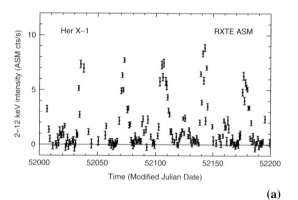

(a)

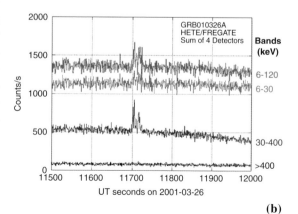

(b)

Fig 12.7. (a) X-ray emission from HZ Her. The intensity is plotted as a function of time into the period. (b) X-ray emission from a burster. [(a) Alan M. Levine, MIT Center for Space Research; (b) MIT Center for Space Research]

The period changes regularly throughout the 1.7 day cycle. This can be interpreted as a Doppler shift (see Problem 12.7). We can think of the X-rays as a signal being emitted with a period of 1.24 s. When the source is moving away from us the period appears longer, and when the source is coming toward us the period appears shorter. The X-ray source also appears to be eclipsed every 1.7 days.

The mass transfer rate, estimated from the X-ray observations, is about 10^{-9} $M_\odot$/yr. The luminosity is about 10^{37} erg/s. The temperature is estimated to be 10^8 K. At this temperature, we can estimate the frequency of a photon with energy kT. The frequency, ν, is kT/h, or 2×10^{18} Hz. This corresponds to a wavelength of 0.14 nm, clearly in the X-ray part of the spectrum.

Theoreticians have speculated on the future evolution of such a system. The mass transfer rate may become so large that the X-ray emission is quenched. The outgoing X-rays are effectively blocked by the infalling material. This system may eventually end up as two compact objects.

Mass transfer in a system with a neutron star can also explain the existence of pulsars with very short periods, as short as a few milliseconds, known as *millisecond pulsars*. One of the intriguing features of these objects is that their periods are not decreasing as rapidly as those for normal pulsars. Some of them have values of $P/(dP/dt)$ as large as 10^{10} per year. That is, in one year, the change in the period is only 10^{-10} (one-ten-billionth) of the period. This means that they are extraordinarily stable clocks (something which had originally been hoped for normal pulsars until their period changes were observed, as discussed in Chapter 11).

To see how mass transfer can explain millisecond pulsars, remember that if transferred material is not aimed directly at the center of the neutron star, it has a large amount of angular momentum. This explains the formation of an accretion disk, as material goes into orbit rather than falling onto the neutron star surface. As material leaks inward from the accretion disk onto a (normal) pulsar, it transfers a lot of angular momentum to the pulsar, causing a large increase in the rotation rate (decrease in the period). This explains how a normal pulsar can be "spun up" into a millisecond pul-

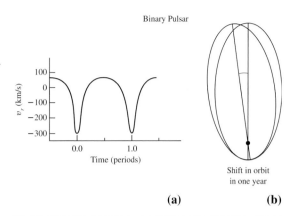

Fig 12.8. Binary pulsar orbit. (a) The radial velocity as a function of time within a period, expressed as a fraction of the period. (b) The shift in the orbit over a one year period. [Joseph Taylor, Princeton University]

sar. Also, as long as there is mass transfer, the rotation rate can be kept very high, explaining why the slowdown of millisecond pulsars is very small.

While discussing neutron stars in binary systems, we should mention one other interesting object. This is a radio pulsar discovered in 1974 by *Joseph Taylor* and *Russel Hulse*, then of the University of Massachusetts. This pulsar has a period of 0.059 s but the period varies periodically, suggesting that the pulsar is in a binary system. The variation in the period is like a Doppler shift (Fig. 12.8), with the period appearing longer when the pulsar is moving away from us and shorter when it is coming towards us. This system provides us with an interesting test of a prediction of general relativity–gravitational radiation. For their studies of this *binary pulsar*, Taylor and Hulse were awarded the Nobel Prize in physics in 1993.

Since the gravitational radiation carries away energy, the total energy of the orbit should be decreasing. In hopes of seeing this, Taylor and his co-workers have monitored this binary pulsar for several years. Some results are shown in Fig. 12.9. They have found that the orbital period is changing by -2.3×10^{-12} s/s. The change in energy of the orbit corresponds to the energy that would be given off by gravitational radiation from the orbiting bodies. This may provide us with the first indirect observational confirmation of gravitational radiation.

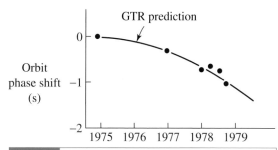

Fig 12.9. A comparison of theory and experiment for the binary pulsar. The plotted quantity keeps track of energy lost in the orbit over the years that it has been observed. The predictions from general relativity include gravitational radiation as the mechanism for energy loss. [Joseph Taylor, Princeton University]

12.4 | Systems with black holes

In Chapter 11, we saw that neutron stars are supported by neutron degeneracy pressure. However, if the neutron star is too massive, neutron degeneracy pressure is insufficient to support the star. We think this limit is about 3 $M_\odot$. We know of no other source of pressure that will stop the collapse of the star. It will collapse right through the Schwarzschild radius for its mass, and will become a black hole. Black holes would be a normal state of evolution for some stars.

How would we detect a stellar black hole? We obviously could not see it directly. We could not even see it in silhouette against a bright source since the area blocked would be only a few kilometers across. We have to detect stellar black holes indirectly. We hope to see their gravitational effects on the surrounding environment. This is not a hopeless task, since we might expect to find a reasonable number of binary systems with black holes. By studying a binary with a suspected black hole, the problem would be to show the existence of a very small object (as inferred from orbits) with a mass in excess of 3 $M_\odot$.

How do we find a candidate binary to study? In Section 12.3, we saw that a neutron star in a binary system can give rise to strong X-ray sources. The importance of the neutron star is that its radius is so small that infalling material acquires enough energy to give off X-rays. A stellar black hole would be smaller than a neutron star, so material falling

in would also emit X-rays (before crossing the event horizon). We could start searching for stellar black holes by looking for irregular X-ray sources.

One interesting possibility is known as *Cyg X-1* (Fig. 12.10), the brightest X-ray source in Cygnus. The Uhuru satellite showed this to have both short and long term variability. Until the Einstein observatory was launched in 1978, the positions of X-ray sources were not accurately determined. However, there is also a radio source associated with the X-ray source. We know that the X-ray and radio sources are associated because they have the same pattern of variability. The position of

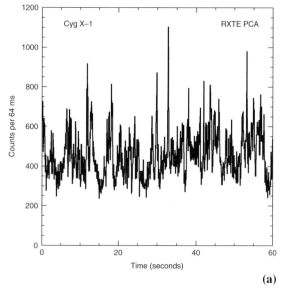

(a)

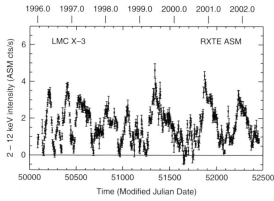

(b)

Fig 12.10. X-ray light curves for suspect stellar black holes in close binaries: (a) Cyg X-1, (b) LMCX-3. [(a) Edward H. Morgan, MIT Center for Space Research; (b) Alan Levine, MIT Center for Space Research]

the radio source is determined accurately from radio interferometry. Once the radio position was known, optical photographs were studied to see if there is an optical counterpart. This optical counterpart was found to be a ninth magnitude star HDE226868. A study of the star's spectrum shows that is an O9.7Ib (blue supergiant) star. This places its mass at about 15 $M_\odot$.

This star is also a spectroscopic binary. Its orbital period is 5.6 days. The star also varies in brightness with this period. The amount of variation is 0.07 mag. The source of the variation is thought to be a distortion in the shape of the blue supergiant due to thee strong tidal effects of the unseen companion. The distortion results in the star appearing as different sizes at different points in its orbit. From the amount of brightness variation, it was concluded that the inclination of the orbit is 30°.

We now look at what can be deduced about the companion. From the analysis of the Doppler shift data, the mass function (discussed in Chapter 5) can be found. Since we know the mass of the blue supergiant and have an estimate of the inclination angle of the orbit, we can derive the mass of the unseen companion from the mass function. The best estimate for this is 8 $M_\odot$. There is some uncertainty, but it seems very likely that the mass of the unseen companion is a least greater than 4 $M_\odot$.

The X-ray observations tell us that the companion must be very small. This is because the X-ray emission varies significantly in intensity on time scales of about 5 ms. This requires that the emitting region be less than 1500 km in extent (see Problem 12.9). Thus, we have an object that is definitely smaller than a normal star, but which has a mass greater than 4 $M_\odot$ (and probably closer to 8 $M_\odot$). It would therefore seem likely that the object is a black hole!

It is interesting that such an astonishing conclusion rests on the foundation of some "standard" observational techniques. These include using the spectrum of HDE226868 to find its mass, and the classical analysis of spectroscopic binary orbits, using the mass function.

Though most astronomers probably accept the black hole explanation for Cyg X-1, the argument is not airtight. For example, our knowledge of the mass of the visible star comes from assuming that it is like another O9.7Ib star. However, we know that it is in a close binary system, and is tidally distorted. We are only now beginning to understand stellar structure and evolution in close binary systems. It may be that the spectrum we classify as O9.7Ib is really produced by a star of a different mass. We are also not sure of the inclination of the orbit, and it enters in the mass function as $\sin^3 i$. It is also possible to avoid an 8 $M_\odot$ companion if we postulate the triple-star system, but searches for a third component have not been successful.

For ten years, Cyg X-1 was the only strong stellar binary black hole candidate. Astronomers have studied large numbers of possible candidates. The starting points are generally X-ray sources for which optical counterparts can be identified. In the last few years, a few more solid possibilities have emerged.

One is an X-ray source in the Large Magellanic Cloud (a companion to our own galaxy), known as *LMCX-3*. Based on Einstein positional observations, an optical counterpart to the X-ray source was identified. It turns out to be a 17th magnitude B3V star. The star is a spectroscopic binary with an orbital period of 1.7 days, and an orbital Doppler shift of 230 km/s. The system does not eclipse, so this puts some constraints on the orbital inclination. An inital analysis put the most likely mass range of the compact object as 4 to 11 $M_\odot$. A more detailed analysis showed that it is more than 6 $M_\odot$. One uncertainty in studies of this object is that it seems that some of the optical light comes from the accretion disk, making analysis of the optical variability more difficult. (This is not a problem in Cyg X-1, where the optical light appears to all come from the visible star.) Even with these uncertainties, it seems that 4 $M_\odot$ is a reliable lower limit.

The third strong candidate that has emerged is different from Cyg X-1 or LMCX-3 in that it is a transient X-ray source. Depending on what cataloge it is found in, it has numerous names: *0620-00, V616 Mon, Nova Mon 1975, 1917*. It has been identified with an optically recurrent nova (with 1917 and 1975 being the two most recent outbursts). The identification of the X-ray and optical novae allowed for the secure identification with the X-ray source with an optical source that could be studied in detail. The optical star is a K5/7 dwarf. The properties of low mass dwarfs are better known than the more massive stars found in Cyg X-1 and LMCX-3. This K dwarf was found to be a spectroscopic binary with a period of 7.8 hr, and a large orbital Doppler shift (470 km/s). Analysis of the mass function places a lower limit of 3 $M_\odot$ on the mass of the compact companion. An analysis of the mass ratio of the two objects placed the mass of the compact object between 4 and 9 $M_\odot$.

Many objects have been studied in detail, but there are always problems in the chain of reasoning. For example, a promising source, LMCX-1, has two possible optical counterparts. Of the three objects mentioned here, Nova Mon 1975 probably has the best evidence for the presence of a black hole. It is amazing to contemplate how far astronomers have come when we can calmly say that the most likely explanation for some observed phenomenon is the presence of a black hole!

12.5 | An unusual object: SS433

To form an idea of the fascinating range of phenomena encountered in close binary systems, we take a look at an object best known by its designation in a particular catalog, SS433. (It was cataloged long before is unusual nature was realized.) The object is a binary at the center of a supernova remnant. It is therefore not surprising that one of the members of the binary system is a neutron star or black hole. The period of the binary system is 13.087 days. The system is also a periodic X-ray source.

Optical observations reveal absorption and emission lines with very large Doppler shifts. The required speeds are up to 0.26c. At any time, both blueshifted and redshifted components are present. The magnitude of the Doppler shift goes through a 164 day cycle, as shown in Fig. 12.11. There is an interesting asymmetry in the Doppler shifts. The redshift is always larger than the blueshift. The maximum redshift corresponds to about 50 000 km/s, and the maximum blueshift corresponds to about 30 000 km/s. If we take an average of the redshifted velocity and blueshifted velocity at any instant (remembering that blueshift corresponds to a negative radial velocity), we obtain a fairly constant value of about 12 000 km/s.

The basic model to explain this behavior is shown in Fig. 12.12. It involves a binary system in

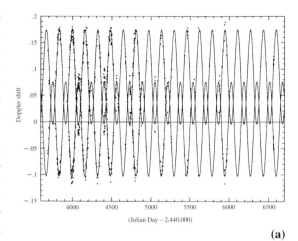

(a)

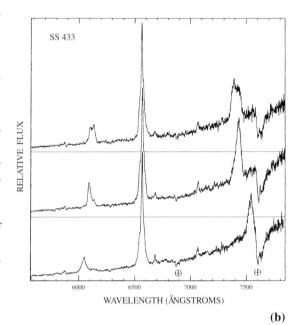

(b)

Fig 12.11. Doppler shift data for SS433. (a) The vertical axis shows the Doppler Shift, $\Delta\lambda/\lambda$, relative to the average velocity of the system. The horizontal axis is in days. The dots represent the actual measurements, and the smooth curves are fits of models to the data. (b) Actual spectra of SS433 at three different times, showing the shifts from the velocities of the Hα emission. (The absorption features marked "⊕" are from the Earth's atmosphere.) [Bruce Margon (University of Washington)/*Margon, B. Astrophys. J. Lett,* **230**, L41, 1979, Fig. 3]

which one member is either a black hole or a neutron star. The period of the binary is 13.087 days. The compact object is a source of two jets, moving in opposite directions. The jets are emitted in a cone with a half-angle of about 20°. The cone is

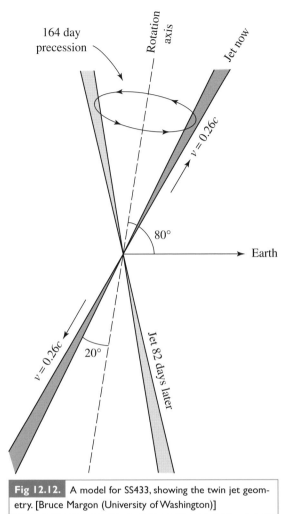

Fig 12.12. A model for SS433, showing the twin jet geometry. [Bruce Margon (University of Washington)]

precession we mean the changing of the orientation of the axis of rotation caused by an external torque. This is like the precession of the Earth, under the combined effects of the Sun and Moon, as discussed in Chapter 23.) This causes the projected angle of the jets to go through a 164 day cycle, giving the variations in the Doppler shifts.

In this model there is a natural explanation for the 12 000 km/s offset in the Doppler shifts. It comes from a transverse Doppler shift arising from the fact that the jet is moving at an appreciable fraction of the speed of light. (We discussed the transverse Doppler effect in Chapter 7; it is essentially a time dilation effect.) At $v/c = 0.26$, $\gamma = 1.036$. According to equation (7.5) this produces a wavelength which is 1.036 times the rest wavelength. If we interpret this shift as arising from a normal Doppler shift, it corresponds to $v/c = 0.036$, or about 11 000 km/s. Since this is just a time dilation effect, this is added in as a constant to any radial Doppler shift. Since it always increases the wavelengths, it make the redshifts larger and the blueshifts smaller.

In this general model, the X-rays come from material falling into an accretion disk around the compact object. The binary is an eclipsing binary, so we can estimate the masses of the components. The best estimate for the mass of the companion is about 4 $M_\odot$. This would make it a black hole. However, there are enough uncertainties in the estimate to allow it to be either a black hole or a neutron star. Theoreticians are still working on the mechanisms for the collimating of the jets, and for generating the energy for getting the jets moving so fast.

inclined by 80° to the line of sight. The compact object precesses with a period of 164 days. (By

Chapter summary

We saw in this chapter how the evolution of stars is altered by placing them in close binary systems. The change in evolution arises mainly from strong tidal effects and from mass transfer.

If the component receiving mass is a compact object – white dwarf, neutron star or black hole – infalling material can acquire enough energy to emit X-rays as it falls in to form an accretion disk.

We saw how mass transfer onto white dwarfs can account for novae. If mass transfer is large enough, it can account for type I supernovae.

We saw how mass transfer onto neutron stars can produce strong X-ray emission, such as from HZ Her. We also saw how mass transfer can spin up the pulsar, making it into a millisecond pulsar. We also saw how studies of a binary pulsar have provided evidence for gravitational radiation.

We saw how studies of Cyg X-1, LMCX-3 and Nova Mon 1975 have provided us with evidence for the existence of a stellar black hole.

Questions

12.1. Explain why surfaces of constant effective potential must be surfaces of constant density.

12.2. Why is the formation of an accretion disk a likely event when we have mass transfer onto a compact object in a binary system?

12.3. Why is the first star in a binary to go supernova usually the more massive one?

12.4. Why is it unusual to find a neutron star in a binary system?

12.5. How does mass transfer explain the existence and stability of millisecond pulsars?

12.6. Why is the binary pulsar so important?

12.7. Explain the steps leading to the conclusion that Cyg X-1 is a black hole. Which are the most suspect? Compare the problems in the reasoning for LMCX-3 and Nova Mon 1975.

Problems

12.1. Show that, for a fixed total angular momentum, the "synchronized" spins situation is the lowest energy state for an orbiting system. (Hint: Consider the sum of the orbital and rotation energies.)

12.2. The radial force F_r is related to a potential $V(r)$ by

$$F_r = -dV(r)/dr$$

Show that the "centrifugal" force can be derived from the term $J^2/2mr^2$ in the effective potential.

*12.3. Suppose we have a pulsar in orbit around another object. The pulse period, as emitted by the pulsar, is P_0. The orbit is circular with a speed v. We are observing in the plane of the orbit. Find an expression for the observed pulse period as a function of the position of the pulsar in the orbit. (Hint: Consider the advance and delay in the arrival time of pulses as the pulsar moves toward us and away from us.)

12.4. For the rate of change in the orbital period for the binary pulsar discussed in Section

12.3, what is the fractional change in the orbital energy $\Delta E/E$ per second?

12.5. Derive the first line of equation (12.1).

12.6. Derive equation (12.2), following the outline given in the discussion.

12.7. For the system HZ Her, how large a shift in the pulsar period would be observed if the visible star has a mass of 3.0 $M_\odot$, the compact object has a mass of 2.0 $M_\odot$, and the orbital period is 1.7 days? The pulsar period is 1.24 s.

12.8. For HZ Her, assuming that the material falls in from far away at the given rate, how close must it come to the star to provide the given X-ray luminosity?

12.9. For Cyg X-1, the most likely value of the mass function is

$$(M_x \sin i)^3/(M_x + M_{opt})^2 = 0.25\ M_\odot.$$

For an inclination angle of 30°, and an optical star mass of 33 $M_\odot$, find the mass of the compact object.

Computer problems

12.1. Draw a graph showing mass accretion luminosity for material falling onto a neutron star (at a rate of $10^{-8}\ M_\odot/yr$) for neutron star masses ranging from 1 to 3 $M_\odot$. For each neutron star mass, assume the radius as given by the mass–radius relation discussed in Chapter 11.

12.2. Construct a table showing how much mass would have to be lost in a supernova explosion to unbind

the system, using equation (12.2), for combinations involving $m_1 = 2, 5, 10, 20$ and 50 $M_\odot$ and $m_2 = 1, 2$ and 5 $M_\odot$.

12.3. Repeat Problem 12.9, showing the effects of changing the inclination angle to 15° and 45°, and the optical star mass to 28 $M_\odot$ and 38 $M_\odot$.

Chapter 13

Clusters of stars

When we look at the spatial distribution of stars in our galaxy, we find that most of the light is concentrated in a thin disk. We are inside this disk, so we see it as a band of light on the sky, called the *Milky Way*. We will discuss this farther in Part III, but we will see in this chapter that location of stars in the galaxy can tell us something about those stars. In particular, some stars are confined to the thin disk of the Milky Way, while others form a more spherical distribution. In this chapter, we will discuss groupings of stars, called *clusters*, and see how they vary in size, content and galactic distribution.

13.1 | Types of clusters

We distinguish between two types of star clusters – galactic clusters and globular clusters.

Galactic clusters are named for their confinement to the galactic disk. A selection of images of galactic clusters is in Fig. 13.1. A familiar galactic cluster, *the Pleiades,* is shown in Fig. 13.1(a). Note the open appearance in which individual stars can be seen. Because of this appearance, galactic clusters are also called *open clusters*. Galactic clusters typically contain $<10^3$ stars, and are less than ~10 pc across. Recent sensitive near IR surveys are showing more members than we had previously thought in many clusters. In the photograph, we see some starlight reflected from interstellar dust. Galactic clusters are sometimes associated with interstellar gas and dust.

Globular clusters are named for their compact spherical appearance (Fig. 13.2). They have 10^4 to 10^6 stars, and are 20 to 100 pc across. They seem to have no associated gas or dust. Some do have planetary nebulae, though. Globular clusters are not confined to the disk of the galaxy. Harlow Shapley used RR Lyrae stars and Cepheids to find the distances to globular clusters. This placed the globular clusters in three dimensions. It was found that the globular clusters form a spherical distribution with the Sun being about 10 kpc from the center. (This is still one of the best techniques for finding the distance to the galactic center.)

Before we look at the properties of the clusters themselves, we will look at an important technique for determining distances to relatively nearby galactic clusters.

13.2 | Distances to moving clusters

Let's assume that we have a star (or cluster) moving through space with a velocity v. The velocity makes an angle A with the line of sight. We can break the velocity into components parallel to the line of sight and perpendicular to the line of sight. The component parallel to the line of sight is the radial velocity v_r and is responsible for the Doppler shift we observe. The component perpendicular to the line of sight is the *transverse velocity* v_T. It is responsible for the motion of the star across the sky, called the *proper motion*.

From the right triangle in Fig. 13.3, we can see that these quantities are related by

$$v^2 = v_r^2 + v_T^2 \tag{13.1}$$

$$v_r = v \cos A \tag{13.2}$$

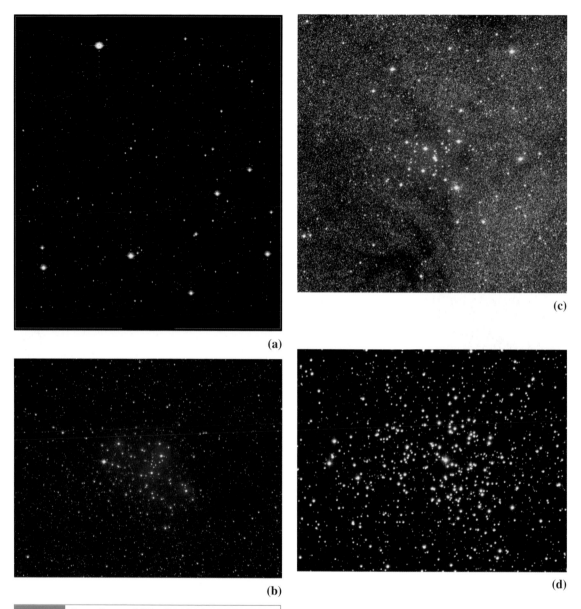

(a)

(b)

(c)

(d)

Fig 13.1. Open clusters. (a) The Pleiades (M45) in Taurus. The nebulosity seen here is starlight reflected from interstellar dust. (b) M6, also known as the Butterfly Cluster. (c) M7 in Scorpius. Its distance is about 300 pc, and it is about 8 pc across. (d) M37 in Aurigi, at a distance of 1.5 kpc. [(a) Courtesy of 2MASS/UMASS/IPAC/NASA/JPL/Caltech; (b)–(d) NOAO/AURA/NSF]

$$v_T = v \sin A \qquad (13.3)$$

$$\tan A = v_T/v_r \qquad (13.4)$$

The relationship between proper motion and tangential velocity is shown in Fig. 13.4. The proper motion μ, expressed in radians per sec-

ond, is just the transverse velocity divided by the distance to the star d:

$$\mu(\text{rad/s}) = v_T(\text{km/s})/d(\text{km}) \qquad (13.5)$$

The greater the transverse velocity, the faster the star will appear to move across the sky. Also, the closer the star is, the greater the motion across the sky. In general, proper motions are very small, amounting to a few arc seconds per year, or less. (The largest proper motion is 10.3 arc sec/yr for Barnard's star.) For this reason, we would like to rewrite equation (13.5), expressing μ in arc seconds per year, v in kilometers per second, and d in

(a)

(b)

(c)

(d)

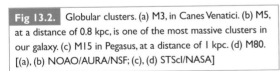

Fig 13.2. Globular clusters. (a) M3, in Canes Venatici. (b) M5, at a distance of 0.8 kpc, is one of the most massive clusters in our galaxy. (c) M15 in Pegasus, at a distance of 1 kpc. (d) M80. [(a), (b) NOAO/AURA/NSF; (c), (d) STScI/NASA]

parsecs. We can then rewrite equation (13.5) as

$$\frac{v_T}{[1 \text{ km/s}]} = \frac{\mu}{[1 \text{ rad/s}]} \frac{d}{[1 \text{ km}]}$$

$$\times \frac{[1 \text{ rad}]}{[2.063 \times 10^5 \text{ arc sec/rad}]}$$

$$\times \frac{1 \text{ yr}}{[3.156 \times 10^7 \text{s/yr}]} \times \frac{[3.086 \times 10^{13} \text{ km}]}{[1 \text{ pc}]}$$

$$v_T(\text{km/s}) = 4.74\mu(\text{arc sec/yr})d(\text{pc}) \qquad (13.6)$$

In general, we can measure the radial velocity (from the Doppler shift) and we can also measure the proper motion. If we know the transverse velocity, we can find the distance from

$$d = v_T/4.74\mu \qquad (13.7)$$

If, instead of v_T, we know A, then we can use equation (13.4) in equation (13.7) to give

$$d = v_r \tan A/4.74\mu \qquad (13.8)$$

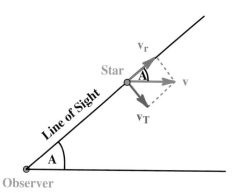

Fig 13.3. Space velocity. The velocity of the star is *v*, which makes an angle *A* with the line of sight. The radial and tangential components of the velocity are v_r and v_T, respectively.

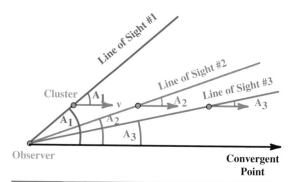

Fig 13.5. Convergent point. As the cluster moves farther away, the angle *A* between **v** and the line of sight approaches zero. That is $A_1 > A_2 > A_3$. As this happens, the cluster approaches one line of sight, the convergent point. Note that this figure is exaggerated to show the effect. Real clusters do not move that much over the times we could observe them.

With a cluster of stars, we can compare the proper motion with the rate at which the angular size of the cluster changes to find *A*. To see how this works, let us consider the case of a cluster moving away from us (positive radial velocity). As the cluster moves away (Fig. 13.5), *A* becomes smaller and approaches zero. As the cluster moves farther away, the proper motion approaches zero (by equation 13.5). Thus, the cluster appears to be heading toward a particular point in the sky. We

call this point the *convergent point* of the cluster. We can see from the figure that the angle between the *current line of sight* and the line of sight to the convergent point is the *current value of A*. Similar reasoning applies to clusters that are moving toward us. They are moving away from their convergent point, so we find it by extrapolating their motion backwards in time.

To apply these ideas, we take a series of images a number of years apart. From the proper motion, and the change in angular size, we can find the convergent point, and therefore we know *A*. This is shown schematically in Fig. 13.5.

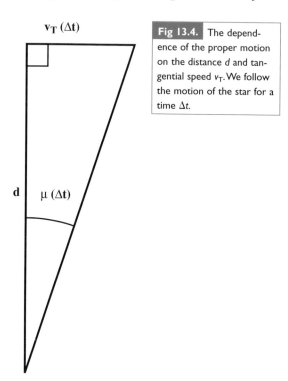

Fig 13.4. The dependence of the proper motion on the distance *d* and tangential speed v_T. We follow the motion of the star for a time Δt.

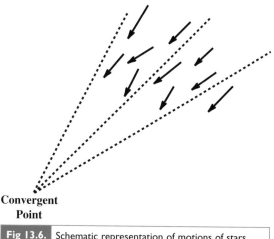

Fig 13.6. Schematic representation of motions of stars within a cluster. Arrows represent total motions.

We measure the radial velocity v_r and proper motion μ. Since we know A, v_r and μ, we can find d from equation (13.8). The best determination of a distance to a moving cluster is the *Hyades*. This determination is an important cornerstone in our determination of distances to more distant objects in our galaxy and in other galaxies.

13.3 | Clusters as dynamical entities

In this section we look at the internal dynamics of star clusters. If the gravitational forces between the stars are sufficient to keep the cluster together, we say the cluster is gravitationally bound. (We have already discussed the idea of gravitational binding when we talked about binary stars, in Chapter 5 and Chapter 12.) However, gravity does more than assure the overall existence of the cluster. As stars move around within the cluster, pairs of stars will pass near each other. The gravitational attraction between the two stars in the pair will alter the motion of each star. The momentum and energy of each star will change in this gravitational encounter. Thus, these encounters alter the distribution of speeds, the number of stars traveling at a given speed. If there has been sufficient time for many encounters to occur, the distribution of speeds will reach some equilibrium. For every star that suffers a collision changing its speed from v_1, there is another collision in which some other star has its speed changed to v_1. (We refer to these gravitational encounters as collisions, even though the stars never actually get close enough for their surfaces to touch.) When a cluster has reached this stage, we say that it is *dynamically relaxed*.

13.3.1 The virial theorem
In a dynamically relaxed system, the kinetic and potential energies are related in a very specific way. This relationship is known as the *virial theorem*. In this section we derive it.

We begin with a collection of N particles. (We can think of each star in a cluster or each atom in a gas cloud being represented by a particle.) To simplify the calculation we assume that all particles have the same mass m. The final result would be the same if we allowed for different masses.

(See Problem 13.10.) We let the position of the ith particle, relative to some origin, be $\mathbf{r}_i$. If we have two particles, i and j, the vector giving their separation is $\mathbf{r}_j - \mathbf{r}_i$, as shown in Fig. 13.7. We let $\mathbf{F}_i$ be the net force on the ith particle. We can therefore write the equation of motion for this particle as

$$\mathbf{F}_i = m\,\frac{d^2\mathbf{r}_i}{dt^2} \qquad (13.9)$$

We are looking for a relationship between various types of energy. The vector dot product between force and distance will give us an energylike quantity. We therefore take the dot product of $\mathbf{r}_i$ with both sides of equation (13.9) and then sum the resulting quantities over all the particles to give

$$\sum_{i=1}^{N} \mathbf{r}_i \cdot \mathbf{F}_i = m \sum_{i=1}^{N} \mathbf{r}_i \cdot \left(\frac{d^2\mathbf{r}_i}{dt^2}\right) \qquad (13.10)$$

We can rearrange this by using the following:

$$\frac{d^2(\mathbf{r}_i^2)}{dt^2} = \frac{d}{dt}\left(2\mathbf{r}_i \cdot \frac{d\mathbf{r}_i}{dt}\right)$$
$$= 2\left(\frac{d\mathbf{r}_i}{dt}\right)^2 + 2\mathbf{r}_i \cdot \left(\frac{d^2\mathbf{r}_i}{dt^2}\right) \qquad (13.11)$$

where we have used the fact that for any vector $\mathbf{x}$,

$$\mathbf{x}^2 \equiv \mathbf{x} \cdot \mathbf{x} = x^2$$

This gives

$$\sum_{i=1}^{N} \mathbf{r}_i \cdot \mathbf{F}_i = \left(\frac{1}{2}\right)\frac{d^2}{dt^2}\left(\sum_{i=1}^{N} m\mathbf{r}_i^2\right) - \sum_{i=1}^{N} m\left(\frac{d\mathbf{r}_i}{dt}\right)^2 \qquad (13.12)$$

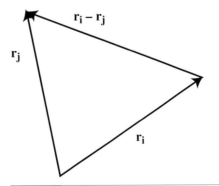

Fig 13.7. Position vectors $\mathbf{r}_i$ from the origin to the ith particle and $\mathbf{r}_j$ from the origin to the jth particle.

We now introduce the velocity of the ith particle, $\mathbf{v}_i$, defined as

$$\mathbf{v}_i = d\mathbf{r}_i/dt \tag{13.13}$$

We also introduce the quantity I, defined as

$$I = \sum_{i=1}^{N} m\mathbf{r}_i^2 \tag{13.14}$$

If $\mathbf{r}_i$ were the distance of particle i from some axis, then I would be the moment of inertia about that axis. However, an axis is a line, and $\mathbf{r}_i$ is measured from a point (the origin). Therefore, I is a quantity similar to, but not the same as, the familiar moment of inertia. (For most normal shaped objects it has a numerical value close to the moment of inertia.) We also set the total kinetic energy K equal to the sum of the kinetic energies of the individual particles,

$$K = \left(\frac{1}{2}\right) \sum_{i=1}^{N} m\mathbf{v}_i^2 \tag{13.15}$$

In terms of these quantities, equation (13.12) becomes

$$\sum_{i=1}^{N} \mathbf{r}_i \cdot \mathbf{F}_i = \frac{1}{2} \frac{d^2 I}{dt^2} - \sum_{i=1}^{N} m\mathbf{v}_i^2 \tag{13.16}$$

$$= \frac{1}{2} \frac{d^2 I}{dt^2} - 2K \tag{13.17}$$

We evaluate $\sum \mathbf{r}_i \cdot \mathbf{F}_i$ for the $\mathbf{F}_i$ being the gravitational forces between particles. We let

$$\mathbf{r}_{ij} = \mathbf{r}_j - \mathbf{r}_i \tag{13.18}$$

be the vector distance between the two particles. The net force on a given particle is the sum of the forces exerted on it by all of the other particles. Therefore

$$\mathbf{F}_i = \sum_{\substack{j=1 \\ i \neq j}}^{N} \frac{m^2 G \mathbf{r}_{ij}}{|\mathbf{r}_{ij}|^3} \tag{13.19}$$

which gives us

$$\sum_{i=1}^{N} \mathbf{r}_i \cdot \mathbf{F}_i = \sum_{i=1}^{N} \sum_{\substack{j=1 \\ i \neq j}}^{N} Gm^2 \left(\frac{\mathbf{r}_i \cdot \mathbf{r}_{ij}}{|\mathbf{r}_{ij}|^3} \right) \tag{13.20}$$

Since the i and j both go over the full range, 1 to N, we could interchange the i and j on the right-hand side of equation (13.20) without really changing anything. If we rewrite the right-hand side, interchanging i and j, and then adding it to the right-hand side, we will have a quantity equal to twice the original right-hand side. This gives us

$$\sum_{i=1}^{N} \mathbf{r}_i \cdot \mathbf{F}_i = \left(\frac{1}{2}\right) \sum_{i=1}^{N} \sum_{\substack{j=1 \\ i \neq j}}^{N} Gm^2 \left(\frac{\mathbf{r}_i \cdot \mathbf{r}_{ij}}{|\mathbf{r}_{ij}|^3} + \frac{\mathbf{r}_j \cdot \mathbf{r}_{ji}}{|\mathbf{r}_{ji}|^3} \right) \tag{13.21}$$

$$= \left(\frac{1}{2}\right) \sum_{i=1}^{N} \sum_{\substack{j=1 \\ i \neq j}}^{N} Gm^2 \left(\frac{\mathbf{r}_i \cdot (\mathbf{r}_j - \mathbf{r}_i) + \mathbf{r}_j \cdot (\mathbf{r}_i - \mathbf{r}_j)}{|\mathbf{r}_{ij}|^3} \right) \tag{13.22}$$

where we have also used the fact that $|\mathbf{r}_{ij}| = |\mathbf{r}_{ji}|$. This procedure actually allows for a simplification. To see this, we first multiply out the numerator on the right-hand side to obtain

$$\sum_{i=1}^{N} \mathbf{r}_i \cdot \mathbf{F}_i = \left(\frac{1}{2}\right) \sum_{i=1}^{N} \sum_{\substack{j=1 \\ i \neq j}}^{N} Gm^2$$
$$\times \left(\frac{\mathbf{r}_i \cdot \mathbf{r}_j - \mathbf{r}_i^2 - \mathbf{r}_j^2 + \mathbf{r}_j \cdot \mathbf{r}_i}{|\mathbf{r}_{ij}|^3} \right) \tag{13.23}$$

We also note that

$$|\mathbf{r}_{ij}|^2 = (\mathbf{r}_i - \mathbf{r}_j) \cdot (\mathbf{r}_i - \mathbf{r}_j)$$
$$= r_i^2 + r_j^2 - \mathbf{r}_i \cdot \mathbf{r}_j - \mathbf{r}_j \cdot \mathbf{r}_i \tag{13.24}$$

Substituting into equation (13.23) gives

$$\sum_{i=1}^{N} \mathbf{r}_i \cdot \mathbf{F}_i = \left(\frac{1}{2}\right) \sum_{i=1}^{N} \sum_{\substack{j=1 \\ i \neq j}}^{N} Gm^2 \left(-\frac{|\mathbf{r}_{ij}|^2}{|\mathbf{r}_{ij}|^3} \right) \tag{13.25}$$

which simplifies to

$$\sum_{i=1}^{N} \mathbf{r}_i \cdot \mathbf{F}_i = -\left(\frac{1}{2}\right) \sum_{i=1}^{N} \sum_{\substack{j=1 \\ i \neq j}}^{N} \left(\frac{Gm^2}{|\mathbf{r}_{ij}|} \right) \tag{13.26}$$

The term on the right-hand side is the sum of the gravitational potential energies of each pair of particles. Note that each pair appears twice in the sum, since the energy of the pair is independent of which particle in the pair we count first. For example, for particle 1 and 2, both the quantities $Gm^2/|\mathbf{r}_{12}|$ and $Gm^2/|\mathbf{r}_{21}|$ appear. This means that the double sum on the right-hand side of equation (13.26) gives us twice the gravitational potential energy, but there is also a factor of one-half in front, so the right-hand side is equal to the gravitational potential energy U. We can therefore rewrite equation (13.17) as

$$\left(\frac{1}{2}\right) \frac{d^2 I}{dt^2} = 2K + U \tag{13.27}$$

If we take the time average of these quantities over a sufficiently long time, the left-hand side approaches zero. This leaves

$$0 = 2\langle K \rangle + \langle U \rangle \tag{13.28}$$

where the $\langle \rangle$ represents the time average of the enclosed quantity. Equation (13.28) is the simplest form of the virial theorem.

The virial theorem applies to any gravitationally bound system that has had sufficient time to come to equilibrium. Even simple systems, like binary stars, obey the virial theorem (see Problem 13.11). If the orbits are circular then $K = -U/2$ at all points. For elliptical orbits, r and v are changing, so K and U are changing, while their sum E is fixed. This means that we have to average over a whole orbit to get $\langle K \rangle = -\langle U \rangle/2$.

Remember, for any system, the total energy is

$$E = K + U \tag{13.29}$$

So for a system to which the virial theorem, in the form of equation (13.28) applies, we set $\langle K \rangle = -\langle U \rangle/2$, to give

$$E = \langle U \rangle/2 \tag{13.30}$$

(We don't have to take the time average of E, since E is constant.) Remember, the gravitational potential energy, defined so that it is zero when the particles are infinitely far apart, is negative. Therefore, the total energy of a bound system is negative. This means that we have to put energy in to break up the system.

13.3.2 Energies

We now look at the kinetic and potential energies of a cluster. In Section 9.1, we saw that the gravitational potential energy for a constant density sphere of mass M and radius R is

$$U = -(3/5)GM^2/R$$

Example 13.1 Potential energy for a globular cluster
Find the gravitational potential energy for a spherical cluster of stars with 10^6 stars each of 0.5 $M_\odot$. The radius of the cluster core is 5 pc.

SOLUTION
We use the above equation to give

$$U = -\frac{(0.6)(6.67 \times 10^{-8}\,\text{dyn cm}^2/\text{g}^2)[(0.5)(2 \times 10^{33}\,\text{g})(10^6)]^2}{(5\,\text{pc})(3.18 \times 10^{18}\text{cm/pc})}$$

$$= -2.5 \times 10^{51}\,\text{erg}$$

If the virial theorem applies, then the total energy is $E = -U/2 = 1.2 \times 10^{51}$ erg.

We now look at the kinetic energy. In a cluster of stars, the kinetic energy is in the random motions of the stars. If the cluster has N stars, each of mass m, the kinetic energy is

$$K = \left(\frac{1}{2}\right)\sum_{i=1}^{N} mv_i^2$$

$$= \left(\frac{1}{2}\right)m\sum_{i=1}^{N} v_i^2$$

The total mass of the cluster is $M = mN$; so

$$K = \left(\frac{1}{2}\right)(mN)\left(\frac{1}{N}\right)\sum_{i=1}^{N} v_i^2 \tag{13.31}$$

If we take the sum of N quantities and then divide by N, the result is the average of that quantity. Therefore $(1/N)\sum v_1^2$ is the average of the quantity v^2. We write this average as $\langle v^2 \rangle$. Remembering that $mN = M$, equation (13.31) becomes

$$K = (1/2)M\langle v^2 \rangle \tag{13.32}$$

If we put this and the potential energy into the virial theorem, we find

$$M\langle v^2 \rangle = (3/5)GM^2/R \tag{13.33}$$

Dividing both sides by M gives

$$\langle v^2 \rangle = (3/5)GM/R \tag{13.34}$$

The quantity $\langle v^2 \rangle$ is the mean (average) of the square of the velocity. If we take the square root of this quantity, we have the *root mean square velocity* or *rms velocity*. It is a measure of the internal motions in the cluster.

Example 13.2 The rms velocity in a cluster
Find the rms velocity for the cluster used in the previous example.

SOLUTION
We use equation (13.34) with the given quantities:

$$\langle v^2 \rangle = -\frac{(0.6)(6.67 \times 10^{-8}\,\text{dyn cm}^2/\text{g}^2)(0.5)(10^6)(2 \times 10^{33}\,\text{g})}{(5\,\text{pc})(3.18 \times 10^{18}\,\text{cm/pc})}$$

$$= 2.5 \times 10^{12}\,(\text{cm/s})^2$$

Taking the square root gives

$$v_{\text{rms}} = 1.6 \times 10^6\,\text{cm/s}$$

$$= 16\,\text{km/s}$$

We can relate the gravitational potential energy to the *escape velocity* v_e, the speed with which an object must be launched from the surface to escape permanently from the cluster. Consider a particle of mass m, moving outward from the surface at speed v_e. If the object escapes, it must get so far away that the potential energy is essentially zero. Since the kinetic energy is always greater than or equal to zero, the total energy of the object far away must be greater than or equal to zero. Since the total energy is conserved, the total energy for an escaping object must be zero or positive when it is launched. The kinetic energy of the particle is

$$KE = (1/2)mv_e^2 \qquad (13.35)$$

Since it is at the surface of the sphere of mass M and radius R, its potential energy is just

$$PE = -GmM/R \qquad (13.36)$$

For the total energy to be zero (the condition that the particle barely escapes), $KE = -PE$, giving

$$v_e^2 = 2GM/R \qquad (13.37)$$

Note that the escape velocity is approximately twice the rms speed. For a gravitationally bound system, we would expect $v_e > v_{rms}$. If it were the other way around, many particles would have speeds greater than the escape velocity and would escape. The cluster would not be gravitationally bound.

13.3.3 Relaxation time

In any given cluster, stars will be in orbits about the center of mass. Pairs of stars can exchange energy and momentum via gravitational encounters. (By exchange, we just mean that some energy and momentum is transferred, not that each one acquires the other's energy and momentum.) As we have said, if there are enough collisions, an equilibrium velocity distribution will be reached. Not all are moving at the average speed. Some move faster and others move slower. In a cluster, there will be some stars with speeds greater than v_e. They will escape. This alters the velocity distribution by removing the highest velocity stars. The remaining stars must adjust, re-establishing the equilibrium velocity distribution. (This is equivalent to the evaporation of a puddle of water on the

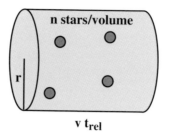

Fig 13.8. Calculation of relaxation time.

Earth. The highest speed molecules escape, leaving the water a little cooler.)

In discussing cluster dynamics, it is important to remember that we are studying the motions of the stars in a cluster relative to the center of mass of the cluster. We are not concerned with the overall motion of the cluster through space.

We call the time it takes to re-establish equilibrium the *relaxation time*, t_{rel}. We can estimate t_{rel} by following a single star as it moves through the cluster (Fig. 13.8). We assume that there are n stars per unit volume in the cluster. We would like to know how long our star will go between collisions with other stars. That depends, in part, on how we define a collision. We would like to define a distance r and say that if two stars pass within this distance we will count it as a collision. We define r so that the potential energy of the star is equal in magnitude to the kinetic energy of our star. If our star is moving with speed v, this means that r is defined by

$$Gm^2/r = mv^2/2 \qquad (13.38)$$

We can think of our star as sweeping out a cylinder in a given time t_{rel}. The radius of the cylinder is r, and the length is vt_{rel}. Therefore, the volume swept out is $\pi r^2 vt_{rel}$. The number of stars in this volume is n multiplied by the volume. If we define t_{rel} so that it is the time for one collision, we have the condition

$$n\left(\pi r^2 v\, t_{rel}\right) = 1 \qquad (13.39)$$

Solving for t_{rel} gives

$$t_{rel} = 1/n\pi r^2 v \qquad (13.40)$$

Substituting for r from equation (13.38) gives

$$t_{rel} = v^3/4\pi G^2 m^2 n \qquad (13.41)$$

The number of stars per unit volume is simply

$$n = \frac{N}{(4\pi/3)R^3}$$

$$= \frac{M/m}{(4\pi/3)R^3}$$

(13.42)

Substituting into equation (13.41) gives

$$t_{rel} = \frac{v^3 R^3}{3G^2 mM}$$

$$= \frac{(R/v)v^4 R^2 (M/m)}{3G^2 M^2}$$

(13.43)

Using the virial theorem to eliminate two factors of v^2, and ignoring numerical factors that are close to unity (since this is just an estimate), this simplifies to

$$t_{rel} \cong (R/v)(M/m)$$

$$= NR/v$$

(13.44)

Equation (13.44) is just an estimate in which the effects of a few close encounters dominate. However, for every close encounter that a star has, it has many more distant encounters. This is because the effective area for an encounter at distance r is proportional to r^2. The effect of any one distant encounter is small. However, because there are so many of them, they actually dominate the relaxation process and speed it up. A more detailed calculation shows that the effect of many distant encounters is to reduce t_{rel} by a factor of $12 \ln(N/2)$.

Example 13.3 Relaxation time
Estimate the relaxation time for the cluster discussed in the two previous examples.

SOLUTION
We use equation (13.44). For the speed v, we use the calculated v_{rms}:

$$t_{rel} = \frac{(10^6)(5 \text{ pc})(3.18 \times 10^{18} \text{ cm/pc})}{1.6 \times 10^6 \text{ cm/s}}$$

$$= 1.0 \times 10^{19} \text{ s}$$

$$= 3 \times 10^{11} \text{ yr}$$

$$= 300 \text{ Gyr}$$

If we apply the correction for distant encounters, $12 \ln(N/2) = 1.6 \times 10^2$, t_{rel} is reduced to 2 Gyr. For comparison, we will see in Chapter 21, that the approximate age of the universe is 15 Gyr. So, if we don't include the correction for distant encounters, then the relaxation time would be longer than the age of the universe, but, with that correction, that is not a problem.

We can also define an *evaporation time*, which is the time for a significant number of stars to leave the cluster. The evaporation time is approximately 100 times the relaxation time. For the cluster in the above example, the evaporation time would be 200 Gyr, much more than the age of the universe. So, that cluster has had enough time to become relaxed, but not evaporate.

Once relaxation takes place, the velocity distribution will evolve toward a Maxwell–Boltzmann distribution, given for a gas in equation (9.16). For a cluster of stars, we let kT become $(1/3)mv^2_{rms}$, the number of stars in the velocity range v to $v + dv$ is given by:

$$n(v) \, dv \propto \left(\frac{v^2}{v^3_{rms}} \right) \exp\left(-\frac{3v^2}{2v^2_{rms}} \right) dv$$

(13.45)

Calculations show that a core denser than the outer parts of the cluster will develop. It has been speculated that, at some point, the core can become so massive that it collapses to form a large black hole. Recent observations of some globular clusters have revealed the existence of a luminous extended object in the center. The size of these objects are in the 0.1 pc range. In each case the object is bluer than the rest of the cluster, meaning that it is not an unresolved group of red stars.

13.3.4 Virial masses for clusters

For dynamically relaxed clusters, we can use the virial theorem to estimate the mass of the cluster. For a uniform cluster, with N stars, each of mass m, and the total mass of the cluster $M = mN$, the cluster potential energy is

$$U = -\left(\frac{3}{5} \right) \frac{GM^2}{R}$$

where R is the radius of the cluster. The kinetic energy is (equation 13.32)

$$K = (1/2)M \langle v^2 \rangle$$

Substituting these into the virial theorem gives

$$M \langle v^2 \rangle = 3 \, GM^2/5R$$

(13.46)

Solving for M, we have

$$M = \left(\frac{5}{3}\right)\frac{\langle v^2 \rangle R}{G} \qquad (13.47)$$

It is important to understand which motions we are talking about. The cluster has some overall motion of its center of mass, shared by all the stars in the cluster. The stars have individual motions within the cluster (with respect to the center of mass of the cluster). The net motion of each star is the vector sum of these two motions, and that is what we observe. In equation (13.47) the quantity $\langle v^2 \rangle$ is the average of the square of the star velocities with respect to the center of mass of the cluster.

The best way for us to measure the velocities of individual stars is through their Doppler shifts. However, this only gives us the component of the velocity along the line of sight. This means that we are measuring $\langle v_r^2 \rangle$ rather than $\langle v^2 \rangle$. However, if the internal motions of the stars are random, we can relate these two quantities.

Suppose we resolve the motion of any star into its components in an (x, y, z) coordinate system. The velocity can then be written in terms of its components as

$$\mathbf{v} = v_x\hat{\mathbf{x}} + v_y\hat{\mathbf{y}} + v_z\hat{\mathbf{z}} \qquad (13.48)$$

where $\hat{\mathbf{x}}$, $\hat{\mathbf{y}}$ and $\hat{\mathbf{z}}$ are the unit vectors in the three directions, respectively. To be definite, we can let the x-direction correspond to the line of sight. The square of v, which is $\mathbf{v} \cdot \mathbf{v}$, is simply the sum of the squares of the components,

$$v^2 = v_x^2 + v_y^2 + v_z^2 \qquad (13.49)$$

If we then take the average of both sides of the equation, we have

$$\langle v^2 \rangle = \langle v_x^2 \rangle + \langle v_y^2 \rangle + \langle v_z^2 \rangle$$

However, if the motions are random, the averages of the squares of the components should be the same for all directions. This means that

$$\langle v_x^2 \rangle = \langle v_y^2 \rangle = \langle v_z^2 \rangle \qquad (13.50)$$

Using this, equation (13.49) becomes

$$\langle v^2 \rangle = 3\langle v_x^2 \rangle$$

and since the x-direction is the one corresponding to the line of sight, $v_r = v_x$, so

$$\langle v^2 \rangle = 3\langle v_r^2 \rangle \qquad (13.51)$$

If we substitute this into the virial theorem (equation 13.47), the mass is given by

$$M = \frac{5 \langle v_r^2 \rangle R}{G} \qquad (13.52)$$

Example 13.4 Virial mass of cluster
Find the virial mass of a cluster with $\langle v_r \rangle = 10$ km/s and $R = 5$ pc.

SOLUTION
From equation (13.52) we have

$$M = \frac{(5)(1.0 \times 10^6 \text{ cm/s})^2(5 \text{ pc})(3.18 \times 10^{18} \text{ cm/pc})}{6.67 \times 10^{-8} \text{ dyn cm}^2/\text{g}^2}$$

$$= 1.2 \times 10^{39} \text{ g}$$

$$= 6 \times 10^5 \, M_\odot$$

When we talked about binary stars (Chapter 5), we noted that the best way to measure the mass of an object, or a group of objects, is to measure their gravitational effects on other objects. The gravitational effects are independent of how bright the objects are; they depend only on how massive they are. Using virial masses is an extension of these ideas. The more massive the cluster, the greater the internal motions that we will observe. To determine $\langle v_r \rangle$, it is not even necessary to measure radial velocities for all the stars in the cluster. We just need a representative sample.

What are the limitations of this method? An important one is that we don't know if any particular cluster is dynamically relaxed, or even gravitationally bound. We may measure large internal motions in an unbound system and mistake them for bound motions in a more massive system. This can introduce errors that are off by as much as a few orders of magnitude. Our calculation of the relaxation time suggests that all but the youngest clusters should be relaxed. We will talk about indicators of age of a cluster in the next section. Another limitation can be from geometric effects. We may observe clusters that are elliptical, rather than spherical. Or we may observe clusters that don't have a uniform density. The most likely variation is having a higher density in the center. The effect of these geometric effects can be to produce errors of order unity (see Problem 13.16). For most applications, astronomers use the virial theorem, knowing that it is a technique that may be off by a factor of two or three. But this can be very useful for measuring the masses of a variety of astronomical systems. We will look more at virial masses when we talk about the masses of interstellar clouds (Chapter 14) and the masses of clusters of galaxies (Chapter 18).

13.4 | HR diagrams for clusters

By studying the HR diagram for a cluster, we are studying a group of stars with a common distance. We can study their relative properties without knowing what their actual distance is. If we do know the distance to the cluster, we plot directly the absolute magnitudes on the HR diagram. If we don't know the distance, we plot the apparent magnitudes. We then see how many magnitudes we would have to shift the diagram up or down to calculate the right absolute magnitudes for each spectral type. The amount of shift gives the distance modulus for the cluster, and therefore the distance. This procedure is known as *main sequence fitting*. It is like doing a spectroscopic parallax measurement, but it uses the information from all of the main sequence stars in the cluster. This is more accurate than studying a single star.

The HR diagram for a group of galactic clusters is shown schematically in Fig. 13.9. Note that the lower (cooler or later) part of the main sequence is the same for all the clusters shown. For each cluster, there is some point at which the main sequence stops. Beyond that point, no hotter stars are seen on the main sequence. The hotter stars all appear to be above the main sequence. The point at which this happens for a given cluster is called the *turn-off point*. Stars of earlier spectral type (hotter than) the turn-off point appear above the main sequence, meaning that they are more luminous, and therefore larger than main sequence stars of the same spectral type. Each cluster has its turn-off point at a different spectral type. Data for one cluster are shown in Fig. 13.10, to see the scatter in the points.

We interpret this behavior as representing stellar aging, in which stars use up their basic fuel supply, as described in Chapter 10. Hotter, more massive, stars evolve faster than the cooler, low mass stars, and leave the main sequence sooner. We assume that the stars in a cluster were formed

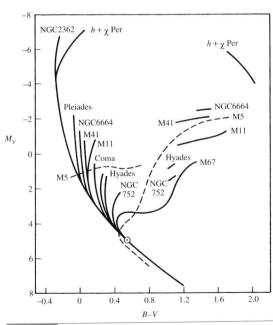

Fig 13.9. Schematic HR diagrams for various galactic clusters.

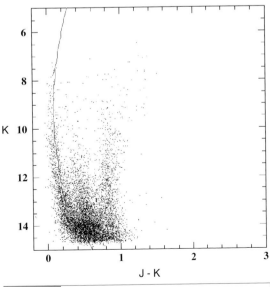

Fig 13.10. Color–magnitude diagram for a galactic cluster H and Chi Persei (the double cluster shown in Fig. 2.4). [Courtesy of 2MASS/UMASS/IPAC/NASA/JPL/Caltech]

at approximately the same time. As a cluster ages, later and later spectral types evolve away from the main sequence. This means the turn-off point shifts to later spectral types as the cluster ages. We can tell the relative age of two clusters by comparing their turn-off points. If we know how long different spectral type stars actually stay on the main sequence, we can tell the absolute age of a cluster from its turn-off point.

We note that there are some galactic clusters, not shown in Fig. 13.9, that are missing the lower (cooler) end of the main sequence. We think that these clusters are very young. The lower mass stars are still in the process of collapse, and have not yet reached the main sequence. (We think that lower mass stars take longer to collapse than higher mass stars.)

An HR diagram for a composite of globular clusters is shown in Fig. 13.11. These appear to be different from the HR diagrams for galactic

clusters. For globular clusters, only the lower (cooler) part of the main sequence is present. All earlier spectral types have turned off the main sequence. This tells us that globular clusters must be very old. Globular clusters contain a large number of red giants. In Chapter 10 we saw that the red giant state is symptomatic of old age in a star.

13.5 | The concept of populations

There is another important difference between the stars in galactic and globular clusters. It concerns the abundances of "metals", elements heavier than hydrogen and helium. Many globular clusters have stars with very low metal abundances, while galactic cluster star are higher in metal abundance. We refer to high metal stars as *population I* stars and low metal stars as *population II* stars. We have a general sense that population I stars represent younger, more recently formed stars. We interpret the metal abundance differences as reflecting the conditions in our galaxy at the time each type of star was formed. When the older stars were formed, our galaxy had only hydrogen and helium. When the newer stars were formed, the galaxy had been enriched in the metals. This enriched material comes from nuclear processing in stars, followed by spreading into the interstellar medium, especially through supernova explosions.

The differences between galactic and globular clusters start us thinking about old and new material in our galaxy. The globular clusters are older, and form a spherical distribution, while the galactic clusters are newer and are confined to the galactic disk. This suggests that, a long time ago, star formation took place in a large spherical volume, but now it only takes place in the disk. This is supported by the fact that globular clusters are free of interstellar gas and dust, the material out of which new stars can form, while galactic clusters are sometimes associated with gas and dust.

The concept of stellar populations is important in our understanding of the evolution of our galaxy. This will be discussed farther in Chapters 14–16.

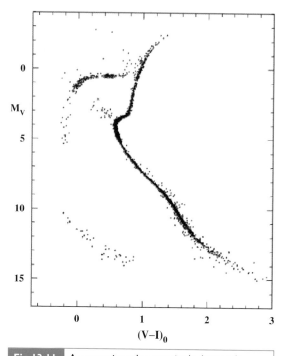

Fig 13.11. A composite color–magnitude diagram for a "metal-poor" globular cluster, constructed from real photometric data from several Milky Way globular clusters including M3, M55, M68, NGC 6397 and NGC 2419. [William Harris, McMaster University, STScI/NASA 10th Anniversary Symposium Proceedings (STScI, Baltimore), May 2000]

Chapter summary

In this chapter we saw what could be learned from studying clusters of stars. Clusters are useful in studying a variety of astronomical problems because they provide us with groupings of stars all at the same distance.

We looked at an important technique for determining the distances to nearby clusters. It is free of any assumptions about luminosities, and serves as an important cornerstone in our distance determination scheme.

We looked at the dynamical properties of clusters. Clusters have internal motions which eventually reach some equilibrium velocity distribution.

This distribution is brought about by star–star gravitational encounters, and the effects of many distant encounters are important. We saw that it is possible for gravitationally bound clusters to evaporate by losing the stars that are moving the fastest.

We developed the idea of stellar populations, signifying old and new material in the galaxy. Globular clusters seem associated with the old material, and galactic (open) clusters seem associated with the newer material. Some of these differences are very evident in comparing HR diagrams, as well as in comparing the metal content of the stars in the two types of clusters.

Questions

13.1. Suppose you had photographs of a cluster taken ten years apart. How would you use those photographs to find the convergent point of the cluster?

13.2. List the distance measurement techniques that we have encountered so far in this book, and estimate the distance range over which each is useful.

13.3. What is the relationship between main sequence fitting and spectroscopic parallax?

13.4. What is the significance of the main sequence turn-off point of a cluster?

13.5. Construct a table, contrasting the properties of globular clusters and galactic clusters.

13.6. If a cluster is moving through space at 50 km/s, how should this motion be included in $\langle v^2 \rangle$, which appears in equation (13.34)?

13.7. In a cluster for which $v_e > v_{rms}$, some stars can still escape. How does this happen?

13.8. What are the advantantages and disadvantages of using the virial theorem to determine the mass of a cluster?

Problems

13.1. The Hyades has a proper motion of 0.07 arc sec/yr and appears 26° from its convergent point. The radial velocity is 35 km/s. (a) How far away is the cluster? (b) What is the actual speed of the cluster?

13.2. Suppose we can detect proper motions down to 0.1 arc sec/yr. How far away can we detect a transverse velocity of 10 km/s?

13.3. Suppose we have two photographs of a cluster, taken ten years apart. In the second photograph the cluster has moved over by 1.0 arc sec. Two stars that were originally 20.0 arc sec apart are now 19.5 arc sec apart. What is the angle between the current line of sight to the cluster and the line of sight to the convergent point?

*13.4. In Chapter 4, we discussed the resolving power of gratings. How is the minimum radial velocity shift we can measure related to this resolving power?

*13.5. For the globular cluster treated in the example in this chapter, estimate the average time between collisions in which the stars actually hit.

13.6. Compare the rms speeds in a typical globular cluster with those in a typical galactic cluster.

13.7. Verify that equation (13.44) can be obtained, as outlined, from equation (13.43).

13.8. The *crossing time* for a cluster is the average time for a star to move from one side of the cluster to the other. (a) What is the relationship between the crossing time and the relaxation time? (b) How would you explain that relationship?

*13.9. Calculate the relaxation time for the typical galactic cluster discussed in this chapter, and compare it with that for the typical globular cluster.

*13.10. Show that the derivation of the virial theorem gives the same result if we allow the masses of the stars to be different.

13.11. Show that binary stars obey the virial theorem.

13.12. Suppose we have a gravitationally bound object in virial equilibrium. Show that if more than half of the mass of the system is lost, with no change in the velocities of the remaining material, then the system will be unbound.

13.13. Find an expression that gives the virial mass in solar masses, when rms velocity is in km/s and the size is in pc.

13.14. (a) Find the virial mass of a cluster for which the rms radial velocity dispersion is 5 km/s and the radius is 15 pc. (b) What are U, K and E for this cluster?

13.15. Compare the evaporation times for typical galactic and globular clusters discussed in this chapter. Include the correction for the effects of distant collisions.

*13.16. Find an expression for the gravitational potential energy for a spherical cluster of mass, M, and radius, R, where the density falls as $1/r^2$. (Hint: Integrate equation (9.6) with a variable density.) How does your result compare with the case of a uniform density sphere?

Computer problem

13.1. Find the virial masses, and U, K and E for clusters with the properties shown in Table 13.1.

Table 13.1.

v_{rms} (km/s)	R (pc)
2	2
2	5
2	10
5	2
5	5
5	10
10	2
10	5
10	10

Part IV

The Milky Way

Most of the light we can see from our galaxy appears as a narrow band around the sky. From its appearance, we think that we are in the plane of a disk, and that this disk looks something like the Andromeda galaxy. However, our location within our own galaxy makes its structure very difficult to study. In this part we will see both how we learn about our galaxy and what we have learned about it so far.

Most of the light that we see comes directly from stars. Among all the objects we can see, the stars provide most of the mass. Averaged over the whole galaxy, the gas and dust between the stars – the interstellar medium – contains only about 1% as much mass as the stars themselves. Of the interstellar medium, 99% of the mass is in the form of gas, and 1% of the mass is in the form of dust. However, this small amount of dust is very efficient at blocking light, making optical observations of distant objects difficult.

We expect that stars form out of interstellar material. Since most of the mass of the interstellar material is in the form of gas, it is the gas that will provide the gravitational attraction for the star formation process. In this part, we will first look at the contents of the interstellar medium. We will then look at how stars are born. Finally, we will see how the stars and interstellar medium are arranged in the galaxy as a whole.

Chapter 14

Contents of the interstellar medium

14.1 | Overview

When we look at photographs of the Milky Way (see Fig. 16.1), we note large regions where no light is seen. We think that these are due to dust blocking the light between us and the stars. We can see the same effect on a smaller scale (Fig. 14.1). Note that there is a high density of stars near the edges of the image. As one moves close to the center, the density of stars declines sharply. Near the center, no stars can be seen. This apparent hole in the distribution of stars is really caused by a small dust cloud, called a *globule*. The more dust there is in the globule, the fewer background stars we can see through the globule. We can use images like this to trace out the interstellar dust. We find that it is not uniformly distributed. Rather, it is mostly confined to concentrations or *interstellar clouds*.

We detect the presence of the gas by observing absorption or emission lines from the gas. By tracing these lines, we find that the gas also has an irregular distribution. Often the gas appears along the same lines of sight as the dust clouds. From this apparent coincidence we form the idea that the gas and dust are generally well mixed, with the gas having about 99% of the mass in a given cloud. In this chapter, we will see how the masses of different types of clouds are determined.

One of the reasons that the interstellar medium is so interesting is that it is the birth-place of stars. How do we know that stars are still forming in our galaxy? We have seen that stars are dying, and we know that there is still a large number of stars in the galaxy. We therefore presume that stars are being created at a rate that approximately offsets the rate at which they are dying. This is not an airtight argument, because it could be that many stars were formed early in the history of the galaxy and we are just seeing the ones that haven't died yet. However, we know that O stars live only about 10^7 years or less on the main sequence. Since we see O stars today, there must have been O star formation in the last 10^7 years. We think that the galaxy is ten billion years old. Compared with this, ten million years is almost like yesterday. If the conditions were right for star formation in the last 10^7 years, they must be right for star formation now. The actual star formation process will be discussed in Chapter 15.

14.2 | Interstellar extinction

If we want to see direct emission from the dust, we have to look in the infrared, as we will discuss in the next section. In the visible part of the spectrum the dust is generally evidenced by its blocking of starlight. The blocking arises from two processes, *scattering* and *absorption*. In scattering, the incoming photon is not destroyed, but its direction is changed. In absorption, the incoming photon is destroyed, with its energy remaining in the dust grain. The combined effects of scattering and absorption are called interstellar *extinction*. In Fig. 14.2 these two processes are depicted schematically. A dark nebula, in which background light

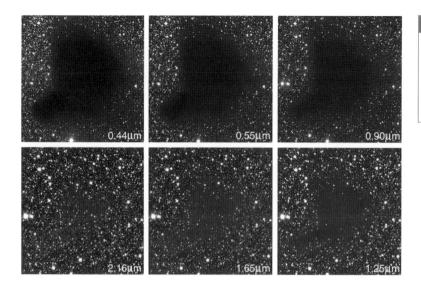

Fig 14.1. An image of a globule at various wavelengths. The globule is the region with the fewest number of stars per unit area. The dust in the globule is blocking the light from the background stars. [ESO]

is being blocked, and a *reflection nebula*, in which scattered starlight is being sent in our direction, are depicted in Fig. 14.3.

14.2.1 The effect of extinction

We quantify interstellar extinction as the number of magnitudes by which a cloud dims starlight passing through it. For example, if a particular star would have an apparent magnitude m without extinction, but its light passes through a cloud with A magnitudes of extinction, then the star will be observed to have a magnitude $m' = m + A$. (Remember, extinction dims the starlight, so the magnitude increases.)

We can relate the extinction, in magnitudes, to the optical depth τ of the dust. This is useful, since it is the extinction in magnitudes that will be directly measurable, but it is the optical depth that is directly related to the dust properties. If we have light of incident I_0 passing through the cloud of optical depth τ, and intensity I emerges, then these are related by (as we saw in equation 6.18)

$$I = I_0\, e^{-\tau} \tag{14.1}$$

From the definition of extinction and the magnitude scale

$$A = m' - m$$

$$= 2.5 \log_{10} (I_0/I) \tag{14.2}$$

Using equation (14.1), this becomes

$$A = 2.5 \log_{10} (e^{\tau})$$

$$= 2.5\, \tau \log_{10} (e)$$

$$= (2.5)(0.4343)\, \tau$$

$$= (1.086)\, \tau \tag{14.3}$$

This means that one magnitude of extinction corresponds approximately to an optical depth of one.

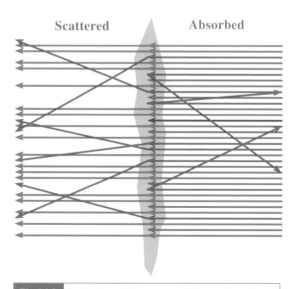

Scattered   Absorbed

Fig 14.2. Scattering and absorption. Light is incident from the right. Light rays that are absorbed stop inside the cloud. Light rays that are scattered change direction. Rays that are scattered or absorbed are color coded as indicated.

(a)

(b)

Fig 14.3. The *Horsehead Nebula*, in Orion's belt, is formed by the dust blocking the light from the glowing gas in the background. In this image, north is to the left. (a) A wider view. The fuzzy blue patch at the lower left of the Horsehead is a reflection nebula, where dust is scattering light from a hidden background star towards us. Just off the left (north) of the image is the southwestern most star in Orion's belt. (b) A closer view from HST. This shows the intricacy in the structure in both the glowing gas and the absorbing dust. [(a) NOAO/AURA/NSF; (b) STScI/NASA]

If we have a star of known distance and spectral type, we can determine the extinction between the star and us. The spectral type gives us the absolute magnitude M. We can measure the apparent magnitude m. In the presence of A magnitudes of extinction, the star will appear A magnitudes fainter than without extinction, so

$$m = M + 5 \log (r/10 \text{ pc}) + A \qquad (14.4)$$

Since we know m, M and r, we can find A. Obviously the presence of interstellar extinction will affect distance measurements by spectroscopic parallax. If we don't correct for extinction, then a star will appear to be farther away than it actually is. You can see that if both r and A are unknown, then equation (14.4) only gives us one equation with two unknowns. We will see below that there is a way of obtaining additional information by observing at different wavelengths.

Example 14.1 Interstellar extinction
Suppose we observe a B5 ($M = -0.9$) star to have an apparent magnitude of 9.2. The star is in a cluster whose distance is known to be 400 pc. What is the extinction between us and the star?

SOLUTION
We solve equation (14.4) for A to give

$$A = m - M - 5 \log (r/10 \text{ pc})$$

$$= 9.2 + 0.9 - 5 \log(40)$$

$$= 2.1 \text{ mag}$$

14.2.2 Star counting

If we record an image of a field which has some interstellar extinction, fewer stars will appear than if the extinction were not present. This is because the light from each star is dimmed by the extinction. Some stars that would have appeared on the image if there were no extinction are now too dim to appear with extinction. We can estimate the extinction in a cloud by comparing the number of stars we can see through the cloud with the number we can see in an unobscured region of the same size. Suppose an image is exposed to a threshold magnitude m_0. All stars with apparent magnitude less than m_0 (that is, stars that are brighter than m_0) will appear on the image. If the light from each star is dimmed by A magnitudes, only stars that have undimmed magnitudes of $m_0 - A$ will appear.

There are two ways of applying this idea. In one, we measure the number of stars per unit area

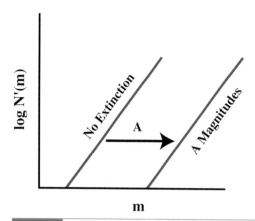

Fig 14.4. Effect of extinction on star counts. The plot gives the number of stars per magnitude interval, as a function of magnitude m. The distribution is such that when the vertical axis is logarithmic, the curve is close to a straight line. The effect of a cloud with A magnitudes of extinction is to make each star A magnitudes fainter, shifting the curve to the right by A magnitudes.

in each magnitude range. In the more common way, we measure just the total number of stars per unit area. We define the function $N'(m)$ such that $N'(m) \, dm$ is the number of stars per unit area with magnitudes between m and $m + dm$ in the absence of extinction. We measure $N'(m)$ for a region we think is partially obscured by dust and for a nearby region we think is unobscured. If we plot graphs of these two quantities, as shown in Fig. 14.4, we see that the two curves look like each other, except that one is shifted by a certain number of magnitudes. The amount of the shift is the extinction in the partially obscured region.

Often we don't have enough stars in each magnitude range to obtain a good measure of $N'(m)$. In that case we must use integrated star counts. We let $N(m)$ be the number of stars per unit area brighter than magnitude m. This is related to $N'(m)$ by

$$N(m) = \int_{-\infty}^{m} N'(m') \, dm' \tag{14.5}$$

If a photographic plate has a limiting magnitude m_0, then the number of stars per unit area, without extinction, is

$$N(m_0) = \int_{-\infty}^{m_0} N'(m') \, dm' \tag{14.6}$$

If we now look at a region with extinction A, only the stars that would have had magnitude $m_0 - A$ without extinction will show up. We therefore count

$$N(m_0 - A) = \int_{-\infty}^{m_0 - A} N'(m') \, dm' \tag{14.7}$$

Therefore, if we know $N'(m)$, we can predict $N(m_0 - A)$ for various values of A. If we use plates with a limiting magnitude of 20, then we can generally obtain good star count data for A in the range 1 to 6 mag. For A much less than 1 mag, the difference between an obscured and an unobscured region is hard to detect. For A much greater than 6 mag, there are very few stars bright enough to shine through, and the obscured region will appear blank, a situation in which 6 mag of extinction is indistinguishable from 20 mag.

14.2.3 Reddening

If we measure interstellar extinction we find that it is not the same at all wavelengths. In general, the shorter the wavelength is, the higher is the extinction. This means that blue light from a star is more efficiently blocked than red light. In the presence of extinction, the images of stars will therefore appear redder than normal, as shown in Fig. 14.5. This is called *interstellar reddening*. You

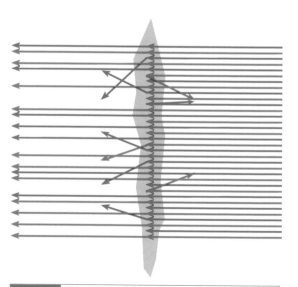

Fig 14.5. Interstellar reddening. More blue light is removed from the incoming beam than red.

can see the effect of reddening in the various wavelength images of the globule in Fig. 14.1. More stars shine through at longer wavelengths.

Suppose we measure the magnitude of a star in two different wavelength ranges, say those corresponding to the B and V filters. Then, from equation (14.4) we have

$$m_V = M_V + 5 \log (r/10 \text{ pc}) + A_V \qquad (14.8a)$$

$$m_B = M_B + 5 \log (r/10 \text{ pc}) + A_B \qquad (14.8b)$$

If we take the difference $m_B - m_V$, the distance r drops out, giving

$$(m_B - m_V) = (M_B - M_V) + (A_B - A_V) \qquad (14.9)$$

In equation (14.9), the quantity on the left-hand side is directly observed. The first quantity on the right-hand side depends only on the spectral type of the star. It is simply the $B - V$ color of the star. We know it because we can observe the star's absorption line spectrum to determine its spectral type. Since the spectral type determination depends on the presence of certain spectral lines, it is not greatly influenced by interstellar extinction. We can therefore determine the quantity $(A_B - A_V)$.

Since both A_B and A_V are proportional to the total dust column density N_D, their difference is also proportional to N_D. If we define a quantity

$$R = A_V/(A_B - A_V) \qquad (14.10)$$

it will not depend on N_D since N_D appears in both the numerator and denominator. We call this quantity the *ratio of total-to-selective extinction*. Extensive observational studies have shown that, in almost all regions, R has a value very close to 3.1. (There are a few special regions where R is as high as 6.) This has a very important consequence. It means that if we can measure $(A_B - A_V)$, we need only multiply by 3.1 to give A_V. We have already seen that the difference can be determined by knowing the spectral type of a star and measuring its B and V apparent magnitudes, and then using equation (14.9). Note that we have not made use of knowing the distance to the star r. We can still go back to equation (14.8a) to find the distance to the star. So, the method of spectroscopic parallax works even in the presence of interstellar extinction. We just need to do an extra observation at a different wavelength.

Example 14.2 Spectroscopic parallax with extinction

Suppose we observe a B5 star ($M_V = -0.9$, $B - V = -0.17$) to have $m_B = 11.0$ and $m_v = 10.0$. What is the visual extinction between us and the star, and how far away is the star?

SOLUTION
From equation (14.9) we have

$$(A_B - A_V) = (m_B - m_V) - (M_B - M_V)$$

$$= 1.00 + 0.17$$

$$= 1.17$$

We can now use the ratio of total-to-selective extinction to convert this to A_V:

$$A_V = R (A_B - A_V)$$

$$= 3.6 \text{ mag}$$

We can now find the distance from

$$5 \log (r/10 \text{ pc}) = m_V - M_V - A_V$$

$$= 7.3$$

This gives

$$r = 280 \text{ pc}$$

14.2.4 Extinction curves

If we study how extinction varies with wavelength, we can learn something about the properties of interstellar dust grains. We try to measure the $A(\lambda)$ in the directions of several stars, to see the degree to which grain properties are the same or different in different directions. Since the dust column densities are different in various directions, we do not directly compare values of A. Instead, we divide by A_V or $(A_B - A_V)$, to get a quantity that is independent of the column density. It is conventional to plot the following function to represent *interstellar extinction curves*:

$$f(\lambda) = \frac{A(\lambda)}{A_V} \qquad (14.11)$$

A typical curve is shown in Fig. 14.6. One general feature is that in the visible part of the spectrum $f(\lambda)$ is roughly proportional to $1/\lambda$. In the ultraviolet there is a broad 'hump' in the curve. The size of this hump varies from one line of

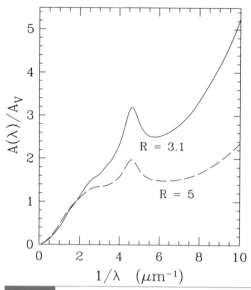

Fig 14.6. Average interstellar extinction curves. The vertical axis is just $[A(\lambda)]/A_V$. This is plotted as a function of $1/\lambda$, since it is approximately linear in $1/\lambda$ near the visible. The V filter would be at $1/\lambda = 1.8$. The solid line is for most normal regions, with R (ratio of total-to-selective extinction, defined in equation 14.10) = 3.1. The dashed line is for regions with an unusually high value of R = 5. [John Mathis, University of Wisconsin]

sight to another. In the infrared there are absorption features of various strengths. We will see that these infrared features tell us a lot about the grain composition.

14.2.5 Polarization

Sometimes the light we receive from celestial objects is polarized. We can detect this polarization by placing a polarizing filter in front of our detector. Such a filter only passes radiation whose electric field vector is parallel to the polarization direction of that filter. As we rotate the filter, different polarizations of the incoming light are passed. If the incoming light is unpolarized, then the amount of light coming through will not depend on the angle through which the filter has been rotated. If the incoming light is completely linearly polarized, there will be one position of the filter for which we see the image at full brightness and another position, 90° away, for which we see no light. If the incoming light is partially polarized, we will see a maximum brightness in one position and a minimum

brightness when the filter is rotated by 90°. The greater the amount of polarization, the greater the contrast between maximum and minimum.

When unpolarized starlight passes through interstellar dust clouds, it can emerge with a slight degree of linear polarization. This means that the polarization must be caused by the dust itself. The amount of polarization is very small, only a few percent at most. We find that there is a weak wavelength dependence on the amount of polarization. We also find that the amount of polarization generally depends on the visual extinction A_V. When A_V is low, the polarization is low. When A_V is high, the polarization can be low or high. Dust is necessary for the polarization, but something else must be necessary. There must be a mechanism of aligning non-spherical dust grains (at least partially) to produce the polarization. We will discuss this in the next section.

14.2.6 Scattering vs. absorption

We have said that extinction is the combined effect of scattering and absorption. The relative importance of these two effects depends on the physical properties of the grains and the wavelength of the incoming light. The fraction of the extinction that results from scattering is called the *albedo* a_λ of the dust grains. If a_λ is the fraction scattered, then $1 - a_\lambda$ must be the fraction absorbed.

The albedo is much harder to measure than the extinction. Studies of reflection nebulae are particularly useful, since they provide us with light that we know is scattered by the dust. It appears that the albedo is quite high, about 50–70%, at most wavelengths. (The albedo is lower in the range of the ultraviolet bump in the extinction curve, meaning that the bump is due to a strong absorption feature.) The high albedo also means that a photon may be scattered a few times before it is actually absorbed.

If a photon is scattered by a dust grain, we would like to know the directions in which it is most likely to travel. Studies indicate that about half of the scattered photons move in almost the same direction as they were going when they struck the grain. The rest of the photons have almost an equal probability for being scattered in any direction.

14.3 | Physics of dust grains

In this section we will see how we use observations and theory to deduce a number of properties of interstellar dust grains. We will also see how grains interact with their environment. Some of the things we would like to know about interstellar dust are:

(1) size and shape;
(2) alignment mechanism for polarization;
(3) composition;
(4) temperature;
(5) electric charge;
(6) formation and evolution.

14.3.1 Size and shape

We try to deduce grain sizes from the observed properties of the interstellar extinction curve, the variation of extinction with wavelength. If the grain size r is much greater than the wavelength λ, then we are close to the situation in which geometric optics applies. The wavelength is unimportant, and $A(\lambda)$ is roughly constant. If $r \ll \lambda$, the waves are too large to 'see' the dust grains, and $A(\lambda)$ is very small (explaining the low extinction at long wavelengths). If r is comparable to λ, then diffraction effects in the scattering process are important. Hence the wavelength dependence is strongest in this range.

We compare the observations with theoretical calculations of scattering and absorption by grains of different sizes and composition to see which gives the best agreement. We find that interstellar grains are not all the same size. There is some spread about an average (just as people are not all the same height, but have some spread of heights about some average value). In fact, the situation is more complicated than that. Observations of extinction curves indicate that there are probably at least two different types of grains with distinctly different average sizes (just as men and women are different types of people with different average heights).

We can deduce something about the shapes of interstellar grains from their ability to polarize beams of light. If the grains were perfect spheres, there would be no preferred direction, and there would be no way of producing the polarization. Therefore, some significant fraction of the grains must either be elongated, like cigars, or flattened, like disks.

There must also be a mechanism for actually aligning the asymmetric grains. This mechanism probably involves the interstellar magnetic field. The grains are probably not ferromagnetic. This means that a collection of grains cannot make a permanent magnet. However, they may be paramagnetic. In a paramagnetic material the individual particles have magnetic moments. These can be aligned by a magnetic field. The tendency to align is offset by the random thermal motions of the grains. We think that a partial alignment arises from a combination of two effects: (1) the tendency of elongated grains, shaped like cigars, to rotate end-over-end rather than about the long axis, since less energy is required for end-over-end motion; and (2) the tendency for the magnetic moment of the grain to align with the rotation axis.

14.3.2 Composition

We can deduce the composition of the larger grains from infrared absorption features. This is not as exact as using optical absorption spectra to tell us about the compositions of stellar atmospheres. Because the dust grains are solids, certain motions of atoms within the grains are inhibited by the close bonding to neighboring atoms, so the spectra consist of a few smeared out features instead of many sharp spectral lines. We observe absorption features at 10 μm and 12 μm, which correspond to the wavelengths for vibrational transitions (stretching of bonds) in silicates (SiO and SiO_2) and water ice. Since silicates are an important component of normal dirt on Earth, we sometimes talk of grains as being 'dirty ice'.

The extinction in the ultraviolet (including the hump) cannot be explained by dirty ice. For that, carbon is probably needed. Therefore, interstellar grains are probably a combination of large dirty ice grains and small graphite grains.

At the smallest end of the size distribution, probably only 1 nm across, are grains consisting of 20 to 100 C atoms in an aromatic hydrocarbon form. These are called *polycyclic aromatic hydrocarbons,*

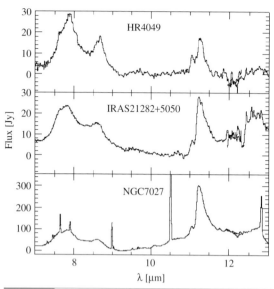

Fig 14.7. Infrared spectra of dust, showing the main spectral lines. The emission features at 3.3, 6.2, 7.7, 8.6 and 11.3 μm are probably from PAHs. [ESA/ISO, SWS, F. J. Molster et al.]

PAHs. They can be identified through some of their emission lines, like those shown in Fig. 14.7. They are stable at high temperatures, and were originally proposed to explain diffuse emission from hot extended clouds.

It also seems likely that many grains, especially the larger ones, are not of uniform composition. Rather, they have layers, like the schematic grain in Fig. 14.7. There could be a core of silicates and carbon. Outside of that are various mantles, one with water and ammonia ice and another with O, N and CO (carbon monoxide) all in solid form. There may even be a thin outer layer of hydrogen.

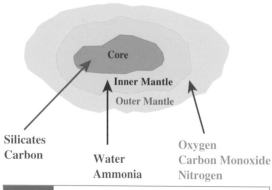

Fig 14.8. A multilayered interstellar grain.

14.3.3 Electric charge

We can deduce the electric charge for grains from theoretical considerations. There are two ways for grains to acquire charge: (1) Charged particles (both positive and negative) from the gas can strike the grains and stick to the surface. (2) Photons striking the grain surface can eject electrons via the photoelectric effect. The grain is left with one unit of positive charge for every electron ejected. In an equilibrium situation the net charge on the grains must be constant.

We first consider the situation in which the photoelectric effect is not important. This would be the case in regions of high extinction. For particles striking the grains, the negative charges are mostly carried by electrons, and the positive charges are mostly carried by protons. At any given temperature the average speed of the electrons, which have lower mass, will be greater than that of the protons. Therefore, electrons will hit grains at a greater rate than protons or C^+, ionized carbon. If the grains are initially neutral, this will tend to build up negative charge. However, once the grains have a small negative charge, the electrons will be slowed down as they approach the grains, while the positive charges will be accelerated. Therefore, if the grains have a net negative charge it is possible to have protons and electrons striking the grain at the same rate, keeping the charge on the grain constant. Note that, if the grains have a net negative charge, the gas must have a net positive charge if the interstellar medium as a whole is to be neutral.

Example 14.3 Charge on dust grains
Estimate the electric charge (in multiples of e) required to keep the charge on dust grains constant. Take the radius of the grain to be 10^{-5} cm and the gas kinetic temperature to be 100 K.

SOLUTION
We estimate the grain charge for which the electric potential energy of an electron at the grain surface is equal in magnitude to the average kinetic energy of the electrons in the gas. If the net charge on the grain is $-Ne$, the potential energy for an electron on the surface, a distance r from the center, is

$$U = Ne^2/r$$

The average kinetic energy is $(3/2)kT$. Equating these and solving for N gives

$$N = \left(\frac{3}{2}\right)\frac{kT\,r}{e^2}$$

$$= \frac{(3)(1.38 \times 10^{-16}\,\text{erg/K})(100\,\text{K})(10^{-5}\,\text{cm})}{(2)(4.8 \times 10^{-10}\,\text{esu})^2}$$

$$= 0.9$$

This says that each grain should have a net charge of about $-e$. However, the actual charge is about a factor of ten larger because we have only considered electrons of the average energy. Electrons moving faster than average contribute significantly to the charge buildup. Also, the charge becomes more negative at higher temperatures.

If the photoelectric effect is dominant, the grains will have a positive charge. There must be a balance between the rate at which electrons are being ejected and the rate at which they strike the grain. For positively charged grains, the electrons in the gas are attracted, meaning the tendency for the electrons to strike the grains at a greater rate than positively charged particles is enhanced.

14.3.4 Temperature

The temperature of a large dust grain is determined by the fact that, on a time average, it must emit radiation at the same rate as it receives radiation. This keeps the temperature constant. The temperature of a dust grain will therefore depend on its environment. If it is very close to a star it will be hot. If it is far from any one star it is cool, receiving energy only from the combined light of many distant stars.

Let's look at the case of a dust grain a distance d from a star whose radius and temperature are R_* and T_*. We will assume that the albedo is the same at all wavelengths. (If the albedo varies with wavelength, as in more realistic cases, the fraction of incoming radiation absorbed is different at different wavelengths, and the calculation is harder. See Problem 14.11.) The luminosity of the star is

$$L_* = 4\pi R_*^2\,\sigma\,T_*^4 \tag{14.12}$$

The fraction of this power striking the grain is the projected area of the grain πr_g^2, divided by the area of a sphere of radius d. That is

$$\text{fraction striking grain } = \pi r_g^2/4\pi d^2 \tag{14.13}$$

If a is the albedo, then $(1 - a)$ is the fraction of incoming radiation absorbed by the grain. Therefore the rate P at which energy is being absorbed by the dust grain is

$$P_{\text{abs}} = \frac{(1 - a)(4\pi R_*^2)(\sigma T_*^4)(\pi r_g^2)}{(4\pi d^2)} \tag{14.14}$$

$$= \frac{(1 - a)\pi\sigma R_*^2\,T_*^4 r_g^2}{d^2}$$

The quantity $\pi R_*^2/d^2$ is the solid angle subtended by the star as seen from the dust grain. We say that the star acts like a *dilute blackbody*. It has the spectrum of a blackbody at a temperature T_*, but the intensity is down by a factor of (solid angle/4π).

We now look at the rate at which the grain radiates energy. Since it can only absorb $(1 - a)$ of the radiation striking it, it can only emit $(1 - a)$ of the radiation that a perfect blackbody would emit. (A perfect blackbody has an albedo of zero, by definition.) If the grain temperature is T_g, the power radiated is

$$P_{\text{rad}} = (1 - a)4\pi r_g^2\,\sigma T_g^4 \tag{14.15}$$

Equating the power radiated and the power received, and solving for T, gives

$$T_g = T_*\,(R_*/2d)^{1/2} \tag{14.16}$$

Note that the final result does not depend on the size of the grain or the albedo. That is because both enter into the emission and absorption processes. (See Problem 14.10 for a discussion of what happens if the albedo is a function of wavelength.) This result is the same as that derived for a planet in Section 23.2.

Example 14.4 Temperature of a dust grain near a star

What is the temperature of a dust grain a distance 5000 stellar radii from a star whose temperature is 10^4 K?

SOLUTION
Using equation (14.16), we have

$$T = (10^4\,\text{K})\left[\frac{R_*}{(2)(5000)R_*}\right]^{1/2}$$

$$= 100\,\text{K}$$

When dust is sufficiently warm $(T_g > 20\,\text{K})$, it is a good emitter in the infrared, and we can

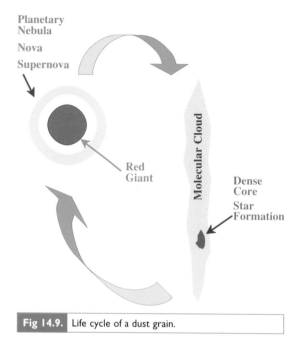

Planetary
Nebula

Nova

Supernova

Red
Giant

Molecular Cloud

Dense
Core

Star
Formation

Fig 14.9. Life cycle of a dust grain.

determine its temperature directly from infrared observations.

14.3.5 Evolution

We still know very little about the evolution of dust grains. Fig. 14.9 indicates, schematically, the life cycle of a typical dust grain. The densities in interstellar clouds are probably too low for the grains to be formed directly where we see them. We think that most dust grains are formed in the envelopes of red giants undergoing mass loss. As material leaves the surface, it is hot enough to be gaseous. However, as it gets farther from the surface, it cools. When the temperature is low enough, about 1000–2000 K, many of the materials, such as silicates, can no longer exist as a gas. They form small solid particles. These particles are blown into the interstellar medium either via the effects of stellar winds, or as part of a planetary nebula. This is another example of the cyclical processing of material between stars and the interstellar medium.

Once the grains are in clouds, they can collect particles from the gas and grow. There are some limits. For example, once a layer of molecular hydrogen (H_2) one molecule thick forms on the grains, no more hydrogen will stick. Grains will be destroyed, or diminished in size by a number of processes. Sometimes molecules can simply sublime from the surface. (Sublimation is a phase change directly from the solid phase to the gas phase.) Collisions with atoms in the gas can break up grains. Collisions between grains can also destroy the grains.

14.4 | Interstellar gas

14.4.1 Optical and ultraviolet studies

Early studies of cold interstellar gas utilized optical absorption lines. When light from a star passes through a cloud, as shown in Fig. 14.10, some energy is removed at wavelengths corresponding to transitions in the atoms and molecules in the cloud. These studies revealed the existence of trace elements such as sodium or calcium. (These elements happen to have convenient spectral lines to study.) In addition to these atoms, some simple molecules were discovered: CH (in 1937), CN (in 1940), and CH^+ (in 1941).

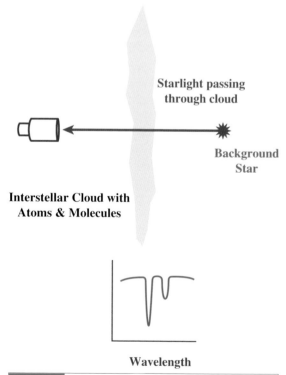

Starlight passing
through cloud

Background
Star

Interstellar Cloud with
Atoms & Molecules

Wavelength

Fig 14.10. On top is the arrangement for doing optical or uv absorption spectroscopy on an interstellar cloud. Below is a schematic absorption spectrum.

These simple molecules are not generally stable in the laboratory. CH^+ is charged and would combine with a negative ion or electron under laboratory conditions. CH and CN have an outer electronic shell with only one electron (as does H), making them chemically reactive. The presence of these unstable molecules in the interstellar gas suggests densities much lower than in the typical laboratory.

In these early studies no hydrogen absorption lines were observed. This is not because there was no hydrogen present. The temperatures in interstellar clouds are generally low, and most of the hydrogen is in the ground state. Therefore, the only H absorption lines possible are the Lyman lines in the ultraviolet. Now that ultraviolet observations are possible from satellites, astronomers can study these absorption lines.

You might wonder how we know that the absorption lines are coming from the interstellar gas and not from the stars themselves. After all, we have already seen the large number of absorption lines present in stars. One distinguishing feature is that the interstellar lines are much narrower than the stellar absorption lines. By narrower we mean they cover a smaller range of frequency (or wavelength). Interstellar lines have Doppler broadenings that correspond to a few kilometers per second. If the Doppler broadening is produced by random thermal motions, this suggests a temperature of about 100 K. Also, systematic studies show that, on the average, the absorption lines are stronger when detected in the light of more distant stars. The more distant the star is, the more interstellar material there is between us and the star. The narrow interstellar lines do not appear in the spectra of all stars. This suggests that the interstellar gas is clumpy, just as the interstellar dust is clumpy.

14.4.2 Radio studies of atomic hydrogen

Much of what we know about the interstellar medium comes from radio observations. We have already seen that supernova remnants, planetary nebulae and pulsars are sources of radio emission. These are generally hot sources, or sources with high energy electrons that produce a high radio luminosity. However, most of the interstellar gas is cool and does not produce a strong radio continuum emission. The cool interstellar gas must be observed via radio spectral lines.

The first interstellar radio line to be observed was from atomic hydrogen, but it was not a transition in which an electron jumps from one orbit to another. As we have said, these transitions are in the visible and ultraviolet parts of the spectrum. For the radio transition, the hydrogen stays in the ground electronic state. This is illustrated in Fig. 14.11. Both the electron and proton have intrinsic angular momentum, called spin. We have already seen that this spin can have two

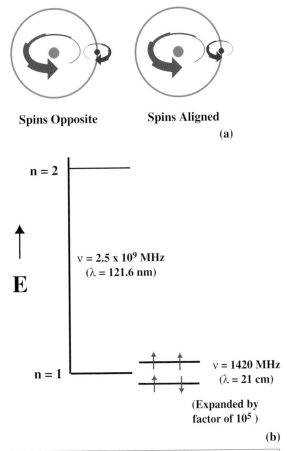

Spins Opposite   **Spins Aligned**

(a)

n = 2

E

$\nu = 2.5 \times 10^9$ MHz
($\lambda = 121.6$ nm)

n = 1

$\nu = 1420$ MHz
($\lambda = 21$ cm)

(Expanded by factor of 10^5)

(b)

Fig 14.11. (a) Origin of the 21 cm line. The splitting comes from a magnetic interaction which depends on the spin directions of the electron and the proton. The energy is higher when the spins are parallel and lower when they are antiparallel. (b) Energy level diagram showing the splitting of the hydrogen ground state (n = 1). The splitting is greatly exaggerated in this figure.

possible orientations. We refer to them as 'up' and 'down'. This means that the electron and proton spins can be either parallel or antiparallel. The relative orientation of the spins affects the magnetic force between the electron and the proton. The state with the spins parallel has slightly more energy than the state with the spins antiparallel. The atom can undergo transitions between these two states. The energy difference corresponds to a frequency of about 1400 MHz, or a wavelength of 21 cm. This is generally referred to as the *21 cm line*.

If we take the energy of the transition $h\upsilon$ and divide by Boltzmann's constant, the quantity $h\upsilon/k$ gives the temperature necessary to see collisional excitation of the hydrogen upper state. This is about 0.07 K. This means that even at the low temperatures of interstellar space there will be sufficient energy to excite transitions between these two states in hydrogen. The 21 cm line is easily observable under interstellar conditions. The possibility of detecting this line was discussed in Leiden (Netherlands) in the early 1940s by *Henk van de Hulst*. After that there was a race among Australian, Dutch and American groups to detect the line. The first detection of the 21 cm line from interstellar hydrogen was in the early 1950s by a group at Harvard, led by *Edwin Purcell*, who won the Nobel Prize in physics for his work.

Since that time there have been extensive observations of the 21 cm line by radio astronomers all over the world. It is probably fair to say that it was the dominant tool for studying the interstellar medium and galactic structure through the 1960s, and continues to be very useful. In these studies, the line was observed in both emission and absorption. The conditions for emission or absorption lines are shown in Fig. 14.12. In order for the line to be in absorption, there must be a background continuum source whose brightness temperature at 21 cm is greater than the excitation temperature of the atoms in the particular cloud being observed. Under most conditions the excitation temperature of the 21 cm line is close to the kinetic temperature of the clouds.

By studying both absorption and emission lines in a given region it is possible to deduce

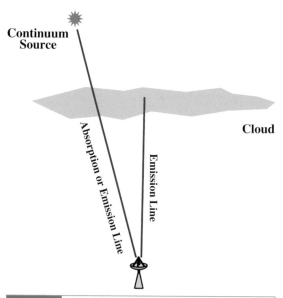

Fig 14.12. Conditions for radio absorption and emission lines. If a radio continuum source is viewed through an interstellar cloud, then radio absorption lines can be seen against the continuum source. (This also requires the continuum source to appear hotter than the cloud at the wavelength of observation.) If there is no background continuum source, then only emission lines can be seen.

both the excitation temperature and the optical depth. The excitation temperature enables us to calculate the kinetic temperature of the gas, an important quantity. The optical depth can be converted into a column density for atomic hydrogen. If we know the column density and size of a cloud, we can also find the average local density of hydrogen. So you can see that the 21 cm line observations provide astronomers with an important tool for studying the physical conditions in many interstellar clouds.

One important feature of the radio observations is that interstellar dust is transparent at radio wavelengths. Therefore we can use radio telescopes to detect objects across the galaxy, far beyond what we can see optically in the presence of dust. Since we can use it to observe clouds anywhere in the galaxy, the 21 cm line is a very useful tool for studying galactic structure. Also, since it is a spectral line, we can observe its Doppler shift and learn about motions throughout our galaxy. We will see how these studies are used in Chapter 16.

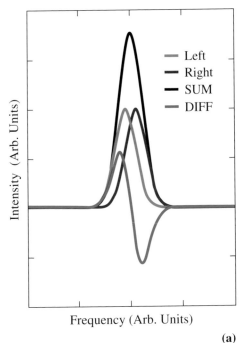

(a)

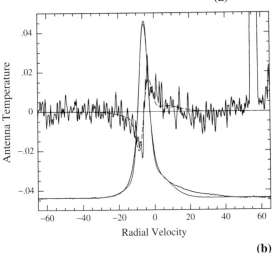

(b)

Fig 14.13. Zeeman effect in atomic hydrogen. (a) Idealized situation. The line shift is a very small fraction of the linewidth. However, opposite polarizations are shifted in opposite directions. When we subtract one polarization from the other, we are left with a very distinctive pattern. (Note that the difference curve is approximately the derivative of the original curve. This follows directly from the definition of a derivative.) (b) Spectra from a real source. In this figure, the two smooth curves that have very little noise are the spectra for opposite polarizations. The difference is shown on an expanded scale, so it looks noisy. The dashed line is the best fit to that difference spectrum. [(b)Carl Heiles (University of California at Berkeley)/Heiles, C., Astrophys. J., **336**, 808, 1989, Fig. 1a]

We have already seen (in Chapter 6) that some energy levels shift in the presence of a magnetic field, the Zeeman effect. The levels involved in the 21 cm line fall into this category. The stronger the magnetic field, the greater the shift. This means that we can use the Zeeman shift in the 21 cm line to measure the strength of interstellar magnetic fields. The experiment is difficult because the Zeeman shift is much less than the width of the normal 21 cm line. However, opposite polarizations are shifted in opposite directions. Since we can detect different polarizations separately, we can subtract one polarization's spectrum from the other, leaving a very small signal, as shown in Fig. 14.13. The experiment is also difficult because a small difference in the response of the telescope to the two polarizations can mimic the effects of a Zeeman shift. Despite these difficulties, recent experiments have succeeded in measuring fields of the order of tens of microgauss in a growing number of interstellar clouds. Fields of this strength may sound very weak, but they are strong enough to influence the evolution of these clouds, as we will discuss in the next chapter.

By making maps of the 21 cm emission astronomers have been able to form a good picture of the cloud structure in the interstellar gas. These maps show an irregular cloud structure, similar to that shown in the dust clouds. Typical clouds have the following physical parameters: temperature, 100 K; hydrogen density $n_H \sim$ 1–10 cm^{-3}, lengths of tens of parsecs; hydrogen column densities up to $\sim 10^{21}$ cm^{-2}. The clouds fill about 5% of the volume of interstellar space, meaning that the average density of atomic hydrogen in interstellar space is of the order of 0.1 cm^{-3}; One interesting recent finding is the presence of large HI shells, which stir up the interstellar medium throughout the galaxy (Fig. 14.14).

The regions between the clouds are not empty. Studies of the line profiles of the 21 cm line show very broad, faint wings. This is interpreted as coming from a small amount of very hot gas. Temperatures of about 10^4 K have been estimated for this low density gas between the clouds. We will see later in this chapter that the low density means it is very hard for the gas to lose energy

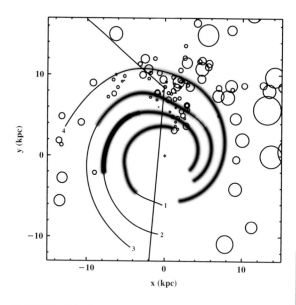

Fig 14.14. Images made from 21 cm observations, showing the large scale structure, projected in a galactic coordinate system, with the plane of the Milky Way acting as the equator. The map shows the locations of the largest HI shells, called 'supershells'. The larger circles indicate larger shells. To give an idea of galactic scale, these are superimposed on estimates of where pieces of spiral arms may lie. We will talk more about the spiral structure of the galaxy in Chapter 16. [Naomi McClure-Griffiths, Australia Tel. National Facility/McClure-Griffiths, N. M. et al., Astrophys. J., **578**, 189, (2002), Fig. 18]

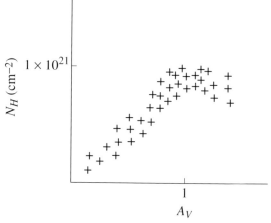

Fig 14.15. Hydrogen column density N_H as a function of visual extinction A_V. Though there is some scatter, there is a good correlation in the two quantities as long as A_V is less than one magnitude.

and cool (a situation somewhat similar to the solar corona). It has been noted that if we compare the pressure within a cloud P_{cl} with the pressure in the intercloud medium P_{ic}, we find

$$\frac{P_{cl}}{P_{ic}} = \frac{n_{cl}T_{cl}}{n_{ic}T_{ic}}$$

$$= \frac{(10 \text{ cm}^{-3})(10^2 \text{ K})}{(0.1 \text{ cm}^{-3})(10^4 \text{ K})}$$

$$= 1$$

The pressure in the clouds and intercloud medium is approximately the same. Some theoreticians have proposed a picture of the interstellar medium known as the *two-phase model*, in which this equality of pressures is not a coincidence, but follows from the ways in which the gas can cool. The two-phase model is now considered overly simplified and has been replaced by more dynamic pictures of the interstellar medium.

For clouds that are near enough to be seen optically, it was found that the 21 cm emission often follows the optical obscuration of the dust. This suggests that the gas and dust are well mixed. This idea was tested in detail by seeing the degree to which the hydrogen column density N_H correlates with the visual extinction A_V. The results of these studies are shown in Fig. 14.15. There is some scatter in the data, but it is clear that N_H and A_V are related. The general ratio N_H/A_V is approximately 10^{21} atoms cm^{-2}/1 mag.

This ratio was found to hold as long as the extinction is less than 1 mag. When the extinction becomes higher, the relationship no longer holds. This was a mystery for many years. Two possible solutions were proposed. One was that with a lot of dust, very little radiation can penetrate to heat the cloud. It is possible that the hydrogen is so cold that the emission lines are just very weak. The other possibility is that under the higher extinction conditions, pairs of hydrogen atoms combine to form molecular hydrogen, H_2. Molecular hydrogen obviously has a very different structure than atomic hydrogen, and has no equivalent of the 21 cm line. In fact, cold molecular hydrogen has no emission or absorption lines in the radio or visible parts of the spectrum. The recent discovery of a large number of interstellar molecules, including H_2, tells us that the latter explanation is correct.

14.5 | Interstellar molecules

14.5.1 Discovery

The discovery of optical absorption lines of CH, CH^+ and CN raised the possibility that molecules might be an important constituent of the interstellar gas. However, it was thought that the densities were too low for chemistry to proceed very far. In fact, the existence of these three unstable species supported that general notion. If chemistry had proceeded very far, these species would have been incorporated into more complex molecules.

To see how these arguments worked, we can estimate the rate at which molecules form in a cloud. Let's take the example of C and O coming together to form the simple molecule CO. The rate of formation of CO per unit volume is given by

$$R_{form} = n_C n_O v \sigma \tag{14.17}$$

where n_C and n_O are the C and O densities, respectively, v is the relative speed of the atoms, and σ is the cross section for a collision. We take σ to be the geometric cross section (the approximate size of an atom, 10^{-16} cm^{-2}) and v to be the average thermal speed at a temperature of 100 K (about 10^5 cm/s). Finally, since both C and O have cosmic abundances of about 10^{-3} that of H, we take each of their densities to be $10^{-3} n_H$. This will give us a factor of n_H^2 in the rate, so the density is very important. If $n_H = 10$ cm^{-3} the rate becomes 10^{-15} cm^{-3}/s.

We have to compare this with the rate at which CO is destroyed. One destruction mechanism is photodissociation. An ultraviolet photon strikes the molecule with a sufficient energy to break it apart. An unprotected CO molecule can live an average of 10^3 years (3×10^{10} s) in the interstellar radiation field. The dissociation rate per molecule is the inverse of the lifetime. The dissociation rate per unit volume is the density of CO molecules divided by the lifetime:

$$R_{dis} = n_{CO}/t_{dis} \tag{14.18}$$

If we equate the formation and destruction rates, we can solve for the equilibrium CO abundance:

$$n_{CO} = (1 \times 10^{-15} \text{ cm}^{-3}/\text{s})t_{dis}$$

$$= 3 \times 10^{-5} \text{ cm}^{-3}$$

Since $n_H = 10$ cm^{-3} the fractional abundance of CO, n_{CO}/n_H, is about 3×10^{-6}. This is low enough that it did not raise the hopes of finding very complex molecules. We have even been very optimistic by assuming that every collision between C and O leads to a CO molecule.

However, radio searches for small molecules were carried out, with some of the initial candidates being chosen by the availability of convenient radio transitions. In the 1960s three simple molecules were found, OH (at a wavelength of 18 cm), H_2O (at a wavelength of 1 cm), and NH_3 (at a wavelength of 1 cm). The abundances of these molecules were surprisingly high, and astronomers were encouraged to carry out searches for other molecules. In 1969 one of the most important molecular discoveries took place. CO was found at a wavelength of 2.6 mm, by a group at Bell Laboratories, led by *Arno Penzias* and *Robert Wilson* (who shared the Nobel Prize in physics for their earlier discovery of the cosmic background radiation, to be discussed in Chapter 21). They used the NRAO millimeter telescope shown in Fig. 4.28(a). This was the first molecule to be found at millimeter wavelengths. Remember, at shorter wavelengths we can produce good angular resolution with modest sized telescopes. (Of course, the telescope surfaces require greater precision and must be placed at dry sites.) The abundance of CO is also very high, with CO densities of about 1 cm^{-3}, much higher than our previous estimate. As we will see, the 2.6 mm line of CO has taken its place alongside the 21 cm line as one of the important tools in studying the cool interstellar gas.

Following these initial discoveries, a large number of interstellar molecules were found. Over 100 have been discovered to date. They are listed in Table 14.1. There are many familiar molecules, such as formaldehyde (H_2CO), methyl alcohol (CH_3OH), and ethyl alcohol (CH_2CH_3OH). There are some unfamiliar molecules. Some of these are charged species, such as HCO^+, and

Table 14.1. Interstellar molecules, arranged by number of atoms.

2	3	4	5	6	7	8	9	10	11	13
AlF	C_3	$c\text{-}C_3H$	C_5	C_5H	C_6H	CH_3C_3N	CH_3C_4H	CH_3C_5N	HC_9N	$HC_{11}N$
$AlCl$	C_2H	$l\text{-}C_3H$	C_4H	C_5O	CH_2CHCN	$HCOOCH_3$	CH_3CH_2CN	$(CH_3)_2CO$		
C_2	C_2O	C_3N	C_4Si	C_2H_4	CH_3C_2H	C_2H_6	$(CH_3)_2O$	NH_2CH_2COOH		
CH	C_2S	C_3O	$l\text{-}C_3H_2$	CH_3CN	HC_5N	C_6H_2	CH_3CH_2OH			
CH^+	CH_2	C_3S	$c\text{-}C_3H_2$	CH_3NC	$HCOCH_3$	C_7H	HC_7N			
CN	HCN	C_2H_2	CH_2CN	CH_3OH	NH_2CH_3		HC_8			
CO	HCO	CH_2D^+	CH_4	CH_3SH	C_3H_4					
CO^+	HCO^+	$HCCN$	HC_3N	HC_3NH^+						
CP	HCS^+	$HCNH^+$	HC_2NC	$HCOCH_2$						
CS	HOC^+	$HNCO$	$HCOOH$	$HCONH_2$						
CSi	H_2O	$HNCS$	H_2CHN	H_2C_4						
HCl	H_2S	$HOCO^+$	H_2C_2O	C_3H_2O						
H_2	HNC	H_2CO	H_2NCN	C_5N						
KCl	HNO	H_2CN	HNC_3							
NH	$MgCN$	H_2CS	SiH_4							
NO	$MgNC$	H_3O^+	CH_3O^+							
NS	NH_2	NH_3								
$NaCl$	NH_2^+	$HOCS^+$								
OH	N_2O	CH_3								
PN	$NaCN$	C_3Si								
SO	OCS									
SO^+	SO_2									
SiN	$c\text{-}SiC_2$									
SiO	CO_2									
SiS	CO_2^+									
CH^+	HN_2^+									
HF	H_2O^+									
HLi	H_3^+									
OH^+										
HS										
N_2^+										

others have unpaired electrons and are chemically active in the laboratory, such as CCH. Even carbon chain molecules of moderate length (such as $HC_{11}N$) have been found. There are even some simple ring molecules. Many of these molecules were discovered by observations at millimeter wavelengths on telescopes such as that shown in Fig. 4.28(a).

The discovery of so many interstellar molecules was obviously a surprise. How could the predictions that molecules could not form have been so wrong? One answer is that the clouds in which the molecules have been found are not the same clouds that were studied at 21 cm. They have higher densities and visual extinctions, and lower temperatures. The higher densities mean that chemical reactions take place faster (remember the formation rate goes roughly as the square of the overall density). The higher visual extinctions provide shielding from the ultraviolet radiation that dissociates the molecules.

We don't see 21 cm emission from these clouds because the atomic hydrogen has been converted to molecular hydrogen. As we have already seen, the molecular hydrogen has no radio or optical spectrum. Since hydrogen is the most abundant element, we classify interstellar clouds by the form in which the hydrogen is found. For example, clouds in which the hydrogen is mostly atomic are called *HI clouds*. Clouds in which the hydrogen is mostly ionized are called *HII regions* (to be discussed in the next chapter). Clouds in which the hydrogen is mostly molecular are called *molecular clouds*.

14.5.2 Interstellar chemistry

Since the discovery of so many interstellar molecules, considerable effort has gone toward a better understanding of interstellar chemistry. It appears that some of the chemical reactions take place on grain surfaces. The grain surface provides a place for two atoms to migrate around until they find each other. They also provide a sink for the binding energy of a molecule. The example of molecular hydrogen is shown in Fig. 14.16. If two H atoms formed in the gas phase, the particular properties of the H_2 molecule would keep it from radiating away the excess energy before the molecule flew apart. On a grain

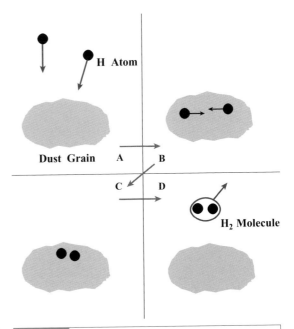

Fig 14.16. Formation of H_2 on a grain surface.

surface, the energy can be efficiently transferred to the grain, resulting in a slight increase in grain temperature. The fact that the dust plays an important role in the formation of H_2 and the protection of H_2 once it is formed, results in an interesting situation. When a cloud has a visual extinction of less than one magnitude, almost all of the hydrogen is atomic. When the extinction is greater than 1 mag, almost all of the hydrogen is molecular. This explains the breakdown in the relationship between N_H and A_V above 1 mag.

Despite the important role that dust plays in the formation of the most abundant molecule H_2, most of the interstellar chemistry cannot proceed in this way. Many of the molecules are formed in the gas. At the beginning of this section we calculated a very low rate for two atoms to collide in the gas to form a molecule. However, the densities in molecular clouds are at least 10^3 times those we used for our estimate, and the reaction rates go as the square of the density. Therefore, the reaction rates in molecular clouds are much faster than our initial calculation suggests. There is also another factor that increases the cross section for collisions if one of the reactants is an ion and the other is a neutral. Such a reaction is called an *ion–molecule reaction*.

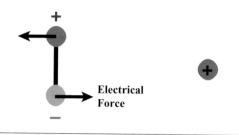

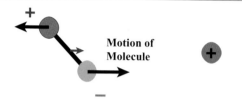

Fig 14.17. Dipole in an electric field. In this case, the electric field is provided by the positive charge, and weakens with distance from that charge. The negative end of the dipole is closer to the positive charge, so an attractive force felt by the negative end is greater than the repulsive force felt by the positive end. The dipole is thus attracted to the charge. (The same thing would happen with a negative charge.)

To see how the rate is enhanced, let's consider the case of a positive ion (as shown in Fig. 14.17). We have already said that the grains must be negatively charged, so the gas must be positively charged. In addition, Table 14.1 shows that many positive ions have been detected. The neutral atoms can still have an electric dipole moment, even though it has no net charge. The dipole will tend to line up with the electric field of the ion. Since the ion is positive, the negative end of the dipole will end up closer to the ion. The negative end of the dipole will therefore feel an attractive force which is slightly greater than the repulsive force felt by the positive end, which is farther away. The dipole will feel a net attractive force. This attractive force significantly increases the effective cross section of the reactants and speeds the reaction.

Theoreticians have tried to identify the chemical reactions that might be important in the interstellar medium. They then carry out model calculations in which they calculate the equilib-

rium abundances of various molecules. These are the abundances for which the rates of destruction and formation are equal. These theories have been quite successful at predicting the abundances of most of the simpler (especially two- and three-atom) molecules. More work is still needed for the heavier molecules. In addition, it may be that many interstellar clouds are not old enough to have reached an equilibrium situation. If that is the case, the abundances should still be changing.

14.5.3 Observing interstellar molecules

When we observe interstellar molecules, we are not observing transitions in which electrons jump from one level to another. Such transitions do exist for molecules, as they do for atoms. However, they require energies of the order of at least a few electron volts and are in the visible part of the spectrum. These transitions are not easily excited in the cool interstellar medium. Another type of transition in molecules, involving lower energies, is vibrational. We can think of a molecule as consisting of a number of balls connected by springs. The springs can stretch and bend at certain frequencies, with certain energies. Transitions between vibrational states are possible. The energies associated with vibrational transitions usually place the resulting photons in the infrared. This is still too energetic for the cool clouds.

There is another type of transition, with even lower energies. It involves the rotation of the molecules. The rotational motion is also quantized and transitions among rotational states can take place. The photons associated with these transitions are generally in the radio part of the spectrum. To see what the energy levels look like in this case, we consider a diatomic molecule (such as CO), rotating end-over-end about its center of mass. The rotational inertia is I. If the molecule is rotating with an angular speed ω, the energy is given by

$$E = (1/2)I\,\omega^2 \tag{14.19}$$

The angular frequency can be expressed in terms of the angular momentum L as

$$\omega = L/I \tag{14.20}$$

Using this, equation (14.19) becomes

$$E = L^2/2\,I \tag{14.21}$$

The condition for the quantization of angular momentum is different than the one we saw for electrons in an atom. If J is an integer, called the *rotational quantum number*, then L^2 is related to J by

$$L^2 = J(J + 1)(h/2\pi)^2 \qquad (14.22)$$

(Note: for large values of J this is not very different from the condition $L = Jh/2\pi$ for orbiting electrons.) If we put equation (14.22) into equation (14.21), the energy becomes

$$E = \frac{J(J + 1)(h/2\pi)^2}{2I} \qquad (14.23)$$

For any given molecule, the energy levels are determined by the rotational inertia. If I is large, the energy levels will be close together. If I is small, the energy levels will be farther apart. The levels for CO and CS are shown in Fig. 14.18. The

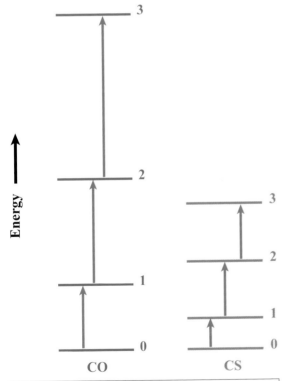

Fig 14.18. Rotational energy levels for two diatomic molecules, CO and CS. The states are designated by the rotational quantum number J. The differences between the two molecules arise from the differences in rotational inertias resulting from different masses for O and S, and different bond lengths for the two molecules.

major difference between I(CO) and I(CS) are from the difference in the masses of the O and S. In addition, the bond lengths are slightly different in the two molecules. Note that the closest spacing is for the first two energy levels ($J = 0$ and $J = 1$). As we go to higher values of J, the energy levels are farther apart. This means that at low temperatures only a few of the lowest energy levels are populated. For example, the 2.6 mm transition in which CO is most commonly observed is the $J = 1 \rightarrow 0$. The values of rotational inertia for many simple molecules are such that the lowest transitions lie in the millimeter part of the radio spectrum. That is why so many molecules were discovered at millimeter wavelengths.

Adding more atoms to a molecule can complicate the spectra. If we still have a linear molecule (for example, in HCN the three atoms are in a line), then the energy levels are essentially the same as the diatomic case, with the appropriate value for I. If molecules are not linear, then the spectra are more complicated, since we have to allow for rotation about more than one axis, but there are similarities to the linear case.

If we want to look at a new interstellar molecule, we need to know the wavelengths at which it can emit. For the most part, we rely on accurate laboratory measurements of molecular spectra. Once the wavelengths of a few transitions have been measured, those of other transitions can be calculated very accurately (using expressions such as equation (14.23)). There are some molecules that have been found in interstellar space without prior laboratory study. These were found accidentally, in the course of searches for other molecules. In some cases the interstellar medium provides us with a unique opportunity to study molecules that are not stable in the laboratory.

The most important feature of interstellar molecules is that they provide us with a way of obtaining information about the physical conditions in the molecular clouds. If we take the energy corresponding to the 2.6 mm photon, and divide by k, we find an equivalent temperature of 5.5 K. This means that rotational transitions in molecules are excited even at low temperatures. Also, the factor $e^{-E/kT}$ in the Boltzmann equation is most sensitive to changes in temperature and density when E is of the order of kT.

	Before	After	

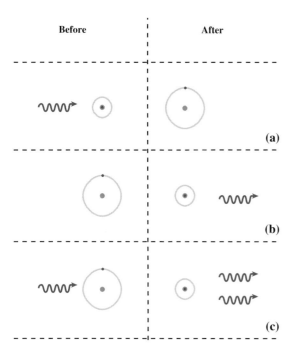

Fig 14.19. Types of interactions between radiation and matter. (a) Absorption. A photon is absorbed, leaving the atom or molecule in an excited state. In this case the excited state is denoted by a larger symbol for the atom or molecule. (b) Emission. The atom or molecule starts in the excited state, and spontaneously makes a transition to the lower state, giving off a photon of the appropriate energy. (c) Stimulated emission. The atom or molecule is in the excited state. A photon whose energy would be right to produce an absorption if the atom or molecule were in the lower state, strikes the atom or molecule, inducing a downward transition and the emission of a photon. There are now two photons. They are at the same frequency, traveling in the same direction and in phase.

To see how we use molecules as probes of the physical conditions in interstellar clouds, we must first know the ways in which a molecule can go from one rotational state to another. One set of processes involves the interaction with radiation, either by emission or absorption of photons. Radiative processes are illustrated in Fig. 14.19. We are already familiar with absorption and emission of photons. For absorption, the photon must have the energy corresponding to a transition between the molecule's initial state to a higher energy state. For emission, the molecule goes from a higher energy state to a lower energy

state. The emitted photon has an energy equal to the energy difference between the states.

There is an additional radiative process that is important – *stimulated emission*. This is emission of a photon, stimulated by the presence of another photon. Suppose we consider only two energy states. The molecule starts in the higher energy state. When the molecule is struck by a photon, whose energy is equal to the energy difference between the states, the molecules cannot absorb the photon, since the molecule is already in the higher state. However, the presence of that photon can cause the molecule to drop to the lower state, emitting a second photon. In the process of stimulated emission one photon comes in and two photons go out. The two photons have the same wavelength, are in phase with each other, and travel in the same direction. As will see below, it is stimulated emission that is responsible for amplification in masers and lasers.

Molecules can also be induced to make transitions by collisions with other particles. In a molecular cloud most of the matter is in H_2, which we don't usually directly observe. However, this H_2 makes its presence felt by forcing transitions in other molecules, as illustrated in Fig. 14.20. The process works in both directions. An H_2 molecule can strike a CO molecule, for example. In the process, the H_2 can lose kinetic energy, while the CO is excited to a higher energy state, or the H_2 molecule can gain kinetic energy with the CO going to a lower energy state.

In order to use observations of interstellar molecules to tell us about cloud physical conditions, we must be able to calculate the rates at which these various processes occur under different conditions. It is then necessary to carry out large calculations to model the conditions in a cloud. In these models we require that the population of each level stays constant. The rate at which molecules can reach any state must equal the rate at which they leave that state. We use the model calculations to predict the strengths of various molecular lines. We then compare those predictions with observations. The models are adjusted until agreement is found. The model is then used to predict the results of new observations, and the process continues.

H₂ Molecule

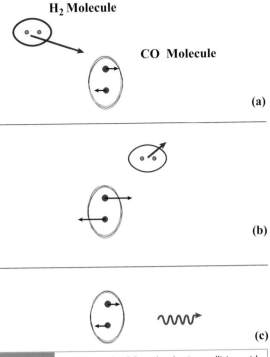

CO Molecule

(a)

(b)

(c)

Fig 14.20. Excitation of a CO molecule via a collision with an H₂ molecule. (a) In this case the H₂ strikes the CO. (b) The H₂ is moving slower, having lost some kinetic energy, and the CO is rotating faster with energy that it gained in the collision. (c) The CO emits a photon (spontaneously) and drops back to the lower state. (A transition to the lower state could have also been caused by a collision with a different H₂ molecule, causing that H₂ molecule to speed up, taking away the rotational energy of the CO.)

In studying molecular lines we can also learn about the velocities within a cloud by studying the line profiles. An interesting feature of molecular lines is that they are almost always wider (cover a wider frequency range) than would be expected for a line in which the only Doppler broadening is from random thermal motions. This implies that we are seeing the effects of other internal motions in the cloud. These motions can include collapse, expansion or rotation. We may also be seeing the effects of turbulent stirring of the gas. This stirring may be driven by mass loss from both old and young stars.

Probably the most usefully studied interstellar molecule is carbon monoxide, CO. It is very abundant. Its abundance is about 10^{-4} to 10^{-5} times

that of hydrogen, while most other molecules have abundances that are about 10^{-9} times that of hydrogen (a part per billion) or less. The CO is easy to excite and to observe. It is particularly useful in tracing out the extent of molecular clouds. Also, observations of CO allow us to estimate cloud masses and kinetic temperatures. Some other molecules that are very useful are carbon sulfide (CS) and formaldehyde (H_2CO). These other molecules are rarer and harder to excite than CO. We only see them in very dense parts of clouds. Therefore, we can use these molecules to tell us about the densities in these clouds.

Determining the masses of molecular clouds is more difficult than for atomic clouds. When we look at atomic clouds through the 21 cm line we are looking at the primary constituent of those clouds – HI. So, when we measure HI column densities from the strength of the 21 cm line, we can convert those column densities directly into cloud masses. Also, since the optical depth of that line is generally small, the strength of the line is close to proportional to the column density. When we look at molecular clouds, say by observing CO, we are observing a trace constituent, something present in a part per million relative to H_2, and that abundance may vary according to local chemical conditions. So, even if we could measure the total mass of CO in a molecular cloud, we wouldn't know by what factor to multiply that to determine the H_2 mass. In addition, for the most part, the transitions that we observe in CO are not optically thin, so it is hard to relate their intensity directly to the CO column density.

Analyzing molecular excitation to give H_2 densities in different parts of a cloud can be useful in studying local conditions within a cloud. These do work, but involve extensive observations, making maps of different lines and different molecules, taking advantage of the fact that different transitions are excited in different temperature and density regimes.

There is one technique that is potentially useful. In the last chapter, we discussed using the virial theorem to determine the masses of clusters of stars. This might also work for molecular clouds. Even though the H_2 is mostly invisible, the CO in the cloud still feels its gravitational effects. So, a more massive cloud would be able to support larger internal motions and still remain bound. We can use the CO emission to measure both the extent of the cloud and the internal motions, from the widths the spectral lines. There is always the problem that we don't know if a cloud is dynamically relaxed, or even bound. After all, we have seen earlier in this chapter that the interstellar medium is a very turbulent place. If we want to use virial masses, we need to find clouds that seem to be in relatively quiet regions.

Example 14.5 Virial mass of a molecular cloud
Find the virial mass of a molecular cloud with $\langle v_r \rangle = 3.0$ km/s and $R = 10$ pc.

SOLUTION
From equation (13.52) we have

$$M = \frac{(5)(3.0 \times 10^5 \text{ cm/s})^2(10 \text{ pc})(3.18 \times 10^{18} \text{ cm/pc})}{(6.67 \times 10^{-8} \text{ dyn cm}^2/\text{g}^2)}$$

$$= 2.1 \times 10^{38} \text{ g}$$

$$= 1.0 \times 10^5 \, M_\odot$$

Objects this massive are called *Giant Molecular Clouds* (GMCs) and will be discussed more in the next chapter.

Another interesting aspect of interstellar molecules involves the substitution of various isotopes. For example, the most common form of carbon is ^{12}C and most of the CO is in this form. However, some CO is formed with the rarer species ^{13}C. Making such a substitution changes the rotational inertia, I, for the molecule, shifting the spectral lines. The shifts are quite large, and are easy to detect. By contrast, the wavelengths of electronic transitions (in C atoms for example) depend on the electron nucleus reduced mass, and the shift is very small when we go from ^{12}C to ^{13}C. We can therefore measure the amounts of different isotopes in the interstellar medium. Since all of the heavy elements come from stars, these measurements can tell us about the ways in which earlier generations of stars have enriched the interstellar medium. It also turns out that changing isotopes changes the chemical reaction rates. This is particularly true for ion–molecule reactions in cool clouds. Therefore, observing the same molecule with different isotopic substitutions can tell us about how that molecule was made.

14.6 | Thermodynamics of the interstellar medium

The temperature of any object is determined by the balance between heating and cooling. There is generally some temperature for which the rates of heating and cooling will be the same, allowing the temperature to stay constant. In this section we will look at the heating and cooling processes. When we talk about a heating process we mean one that tends to increase the kinetic energy of the gas. When we talk about a cooling process we mean one that tends to decrease the kinetic energy.

One way for an interstellar cloud to be heated is by the absorption of photons. These photons come from a nearby star, or from the combined light of many distant stars. A photon entering a cloud is not, by itself, a mechanism for heating the cloud. We must have a way of converting the energy of the photon into kinetic energy in the gas. The most important mechanisms for photon heating are as follows.

(1) Heating the dust. A photon strikes a dust grain with the photon energy going towards increasing the grain temperature. The hot grain is then struck by an atom or molecule in the gas, and it transfers some of its energy to that atom or molecule.

(2) Excitation of atoms or molecules. A photon strikes an atom or molecule, leaving it in an excited state. The excited atom or molecule undergoes a collision with another atom or molecule. The first atom or molecule drops back to the lower energy state, and the energy shows up as an increased kinetic energy for the second atom or molecule. It is important that the collision takes place before the first atom or molecule has had time simply to emit a photon and drop to the lower state, since the photon can escape, leaving the cloud with no additional energy.

(3) Ionization. The incoming photon strikes an atom or molecule, ejecting an electron. The electron can then transfer its kinetic energy to the rest of the gas through collisions. These collisions must take place before the electron recombines with an ion, releasing a photon.

(4) Photoelectric effect. In the process, an incoming photon strikes a grain surface, causing the ejection of an electron. The electron's kinetic energy is then available to heat the gas.

The above processes also work for heating by streams of high energy particles, known as *cosmic rays*, which permeate the interstellar medium. The sources of cosmic rays will be discussed in Chapter 19.

The interstellar medium can also be heated by the direct injection of mechanical energy from high velocity flows. For example, a supernova remnant, traveling at high speeds, will transfer some of its kinetic energy to material it overruns. Stellar winds can accomplish the same things. When we look at large scale maps of interstellar clouds, we see evidence for many loops and 'bubbles'. These suggest that the interstellar medium is constantly being stirred up by processes such as supernova explosions.

We now look at cooling processes. We must remember that a cooling process must take kinetic energy from the gas and remove that energy from the cloud. These processes are just the inverse of the heating processes.

(1) Emission from grains. An atom or molecule strikes a grain, with the atom or molecule losing kinetic energy and the grain becoming hotter. The grain can then radiate this excess energy away. It must radiate away before it is struck by another atom or molecule that might take back the energy.

(2) Excitation. One atom or molecule strikes another, with the first losing kinetic energy and the second being driven into an excited state. The second one then emits a photon and drops back to its lower state. Of course, the emission of the photon must take place before another collision forces the second one back to the lower state.

(3) Ionization. One atom or molecule strikes another, with the second being ionized. The electron then recombines, accompanied by the emission of a photon, before it can collide with another particle in the gas.

In the heating and cooling processes, different atoms and molecules play important roles in different density and temperature regimes. For example, in the cool molecular clouds, much of the cooling comes from radiation by CO. In very hot regions the cooling can come from unusual emission lines in certain ions (discussed in Chapter 15). This multitude of processes allows us to have a wide variety of temperatures in the interstellar medium, ranging from cool (10 K) high density regions to hot (10^4 K) low density regions.

Chapter summary

In this chapter we looked at various components of the interstellar medium. We saw how they are observed, and we looked at the physical processes that are important in their current state and evolution.

Though only 1% of the interstellar mass, the dust is the most easily visible part of the interstellar medium. We detect dust by its blocking of starlight, known as extinction. Warm dust can be detected by emission in the infrared. The extinction consists of both scattering and absorption. The

extinction is wavelength dependent, producing interstellar reddening. We saw what could be deduced about grain sizes from extinction curves. We obtain information on grain composition from infrared spectra. We saw how the equilibrium temperatures of grains are determined by a balance between radiation absorbed and radiation emitted.

The interstellar gas can be observed in the optical and ultraviolet parts of the spectrum, but radio observations are most useful in studying the cool gas. Extensive studies have been made using the 21 cm line of hydrogen. However, the clouds revealed by these studies do not have high enough densities for them to be the likely sites of star formation.

Star formation probably takes place in cooler, denser, molecular clouds. In these clouds most of the hydrogen is in the form of H_2. We cannot observe the H_2 from the ground, except when it is heated in a few small regions. Instead, we study molecular clouds by emission from trace molecules, such as CO. At these low temperatures (10–50 K), we are usually observing transitions from one rotational state to another. At these low temperatures and densities, most of the chemical reactions are probably between ions and neutral species.

We also looked at how interstellar clouds are heated and at how they cool. In heating, any energy input must eventually be converted into kinetic energy in the gas. In cooling, the kinetic energy of the gas must be converted into energy, such as radiation, that can leave the cloud.

Questions

14.1. When we look at an image like Fig. 14.1, how do we know that something is blocking the light of distant stars, rather than there simply being fewer stars in one direction?

14.2. How will scattering affect the temperature of the struck dust grain? How will absorption?

14.3. If we look at a bright nebula, how might we know that it is a reflection nebula?

14.4. What does counting stars tell us about the extinction in a cloud?

14.5. When we use star counts, what are the relative advantages of looking in magnitude ranges and taking total numbers of stars?

14.6. What is the significance of the constancy of the ratio of total-to-selective extinction?

14.7. What is the evidence that there is more than one type of interstellar dust grain?

14.8. Why is it not likely that interstellar dust grains are all spherical?

14.9. What do we mean by a dilute blackbody?

*14.10. How would the equilibrium temperature of a dust grain change if the albedo were less in the infrared than in the visible?

14.11. Why do we not see Hα absorption from interstellar HI clouds?

14.12. What is the explanation for the disappearance in the correlation between HI column density and visual extinction above one magnitude?

14.13. What is the evidence that interstellar gas and dust are well mixed?

14.14. How can we study the Zeeman shift in interstellar HI when it is only a small fraction of the linewidth?

14.15. What are the advantages of using radio observations (as opposed to optical) in studying the interstellar gas?

14.16. What are the mechanisms that we have discussed for broadening interstellar lines?

14.17. What does the two-phase model of the interstellar medium try to explain?

14.18. Compare the advantages of studying the 21 cm line of H with the 2.6 mm line of CO. Discuss both the observational advantages and the differences in the physical information that we obtain.

14.19. Why did early molecular discoveries (CH, CH^+, CN) discourage further searches?

14.20. How do ion–molecule reactions help in interstellar chemistry?

14.21. Why is it hard to make H_2 directly in the gas phase?

14.22. As you increase the mass of the atoms in a diatomic molecule, what happens to the rotational energy levels?

14.23. Explain how we use observations of a molecule such as CO to tell us about conditions in molecular clouds.

14.24. In studying molecular clouds, what are the advantages of studying a variety of molecules?

14.25. Explain how CO can act to cool molecular clouds. Go through the steps of how the energy would transform from the kinetic energy of the gas to a form where it can escape the cloud.

14.26. Why do low density gases have difficulty cooling?

Problems

(For all of these problems, where necessary assume the normal ratio of total-to-selective extinction.)

14.1. Suppose we observe an A0 V star to have an apparent (visual) magnitude of 15.7. The star is in a cluster whose distance is known to be 950 pc. (a) What is the extinction between us and the star? (b) If we had not known the distance to the cluster, and not taken the extinction into account, by how much would our distance calculation have been off?

14.2. We observe a star that is 1000 pc away. How much extinction would there be if we calculated a distance of 2000 pc when we didn't take extinction into account?

14.3. Suppose we observe an A0 V star to have m_V = 16.0, m_B = 17.0. (a) What is the extinction between us and the star, and how far away is the star? (b) What are the extinction and distance if m_B = 18.0?

14.4. Suppose we observe a K0 V star to have m_V = 16.0, m_B = 17.0. (a) What is the extinction between us and the star, and how far away is the star? (b) What are the extinction and distance if m_B = 18.0?

14.5. We are observing a cluster whose distance is known to be 850 pc. The visual extinction between us and the cluster is 2 mag. What would be the m_V and m_B for (a) an O5 star and (b) a G5 star?

14.6. Prove that, for the same angular momentum, end-over-end rotation for a cigar-shaped object has a lower energy than rotation about the long axis.

14.7. Suppose we have a grain with a radius of 10^{-5} cm and a charge of $+e$. An electron starts from far away at rest. It is attracted and eventually hits the surface. (a) How fast is it going when it hits the surface? (b) How does this compare with the the average thermal speed of the electrons if they are at 100 K?

14.8. Use the results of Example 14.4 to give a scaling relationship that can be used to calculate the dust temperature (in Kelvin) when the stellar temperature is given in units of 10^4 K, and the distance from the star is given in stellar radii.

14.9. How far must a dust grain be from a 10^4 K star for it to have a temperature of 1000 K?

*14.10. Suppose the albedo of a dust grain is constant in the infrared (where it will emit most of its energy), at a value a(IR), and is constant over the visible and ultraviolet (where it will absorb most of its energy), with a value a(V). It is near a star whose spectrum is that of a blackbody at temperature T_*, and whose radius is R_*. The grain is a distance d from the star. Derive an expression for the dust temperature.

*14.11. Suppose a dust grain has an albedo $a(\lambda)$. It is near a star whose spectrum is that of a blackbody at temperature T_*, and whose radius is R_*. The grain is a distance d from the star. Derive an expression for the dust temperature. (You will have to leave your answer in terms of an integral, since the form of $a(\lambda)$ is not given.)

14.12. What is the range of distances from a B0 V star for which the dust temperature is between 50 and 1000 K?

*14.13. In our discussion of the temperature of a dust grain near a star, we did not account for the fact that the dust grains near the star would block some light from reaching dust grains far from the star. Show how this effect modifies the results.

14.14. Show that, in a gas, the number of particles hitting a surface per second per unit surface area is nv, where n is the number density of particles per volume and v is the speed of the particles.

14.15. Compare the width (both in km/s and nm) of the Na D line (589.6 nm) for thermal broadening in an interstellar cloud (50 K) and a stellar atmosphere (5000 K).

14.16. What frequency resolution would be needed to observe the 21 cm line with a velocity resolution of 0.1 km/s?

14.17. For an excitation temperature of 100 K, what is the ratio of populations for the two levels in the 21 cm transition? (Take the statistical weight of the lower level to be 1 and the upper level to be 3.)

14.18. How does the angular resolution of a 100 m diameter telescope at 21 cm compare with that of a 12 m diameter telescope at 2.6 mm? What is the significance of these numbers?

*14.19. Estimate the rate at which two H atoms can form an H_2 molecule on a grain surface. Assume that all atoms hitting a grain stick, and that as soon as two H atoms are on the grain surface, they immediately form a molecule.

14.20. Suppose we have an electric dipole made up of two charges, $+q$ and $-q$, a distance d apart. The dipole is placed a distance r from a charge $+Q$. (a) Find an expression for the net force on the dipole. (b) How is this related to chemical reactions in the interstellar medium?

14.21. For the CO, $J = 1 \rightarrow 0$ transition (just considering the two lowest levels), what is the ratio of the populations of the two levels for temperatures of 10, 20, 30 K? The statistical weights of the two states are 1 and 3.

14.22. For scattering of light by interstellar grains we define the *phase function* as the average value of cos θ, where θ is the angle through which the light is scattered, and $\theta = 0$ for forward scattering. Find the value of the phase function in the following limiting cases: (a) all forward scattering, (b) all rearward scattering, (c) random scattering with all directions equally likely.

14.23. For rotational energy levels of diatomic molecules the most likely transitions are those for which J changes by ± 1. Find an expression for the energies of these allowed transitions, as a function of J and I.

14.24. Calculate the mass of an HI cloud with a circular appearance, with a radius of 10 pc and an average H column density of 10^{21} cm^{-2}.

14.25. Calculate the virial mass for a molecular cloud with $\langle v_r \rangle = 5$ km/s and $R = 20$ pc.

14.26. For a 30 pc radius molecular cloud to have a virial mass of 10^6 $M_\odot$, what must be the value of $\langle v_r \rangle$?

Computer problems

14.1. If we are far from an individual star, we can treat the interstellar radiation field as a dilute blackbody of temperature 10^4 K, with a dilution factor of 10^{-14}. This means that the spectrum of the radiation looks like that of a 10^4 K blackbody, with the intensity reduced by a factor of 10^{-14}. Use this to estimate the temperature of an interstellar grain in such an environment.

14.2. Plot the graph of the temperature of a dust grain vs. distance from an O5 star, with the graph covering distances from 1000 AU to 1 pc.

14.3. Draw diagrams like Fig. 14.13(a) where the splitting between the two components is a smaller and a larger fraction of the linewidth. (The red and blue curves are both gaussians with the same peak intensity and linewidth.)

14.4. Assume that the populations of the 21 cm transition, in a gas of temperature T, are described by the Boltzmann equation with statistical weights 1. Plot the graph of the fraction in the upper state vs. T for T ranging from 0 to 1000 K.

14.5. For CO and CS, what rotational transitions would be observable in the visible part of the spectrum?

14.6. For the CO, $J = 1 \rightarrow 0$ transition (just considering the two lowest levels), make a graph of the ratio of the populations of the two levels for temperatures ranging from 0 to 1000 K. The statistical weights of the two states are 1 and 3.

14.7. For the lowest three energy levels of CO, draw a graph of the fraction in the $J = 1$ state vs. T for T ranging from 0 to 1000 K. The statistical weights of the three states are 1, 3 and 5.

14.8. Calculate virial masses for molecular clouds with $R = 1, 3, 5$ and 10 pc, and $\langle v_r \rangle = 1, 2$ and 5 km/s.

Chapter 15

Star formation

In Chapter 14, we discussed the contents of the interstellar medium, the material out of which new stars must be formed. In this chapter, we will identify those parts of the interstellar medium that are involved in star formation, and see what we know, and what we have to learn, about the star formation process.

15.1 | Gravitational binding

In Chapter 13, we talked about gravitational binding for clusters of stars. The same concepts apply to interstellar clouds, with the stars in the cluster being replaced by the particles that make up the cloud (either H or H_2). The gravitational potential energy is now due to the interaction among all of the particles in a cloud. For a uniform spherical cloud, the gravitational potential energy is $-(3/5)GM^2/R$. The kinetic energy is still related to the rms velocity dispersion, but with a large number of particles, which can easily be related to the cloud temperature, so the kinetic energy is $(3/2)(M/m)kT$, where M is the total mass of the cloud and m is the mass per particle.

The clouds are kept together by the gravitational attraction amongst all of the particles in the cloud. If the gravitational forces that hold the cloud together are greater than the forces driving it apart, we say the cloud is gravitationally bound. We can think of the random thermal motions in the gas as resisting the collapse.

The condition for gravitational binding (total energy negative) is then

$$(3/5)GM^2/R \geq (3/2)(M/m)kT$$

Dividing both sides by GM and multiplying by $(5/3)$ gives

$$(M/R) \geq (5/2)(kT/Gm) \tag{15.1}$$

The mass and radius of a cloud are not independent, since they are related to the density $\rho = M/(4\pi/3)R^3$. We might therefore like to use equation (15.1) to estimate the *smallest size cloud* of a given ρ, m and T for which the cloud is gravitationally bound. This quantity is called the *Jeans length*, R_J. James Jeans obtained essentially the same result with a more sophisticated analysis. We therefore eliminate M in equation (15.1), and change the inequality to an equality, since we are looking for the value of R that is on the boundary between bound and unbound. This gives

$$(4\pi/3)R_J^3\rho/R_J = 5kT/2Gm$$

Solving for R_J,

$$R_J = (15kT/8\pi Gm\rho)^{1/2} \tag{15.2}$$

Note that $(15/8\pi)^{1/2} = 0.77$, which is close to unity. As the geometry of the cloud changes, the exact value of the constant will change, but it will still be close to unity. We then write

$$R_J \cong (kT/Gm\rho)^{1/2} \tag{15.3}$$

We can rewrite this in terms of n, the number of particles per unit volume ($n = \rho/m$), as

$$R_J \cong (kT/Gm^2n)^{1/2} \tag{15.4}$$

We can also use equation (15.1) to give us the minimum mass for which a cloud of given ρ, T and m will be bound. This minimum mass is called the *Jeans mass*. It is the mass of an object whose radius is R_J, so

$$M_J = (4\pi/3)R_J^3\rho$$

$$= (4\pi/3)(kT/Gm\rho)^{3/2}\rho$$

$$\cong 4(kT/Gm)^{3/2}\,\rho^{-1/2}$$

$$= 4(kT/Gm)^{3/2}(nm)^{-1/2} \tag{15.5}$$

Example 15.1 Jeans length and mass
Find the Jeans length and mass in a cloud with 10^5 H atoms per centimeter cubed and a temperature of 50 K.

SOLUTION
We use equation (15.4) to find R_J:

$$R_J = \left[\frac{(1.38 \times 10^{-16}\,\text{erg /K})(50\,\text{K})}{(6.67 \times 10^{-8}\text{dyn cm}^2/\text{g}^2)(1.67 \times 10^{-24}\text{g})^2(10^5\text{cm})}\right]^{1/2}$$

$$= 6.1 \times 10^{17}\,\text{cm}$$

$$= 0.2\,\text{pc}$$

We find the mass by multiplying the density by the volume:

$$M_J = (4\pi/3)(1.67 \times 10^{-24}\,\text{g})(10^5\,\text{cm}^{-3})(6.1 \times 10^{17}\,\text{cm})^3$$

$$= 1.5 \times 10^{35}\,\text{g}$$

$$= 76\,M_\odot$$

We could have obtained the same mass directly from equation (15.5).
As we will see below, not all the mass will end up in the star.

Once a cloud becomes gravitationally bound, it will begin to collapse. We would like to be able to estimate the time for the collapse to take place. We begin by considering a particle a distance r from the center of the cloud. It will accelerate toward the center under the influence of the mass closer to the center than r. The acceleration is given by

$$a(r) = GM(r)/r^2$$

$$= G(4\pi/3)r^3\rho/r^2$$

$$= (4\pi/3)Gr\rho \tag{15.6}$$

If the acceleration of this particle stayed constant with time, then the *free-fall time*, the time for it to fall a distance r, would be

$$t_{ff} = \left[\frac{2r}{a(r)}\right]^{1/2} \tag{15.7}$$

$$= \left[\frac{2r}{(4\pi/3)(Gr\rho)}\right]^{1/2}$$

Note that the constant $(3/2\pi)^{1/2} = 0.7$, which we can approximate as unity, since we are making an estimate of the time. This gives

$$t_{ff} \cong 1/(G\rho)^{1/2} \tag{15.8}$$

The free-fall time is independent of the starting radius. Therefore, all matter in a constant density cloud has approximately the same free-fall time. However, as the cloud collapses, the density increases. The collapse proceeds faster. The free-fall time for the original cloud is therefore an upper limit to the actual collapse time. However, the result is not very different, since most of the time will be taken up in the early stages of the collapse, when the acceleration is not appreciably different from the one we have calculated. Therefore, we use the free-fall time as a reasonable estimate of the time it will take a cloud to collapse.

There is one important difference between our idealized cloud and a real cloud. A real cloud will probably have a higher density in the center. We can see this as follows. If the cloud is initially of uniform density, all points will have the same inward acceleration. This means that all particles will cover the same inward distance dr (where $dr < 0$), in some time interval dt. We can see how this changes the density for different volume spheres. If the initial density is ρ_0, then the density of a constant mass collapsing sphere that shrinks from r_0 to r is

$$\rho = \rho_0 (r_0 / r)^3$$

The change in density $d\rho$ is found by differentiating to give

$$d\rho = -3\,\rho_0 (r_0^3 / r^4)\,dr$$

The fractional change in density, $d\rho/\rho$, is

$$d\rho/\rho = -3 \, dr/r \qquad (15.9)$$

This means that the smaller the initial sphere we consider, the faster its density will grow.

With a higher density at the center, the free-fall time for material near the center will be less than for material near the edge. The material from the edge will lag behind the material closer in. This will enhance the density concentration in the center. The net result is that we end up with a strong concentration in the center. The concentration will eventually become the star, but the material from the outer parts of the cloud will continue to fall in on the star for quite some time.

Example 15.2 Free-fall time
Calculate the free-fall time for the cloud in the above example.

SOLUTION
Using equation (15.8) gives

$$t_{ff} = \frac{[(6.67 \times 10^{-8} \, \text{dyn cm}^2/\text{g}^2)(1.67 \times 10^{-24} \, \text{g})}{(10^5 \, \text{cm}^{-3})]^{-1/2}}$$

$$= 9.5 \times 10^{12} \, \text{s}$$

$$= 3 \times 10^5 \, \text{yr}$$

While almost a million years might sound like a long time, it is short compared with the main sequence lifetime of the star that will be formed, or the age of the galaxy.

If the cloud is rotating, then the collapse will be affected by the fact that the cloud's angular momentum must remain constant. The angular momentum L is the product of the moment of inertia I and the angular speed ω,

$$L = I\omega \qquad (15.10)$$

For a uniform sphere, the moment of inertia is

$$I = (2/5)Mr^2 \qquad (15.11)$$

If I_0 and ω_0 are the original moment of inertia and angular speed, and I and ω are their values at some later time, conservation of angular momentum tells us that

$$I_0 \omega_0 = I\omega \qquad (15.12)$$

Using equation (15.11) to eliminate I and I_0, we have

$$(\omega/\omega_0) = (r_0 / r)^2 \qquad (15.13)$$

(This explains why a figure skater rotates faster as she brings her arms in. The $1/r^2$ dependence of the angular speed has a dramatic effect.)

To see what effect this has on collapse, we again look at a particle a distance r from the center of a collapsing cloud. The acceleration at that point is still $GM(r)/r^2$. However, the radial acceleration now has two parts: (1) $a(r)$ is associated with the change in magnitude of the radius, and (2) the acceleration associated with the change of direction, $r\omega^2$. Therefore,

$$GM(r)/r^2 = a(r) + r\omega^2 \qquad (15.14a)$$

Solving for $a(r)$ gives

$$a(r) = GM(r)/r^2 - r\omega^2 \qquad (15.14b)$$

In comparing this with equation (15.6), we see that the acceleration $a(r)$ is less for a rotating cloud than for a non-rotating cloud. The effect of the rotation is to slow down the collapse perpendicular to the axis of rotation.

The effects of rotation will be most significant when the second term on the right-hand side of equation (15.14a) is much greater than the first term, in which case

$$GM(r)/r^2 = r\omega^2$$

Multiplying both sides by r^2 gives

$$GM(r) = r^3 \omega^2$$

$$= r^3 \omega_0^2 \, (r_0/r)^4$$

$$= (\omega_0 r_0)^2 \, (r_0/r) \, r_0 \qquad (15.15)$$

Noting that $v_0 = \omega_0 \, r_0$, where v_0 is the speed of a particle a distance r_0 from the center,

$$GM(r) = v_0^2 \, r_0 \, (r_0/r)$$

We now solve for r/r_0, the amount by which the cloud collapses before the rotation dominates:

$$r/r_0 = v_0^2 \, r_0 / GM(r) \qquad (15.16)$$

For the cloud given in the two previous examples, with an initial rotation speed $v_0 = 1$ km/s, $r/r_0 = 0.6$. This means that, by the time the cloud

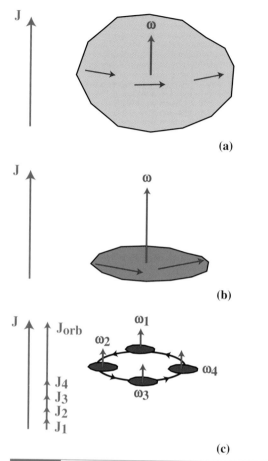

Fig 15.1. Fragmentation of a collapsing interstellar cloud. (a) The cloud is initially rotating as shown. As it collapses, the angular momentum **J** is conserved. (b) As the cloud becomes smaller, its angular speed ω must increase to keep the angular momentum fixed. The rotation inhibits collapse perpendicular to the axis of rotation, and the cloud flattens. (c) Unable to collapse any further, the cloud breaks up, with the total angular momentum being divided between the spin and orbital angular momenta of the individual fragments.

reaches half its initial size, the rotation can completely stop the collapse perpendicular to the axis of rotation. Motions parallel to the axis of rotation are not affected by this, so collapse parallel to the axis of rotation can proceed unimpeded, and the cloud will flatten. (We will see that the tendency of rotating objects to form disks will reappear in many astrophysical situations.) Since the collapse is then only in one dimension, it is harder to reach stellar densities. Thus, the effect of rotation is to keep much of the material from becoming a star.

More of the material can end up in stars if the cloud breaks up into a multiple star system. The angular momentum can be taken up in the orbital motion, but individual clumps can continue contracting. This fragmentation process is probably responsible for the high incidence of binary systems. If a cloud shrinks to half its initial size, the average density will go up by a factor of eight. (The density is proportional to $1/\text{volume}$, and the volume is proportional to r^3.) From equation (15.5), we see that the Jeans mass of the denser cloud will be approximately one-third of the original Jeans mass. This means that it is possible for the less massive clumps to be bound and continue their collapse. The fragmentation process (Fig. 15.1) may be repeated until stellar mass objects are reached.

15.2 | Problems in star formation

We would like to know the conditions under which stars will form. We would like to know which types of interstellar clouds are most likely to form stars, and which locations within the clouds are the most likely sites of star formation. We would also like to know whether star formation is spontaneous or whether it needs some outside *trigger*. When we say a trigger is necessary, we mean that the conditions in a cloud are right for star formation, but something is necessary to compress the cloud somewhat to get the process started. Once started, it continues on its own. Sources for triggering star formation that have been suggested are the passage of a supernova remnant shock front, or the compression caused by a stellar wind. (Later in this chapter we will see how expanding HII regions might act as triggers, and in Chapter 17 we will discuss density waves associated with galactic spiral structure. These might also induce star formation.)

Once the collapse to form stars starts, we would like to know how it proceeds, and what fraction of the cloud mass ends up in stars. This is sometimes referred to as the *efficiency of star formation*. We would also like to know how much of the mass that goes into stars goes into stars of various masses. This is called the *initial mass function*. By "initial" we mean the distribution of stellar masses at the time that a cloud gives birth to stars. The actual mass distribution in the galaxy is altered by the fact that stars of different masses have different lifetimes.

An important problem in understanding the evolution of star forming clouds comes from the angular momentum of the cloud. In the previous section, we saw that the collapse can be slowed down or even stopped in a rotating cloud. How a cloud distributes and loses its angular momentum probably affects the efficiency of star formation and the initial mass function. In addition, it may account for the high fraction of multiple star systems and for the formation of planetary systems.

An interesting set of problems is posed by groupings of stars called *OB associations*. These are groups of stars in which it has been suggested that all O stars form. We refer to these groupings as associations rather than clusters because associations are not gravitationally bound. They are expanding, and eventually dissolve into the background of stars. We would like to know how an initially bound cloud can give birth to an unbound grouping of stars. In Chapter 13 we saw that, if a system in virial equilibrium loses more than half of its mass without the velocity distribution changing, then the system becomes unbound. It is clear that the clusters have lost more than half their mass.

Another interesting feature of OB associations is the existence of *subgroups*. Some associations have as many as three or four distinct groupings of stars. The subgroups have different ages, as determined from their HR diagrams. Also, the older subgroups seem to be larger, which makes sense if they are expanding. A major question in star formation is explaining what appears to be a sequential wave of star formation through an association. It is in this context that triggers have been most actively discussed. OB associations are often near molecular clouds, as shown in Fig. 15.2. The younger subgroups tend to appear closer to the molecular clouds.

We would also like to know whether low and high mass star formation take place in different ways or in different environments. It has been suggested, for example, that low mass stars are being made all the time, whereas high mass star formation takes place in bursts. We would also

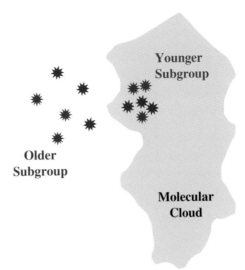

Fig 15.2. OB associations and molecular clouds. These associations often have a few subgroups. The older subgroups are more extended, since the associations are unbound and are expanding. Younger subgroups tend to be more closely related to molecular clouds.

like to know how an interstellar cloud knows what distribution of stellar masses it is supposed to make.

15.3 | Molecular clouds and star formation

We discussed the properties of molecular clouds in Chapter 14. They are important for star formation because they are both cool and dense, relative to the rest of the interstellar medium. In Section 15.1, we discussed the conditions under which an interstellar cloud is gravitationally bound, and expressed the result as a Jeans length. For a cloud of a given temperature T and number density n, the Jeans length is the minimum size of a gravitationally bound cloud. We approximated the Jeans length (equation 15.4) as

$$R_J \cong (kT/Gm^2 n)^{1/2}$$

Example 15.3 Jeans length for atomic and molecular clouds
Compare the Jeans length for an atomic cloud ($T = 100$ K, $n = 1$ cm^{-3}) and a molecular cloud ($T = 10$ K, $n = 10^3$ cm^{-3}).

SOLUTION

We simply take the ratios, noting also that the mass per particle in molecular clouds is twice that in atomic clouds.

$$R_J(at)/R_J(mol) = [(10^3)(10)(4)]^{1/2}$$

$$= 200$$

This means that a much smaller piece of a molecular cloud can become gravitationally bound than for an atomic cloud. It is therefore much easier to get a bound section in a molecular cloud than in an HI cloud. This comes about because of both the higher density and lower temperature.

We find a number of different types of molecular clouds. Their basic properties are summarized in Table 15.1. The simplest are the globules, as shown in Fig. 14.1. These are sometimes called *Bok globules*, after *Bart Bok*, who suggested that they were potential sites for star formation. Globules are typically a few parsecs across. They generally have a simple, round appearance. This simplicity makes them attractive to study. Their visual extinctions fall in the range 1−10 mag, which can be determined by star counting. From CO observations, we find their kinetic temperatures are about 10 K. From observations of CO and CS we estimate their densities at 10^3 cm^{-3} and up to 10^4 cm^{-3}, and masses in the range 10 to 100 $M_\odot$. We think that they are in a state of slow gravitational contraction.

The *dark clouds*, such as those shown in Fig. 15.3, have local conditions (density, temperature) similar

Fig 15.3. Dark clouds blocking the light from background stars. Note the intricate shapes. This is a near infrared image of the star forming region RCW108. We see a few bright young stars, but mostly irregular dark clouds. (Compare this with the simple shape of the globule in Fig. 14.1.) [ESO]

to those in globules, but the dark clouds are larger. Typical sizes for the dark clouds are in the tens of parsecs range. Often a size is hard to define because dark clouds appear to consist of a number of small clouds in an irregular arrangement. There is evidence that they contain low mass stars.

The largest molecular clouds are called *giant molecular clouds* or *GMCs*. They are generally elongated, with a length of about 50 to 100 pc. Their

Table 15.1. Interstellar molecular clouds.

Type	T_k (K)	$n(H_2)$ (cm^{-3})	R (pc)	M ($M_\odot$)	A_V (mag)	Probes
Dark cloud	10	10^3	5–10	10^4	1–6	CO
Globule	10	10^3–10^4	1	10^2–10^3	5–15	CO, CS
GMC envelope	15	300	50	10^5	1–5	CO
Dense core	30–100	10^5–10^6	1	10^3	100	CO, CS, H_2CO, NH$_3$ and many others
Protostellar cores	100–200	10^7				high J transitions of various molecules
Energetic flows	1000					CO, SiO, H_2
Envelope of evolved stars	2000					SiO masers

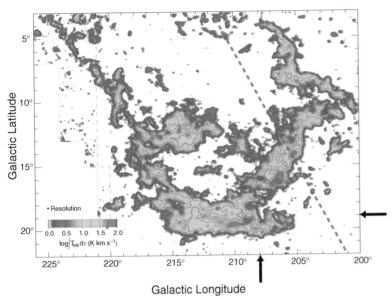

Fig 15.4. Molecular clouds in the Orion region. Giant molecular clouds are indicated by their contours of emission from the CO molecule at 2.6 mm. This was done with a 1 m telescope, providing an angular resolution of 12 arc min, which, as the black dot at lower left shows, is quite adequate for this map. This was part of a survey of CO emission in our galaxy (which we will talk about more in the next chapter), so this is plotted in galactic coordinates, which are tilted with respect to the celestial equator, which is shown in a red dashed line in this figure. The Orion Nebula (shown in Fig. 15.28) is at the intersection of the two black arrows in the lower right corner of the figure. At the 500 pc distance of Orion, 100 pc would subtend 11 degrees. So, the cloud containing the Orion Nebula is some 80 pc long. There is an OB association around the region of the nebula. There is another equally large cloud that extends to the north across the celestial equator. This contains the association connected to Orion's belt and includes (as a very small part) the Horsehead Nebula (Fig. 14.3). Note that there are some other clouds which extend to the west, into Monoceros, containing yet another OB association. [Thomas Dame, CFA]

densities are about 300 cm^{-3}, a little lower than for globules or dark clouds. They are also warmer, with $T = 15$ K. Their extent can be traced using CO observations, like those shown in Fig. 15.4. By observing CO in nearby GMCs, where we can see the dust, we gain confidence in the fact that the CO tells us where the dust (and molecular hydrogen) is. We therefore use the CO to trace out GMCs that are so far away that foreground dust

blocks our view of them. GMCs typically have masses of a few times 10^5 $M_{\odot}$, and seem to come in complexes whose masses exceed 10^6 $M_{\odot}$. These complexes are among the most massive entities in the galaxy. There seems to be a close connection between giant molecular clouds and OB associations. It therefore appears that O and B stars form in GMCs. GMCs also have lower mass stars in addition to the O and B stars.

Within the giant molecular clouds we find denser regions, called *dense cloud cores*. These are denser and warmer than the surrounding cloud. Their temperatures are above 50 K. Their densities, determined from studies of a number of different molecules, are in the range 10^5 to 10^6 cm^{-3}. (Even though we call these clouds "dense", their densities are comparable to the best vacuums we can obtain in the laboratory!) These cores are small, only a few tenths of a parsec across, and have masses of a few hundred $M_{\odot}$. Our ability to study them is limited by the angular resolution of our telescopes, but that is helped by the development of interferometers working at millimeter wavelengths. We think that these cores are the places in the GMCs where the star formation is taking place. (Some dense cores are also found in dark clouds and globules.)

One of the observational challenges in studying dense cloud cores is to find cores in which there is unambiguous evidence for collapse. After all, if these are in the process of forming stars, there should be large inward motions that should be detectable via Doppler shifts. Material on the side of the core closest to us should be moving away from us, and we should see red-shifted spectral lines. Material on the far side of the core should be moving towards us, and we should see blueshifted spectral lines. We do see objects with both large red- and blueshifts. The problem is that the Doppler shift measurements cannot tell us which side of the cloud each part of the emission is coming from. So, an expanding cloud could have the same spectra as a collapsing cloud. The trick lies in having high angular resolution and seeing how the Doppler shift changes with position. Studies of possible collapsing clouds using millimeter interferometers are just beginning to yield results.

In our discussions, it is important to remember that, while we think in terms of spherical clouds for simplicity, real clouds have irregular shapes. Most of the larger clouds (GMCs) appear elongated, part of larger filamentary structures. It has even been suggested that the geometry of interstellar clouds may be better represented by fractals. However, we can still form a good insight into the physical processes that govern star formation using our simplified models.

15.4 | Magnetic effects and star formation

Astronomers are becoming increasingly aware of the fact that magnetic fields can have an important effect on the star formation in an interstellar cloud. Work in this area has been slow for two reasons. (1) As we have seen, measurements of the interstellar magnetic fields are very difficult. Until we have a good idea of field strengths, it is hard to estimate their effects. As we mentioned in the last chapter, observations of the Zeeman effect in HI yield field intensities of tens of microgauss in a number of clouds. (2) Theories that include magnetic fields are much harder to work out than those that don't. However, computer simulations of gravitational collapse in clouds with substantial magnetic fields are being carried out more routinely.

We would expect the magnetic effects to be important when the energy associated with the presence of the magnetic field is comparable to the gravitational energy in magnitude. In cgs units, the energy density (erg/cm^3) associated with a magnetic field B is

$$u_B = B^2/8\pi \qquad (15.17)$$

Example 15.4 Magnetic energy
For what magnetic field strength B does the magnetic energy of a cloud equal the absolute value of the gravitational potential energy? Assume a spherical cloud with a radius $R = 10$ pc, and a density of molecular hydrogen $n(H_2) = 300$ cm^{-3}.

SOLUTION
The magnetic energy U_B is the energy density, multiplied by the volume of the cloud:

$$U_B = \left(\frac{B^2}{8\pi}\right)\left(\frac{4\pi}{3}\right)R^3$$

$$= \frac{B^2 R^3}{6}$$

The magnitude of the (negative) gravitational potential energy is

$$-U_G = \frac{3}{5}\frac{GM^2}{R}$$

$$= \frac{3}{5}G\frac{\left[\left(\frac{4\pi}{3}R^3\right)n(H_2)(2m_p)\right]^2}{R}$$

$$= \left(\frac{16\pi^2}{15}\right)GR^5(n(H_2)(2m_p))^2$$

$$= 10\,GR^5\,(n(H_2)(2m_p))^2$$

Equating these and solving for B, we have

$$B = (60\,G)^{1/2}\,(2n(H_2)m_p\,R)$$

$$= 6 \times 10^{-5}\text{ gauss}$$

$$= 60\ \mu\text{G}$$

This is of the order of strengths of fields that have been measured from HI Zeeman measurements.

As a molecular cloud collapses, the magnetic field strength will increase, as illustrated in Fig. 15.5. This is because of the flux freezing, discussed in Chapter 11. (Remember, Faraday's law

Magnetic Field

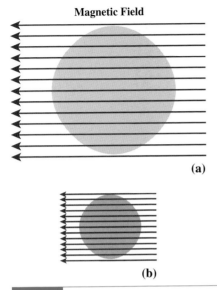

(a)

(b)

Fig 15.5. Flux freezing in a collapsing interstellar cloud. Graphically, Faraday's law tells us that the number of field lines crossing the cloud's surface stays constant as the cloud collapses. This means that the field lines are closer together, signifying a stronger field in (b).

requires that the flux through a conducting surface be constant.) This only takes place if the cloud is a good conductor. Most interstellar clouds have sufficient ionization for this to be the case. The ionization in cold clouds probably results mostly from cosmic rays. Most of the mass of the cloud is in the form of neutral atoms or molecules. There are roughly 10^7 neutrals for every ion. However, as the cloud collapses, these neutral particles carry the charged particles along with them. The charged particles, in turn, provide the conductivity to insure the flux freezing. This process allows the magnetic field effectively to exert a pressure which can inhibit the collapse.

Example 15.5 Flux freezing

For a spherical cloud with the magnetic flux constant as the cloud collapses, find how the magnetic energy varies with the cloud radius R. Compare this with the gravitational energy.

SOLUTION

For a uniform cloud, the magnetic flux is the product of the field B and the projected area of the cloud πR^2. This means that the flux is proportional

to BR^2. If the flux is constant, then BR^2 must be constant. This means that

$$B \propto 1/R^2$$

From the previous example, we see that the magnetic energy U_B is proportional to B^2R^3. This means that

$$U_B \propto (1/R^2)^2 \, R^3$$

$$\propto 1/R$$

The gravitational potential energy is

$$U_G \propto GM^2/R$$

Since the mass of the cloud stays constant as it collapses,

$$U_G \propto 1/R$$

Therefore, the magnetic and gravitational energies have the same dependence on R as the cloud collapses. If the magnetic field cannot prevent the initial collapse, then it cannot prevent the further collapse. However, if the magnetic field is important in the initial collapse, it will continue to be important.

As a cloud evolves, the ions and neutrals do not always stay perfectly mixed. The ions drift with respect to the neutrals. If this happens, some of the magnetic flux will escape from the cloud, meaning that the field is not as high as one would calculate from flux freezing. The process, called *ambipolar diffusion*, has another effect. As the ions move past the neutrals some collisions occur. This converts some of the drift motion into random motions of the neutrals, meaning an increase in the cloud temperature. Therefore, ambipolar diffusion can serve as a general heat source in a cloud.

The current picture that has emerged suggests that there are two ways in which the magnetic support of clouds is overcome. One is by ambipolar diffusion. This occurs in clouds where the magnetic energy is comparable to the gravitational energy. Ambipolar diffusion allows for the gradual contraction of the cloud. It is thought that this process produces low mass stars at a roughly steady rate throughout the galaxy. In the alternative situation enough material is gathered together so that the absolute value of the gravitational energy is much greater than the magnetic energy, and star formation takes place rapidly. It is thought that this process makes a mixture of high and low mass stars.

15.5 | Protostars

15.5.1 Luminosity of collapsing clouds

As a cloud collapses the gravitational potential energy decreases. This is because the particles within the cloud are moving closer to the center. The decrease in potential energy must be offset by energy radiated away or by an increase in the kinetic energy. This increased kinetic energy can show up in two forms: (1) it can go into the faster infall of the particles in the collapsing cloud; or (2) it can go into heating the cloud.

Let's see what happens to the energy in a collapsing cloud. From the virial theorem, we know

$$E = - \langle K \rangle \tag{15.18}$$

This tells us that as the cloud collapses its internal kinetic energy K will increase. However, only half the potential energy shows up as increased kinetic energy. We can therefore see that the total energy of the collapsing cloud is decreasing. This means that the cloud must be radiating energy away. The virial theorem tells us that half of the lost potential energy shows up as kinetic energy, and half the energy is radiated away.

We can relate the luminosity of a contracting cloud to its total energy. The total energy is

$$E = (-3/10)GM^2/R \tag{15.19}$$

The energy lost in radiation must be balanced by a corresponding decrease in E. The luminosity, L, must therefore be equal to dE/dt. Differentiating equation (15.19) gives

$$\frac{dE}{dt} = \frac{3}{10}\left(\frac{GM^2}{R^2}\right)\left(\frac{dR}{dt}\right) \tag{15.20}$$

We can solve for dR/dt to find the collapse rate for a given luminosity:

$$\frac{dR}{dt} = \frac{10}{3}\left(\frac{R^2}{GM^2}\right)\left(\frac{dE}{dt}\right) \tag{15.21}$$

(Remember, for a collapsing cloud, both dR/dt and dE/dt are negative numbers.) If we solve

equation (15.19) for R, and substitute that solution for one of the R's in equation (15.20) or (15.21), we find that

$$\frac{1}{E}\frac{dE}{dt} = \frac{1}{R}\left(\frac{dR}{dt}\right) \tag{15.22}$$

This tells us that, in any time interval dt, the fractional change in the energy dE/E is equal to the fractional change in the radius dR/R. These results tell us that the rate of collapse can be limited by the rate at which energy can be radiated.

We now look at the luminosity in various stages of the collapse. As the collapsing cloud heats, it is still well below normal stellar temperatures, so most of the radiation is given off in the infrared. Therefore, the opacity of the cloud in the infrared plays an important role in determining the nature of the collapse.

When the collapse begins, the material is mostly atomic and molecular hydrogen and atomic helium. As the collapse continues, half the liberated energy goes into the internal energy of the gas. However, this doesn't increase the temperature. Instead, the energy goes into the ionization of these neutral species. Following this, the liberated energy goes into heating the gas, and the gas pressure can eventually slow the collapse. For a $1\,M_\odot$ protostar, the free-fall phase ends when the radius is about $500\,R_\odot$. (The radius varies approximately with mass.) During the free-fall stage, the luminosity increases and $|dR/dt|$ increases.

Example 15.6 Luminosity of a collapsing cloud
For a $1\,M_\odot$ protostar that has collapsed to a radius of $500\,R_\odot$, (a) calculate the energy that has been liberated to this point; (b) use this to calculate the average luminosity if most of the energy is liberated in the last 100 years of the collapse.

SOLUTION

From the virial theorem, the energy radiated will be one-half times the current gravitational potential energy:

$$E = -\left(\frac{1}{2}\right)\left(-\frac{3}{5}\right)\frac{GM^2}{R}$$

$$= \left(\frac{3}{10}\right)\frac{(6.67 \times 10^{-8}\,\text{dyn cm}^2/\text{g}^2)(2 \times 10^{33}\,\text{g})^2}{(500)(7 \times 10^{10}\,\text{cm})}$$

$$= 2 \times 10^{45}\,\text{erg}$$

The average luminosity is this energy divided by the time over which it is radiated:

$$L(\text{avg}) = \frac{2 \times 10^{45}\,\text{erg}}{(100)(3 \times 10^7\,\text{s})}$$

$$= 7 \times 10^{35}\,\text{erg/s}$$

$$= 170\,L_\odot$$

This is the average luminosity over the 100 year period, but the actual luminosity at the end of the period is higher, since $|dR/dt|$ is greatest then.

Once a cloud is producing stellar luminosities by gravitational collapse, we call it a *protostar*. Once the cloud becomes opaque the radiation can only escape from near the surface. (When the opacity is low a photon can escape from anywhere within the volume.) Since energy escapes slowly, the temperature rises quickly. Also, a large temperature difference can exist between the center and the edge. Under these conditions, the most efficient form of energy transport from the center to just outside is by convection. This point was first realized in 1961 by the Japanese astrophysicist Chushiro Hayashi. During this stage the surface temperature stays roughly constant at about 2500 K. Since the radius is decreasing, and the temperature is approximately constant, the luminosity decreases.

During this stage the central temperature is still rising. When it is high enough, nuclear reactions start. The contraction goes on for some time in the outer parts, as the pressure builds up in the core. Eventually the pressure in the core is sufficient to halt the collapse, and the star is ready to settle into its main sequence existence.

For a protostar, the continuous spectrum peaks in the near infrared. The dust in the collapsing cloud surrounding the protostar will absorb some of the radiation. The dust will be heated, but will not be the same temperature as the star. The emission from the dust will be in the far infrared. From this we see that protostars are best observed in the infrared part of the spectrum.

15.5.2 Evolutionary tracks for protostars

When we plot an HR diagram with stars we see now, we are plotting the distribution of L and T as they are now. However, as a star evolves, its luminosity and temperature change. Therefore, its location on an HR diagram changes. If $L(t)$ is the luminosity of a star as a function of time and $T(t)$ is the temperature as a function of time, we can plot a series of points and connect them to follow the evolution of a star. Such a series of points is called an evolutionary track. Stars evolve so slowly compared with human lifetimes that we cannot deduce the evolutionary track by observing one star. However, by observing many stars, each at a different stage, we can infer the evolutionary tracks. (We have already used evolutionary tracks in our discussion of post main sequence evolution, in Chapters 10 and 11.)

We can also predict evolutionary tracks from theoretical models of protostars and stars. We use basic physics to calculate the physical conditions, and see how the star's radius and temperature change with time. Since the luminosity is given by $L = (4\pi R^2)(\sigma T^4)$, we can relate changes in R and T to changes in L and T. When we calculate model tracks, we find that the evolutionary track of a protostar depends on its mass. This is not surprising, since we have already seen that the mass determines where a star will appear on the main sequence.

Some evolutionary tracks for protostars and pre-main sequence stars are shown in Fig. 15.6. Note that the protostars appear above the main sequence. This means that for a given temperature, T, protostars are more luminous than main sequence stars of the same temperature. Protostars are also larger than main sequence stars of the same temperature. This is not surprising since protostars are still collapsing. Once the accretion phase stops, but before the main sequence is reached, we call these objects *pre-main sequence stars*.

Fig. 15.7 shows a model for the collapse of an interstellar cloud into a 1 $M_\odot$ protostar. At first the cloud is cool, and then it contracts and heats. As discussed above, the T^4 increase is greater than the R^2 decrease, and the luminosity of the protostar increases. The peak luminosity is reached when the temperature reaches 600 K. As the protostar becomes denser, its opacity increases. Eventually, it is harder for the radiation from the center to escape, and the luminosity begins to decrease. During this stage energy transport in

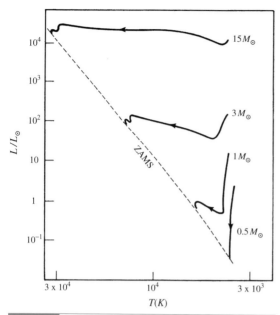

Fig 15.6. Evolutionary tracks for pre-main sequence stars on an HR diagram. Tracks are marked by the mass used in the model. The dashed line represents the zero age main sequence (ZAMS), the place where stars first join the main sequence.

the star is mostly by convection. The part of the evolutionary track at which the luminosity is decreasing quickly while the temperature increases slightly is called the *Hayashi track*. After this collapse slows, the star begins to approach the main sequence. Eventually, it reaches the luminosity of a main sequence star, though it may vary somewhat before settling down.

15.6 | Regions of recent star formation

When we study star formation, we find that there are some very obvious signposts of recent or ongoing star formation. Regions of recent star formation are important for a number of reasons. First, they call our attention to places where star formation might still be taking place. Second, the newly formed stars have some effect on their immediate vicinity, which might promote or inhibit further star formation. In this section we will look at some of the most prominent: (a) HII regions, (b) masers, (c) energetic flows, and (d) protostellar cores. In each case the object becomes prominent either because of the unique conditions that accompany star formation or because of the effect of newly formed stars on the cloud out of which they were born.

15.6.1 HII regions

When a massive star forms it gives off visible and ultraviolet photons. Photons with wavelengths shorter than 91.2 nm, in the ultraviolet, have enough energy (> 13.6 eV) to ionize H. The stars that give off sufficient ultraviolet radiation to cause significant ionization are the O and early B stars. When most of the hydrogen is ionized, we call the resulting part of the cloud an *HII region*, as shown in Fig. 15.8.

In equilibrium in an HII region there is a balance between ionizations and recombinations.

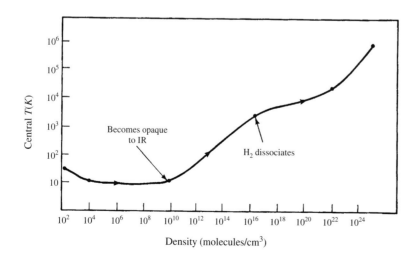

Fig 15.7. Model for the collapse of an interstellar cloud into a protostar and a pre-main sequence star.

Fig 15.8. HII regions. (a) The Lagoon Nebula (M8), in Sagittarius, at a distance of 2 kpc. It is 20 pc across. Notice the cluster of bright blue-white stars, which produce ionizing radiation. The ionized gas glows red. The name comes from the dust lane that cuts across the front, blocking our view of the gas behind. (b) HST image view of M8. (c) The Eagle Nebula (M16), in Serpens. (d) HST image of the dust lanes in M16. The bright edges are regions of recent ionization. (e) HST image of the Omega Nebula (M17), in Sgr, at a distance of 2 kpc. Here the ionizing stars are not as obvious, and are embedded deep within the nebula.

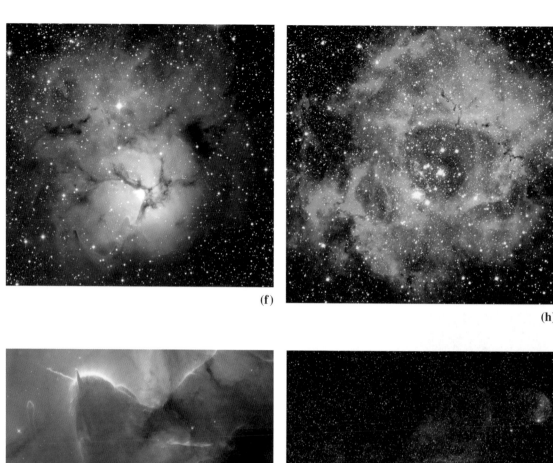

(f)

(h)

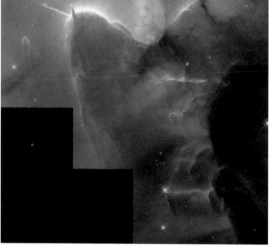

(g)

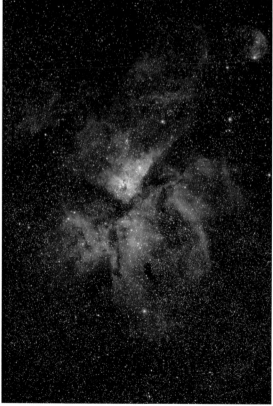

(i)

Fig 15.8. (*Continued*) (f) The Trifid Nebula (M20), in Sgr. It is named for the three-part appearance produced by the dust lanes. The blue part on top is starlight reflected from associated dust, a reflection nebula. (g) HST image of M20 (h) The Rosette Nebula (NGC 2244) in Monoceros, named for its red color and petal-like appearance. The cluster of blue stars in the center has created a cavity in the center of the cloud. It is 1.3 kpc away and 15 pc across. (i) The Eta Carina Nebula (NGC 3372), named for the bright star that illuminates it. It is 3 kpc from Earth.

Fig 15.8. (Continued) (j) The central region of the Eta Carina Nebula. (k) HST image of the immediate vicinity of Eta Carina. [(a), (c), (f), (h)–(j) NOAO/AURA/NSF; (b), (d), (g), (k) STScI/NASA; (e) ESO]

(j)

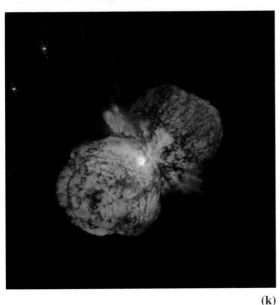

(k)

Free electrons and protons collide, forming neutral hydrogen atoms. However, the ultraviolet photons from the star are continuously breaking up those atoms to form proton–electron pairs. The balance between these two processes determines how large a particular HII region can be. Within the HII region, almost all of the hydrogen is ionized. There is a rapid transition at the edge, from almost entirely ionized gas to almost entirely neutral gas. The theoretical reasons for this sharp transition were first demonstrated by the Swedish astrophysicist, *Bengt Stromgren*. For

this reason, HII regions are often referred to as *Stromgren spheres*, and the radius of an HII region is called the *Stromgren radius, r_S.*

We can see how the balance between ionizations and recombinations determines the Stromgren radius. If N_{uv} is the number of ultraviolet photons per second given off by the star capable of ionizing hydrogen, then this is the number of hydrogen atoms per second that can be ionized. That is, the rate of ionizations R_i is given by

$$R_i = N_{uv} \tag{15.23}$$

The higher the density of protons and electrons, the greater the rate of recombinations. The recombination rate is given by

$$R_r = \alpha n_e n_p V \tag{15.24}$$

where V is the volume of the HII region and α is a coefficient (which depends on temperature in a known way). For the volume, we can substitute the volume of a sphere with radius r_S. If the only ionization is of hydrogen, the number density of electrons must equal that of protons, since both come from ionizations of hydrogen. Equation (15.24) then becomes

$$R_r = \alpha n_p^2 (4\pi r_S^3/3) \tag{15.25}$$

Equating the ionization and recombination rates gives

$$N_{uv} = \alpha n_p^2 (4\pi r_S^3/3) \tag{15.26}$$

Solving for r_S gives

$$r_S = (3/4\pi\alpha)^{1/3} (N_{uv})^{1/3} n_p^{-2/3} \qquad (15.27)$$

From equation (15.27) we can see that the size of an HII region depends on the rate at which the star gives off ionizing photons and the density of the gas. If the gas density is high, the ionizing photons do not get very far before reaching their quota of atoms that can be ionized. The rate at which hydrogen ionizing photons are given off changes very rapidly with spectral type, as indicated in Table 15.2, so the HII region around an O7 star is very different from that around a B0 star. Often, O and early B stars are found in very small groupings. In these groupings, the HII regions from various stars overlap, and the region appear as one large HII region.

The ultraviolet radiation from stars can also ionize other elements. For example, after hydrogen, the next most abundant element is helium. However, the ionization energy of helium is so large that only the hottest stars produce significant numbers of photons capable of ionizing helium. On the other hand, the ionization energy of carbon (for removing one electron) is less than that of hydrogen. There are many photons that are capable of ionizing carbon that will not ionize hydrogen. This, combined with the lower abundance of C relative to H, means that CII regions are generally much larger than HII regions (see Problem 15.20).

There are actually two conditions under which the boundary for an HII region can exist. One is that which we have already discussed. The cloud continues beyond the range of the hydrogen-ionizing photons. When this happens, we say that the HII region is *ionization bounded*. The other possibility is that the cloud itself comes to an end while there is still hydrogen-ionizing radiation. In this case, we say that the HII region is *density bounded*, since its boundary is determined by the place where the density is so low that we no longer think of the cloud as existing. When an HII region is density bounded, hydrogen-ionizing radiation can slip out into the general interstellar radiation field. This is an important source of ionizing radiation in the general interstellar medium (i.e. not near HII regions).

The temperature of HII regions is quite high – about 10^4 K. HII regions are heated by the ionization

Table 15.2. Rates of H-ionizing photons for main sequence stars.

Spectral type	Photons/s ($\times 10^{48}$)
O5	51
O6	17.4
O7	7.2
O8	3.9
O9	2.1
B0	0.43
B1	0.0033

of hydrogen. When an ultraviolet photon causes an ionization, some of the photon's energy shows up as the kinetic energy of the free proton and electron. Cooling in an HII region is inefficient, since there are no hydrogen atoms and no molecules. Cooling can only take place through trace constituents, such as oxygen. Transitions within these constituents are excited by collisions with protons and electrons. The collisions transfer kinetic energy from the gas to the internal energy of the oxygen. The oxygen then radiates that energy away. Since the heating is efficient and the cooling is inefficient, the temperature is high.

HII regions can give off continuous radiation, which can be detected in the radio part of the spectrum. This radiation results from collisions between electrons and protons in which the two do not recombine. Instead, the electron scatters off the proton. In the process the electron changes its velocity. When a charged particle changes it velocity, it can emit or absorb a photon. This radiation is called *Bremsstrahlung* (from the German for "stopping radiation"). It is also called *free–free radiation*, because the electron is free (not bound to the proton) both before and after the collision. The spectrum of free–free radiation (Fig. 15.9) is characterized by the temperature of a gas. The spectrum is not that of a blackbody because the gas is not optically thick. The spectrum is a blackbody curve multiplied by a frequency dependent opacity. Because the radiation can be described by the gas temperature, it is also known as *thermal radiation*. This radiation is strongest in the radio part of the spectrum. Therefore, we can use radio continuum observations to see HII regions anywhere in our

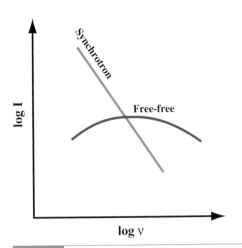

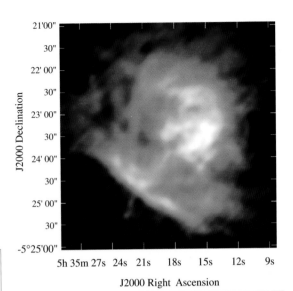

J2000 Right Ascension

Fig 15.10. Radio image (made with the VLA) of free–free emission from an HII region, the Orion Nebula (for which optical images appear in Fig. 15.28). This is a higher resolution image than the single dish version in Fig. 4.25. It shows the fine scale structure in the core of the nebula. The image was made with the VLA in D (smallest) configuration at 8.4 GHz, providing 8.4 arc sec resolution. This is a nine-field (3 × 3) mosaic. The interferometer picks up less than one-half of the total flux density because it is insensitive to the extended emission. Of course, it also gives beautiful detail of the structure in the nebula. [D. Shephard, R. Maddalena, J. McMullin, NRAO/AUI/NSF]

galaxy. A map of continuum emission from an HII region is shown in Fig. 15.10. (Note: In the encounter between the electron and proton, the proton also accelerates and gives off radiation. However, the acceleration of the proton is much less than that of the electron, by the ratio of their masses. This means that the radiation given off by the protons is not very important.)

HII regions also give off spectral line radiation, called *recombination line* radiation. When an electron and proton recombine to form a hydrogen atom, the electron often ends up in a very high state. The electron then starts to drop down. It usually falls one level at a time. Larger jumps are also possible, but less frequent. With each jump, a photon is emitted at a frequency corresponding to the energy difference for the particular jump. (The energies are given by equation 3.6.) For very high states, the energy levels are close together and the radiation is in the radio part of the spectrum. As the electron jumps to lower states the lines pass through the infrared and into the visible. Generally, the electron can go all the way down to the ground state before the atom is re-ionized. We even see Hα emission as part of this recombination line series. This gives HII regions a red glow. (This red glow allows us to distinguish HII regions from reflection nebulae, which appear blue.)

HII regions expand with time. When an HII region first forms (Fig. 15.11), it must grow to its equilibrium radius. Even after it reaches this equilibrium size, it will continue to expand. This is because the pressure in the HII region is greater than that in the expanding cloud. The higher pressure results from the higher temperature in the HII region. Remember, the temperature in an HII region is about 10^4 K, while that in the surrounding cloud is less than 100 K. The densities in the HII region and surrounding cloud are similar.

As the HII region expands, it can compress the material in the surrounding cloud, possibly initiating a new wave of star formation, as illustrated in Fig. 15.12. This is one possibility that has been discussed for the triggering of star formation. The gas compressed by an expanding HII region will not automatically form stars. That is because the gas will be heated as it is compressed. If that heat is not lost, the temperature of the cloud will increase.

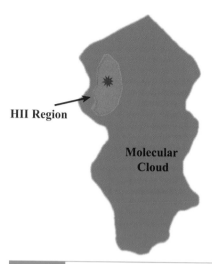

Fig 15.11. HII region in a molecular cloud. HII regions usually form near the edge.

The pressure will increase and the gas will expand again. The re-expansion of compressed gas can only be avoided if the gas can cool as it is compressed. Radiation from molecules such as CO in the surrounding cloud can help with this cooling process.

15.6.2 Masers

We have already seen how the process of stimulated emission can lead to a multiplication – or

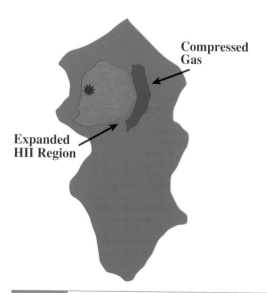

Fig 15.12. The HII region expands, compressing gas deeper within the cloud. If this gas can cool quickly, then it can collapse to form more stars.

amplification – in the number of photons passing through a material. In the stimulated emission process, one photon strikes an atom or molecule, and two photons emerge. The two photons are in phase and are traveling in the same direction. The fact that they are in phase means that their intensities add constructively. Stimulated emission can only take place if the incoming photon has an energy corresponding to the difference between two levels in the atom or molecule, and the atom or molecule is in the upper of the two levels.

If only a few atoms or molecules are in the correct state, there will not be a significant increase in the number of photons. Suppose we designate the two states in the transition as 1 and 2. The population of the lower state is n_1 and the population of the upper state is n_2. The requirement for amplification is that n_2/n_1 be greater than g_2/g_1, where the g are the statistical weights. The situation is called a *population inversion*, since it is the opposite to the normal situation. Formally, it corresponds to a negative temperature in the Boltzmann equation (see Problem 15.22). This is clearly not an equilibrium situation. The population inversion in a particular pair of levels must be produced by a process, called a *pump*. The pump may involve both radiation and collisions. The net effect of the pump process is to put energy into the collection of atoms or molecules so that some of that energy can come out in the form of an intense, monochromatic, coherent (in phase) beam of radiation.

This was first realized in the laboratory, in the 1950s, by *Charles Townes* (then at Columbia University). Townes won the Nobel Prize in physics for this work. Since microwaves were being amplified in the process, the device was called a *maser* (Fig. 15.13), an acronym for *microwave amplification by stimulated emission of radiation*. Subsequently, *lasers* were developed for the amplification of visible light. In any laser or maser, two things are necessary: (1) a pump to provide the population inversion, and (2) sufficient path length to provide significant amplification. In interstellar space, the path length is provided by the large side of interstellar clouds. In the laboratory that path length is provided by mirrors. (Laboratory masers are used as amplifiers in some radio telescope receivers.)

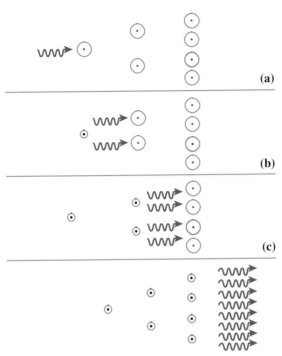

Fig 15.13. Maser amplification. In each frame, a molecule in the upper level of the maser transition is indicated by a large circle, and one in the lower level is indicated by a small circle. (a) All of the molecules are in the upper state, and a photon is incident from the left. (b) The photon stimulates emission from the first molecule, so there are now two photons, in phase. (c) These photons stimulate emission from the next two molecules, resulting in four photons. (d) The process continues with another doubling in the number of photons.

Shortly after the development of laboratory masers, an interstellar maser was discovered. It involved the molecule OH. Four emission lines of OH were observed, but their relative intensities were wrong for a molecule in equilibrium. As radio telescopes were developed with better angular resolution, the emission was observed to become stronger and stronger. This means that the emission is probably very intense, but coming from a very small area. This behavior was suggestive of an interstellar maser. The next maser discovered was in the water (H_2O) molecule, at a wavelength of 1 cm. As observations with better resolution became possible, it was clear that the objects were giving off as much energy as a 10^{15} K blackbody over that narrow wavelength range in which the emission was taking place.

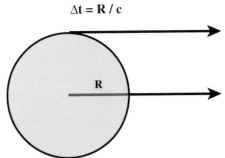

$$\Delta t = R / c$$

Fig 15.14. Time variability and source size. The signal from the farthest point the eye can see must travel an extra distance R over that from the nearest point the eye can see.

A small size for these sources was also deduced from rapid variations in their intensity. Suppose we have a sphere of radius R, as shown in Fig. 15.14. If the sphere were suddenly to become luminous, then the first photons to leave each point on the surface would not reach us simultaneously. The photons from the edge of the sphere have to travel a distance R farther than the photons from the nearest point. These photons will arrive a time $\Delta t = R/c$ later than the first photons. Therefore, it will take this time for the light we see to rise from its initial low level to the final high value. A similar analysis holds for the time it would take for us to see the light turning off.

The above analysis tells us that an object's brightness cannot vary on a time scale faster than the size of the emitting region, divided by c. If we see variations in intensity over a time scale of a year, the source cannot be larger than a light year across. Interstellar masers were found to vary in intensity on an even shorter time scale, of the order of a month, indicating an even smaller size.

Example 15.7 Maser size
Estimate the maximum size of a maser that varies on the time scale of one month. What is the angular size of this object at a distance of 500 pc?

SOLUTION
The time scale for the variations is

$$\Delta t = (24 \text{ h/day})(3600 \text{ s/h})(30 \text{ day})$$

$$= 2.6 \times 10^6 \text{ s}$$

This corresponds to a size of

$R = c\Delta t$

$\qquad = 7.8 \times 10^{16}$ cm

$\qquad = 5.2 \times 10^3$ AU

The angular size (in arc seconds) is related to R (in AU), and the distance d (in pc) by (equation 2.16)

$\theta('') = R(\text{AU})/d(\text{pc})$

$\qquad = 5.2 \times 10^3 / 500$

$\qquad = 10$ arc sec

In fact, masers are even smaller than this size, and have angular extents much less than 1 arc second. This means that we need radio interferometers to study masers. Very long baseline interferometry has been used to study masers.

When we try to understand interstellar masers we must explain both the pump and the path length for the gain. Many of the theories require very high densities. For example, we think that the presence of water masers suggests densities in excess of 10^8 cm^{-3}. This is much denser than even the dense cores that we normally see in molecular clouds. We therefore think that masers are associated with objects collapsing to become protostars. We take the presence of H_2O or OH masers in a region to indicate the possibility of ongoing star formation.

When we observe masers, we often see them in clusters, such as that depicted in Fig. 15.15. With radio interferometry, we can measure the positions of the masers very accurately. We can even measure their proper motions. We can use Doppler shifts to measure their radial velocities. However, we expect the motions of a cluster to be random, so the average radial velocity should

equal the average transverse velocity v_T. From equation (13.6) we see that the distance is related to the proper motion and transverse velocity by

$d(\text{pc}) = v_T(\text{km/s})/4.74\mu(\text{arc sec/yr})$

Therefore, an accurate study of the motions of masers allows us to determine the distance to a cluster of masers. It is hoped that this will develop into a very powerful distance measuring technique. This technique works equally well for random motions or for the masers being in an expanding shell. The only requirement is that the average velocity along the line of sight is the same as the average velocity perpendicular to the line of sight.

Maser emission is also observed in the molecule SiO (silicon monoxide). From the regions in which it is observed, it seems that SiO maser emission is associated with mass outflow from evolved red giant stars. Also, some OH masers are associated with similar regions.

15.6.3 Energetic Flows

A major recent discovery is that many regions of star formation seem to be characterized by strong outflows of material. One piece of evidence for such flows comes from the observation of very broad wings on the emission lines of CO (Fig. 15.16). The widths of these wings range from 10 to 200 km/s. The broad wings are usually seen only over a small region where the CO emission is strongest. A peculiar feature of this emission is that the redshifted wing and blueshifted wing seem to be

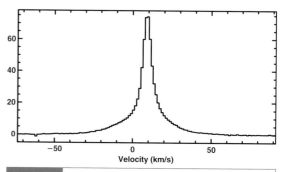

Fig 15.16. Spectrum of the 2.6 mm CO emission line from the core of the molecular cloud behind the Orion Nebula (Fig. 15.28.) The broad wings extend many tens of kilometers per second on both sides of the line center. [Jeffrey Mangum, NRAO/AUI/NSF]

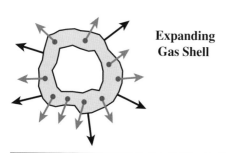

Expanding Gas Shell

Fig 15.15. Cluster of masers in an expanding shell.

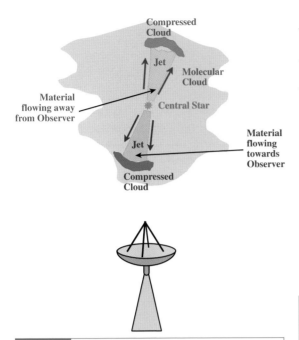

Fig 15.17. Model for a bipolar flow. Material coming towards us on the near side of the cloud is blueshifted. Material going away from us on the far side of the cloud is redshifted. If the flow is not aligned along the line of sight, the redshifted and blueshifted emission will appear in different locations on the sky.

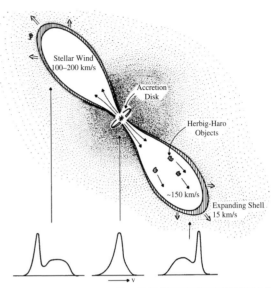

Fig 15.18. A model for sources with bipolar flows and HH objects. The stellar wind comes out in all directions but is blocked in most directions by a disk around the star. The wind emerges mostly at the poles of the disk. This drives material in the surrounding cloud away. Below, the effects of the motion on the CO line profiles are shown, assuming that the wind to the upper left moves away from the observer and the wind to the lower right moves towards the observer. [Ronald Snell (University of Massachusetts) Snell, R. A. et al., Astrophys. J. Lett., **239**, L17, 1980, Figs. 2 & 5]

coming from different parts of the cloud. This suggests that we are seeing two jets of gas, one coming partially towards us and the other partially away from us, as shown in Fig. 15.17. Because of this structure, we call these objects sources of *bipolar flows*.

Actually, we could also envision a model in which we are seeing infall rather than outflow. However, there is evidence we are seeing the effects of a wind striking the surrounding cloud, heating a small region. These small heated regions show emission in the infrared from H_2, requiring temperatures of about 10^3 K. Current theories of these sources involve strong stellar winds, as shown in Fig. 15.18. The star is also surrounded by a dense disk of material. This disk blocks the wind in most directions, but allows it to escape along the axis, explaining the bipolar appearance. Remember, we saw earlier in this chapter that disks are likely to form around the collapsing star.

Evidence for collimated winds is present in another interesting class of objects that we think are associated with pre-main sequence stars, the *Herbig–Haro (HH) objects* shown in Fig. 15.19; see also Fig. 15.20. They were discovered independently by *George Herbig* of the Lick Observatory and *Guillermo Haro* of the Mexican National Observatory. HH objects appear as bright nebulosity on optical photographs. Their spectra resemble those of stars, and usually show emission lines, but no star is present in the nebulosity (Fig. 15.21). We now think that the wind from a pre-main sequence star clears a path through the cloud. The part of the cloud where the wind runs into the cloud is heated, and glows. We also see starlight reflected from the dust, explaining the stellar spectrum. The exciting star is deep within the cloud and is not seen directly.

These observations indicate that winds are an important feature of protostellar evolution for most stars. For low mass stars such as the Sun,

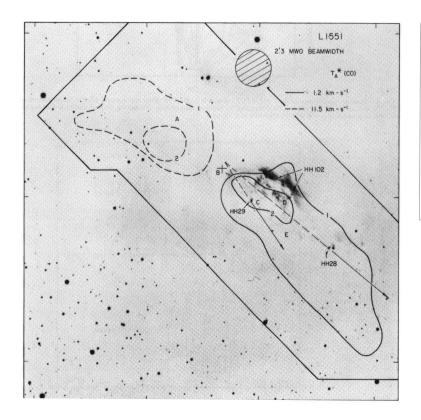

Fig 15.19. A negative optical image of a region, containing a bipolar flow and a number of Herbig–Haro objects. The solid contours are the CO emission that is blueshifted with respect to the average velocity. The dashed contours show the redshifted CO. The plus (+) marked B is the location of the suspected source for the flow. [Snell, R. A. et al., Astrophys. J. Lett., 239, L17, Figs. 2 & 5]

this wind is relatively gentle, and can clear some of the debris from around the forming star, leaving any planetary system intact. For massive (O and B) stars the strong winds drive away a large mass. It has been suggested that the combined effects of winds in OB associations can drive off enough mass to unbind the association, explaining why associations are not gravitationally bound. Winds can also carry away some of the angular momentum in a cloud, allowing the collapse of the remaining material to continue.

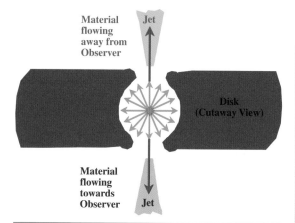

Fig 15.20. How an isotropic flow can be converted into two jets. The source of the isotropic outflow is in a hole in a disk. In most directions the disk can block the flow, but not in the direction where the disk is thinnest. Jets emerge in that direction.

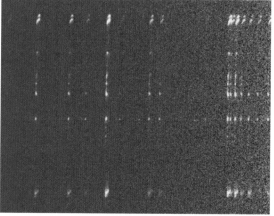

Fig 15.21. IR spectrum of Herbig–Haro object number 212. Notice the bright emission lines. [ESO]

(a)

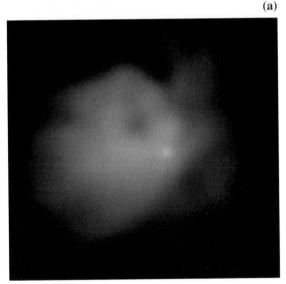

(b)

Fig 15.22. Images of two T Tauri stars. (a) T Tauri. The star is unremarkable, but there is a small nebula to the right. The star is in a dust cloud. The presence of the dust can be deduced from the fact that it blocks light from the background stars. There are therefore fewer background stars in the center of the photograph than around the edges. (b) HL Tau. Again, there is some nebulosity near the star and the star is in a dust cloud. [(a) Courtesy of 2MASS/UMASS/IPAC/NASA/JPL/Caltech; (b) Laird Close, (University of Arizona)/Close, L., *Astrophys. J.*, **486**, 766, 1997]

15.6.4 T Tauri stars and related objects

Another group of pre-main sequence objects are called *T Tauri stars* (Fig. 15.22). T Tauri is a variable star in the constellation Taurus, and T Tauri stars are variables with properties like those of T Tauri. Light curves are shown in Fig. 15.23, and a spectrum is shown in Fig. 15.24. These stars are spectral class K, and appear above the main sequence on the HR diagram. These show an irregular variability. Their spectra are also characterized by the presence of emission lines.

There are three possible sources of the variability we see:

(1) The variability could arise in a photosphere. One model for this involves star spots. These are dark areas, like sunspots, only larger. As the star rotates, a different fraction of the

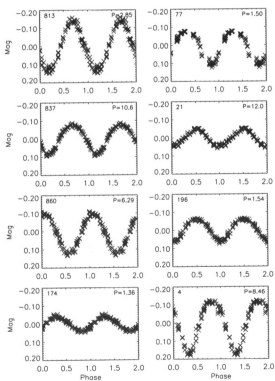

Fig 15.23. Light curves for a selection of T Tauri stars in the Orion Nebula cluster. The horizontal axis is fractions of a period (which is given in days in the upper right-hand corner of each curve). Variability is caused by rotation of stars with large, cool spots. [William Herbst (Wesleyan University)/Herbst, W. et al., *Astrophys. J. Lett*, **554**, L197, 2001]

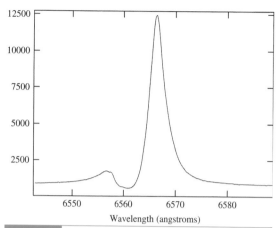

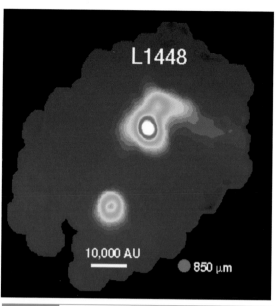

Fig 15.24. Strong Hα emission line from a T Tauri star. [George Herbig, IFA, Hawaii]

observed surface is covered by the spots, and the brightness changes.

(2) The variations may arise in the chromosphere.

(3) The variations may actually result from changes in the opacity of the dust shell surrounding the star.

The emission lines show Doppler-shifted absorption wings, like those in Fig. 15.24. This suggests material both falling into the star and material coming off the star. The infall may be close to the star as the final stage of collapse, while the outflow is a wind (like the solar wind, but stronger) farther away from the star. Alternatively, the infall may be in the form of a disk around the star's equator, while the outflow is along the polar axes.

Fig 15.25. Far IR image of protostellar core. This is a ground-based image from Mauna Kea, at 850 μm. The beam size is shown in the red circle to the lower right, so you can see that the sources are barely resolved. Notice the separation into two sources. The irregular edges of the image are due to problems with the detectors near the edge. This is an example of what is called a Class 0 protostar, which is thought to be the youngest stage, where there is a strong outflow but the surrounding cloud has not been driven away. [Yancy Shirley, University of Texas, Austin, made with SCUBA on the JCMT]

From studies of the spectral lines, we think that the winds may have speeds of about 200 km/s. The mass loss in the wind, dM/dt, is about 10^{-7} $M_\odot$/yr. The total luminosity in the

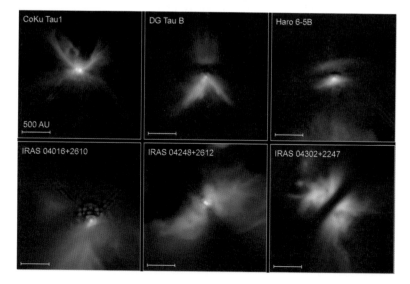

Fig 15.26. HST images of infrared emission from selected disks around forming stars. All six objects are in Taurus, at a distance of 150 pc. [STScI/NASA]

wind is that rate at which kinetic energy is carried away in the wind,

$$L_w = (1/2)(dM/dt)v^2 \qquad (15.28)$$

Using the numbers given, we find a wind luminosity of about 1 $L_\odot$. That is, the star gives off as much energy per second in its wind as the Sun gives off at all wavelengths. However, the wind phase is a short lived one. The wind does sweep away some of the dust that has collected around the star. We think that a similar wind from the Sun was important in clearing debris out of the early Solar System.

It is now possible to make far IR images of protostellar cores, like that in Fig. 15.25. Far IR emission HST and IR satellites have provided us with some images, such as those in Fig. 15.26, which may be the result of infrared emission from dust shells around recently formed stars. Molecular spectral line observations of these disks will require resolutions achievable using interferometers.

15.7 | Picture of a star forming region: Orion

The Orion region (Fig. 15.27) is one of the most extensively studied star forming regions. It is relatively nearby, only about 500 pc from the Sun. It is away from the plane of the Milky Way, so there is little confusion with foreground and background stars. There is also an interesting variety of activity in this region.

The region contains a large OB association. There are four distinct subgroups. The two oldest are near Orion's belt, and the two youngest are near the Orion Nebula (Fig. 15.28a) in Orion's sword. The Orion Nebula is an HII region powered by the brightest stars in the youngest subgroup. Images of the Orion Nebula and its cluster are in Figs. 15.28(b)–(e).

The region also contains two giant molecular cloud complexes, also shown in Fig. 15.4. A far IR image of the GMCs is shown in Fig. 15.27(a). The northern complex is associated with the belt region, which contains the two oldest subgroups, and a strong HII region just north of the Horsehead Nebula. The southern complex is associated with the sword region, which contains the two youngest subgroups. The southern complex

contains the Orion Nebula and several smaller HII regions. The Orion Nebula is actually on the front side of the molecular cloud. Behind the HII region is a dense molecular core. It is totally invisible in the optical part of the spectrum because of the foreground material. However, it can be studied in detail using radio observations of molecules. In addition, it is a source of infrared emission.

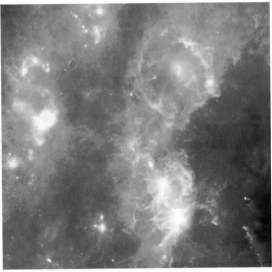

(a)

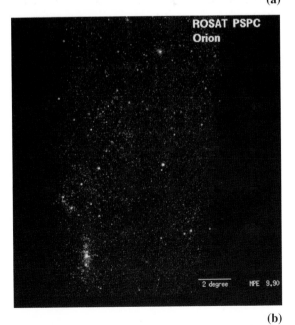

(b)

Fig 15.27. Large-scale images of the Orion region, mostly at a distance of 500 pc. (a) Far IR image from IRAS. (b) X-ray image from ROSAT. [(a) NASA, (b) NASA/MPI]

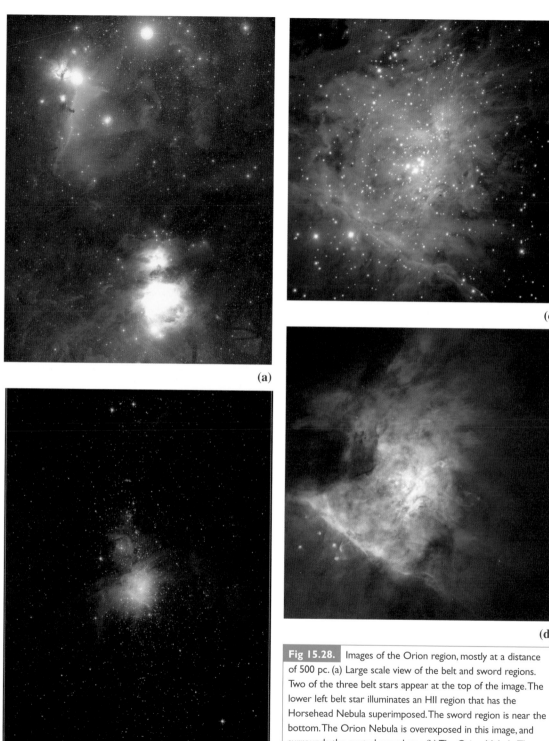

(a)

(b)

(c)

(d)

Fig 15.28. Images of the Orion region, mostly at a distance of 500 pc. (a) Large scale view of the belt and sword regions. Two of the three belt stars appear at the top of the image. The lower left belt star illuminates an HII region that has the Horsehead Nebula superimposed. The sword region is near the bottom. The Orion Nebula is overexposed in this image, and surrounds the central sword star. (b) The Orion Nebula. There is a small cluster of O and B stars that ionize this nebula. A dense dust cloud provides over 100 mag of visual extinction in the central part. The object near the top of the photo is a separate small HII region illuminated by a single star. Compare its simple appearance to the complex structure of the big nebula.

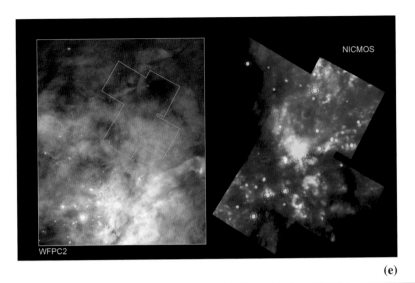

(e)

Fig 15.28. *(Continued)* (c) Central part of the Orion Nebula. This gives a better view of the O and B stars. The brightest four stars in the central part of the cloud are called the Trapezium, because of their arrangement. Notice the dense dark cloud intruding from the upper left. (d) A larger field HST image of the Orion Nebula, assembled from 15 smaller field images. This image is approximately 1 pc across. (e) HST image of filaments in the nebula. The diagonal length of this image is 0.5 pc. The plume of gas at the lower left is the result of a wind from a newly formed star. Red light depicts emission in nitrogen; green is hydrogen; blue is oxygen. [(a) © Anglo-Australian Observatory/Royal Observator, Edinburgh. Photograph from ULK Schmidt plates by David Malin; (b) Courtesy of 2MASS/UMASS/IPAC/NASA/JPL/Caltech; (c) ESO; (d), (e) STScI/NASA]

The general picture of the southern complex that has emerged has the HII region expanding into the molecular cloud, compressing it. This compression probably triggered a new generation of star formation. This new cluster of stars now appears as a series of small infrared sources and masers. There is evidence for an energetic flow. We see very broad (over 100 km/s) lines in CO and other molecules. These wings have the characteristics of the bipolar flows discussed in the preceding section. We also see evidence for small regions of gas heated to high temperatures in the regions where the wind strikes the surrounding cloud. Infrared line emission from H_2 is seen from this 2000 K gas. We can also study the proper motions of the maser and see that the region is expanding. (The best measurement of the distances to this region comes from studying the proper motions of masers, as discussed in the preceding section.) It is likely that this dense region will appear as another OB subgroup when there has been sufficient time to clear the interstellar material away.

Chapter summary

In this chapter we saw how the various components of the interstellar medium, discussed in Chapter 14, are involved in the process of star formation. We first looked at the conditions for gravitational binding, and found that molecular clouds are the likely sites of star formation. Current problems in star formation include how the collapse is actually initiated, what fraction of the mass is converted into stars, how planetary systems form, and how OB associations form. We saw how rotation can slow or stop the collapse, and lead to the formation of binaries, planetary systems and disks.

We saw how the magnetic field in a cloud can affect its collapse if the magnetic energy is comparable to the gravitational potential energy. As a cloud collapses, we expect that flux freezing will lead to an increase in the magnetic field strength within the cloud. We saw that there are two ways to overcome the support the magnetic field provides. One produces a steady stream of low mass stars, and the other produces waves of high mass (as well as low mass) star formation.

We saw how the molecular clouds, being cool and dense, are the most likely sites of star formation. The most massive stars seem to be born in the giant molecular clouds. These clouds have masses in excess of 10^5 $M_\odot$ and are found in complexes with masses in excess of 10^6 $M_\odot$.

We looked at how collapsing clouds eventually become protostars, glowing mostly in the near infrared. Their radiation is provided by the decrease in the gravitational potential energy as the cloud collapses.

Once an O or early B star forms in a molecular cloud, the gas around the cloud is ionized, producing an HII region. The size of the HII region is determined by a balance between ionizations and recombinations. The continuous radiation from HII regions comes from free–free scattering of electrons. Recombinations lead to electrons passing through many energy levels, giving off recombination line radiation.

Other indicators of recent star formation are masers and energetic flows. The bipolar flows provide indirect evidence for the existence of disks that collimate the flows.

We looked at the Orion region as an nearby example of ongoing high and low mass star formation.

Questions

15.1. Explain why cool dense regions are the most likely sites of star formation.

15.2. Suppose we have a cloud of 2 pc radius, formed of a material whose Jeans length is 10 km. Will the cloud collapse?

15.3. As a cloud collapses, what happens to the Jeans mass of the new individual fragments?

15.4. If a cloud has a higher density in the center than at the edge when its collapse starts, explain what happens to the density contrast between the center and the outer part as the collapse continues. (Hint: Think about free-fall time.)

*15.5. As a cloud collapses, the acceleration of the particles increases. Therefore, the free-fall time will be less than we would calculate on the basis on constant acceleration. However, we still use the initial free-fall time as an estimate of the total free-fall time. Why does this work? (Hint: Think in terms of the time for the radius to halve, and halve again, and so on.)

15.6. Explain why a rotating cloud flattens as it collapses.

15.7. How does the formation of multiple star systems help in the star formation process?

15.8. As a cloud collapses, is it likely that it will rotate as a rigid body (ω independent of r)?

15.9. What do we mean by a "trigger" for star formation?

15.10. How is it possible for a gravitationally bound cloud to give birth to a gravitationally unbound association?

15.11. (a) What is the difference between ionization bounded and density bounded HII regions? (b) Why are density bounded HII regions important?

15.12. Explain why the CII zone around a star is much larger than the HII region. Consider (a) the relative ionization energies, and (b) the relative abundances.

15.13. Explain why almost all carbon ionizations result from photons with energies between the ionization energy of carbon and that of hydrogen, even though photons that can ionize hydrogen can also ionize carbon.

15.14. Why is the temperature of an HII region so high?

15.15. (a) How can an expanding HII region trigger star formation? (b) Why is the rate at which the gas can cool important to the process?

15.16. Why is coherence important for a maser?

15.17. How do we know that masers are small?

15.18. What features of T Tauri stars lead us to believe that they are still in the formative process?

Problems

15.1. Compare the Jeans mass and radius for typical HI and molecular clouds.

15.2. Suppose we made an interferometer with one telescope on the Earth and the other on the Moon. What would be its angular resolution at a wavelength of 1 cm?

15.3. Suppose that a cloud contracts to one-tenth of its initial size. How do the Jeans length and mass compare with those in the original cloud?

15.4. Compare the density and pressure of a dense interstellar cloud with those of the gas in your room.

15.5. (a) Find the Jeans length for a cloud with a density 10^6 H_2 molecules/cm^3 and $T = 100$ K. (b) What angle does this subtend at a distance of 1 kpc from the Earth?

15.6. Use the results of Example 15.2 to give the free-fall time (in years) for objects whose density is given as a fraction of 10^5 atoms/cm^3.

15.7. An interstellar cloud is found to be rotating such that a velocity shift of 1 km/s/pc is observed across it. (a) What is the angular speed (rad/s)? (b) What is the rotation period?

15.8. For an object on the surface of the Sun, compare the two terms on the right-hand side of equation (15.14b).

15.9. For a spherical interstellar cloud of mass 1000 $M_\odot$ and radius 10 pc, rotating once every 10^7 years, compare the two terms on the right-hand side of equation (15.14b).

15.10. For a spherical interstellar cloud of mass 1000 $M_\odot$ and radius 10 pc, rotating once every 10^7 years, compare the gravitational potential energy with the kinetic energy of rotation.

15.11. For a uniform sphere, what does it mean for d^2I/dt^2 to be small compared with the gravitational potential energy?

15.12. For the protostar considered in Example 15.6, how much more energy will be released as the star contracts by another factor of five? If this takes another 100 years, what is the average luminosity over that period?

15.13. For a 10 $M_\odot$ protostar that has collapsed to a radius of 1000 $R_\odot$, (a) calculate the energy that has been liberated to this point, and (b) use this to calculate the average luminos-

ity if most of the energy is liberated over the last 50 years of the collapse.

15.14. From the evolutionary track in Fig. 15.6, calculate the radius of a 3 $M_\odot$ protostar when it has reached a temperature of 10^4 K.

15.15. (a) For a star of radius R, emitting like a blackbody of temperature T, find an expression for the number of photons per second emitted capable of ionizing hydrogen. (b) For $h\nu \gg kT$, evaluate the integral. (c) Why is the approximation in (b) valid for an ordinary star?

15.16. Compare the radii of HII regions around an O5 and an O7 star if the density is $n_H = 10^3$ cm^{-3}.

15.17. What is the Doppler width of the H110α line at a wavelength of 6 cm, for a temperature of 10^4 K?

15.18. What is the shift from the H110α to the C110α line wavelength, if it just depends on the nucleus–electron reduced mass?

15.19. Suppose an HII region is formed in a cloud with a density of 10^4 H_2 molecules/cm^3. If the temperature in the HII region is 10^4 K and the temperature of the surrounding molecular cloud is 10^2 K, how much will the HII region expand before the pressures equalize?

*15.20.(a) Derive an expression, analogous to equation (15.27), for the radius r_S (He) of an ionized helium region around a star in terms of N_{uv} (He), the number of photons per second emitted by the star capable of ionizing helium. Give your answer in terms of n_p and n_{He}. Assume that α is the same for H and He. Assume that anywhere that He is ionized, all of the H is ionized, so that $n_e = n_p$. (b) Repeat the calculation for carbon, giving your answer in terms of N_{uv} (C), n_p and n_C. For carbon, assume that it is ionized over a much larger region than H because of its lower ionization energy, so that $n_e = n_C$.

15.21. From observations of the masers in Orion, we find an average radial velocity of 24.0 km/s, and proper motions of 0.01 arc sec/yr. How far away are these masers?

15.22. (a) For a maser, show that a population inversion corresponds to a negative excitation temperature. (b) The negative excitation temperature gives a negative optical depth. Show from the radiative transfer equation that this implies amplification.

15.23. If the Sun turns off, how long will it take the light to go from full intensity to zero?

15.24. (a) If an object dims in a day, how large can it be? (b) What angle would it subtend at a distance of 1 kpc?

15.25. Calculate the rate at which energy is delivered to a cloud by a 10^{-8} $M_\odot$/yr flow with a speed of 100 km/s.

15.26. For the flow in the previous question, what is the rate at which interstellar material is being swept up if the sweeping occurs as long as the wind speed is greater than 5 km/s?

15.27. (a) For the typical T Tauri wind described in this chapter, what is the momentum per second carried away by the wind? (b) If the wind drives away dust, and slows by conservation of momentum, and the wind is effective at driving dust away until it slows to 5 km/s, what is the rate at which dust can be driven away?

15.28. By how much is the Hα line shifted by the 200 km/s wind in T Tauri?

15.29. Calculate the energy used up in ionizing 1 $M_\odot$ of atomic H. To what radius must a 1 $M_\odot$ protostar collapse for this much energy to be released in the change in gravitational potential energy?

Computer problems

15.1. Make a table showing the Jeans mass and length for clouds with densities of 1, 10^3 and 10^5 H atoms/cm^3, and kinetic temperatures of 10, 30 and 100 K.

15.2. Add a column to Table 15.2, giving the Stromgren radius in each case. Assume that n_p is 10^4 cm^{-3}. Take a constant value of $\alpha = 2.6 \times 10^{-13}$ cm^3/s.

15.3. Suppose we start with an interstellar cloud with a density of 10^5 H atoms/cm^3. Using the variable acceleration in equation (15.6), find the actual collapse time, and compare it with the estimate for the free-fall time, given in equation (15.8).

15.4. Reproduce Table 15.2, using the radii and temperatures of the spectral types given.

Chapter 16

The Milky Way galaxy

16.1 | Overview

Throughout this book we have discussed the components of our galaxy: stars, clusters of stars, interstellar gas and dust. We now look at how these components are arranged in the galaxy. The study of the large scale structure of our galaxy is difficult from our particular viewing point. We are in the plane of the galaxy, so all we see is a band of light (Fig. 16.1). The interstellar dust prevents us from seeing very far into the galaxy. We see a distorted view.

The first evidence on our true position in the galaxy came from the work of *Harlow Shapley*, who studied the distribution of globular clusters (Fig. 16.2). He found the distances to the clusters from observations of Cepheids and RR Lyrae stars. Shapley found that the globular clusters form a spherical distribution. The center of this distribution is some 10 kpc from the Sun. Presumably, the center of the globular cluster distribution is the center of the galaxy. This means that we are about 10 kpc from the galactic center.

In Chapter 13, when we studied HR diagrams for clusters, we introduced the concept of stellar populations I and II. The distribution of these populations in the galaxy can help us understand how the galaxy has evolved. Population I material is loosely thought of as being the young material in the galaxy. Population I stars are found in galactic clusters, and are characterized by high metalicity. Some are also associated with interstellar gas and dust, suggesting that they are young enough to have some of their parent cloud around them. Population I stars are confined to the galactic plane.

Population II stars are thought of as being the "old" component of the galaxy. They are found in globular clusters and are characterized by low metalicity. They have no gas and dust around them. Their galactic distribution is very different from that of population I stars. The population II stars form a spherical distribution, as opposed to a disk. This spherical distribution is sometimes called the *halo*. When we talk about a spherical distribution, we do not mean just a spherical shell around the galaxy. Instead, we mean the spherically symmetric distribution whose density falls off with increasing distance from the galactic center. Population II objects also seem to have a larger velocity spread in their motions than do population I objects. Table 16.1 shows the characteristic thicknesses and velocity dispersions for some components of the galaxy.

The schematic arrangement of these components is shown in Fig. 16.3. First we see the disk and the halo. Note that the halo has a spherical distribution with a density of material that falls off radially. We then look at the disk. First there is an overhead view, showing the location of the Sun. The best estimates, which we will discuss later in this chapter, place the Sun 8.5 kpc from the galactic center, about halfway out to the edge of the disk. We then look at a side view of the disk. The inner part of the disk (closer to the center than the Sun) is relatively thin and flat. The outer part of the disk is *warped* and also gets thicker, that is, it *flares*. We will also discuss the evidence for this in this chapter. Surrounding the center region is a *bulge*.

Fig 16.1. Images of the Milky Way at different wavelengths. (a) Radio continuum (408 MHz), from Bonn, Jodrell Bank and Parkes. (b) HI column density, derived from 21 cm emission (Leiden–Dwingeloo survey). (c) Radio continuum (2.4–2.7 GHz), from Bonn and Parkes. (d) H2 column densities based on CO emission at 115 GHz (CFA at Harvard and in Chile). (e) Composite mid and far infrared from IRAS (1.25, 2.2 and 3.5 μm). (f) Mid infrared (6.8–10.8 μm), data from the SPIRIT III instrument on the Midcourse Space Experiment (MSX) satellite. (g) Near infrared from COBE (12, 60 and 100 μm). (h) Optical from wide field photographs (compiled at ESO). (i) X-ray (0.25, 0.75 and 1.5 keV) from ROSAT. (j) Gamma-ray (CGRO). All of the images (except the optical) have angular resolutions of between one-half and two degrees. All images are centered on the galactic center, and cover 10° above and below the galactic plane. [Courtesy of NASA, provided by the GSFC Astrophysics Data Facility (ADF) (a), (e), (g), (i), (j) ADF; (b) Dap Hartmann, "The Leider/Dwingeloo Survey of Galactic Neutral Hydrogen", Ph.D. Thesis, University of Leider, 1974; (c) A. R. Duncan, University of Queensland, Australia; (d) Thonas Dame, CFA; (f) Stephan D. Price, Hanscom AFB; (h) Axel Mellinger, University of Potsdam, Germany]

16.2 | Differential galactic rotation

All of the material in the galaxy orbits the galactic center. If the galaxy were a rigid body, all of the gas and dust would orbit with the same period. However, material closer to the galactic center orbits with a shorter period than material farther out. This is not an unusual situation. After all, the planets in the Solar System exhibit the same behavior: Mercury takes less time to orbit the Sun than does the Earth, and so on. When the orbital period depends on the distance from the center, we say that the material is exhibiting *differential rotation*.

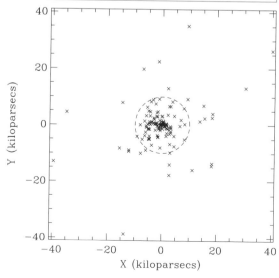

(a)

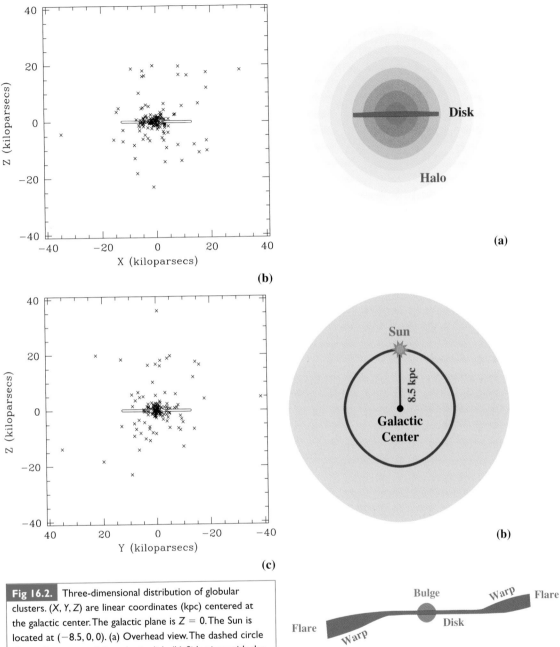

Fig 16.2. Three-dimensional distribution of globular clusters. (X, Y, Z) are linear coordinates (kpc) centered at the galactic center. The galactic plane is Z = 0. The Sun is located at (−8.5, 0, 0). (a) Overhead view. The dashed circle shows the extent of the galactic disk. (b) Side view, with the solid line indicating the extent of the disk. (c) The other side view, with the solid line indicating the extent of the disk. [William Harris, McMaster University]

Fig 16.3. Schematic arrangements of major components of the galaxy. (a) The relationship between the halo and the disk. The lighter shading as one moves farther out in the halo is to suggest a decrease in density with distance from the center. (b) Top view of the disk, showing the location of the Sun. (c) Side view of the disk, showing the warp and flare in the outer part of the disk, and the bulge near the galactic center.

16.2.1 Rotation and mass distribution

The orbital period of any particle will depend on the mass about which it is orbiting and the radius of the orbit. Just as we use the period

Table 16.1. Scale heights[a] and velocity dispersions.

Constituent	Scale height (pc)	Velocity dispersion (km/s)
O Stars	50	5
GMCs	60	7
Galactic clusters	80	
HI clouds	120	
F stars	190	20
Planetary nebulae	260	20
M stars	350	
RR Lyrae stars:		
Short period	900	25
Long period	2000	30
Globular clusters	3000	

[a]Scale height is the distance over which a quantity falls to 1/e of its maximum value.

and size of the Earth's orbit to tell us the mass of the Sun, we can use the orbital periods at different distances from the galactic center to tell us about the distribution of mass in the galaxy.

To see how this works, we assume that all matter follows circular orbits. At a distance R from the center, the orbital speed is $v(R)$, and the angular speed is $\Omega(R)$. The mass interior to radius R is

$$M(R) = \int_0^R \rho(r)\, dV \tag{16.1}$$

where $\rho(r)$ is the density at radius r, and dV is a volume element. For a spherical mass distribution, the motion of an object at R depends only on $M(R)$. Furthermore, the mass $M(R)$ behaves as if it were all concentrated at the center. (This also works for particles in the plane of a thin disk.)

For a particle of mass m orbiting at a radius R, the gravitational force is $GM(R)m/R^2$, and this must provide the acceleration for circular motion, so

$$\frac{GM(R)m}{R^2} = \frac{mv^2(R)}{R} \tag{16.2}$$

Solving for $M(R)$ gives

$$M(R) = \frac{v^2(R)R}{G} \tag{16.3}$$

Therefore, if we can measure $v(R)$, we can deduce $M(R)$, the mass distribution in the galaxy. Equivalently, we can use $\Omega(R)$, since

$$\Omega(R) = v(R)/R \tag{16.4}$$

Substituting into equation (16.3) gives

$$M(R) = \Omega^2(R)R^3/G \tag{16.5}$$

If all of the mass is, indeed, concentrated at the center of the galaxy, then $M(R)$ is a constant, so equation (16.5) gives Kepler's third law (mentioned in Chapter 5). (We speak of the orbits as being "Keplerian".) The function $v(R)$, or $\Omega(R)$, is called the *rotation curve* for the galaxy.

16.2.2 Rotation curve and Doppler shift

The differential galactic rotation produces Doppler shifts in spectral lines that we observe from gas at different distances from the galactic center than the Sun. This is illustrated in Fig. 16.4. In Fig. 16.4(a), we look at five test particles at different distances along the line of sight. In each case the Doppler shift depends on the relative radial velocity of the test particle and the Sun. That is, we take the line of sight component of the particle's motion and subtract the line of sight component of the Sun's motion.

In Fig. 16.4(b) we look at the Doppler shifts for each test particle. Point 1 is slightly closer to the center than the Sun. It is moving slightly faster than we are, so there will be a small Doppler shift. It is moving away from us so that the shift will be to longer wavelength (redshift). Point 2 is where our line of sight crosses the same circle. The speed is the same as at point 1, and the angle with the line of sight is the negative of that at point 1. Since the line of sight component depends on the cosine of that angle, that component is the same. The Doppler shift for point 2 is therefore the same as for point 1. Point 3 is where the line of sight passes closest to the center. Material is moving fastest around that circle. It is also moving directly away from us, so that point has the largest Doppler shift to the red. Point 4 is in the same orbit as the Sun. Our distance from that object is constant, so our relative radial velocity must be zero, meaning that the Doppler shift is zero. Point 5 is farther from the center than the Sun, and is moving

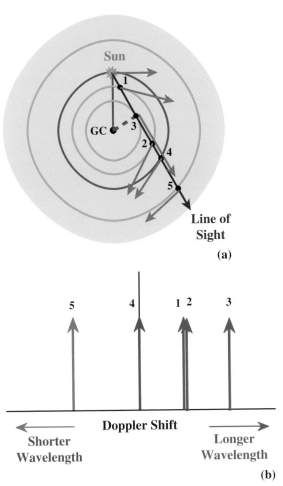

(a)

(b)

Fig 16.4. Doppler shift produced by galactic differential rotation by material at different locations along a given line of sight. (a) The locations of five test particles in an overhead view. Arrows for each particle indicate the velocity (magnitude and direction). For each particle the Doppler shift will depend on the relative radial velocity of that particle and the Sun. (b) For each particle, the position of the arrow shows the amount of Doppler shift.

measured around the galactic plane, starting in the direction of the galactic center.

We also define a convenient reference frame for measuring velocities, called the *local standard of rest*, LSR. If the only motion the Sun had were its orbital motion about the galactic center, then the local standard of rest would coincide with the Sun's motion. That is, we could simply measure motions with respect to the Sun (so-called heliocentric velocities). However, because of gravitational interactions with its nearest neighbors, the Sun has a small motion superimposed on its circular orbital motion, so it is not a convenient reference point for velocities. There are actually two ways of defining the LSR:

(1) *Dynamical.* The origin of the coordinate system orbits at a distance R_0 from the galactic center, and R_0 is the distance of the Sun from the galactic center. The coordinate system moves with a velocity $v(R_0) = v_0$, or $\Omega(R_0) = \Omega_0$, appropriate for circular motion at R_0. Defined

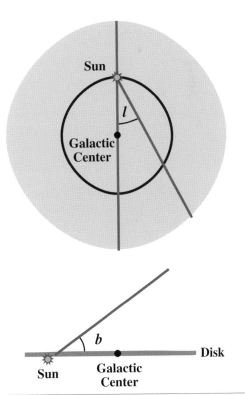

Fig 16.5. Galactic coordinates.

slower than us. We are therefore overtaking it, so there is a Doppler shift to shorter wavelength (blueshift).

In determining the rotation curve for our galaxy, it is convenient to introduce a set of coordinates, known as *galactic coordinates*, measured from our viewing point at the Sun. They are shown in Fig. 16.5. The *galactic latitude, b,* measures the angle above or below the galactic plane of an object. The *galactic longitude, ℓ,* is an angle

this way, the motion of the LSR depends only on $M(R_0)$.

(2) *Kinematic*. The origin of the coordinate system moves with the average velocity of all the stars in the vicinity of the Sun. This averages out the effects of the random motions of these stars.

The two definitions should result in the same velocity system. However, there are small differences, which we will ignore. (The difference tells us about the dynamical properties of the galaxy.) With respect to the LSR, the Sun is moving at about 20 km/s towards a right ascension of 18 h, and a declination of 30°. (In galactic coordinates, this is $\ell = 56°$, $b = 23°$.) Once we know the Sun's motion, we can use it to correct Doppler shift measurements to give us the radial velocity of the object with respect to the LSR.

We now look at the Doppler shifts we will observe for some material a distance R from the galactic center, moving in a circular orbit with a speed $v(R)$. The situation is shown in Fig. 16.6 for $R < R_0$, but the result holds for $R > R_0$ (see Problem 16.12). The relative radial velocity is given by

$$v_r = v(R) \cos (90° - \theta) - v_0 \cos (90° - \ell) \quad (16.6)$$

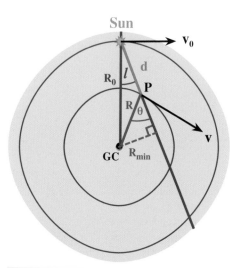

Fig 16.6. Differential rotation and radial velocities. The sun is a distance R_0 from the galactic center. We observe an object at point P, along a line making an angle ℓ with the line of sight from us to the galactic center. P is a distance R from the galactic center.

Using the relationship between sines and cosines, gives

$$v_r = v(R) \sin \theta - v_0 \sin \ell$$
$$= R\Omega(R) \sin \theta - R_0\Omega_0 \sin \ell \quad (16.7)$$

We can measure ℓ, but not θ, so we must eliminate it using the law of sines:

$$\sin(180° - \theta)/R_0 = \sin \ell/R \quad (16.8)$$

Simplifying the left-hand side gives

$$\sin \theta/R_0 = \sin \ell/ R \quad (16.9)$$

Substituting into equation (16.7) gives

$$v_r = R_0\Omega(R)\sin \ell - R_0\Omega_0 \sin \ell \quad (16.10)$$

Factoring out the $R_0 \sin \ell$ gives

$$v_r = [\Omega(R) - \Omega_0]R_0\sin \ell \quad (16.11)$$

In tracing the behavior of v_r, it is convenient to divide the galaxy into *quadrants*, based on the value of ℓ. This is illustrated in Fig. 16.7.

Let's look along a line of sight at some galactic longitude ℓ and see how the radial velocity changes with increasing distance d from the Sun. This is illustrated in Fig. 16.8. We first look at the case for $\ell < 90°$ (first quadrant). As we look at material closer to the galactic center, the quantity $[\Omega(R) - \Omega_0]$ becomes larger. This means that v_r increases. We see that this line of sight has a

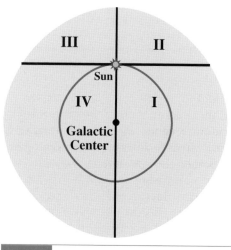

Fig 16.7. Galactic quadrants.

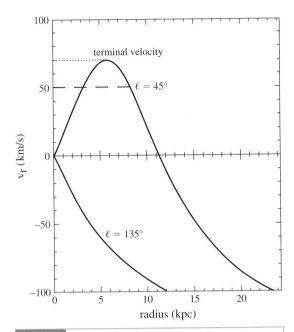

Fig 16.8. Radial velocity as a function of distance from the Sun, d, along a given line of sight. The upper curve is typical of ℓ between 0 and 90°. The maximum v_r corresponds to the point at which the line of sight passes closest to the galactic center. The close point with $v_r = 0$ corresponds to local material, and the far point with $v_r = 0$ corresponds to our line of sight crossing the Sun's orbit. Inside the Sun's orbit, each v_r (except for the maximum) occurs twice. The lower curve is typical of values of ℓ between 90° and 180°. All of these points are outside the Sun's orbit, so each circle is crossed only once, and each v_r is reached only once. [Francoise Combes, Observatoire de Meudon]

point of closest approach to the galactic center. This point is called the *subcentral point*. Of all the material along this line of sight, the material at the subcentral point produces the largest v_r. As we go beyond the subcentral point, we are recrossing orbits, and v_r becomes smaller. Eventually, v_r reaches zero, when the Sun's orbit is crossed again. For points beyond the Sun's orbit, $\Omega(R) < \Omega_0$. This means that v_r is negative and increases in absolute value as d increases.

For $\ell > 270°$ (the fourth quadrant), the behavior of v_r is similar to that in the first quadrant, except that when v_r is positive in the first quadrant it is negative in the fourth quadrant, and

when it is negative in the first quadrant it is positive in the fourth quadrant.

In the second quadrant ($90° < \ell < 180°$), all lines of sight pass only through material outside the Sun's orbit. There is no maximum v_r; it just increases with d. The behavior in the third quadrant is the negative of that in the second quadrant.

We can also find an expression for the relative transverse velocity. This will produce proper motions. The relative velocity is given by

$$v_T = v(R) \sin(90° - \theta) - v_0 \sin(90° - \ell)$$

$$= v(R) \cos \theta - v_0 \cos \ell$$

$$= R\Omega(R) \cos \theta - R_0\Omega_0 \cos \ell \qquad (16.12)$$

From Fig. 16.6 we see that

$$R_0 \cos \ell = d + R \cos \theta \qquad (16.13)$$

This gives

$$R \cos \theta = R_0 \cos \ell - d \qquad (16.14)$$

Substituting this into equation (16.12) gives

$$v_T = \Omega(R)[R_0 \cos \ell - d] - R_0\Omega_0 \cos \ell \qquad (16.15)$$

Grouping the terms with $\cos \ell$ gives

$$v_T = [\Omega(R) - \Omega_0]R_0 \cos \ell - \Omega(R) d \qquad (16.16)$$

The quantity R_0, our distance from the galactic center, is determined from studies of the distribution of globular clusters, and more recently from the studies of clusters of masers near the galactic center. For approximately 20 years prior to 1985, the generally accepted value was 10.0 kpc. However, data accumulated by 1985 suggest a smaller value. As of 1985, the International Astronomical Union started recommending the value

$$R_0 = 8.5 \text{ kpc}$$

By using an agreed upon value, astronomers can be sure that they are using the same values when they compare their studies of various aspects of galactic structure. Prior to 1985, the adopted value for v_0, the orbital speed about the galactic center, was 250 km/s. The value recommended in 1985, to go with the new R_0, is

$$v_0 = 220 \text{ km/s}$$

Example 16.1 Galactic rotation

For the values of the galactic rotation constants, find the time it takes the Sun to orbit the galactic center and the mass interior to the Sun's orbit.

SOLUTION

The time for the Sun to orbit is simply the circumference, $2\pi R_0$, divided by the speed v_0:

$$t = \frac{2\pi R_0}{v_0}$$

$$= \frac{2\pi (8.5 \times 10^3 \text{ pc})(3.1 \times 10^{13} \text{ km/pc})}{(220 \text{ km/s})}$$

$$= 7.5 \times 10^{15} \text{ s}$$

$$= 2.4 \times 10^8 \text{ yr}$$

The Sun orbits the galactic center in 240 million years. We find the mass interior to R_0 from equation (16.3):

$$M(R) = \frac{v_0^2 R_0}{G}$$

$$= \frac{(220 \times 10^5 \text{ cm/s})^2 (8.5 \times 10^3 \text{ pc})(3.1 \times 10^{18} \text{ cm/pc})}{(6.67 \times 10^{-8} \text{ dyn cm}^2/\text{g}^2)}$$

$$= 2.0 \times 10^{44} \text{ g}$$

$$= 1.0 \times 10^{11} \text{ M}_\odot$$

16.3 | Determination of the rotation curve

The rotation curve for material within the Sun's orbit can be determined reasonably well from 21 cm line observations. In determining the rotation curve, we make two important assumptions: (1) The orbits are circular. This means that we need to determine $v(R)$ at only one point for each value of R. (2) There is some atomic hydrogen all along any given line of sight. It is especially important that there be some hydrogen at the subcentral point of each line of sight.

The method takes advantage of the fact that, for lines of sight through the part of the galaxy interior to the Sun's orbit, there is a maximum Doppler shift. It is easy to inspect the 21 cm spectrum at each longitude and determine the maximum Doppler shift. We then assign that Doppler shift to material at the subcentral point (the point of closest approach to the galactic center) for that particular longitude. We can see from Fig. 16.6 that the distance of the subcentral point to the galactic center, R_{min}, is

$$R_{min} = R_0 \sin \ell \tag{16.17}$$

From equation (16.11), we see that if v_{max} is the maximum radial velocity along a given line of sight, then the angular speed $\Omega(R_{min})$ for that line of sight is given by

$$\Omega(R_0 \sin \ell) = (v_{max}/R_0 \sin \ell) + \Omega_0 \tag{16.18}$$

By studying lines of sight with longitudes ranging from 0° to 90°, the corresponding value of R_{min} will range from zero to R_0. This means that we measure $\Omega(R)$ once for each value of R from zero to R_0. However, we have already said that, if the material is moving in circular orbits, one measurement per orbit is sufficient to determine the rotation curve.

There are some limitations to this technique. We have already seen that the distribution of interstellar gas is irregular. If there happens to be no atomic hydrogen at the subcentral point for some line of sight, we will see a v_{max} which is less than the value that we would see if there were material at the subcentral point. There are also problems arising from non-circular orbits. The effect of both of these problems can be reduced by repeating the procedure for the fourth quadrant. Because of the inclination of the galactic plane relative to the celestial equator, the fourth quadrant studies must be performed in the southern hemisphere, and have been done in Australia. For a number of years it seemed that there were disagreements between the first and fourth quadrant rotation curves, but we now think we understand them in terms of non-circular motions. There is further evidence for such motions in the large radial velocities observed for ℓ close to 0° and 180°.

Another problem is that we cannot really cover the full range from zero to R_0. For ℓ close to zero, $\sin \ell$ is close to zero, and the Doppler shift is very small. Random motions of clouds are much larger than the radial velocity due to galactic rotation. Similarly, for ℓ near 90°, Ω is close to Ω_0, providing a small radial velocity due to galactic rotation.

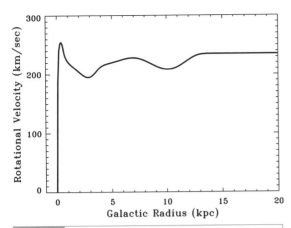

Fig 16.9. Rotation curve for the Milky Way. The curve inside the Sun's orbit is well determined from 21 cm studies. Outside the Sun's orbit, we use HII regions in molecular clouds. We use spectroscopic parallax on the exciting stars to give the distance to the HII regions, and then CO observations to measure the Doppler shift. [Daniel Clemens (Boston University) Clemens, D., Astrophys. J., **295**, 422, 1985]

When we eliminate these ends, a reasonable rotation curve has been derived for R in the range 3 to 8 kpc. This is shown as part of the curve in Fig. 16.9.

Using HI to determine the rotation curve outside the Sun's orbit is more difficult. There is no maximum Doppler shift along any line of sight. It is therefore necessary to measure independently v and d, the distance from the Sun to the material being studied. From d and ℓ, we can deduce R (see Problem 16.6). Until molecular observations, there was no reliable rotation curve for $R > R_0$. The best astronomers could do was to derive the mass distribution for $R < R_0$, and then make some assumptions about how the mass distribution would continue for $R > R_0$. From the assumed mass distribution, a rotation curve could be derived. It was assumed that there was relatively little mass outside the Sun's orbit, so the rotation curve was characterized by a falloff in $v(R)$ that was close to that predicted by Kepler's third law.

However, recent observations of molecular clouds have provided a direct method of measuring the rotation curve outside the Sun's orbit. Molecular clouds associated with HII regions were studied. The radial velocities were determined from observations of the carbon monoxide (CO) emission from the clouds. The distance to the stars exciting the HII regions were determined by spectroscopic parallax. This gives a reliable rotation curve at least out to about 20 kpc. The combined rotation curve (using HI inside the solar circle and CO outside, is shown in Fig. 16.9. There is no falloff in $v(R)$ out to 20 kpc, and there may even be a slight rise. This means that there is much more mass outside the Sun's orbit than previously thought!

We can see from equation (16.3) that if $v(R)$ is constant from 8 to 16 kpc, then $M(16 \text{ kpc})$ will be twice $M(8 \text{ kpc})$. This means that there is as much mass between 8 and 16 kpc as there is out to 8 kpc. However, the luminosity of our galaxy is falling very fast as R increases. Since the luminous part of the matter is mostly in the disk, it would seem that this extra mass cannot be part of the disk. Current thinking places the extra mass in the halo of the galaxy. We still have little idea of what form this matter takes. It has been suggested that it can be in the form of faint red stars, but recent results make this seen unlikely. This is our first encounter with *dark matter*, matter whose gravitational effects are felt, but which is not very luminous. (Astronomers used to call this "missing" matter, but it is not missing. We can tell that it is there by its gravitational effects. We just can't see it.) We will see that there is strong evidence for dark matter in other galaxies, and we will discuss it farther in Chapter 17.

Example 16.2 Galactic mass distribution
For what (spherically symmetric) mass distribution is $v(R)$ constant?

SOLUTION
From equation (16.3), we know how $M(R)$ is related to $v(R)$, and from equation (16.1) we know how $\rho(r)$ is related to $M(R)$.
For a spherical coordinate system, the volume element is $4\pi r^2 dr$, so equation (16.1) becomes

$$M(R) = \int_0^R \rho(r)\, 4\pi r^2\, dr$$

Differentiating both sides tells us that

$$\frac{dM(r)}{dr} = \rho(r)\, 4\pi r^2$$

or

$$\rho(r) = \frac{1}{4\pi r^2}\frac{dM(r)}{dr}$$

If $v(R)$ is a constant, say v_0, then equation (16.3) tells us that

$$M(r) = \frac{v_0^2 r}{G}$$

$$\frac{dM(r)}{dr} = \frac{v_0^2}{G}$$

Equating the two expressions for $dM(r)/dr$ gives

$$\rho(r) = \frac{v_0^2}{4\pi r^2 G}$$

This means that the density falls as $1/r^2$. While this sounds like a rapid falloff, remember that the volume of a shell of thickness dr and radius r is $4\pi r^2\,dr$. So, in calculating the mass of each shell, the r^2 factors cancel. This means that, as we go farther out, the mass of each shell stays constant. So, as far as we continue to add shells with this $1/r^2$ density falloff, the mass of the galaxy grows by the same amount with every shell we add.

Once we have a rotation curve for our galaxy, it is possible to use measured Doppler shifts to determine distances to objects. Since these distances are determined from the motions of the objects, they are called *kinematic distances*. For any particular object, we measure v_r and ℓ. We then determine the angular speed Ω from

$$\Omega = (v_r/R_0 \sin \ell) + \Omega_0 \qquad (16.19)$$

It is assumed that we know R_0 and Ω_0. Once we know Ω we can use the known rotation curve to find the value of R to which the Ω corresponds.

There are a few limitations to this technique. It does not work for material whose radial velocity due to galactic rotation is less than that due to the random motions of the clouds. This rules out material near longitudes of $0°$ and $180°$, as well as material close to the Sun.

Another problem arises for material inside the Sun's orbit. There are two points along the same line of sight that produce the same radial velocity. (The one exception is the subcentral point.) Both of these points are the same distance from the galactic center, but they are different distances from us. This problem is called the *distance ambiguity*. We can use the rotation curve to say that the object is in one of two places, and we must then use other information to resolve the

ambiguity. Remember, there is no distance ambiguity for material outside the Sun's orbit, making this an interesting part of the galaxy to study.

16.4 | Average gas distribution

To understand star formation on a galactic scale, we must know how the interstellar gas, out of which the stars will be formed, is distributed in the galaxy. We are interested in the average distributions of various constituents. By "average" we mean that we are interested only in the large-scale structure. We would like to know the radial distribution of interstellar gas. (Remember, this is not the same as $M(R)$, which includes mass in all forms.) We would also like to know the degree to which the gas is confined to the disk. We can express this as a thickness of the disk, as determined from various constituents. We would also like to know whether the thickness is constant, or whether it varies with position in the galaxy. Finally, we would like to know if the plane of the galaxy is truly flat, or if it has some large-scale bumps and wiggles.

We first look at HI. The amount of HI doesn't fall off very quickly as one goes to larger R. For example, the mass of HI interior to R_0 is about $1 \times 10^9\ M_\odot$, and the mass exterior to R_0 is about

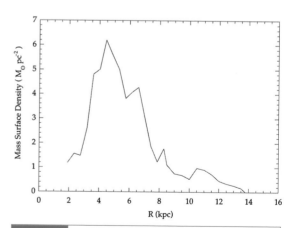

Fig 16.10. Radial distribution of H_2, with the H_2 deduced from CO observations, assuming a constant conversion from CO luminosity to mass. There is growing evidence that this conversion factor actually changes with environment, and this curve may underestimate the mass in the outer galaxy by a factor of three. Also, remember that there is a larger volume, so even a lower density of material can still translate into a significant mass. [Thomas Dame, CFA]

$2 \times 10^9 \, M_\odot$. (Of course, this larger mass is spread out over a larger volume. See Problem 16.5.) Note that the mass of HI is only about 1% of the total mass interior to a given radius. This means that the gas does not provide most of the large-scale gravitational force in the galaxy. It just responds to the gravitational effects of the stars, and whatever dark matter there is in the halo.

The abundance of H_2 (Fig. 16.10) falls off more rapidly with R than does that of HI. Inside the Sun's orbit, the mass of H_2 is approximately equal to that of HI, about $1 \times 10^9 \, M_\odot$. Outside the Sun's orbit, the mass of H_2 is about $5 \times 10^8 \, M_\odot$, about one-quarter that of HI. There appears to be a peak in the H_2 distribution about 6 kpc from the galactic center. This is sometimes called the *molecular ring*. It appears that most of the H_2 seems to be concentrated into a few thousand giant molecular clouds, rather than a large number of small clouds. The distribution of molecular hydrogen H_2 is generally deduced indirectly from observations of CO. There are still disagreements over how to derive the H_2 abundance from the intensity of the CO emission. There is growing evidence that the conversion factor changes with galactic environment, and that the mass of the outer galaxy (like the derived mass distribution in Fig. 16.10) are underestimated by as much as a factor of three.

We generally express the thickness of the disk by finding the separation between the two points, one above and one below, at which the HI density falls to half of its value in the middle of the plane. This is called the full-width at half-maximum or FWHM. At the orbit of the Sun, the thickness of the HI layer is about 300 pc. At $R = 15$ kpc, the thickness is about 1 kpc. This trend is shown in Fig. 16.11. This means that the plane becomes thicker as one goes farther out from the center of the galaxy (as depicted schematically in Fig. 16.3c). This is called the "flaring" of the galactic disk. In addition, we find that the disk isn't flat. It has a warp to it, like the brim of a hat (Fig. 16.12). This is also shown schematically in Fig. 16.3(c). The warp is most prominent outside the Sun's orbit. The bend is upward in the first and second quadrants and downward in the third and fourth quadrants.

It is also interesting to see the degree to which various constitents (atomic, molecular, ionized) are

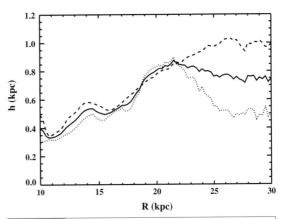

Fig 16.11. Thickness of the plane of the Milky Way as determined from 21 cm observations. The plot shows the layer thickness (measured as a gaussian width), with dotted lines for range $20° < \theta < 160°$, the dashed for $200° < \theta < 340°$, and the solid line the average of both ranges. θ is galactocentric azimuth, where $\theta = 0°$ is the direction $\ell = 0°$. [Butler Burton, Sterrewacht, Leiden University]

confined to the plane. Fig. 16.13 shows the thickness of the plane as measured by various tracers. In general, we think that components that are more closely tied to the plane are younger. This means that HII regions and molecular clouds have formed more recently than HI clouds or globular clusters.

Molecular hydrogen is more closely confined to the galactic plane than atomic hydrogen. The

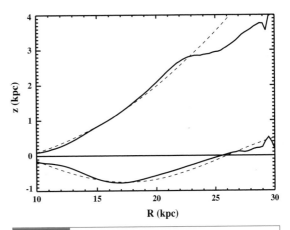

Fig 16.12. Height of the plane of the Milky Way, as determined from 21 cm observations. The solid lines trace the warp heights in the first and second quadrants (upper) and in the third and fourth (lower). The dotted lines are fits used in modelling. [Butler Burton, Sterrewacht, Leiden University]

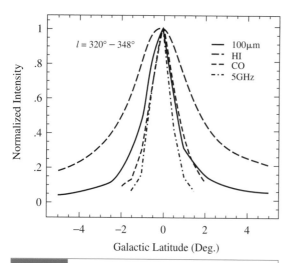

Fig 16.13. Thickness of the plane as measure by four tracers. In general, components that are more concentrated to the plane are thought of as being younger, in the sense that they have not had as much time to spread out. [Eli Dwek, NASA/Sodroski, T. et al., Astrophys. J., **322**, 101, 1987]

thickness of the plane in H_2 is about half that in HI. We have seen that one feature of population I material is its confinement to the galactic plane. We therefore think of molecular clouds as representing a more extreme population I than that represented by the HI clouds. This might indicate that molecular clouds, as a whole, are more recently formed than the HI clouds. Finally, the H_2 shows the same flaring and warp as the HI.

Additional information on the total gas distribution (HI plus H_2) comes from observations of gamma-rays. These gamma-rays are created when cosmic rays strike protons. It doesn't matter what types of atoms or molecules the protons are in. The results are somewhat uncertain, but provide an additional constraint. The general conclusion from such observations supports the conclusions from the CO observations.

16.5 | Spiral structure in the Milky Way

Early photographic surveys revealed a number of nebulae with a spiral appearance. In Chapter 17 we will discuss the reasons for believing that these spiral nebulae are distant galaxies, distinct from our own, rather than part of the interstellar medium in our galaxy. Some examples of spiral galaxies are shown in Fig. 17.3. Given that many other galaxies seem to have a spiral appearance, it is reasonable to wonder whether our galaxy is also a spiral. Unfortunately, from our vantage point within the galaxy, it is very difficult to see the overall pattern. In other galaxies, the spiral pattern appears in the disk.

However, we can detect certain similarities between our galaxy and spiral galaxies. In particular, spiral galaxies have a significant amount of gas and dust. The amount of gas and dust in our galaxy is comparable to that in other spirals. For this reason, astronomers have been encouraged in trying to unravel the spiral structure of our galaxy.

16.5.1 Optical tracers of spiral structure

When we look at spiral galaxies we see that the spiral arms are not continuous bands of light. Rather, they appear to contain knots of bright stars and glowing gas. For example, it appears that HII regions and OB associations trace out spiral arms in other galaxies. For this reason, we have tried to see if the HII regions we can see optically in our galaxy form any distinct pattern. By using optical observations we can rely on distances determined by spectroscopic parallax for the stars exciting the HII regions. Similarly, we can also look at the distribution of OB associations. A drawback is that, with optical observations, we cannot see very far along the plane of the galaxy. When we study the distribution of HII regions and OB associations, it is clear that the placement is not entirely random. We seem to see at least pieces of connected chains of HII regions and OB associations. These pieces have been identified as a series of named "arms", identified by the constellation in which they are most prominent. This is all a tantalizing hint of spiral structure, but is not a definitive picture.

16.5.2 Radio tracers of spiral structure

We can view the situation on a larger scale by using radio observations to look at the distribution of interstellar gas. This allows us to see across the whole galaxy. We can utilize kinematic distances,

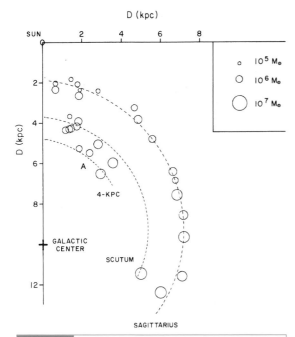

studying the spiral structure in inner parts of our own galaxy, especially with the distance ambiguity. Or, it may be telling us that the spiral pattern in the inner galaxy is not that well defined. We will see in Chapter 17 that, in many other spirals, the pattern gets stronger as one moves farther out.

Outside the Sun's orbit, the approach is more direct. Since we have a rotation curve for the outer part of the galaxy and there is no distance ambiguity, it is easier to trace out the large-scale structure. It is in the regions outside the Sun's orbit that we see the best evidence for spiral structure, with some features being traced over at least a quarter the circumference of the galaxy (Fig. 16.15). There is a growing confidence that the outer part of our galaxy is a four arm spiral. This work is still going on.

Much of our understanding of spiral structure comes from comparing our galaxy to other but must still deal with the distance ambiguity. Initial radio studies of the interstellar gas and spiral structure involved the 21 cm line. Again, long connected features have been identified.

With the discovery of molecular clouds, it was hoped that they would reveal the spiral structure of our galaxy. This is because the optical tracers of spiral structure we see in other galaxies – OB associations, HII regions, dust lanes – are all associated with giant molecular clouds. A number of groups have carried out large-scale surveys of emission from CO throughout the galaxy. We look at some results in Fig. 16.14. Most of the work involved material inside the Sun's orbit. The problem is that the distance ambiguity makes it difficult to place uniquely all of the emitting regions. One approach has been to take specific models for spiral arms and predict the outcome of CO observations. Again, pieces of arms have been identified, and again we see pieces of arms. This may be telling us about the difficulties of

Locations of Mapped Clouds

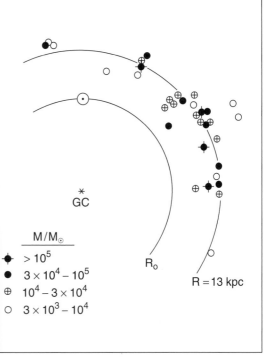

(a)

Fig 16.15. Molecular clouds outside the Sun's orbit and spiral structure. (a) First and second galactic quadrants. The cloud masses are denoted by the symbols, shown at the lower left. The circle at 13 kpc is drawn in for reference.

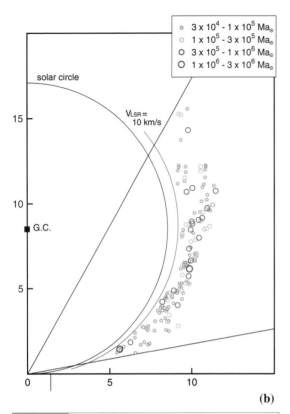

Fig 16.15. (Continued) (b) The fourth quadrant, in a galactic longitude range 280° to 330°. These clouds form part of what is called the Carina arm. [(a) Kathryn N. Mead; (b) Yasuo Fukui, Nobeyama Radio Observatory]

galaxies. Therefore, we will leave further explanation of spiral structure until Chapter 17, when we look at other galaxies.

16.6 | The galactic center

16.6.1 Distribution of material near the center

Ever since it was realized that the galactic center is someplace else from where we are, astronomers have wondered about its nature. Is it simply a geometric location, or is it the site of unusual activity? We will see in Chapter 19 that the centers of many other galaxies are the sites of unusual activity. This makes the search for activity in our galactic center a natural focus. When we talk about the center region, we are talking about roughly the inner 500 pc of the galaxy.

Visual extinction in the galactic plane makes optical studies of the galactic center virtually impossible. We are able to observe the galactic center in the radio and infrared parts of the spectrum. In the radio, we detect continuum emission from ionized gas and line emission from molecular clouds. In the infrared, continuum observations tell us about the dust temperature and opacity, and spectral line observations tell us about the neutral and ionized gas. Near infrared observations can also be used to study rich star clusters near the center. Millimeter observations of interstellar molecules tell us about the cold dense component. Because of the great distance to the galactic center, many observations are limited by poor angular resolution. In the past several years, a major breakthrough has resulted from VLA and VLBI observations. The development of sensitive infrared array detectors has also been very important. Maps of radio emission and X-ray emission are shown in Fig. 16.16.

Studies of the ionized gas show a bent arc of emission perpendicular to the galactic plane. This structure is about 15 arc min in extent. It shows a filamentary structure. This feature is also seen in the infrared. The emission seems to be a combination of thermal (free–free) radiation and non-thermal (synchrotron) radiation. The presence of extended X-ray emission (Fig. 16.16b), suggests temperatures as high as 10^7 to 10^8 K. It is speculated that these high temperatures result from a past explosive event. This event may have been a number of supernova explosions following an unusual wave of star formation. (We will discuss these "starbursts" more in Chapter 19.) At the higher end of that range, the gas would not be bound to the galaxy.

The molecular material in the inner 200 pc is remarkably hot and dense, compared with giant molecular clouds elsewhere in the galaxy (Chapter 15). Typical temperatures are 70 K, and densities are greater than 10^4 cm^{-3}. These are conditions found in normal molecular cloud cores, but they are found over the full extent of the clouds. The internal velocity dispersions are up to 50 km/s, much greater than even in molecular cloud cores. The amount of molecular material may be as high as 10^8 $M_{\odot}$. It has been suggested that the ionized

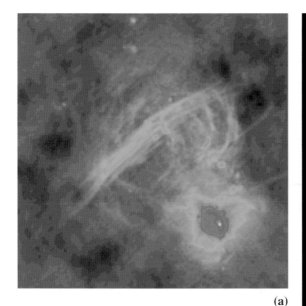

(a)

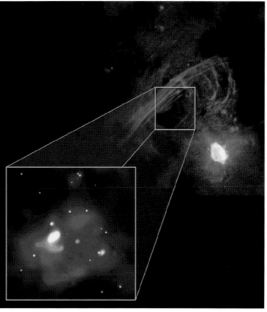

(c)

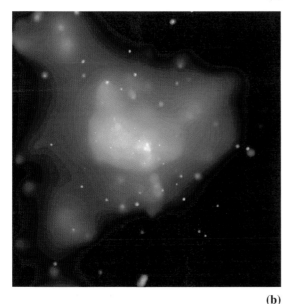

(b)

Fig 16.16. Maps of the galactic center. (a) VLA map of radio emission at a wavelength of 6 cm. An arc like structure is visible. Much of this pattern is viewed as coming from a tilted ring of ionized gas. (b) Chandra image of X-ray emission from hot gas hear the galactic center. (c) The blue image shows the X-ray emission relative to the radio emission, which is shown in red. [(a) NRAO/AUI/NSF; (b), (c) NASA]

gas filaments are the boundaries of dense molecular clouds, which have been exposed to the intense ionizing radiation near the center.

If we are to understand the dynamics of this unusual environment, we must also know the magnetic field strength (and arrangement). As we discussed in Chapter 14, magnetic fields are generally hard to measure. For the galactic center, even the Zeeman effect in HI is hard to measure because the spectral lines are so much wider than in the normal interstellar medium. However, a variety of techniques suggest very high field strengths, of the order of a few milligauss. (Remember, typical interstellar fields are tens of microgauss.)

The galactic center provides a totally unique environment for star formation. There is a high density to help, but it is inhibited by higher temperatures, velocity dispersions, magnetic pressure, and tidal effects. Observations (such as those shown in Fig. 16.17) tell us that the central parsec is very rich in past and current star formation. The star formation rate in the central region is estimated to be as high as $1 M_\odot/\text{yr}$. The central cluster shows evidence for a new $10^4 M_\odot$ superimposed on an older $10^6 M_\odot$ cluster. It has been suggested that the formation of the newest cluster triggered the formation of at least two other rich nearby clusters. (We discussed triggering of star formation in Chapter 15.)

Arches Cluster

Quintuplet Cluster

1 light year

2 light years

Fig 16.17. HST near IR images of the galactic center, showing rich star clusters. [STScI/NASA]

16.6.2 A massive black hole?

There has been considerable speculation on the nature of the central object. It has been suggested that it might be a few million $M_\odot$ black hole. In establishing the existence of such a black hole, there are two observational steps that must be made. First, we must accurately determine the mass of the object. The best way to do this is by its gravitational influence on nearby surrounding stars or gas. Second, we must show that the object is smaller than the Schwarzschild radius for that mass.

You can get an idea of how difficult this is. By equation (8.10), the Schwarzschild radius for a $10^6 M_\odot$ black hole is 3×10^6 km, which is about 10^{-7} pc, and would subtend an angle of about 10^{-6} arc sec.

One interesting approach to this problem was started in 1995 by *Andrea Ghez* (Fig. 16.18) of UCLA, and co-workers, using the Keck 10 m telescope (described in Chapter 4). Their goal was to measure the proper motions of stars in the direction of SgrA*, the source in the center of the galaxy. Ghez's group observed in the near infrared (2.2 μm), which provided good angular resolution, but still had less extinction to the galactic center than they would have had in the visible. To obtain the best possible resolution, they took a number of short exposures, so there was little atmospheric smudging in each image. In adding the images together, they shifted the image to remove changes in atmospheric refraction from image to image. The final images had a resolution of about 0.5 arc sec. If the diffraction pattern is very clean (and it is for their system) then it is possible to measure the position of each star to much greater accu-

racy, in their case to within 0.002 arc sec. When we have previously talked about using velocities to measure the mass of a system, in using the virial theorem for example, we used radial velocities as measured from Doppler shifts. We could have used proper motions, but they are generally too small to measure over a period of a few years. Velocities near a massive black hole would be large

Fig 16.18. Andrea Ghez. [Andrea Ghez, UCLA]

enough to produce measurable proper motions. Measuring Doppler shift for these stars would be hard, because they required such short exposures to achieve good angular resolution. Their results are summarized in Fig. 16.19.

With three observing runs, a year apart, Ghez's group were able to identify 90 stars with proper motions large enough to measure. The largest proper motion corresponded to a tangential velocity of 1400 km/s. They found that this motion is organized about the position of SgrA* to within 0.1 arc sec. The projected positions and velocities are consistent with Keplerian motion. This means that essentially all of the mass is closer to the center than any of the stars measured. The best estimate of the mass in the central object is $(2.6 \pm 0.2) \times 10^6 \, M_\odot$. The observations don't

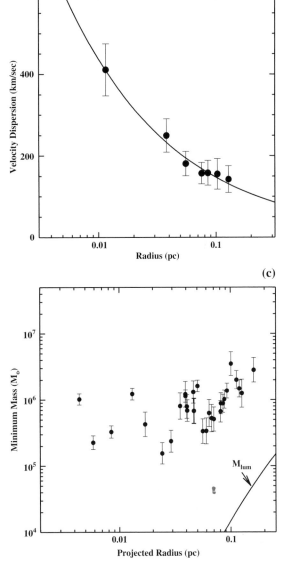

Fig 16.19. (a) Measured positions for three observing years near the position of SgrA*. (b) Positions of the 90 stars whose proper motions were measured, with the size of the symbol scaled to the size of the motion. (c) Projected stellar velocity dispersion as a function of distance from SgrA*. The solid line is the value for Keplerian orbits (still outside all of the mass). (d) Enclosed mass as a function of projected distance for 30 stars.

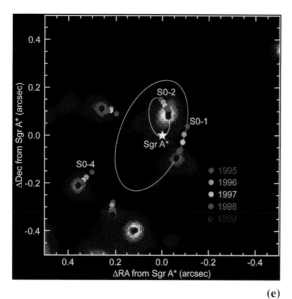

(e)

Fig 16.19. (*Continued*) (e) Curvature in the paths of three stars. [Andrea Ghez, UCLA; (a)–(d) Ghez, A. M. *et al.*, *Astrophys. J.*, **509**, 678, 1998, Figs. 3–6; (e) Ghez, A. M. *et al.* Reprinted by permission from *Nature*, **407**, 307, Fig. 1, Copyright (2000) Macmillan Publishers Ltd.]

By following these stars for another two years (through 1999), the group found three stars whose paths had a measurable curvature about the SgrA* position (Fig. 16.19e). These stars had projected positions on the sky ~0.005 pc from SgrA*. From this curvature, they could directly measure the acceleration in the circular orbit. It turns out to be numerically close to the value of the Earth's acceleration in its motion about the Sun. The shortest orbital period is 20 years, so we have the prospect of being able to watch something make a complete revolution about the galactic center in our lifetimes. Acceleration vectors don't allow for a better mass estimate, but if you use the mass estimate from the proper motions, the volume the mass is constrained to is decreased by an order of magnitude. This greatly strengthens the case for that mass to be included within its Schwarzschild radius.

There is also indirect evidence for explosive activity in the galactic center region. For example, there is an armlike feature in our galaxy, some 3 to 4 kpc from the galactic center, called the "3 kpc arm", which appears to be expanding at about 50 km/s. Speculation is that this expansion is due to some explosion in the relatively recent past (see Problem 16.14). We see other similar features closer in, suggesting that this activity has taken place on a continuing basis. We will see in Chapter 19 that the activity in our galactic center is small compared with that in many galaxies. However, it gives us our best opportunity to study such activity "close up". For this reason, study of the galactic center is a very active field.

conclusively show that this mass is contained within R_S, but alternative explanations, such as a very rich cluster, don't seem very likely. These masses are consistent with previous determinations using other techniques, but the new measurements place much tighter constraints on the confinement of the matter, and the coincidence with SgrA*.

Chapter summary

In this chapter we saw how stars and interstellar material are arranged in the Milky Way.

We saw how the rotation curve tells us about the mass distribution in the galaxy, and how that rotation curve is determined. Different techniques are needed for material inside and outside the Sun's orbit about the galactic center. We saw how the rotation provides evidence for dark matter. Once the rotation curve is known, velocities of objects can be used to estimate distances. This works better outside the Sun's orbit than inside, because of the distance ambiguity.

In looking at the average gas distribution, we found that the HI is extended beyond the Sun's orbit, while the number of molecular clouds falls off more sharply with distance from the center of the galaxy. The molecular clouds are also more tightly concentrated toward the galactic plane than the HI clouds. This concentration toward the plane suggests that the molecular clouds are younger.

We discussed evidence for spiral structure in the Milky Way and the difficulties in tracing out spiral arms in our galaxy. We saw that tracers for spiral structure include HII regions, OB associations and molecular clouds.

In looking at the galactic center, we found that it contains a small, active region. The center has a mass concentration which may be a few million $M_\odot$ black hole.

Questions

16.1. How do we know that $v(R)$ tells us the total mass interior to R?

*16.2. Why should the kinematic and dynamical definitions of the local standard of rest give the same rest frame?

16.3. Why is it sufficient to measure the rotation curve only for subcentral points for $R < R_0$?

*16.4. Why can't we use transverse velocities to measure the rotation curve of our galaxy?

16.5. Contrast the method of measuring the rotation curve inside the Sun's orbit with that for measuring it outside the Sun's orbit.

16.6. Describe the difficulties in studying spiral structure in our galaxy.

16.7. Compare the methods for using GMCs and HII regions to study spiral structure.

16.8. Why is it easier to study spiral structure in the outer part of our galaxy than in the inner part?

Problems

16.1. Show that the radial velocity equation (16.11), derived for the first quadrant of the galaxy, holds for the second quadrant.

*16.2. For material observed at a radial velocity v_r along a line of sight at galactic longitude ℓ, find an expression for the separation between the near and far points producing that v_r.

*16.3. Calculate $\Omega(R)$ and $v(R)$ for the following density models: (a) all the mass M at the center of the galaxy; (b) a constant density, adding up to a mass $M(R_0)$ at the Sun's orbit and no mass beyond.

16.4. Suppose the rotation curve of the Milky Way is flat out to $2R_0$. What mass does that imply out to that distance?

16.5. Convert the HI and H_2 masses given in the chapter into: (a) average volume densities and (b) average surface densities for the regions $R < R_0$ and $R_0 < R < 2R_0$ (assuming that all of the mass outside R_0 is between R_0 and $2R_0$).

*16.6. For material outside the Sun's orbit, derive an expression to convert observed v_r and d into $\Omega(R)$ and R.

16.7. (a) Estimate the age of the 3 kpc arm from its radius and expansion speed. (b) Is this an upper or lower limit to its age? Explain.

16.8. Draw a diagram showing two points with the same radial velocity along the same line of sight. Show that the component of their velocity along the line of sight is the same.

16.9. Show that the radial velocity of a point in the fourth quadrant is the negative of that of the corresponding point in the first quadrant.

16.10. If we can only measure kinematic distances for points with $v_r > 10$ km/s, what range of R can we study?

16.11. For $\ell = 45°$, we observe $v_r = +30$ km/s. What are R and d?

16.12. If $v(R) = 220$ km/s for $R_0 < R < 2R_0$, find an expression for v_r as a function of (ℓ, R).

16.13. When the constants R_0 and v_0 changed from 10.0 kpc and 250 km/s to 8.5 kpc and 220 km/s, respectively, by what factor did the mass inside the Sun's orbit change?

16.14. What is the Schwarzschild radius for a 5×10^6 $M_\odot$ black hole? How does this compare with the sizes of the structures found in the galactic center?

*16.15. The mass distribution for material near the galactic center can be determined from studies of the rotation curve close to the center. When the rotation velocity is plotted vs. log R for the inner 10 pc, the result is approximately a straight line. $v(R)$ is 200 km/s at $r = 2$ pc and 70 km/s at $R = 10$ pc. (a) Use this data to find $M(R)$ for 2 pc $< R < 10$ pc. (b) Assuming a spherical distribution, find $\rho(r)$.

Computer problems

16.1. Plot a graph of Doppler shift as a function of distance along a line of sight for $\ell = 45°$ and for $\ell = 135°$.

16.2. Plot a graph of v_{max} vs. ℓ for ℓ ranging from 0 to 90 degrees.

16.3. Complete the following table by giving kinematic R and d for the indicated ℓ and v_r. (In case of a distance ambiguity, give both values.)

ℓ	v_r
30	+30
30	−30
60	+30
60	−30
120	−40
240	+40
330	+30
330	−30

Part V

The universe at large

To this point we have been studying the stellar life cycle and how stars and other material are arranged in the Milky Way Galaxy. We will now turn to studies on a much larger scale. We will first look at other galaxies, and see that some of them tell us more about our own galaxy, which is so hard to observe. When we talk about how the universe is put together, each galaxy has only as much importance as a single molecule of oxygen has in describing the gas in your room.

As we go to larger scales, we will look at how galaxies are distributed on the sky, and how they move relative to one another. We will also see how the problem of dark matter becomes more important as we go to larger and larger scales.

As we go to larger scales, increasing the number of galaxies that we observe, we also find a variety of interesting phenomena associated with galaxies. In Chapter 19 we will discuss aspects of galactic activity, particularly as evidenced by radio galaxies and quasars.

In Chapter 20 and 21 we will turn to cosmology, the study of the universe on the largest scales. This also includes the past and future evolution of the universe. It is in the study of the past that we encounter one of the most fascinating aspects of modern astrophysics research, the merging of physics on the smallest (elementary particles) and largest (structure of the universe) scales.

Chapter 17

Normal galaxies

Our study of the Milky Way has been aided greatly by studies of other galaxies. However, for a long time it wasn't clear that the spiral nebulae we see in the sky are really other galaxies. From their appearance, it might just be assumed that these nebulae are small nearby objects, just as HII regions are part of our galaxy.

The issues were crystallized in 1920 in a debate between *Harlow Shapley* and *Heber D. Curtis*. Curtis argued that spiral nebulae were really other galaxies. His argument was based on some erroneous assumptions. First, he confused novae in our galaxy with supernovae in other galaxies. Shapley thought the spiral nebulae were part of our own galaxy, partly based on an erroneous report of a measurable proper motion for some nebulae.

The issue was settled in 1924 by the observational astronomer *Edwin Hubble* (after whom the Space Telescope is named). Hubble studied Cepheids in three spiral nebulae (including the Andromeda Galaxy), and clearly established their distance as being large compared with the size of the Milky Way. There is some problem with Hubble's analysis, involving type I vs. type II Cepheids. However, even this factor of 2 error in the distance was not enough to alter the basic conclusion that spiral nebulae are not part of our own galaxy. Following this work, Hubble made a number of pioneering studies of other galaxies, essentially opening up the field of extragalactic astronomy.

17.1 | Types of galaxies

In his studies Hubble realized immediately that not all spiral galaxies have the same appearance.

Furthermore, he found galaxies that do not have a spiral structure. Hubble classified the galaxies he studied according to their basic appearance. It was originally thought that the different types of galaxies represented different stages of galactic evolution. (Similarly, some astronomers thought that different spectral type stars along the main sequence were evolutionary states of the same star.) We now know that this is not the case. However, Hubble's classification scheme, depicted in Fig. 17.1, is still quite useful.

17.1.1 Elliptical galaxies

Elliptical galaxies have, as their name suggests, simple elliptical appearances. Some examples of ellipticals are shown in Fig. 17.2. The ellipticals are classified according to their degree of eccentricity. The ones that look spherical (zero eccentricity) are called E0, and the most eccentric are called E7.

The most common type of elliptical galaxies are called *dwarf ellipticals*, since they are also the smallest. Their sizes are typically a few kiloparsecs and their masses are a few million solar masses. More spectacular are the *giant ellipticals*, with extents up to 100 kpc and masses of about $10^{12}M_\odot$, with some with masses up to a factor of ten higher.

The gas content of ellipticals is low. Studies of HI, using the 21 cm line emission, as well as IRAS observations of weak emission from their dust, suggest that the mass of the interstellar medium may be up to about 1% of the mass of the stars that we see. The low gas content rules out the possibility that ellipticals eventually flatten to form spirals. The continuing process of star formation in a galaxy depletes its supply of interstellar matter,

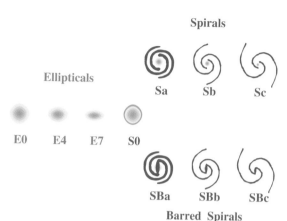

metal abundances are not low. Giant ellipticals have metal abundances that are quite high, about twice the solar value.

Some ellipticals are rotating very slowly. They have a higher ratio of random velocities to rotational velocities than do spirals. We think that their slow rotation means that they could collapse without much flattening. Remember, when we discussed collapsing interstellar clouds,

Fig 17.1. Hubble classification of galaxies. Ellipticals range from E0 (round) to E7 (the most oblate). The regular spirals are divided according to the relative size of the nucleus and the disk, and the tightness of the spiral arms. The Sa have the largest nuclei and the most open arms. The barred spirals, SB, follow the same classification as the normal spirals. S0 galaxies have nuclei and small disks but no spiral arms.

so if spirals are merely evolved ellipticals, we have no way of understanding why spirals have more gas and dust than ellipticals.

Ellipticals generally contain an evolved stellar population, with no O or B stars. However, their

(a)

(b)

(c)

Fig 17.2. Elliptical galaxies. (a) M87, in Virgo, which is a giant elliptical, type E0. The fuzzy patches visible near the edge of the galaxy are globular star clusters. The inset shows a blow-up of the center. (b) M49, in Virgo, type E1. It is about 15 Mpc away, and about 15 kpc across. (c) M32 in Andromeda, which is a dwarf elliptical companion to the Andromeda Galaxy (M31, Fig. 17.3b) and is type E2. It is only 800 pc across. [NOAO/AURA/NSF]

we found that rapid rotation retards collapse perpendicular to the axis of rotation, resulting in the formation of a disk.

We can use photometry to study the brightness distribution in ellipticals. Since we see the galaxy projected as a two-dimensional object on the sky, it is convenient to speak of the luminosity per unit surface area $L(r)$, where r is the projected distance from the center of the elliptical. Studies show that the light from most ellipticals can be described well by a simple relationship (known as de Vaucouleur's law):

$$L(r) = L(0)e^{-(r/r_0)^{1/4}} \tag{17.1}$$

In this expression $L(0)$ and r_0 are constants. The values of $L(0)$ are found not to vary very much, with a typical value of about $2 \times 10^5 \, L_\odot/\text{pc}^2$. The values of r_0, however, show a very large spread.

17.1.2 Spiral galaxies

Spirals make up about two-thirds of all bright galaxies. They are subdivided into classes Sa, Sb and Sc. The two important features of the classification are (1) the openness or tightness of the winding of the spiral pattern, and (2) the relative importance of the central bulge and the disk of the galaxy. Sa galaxies have the largest bulges and the most tightly wound arms. Sc galaxies have the smallest bulges and the most open arms. We think that the Milky Way is between Sb and Sc. Different types of spirals are shown in Fig. 17.3.

Some spirals have a bright bar running through their center, out to the point where the arms appear to start. These are called *barred spirals*. Some examples are shown in Fig. 17.4. The barred spirals are also subclassified into SBa, SBb and SBc, according to the same criteria as Sa, Sb and Sc. In general, the spiral pattern in barred spirals is quite well defined.

When the spiral pattern is well defined, we call the galaxy a *grand design spiral*. They have a continuous pattern running throughout the galaxy. Other spirals have a less organized appearance. These are called *flocculent spirals*. A comparison is shown in Fig. 17.5.

It has been realized that spirals of the same type can have different luminosities. This point is important in trying to determine the distances to galaxies that are so far away that we cannot distinguish individual stars of HII regions. As suggested by *Sydney Van den Bergh*, astronomers have taken to adding a *luminosity class* to the spiral

(a)

(b)

Fig 17.3. Various types of spiral galaxies. (a) NGC 7217, an Sa galaxy. (b) The Andromeda Galaxy (M31), type Sb, one of our nearest neighbors. It is at a distance of about 700 kpc, and is more than 20 kpc across. Notice the two companions, including M32. We think that it is very similar to our own Milky Way.

Fig 17.3. (*Continued*) (c) NGC 4622 in Centaurus, at a distance of 60 Mpc. It is type Sab. (d) M101, in Ursa Major, at a distance of 5 Mpc, type Sc. (e) A section of M101. Note the spiral arms of collections of bright patches, probably HII regions and OB associations, with dust lanes superimposed. (f) M33, in Triangulum, is type Sc. (g) NGC 1232 (Sc).

Fig 17.3. (*Continued*) (h) M83 (the Southern Pinwheel) in Hydra, is 3 Mpc away, and is type Sc. (i) M104, (Sab) the Sombrero Galaxy, looks like a Mexican hat viewed almost edge-on. This is an edge-on spiral with a prominent bulge. The dark lane is the collective extinction of dusty molecular clouds in the disk. (j) The edge-on spiral NGC 4565, in Coma Berenices, is type Sb. [(a) Image from the OSU Bright Spiral Galaxy Survey; (b)–(f), (h), (j), (k) NOAO; (g), (i) ESO]

(h)

(i)

(j)

classification. This is done by adding a I through V following the Hubble classification, with I being the brightest (just as for stars). Efforts are still underway to find other properties of spirals that correlate with luminosity class. In this way, the luminosity of a galaxy can be determined without needing to know its distance. (Similarly, the luminosity class of a star can be determined from the shapes of certain spectral lines, allowing us to know the absolute magnitude of a star without knowing its distance.) Once the absolute

magnitude of a galaxy is known, and its apparent magnitude is observed, its distance can be determined.

An important feature of spirals is the obvious presence of an interstellar medium – gas and dust. Even when a spiral is seen edge-on, we can tell that it is a spiral by the presence of a lane of obscuring dust in the disk of the galaxy. The light from spirals contains an important contribution from a relatively small number of young blue stars, suggesting that star formation is still

(a)

(a)

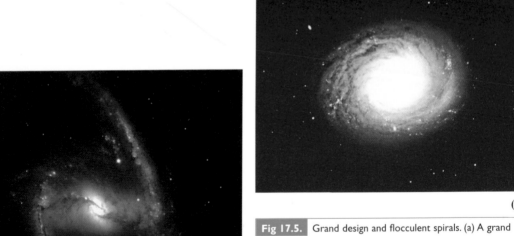

(b)

Fig 17.5. Grand design and flocculent spirals. (a) A grand design galaxy, M81, type Sb. (b) A flocculent spiral, M94, also a type Sb. [NOAO/AURA/NSF]

(b)

Fig 17.4. Various types of barred spirals. (a) NGC 1530, in Cameleopardalis, is type SBb. (b) NGC 1365, in Fornax, type SBc. [(a) NOAO/AURA/NSF; (b) ESO]

taking place in spirals. Where galaxies are found in a cluster, to be discussed in Chapter 18, approximately 80% of the galaxies are ellipticals. Outside of clusters, approximately 80% are spirals. Typical radii for the luminous part of the disk in spirals are about 10 to 30 kpc. Stellar masses of the galaxies we can see range from 10^7 to 10^{11} $M_\odot$.

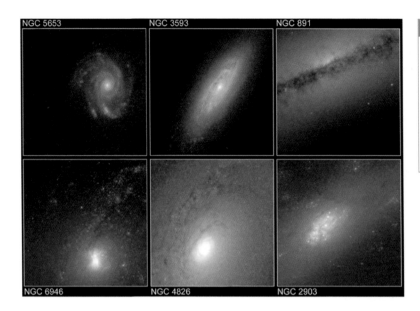

NGC 5653   NGC 3593   NGC 891

NGC 6946   NGC 4826   NGC 2903

Fig 17.6. HST images of six spiral galaxies, showing regions of star formation. These are false color images made through three separate filters. The red represents the Paschen line from H at 1.87 μm. Blue shows near IR emission (1.4 to 1.8 μm). Green is a mixture of the two. [STScI/NASA]

Fig. 17.6. shows the distribution of tracers of young stars in a selection of spirals. In studying general trends, we can get a good idea of the distribution of young stars as we go farther out in the disks of spiral galaxies. The luminosity of the disk falls off sharply with r, the distance from the center. We can approximately fit the observed falloff with an exponential expression. That is, if L_0 is the luminosity at the center, the $L(r)$, the luminosity at radius r, is given by

$$L(r) = L_0\, e^{(-r/D)} \qquad (17.2)$$

In this expression, D is called the *luminosity scale length* and gives a measure of the characteristic radius of the galaxy as seen in visible light. Typical values of D are about 5 kpc. This means that the luminosity of the disk of a spiral falls to $1/e$ of its peak value at $r = 5$ kpc.

17.1.3 Other types of galaxies

There is an additional type of galaxy that has certain features in common with spirals, but does not show spiral arms. This type is called *S0* ('S-zero') (Fig. 17.7). The bulge in an S0 is almost as large as the rest of the disk, giving the galaxy an almost spherical appearance. Some S0 galaxies also contain gas and dust, suggesting that they belong in the spiral classification. However, most

Table 17.1. Properties of spirals and ellipticals.

Property	Spirals	Ellipticals
Gas	yes	some
Dust	yes	some
Young stars	yes	none
Shape	flat	round
Stellar motions	circular rotation	random
Color	blue	red

Fig 17.7. M102, type S0. [NOAO/AURA/NSF]

Fig 17.8. The Magellanic Clouds. (a) The Large Magellanic Cloud (LMC) is 50 kpc from us. (b) The Small Magellanic Cloud (SMC), which is 65 kpc from us. [NOAO/AURA/NSF]

Fig 17.9. This galaxy, M82, is the scene of very unusual activity. At first it was thought that this galaxy was exploding. However, it just seems to have undergone a rapid wave of star formation. Galaxies like this are known as starburst galaxies, and we will discuss them more in Chapter 19. [NOAO/AURA/NSF]

S0 galaxies have no detectable gas. The role of S0 galaxies is still not well understood.

Some galaxies have no regular pattern in their appearance. These are called *irregular galaxies*. The Magellanic Clouds, shown in Fig. 17.8, are irregular companions to our own galaxy. Irregulars make up a few percent of all galaxies. We distinguish between two types of irregulars: Irr I galaxies are resolved into stars and nebulae; Irr II galaxies just have a general amorphous appearance. *Lenticular galaxies* have an irregular elongated structure. *Ring galaxies* have prominent bright rings around their centers.

Peculiar galaxies have a general overall pattern, but also have some irregular structure indicative of unusual activity in the galaxy. An example is shown in Fig. 17.9.

There are also types of galaxies that are characterized by a very bright nucleus. *Seyfert galaxies* are named after their discoverer *Carl Seyfert*, who

reported their existence in 1943. These are spiral galaxies with a bright small nucleus. The spectra show broad emission lines, an indication of a very hot or energetic gas. Seyferts make up about 2 to 5% of all spiral galaxies. An example of a similar phenomenon in ellipticals is found in *N galaxies* (where the N stands for 'nucleus'). There is also a class of galaxies that give off very strong radio emission, *radio galaxies*. All of these *active galaxies* will be discussed in Chapter 19.

17.2 | Star formation in galaxies

An important opportunity provided by other galaxies is the opportunity to study star formation in a variety of environments. We can see how various factors, such as the type of galaxy, its metallicity, and the interstellar radiation field within the galaxy, affect star formation. We discussed the basic idea of star formation in the interstellar medium in Chapters 14–16. We would like to apply the ideas we developed when studying star formation in our galaxy to help us understand other galaxies. In turn, our understanding of other galaxies will help our analysis of our galaxy. We can ask a number of questions about star formation in galaxies.

(1) Does star formation take place in molecular clouds, as it does in the Milky Way? If so, are

the properties of those clouds similar to the ones we find in our galaxy?

(2) What is the large-scale distribution of star-forming material within the galaxy?

(3) What is the distribution of newly formed stars?

(4) Does the mix of stellar masses (the initial mass function) appear to vary from galaxy to galaxy, or within galaxies?

(5) How does the star formation rate vary from galaxy to galaxy and within galaxies?

To study molecular material, we still need to observe trace constituents, such as CO, using millimeter telescopes. For more distant galaxies, angular resolution with single dishes is a limiting factor, but millimeter interferometers are becoming more powerful, and give much more useful resolutions. To study hot cores and protostars, near infrared observations are useful, and angular resolution is not a serious problem. To study recent sites of massive star formation, we look for HII regions, either by their Hα emission (through Hα filters), or by observing their radio continuum emission, at centimeter wavelengths. The latter requires interferometers for good angular resolution.

17.2.1 Star formation in the Large Magellanic Cloud

The Magellanic Clouds (Fig. 17.8) are the closest galaxies in which we can study star formation. At the 50 kpc distance of the LMC, a 1 parsec extent, subtends an angle of 4 arc sec. So the 20 arc sec resolution of typical millimeter telescopes corresponds to a size of 5 pc. This is adequate for studying giant molecular clouds with extents of tens of parsecs (though ultimately arrays like ALMA, shown in Fig. 4.32, will be needed to study small clouds and cloud cores). The Swedish-ESO-Submillimetre Telescope (SEST, Fig. 4.28b), placed in Chile, has been particularly well suited to study the Magellanic Clouds.

As Fig. 17.10 shows, the LMC is the site of many HII regions. As we discussed in Chapter 15, this implies the existence of recent massive (O and B) star formation. Fig. 17.10(a) shows an optical image of the brightest HII region in the LMC. Because of its appearance, it is called the Tarantula Nebula. The exciting star is 30 Dorado, and this is some-

(a)

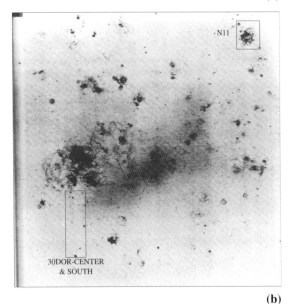

(b)

Fig 17.10. HII regions in the LMC. (a) Optical image of the Tarantula Nebula, 30 Dorado. (b) Image of the LMC, taken with an Hα filter, shows the locations of the HII regions. The two rectangular boxes show the regions of detailed study in Fig. 17.11. [(a) STScI/NASA; (b) Monica Rubio, University of Chile]

times referred to as the 30 Dor complex. An Hα image of the whole LMC is shown in Fig. 17.10(b), showing HII regions all over the LMC. We will look in a little detail at the large dark cloud (visible in Fig. 17.8b) that runs south from 30 Dor, and an isolated HII region in the northeast corner of the image, known as N11. Both of these regions are indicated by the rectangular boxes on Fig. 17.10(b)

Maps of the CO emission from these regions of detailed study are shown in Fig. 17.11, both by

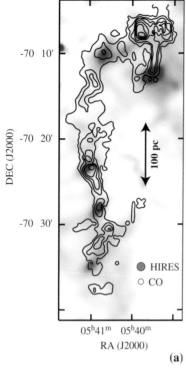

Fig 17.11. CO images (contours) of star forming regions in the LMC, with observations done on the SEST, on La Silla, Chile. These are superimposed on FIR images from IRAS (the gray scale). (a) The 30 Dor Complex. Notice the complex of giant molecular clouds. More detailed maps show two dozen GMCs in this complex, with properties very much like Milky Way GMCs. FIR peaks all have associated strong CO emission, suggesting there are dense cores there. There are also CO peaks away from FIR emission. Perhaps these are clouds that are not as far along the star formation process. (b) The N11 region. In this more open ring, the individual clouds are more easily seen. HIRES refers to a type of image processing, which enhances the angular resolution [author].

(determined from the virial theorem) of a few times $10^5 M_\odot$. These are very much like giant molecular clouds, and this whole long dark cloud is like a GMC complex in the Milky Way. The FIR image shows a number of embedded regions where dust is being heated by ongoing or recent star formation, just as for massive star forming regions in the Milky Way. It appears that the Tarantula Nebula is at the northern end of this complex, and is in a region where there are more young stars, but less molecular clouds. It has been suggested that this is a site of sequential star formation, also similar to situations found in the Milky Way.

The N11 region, Fig. 17.11(b), has a more open appearance, so it is easier to see the structure. Here we see a ring of clouds (with an extension to the northwest). These clouds also look like Milky Way GMCs, and they also have masses of a few times $10^5 M_\odot$. All of these clouds have internal velocity dispersions comparable to Milky Way GMCs. This grouping has a similar appearance to the Orion region (partly shown in Fig. 15.4). So, it appears that star forming regions in the LMC are very similar to those in the Milky Way, despite the many differences between the LMC and the Milky Way.

17.2.2 Star formation in spiral galaxies

When we look at the various images of spirals (Fig. 17.3), we see that the spiral arms are traced by strings of bright HII regions. This suggests that the spiral arms are sites of enhanced massive star formation. We would like to apply the ideas we developed when studying star formation in our galaxy to help us understand other spiral galaxies. In turn, our understanding of other galaxies will help our analysis of our galaxy. We can ask a

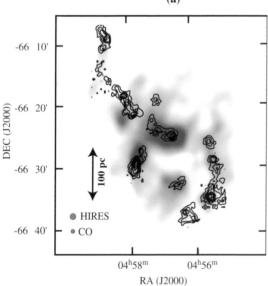

themselves, and superimposed in the FIR images. We first look at the 30 Dor complex (Fig. 17.11a). The CO emission shows a complex of molecular clouds that extend over part of an arc for about 600 pc. A more detailed picture shows that this is composed of some two dozen clouds, each with an extent of tens of parsecs, and each with a mass

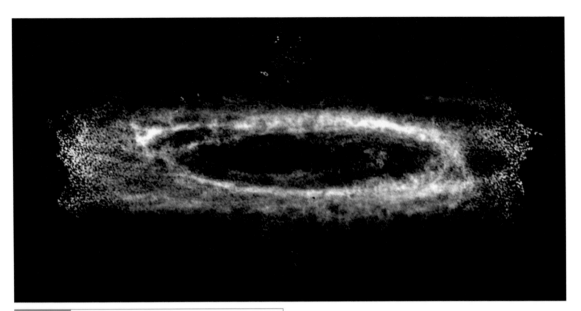

Fig 17.12. 21 cm map of the Andromeda Galaxy, M31, made using the Westerbork interferometer in the Netherlands. [Elias Brinks, Sterre wacht, Leiden University]

number of questions about star formation in spiral galaxies.

(1) What is the large-scale distribution of star forming material? How does it vary in the disk with distance from the center? How does it vary with distance from the central plane of the disk?
(2) Is the interstellar medium concentrated into the spiral arms?
(3) How do the sizes of molecular clouds compare with those in our galaxy? Are the physical conditions within the star forming clouds the same?

In studying the interstellar medium of our galaxy, or any other, radio observations play an important role. Except for the nearest galaxies, single-dish radio observations do not provide much spatial detail. However, with the extensive use of interferometers we have now obtained very detailed maps of many galaxies. Continuum observations can be used to study the positions of HII regions and young supernova remnants, both signs of relatively recent star formation. Studies of spiral structure have been limited by poor angular resolution for single-dish studies. However, sufficient resolution is available to study nearby galaxies.

M31 is the nearest spiral, so it provides the best opportunity for studying the interstellar medium in detail. At 700 kpc distance, 1 arc sec corresponds to a linear size of 3 pc, so a 100 pc long giant molecular cloud would subtend an angle of 30 arc sec. This corresponds to the resolution of typical single-dish millimeter telescopes. To study the large-scale distribution of molecular material, we could use single-dish observations, but to study individual clouds we have to use millimeter arrays. To study the HI, single-dish observations at 21 cm do not give sufficient resolution, so we must use arrays.

Fig. 17.12. shows an interferometer map of the large-scale distribution of HI in M31. The large-scale molecular distribution is shown in a single-dish CO map in Fig. 17.13. One of the problems in studying the spiral structure in M31 is that it is tilted so it is hard to trace accurately the spiral arms as they would appear if we were looking from overhead. The large-scale distribution of star forming regions in M31 is shown by the FIR image in Fig. 17.14.

The single-dish CO observations of spirals reveal a sharp falloff in brightness with radius, similar to that of the visible light. The falloff in CO emission may indicate the true gas distribution. However, it may be due to the fact that the gas cools, and therefore radiates less strongly where there are fewer stars to heat it. There may

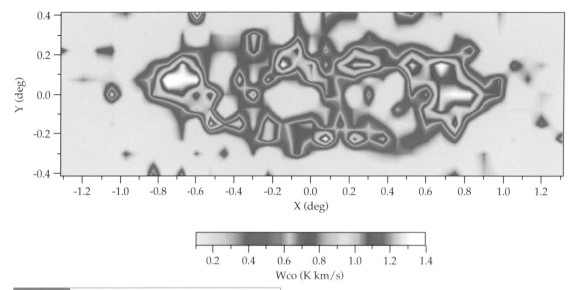

Fig 17.13. Large-scale distribution of CO emission in the galaxy M31. This map was made using a 1 m diameter telescope at the Center for Astrophysics (Harvard). [Thomas Dame, CFA/Dame, T. et al., Astrophys. J., **418**, 730, 1993, Fig. 1(a)]

also be significant amounts of molecular gas in the outer parts of spiral galaxies.

In studies of several spirals it is found that the relative amounts of molecular and atomic hydrogen vary significantly. These variations occur both within a galaxy and from galaxy to galaxy. Within a galaxy, the general trend is to have the molecular hydrogen abundance fall off faster with radius than the atomic hydrogen abundance. We find some galaxies in which the molecular hydrogen makes up over half of the interstellar medium, and others in which it seems to be less than 10%, as determined from a deficiency of CO emission. In galaxies that seem to be deficient in molecular hydrogen, we still don't have observations with sufficient resolution to tell us whether this is because they contain fewer molecular clouds than the other galaxies, or whether the clouds are less dense. Also, even galaxies that appear to be deficient in molecular clouds have O stars. This tells us that we still do not fully understand the connection between molecular clouds and massive star formation.

17.3 | Explanations of spiral structure

It is actually quite surprising that we see any spiral structure in galaxies. The differential rotation of a galaxy should smear out any pattern on a time scale comparable to the orbital period. For the Sun, the orbit period is about 200 million years. We think that the age of the galaxy is about 10 billion years. This means that any initial spiral

Fig 17.14. IRAS map of M31. [NASA]

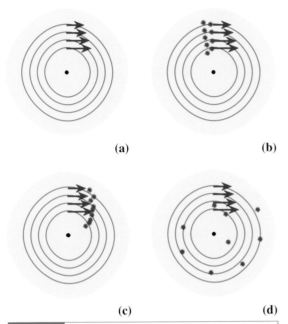

(a)   **(b)**

(c)   **(d)**

Fig 17.15. Scenario for temporary spiral structure. (a) The dark area is the center of the galaxy, and the arrows indicate the orbital speeds of material at different distances from the center. (b) A large-scale burst of star formation takes place. (c) The differential rotation stretches the stars out, producing part of a spiral arm. (d) After a few rotations the arm is stretched out so much that we can no longer detect its presence.

A scenario for temporary arm formation is shown in Fig. 17.15. For one reason or another, star formation starts in one region. It may even spread via some of the triggering mechanisms discussed in Chapter 15. The region with new stars is then stretched out by the differential rotation into a piece of a spiral pattern. This may explain why we see pieces of spiral features but no complete arms in our galaxy. In this scenario each 'arm' lives for a short time, and new ones are always forming.

If spiral patterns persist for many revolutions of a galaxy, then it may be that the pattern and the matter itself are moving at different speeds. At first this may seem strange, but we can use an analogy to see how it might work (Fig. 17.16). Suppose we have cars moving along a two-lane highway, and we are looking from above, in a traffic helicopter. A truck breaks down in the right lane, causing a traffic jam. If we look from above, we see cars backed up for some distance behind the truck. The density of cars is higher for this region. Far behind the truck, the cars are still moving at their normal speed, and after the cars squeeze past the truck, they will resume their normal speed. If we come back a few minutes later, we will see the same pattern of cars. However, the specific cars involved in the buildup will be different. The cars that we saw originally will be far down the road. In this case, the cars are moving along, but the pattern stays in the same place because the truck stays in the same place.

pattern would have had ample time to smear out. Therefore, spiral arms must be temporary, or there must be a way of perpetuating them.

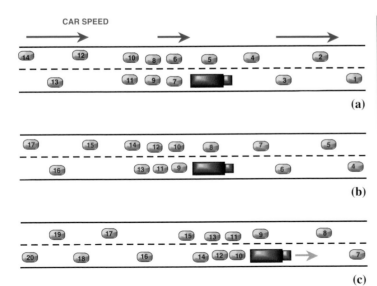

CAR SPEED

(a)

(b)

(c)

Fig 17.16. Car analogy to density wave. (a) A truck (shaded) is broken down in the right lane. Far in front and behind the truck, cars have the normal speed and low density. Just behind the truck, the density of cars goes up, and their speed goes down. (b) As time goes on, cars slowly squeeze by the truck. The basic pattern is retained. However, as the numbers on the cars show, different cars are stuck behind the truck than in the earlier frame. We therefore have a density concentration along the highway while the individual cars are not permanently attached to the concentration. (c) This concentration can even move. Instead of being stuck, suppose the truck is moving slowly. The pattern moves along with the truck, while the individual cars move at higher speeds.

Now suppose the truck is moving at a slow speed. Again, there will be a buildup behind the truck as cars squeeze into one lane to move past the truck. As in the case of the stationary truck, we see the cars moving at their normal speed. However, now the pattern moves. The speed of the pattern is not related to the speed of the cars – it is determined by the speed of the truck. The truck is responsible for the pattern. The cars simply respond to the presence of the truck. This is the type of situation in which the pattern (the traffic jam) can move at one speed, and the matter (the cars) at another.

There is a theory that the same type of situation can occur in spiral galaxies. Since the matter moves at a different speed from a density buildup, the theory is called the *density wave* theory. In a galaxy, the dynamics is controlled by the halo, which contains most of the mass. The bright spiral arms contain a small fraction of the mass of the galaxy, and represent material which is orbiting at its normal speed, but responding to the gravitational effects of the asymmetric distribution of the stars in the halo. The mathematician *C. C. Lin* has shown that once a spiral pattern is established in a galaxy, it can sustain itself for a long time in this type of wave. Eventually, the wave will die out and a new one must be generated.

In the density wave picture, the visible arms are a result of a gathering of interstellar matter. When high enough densities are reached, star formation may take place. One scenario for this is illustrated in Fig. 17.17. A large HI cloud, or a group of small clouds, approaches an arm at a speed of about 100 km/s relative to the arm. (In this case, the arm may be moving at 100 km/s, and the matter overtaking it at 200 km/s.) The arm acts like a gravitational potential well, causing material to take more time to traverse the arm than a similar distance between arms. The matter entering an arm will leave its circular path, and have some motion along the arm, before finally emerging. It should be noted that, even if the density waves don't cause strong visible arms, they alter the orbits, resulting in noncircular motions. Some of the results of critical calculations of density waves are shown in Fig. 17.18.

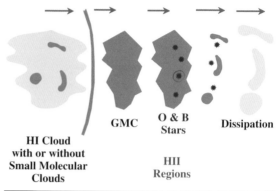

HI Cloud with or without Small Molecular Clouds   **GMC**   **O & B Stars**   **HII Regions**   **Dissipation**

Fig 17.17. Scenario for molecular clouds tracing spiral arms. If the gas is circulating faster than the spiral pattern, as with the density wave picture, the gas can overtake the spiral arms from behind. Before reaching the arms, the gas is in the form of low density HI cloud, possibly containing some small molecular clouds. The entry into the arm slows the HI clouds down and compresses them. It may also gather together small molecular clouds. In any event, giant molecular clouds form. These clouds give birth to O and B stars (as well as to all the other types). The radiation from the O and B stars disrupts the clouds.

As the front of the cloud enters the arm, it slows down. The only way for the back of the cloud to know that this has happened is for a pressure wave (sound wave) to travel from the front of the cloud to the back. However, the speed of the cloud is greatly in excess of the speed of sound within the cloud (a few kilometers per second). The back of the cloud doesn't receive the message to slow down until it has almost overtaken the front. The cloud has been compressed. (This may also gather small molecular clouds into giant molecular clouds.) At this point, we see the cloud as part of a dust lane, explaining why dust lanes appear at the back of spiral arms. Since the cloud has been compressed, star formation is initiated. Massive (as well as low mass) star formation takes place. The bright stars don't live very long, so we see them only over a small range, forming the bright chains that mark the fronts of the arms.

The massive stars also have the effect of driving the clouds apart. This can be through the effect of stellar winds, expanding HII regions, and supernova explosions. The clouds dissipate. The material again resembles that which originally entered the back of the arm. It remains this way

Fig 17.18. The results of theoretical calculations demonstrating the ability of galaxies to amplify small perturbations into a well defined spiral pattern. [Alar Toomre, MIT]

until it overtakes the next arm. According to this picture, giant molecular clouds should be seen almost exclusively in spiral arms. It is a good observational test of the theory, but, as we have seen, the limited angular resolution of single dishes makes these observations difficult.

We have already seen that spirals seem to fall into two categories: grand design and flocculent. It may be that the underlying cause of the two types of spiral structure is different. One observation to support this is that grand design spirals seem to occur in galaxies with internal bars or with nearby neighbors. It is suspected that the tidal interactions between the galaxy and the neighbors set up the spiral density wave in the mass distribution of the galaxy. As the gas streams through the density wave, it is compressed into giant molecular clouds, which give birth to stars and then dissipate, as shown in Fig. 17.17. In flocculent spirals, the spiral structure may be a series of temporary patterns, in which the results of local bursts of star formation are drawn into a spiral by the differential rotation of the galaxy.

The density wave theory seems to be best applied to grand design galaxies. We will briefly look at one that has been studied in detail, M51 (the 'Whirlpool'). This is shown in Fig. 17.19. In

Fig 17.19. Optical image of the Whirlpool Galaxy, M51, 9.6 Mpc distant [NOAO/AURA/NSF].

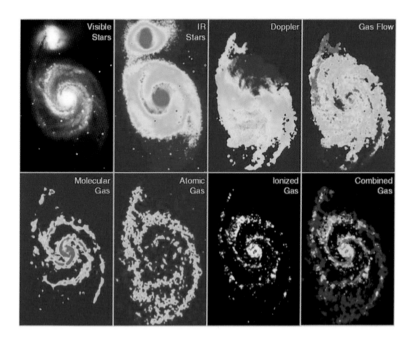

Fig 17.20. Tracers of the spiral pattern in M51. 'Visible' is R band taken at Mt Palomar. 'IR' is K band, taken at Kitt Peak. 'Doppler' is Hα velocities, taken at Mt Palomar. 'Gas Flow' are the residual velocities from 'Doppler' with the average rotation taken out. This gives the streaming motions. 'Molecular Gas' is CO emission from BIMA. 'Atomic Gas' is HI emission with the VLA. 'Ionized Gas' is Hα from Kitt Peak. 'Combined Gas' is HI in blue, Hα in green and CO in red, to compare the locations of these components. [Stuart Vogel, University of Maryland; HI emission, Arnold Rots, CFA]

M51 we can see clearly the location of the dust lanes with respect to the bright arms. The lanes are on the inside edges of the bright arms. Remember, in the density wave picture, the interstellar gas is swept up, forming giant molecular clouds. These giant molecular clouds are visible as the dust patches. Eventually, these clouds give birth to O stars, which illuminate HII regions. The O stars and HII regions trace out bright arms. The density wave picture therefore makes a specific prediction of the relative positions of the dust and bright arms. Radio continuum maps of M51 show that the synchrotron emission is strongest in the direction of the dust lanes. At first we might expect the synchrotron emission to be strongest where there are the most supernovae, on the bright side of the arm. However, the compression of the interstellar medium on the dark dust side of the arms produces relatively strong magnetic fields, and the synchrotron emission becomes stronger when the magnetic field strengthens.

Fig. 17.20. shows the results of a detailed study of the spiral structure made by Stuart Vogel (University of Maryland). One of the features that we see is that the CO emission comes from the inside of the arms, consistent with predictions of density wave theory. Also note

that the molecular gas dominates the inner disk and the atomic gas dominates the outer part of the disk.

The face-on appearance of M51 means that we can easily see the relative placement of features. However, we have very little velocity information, since all of the galactic rotation is perpendicular to our line of sight. We see that the largest Doppler shifts are in edge-on spirals, but we cannot make out any spatial structure from their edge-on appearance. Studies of more inclined spirals, such as M81, shown in Fig. 17.5(a), have been useful in testing velocity shift predictions of the density wave theory.

17.4 | Dark matter in galaxies

When we look at a galaxy in visible light we obviously see the most luminous objects. However, some of the mass may not be luminous. It could be there but hard to detect. The only sure way to trace out the total mass, whether it is bright or dark, is to study its gravitational effects. In a galaxy, the easiest way to study the gravitational forces is to measure the rotation curve. We have already discussed the rotation curve in our galaxy in Chapter 16.

that $v(r)$ stays roughly constant out as far as we see luminous material. This immediately tells us that the mass doesn't fall off as fast as the luminosity. The masses that are found are as high as $2 \times 10^{12} M_\odot$. In many galaxies, no edge has yet been found. The rotation curves are still flat out to radii where the interstellar medium can no longer be detected, even using 21 cm observations, which show material farther out than the Hα emission.

Where can this matter reside? One possibility is that it is part of the disk. However, theoretical models show that such a large mass would gather the disk into a bar. The disks that we see would

(a)

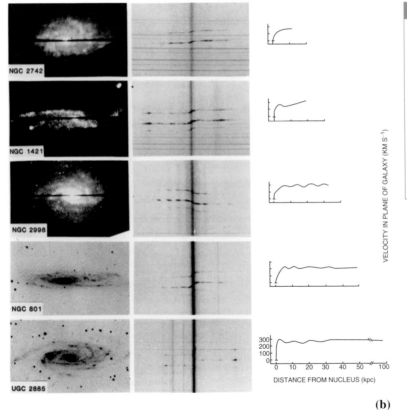

(b)

> **Fig 17.21.** (a) Vera Rubin.
> (b) Rotation velocity as a function of distance from the center of a galaxy, for five spiral galaxies. The fact that the rotation velocity is large at great distances from the center of the galaxy is taken as evidence for the existence of dark matter. [Vera Rubin, Department of Terrestrial Magnetism; (b) Reprinted with permission from Rubin, V., *Science*, **220**, 1339, Fig.1 © (1983) AAAS.]

Rotation curves can be determined from the measurement of Doppler shifts in spectral lines. This can be done with optical lines, such as Hα. Extensive studies have been carried out by *Vera Rubin* (Fig. 17.21 a), at the Carnegie Institution. Typical results are shown in Fig. 17.21 (b). We find

not be stable. The galaxies are more stable if the dark mass has a spherical distribution. This would mean that the dark matter is most likely in the halo. Remember, the halo is not a ring around the galaxy. It is a spherically symmetric mass distribution.

We can use the rotation curve to give us the mass distribution in the halo. If $M(r)$ is the mass interior to radius r, then $v(r)$ is related to it by the fact that the acceleration of gravity must provide the acceleration for a circular orbit (v^2/r), so

$$\frac{GM(r)}{r^2} = \frac{v^2(r)}{r}$$

Solving for $M(r)$ gives

$$M(r) = \frac{rv^2(r)}{G} \tag{17.3}$$

We can relate $M(r)$ to the density distribution $\rho(r)$ by the equation of mass continuity (equation 9.33), which was one of the equations of stellar structure:

$$dM/dr = 4\pi r^2 \rho(r) \tag{17.4}$$

Solving for $\rho(r)$ gives

$$\rho(r) = \left(\frac{1}{4\pi r^2}\right)\left(\frac{dM(r)}{dr}\right) \tag{17.5}$$

If we take $v(r) = v_0$, a constant, then differentiating equation (17.3) with respect to r gives

$$\frac{dM(r)}{dr} = \frac{v_0^2}{G} \tag{17.6}$$

Finally, substituting equation (17.6) into equation (17.5) gives

$$\rho(r) = \frac{v_0^2}{4\pi G r^2} \tag{17.7}$$

The density in the halo therefore falls off as $1/r^2$. It is not nearly as fast as the exponential falloff in the light of the disk.

With a $1/r^2$ falloff in density, it might seem that there is not much matter very far out. However, if we divide the galaxy into spherical shells, each of thickness dr, the volume of each shell is $4\pi r^2 dr$. This means that the mass of each shell is constant! As far out as the rotation curve remains flat, we are adding significant amounts of matter to the galaxy. This is particularly important, since the rotation curves seem flat as far out as we can measure. It may also be that there is still a significant amount of matter beyond those points.

What is the dark matter in the halo? We can get a clue by looking at *mass-to-light ratios* of various objects. This is the ratio of mass, expressed in solar masses, to the luminosity, expressed in solar luminosities. By definition, the mass-to-light ratio of the Sun is one. The mass-to-light ratios of main sequence stars are given in the mass–luminosity relationship discussed in Section 5.5. If we know the mass-to-light ratio of a galaxy, we can see what types of objects have similar mass-to-light ratios. For spiral galaxies, the mass-to-light ratio is 1:3 near the center. This means that most of the mass near the center probably comes from normal stars. However, near the edge of the visible disk, the ratio climbs to 20:1, and is above 100:1 for the farthest points to which rotation curves have been measured.

It has been suggested that the halos might consist of faint, old red stars. These stars would have masses less than $1M_\odot$. There is even some direct observational evidence for such stars in the halos of nearby galaxies. The mass-to-light ratio of such stars is about 20:1. They might therefore provide the dark matter out to the edge of the visible disk, but something else is needed beyond that. More recent observations seem to rule out all nuclear burning material as a significant part of the halo. Some astronomers have suggested that a lot of mass could be hidden in Jupiter-sized objects, which are obviously not very luminous.

Recently a new technique has been employed to search for *massive compact halo objects (MACHOs)*. If these objects are massive and small then they should bend starlight passing close to their surface. The bending of light is a prediction of general relativity, as discussed in Chapter 8. When a MACHO passes in front of a star in another nearby galaxy (such as the LMC) there will be a very brief change in the intensity of that light. This is known as *microlensing*. Systematic searches for microlensing events have been carried out. There have been some detections of such events, in which light from stars in the LMC shows variations suggestive of microlensing. However, there is still a question about whether we are seeing microlensing due to a star in the halo of our galaxy, or due to one near the target star in the LMC. (The LMC is used for such studies because it provides us with a large number of stars to study, and it is far from our galactic plane, so, if we see lensing events, it is more likely that they are caused by objects in the halo of our galaxy, than the disk.)

It has been suggested that the dark matter could be in the form of neutrinos, if neutrinos have a small rest mass. Remember, we saw in Chapter 9 that there is growing evidence that the neutrino mass is in about the 10^{-2} eV range This is small enough to have been overlooked in previous experiments. However, there are so many neutrinos in the universe that even a small mass for each neutrino can add a lot of mass to a galaxy. The idea of neutrinos as dark matter in galaxies is still not generally accepted.

The rotation curves of galaxies are our first evidence for the existence of dark matter. We know it is there because we can measure its gravitational effects, but we cannot see it. When astronomers were not sure the matter was there, there was a 'missing mass problem'. However, the mass isn't missing; it is just nonluminous. As we go to larger scales in the universe, we will see that we will find evidence that dark matter becomes more and more important.

Chapter summary

In this chapter we looked at the properties of various types of galaxies.

Elliptical galaxies have no evidence for recent star formation. However, the metal abundances are high. Ellipticals are classified according to the eccentricity of their appearance. Spirals have an evident interstellar medium, as well as O and B stars, meaning that star formation is still taking place. Spirals are classified according to the tightness of the spiral arms and the relative sizes of the nucleus and disk.

We applied some of the ideas of star formation, discussed in previous chapters, to look at star formation in the LMC and in spirals. We also looked at how the density wave theory might explain spiral structure.

In studying rotation curves of galaxies we found that the masses of galaxies are greater than would be determined from luminous material. It has been suggested that most of the matter is distributed in a spherically symmetric halo.

Questions

17.1. If we see a spiral galaxy edge-on, how do we know that it is a spiral?

17.2. (a) Why is it not likely that single spirals formed from single ellipticals? (b) Why is it not likely that ellipticals formed from spirals? (Hint: Think of the effects of rotation.)

17.3. Compare the properties of dwarf ellipticals with the properties of globular clusters in our own galaxy.

17.4. What features of S0 galaxies make them similar to (a) spirals, (b) ellipticals?

17.5. Assume there is some way, either by spectroscopy or by some aspect of the shape of the galaxy, to determine the luminosity class for a galaxy. How can this information be used, along with other observations, to determine the distance to a galaxy?

17.6. How do the relative abundances of atomic and molecular hydrogen vary within a spiral galaxy?

17.7. What parts of the interstellar medium would you expect to best trace out spiral arms? Explain your answer.

17.8. What are the differences between grand design and flocculent spiral galaxies (a) in their appearance and (b) in the scenario by which the spiral arms are formed and maintained?

*17.9. In equation (17.4) we used the relationship between mass and density for a spherically symmetric system. However, spiral galaxies have a disklike appearance. Why is the use of equation (17.4) valid?

17.10. How does the density wave theory help us explain spiral structure?

Problems

17.1. Given the luminosity profile in equation (17.1), what would the luminosity be at (a) $r = r_0$, (b) $2r_0$?

17.2. Given the luminosity profile in equation (17.2), what is the luminosity at (a) $r = D$, (b) $r = 2\,D$? (Express your answer in terms of L_0.)

17.3. Given the luminosity profile in equation (17.2), for a given galaxy how far out would you have to go before you have 95% of the galactic luminosity? Express your answer as a multiple of D.

17.4. Given the luminosity profile in equation (17.2) and the density profile in equation (17.7), find an expression for the mass-to-light ratio as a function of distance from the center of the galaxy r. Assume that the ratio is unity at $r = D$.

17.5. The Andromeda Galaxy is at a distance of 700 kpc. (a) Suppose we observe the CO(2−1) transition at a wavelength of 1.3 mm, with a 30 m diameter telescope. What is the linear resolution?
(b) Suppose we use a millimeter interfero-meter with a maximum baseline of 1 km?
(c) Repeat these calculations for the Whirlpool Galaxy, at a distance of 9.6 Mpc.

17.6. Suppose we are observing a galaxy to measure its rotation curve. By how much would the wavelength of the 21 cm line shift as we moved from the center to the edge of the galaxy, which has an orbital speed of 200 km/s (a) if we are in the plane of the galaxy, (b) if the galaxy is tilted at a 45° angle to our line of sight?

17.7. If we measure the rotation curve of a galaxy, how far out would the orbital speed have to be 300 km/s, to give the largest measured mass, $2 \times 10^{12} M_\odot$?

17.8. For a galaxy, with a flat rotation curve, whose density distribution is given by equation (17.7) out to a radius of 20 kpc, calculate (leaving your answer in terms of v_0) (a) the total mass out to 20 kpc, (b) the gravitational potential energy, (c) the rotational kinetic energy. (d) Compare (b) and (c) and note if they obey the virial theorem.

17.9. Calculate the mass of a galaxy with a flat rotation curve, with $v = 300$ km/s, out to $r = 20$ kpc. (Express your answer in $M_\odot$.)

17.10. If neutrinos have a rest mass equal to 10^{-4} of the electron rest mass, how many neutrinos do you need to give $10^{12} M_\odot$?

Computer problems

17.1. Use the mass–luminosity relationship discussed in Chapter 5 to draw a graph of the mass-to-light ratio for main sequence stars as a function of spectral type.

17.2. Draw a graph of the luminosity profile, $L(r)/L(0)$ vs. r, for elliptical galaxies (equation 17.1), for $r_0 = 5$, 10, 20 kpc.

17.3. Draw a graph of the luminosity profile, $L(r)/L_0$ vs. r, for spiral galaxies (equation 17.2), for $D = 2.5$, 5, 10 kpc.

17.4. Use the rotation curves in Fig. 17.21(b) to determine the mass interior to the largest radius out to which the rotation curve has been measured. (Express your answer in $M_\odot$.)

Chapter 18

Clusters of galaxies

18.1 | Distribution of galaxies

If we look at the distribution of galaxies, such as that shown in Fig. 18.1, we see that the galaxies are not randomly arranged on the sky. Among the patterns we see distinct groupings, called *clusters of galaxies*.

Clusters are interesting for a number of reasons. They may provide us with clues on the formation of galaxies themselves. This is especially true if, as many think, cluster-sized objects formed first and then broke into galaxy-sized objects. (The alternative view is that galaxies formed first and then gathered into clusters.) Clusters also pose us with interesting dynamical problems, including a dark matter problem of their own. Finally, when we reach the scale of clusters of galaxies, we are beginning to reach a scale which has some significance in the overall structure of the universe.

The cluster of galaxies to which the Milky Way belongs is called *the Local Group*. As clusters go, it is not a very rich one. Besides the Milky Way, it contains several irregulars, including our companions, the Large and Small Magellanic Clouds, the spiral galaxies M31 and M33, and a number of dwarf ellipticals. Other nearby clusters are named by the constellation in which they are centered. For example the *Virgo, Coma, Hercules* and *Centaurus clusters* are shown in Fig. 18.2.

18.2 | Cluster dynamics

Just as with clusters of stars, clusters of galaxies may be isolated collections of masses interacting

gravitationally. As such, they are interesting systems to understand. In addition, by studying the gravitational interactions, we learn about the masses of the individual galaxies and the cluster as a whole. It has been found that the number of galaxies per unit area in a cluster falls off approximately as $\exp[-(r/r_0)^{1/4}]$. This is the same as the surface brightness in elliptical galaxies. We know that ellipticals are dynamically relaxed systems. If clusters and ellipticals have similar density distributions, then this suggests that some of the clusters are dynamically relaxed also.

Example 18.1 Crossing time for a cluster of galaxies

The time for a galaxy to cross from one side of a cluster to the other is called the crossing time. Find the crossing time for a cluster of galaxies with an extent of 1 Mpc, and galaxies moving at 10^3 km/s.

SOLUTION

The time for a galaxy to cross is the diameter divided by the speed, so

$$t_{\text{cross}} = (10^6 \, \text{pc})(3 \times 10^{18} \, \text{cm/pc})/(10^8 \, \text{cm/s})$$

$$= 3 \times 10^{16} \, \text{s}$$

$$= 1 \times 10^9 \, \text{yr}$$

We think that clusters of galaxies have been around for over 10^{10} years (the age of our galaxy as determined from globular cluster HR diagrams). If they were not gravitationally bound they would have had many crossing times to evaporate. We therefore think that clusters are gravitationally bound. They have also had sufficient time to become relaxed, so we can apply the virial theorem to analyze their internal motions (Fig. 18.3).

Fig 18.1. Distribution of galaxies. This is a two-dimensional view as seen from the Earth. It is from the APM Galaxy Survey, which detected over two million galaxies, covering approximately one-tenth of the whole sky. This image covers a region of about $100° \times 50°$ about the South Galactic Pole. The intensities of each pixel are scaled by the number of galaxies in that pixel. [Steven Maddox, Nottingham University]

In terms of the internal motions, the virial mass is given by equation (13.47):

$$M = \left(\frac{5}{3}\right)\frac{\langle v^2 \rangle R}{G}$$

In terms of observed Doppler shifts, the virial mass is given by equation (13.52)

$$M = \frac{5 \langle v_r^2 \rangle R}{G} \tag{18.1}$$

Example 18.2 Virial mass of cluster

For the Coma cluster we have $v_{rms} = 860$ km/s and a cluster radius of 6.1 Mpc. Find the virial mass of the cluster.

SOLUTION

From equation (18.1) we have

$$M = \frac{(5)(8.6 \times 10^7 \text{ cm/s})^2 (6.1 \text{ Mpc})(3.1 \times 10^{24} \text{ cm/Mpc})}{(6.67 \times 10^{-8} \text{ dyn cm}^2/\text{g}^2)}$$

$$= 1 \times 10^{49} \text{g}$$

$$= 5 \times 10^{15} M_\odot$$

(a)

(b)

Fig 18.2. Nearby clusters of galaxies. (a) Virgo. This is a rich cluster, with a few thousand members, but it is not very strongly concentrated towards the center. (b) Coma Berenices. This cluster contains more than 1000 galaxies, with a large number of types E and S0.

(c)

(d)

Fig 18.2. (*Continued*) (c) Hercules. This is a small irregular cluster at a distance of 120 Mpc. (d) Centaurus, in the southern hemisphere. [NOAO/AURA/NSF]

When we add up all the mass that we can see in the cluster, we find that it does not add up to the amount required by the virial theorem. This was originally done by using just the mass of the luminous matter in the galaxies. However, we saw in Chapter 17 that the halos of galaxies may contain dark matter. Even if we don't know what that dark matter is, we know it is there, and can add its mass to that of the luminous matter in each galaxy. However, clusters have many ellipticals and S0 galaxies which may not have massive

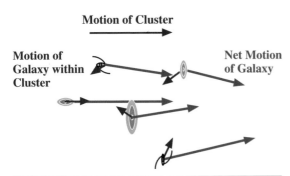

Fig 18.3. Motions of a galaxy in a cluster. The blue arrow shows the overall motion of the cluster. The green arrows show the motions of the galaxies within the cluster. The net motion for each galaxy (shown in red) is the vector sum of its internal motion with the overall motion of the cluster.

halos. We should only add the dark matter that we know is there, so we only add enough to account for the observed rotation curves in different types of galaxies. (We suspect that there may be more dark matter beyond the points where the rotation curves have been measured, because there is no evidence of the rotational velocities beginning to fall off.) Even this amount of dark matter is not enough to account for the virial mass.

Some of the mass may be in the form of low density gas within the cluster, but between the galaxies. This gas has either been ejected from the galaxies, or has fallen into the cluster. In either case, we would expect this gas to be very hot, about 10^7 K. It should be hot enough to give off faint X-ray emission. In fact, such emission is observed. In Figs. 18.4 and 18.5, we see X-ray images of two clusters. The hot gas contributes a significant amount of mass, but doesn't completely solve the problem.

There are still two possible solutions. One is that the individual galaxies have, as some have suspected, halos that go out even farther than the rotation curves can be measured. There is evidence to support this in studies of the interactions of binary galaxies. The advantage of a binary galaxy over the rotation curve studies is that the galaxies in a binary system are far enough apart to sample the full mass of each other. The other possibility is that the clusters contain their own dark matter. This matter may

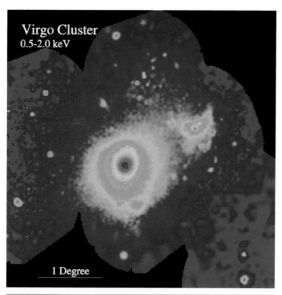

Fig 18.4.   ROSAT X-ray image of the Virgo cluster (Fig. 18.2a) shows the hot intracluster, intergalactic gas. [NASA/MPI]

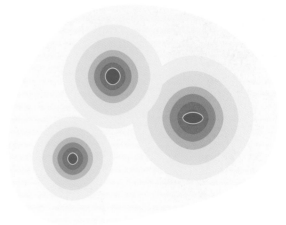

Luminous Matter

Dark Matter

Fig 18.6.   Possible distribution of dark matter in a cluster of galaxies. Each blue patch indicates the position of luminous matter within the cluster. The red areas indicate the locations of the dark matter. The darker the areas are, the greater the concentration of dark matter.

be the same as that in the halos of galaxies, but there just may be additional amounts in the cluster, not bound to any one galaxy. If a rich cluster has the mass implied by the virial theorem, then the mass-to-light ratio is about 200. This would be consistent with either the extended halos in

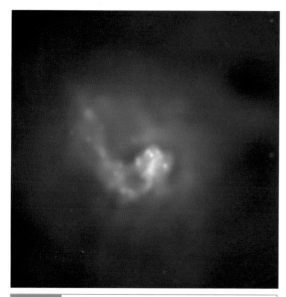

Fig 18.5.   Chandra X-ray image of the Centaurus cluster (Fig. 18.2d), showing more detail in the hot gas [NASA].

individual galaxies or the generally distributed dark matter. A possible distribution of dark matter in a cluster of galaxies is shown schematically in Fig. 18.6.

Another interesting feature about clusters of galaxies is that giant elliptical galaxies are found near the centers of some clusters (such as M87, in Fig. 17.2a). These galaxies are *central dominant* or *cD* galaxies. Some cD galaxies also seem to have multiple nuclei. It has been noted that the center of a cluster is the most likely place for galaxies to pass near each other. Some galaxy collisions result in galaxy mergers. Once a few galaxies have merged, they can swallow galaxies that pass too close. The process is called *galactic cannibalism*.

The whole subject of galaxy encounters is under active study. Numerical simulations have been carried out to find out what happens to the stars and gas in each of the two colliding galaxies, both for very close encounters and for direct collisions. The result of one such calculation is shown in Fig. 18.7. Some examples of interacting galaxies are shown in Fig. 18.8. As you can see, the calculations produce results that look like objects that are actually observed.

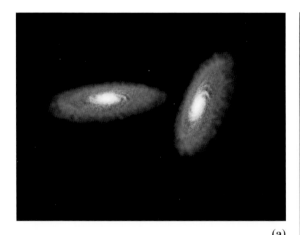

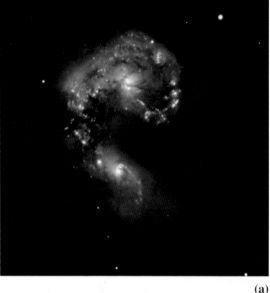

(a)

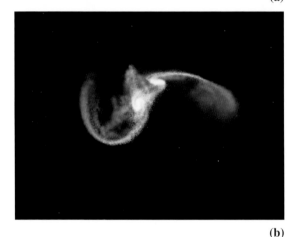

(b)

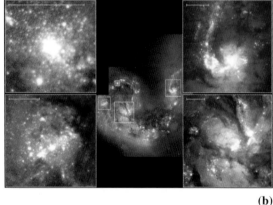

(b)

Fig 18.7. Interacting galaxies. Steps in the computer simulation. (a) The galaxies are far apart. (b) They are closer together, and the effects of the interaction are showing. Note that the tidal effects, tending to stretch the structures, are very important. In encounters, individual stars never actually touch. [Visualization by Frank Summers (STScI); simulation by Chris Mihos (Case Western Reserve University) and Lars Hemquist (Harvard University)]

Fig 18.8. Observations of interacting galaxies. (a) A pair of galaxies, NGC 4038 and 4039, which have an appearance similar to the simulation. Because of their appearance, these are called 'The Antennae'. (b) HST image of the nucleus of the elliptical galaxy NGC 1316. There is an unusually large number of bright young clusters in this image, suggesting that they were formed in some encounter with another galaxy. [STScI/NASA]

18.3 | Expansion of the universe

18.3.1 Hubble's law

In Hubble's study of galaxies, he found that all galaxies have redshifted spectral lines. The redshift means that they are all moving away from us. Furthermore, the rate at which any galaxy is receding from us is proportional to its distance from us. We can write this in the simple form:

$$v = H_0 d \tag{18.2}$$

where v is the speed of the galaxy, d is the distance, and H_0 is a constant, called the *Hubble constant*. (The subscript zero on the H indicates that this is the current value. As we will see in Chapter 20, H is constant in the sense that it is the same at every place, but can change with time.) The relationship given by equation (18.2) is called *Hubble's law*. Results of more recent studies are shown in Fig. 18.9.

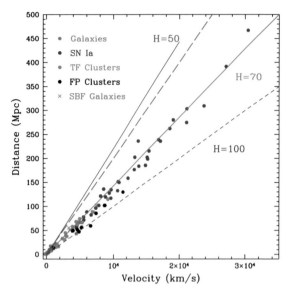

Fig 18.9. Hubble's law. The distance is plotted on the vertical axis, and the radial velocity (from the redshift) is plotted on the horizontal axis. This is the result of the HST Key Project to study Hubble's law, and data using different distance indicators are shown with different colored symbols. The labelled lines show lines corresponding to different values of H. [John Huchra, CFA]

Suppose the rate of expansion has been constant over the age of the universe. If all objects started very close together at $t = 0$ (whatever that time means), then the current distance between any two objects would be

$$d = vt_0$$

where t_0 is the current age of the universe. Solving for v gives

$$v = (1/t_0)d \qquad (18.3)$$

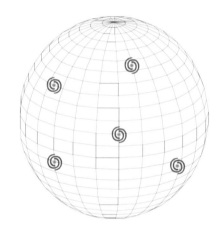

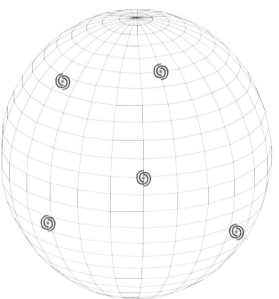

Fig 18.10. The universe as an expanding balloon. The galaxies are painted on the surface of the balloon. As the balloon expands, each galaxy moves away from every other galaxy. This is a two-dimensional analogy to help us with the visualization.

At first it might seem unusual that we are in some special part of the universe, so that all things are moving away from us in a very particular way. However, we interpret Hubble's law as telling us that all galaxies are moving away from each other. This motion represents the overall expansion of the universe. (We will discuss this in more detail in Chapter 20.) To visualize this, we can imagine the galaxies as dots on the surface of a balloon, as shown in Fig. 18.10. As the balloon is blown up, all the dots move away from all the other dots. In Fig. 18.11, we see that the separations between any pair of galaxies increases, and that the larger separations increase faster. This means that we could observe from any of the galaxies, and we would still obtain Hubble's law.

Suppose that in some time Δt, the balloon expands so that all distances are multiplied by a factor $(1 + f)$. If two objects were initially a distance d apart, their distance at the end of the interval is $(1 + f)d$. The change in the distance between the two objects is fd, so the average relative velocity of the two objects is $fd/\Delta t$. This is the same form as Hubble's law.

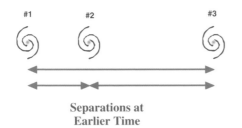

Separations at
Earlier Time

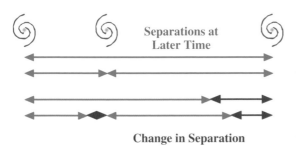

Separations at
Later Time

Change in Separation

Fig 18.11. The effect of all galaxies moving away from each other. The two frames show the positions of the galaxies at different times, with the bottom frame being later. In the top frame the green arrows show the separations at the earlier time. In the bottom frame, the blue arrows show the separations at the later time. The red arrows show the change in separation between the two times for each pair of galaxies (the difference between the blue and green arrows). You can see that the galaxies that were initially farther apart have the greatest change in separation, and the galaxies that were closest have the least change in separation.

This is the same as Hubble's law, if we make the identification

$$H_0 = 1/t_0 \qquad (18.4)$$

Then, $1/H_0$, called the *Hubble time,* is the age of the universe if the expansion has been constant. Actually, as we will discuss in Chapter 20, the expansion is not constant. If the expansion is slowing down, the actual age of the universe is less than the Hubble time. If the expansion is speeding up, the actual age of the universe is greater than the Hubble time.

The value that Hubble obtained for H_0 was 500 km/s/Mpc. Note that the units of the Hubble constant may seem strange, but there is a distance in the numerator (km) and in the denominator (Mpc), so the units really work out to 1/time. We use km/s/Mpc because with it we can express both distance and velocity in convenient units. A value of 500 km/s/Mpc works out to a Hubble time of 2×10^9 yr (2 Gyr; see Problem 18.6). This was a cause for concern, since our understanding of stellar evolution and the HR diagrams of globular clusters in our galaxy tell us that these clusters are about 10 Gyr old. (In fact, radioactive dating places the age of the Solar System at over 4 Gyr.) It is disconcerting to have the universe younger than things in it. However, there is an immediate error in Hubble's value due to confusion between type I and type II Cepheids as distance indicators. Over the years, other refinements have been made. As we will see below, the currently accepted value for the Hubble constant is in the range 50–100 km/s/Mpc.

Apart from telling us something interesting about the universe, Hubble's law is also of great value when determining distances to distant objects. It is important that this only be used for objects that are far enough away that their velocities relative to us are dominated by the expansion of the universe, as shown in Fig. 18.12. We say that objects must be far enough away to be in the *Hubble flow*. Objects within our own cluster of galaxies are not in the Hubble flow. Their motions are dominated by the dynamics of our cluster. Even nearby clusters have random velocities relative to us that are a significant fraction of their Hubble law velocity.

Example 18.3 Hubble's law and distance
For some cluster, we measure $v = 10^3$ km/s. What is the distance? Take $H_0 = 65$ km/s/Mpc.

SOLUTION
We find the distance from

$$d = v/H_0$$

$$= \frac{(10^3 \text{ km/s})}{(65 \text{ km/s/Mpc})}$$

$$= 15 \text{ Mpc}$$

18.3.2 Determining the Hubble constant

Of course, if we are going to apply Hubble's law to determine distances, we need an accurate value for the Hubble constant. This means that we need an independent way of measuring distances to objects that are far enough away to be in the

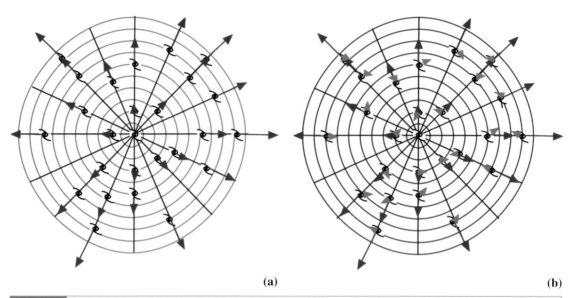

(a)   (b)

Fig 18.12. Expansion of the universe and random velocities of galaxies. (a) We see just the motion due to the expansion of the universe. The center can be any reference galaxy, such as our own. The red arrows show the motions of galaxies at different distances from us, solely due to the expansion of the universe. Notice that all the red arrows are pointed directly away from us, and that they are longer as we look at more distant galaxies. (b) We add in the random motions of the galaxies (green arrows). The random motions point in any direction, and they are the same throughout the universe. The net motion of any galaxy is the sum of its random and expansion motions.

Hubble flow. The importance of an object being far away is illustrated in Fig. 18.13. Suppose an object has a velocity of H_0d from the Hubble flow and Δv from other sources. This Δv probably results from random motions of the galaxies, just as the gas in the room has random motions superimposed on any regular flow. Note that Δv can be positive or negative. The actual radial velocity that we measure will be

$$v_r = H_0d + \Delta v \qquad (18.5)$$

In general, Δv is independent of d. Thus, for more distant objects, H_0d increases while Δv stays the same. For more distant objects, Δv represents a smaller fraction of H_0d, and introduces a smaller fractional error into the determination of H_0.

It would seem a simple task to get around this problem. All we have to do is use the most distant objects we can observe. Unfortunately, our distance indicators work best for nearby galaxies, where we can still see individual stars such as Cepheids. Therein lies the problem. We can measure distances more accurately for nearby objects, but we are not sure if they are in the Hubble flow. Arguments over the proper value of H_0 center

around these two points: (1) what are the correct distance indicators, and (2) where does the Hubble flow start?

We now turn to the problem of measuring distances to distant clusters of galaxies. The procedure involves using our most secure distance indicators to measure the distances to nearby galaxies, and then building up a series of distance indicators, useful at greater and greater distances. The problem can be involved, as shown in Table 18.1.

We start the process by looking at Cepheids to find the distances to nearby galaxies. Of course, we must calibrate the Cepheid period–luminosity relationship within our own galaxy. This involves starting with trigonometric parallax observations of nearby stars and moving cluster observations of nearby star clusters to produce a calibrated HR diagram. The calibrated HR diagram allows us to use spectroscopic parallax for individual stars, and for main sequence fitting for globular clusters containing Cepheids. This gives a calibrated period–luminosity relation for Cepheids as well as RR Lyrae stars. We can then use the Cepheids and RR Lyrae stars as distance indicators for galaxies that are close enough for us to see these stars individually.

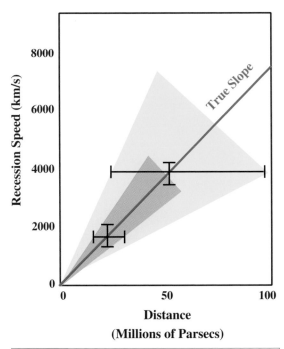

Fig 18.13. Sources of error in measuring the Hubble constant. For both nearby and distant galaxies, uncertainties are indicated schematically by error bars. Straight lines from the origin are then drawn at the widest angle that can still lie within one of the error bars. The error bar that determines this widest angle is the one that generates the largest uncertainty in H_0 since H_0 would be determined by the slope of a line from our measured point to the origin.

Table 18.1. | Distance indicators.

Method	Distance range (Mpc)
Cepheids	0–100
Novae	0–20
RR Lyrae stars	0–0.2
W Virginis stars	0–1
Eclipsing binaries	0–1
Red giants	0–1
Globular clusters	0–50
Supergiants	0–1
Stellar luminosity function	0–1
HII region diameters	0–25
HII region luminosities	0–100
HII loop diameters	0–4
Brightest blue star	0–25
Brightest red stars	0–15
Supernovae	0–400
21 cm linewidths	0–100
Disk luminosity gradients	0–100
U—B colors	0–100
Luminosity classification	0–100
Brightest elliptical in cluster	50–5000

HST has greatly expanded the distance over which we can study Cepheids up to 100 Mpc.

For galaxies that are somewhat farther away, we can still use individual objects within the galaxy, but they have to be brighter than Cepheids. We can, for example, measure the angular sizes of HII regions. Since we think we know what their linear sizes should be, this gives us a measure of distance. One promising technique is the use of supernovae. From the light curve we can tell whether the supernova is a type I or type II. Type I supernovae appear to have similar peak luminosities. By measuring their apparent magnitude at peak brightness, we can find the distances.

Eventually we reach a point where individual objects within galaxies cannot be measured. We must rely on being able to know the total luminosity of the galaxy. If we know the absolute magnitude of a given type of galaxy, and we measure

the apparent magnitude, we have a measure of the distance. The problem is to come up with independent indicators of galactic luminosity. Hubble made the simple assumption that all galaxies have the same absolute visual magnitude. We know that this is not the case. Instead, it has been suggested that the brightest ellipticals in each cluster have the same absolute magnitude. The luminosities of the brightest ellipticals seem to vary, however. For this reason, rather than using the brightest elliptical in a cluster, we use the second or third brightest elliptical in a cluster.

With the recent recognition that galaxies have luminosity classes, much effort has gone into finding luminosity class indicators. In this way, we can observe a galaxy, determine its morphological type (Sa, Sb, etc.), and then have some indicator of its luminosity class. If we know absolute magnitude as a function of morphological type and luminosity classes, and we measure the apparent magnitude, we can convert the difference into a distance.

Another recent discovery, indicated in Fig. 18.14, is that the width of the 21 cm line in a galaxy seems

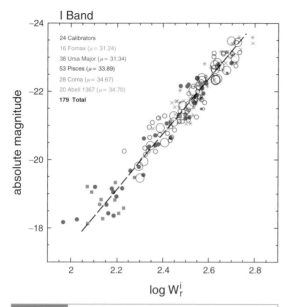

Fig 18.14. Tully–Fisher relation. For various selections of test galaxies, each plotted with a different symbol and color, we plot the 21 cm line width on the horizontal axis, and absolute magnitude in the I band on the vertical axis, for galaxies whose distances are determined by other methods. [Brent Tully, Institute for Astronomy, Hawaii]

calibration is complicated by the fact that the relationship may depend on the type of galaxy.

Even with these methods, there is another nagging problem. When we look at a distant galaxy, we are seeing it as it was a long time ago. However, we know that galaxies evolve. As they evolve, their luminosity changes. Luminosity calibrations on nearby galaxies might not apply to distant galaxies.

More recently supernovae have been an important tool. We can see supernovae far away, especially with HST. One is shown in Fig. 18.15(a). From

to correlate with its absolute infrared luminosity. This is called the *Tully–Fisher relation*. Therefore, we can observe the 21 cm line, measure its width, and know the absolute magnitude of the galaxy. Of course, this relation must be calibrated. The

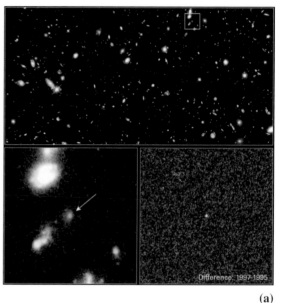

(a)

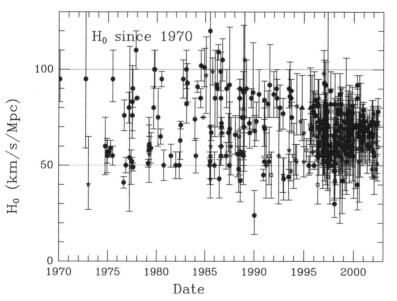

(b)

Fig 18.15. (a) HST image of a supernova in a distant galaxy. The larger image shows the cluster of galaxies in which this event took place. The box on the left shows the location of the individual galaxy in which the supernova took place, and the blowup on the right shows the supernova. The brightness of the supernova places the distance to this cluster at 3 Gpc. (b) Measurements of H_0 since 1970, with error bars shown for each measurement. [(a) STScI/NASA; (b) John Huchra, CFA]

the light curve, we can tell what type of supernova it is. It turns out that type I supernovae (which occur in close binary systems, discussed in Chapter 12), have a reasonably narrow range of peak absolute magnitudes. We can therefore use them as 'standard candles' at great distances.

So you see that there is a problem at every step of the determination of the Hubble constant. For this reason, no one indicator is used at any step along the way. For example, when the distance to the Hyades, as determined from the moving cluster method, was revised by 10%, it only affected the extragalactic distance scale by 5%.

Currently accepted values of the Hubble constant fall in the range 50 to 100 km/s/Mpc. The corresponding Hubble times are 20 Gyr to 10 Gyr, the latter time being small enough to be somewhat worrisome. Measurements since 1970 are shown in Fig. 18.15(b). A number of recent studies suggest values of 70 km/s/Mpc with an uncertainty of 10 km/s/Mpc.

In Chaper 20, when we begin to look at theories of the large-scale structure of the universe, we will see that H_0 shows in various equations. It is useful to be able to do calculations using a value of H_0, but for which the results can be scaled for a different value. We therefore define a dimensionless parameter, h, such that

$$H_0 = h \, [100 \text{ km/s/Mpc}] \qquad (18.6)$$

So, if H_0 turns out to be 70 km/s/Mpc, then $h = 0.7$.

18.4 | Superclusters and voids

Now that we have seen that galaxies are gathered into clusters, we might ask if the clusters are gathered into larger groupings, called *superclusters*. The answer is that they are. The first supercluster identified (in the 1950s) is the one in which we live, called the *local supercluster*. The Virgo cluster of galaxies is near the center of the local supercluster. The local group, our cluster of galaxies, is near the edge. The local supercluster contains 10^6 galaxies in a volume of about 10^{23} cubic light years!

Studies of more distant superclusters have been made difficult by a lack of extensive data on distances to galaxies. After all, we only see two dimensions projected on the sky. We can get a better idea of clustering if we also know distances to galaxies. We obtain the distances from measuring the redshifts of clusters.

It might seem that we can determine the mass of a supercluster by using the virial theorem, in much the same way as we did for clusters of galaxies. However, the crossing time for a cluster in a supercluster is greater than the current age of the universe (as estimated by the Hubble time). This means that superclusters are not dynamically relaxed, and the virial theorem should not apply. It is even possible that superclusters are not gravitationally bound.

In addition to superclusters, there appear to be *voids* of comparable size. As their name implies, voids are large regions of space that are essentially devoid of galaxies.

One of the major breakthroughs in these studies has come from the ability to measure a large number of redshifts in a relatively short time. These *redshift surveys* are carried out at radio (21 cm) and optical (Hα) wavelengths. They cover large sections of the sky, and also cover a large range of redshifts. Having surveys in both the optical and radio parts of the spectrum provides an important check on the results.

Presenting the results of these redshift surveys can be difficult. Not only are there millions of data points, we have the location of each galaxy as a function of three coordinates – two for the position on the sky, and one for the redshift. We can leave the third coordinate in terms of redshift, or convert it to a distance, using Hubble's law. In any case, we are still stuck with trying to plot a three-dimensional distribution of galaxies. One way of doing this is to make slices through our data, and then for each slice make a two-dimensional picture of the resulting distribution of galaxies. In Fig. 18.16, we show how we might uses slices through a three-dimensional Earth to produce a series of two-dimensional images, which, taken together, give us a feel for the three-dimensional structure. For the redshift surveys, we can make our slices showing distribution on the sky in a series of redshift ranges, or showing distribution in one sky coordinate and redshift for a range of the other sky coordinate.

Some representative survey results are shown in Fig. 18.17. In these figures, each dot represents a galaxy; a concentration of dots is a cluster; a

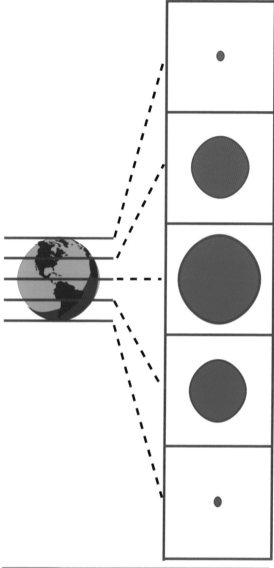

Fig 18.16. How we might use slices through a three-dimensional object to let a series of two-dimensional images represent our three-dimensional structure.

even more complicated. Some have made analogies with sponges and swiss cheese, with the typical sizes of the holes being tens to a hundred million parsecs across. This shows us that the galaxies are distributed in a much more complicated way than is implied by simply talking about superclusters and voids.

From these surveys, we can also deduce the motions of the galaxies with respect to their neighbors. Remember, earlier in this chapter we saw that any galaxy has motion associated with the expansion of the universe, and local motions in response to the gravitational attraction of its neighbors. When we analyze these local motions of the galaxies, we find that they are are not completely random. They are organized, with galaxies in some part of a shell having similar motions. This suggests that there are large amounts of matter attracting the galaxies to produce the organized motion.

We can also see a similar effect in the motion of our galaxy (or the Local Group) through space. How do we detect that motion? Think of the following analogy. Imagine that you are in a room full of people, all standing still. You start to walk through the room. All of a sudden, it appears that people on one side of the room are coming towards you, while people on the other side of the room are moving away from you. Your first thought might be that this is the way that the people are actually moving. On farther thought, you realize that you are seeing your own motion reflected in the apparent motions of the people.

The situation is a little more complicated with galaxies, because the universe is expanding. We start with all of the galaxies moving away from us. If we then start moving in some direction, we will be overtaking some galaxies, and moving away from others in the opposite side of the sky. On average, galaxies in one half of the sky will appear to be moving away from us slightly faster than in the other half of the sky. This effect is actually observed. By seeing which half of the sky appears to be moving away from us a little faster, and which half is moving a little slower, we can determine how fast we are moving and in what direction. What we actually measure is the total motion of the Earth. We must then correct for the motion of the Earth around the Sun, the

concentration of clusters is a supercluster, and a lack of dots is a void. From these figures, we see that superclusters and voids are quite common. There is a very distinctive feature to the distribution. It looks like the galaxies are concentrated on the surfaces of various shapes. The first astronomers studying these distributions thought that the surfaces may be like soap or beer foam bubbles, but it now seems that the structures are

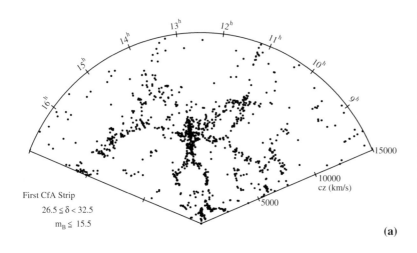

Fig 18.17. The results of red- shift surveys. (a) The first Center for Astrophysics slice, based on optical observations. Around the outside are markings indicating right ascension (in hours) of the galaxy. All of the galaxies in the indicated declination range, covering six degrees, are plotted in this single slice. The redshifts are plotted on the straight axis moving away from the point. (b) Two additional slices, showing different declination ranges. [John Huchra, CFA]

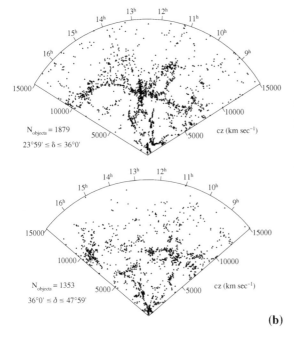

18.5 | Where did all this structure come from?

One of the great questions that we try to answer in astronomy is, 'How did we get here?' In this book, we have already talked about how star formation takes place in various types of galaxies. Therefore, the next step is to ask, 'Where do the galaxies and clusters (and other such structures) come from?'

For many years, attempts to answer this question have been quite vague. Astronomers loosely talked about two general scenarios, a 'top-down' and a 'bottom-up'. These are indicated in Fig. 18.18. In the top-down scenario, the largest-scale structures form first, and then everything else fragments. So, somehow, in the early universe, objects with the masses of superclusters separated out and began to contract gravitationally. Eventually, a density was reached where less massive structures, with the masses of clusters of galaxies, could contract. Finally, these objects fragmented into the galaxies themselves. In the bottom-up picture, the galaxies form first. Galaxies near each other would then attract themselves into clusters, and clusters near each other would then attract themselves into superclusters.

The thing that has changed recently is that the redshift surveys have provided a large body of data. In addition, the advent of supercomputers

motion of the Sun around the galactic center, and the motion of the galaxy within the Local Group. When we do this, we find that our cluster is moving in the general direction of the Virgo cluster, at a speed of approximately 300 km/s. Other clusters, including the Virgo cluster, show a similar motion. This suggests that there is a very massive object out there beyond the Virgo cluster, providing a strong pull. We have no idea what it is, and, for lack of a better name, it is called the *great attractor*.

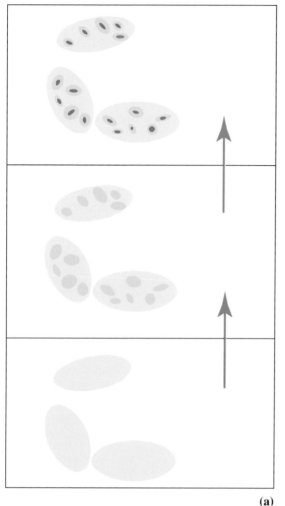

(a)

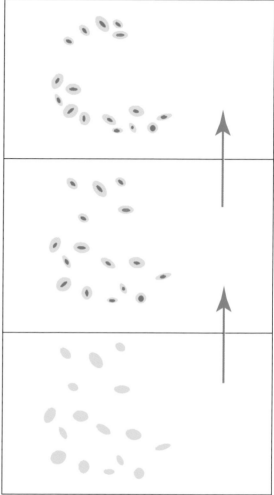

(b)

Fig 18.18. Diagram showing (a) top-down and (b) bottom-up scenarios for galaxy formation.

has made it possible to simulate the evolution of the universe and galaxy formation under various conditions. The results of the computer simulations can then be compared with the large body of data. If a particular simulation can reproduce the distributions of galaxies and their large-scale motions, then it is possible that the physical ideas that went into that simulation might be important in the real universe. However, to date, no single simulation has been able to produce structure on all of the scales that we see.

Part of the problem involves dark matter. We have already seen that there is more dark matter than luminous matter around us. In clusters of galaxies, the domination of dark matter is even greater. Since galaxy formation is initiated by gravitational attraction, most of that attraction will be provided by most of the matter. This means that, even though we see galaxies and clusters as luminous objects, their formation is governed by dark matter. Therefore, to make a successful computer simulation of galaxy formation, we must start with the right type of dark matter. However, we have said that we don't know what the dark matter is. We don't even know if the dark matter in individual galaxies is the same as that in clusters of galaxies.

While this may seem like an insurmountable problem, theoreticians have managed to turn it around. They first realized that, even if we don't know the details of the dark matter, there are probably classes that can be treated as a whole. For example, if the dark matter in individual galaxies is in the form of Jupiter mass objects, then, from the point of view of gravity, it doesn't matter whether we have a certain number of Jupiter mass objects or ten times as many objects whose mass is one-tenth that of Jupiter. The point would be that we were dealing with ordinary matter, but not in quantities large enough to form luminous stars.

By analyzing the various possibilities, theoreticians have been able to group the dark matter into two general types, according to how it behaves. The two types are called *cold dark matter* and *hot dark matter*. These don't have to do with the temperature, but with how the material behaves. An example of hot dark matter would be neutrinos, if they had even a very small rest mass. We will talk about what particles these might actually be in Chapter 21, when we talk about the earliest times in the universe. For now, we only need to know that these produce different types of structures.

So we have gone from vague notions of top-down and bottom-up scenarios to asking a very specific question: *What is the dark matter that dominates galaxy formation?* The hope is that we can answer that question by comparing the simulations with the data. Unfortunately, at this point, neither one works completely. The cold dark matter is good at producing the small-scale structures (the galaxies) but not the large-scale structures (the clusters and superclusters, or the 'bubbles'). The hot dark matter is just the opposite. It appears that cold dark matter does a better job of describing the structures we see, provided we add some modifications that come from general relativity, which we will discuss in Chapter 20.

Even when we correctly identify the dark matter, and show how galaxies and clusters can grow from some small density enhancements in the early universe, we still haven't addressed the issue of where those initial enhancements came from. To do that we must look at the history of the universe, the field called cosmology, and we do that in Chapters 20 and 21.

18.6 | The Hubble Deep Field

With the sensitivity and imaging quality of the refurbished HST, it was decided to devote a large amount of observing time to a small region of the sky to look for the faintest (presumably) most distant objects that could be detected. This resulted in the *Hubble Deep Field*. This field was chosen to be far from any local bright sources (like the galactic plane). The image was released so that any astronomer could study it. This also spurred coordinated observations of this field and objects within it, using a wide range of telescopes.

The field is approximately 1 arc min across. An image of a one-quarter of that field is shown in Fig. 18.19(a). This image contains objects as faint as 30th mag. (The diffraction spikes are around a 20th mag star.) This picture shows a large number of spiral and elliptical galaxies.

(a)

Fig 18.19. Images of the Hubble Deep Field. (a) A true color image of one-quarter of the field.

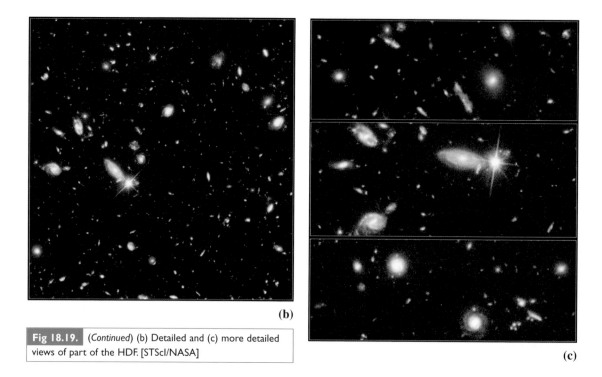

(b)

Fig 18.19. (*Continued*) (b) Detailed and (c) more detailed views of part of the HDF. [STScI/NASA]

(c)

Chapter summary

Galaxies have a very irregular distribution on the sky. The highest concentrations are called clusters. We think that the clusters are gravitationally bound. However, the galaxies that we see in a cluster do not appear to have enough mass to hold the cluster together, even if we allow for the dark matter that we know to be in the galaxies. This means that there must be dark matter in the clusters that is not associated with the individual galaxies.

Virtually all of the galaxies that we see are moving away from us. The more distant galaxies are moving away faster. There is a simple relationship between how far away a galaxy is and how fast it is moving away. This relationship is called Hubble's law. This simple relationship results from the expansion of the universe. We can also use Hubble's law to tell us the distance to galaxies whose redshift we can measure.

The quantity that currently tells us the actual distance to a galaxy with a given redshift is called the Hubble constant. It is constant in that it has the same value everywhere in the universe; how-ever, it can change with time. Measuring the Hubble constant is very difficult. We need measurements of the distances to galaxies that don't involve using the redshift. These are more accurate for nearby galaxies. However, it is only for the distant galaxies that the effect of the expansion of the universe is much greater than the effect of the random motions of the galaxies. The most likely values of the Hubble constant are between 50 and 100 km/s/Mpc, with much recent work suggesting a value of 70 km/s/Mpc.

On larger scales, the clusters are gathered into superclusters, which are tens of megaparsecs in extent. Of comparable size are large volumes with no galaxies, called voids. We learn about this large-scale structure from redshift surveys, in which we measure the redshifts of thousands of galaxies. (These are done both in the visible and radio parts of the spectrum, depending on the types of galaxies we are looking at.)

Efforts to understand all of this structure are just getting under way. An important problem is that we don't know how much dark matter there is on the largest scales, and we don't know the nature of that dark matter. Some have even tried to turn the problem around, using the distribution of galaxies to tell us something about the nature of the dark matter. No matter what we try, no single theory (or type of dark matter) can explain the large- and small-scale structures that we see, but CDM appears to be doing a better job.

Questions

18.1. Discuss the following statement: 'For Hubble's law to be true, we must be at the center of the universe.'

18.2. Draw a diagram, like Fig. 18.11, showing that the argument presented works even if the three galaxies are in a triangle, rather than in a straight line.

18.3. Why do we call the Hubble time an 'upper limit' or a 'lower limit' to the age of the universe?

18.4. What are the main problems in the accurate determination of the Hubble constant?

18.5. What do we mean by 'Hubble flow'?

18.6. Can we use Hubble's law to determine distances within our own galaxy?

18.7. Of the distance measurement techniques listed in Table 18.1, which ones depend on knowing the intrinsic luminosity of some type of object?

18.8. What is the value of supernovae as distance indicators?

18.9. When we look at a distant galaxy, we see it as it was a long time ago. How does this make the use of distance indicators more difficult?

18.10. What is the evidence for the existence of clusters and superclusters of galaxies?

18.11. What makes us think that there is dark matter in the Virgo cluster?

18.12. Why do we expect a cluster of galaxies to obey the virial theorem but not a supercluster?

18.13. In adding up the 'visible' mass in clusters, what is the problem in accounting for the masses of the galaxies that we can see?

18.14. Why would we expect intergalactic gas to be able to emit X-rays?

18.15. What are the current possibilities for the dark matter in clusters?

18.16. If we cannot see the dark matter in clusters of galaxies, how do we know that it is there?

18.17. Explain how we might determine the mass of a binary galaxy system. Why is this important, since we already have masses determined from rotation curves?

18.18. Why are redshift surveys important in studying clustering of galaxies?

18.19. How are hot and dark cold matter related to various scenarios of galaxy formation?

18.20. What type of structure is hot dark matter best at explaining? What types for cold dark matter?

Problems

For all problems, unless otherwise stated, use $H_0 = 70$ km/s/Mpc.

18.1. For the cluster in Example 18.2, what is the total kinetic energy?

18.2. For some cluster of galaxies, the radius is 500 kpc, and the rms radial velocity is 300 km/s. What is the mass of the cluster?

18.3. Rewrite equation (18.1) so that if velocities are entered in km/s and distances in Mpc, the mass results in solar masses.

18.4. Suppose that one-half of the mass of the cluster in Example 18.2 is in the form of hot intergalactic gas, spread out uniformly over the whole cluster. What density of gas would this require?

18.5. For the galaxies represented in Fig. 18.11, draw a graph of the length of the red line vs. distance. Do this three times, each time using a different galaxy as your reference point.

18.6. Find a relationship between the Hubble constant, expressed in km/s/Mpc, and the Hubble time, expressed in years. Use this to find the Hubble times corresponding to Hubble constants of 50, 65, 100 and 500 km/s/Mpc.

18.7. Suppose we have a universe whose size increases by 1% in 1 Gyr. Show that the average rate of separation between any two points is proportional to their distance, and find the proportionality constant.

18.8. For some galaxy, we measure a recession velocity of 2000 km/s. How far away is the galaxy?

18.9. For some galaxy, we measure a recession velocity of 3000 km/s. How far away is that galaxy for Hubble constants of: 50, 65, 70 and 100 km/s/Mpc?

18.10. If the typical random velocity in a cluster is 300 km/s, how far out must we go before this is only 1% of the expansion speed at that distance?

18.11. Suppose that we can use supernovae to measure distances out to 200 Mpc. What is the recession speed at that distance?

18.12. What is the crossing time for a cluster moving at 10^3 km/s through a typical supercluster-sized object?

*18.13. What are the Jeans mass and length if we have 10^{16} $M_\odot$ worth of H spread out over 10 Mpc, with random internal motions of 1000 km/s?

18.14. What is the density of galaxies (galaxies/ly^3) in the local supercluster?

Chapter 19

Active galaxies

In this chapter we look at galaxies with unusual activity within and around them. For many years astronomers thought of these various types of activity as being distinct. We now realize that many of them have similar origins, but differ in the specific conditions within the galaxy or its environment. We realize that all of this activity takes place in the nucleus of the galaxy, or is driven by activity in the nucleus. We say that these phenomena are associated with *active galactic nuclei (AGN)*.

19.1 | Starburst galaxies

Some galaxies appear to be giving out excessive amounts of radiation in the infrared. When we studied star formation in Chapter 15, we saw that regions with recent star formation give off a lot of infrared radiation. The energy comes from the newly formed stars, and heats the dust (from the parent cloud) surrounding the stars. The dust then glows in the infrared. The more energy the young stars put into the cloud, the more infrared radiation is released. The excess infrared radiation from some galaxies suggests that those galaxies have very high rates of star formation. The rate is so high that it cannot be sustained for very long, or it would use up all of the interstellar material. This leads to the idea that this excessive star formation is a short-lived phenomenon. We therefore call such galaxies *starburst galaxies*.

Fig. 19.1 shows a typical starburst galaxy. In Fig. 19.1 (a), we see an optical image of the whole galaxy. Notice that it doesn't look very unusual in this optical image. There does appear to be a lot

of obscuration from dust near the center. In Fig. 19.1 (b), we have a combined image, which shows various tracers of massive star formation (including UV, IR and Hα). We see a small ring of bright young stars, close to the center. The dust lanes that we see correspond to giant molecular clouds.

In Fig. 19.2, we see a spectrum of a starburst galaxy M82. Notice the large number of emission lines. These generally signify the presence of hot gas.

In Chapter 15, we saw that star formation takes place in molecular clouds. We detect molecular clouds by emission from carbon monoxide (CO) in the millimeter part of the spectrum. If the strong infrared emission is really telling us that there is a lot of star formation taking place, then we should be able to 'see' the molecular clouds in which the stars are forming. We 'see' them by looking for the CO emission from those clouds. Indeed, this CO emission is observed. By comparing the CO maps with far infrared maps, we see that the CO and far infrared emissions are strongest in the same place. This is basically what we see in our galaxy, when we have star formation in a molecular cloud (as depicted in Fig. 15.2). This suggests that starburst galaxies really do have a lot of star formation in complexes of molecular clouds.

We can form a better picture of the individual star forming regions by looking at near infrared images. The near infrared image has two advantages: (1) since the wavelength is shorter than the far infrared, it corresponds to hotter temperatures, and allows us to isolate hotter objects; (2) because of the shorter wavelength, we

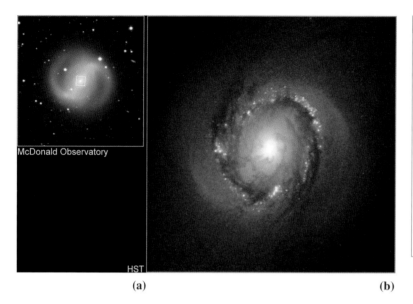

Fig 19.1. Images of the starbust galaxy NGC 4314, at a distance of 13 Mpc. (a) A normal image of the whole galaxy, a barred spiral. (b) An HST image of the central region. This is a composite of images taken in through ultraviolet, blue, visible, infrared and Hα filters. It shows that most of the recent star formation is in a small ring about the center of the galaxy. We also see dust lanes inside the ring, showing the locations of the giant molecular clouds. [(a) McDonald Observatory; (b) STScI/NASA]

(a)   (b)

can produce images with better angular resolution than at longer wavelengths. These near infrared images show clusters of recently formed massive stars.

How do we know that the stars that have formed are truly massive? That is, how do we know that the emission comes from some number of massive stars, rather than a larger number of less massive stars? We know that after a relatively short time, massive stars must undergo supernova explosions (Chapter 11). This means that some number of the stars that have formed should have had time to go supernova. These should no longer be visible as stars, but as supernova remnants. As we saw in Chapter 11, super-

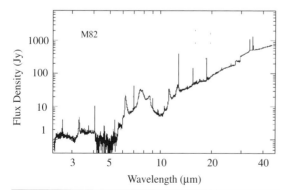

Fig 19.2. A spectrum of a starburst galaxy, M82 (also shown in Fig. 17.9). [ESA/ISO, SWS, R. Gerzel and D. Lutz]

nova remnants are detected by their radio emission. The centers of starburst galaxies also have radio emission that is characteristic of supernova remnants. This tells us that the star formation must have included high mass stars.

When we look at the amount of molecular material out of which the stars could have formed, and we look at the stars that have formed, we come to another interesting conclusion. The star formation must have been very efficient. That is, a large fraction of the available molecular cloud mass was converted into stars in a short period of time. This is in contrast to what we found in the nearby star forming regions, like Orion. In those cases, a very small fraction of the available cloud mass has been converted into stars. We discussed the fact that, in general, in our galaxy, star formation has a low efficiency. That is, something appears to be supporting the clouds until the conditions are just right for star formation. In starburst galaxies, the clouds are not as well supported, and more easily form massive stars (as discussed in Chapter 15). There is some evidence that the density of molecular material is much higher than in normal GMCs near us. They are more like the GMCs in our galactic center, which also have high densities.

What creates the conditions that favor a starburst? This is a topic of current research. Some

Motions of Galaxies

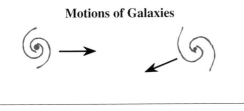

Gas Clouds

Fig 19.3. Diagram showing a model for how a starburst might occur when two galaxies pass close together.

astronomers think that starbursts occur in galaxies that have nearby neighbors, and can steal interstellar gas from those neighbors. This is illustrated, schematically, in Fig. 19.3. The galaxies pass close to each other, and one galaxy manages to draw a large quantity of interstellar material from the other. Most of this new material finds its way to the center of the galaxy. This provides a lot of material for star formation. In addition, that material is denser than the typical molecular clouds that we have encountered. This means that it is more likely to give birth to stars. The result is very efficient star formation – a starburst.

19.2 | Radio galaxies

Many strong radio sources are not objects in our own galaxy. When a radio source is found, it is interesting to see if there is an identifiable optical object at the same position. In the past this could not always be done because the poor angular resolution in single-dish observations left radio astronomers with uncertain positions for their sources. With the development of interferometers, with better angular resolution, the finding of optical counterparts has been easier. Many of the strong radio sources appear in the directions of other galaxies. Galaxies with strong radio emission are called *radio galaxies*.

19.2.1 Properties of radio galaxies

The radio energy output of the radio galaxies is enormous, typically 10^6 times the total output of a normal galaxy. The radio spectrum of a typical radio galaxy is shown in Fig. 19.4. The shape has the power law characteristic of synchrotron radiation. The radiation is also found to be polarized, another signature of synchrotron radiation. Synchrotron radiation requires a strong magnetic field and high energy particles, most likely electrons.

The actual synchrotron spectrum depends on the energy distribution of the electrons. The greater the proportion of the highest energy electrons, the greater is the proportion of high energy photons. As the electrons radiate, they lose energy. Therefore the synchrotron spectrum evolves over time. The most energetic electrons lose their energy the fastest, so the proportion of

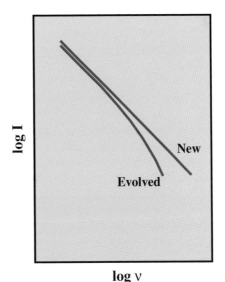

Fig 19.4. Schematic evolution of a radio galaxy spectrum. As the more energetic electrons give off energy, they become less energetic, so the evolved spectrum has proportionately less energy at higher frequencies.

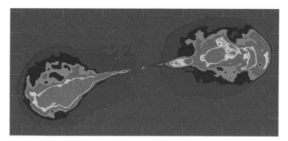

Fig 19.5. VLA radio image of Hercules A ($z = 0.15$). The resolution of the image is 1.4 arc sec; the image itself is a stacked image of four frequencies ($18+21+22+ 23$ cm) and three configurations (A + B + C). It is of the highest sensitivity of any map of any radio source. Notice the narrowness of the jet, and then the much larger extent of the lobes. [Nectaria A. B. Gizani, Observatorio Astronomico de Lisboa, Portugal]

high energy electrons decreases with time. This means that the spectrum evolves, with more and more of the radiation coming out at long wavelengths as shown in Fig. 19.4. By studying the synchrotron spectrum of a radio galaxy, we can estimate how long it has been radiating.

Maps of radio galaxies, such as those shown in Figs. 19.5–19.7, show us what truly amazing objects they are. (In order to determine the linear size, we must know the distance to the radio galaxy. This is found from the redshift of the visible associated galaxy, using Hubble's law.) A typical radio galaxy has a small source near the center of the galaxy, and then two large sources far beyond the optical limits of the galaxy. The optical galaxy is generally a giant elliptical. The two sources, or *radio lobes*, may be separated by up to 10^7 pc (10 Mpc), and be as wide as 10^6 pc (1 Mpc). Sometimes, multiple pairs of lobes are seen. The structure of these sources is suggestive of matter being ejected from the galaxy. We sometimes see

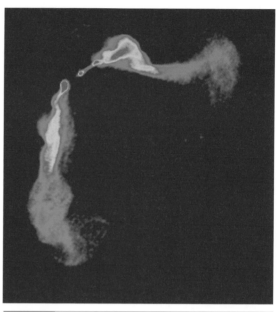

Fig 19.7. VLA radio image of 3C465 ($z = 0.03$) at 327 MHz . Notice the bent appearance. It is not due to the galaxy motion (we know it's moving too slowly). Probably it is due to flows in the cluster atmosphere, relative to the central cD. Such flows can be induced by cluster mergers. [Jean Eilek (NMTech) and Frazer Owen (NRAO/AUI/NSF, STScI/NASA)]

bending of the lobes. Higher resolution observations show narrow structures, called *radio jets*, pointing from the center of the galaxy to the larger radio lobes. Some images of jets are shown

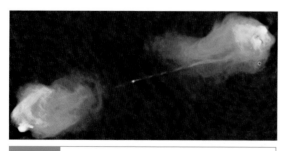

Fig 19.6. VLA radio image of Cygnus A ($z = 0.06$). [Courtesy C. Carilli and R. Perley, NRAO/AUI/NSF]

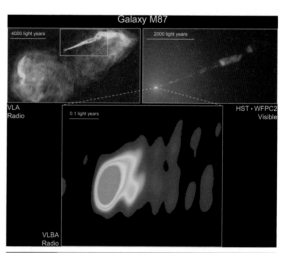

Fig 19.8. M87 on different scales. At upper left is a VLA image of the large-scale structure. To the right is an optical (HST) image of the jet. Below is a VLBA radio image of the center and start of the jet. An optical image of M87 is shown in Fig. 17.2(a). [NRAO/AUI/NSF, STScI/NASA]

in Fig. 19.8. An optical image of Cen A is shown in Fig. 19.9.

19.2.2 Model for radio galaxies

The generally accepted picture of what is going on in radio galaxies is shown in Fig. 19.10. We have interesting phenomena on a wide variety of length scales. On the largest scale we have the radio lobes themselves. The density in these lobes is very low, 10^{-4} to 10^{-3} electrons/cm^3. Because of this low density, collisions are infrequent. Therefore, when energy is deposited into this gas, it takes a long time for collisions to establish an equilibrium (Maxwell–Boltzmann) velocity distribution. The time is longer than the age of the universe, so this never happens. This is how we can have an unusually large number of high energy electrons. The high energy electrons lose their energy by synchrotron radiation, rather than by collisions with lower energy electrons. The magnetic fields in the lobes are thought to be in the range 10–100 μG. The lobes contain smaller-scale structures with extents of 10^3 pc and smaller.

The jets are highly collimated, being some 10^5 pc long by 10^4 pc wide. The flow velocities in the jets are hard to measure, but estimates range from 10^3 to 10^5 km/s. (Some jets are moving close to the speed of light, as discussed in Section 19.2.3.) The densities are about 10^{-2} electrons/cm^3. The magnetic fields are comparable to those in the lobes.

Figure 19.11 shows one mechanism by which the jets can be collimated. We assume that there is material flowing outward in all directions. However, the source of the flow is surrounded by denser gas. This denser gas has a hole in it, and the outflowing material just follows the path of least resistance. Effectively, the material forms itself into a nozzle. One problem with this picture is that we would expect the outflowing material to drive the confining material outward.

(a)

WFPC2

NICMOS

Fig 19.9. (a) Optical image of the radio galaxy Centaurus A. (b) Optical HST image of Centaurus A, showing the inner region with the suspected black hole. [(a) ESO; (b) STScI/NASA]

(b)

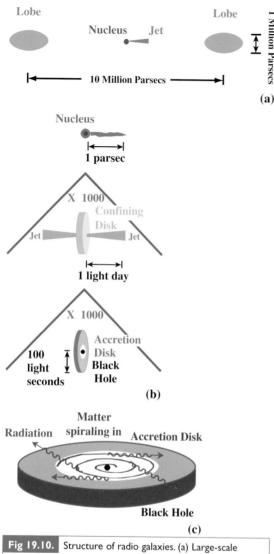

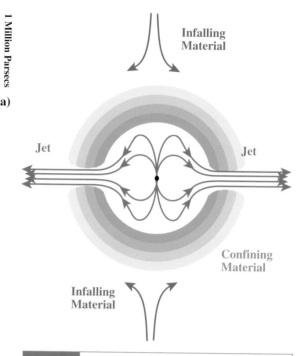

Fig 19.11. Jet collimation mechanism. Material flows out of the small source in the center. It is confined by the ring of material, forming the nozzles for the jets. The shading of the confining material indicates that the density is higher on the inner side. The confining material is stabilized by material falling in from farther out.

Fig 19.10. Structure of radio galaxies. (a) Large-scale schematic. The shaded areas are the large lobes. Closer to the center is the jet pointing to one lobe or the other. (b) Small-scale schematic. A series of three frames, each blown up by a factor of 1000. (c) Schematic of central region.

The confining material may be held in place by material falling in from even farther out, as shown in the figure.

It is possible to have two openings on opposite sides. However, higher resolution observations show that clumps on opposite sides of the center were not ejected at the same time. We also see single jets, but two lobes, in many radio galaxies. It therefore seems that only one nozzle operates at a time. Some flip-flopping of the nozzles is needed to produce the two lobes.

We now turn to the source of the energy in the nucleus of the galaxy. The energy requirements are enormous, since any of the energy ultimately given off in radio waves has its source in the galactic nucleus. At first we might think that nuclear reactions are the most efficient possible energy source. After all, they convert about 0.7% of the available mass into energy. However, if we look at the mass available in the nucleus of a radio galaxy, a higher fraction is being converted into energy. What energy source can be more efficient (by a factor of almost 100) than nuclear reactions?

The answer is mass falling into a black hole. As we saw in Chapter 8, a black hole is important because it allows us to have a large mass in a small radius, and hence strong gravitational forces. We can estimate the amount of energy available in dropping a particle of mass m from far away to the Schwarzschild radius. (Once the mass has passed the Schwarzschild radius, we can no longer get any energy out.) The energy gained by this mass is the negative of the potential energy at R_S, so the maximum energy we can extract is

$$E_{max} = GMm/R_S \qquad (19.1)$$

Substituting $R_S = 2GM/c^2$ gives

$$E_{max} = (1/2)mc^2 \qquad (19.2)$$

This tells us that we can take out up to half the rest energy of the infalling mass. The rate at which energy is generated then depends on the rate at which mass falls into the black hole, dm/dt. The maximum luminosity, dE/dt, is given by

$$L_{max} = (1/2)(dm/dt)c^2 \qquad (19.3)$$

Example 19.1 Luminosity for mass falling into a black hole
Calculate the energy generation rate for mass falling into a black hole at the rate of 1 $M_\odot$/yr.

SOLUTION
Using equation (19.3) gives

$$L = \left(\frac{1}{2}\right)\frac{(2 \times 10^{33}\,\text{g})(3 \times 10^{10}\,\text{cm/s})^2}{(3 \times 10^7\,\text{s})}$$

$$= 3 \times 10^{46}\,\text{erg/s}$$

Remember, the luminosity of the Sun is 3.8×10^{33} erg/s, so this is almost 10^{13} solar luminosities!

However, extracting energy is not a simple as dropping mass into any black hole. If the mass is dropped straight in, most of the energy will be sucked into the black hole. In order to have most of the energy escape, it is necessary for the infalling matter to be in orbit around the black hole, slowly spiraling in. In this case approximately 40% of mc^2 is available to power the galaxy. This 40% is very close to the limit of one-half that we found in our simple calculation, equation (19.3).

In equation (19.3), we see that the luminosity does not depend on the mass of the black hole. However, when we take into consideration the spiraling trajectory for extracting most of the energy, the mass of the black hole, becomes important. The more massive the black hole, the greater the rate at which we can drop in material.

We can think of a more massive black hole as having a larger surface area. Calculations show that, in order to produce the luminosities we see in radio galaxies, black holes with masses of about $10^7\ M_\odot$ are needed!

19.2.3 The problem of superluminal expansion

An interesting problem with some radio sources is that they appear to have small components that are moving faster than the speed of light! This is called the problem of *superluminal expansion*. Of course, we do not actually observe the velocities of these components. We observe the rate of change of the angular separation from the center of the source, $d\theta/dt$, as shown in Fig. 19.12. We can convert this to a velocity only if we know the distance d to the source. If θ is measured in radians, then the speed is

$$v = (d\theta/dt)d$$

If our derived velocity is greater than c, then either (1) the sources are much closer than Hubble's law suggests, or (2) the apparent velocity doesn't represent a true physical velocity.

One explanation is based on the premise that we are not seeing one source moving. Instead, we are seeing a series of sources. Each source turns on as the previous one fades. This creates the illusion of motion, much as do the lights on a movie marquee. Unfortunately, this doesn't solve the problem. It is unlikely that the individual sources will turn on in sequence by chance. Some signal must be coordinating the time when each turns on. In order for us to see the superluminal expansion, the coordinating signal must be traveling faster than the speed of light. Having a signal traveling faster than the speed of light is just as bad as having an object travel faster than c.

There is an alternative explanation, involving a special relativistic effect. The situation is illustrated in Fig. 19.13. Suppose we have an object starting at point O and traveling to P, a distance r away. The object has a speed v, making an angle θ with the line of sight. In this arrangement, we take the x-direction to be along the line of sight. The y-direction is perpendicular to the line of sight, and the motion along the y-direction will

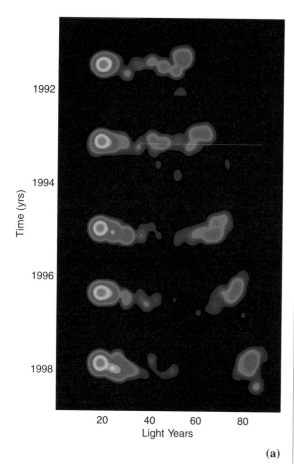

Time (yrs)

1992

1994

1996

1998

20   40   60   80
Light Years

(a)

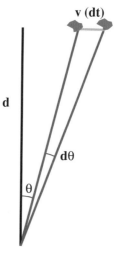

v (dt)

d

dθ

θ

Observer

(b)

Fig 19.12.   Superluminal expansion. (a) VLBA image of the radio jet 3C279, which shows apparent superluminal flows. Superluminal motion is shown in a 'movie' mosaic of five radio images made over seven years. The stationary core is the bright red spot to the left of each image. The observed location of the rightmost blue/green blob moved about 25 light years from 1991 to 1998, hence the changes appear to an observer to be faster than the speed of light, or 'superluminal'. The blue/green blob is part of a jet pointing within 2° to our line of sight, and moving at a speed of 0.997 times the speed of light. (b) The geometry of the problem. We measure a change in angular position $d\theta$ and relate that to a tangential velocity v by knowing the distance d. [(a) Glenn Piner, Whittier College/Wehrle, A. et al., Astrophys. J. Suppl., **133**, 297, 2001]

be detected as the proper motion. The distances x and y traveled along these two directions are

$$x = r \cos \theta$$

$$y = r \sin \theta \tag{19.4}$$

The time for the object to move the distance r is

$$t = r/v \tag{19.5}$$

However, a light or radio wave emitted from P has to travel a shorter distance than one emitted from O before reaching the observer. The path from P to the observer is shorter than the path from O to the observer by a distance x. This means that a light wave emitted from P takes x/c less time to reach us than one emitted at O. Therefore, from the point of view of the observer, the apparent time, t_{app}, for the object to travel from O to P is

$$t_{app} = t - x/c$$

Substituting from equations (19.4) and (19.5), we have

$$t_{app} = (r/v) - (r/c) \cos \theta$$

$$= (r/v) (1 - \beta \cos \theta) \tag{19.6}$$

where we have set $\beta = v/c$. The apparent velocity across the sky, v_{app}, is then

$$v_{app} = \frac{y}{t_{app}}$$

$$= \frac{r \sin \theta}{(r/v)(1 - \beta \cos \theta)}$$

Eliminating r gives

$$v_{app} = \frac{v \sin \theta}{(1 - \beta \cos \theta)} \tag{19.7}$$

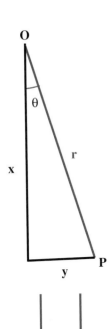

Fig 19.13. Apparent superluminal motion as a clump of material moves from O to P.

Substituting back into equation (19.7), we have

$$\left(\frac{v_{app}}{v}\right)_{max} = \frac{1}{(1 - \beta^2)^{1/2}} \qquad (19.8)$$

This is just the quantity γ that appears in the Lorentz transformations. We know that this quantity can become quite large as v approaches c.

Example 19.2 Superluminal expansion
For an object moving away from the nucleus of a galaxy at $v = 0.95c$, find the maximum value of v_{app} and the angle at which it must be moving to reach this maximum.

SOLUTION
We have

$$\cos \theta = \beta$$

so $\theta = 18.2°$. From equation (19.8) we have

$$v_{app} = \frac{0.95c}{(1 - 0.95^2)^{1/2}}$$

$$= 3c$$

There is a way to test this explanation. For a given speed and direction of motion, there should be a specific Doppler shift for radiation from the moving object. Unfortunately, these radio sources do not have any lines in their spectra. The Doppler shift alters the shape of the synchrotron spectrum, but the interpretation is difficult. Studies of the spectra of these sources are continuing.

19.3 | Seyfert galaxies

Seyfert galaxies are characterized by having nuclei that strongly dominate the total light from the galaxy. On a short exposure, they look like stars with a fuzzy patch around them and like a spiral galaxy on longer exposures (Fig. 19.14). For comparison, normal spirals never look like a star, even on a short exposure. This suggests that we are not resolving the nuclei of Seyferts in most of our images of them. (There are a few Seyferts for which the source of narrow emission lines has been resolved.) When we can study the fuzzy patch around the nucleus, it seems like the

For $v \ll c$, β is close to zero, and $v_{app} \cong v \sin \theta$, the expected result. However, for v close to c, v_{app} can be greater than v. In fact, v_{app} can be so much greater than v that it exceeds c.

To see this, we can find the angle that gives the maximum v_{app} for a given v. Taking equation (19.7), dividing by v, differentiating the result with respect to θ, and setting the result equal to zero, gives

$$\frac{\cos \theta}{(1 - \beta \cos \theta)} - \frac{\beta \sin^2\theta}{(1 - \beta \cos \theta)^2} = 0$$

Multiplying through by $(1 - \beta \cos \theta)^2$ gives

$$\cos \theta - \beta \cos^2 \theta - \beta \sin^2\theta = 0$$

Remembering that $\sin^2\theta + \cos^2\theta = 1$, this simplifies to

$$\cos \theta - \beta = 0$$

or

$$\cos \theta = \beta$$

We find $\sin \theta$ from

$$\sin \theta = (1 - \cos^2\theta)^{1/2}$$

$$= (1 - \beta^2)^{1/2}$$

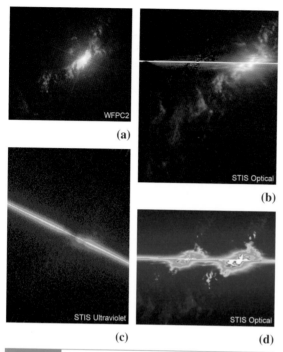

WFPC2
(a)

STIS Optical
(b)

STIS Ultraviolet
(c)

STIS Optical
(d)

Fig 19.14. The Seyfert galaxy NGC 1566. This is at a distance 15 Mpc. The active region in the center is found to vary on a time scale of less than a month. (a) HST Wide Field Planetary Camera 2 (WFPC2) image of the oxygen emission (5007 Å) from the gas at the heart of NGC 4151. Though the twin cone structure can be seen, the image does not provide any information about the motion of the oxygen gas. (b) In this STIS (imaging spectrometer) spectral image of the oxygen gas, the velocities of the knots are determined by comparing the knots of gas in the stationary WFPC2 image to the horizontal location of the knots in the STIS image. (c) This STIS spectral image shows the velocity distribution of the carbon emission from the gas in the core of NGC 4151. It requires more energy to make the carbon gas glow (CIV at 1549 Å) than it does to ionize the oxygen gas seen in the other images. (d) In this false color image the two emission lines of oxygen gas (the weaker one at 4959 Å and the stronger one at 5007 Å) are clearly visible. The horizontal line passing through the image is from the light generated by matter falling into the black hole at the center of NGC 4151. [STScI/NASA]

disk of a spiral galaxy. This suggests that we are seeing spiral galaxies with unusually bright nuclei.

The optical spectra (Fig. 19.15) are characterized by strong, broad emission lines. The lines are more than 10^3 km/s wide. This is more than ten times the width of lines in normal galaxies. If the broadening is thermal Doppler broadening, this would imply a temperature in excess of 10^7 K (see problem 19.10).

There are some differences among Seyferts in the appearance of lines called 'forbidden lines'. These are spectral lines that are not strong under normal circumstances. In some Seyferts, the forbidden lines are broad, and in others they are narrow. The ones with narrow lines are called type I, and the others are called type II. In addition, infrared emission from type I Seyferts is non-thermal, while that from type II Seyferts is thermal emission from dust. Also, type I Seyferts are weak radio sources, while half of the type II have moderate radio emission. Type I Seyferts also have strong X-ray emission, with a correlation between optical and X-ray luminosity. In Fig. 19.15, we see that the Seyfert spectra have similarities with other types of active galaxies.

The brightest Seyfert is shown in Fig. 19.14. It is an 11th magnitude galaxy in Coma Venatici. Ultraviolet emission lines from this galaxy show rapid variations in strength and width. To explain this phenomena, a nucleus with a black hole of mass $10^9 M_\odot$ has been proposed.

Another type of active galaxy is called a BL Lac object, named after the first one of this type observed. A spectrum of a BL Lac object is shown in Fig. 19.16.

19.4 | Quasars

19.4.1 Discovery of quasars

In our discussion of radio galaxies we mentioned the importance of predicting accurate radio positions for the purposes of finding optical counterparts. Before the use of interferometry, some sources were studied by lunar occultation. In such an experiment, the source is observed as the Moon passes in front of the source. Since we know the position of the edge of the Moon very accurately, we can determine the location of a radio source by noting the times at which the source disappears and reappears.

This technique was used at the Parkes radio telescope in Australia to study the radio source 3C273. (The designation means that it is the 273rd

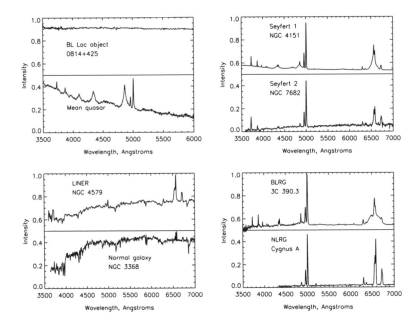

Fig 19.15. Line spectra of various types of active galaxies. [William Keel, University of Alabama]

source in the 3rd Cambridge catalog of radio sources.) When photographs of the area around the radio source were examined, a faint (13th magnitude) star was noted. This was an interesting discovery, since no radio stars were known at the time. Closer observations, producing photographs like that in Fig. 19.17, have shown that 3C273 is not really a star. It looks like a star with a fuzzy patch around it. The most detailed photographs show a jet extending from the core, just as in radio galaxies.

The optical spectrum of 3C273, shown in Fig. 19.18, is quite unusual. The spectrum puzzled astronomers for some time. It was finally noted that a series of lines looked like the Balmer series, but with a very large redshift. For example, the

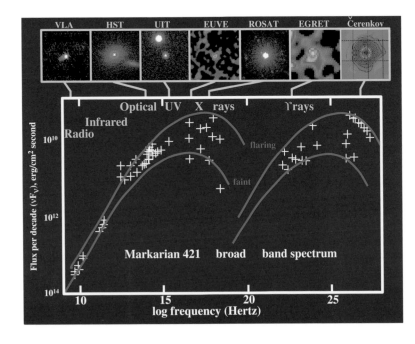

Fig 19.16. A BL Lac object, Markarian 421, showing its full spectrum over all wavelength ranges. [William Keel, University of Alabama]

Fig 19.17. An optical image of the quasar 3C273. Notice the jet to the lower right. [H.-J. Roeser, ESO, and William Keel, University of Alabama]

Hα line, whose rest wavelength is 656.3 nm, was observed at about 760.0 nm. This is a 15% shift in wavelength.

In Chapter 20 we will see that Hubble's law, as presented in Chapter 18, is not a useful description for finding the distance modulus to an object with a very large redshift. This is because it

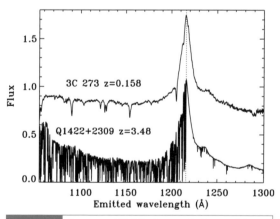

Fig 19.18. Spectrum of 3C273, showing the Lyman alpha line in emission and a number of other lines in absorption. The lower spectrum is of a quasar with z = 3.48, in which there are many absorption lines at lower redshift, the Lyman-alpha forest, corresponding to intervening material. [William Keel, University of Alabama, NASA, and L. Lu, Keck Observatory]

is not proper to think of the redshift as being a Doppler shift. We will see in Chapter 20 that a wavelength shift of 15% means that, in the time it takes light to get from 3C273 to us, the separations between all pairs of objects have increased by 15%. If we approximate the expansion rate as constant, then this means that the light has had to travel for 15% of the Hubble time to reach us. If the distance is d, then the travel time is d/c, so

$$d/c = 0.15 \, (1/H_0)$$

So

$$d = \frac{(0.15)c}{H_0}$$

$$= \frac{(0.15)(3 \times 10^5 \text{ km/s})}{(70 \text{ km/s/Mpc})}$$

$$= 640 \text{ Mpc}$$

We find the distance modulus, $m - M$,

$$m - M = 5 \log_{10} (d/10 \text{ pc})$$

$$= 5 \log_{10} (6.0 \times 10^7)$$

$$= 39.0$$

Since the apparent magnitude is 13, the absolute magnitude is -26. For comparison the absolute visual magnitude of the Sun is $+5$. A difference of 31 magnitudes corresponds to a brightness ratio of a little more than 10^{12}. This means that 3C273 gives off 10^{12} times as much visible light as does the Sun. What makes this even more remarkable is that 3C273 gives off more energy in the radio part of the spectrum than in the visible!

The proposal that the spectrum of 3C273 could be explained with a large redshift was made in 1963 by the Caltech astronomer *Maarten Schmidt*. At that time, the existence of a similar object, 3C48, was known. It had been noted in 1960 that there is a possible correspondence between the radio source and a 16th magnitude star. The spectrum of this star showed an even greater redshift than 3C273, corresponding to a 37% shift. These objects were given the name *quasi-stellar radio sources*, because of the starlike appearance on short-exposure photographs. The name was shortened to *QSR* or *quasar*. Further studies have revealed a class of objects that are like QSRs in optical photographs, and also have large redshifts, but don't have any radio emission. These are called *quasi-stellar objects,* or *QSOs.* We now loosely call both types of objects quasars.

19.4.2 Properties of quasars

The spectra of many quasars show a large ultraviolet excess. This means that they are brighter in the ultraviolet than one would expect from just knowing their visual brightness. This provides us with a way of searching for quasars. We cannot take spectra of all stars to see if they have large redshifts. Since quasars are so faint, it takes a long time to observe a spectrum. We can study radio sources, but not all quasars are radio sources. However, we can compare visible, blue and (now with space observations) ultraviolet images of large fields to find objects with a large ultraviolet excess. Spectra of these objects can then be taken to see if they have large redshifts.

Some quasars are quite variable in their energy output (Fig. 19.19). We have good records of the visible and radio variability of quasars for the past 30 years as a result of specific studies. However, we have optical records going back even farther, since observatories save photographic plates. A quasar may be on a plate exposed for an entirely different purpose. Once the quasar is discovered, an astronomer can go back through plate archives and find its image as far back as 100 years ago.

An important feature of this variability is that it allows us to place an upper limit on the size of the emitting region. We discussed this idea in our study of masers, in Chapter 15. If a significant fraction of the total power varies on a time scale t, then the emitting region can be no larger than ct (as long as motions close to the speed of light are not involved). In the case of quasars, variations on a time scale of a few months limit the size of the emitting region to about 10^{12} km. (This is only about 10^4 AU.)

The spectra of quasars, such as that in Fig. 19.20, show both emission lines and absorption lines. Generally, all of the emission lines can be explained by a single redshift z, but a few groups of absorption lines appear with different redshifts, always less than or equal to that of the emission lines. Often the spectrum is dominated by the Lyman-alpha line at various redshifts. (Remember, Lyα is the lowest transition and is often the most easily excited.) We sometimes refer to this as the *Lyman-alpha forest.* Over the years, extensive absorption line surveys have been carried out, providing good statistical information on absorption line properties. The absorption lines are generally narrow, less than 300 km/s

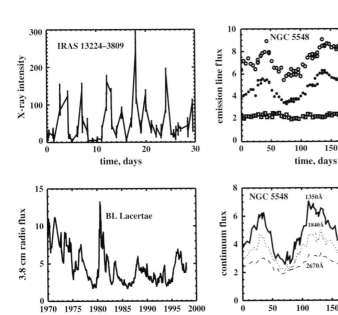

Fig 19.19. Light curves for various active galaxies. [William Keel, University of Alabama]

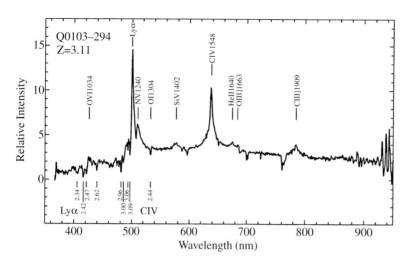

Fig 19.20. Spectrum of the quasar Q0103-294, at $z = 3.11$. [ESO]

wide. They are also mostly from the ground states of atoms, indicating a low temperature. All of this suggests that the absorption lines arise in material between us and the quasar.

At the time of this writing, the largest observed quasar redshift is 6.28. The light from the most distant quasars has been traveling over 90% of the age of the universe to reach us. This means that we are seeing the universe as it was before our own galaxy formed. Therefore, quasars provide us with an important link to our past.

19.4.3 Energy–redshift problem

The immediate problem that astronomers recognized with quasars was explaining their enormous energy output. What makes the problem even more difficult is the fact that the energy must be generated in a small volume. One way out of the problem is to say that quasars are not as far away as we think they are. After all, we only observe their apparent brightness. We infer their absolute brightness by knowing their distances, determined from the redshifts and Hubble's law. (It should be noted that even the factor of two uncertainty in the Hubble constant has no real bearing on this particular problem.) If we are saying that the distances are wrong, we are saying that the quasars do not obey Hubble's law. If that is the case, we must explain the large redshifts. This is the basic problem. Either we have to explain the large energy output, or we have to come up with another redshift mechanism. For this reason, we can think of this as the *energy–redshift problem*.

One possible source of redshift is gravitational. We have already seen that photons are redshifted as they leave the surface of any object. One problem with this explanation arises from the limited range of redshifts seen in the emission lines. This tells us that the emitting gas would have to be in a thin shell around some massive object. An analysis of such systems shows that quasars would have to be so close or so massive that our local part of the galaxy would be greatly affected by their presence. Even with a mass of 10^{11} $M_\odot$, the objects would still be within our galaxy.

There is an additional problem. For the larger redshifts, we need black holes. Even with a black hole to get a redshift greater than 100%, the photons must be emitted from very close to the Schwarzschild radius. We cannot think of any way to come up with a hot radiating gas close to R_S, especially given the narrow range of redshifts in a given quasar. The accretion disks responsible for X-ray emission around objects like Cyg X-1 are generally outside the photon sphere (defined in Section 8.4.2), which is at 1.5 R_S.

Another possible source of redshift is called kinematic. This means that we are observing a high velocity due to something other than the expansion of the universe. For example, they might be shot out of galaxies. However, if this were the case, we might expect to see some blueshifts comparable to the redshifts seen in quasars. To get them all moving away from us, we might think that they have been ejected from our galaxy, but the kinetic energy becomes quite large.

Also, if there were nearby objects moving at high speeds, we would expect to see some proper motion, and we don't.

In an effort to see if some kinematic explanation is possible, some effort has been made to search for galaxies and associated quasars, where the quasar and galaxy have very different redshifts. If it can be shown that a galaxy and a quasar are directly connected, then they must be at approximately the same distance from us. Presumably the distance is accurately indicated by the redshift of the galaxy. If the quasar has a different redshift, then it doesn't obey Hubble's law. Examples have been found of galaxies and quasars that appear near each other on the sky, but have different redshifts. The problem, however, is in proving that the quasar and galaxy are associated. Just because two objects appear along almost the same line of sight doesn't mean that they are at the same distance. A galaxy and a quasar don't have to be at the same distance any more than two stars in the same constellation have to be at the same distance.

Sometimes arguments over whether galaxies and quasars that appear near each other are associated come down to questions of probability and statistics. For example, we might ask, given a quasar, what is the probability that there will be an unassociated galaxy along the same line of sight? Arguments such as this are difficult to frame, and also can lead to fallacious conclusions, akin to a person who takes a bomb on board an airplane to safeguard his passage, on the assumption that if the probability of one person having a bomb on a plane is low, the probability of two people having bombs is much lower. To date, the statistical arguments seem to allow all of the cases of galaxies and quasars in the same direction as being coincidences, with the two being unassociated.

Some effort has also gone into trying to observe direct connections between the galaxy and quasar in such situations. Connections would be in the form of a detectable trail of luminous material between the galaxy and the quasar. Even this problem is not easy, because there are photographic artifacts which can resemble such a bridge between two objects. To date, no convincing evidence has been presented to demonstrate a bridge between a quasar and a galaxy.

In addition, there are direct pieces of evidence that quasars are at cosmological (Hubble's law) distances. The absorption lines in quasar spectra are one example. As we have said they are generally at redshifts that are less than the emission lines, and the most natural explanation is that they are from material between us and the quasar. If the intervening material obeys Hubble's law, then the quasars must be at cosmological distances. Also, the fact that Seyferts appear to be galaxies supports the belief that quasars are also galactic in scale and far away, rather than small and nearby. Finally, the observation of the gravitational lensing of quasars by massive intervening galaxies (discussed in the next section) means that the quasars must be more distant than the intervening galaxies.

Much of the concern over the energy–redshift problem faded in the 1990s. Astrophysicists now feel that the energy generation requirements in quasars are not as outrageous as they seemed in the 1970s. This is partly because of the number of high energy phenomena that have been observed and understood, at least partially. More specifically, our understanding of radio galaxies has reached the point where we think that quasars may be a different manifestation of the same phenomenon. The energy requirements for quasars are large, but not much larger than those for the most luminous radio galaxies. This point will be discussed farther in Section 19.6.

The quasar problem is in many ways typical of problems in astronomy. On the observational side, it provides an excellent example of the interplay among observations in various parts of the spectrum. Radio and optical observations were important in the discovery of the phenomenon. Radio, optical, infrared, ultraviolet and X-ray observations have been important in understanding the phenomenon. In addition, our understanding of quasars has required the results of extensive systematic surveys of quasar properties. Such surveys don't receive the same glory as the initial discovery, but they provide the data on which a solution can ultimately be based.

The quasar problem is also typical in another way. Many problems in astronomy come upon the scene in a spectacular fashion, often with an unexpected observation of a new phenomenon. The solutions, however, do not come in a single, spectacular step. For our understanding of quasars, it has been important to have the problem around long enough for astrophysicists to become more comfortable with the energies required. Part of the process is the discovery of other high energy phenomena. For example, when it was suggested in the 1980s that quasars might involve some form of black hole, the ideas was not taken seriously by most. After all, black holes were theoretic playthings. However, the observations of objects such as Cyg X-1 have made the existence of black holes seem more likely. It is a large jump from a few $M_\odot$ black hole to a $10^8 M_\odot$ one, but it is not as big a jump as from no black holes to a few $M_\odot$ black hole. In addition, the success in explaining many of the details of radio galaxies has made astrophysicists more confident of the model that places a massive black hole in the center of a galaxy. (We may even have one at the center of our galaxy, as noted in Chapter 16.)

19.5 | Gravitationally lensed quasars

When we discussed general relativity and black holes in Chapter 8, we saw that a strong gravitational field can bend the path of a light ray. In Chapter 4, we saw that when we allowed glass (a lens) to bend light rays, images were moved or altered. Is it possible that gravitational bending of light can also produce a lens effect?

By looking at Fig. 19.21, we can see how this might happen. In this case we imagine that the light is coming from a distant quasar, and the bending is caused by a galaxy midway between us

and the quasar. (Clearly, the alignment of the quasar and galaxy are completely accidental. The two have no physical connection.) Let's first look at the ray from the quasar that passes over the top of the galaxy. It is bent slightly downward. When that ray reaches us, we would see an image of the quasar back along the line of the ray reaching us. The image would appear higher than the quasar itself. We can also look at the ray going under the quasar. It is bent upward, and results in the image of the quasar being below the actual quasar. In fact, if we analyzed rays that go all around the galaxy, each produces an image in a slightly different position. When you look at the quasar, you will see a ring of images around the galaxy. This is called an *Einstein ring.*

In real situations we are not likely to find the quasar exactly in line with the galaxy. If it is slightly off line, the ring breaks up into a group of distinct images. Also, if it is off line, the various rays will travel different distances to reach us. This means that each image of the quasar is a 'snapshot' of the quasar at a slightly different time. If the quasar is varying in brightness, the variations will appear in all of the images, but there will be a delay from one image to the next.

Gravitational lensing of quasars has been observed, but the first detection was quite accidental. Astronomers simply noticed two quasars that appeared virtually identical. They had the same spectra and redshifts. The two quasars had the same variations in brightness, but with the variations in one quasar following those in the other by a fixed time delay. It was not obvious that this was a lensing situation, because the intervening galaxy was not bright enough to be seen in the original images. Further observations revealed its presence.

Now that we know that such lensing situations exist, astronomers have searched systematically for more of them. These searches have been done both in the optical and radio parts of the spectrum. Some sample optical images of lensed quasars are shown in Fig. 19.22. The observation of gravitationally lensed quasars is important for a number of reasons. Probably most important, it is a demonstration that the quasars must be at

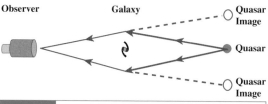

Fig 19.21. Diagram of geometry for the gravitational lensing of a quasar. The size of the shift is hot to scale.

Fig 19.22. Observations of lensing of quasars. (a) Lensing in a rich galaxy cluster Abell 2218. In this HST image, some background objects appear as arcs. (b) PG115 + 080. This ring is the result of gravitational lensing of a quasar. The original image is on the left. The image on the right is processed so that the quasar is removed and the ring remains (See also Fig. 8.1). [STScI/NASA]

(a)

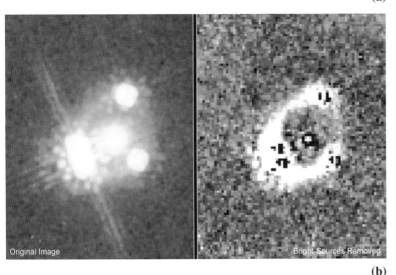

Original Image   Bright Sources Removed

(b)

the distances indicated by their redshifts. This is because the quasar must be at least as far away as the intervening galaxy. Usually the lensing will brighten the image of the quasar. This allows us to study quasars at larger redshifts than we might normally be able to detect. Also, it provides a way of measuring the mass of the intervening galaxy. This allows us to tell how much dark vs. luminous matter there is in that galaxy.

Example 19.3 Mass of lensing galaxy
Suppose we observe the image of a quasar to be shifted by 5 arc sec on the sky. What is the mass of the intervening galaxy? Assume that the light passes past the edge of the galaxy's disk, which has a radius of 50 kpc.

SOLUTION
From equation (8.1) we have

$$\theta = 4\,GM/bc^2$$

where θ is the angle of deflection in radians and b is the impact parameter. Solving for M:

$$M = \frac{\theta\, bc^2}{4\,G}$$

$$= \frac{(5\text{ arc sec})(50\text{ kpc})(3.08 \times 10^{21}\text{cm/kpc})(3 \times 10^{10}\text{cm/s})^2}{(2.06 \times 10^5\text{arc sec/1 rad})(4)(6.67 \times 10^{-8}\text{dyn cm}^2/\text{s}^2)}$$

$$= 1 \times 10^{46}\text{ g}$$

$$= 6 \times 10^{11}M_{\odot}$$

19.6 | A unified picture of active galaxies?

19.6.1 A common picture

When we look at the various examples of AGNs, we see many features in common: large energy output, rapid variability meaning small sources of energy, and jets. The explanations involve massive black holes fed by material passing through an accretion disk. There are also uniform outflows that can be collimated into narrow jets. This leads us to ask whether radio galaxies, Seyferts and quasars are different manifestations of the same basic phenomenon.

The common features would be supermassive black holes in the centers of galaxies. These black holes would be surrounded by accretion disks, with material being fed in, converting gravitational potential energy into kinetic energy and, ultimately, radiation. There are also outflows that are collimated into jets. There are also dense clouds close to the center which give rise to the emission lines that we see.

The particulars of any AGN phenomenon then depend on various parameters. For example, the total luminosity would depend on the mass of the black hole (the more massive, the greater the rate at which we can drop in material and efficiently get energy out) and the amount of material that is available to spiral into the black hole (the amount of fuel that is available for the engine). Some particulars of what we observe will depend on the orientation of the disk with respect to the line of sight. The existence of radio lobes would depend on the AGN having been on long enough for the material to accumulate at the end of the jets.

In this picture massive black holes are present in the nuclei of most large (i.e. not irregular or dwarf) galaxies. Where did these black holes come from? Their universality, and the fact that we see quasars as being quite prominent in the early universe, suggest that these massive black holes must be a natural byproduct of galaxy formation. We will discuss the issue of galaxy formation and dark matter in Chapter 21, but for now we will note that some computer simulations on how galaxies might have formed in the early universe indicate that black holes with masses in the range 10^6 to $10^9\,M_\odot$ were formed in the process, and are naturally found at the centers of galaxies. These black holes are probably rotating, so they will not have the simple structure discussed in Chapter 8. These black holes would become more massive in their active phase. We might wonder about the fate of these black holes. If their host galaxies are isolated, then we might expect the black holes to sit quietly in the centers, long after the fuel to feed the engine has run out. However, if there are other galaxies nearby, then black holes might be ejected in three-galaxy encounters. So, in the present (nearby) universe, we might expect to find some galaxies with black holes and others without. We describe below searches for supermassive black holes in galactic nuclei.

We would expect that the gas surrounding the black hole would settle into a disk. Material spirals in and liberates energy as it becomes closer. The energy can be in the form of radiation, but there are also some mechanical ways in which the energy can be extracted. The simplest models suggest a thin disk, but theoreticians are now able to make more complicated models, and the possibility of thicker disks has been suggested. It may also be that the structure of the disk is affected by magnetic forces.

The continuum emission comes mostly from a region that is several Schwarzschild radii across. This region would have a temperature of a few times 10^5 K, and would therefore give off blackbody radiation that peaks in the ultraviolet, as observed. Other processes also contribute to the continuum, so we don't see exactly a blackbody spectrum. The radiation that we see also depends on the optical depth of the disk and its orientation relative to the line of sight.

In the simplest version of the model, the broad emission lines come from a number of small dense clouds (densities of about 10^9 cm^{-3} and sizes of about 1 AU). These clouds are in random orbits about the central source. They are at a distance from the central source where the density of ionizing photons is about 10^8 cm^{-3} (see Problem 19.17 for the energy density). More detailed analyses indicate that this gas can exist over a larger range of distances from the center, so there is a wider range of physical conditions. Also, recent studies have shown that there is a lot of molecular material close to the centers of AGN. This molecular material can be quite opaque, and, if viewed from the right angle, can hide material on the far side of the center, resulting in asymmetric line profiles.

A further suggestion of how a massive black hole could exist in the nucleus of a starburst galaxy is that the starburst could provide a very rich cluster near the center. The total mass of the stars in the cluster could be in excess of $10^6 \ M_\odot$. Of course, when the cluster is formed, it will not be contained within its Schwarzschild radius. Initially, the motions of the stars within the cluster will keep it from collapsing farther. However, there may be dissipative processes which allow the stars to lose some of their kinetic energy, and settle closer together. If enough energy is disspipated, the cluster can eventually contract to its Schwarzschild radius and become a massive black hole.

19.6.2 Black holes in galactic nuclei?

Obviously, if this general picture is to work then there must be supermassive black holes in the centers of active galaxies. This suggests that we search for observational evidence of those black holes. Also, since quasars seem to have been more prevalent in the past (higher redshift) than now,

we might infer that the AGN phase is something many galaxies went through in the past. If this is the case, then even nearby galaxies with no nuclear activity might have black holes left over from an earlier active phase. Presumably there is currently no activity because there is no material to fall in (no fuel to feed the engine). We might therefore find supermassive black holes in nearby galaxies. The study of nearby galaxies is important since we can study them in the most detail.

The presence of a massive object in the center of a galaxy can be inferred from its gravitational effects on the motions of surrounding objects. In the nuclei of galaxies we have many stars, and, depending on the type of galaxy, we may or not have a lot of gas. Both the stars and the gas will move in response to the presence of a massive object near the center of the galaxy. From the motions (essentially v_r vs. distance from the center) we can learn about the mass of the central object. There is a slight complication because the motions that we are studying are due to rotation and to random motions. In this case we can write the mass interior to radius r as

$$M(r) = (r / G) [v_{rot}^2 + \sigma^2]$$

where v_{rot} is the rotational speed and σ is the rms velocity dispersion for the random motion. When possible, studying the motions of the stars is more reliable than the gas, since the stars are less

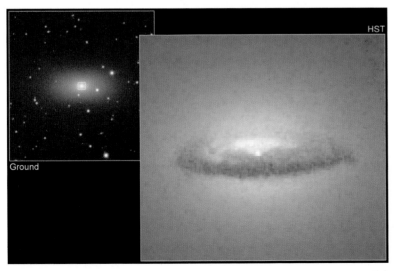

Fig 19.23. Evidence for massive black holes in galactic nuclei. (a) HST image of NGC 7052, showing what may be a massive dusty disk, edge-on.

HST

Ground

(a)

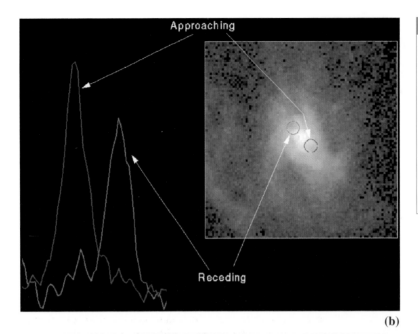

Fig 19.23. (*Continued*) (b) Spectra on opposite sides of the center of M87. These show a very large velocity shift across the center. (c) Imaging spectrometer (STIS) map of M84. The image on the left is a regular (WFPC2) image, with the blue outline showing the line along which spectra were taken. On the right we see the spectrum, showing a very large velocity shift across the nucleus. [STScI/NASA]

(b)

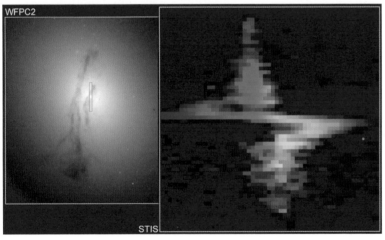

(c)

likely to have their velocities affected by other processes (such as supernova explosions).

So, the first step in looking for a black hole is to use the motions of surrounding objects to measure the mass of the central object. However, this is not enough. We must also show that the supermassive object is contained within the Schwarzschild radius (3×10^9 km for a $10^9\ M_\odot$ black hole). This is hard, because, even with the resolution of HST or VLBI on the nearest galaxies (see Problem 19.18), we can only reach about $10^5\ R_S$.

So, we can hope to show that there are condensed supermassive objects in AGNs, but to this point we cannot conclusively show that such objects are black holes.

To date, studies of motions near the centers of 14 galaxies have produced evidence for a massive compact object. Some of these are AGNs and others are 'normal' nearby galaxies. For example, one of the best cases is M31, the Andromeda galaxy, with a central object mass of $3 \times 10^7\ M_\odot$. As discussed in Chapter 16, there is evidence that the Milky Way contains a $3 \times 10^6 \ M_\odot$ condensed object. The most massive object found to date is $3 \times 10^9 \ M_\odot$ in the well studied active galaxy, M87 (Fig. 19.8). Enough of these objects have been studied to begin to look for correlations between the mass of the central object and other properties of the galaxy. The strongest correlation is with the brightness of the bulge. There is no correlation with the total mass of the galaxy.

Chapter summary

We find a number of galaxies with large amounts of energy coming from a relatively small region in the center of the galaxy. While these used to be thought of independently, we now realize that there are many common physical features of these systems, and we discuss them collectively as active galactic nuclei, or AGNs.

Starburst galaxies appear to have recently undergone a large episode of massive star formation near their centers. While many of them look quite normal in the visible part of the spectrum, since dust blocks the light from most of the stars, that dust is heated and glows strongly in the infrared. In relatively nearby starburst galaxies, we can make CO maps of the molecular clouds out of which the stars are forming. We think that starbursts can occur when two galaxies pass close together, and interstellar matter from one galaxy is transferred to the center of the other. This gas provides the fuel for the starburst.

Some galaxies give off tremendous amounts of radiation in the radio part of the spectrum. These are the radio galaxies. They typically give off a million times as much energy as a normal galaxy. The radiation is polarized, and much more energy is given off at longer wavelengths than shorter wavelengths. These two features suggest that we are observing synchrotron radiation, which means a strong magnetic field and a good supply of high energy electrons. For most galaxies we see the emission coming from two large lobes on either side of the optical part of the galaxy. The lobes are typically 10 Mpc long by 1 Mpc wide. Narrow jets point from the center of the galaxy out to the lobe.

We think that the jets can be produced by a flow in all directions at the center, but collimated by a disk close to the center. The most likely source of energy for the jets is a massive black hole (up to 10 million solar masses of material) into which matter falls. Up to 40% of the rest energy of that matter is converted into energy flowing out.

Seyferts appear to be spiral galaxies with active nuclei. They have spectra that are characterized by strong broad emission lines.

Quasars are also galaxies that seem to have many properties in common with radio galaxies. Many of them have large redshifts, suggesting that they are at great distances. For quasars to appear as bright as they do and be so far away, they must be giving off large amounts of energy. Also, the energy output varies rapidly, meaning that the source of that energy must be very small. We now think that quasars are powered by matter falling into very massive black holes near their centers. If quasars are as far away as their redshifts indicate, then we are seeing them as they were a long time ago, and they give us a way of looking back to see how the universe was at a much earlier time.

Quasars have both emission and absorption lines. We think that the absorption lines are produced by material between us and the quasar. We have now also detected quasar images that are produced by gravitational lensing by intervening material.

Searches for supermassive black holes in galactic nuclei have produced more than a dozen objects with masses ranging from $3 \times 10^7 \ M_\odot$ to $3 \times 10^9 \ M_\odot$. We still do not have sufficient angular resolution to say definitively that these central objects are contained within the Schwarzschild radii.

Questions

19.1. What is the evidence that the Lyman alpha forest arises in material between us and the quasar?

19.2. In this chapter, we saw various examples of active galactic nuclei. List these types. What are the similarities among them? What are the differences that distinguish one type from another?

19.3. What makes us think that starburst galaxies are sites of very active star formation?

19.4. Why is it unlikely that the current rate of star formation in a starburst galaxy can be sustained for a long time?

19.5. What does the detection of many supernova remnants tell us about starburst galaxies?

19.6. In studying starburst galaxies, we make observations in various parts of the spectrum. List them, and state briefly what we learn from each.

19.7. What is the suspected role of galaxy interactions in creating starbursts?

19.8. How do we know that the radio emission from radio galaxies is synchrotron? What does this tell us about the regions that are emitting?

19.9. In a radio galaxy, emission comes from a very large area. What makes us think that the ultimate source of this energy is the nucleus of the galaxy?

19.10. In radio galaxies, high energy particles are transported from the inner parts of the galaxy to the lobes. Once at the lobes, why don't they quickly lose their energy?

19.11. In a radio galaxy, how do we think that the observed jets are formed?

19.12. Why do we think that the ultimate source of energy in a radio galaxy is matter falling onto a black hole? Why is it important for the black hole to be very massive?

19.13. In calculating the energy we can extract by dropping matter into a black hole, we only considered the energy gained in falling to the Schwarzschild radius. However, matter will still accelerate after crossing R_S. Why don't we add this extra energy to what we can extract?

19.14. If we have a particle and an antiparticle annihilating and producing energy, what is the efficiency of that reaction (in terms of the fraction of the mass converted into energy)?

19.15. What are the similarities between radio galaxies and quasars?

19.16. Why were lunar occultations important in the discovery of quasars? How might the discovery of quasars have come sooner if radio interferometry had come sooner?

19.17. What is the problem that was posed by quasars being at the distances indicated by their redshifts (that is, at cosmological distances)?

19.18. Why is the variability of quasars important?

19.19. How would you carry out a search for radio-quiet quasars?

19.20. What do we mean by the 'energy–redshift problem'?

19.21. What evidence is there that quasars are truly at the distances implied by their redshifts?

19.22. How do quasars allow us to see the universe as it was in the distant past?

19.23. Why is the discovery of gravitationally lensed quasars important?

Problems

[For all problems, unless otherwise stated, use $H_0 = 70 \ km/s/Mpc$.]

19.1. (a) To the extent that we can ascribe a temperature to the large lobes of radio galaxies, it is about 10^6 K. For the numbers given in this chapter, estimate the thermal energy stored in one of these lobes. (b) Calculate the magnetic energy, and compare it with the answer in (a).

19.2. For the sizes, densities and speeds given in the chapter, estimate the kinetic energy in the flow of the jets of radio galaxies.

19.3. A radio galaxy has an angular extent of $2°$, and the associated optical galaxy has its spectral lines shifted by 15% of their rest wavelength. What is the diameter of the radio galaxy?

19.4. What are wavelengths of the Hβ and Hγ lines in 3C273?

19.5. What are the wavelengths of the Hα and Hβ lines in 3C48?

19.6. Using the same reasoning as was used for 3C273, estimate the distance to 3C48.

19.7. A quasar is observed with its Hα line at 800.0 nm. Estimate its distance.

19.8. Estimate the redshift of a quasar whose light has been traveling 1 Gyr to reach us.

19.9. What redshift would be needed to shift the Lyα line into the visible part of the spectrum?

19.10. For the 10^3 km/s wide emission lines in Seyferts, what temperature would be required for thermal Doppler broadening?

19.11. For the superluminal source discussed in Example 19.2 (with β = 0.95), what is the range of angles about 18.2° for v_{app} to be greater than c?

19.12. What β is required , assuming the optimal angle, to produce a superluminal source with $v_{app} = 10c$?

19.13. If a quasar varies on a time scale of four months, how does the maximum size of the emitting region compare with the Schwarzschild radius for a 10^9 $M_{\odot}$ black hole?

19.14. At what rate must mass be falling into a black hole to produce the luminosity of a typical radio galaxy?

*19.15. Suppose we try to explain the redshift of 3C273 as a gravitational redshift, rather than cosmological. (a) From what radius, relative to R_S, is the radiation emerging? (b) If the width of the lines is 0.1% of their wavelength, what is the range of radii (relative to R_S) from which the radiation can be emitted?

19.16. What are the Schwarzschild radius and density for a 10^9 $M_{\odot}$ black hole?

19.17. What is the energy density if the density of photons is 10^8 cm^{-3} ? Assume that the photons are all at the peak wavelength of a 10^5 K blackbody.

19.18. Compare the angular resolution for HST observations and for VLBI observations with the Schwarzschild radii for the M31 and M87 central objects at their respective distances.

19.19. For the galaxy NGC 3115 we find a rotational speed of 150 km/s and the dispersion for random motions of 270 km/s at an angular radius of 1.0 arc sec. What is the mass of the central object?

19.20. For the galaxy M87 we find negligible rotation and the dispersion for random motions of 350 km/s at an angular radius of 1.5 arc sec. What is the mass of the central object?

Computer problems

19.1. You drop an object of mass m into a black hole. Calculate the energy it has acquired when it reaches 2 R_S, 1.5 R_S, 1.1 R_S and 1.01 R_S. Express your answer as a fraction of the maximum energy (when reaching R_S).

19.2. Plot a graph of v_{app} vs. θ for v/c = 0.50, 0.90, 0.95, 0.99.

19.3. Estimate the angular sizes of the R_S for 10^6 $M_{\odot}$, 10^7 $M_{\odot}$, 10^8 $M_{\odot}$, 10^9 $M_{\odot}$ black holes for an object in the Virgo cluster and for an object at redshift 1.

19.4. What are the orbital speeds at 2 R_S, for objects in circular orbits around 10^6 $M_{\odot}$, 10^7 $M_{\odot}$, 10^8 $M_{\odot}$ and 10^9 $M_{\odot}$ black holes?

Chapter 20

Cosmology

Einstein once said that the most incomprehensible thing about the universe is that it is comprehensible. It is amazing that we can apparently describe the universe with what are very simple theories. We can ask truly fundamental questions of where we have come from and where we are going and expect scientific answers. In this chapter and the next, we will study *cosmology*, the large-scale structure of the universe. We can learn a great deal using only the physics we have introduced in this book. With the introduction of some more physics, namely elementary particle physics (in Chapter 21), we will see that even more fascinating concepts are within our grasp.

20.1 | The scale of the universe

When we study the gas in a room, we must deal with it as a collection of molecules. We don't care about the fact that the molecules are made up of atoms or that the atoms are made up of protons, neutrons and electrons; or that the protons and neutrons are made up of other particles. All we care about is how the molecules interact with one another, and how that affects the large-scale properties of the gas. When we study the universe, we also treat it as a gas. The molecules of the gas are galaxies. In the big picture, stars, planets, etc., don't matter. Of course these smaller objects can still contain some hidden clues for us to learn about the larger structure. They simply don't affect the larger structure themselves.

How do we study cosmology? On the theoretical side, we look at the universe as large-scale fluid and ignore the lumps. The only force that currently affects the large-scale structure is gravity. We can apply gravity as described by general relativity, though for many things Newtonian gravitation is a sufficiently accurate approximation. The electromagnetic force is important in that electromagnetic radiation carries information, but it doesn't affect the large-scale structure. We will see in the next chapter that there was a time when radiation was dominant and all of the forces we know had an important effect on the structure of the universe.

Until recently, we have had very few observational clues about the large-scale structure of the universe. We will see that recent experiments, some characterized by great difficulty and resourcefulness, have greatly added to our knowledge. The field of observational cosmology is a growing one. In studying cosmology, as with other fields of astrophysics, we combine theory and observations to increase our insight into what is happening.

In making theoretical models of the universe, we start with an assumption, called the *cosmological principle*. It says that *on the largest scales the universe is both homogeneous and isotropic*. By homogeneous, we mean that, at any instant, the general properties, such as density and composition, are the same everywhere. By isotropic, we mean that, at any instant, the universe appears the same in all directions. We know that our everyday world is neither homogeneous nor isotropic, but for the universe, on the scales of many superclusters, this is a very good description.

There is another assumption that we might be tempted to make, namely that the universe is the same at all times. Cosmological theories which incorporate this assumption are called *steady-state theories*. Until the early 1970s, there were vigorous debates about whether or not the universe is steady-state. The theory was favored by many because it had a certain philosophical simplicity. If the universe had no beginning, then there is no need to worry about what went on "before" the beginning.

However, a long chain of observational evidence has been amassed against the steady-state theory, and very few hold it today. For example, we will see that the relative abundance of hydrogen and helium can easily be explained in evolving theories, but not in steady-state theories. In addition, the existence of many quasars some ten billion years ago, but very few now, argues for the conditions in the universe changing over that past ten billion years.

You might think that the steady-state theory would have died with the observation of the expansion of the universe (Hubble's law). If the universe is expanding, its density must be decreasing. If the density is not constant with time, we cannot have a steady-state universe. However, proponents of the steady-state theory pointed out that it might be possible to create matter out of *nothing*. We do not mean creation of mass from energy, but from nothing. This theory, called *continuous creation*, calls for the violation of conservation of energy. (This violation would be permanent, not for short times for the temporary particle–antiparticle creation discussed in Chapter 8.) However, it requires a violation at a level well below that to which conservation of energy has been verified experimentally. We know of no mechanism to create this matter, but proponents of this idea say that there is no experimental evidence to rule it out. It would require the creation of the mass of one proton per year in a box of side 1 km (Problem 20.4).

Evolving theories, called *big-bang theories* finally took the upper hand with the discovery, in 1965, of the *cosmic background radiation (CBR)*. This radiation is a relic of a time when the universe was much hotter and denser than it is now. We will discuss this radiation in detail in Chapter 21. For now, we note that this discovery ended, for most astronomers, the debate over whether the universe is steady-state.

20.2 | Expansion of the universe

20.2.1 Olbers's paradox

There is a very simple observation that we can make every day that tells us about the history and structure of the universe. This simple observation is that the sky is dark at night. The mystery in this observation was noted by *Heinrich Olbers* in 1823. We refer to the problem as *Olbers's paradox*. Qualitatively, Olbers said that, for an infinite universe, every line of sight should eventually end up on the surface of a star. Therefore, the whole night sky should look like the surface of a star. In fact, the whole day sky should appear the same, since the Sun would just be one of a large number of stars.

It may be easier to understand in terms of the quantitative argument, illustrated in Fig. 20.1. Suppose we divide the universe into concentric

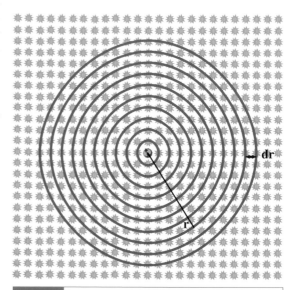

Fig 20.1. Olbers's paradox. We consider the contribution to the night sky from a shell of radius r and thickness dr. (The regular distribution of the stars is a simplification to help view how the number of stars per shell goes up with r. In reality the stars have an irregular distribution, as we have seen in earlier chapters.)

spherical shells, centered on the Earth, each of thickness dr. The volume of each shell is then

$$dV = 4\pi r^2 \, dr$$

If there are n stars per unit volume, the number of stars in each shell is

$$N = 4\pi r^2 n \, dr \qquad (20.1)$$

The number of stars per shell goes up as r^2. However, the brightness we see for each star falls as $1/r^2$. Therefore, the r^2 and $1/r^2$ will cancel, and the brightness for each shell is the same. If there are an infinite number of shells, the sky will appear infinitely bright after we add up the contributions from each shell. Actually, this isn't quite the case. The stars have some extent, and eventually the nearer stars will block the more distant stars. However, this will not happen until the whole sky looks as if it is covered with stars, with no gaps in between.

There might appear to be some obvious ways out of this. You might say that our galaxy doesn't go on forever. Most lines of sight will leave the galaxy before they strike a star. Unfortunately, the argument can be recast in terms of galaxies instead of stars, and the same problem applies. Another possible solution is to invoke the absorption of distant starlight by interstellar dust. However, if the universe has been around forever (or even a very long time), the dust will have absorbed enough energy to increase its temperature to the same as that of the surface of a star. If the dust became any hotter than that, the dust would cool by giving off radiation which the stars would absorb. Even if the sky was not bright from the light of stars, it would be bright from the light of dust. (Scattering by dust wouldn't help because it would make the sky look like a giant reflection nebula, again like the surface of a star.)

The redshift due to the expansion of the universe is of some help. The energy of each photon is reduced in proportion to the distance it travels before we detect it. This adds an additional factor of $1/r$ to the apparent brightness of each shell, meaning that the brightness of each shell falls off as $1/r$, instead of being constant. Note that this doesn't completely solve the problem. If we add up the contributions from all of the shells, we

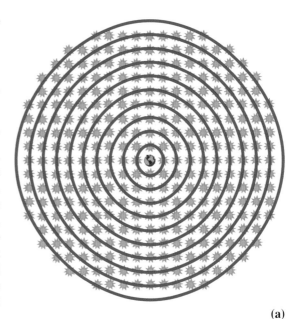

(a)

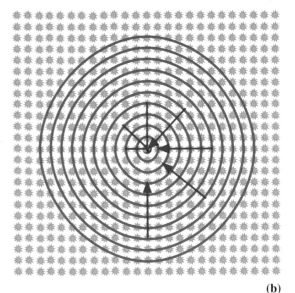

(b)

Fig 20.2. (a) How a finite size for the universe helps with Olbers's paradox. (b) How a finite age for the universe helps with Olbers's paradox.

have to take the integral of dr/r (see Problem 20.3). This gives us $\ln(r)$, so if r can be arbitrarily large the brightness can be also.

It is possible to get out of the problem if the universe has a finite size. This is illustrated in Fig. 20.2(a). If there is a finite size, then we cut off our integral at whatever that size is. There is another way to achieve the same effect. This is if the universe has a finite age t_0, as illustrated in Fig. 20.2(b). We can only see stars that are close enough for their light to have reached us in this time. That is, we can cut off r at ct_0. There may even be a cutoff before this because it took a certain amount of time for stars and galaxies to form. So, there is a finite cutoff to the number of shells that can contribute to the sky brightness, and the problem is solved. It is amazing that this simple observation – that the night sky is dark – leads to the conclusion that the universe has a finite size or age (or both).

20.2.2 Keeping track of expansion
We can show that Hubble's law follows from the assumption of homogeneity. In Fig. 20.3, suppose that P observes two positions, O and O', with distance vectors from P being $\mathbf{r}$ and $\mathbf{r}'$, respectively. The vector from O' to O is $\mathbf{a}$, so that

$$\mathbf{a} = \mathbf{r} - \mathbf{r}' \tag{20.2}$$

We let $\mathbf{v}$ be a function to give the rate of change of length vectors ending at O and $\mathbf{v}'$ the corresponding function for length vectors ending at O'. We then have

$$\mathbf{v}'(\mathbf{r}') = \mathbf{v}(\mathbf{r}) - \mathbf{v}(\mathbf{a}) \tag{20.3}$$

Using equation (20.2) to eliminate $\mathbf{r}'$, this becomes

$$\mathbf{v}'(\mathbf{r} - \mathbf{a}) = \mathbf{v}(\mathbf{r}) - \mathbf{v}(\mathbf{a}) \tag{20.4}$$

The homogeneity of the universe means that the functional form of $\mathbf{v}$ and $\mathbf{v}'$ must be the same. (The functional form of the velocity cannot depend on where you are in the universe.) This means that

$$\mathbf{v}'(\mathbf{r} - \mathbf{a}) = \mathbf{v}(\mathbf{r} - \mathbf{a}) \tag{20.5}$$

Using this, equation (20.4) becomes

$$\mathbf{v}(\mathbf{r} - \mathbf{a}) = \mathbf{v}(\mathbf{r}) - \mathbf{v}(\mathbf{a}) \tag{20.6}$$

This means that $\mathbf{v}(\mathbf{r})$ must be a linear function of $\mathbf{r}$. The only velocity law that satisfies this relationship is

$$\mathbf{v}(\mathbf{r}) = H(t)\,\mathbf{r} \tag{20.7}$$

Note that we haven't required the expansion (H could be zero). However, if there is an expansion, it must follow this law, if the cosmological principle is correct.

When we want to keep track of the expansion of the universe, it is not convenient to think about the size of the universe. Instead, we introduce the *scale factor* which will keep track of the ratios of distances. We let t_0 be the age of the universe at some reference epoch. (It doesn't matter how this reference is chosen.) We let $\mathbf{r}(t)$ be the distance between two points as a function of time. (The points must be far enough apart so that their separation is cosmologically significant.) We define

$$\mathbf{r}_0 = \mathbf{r}(t_0) \tag{20.8}$$

The scale factor $R(t)$ is a scalar, defined from

$$R(t) = \mathbf{r}(t)/\mathbf{r}(t_0)$$
$$= \mathbf{r}(t)/\mathbf{r}_0 \tag{20.9}$$

Note from this definition that $R(t_0) = 1$. If the universe is always expanding, $R < 1$ for $t < t_0$, and $R > 1$ for $t > t_0$.

We can rewrite Hubble's law in terms of the scale factor. We start by writing Hubble's law as

$$d\mathbf{r}/dt = H(t)\,\mathbf{r} \tag{20.10}$$

Using $\mathbf{r}(t) = R(t)\,\mathbf{r}_0$ (equation 20.9) makes this

$$\mathbf{r}_0\,(dR/dt) = H(t)R(t)\,\mathbf{r}_0$$

Dividing by $\mathbf{r}_0$ gives

$$dR/dt = H(t)\,R(t) \tag{20.11}$$

Note that we now only have to deal with a scalar equation, instead of a vector equation. We

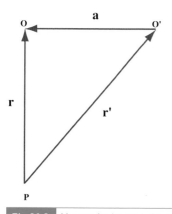

Fig 20.3.   Vectors for locating objects in the universe.

can solve equation (20.11) to give the Hubble parameter in terms of the scale factor:

$$H(t) = [1/R(t)][dR/dt] \qquad (20.12)$$

20.3 | Cosmology and Newtonian gravitation

We can learn a lot about the evolution of an expanding universe by applying Newtonian gravitation. In the next section we will see how the Newtonian results are modified by general relativity.

The assumption of isotropy is equivalent to saying that the universe appears to be spherically symmetric from any point. This means that any spherical volume evolves only under its own influence. The gravitational forces exerted on the volume by material outside the volume sum (vectorially) to zero. If the volume in question has a radius r, and mass $M(r)$, the equation of motion for a particle of mass m on the surface of the sphere at position r is

$$m\ddot{\mathbf{r}} = -\frac{GM(r)m\hat{\mathbf{r}}}{r^2} \qquad (20.13)$$

In this case, we let $\hat{\mathbf{r}}$ be a unit vector in the r-direction (radially outward). The assumption of homogeneity means that the density ρ is the same everywhere (though it can change with time). The mass $M(r)$ is given by

$$M(r) = \frac{4\pi}{3}r^3\rho$$

Substituting into equation (20.13) gives

$$\ddot{\mathbf{r}} = -\frac{4\pi}{3}G\rho\,\mathbf{r} \qquad (20.14)$$

We use equation (20.9) to eliminate $\mathbf{r}$ by using the scale factor, to get

$$\ddot{R} = -\frac{4\pi}{3}G\rho R \qquad (20.15)$$

As the universe expands, any given amount of mass occupies a larger volume. The density goes as 1/volume. The volume is proportional to R^3, so we have

$$\rho(t) = \rho_0\frac{R_0^3}{R^3(t)}$$

where we have used $\rho_0 = \rho(t_0)$, also, since $R_0 = 1$, we have

$$\rho(t) = \frac{\rho_0}{R^3(t)} \qquad (20.16)$$

Substituting this into equation (20.15), we have

$$\ddot{R} = -\frac{4\pi}{3}\frac{G\rho_0}{R^2} \qquad (20.17)$$

Note that if ρ_0 is not zero then $\ddot{R}$ cannot be zero. A universe with matter cannot be static. It must be expanding or contracting. This is like saying that if you throw a ball up, and the Earth's mass is not zero, then the ball must be moving up or down; it cannot be forever stationary.

To integrate the equation of motion (20.17), we first multiply both sides by $\dot{R}$, to give

$$\ddot{R}\dot{R} + \frac{4\pi}{3}\frac{G\rho_0}{R^2}\dot{R} = 0$$

Noting that

$$\frac{d(\dot{R}^2)}{dt} = 2\dot{R}\ddot{R}$$

this becomes

$$\frac{1}{2}\frac{d(\dot{R}^2)}{dt} + \frac{4\pi}{3}\frac{G\rho_0}{R^2}\frac{dR}{dt} = 0$$

Multiplying through by two, and using the fact that $(1/R^2)(dR/dt) = -d(1/R)/dt$, we have

$$\frac{d}{dt}\left[\dot{R}^2 - \frac{8\pi}{3}\frac{G\rho_0}{R}\right] = 0 \qquad (20.18)$$

Since its time derivative is zero, the quantity in the square brackets must be a constant. We set it equal to some arbitrary constant k, so that

$$\dot{R}^2 = \frac{8\pi}{3}\frac{G\rho_0}{R} - k \qquad (20.19)$$

Farther integration of equation (20.19) depends on whether k is zero, positive or negative. We consider each case separately. The schematic behavior of $R(t)$ for each case is shown in Fig. 20.4(a).

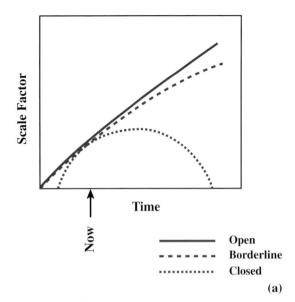

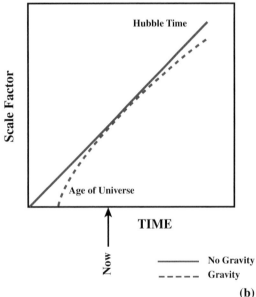

Fig 20.4. (a) The scale factor R as a function of time for borderline, closed and open universes. (b) How the presence of gravitation means that the Hubble time is greater than the real age of the universe.

We first look at the case $k = 0$. Equation (20.19) then becomes

$$\dot{R}^2 = \frac{8\pi}{3} \frac{G\rho_0}{R}$$

(20.20)

It should be noted that $\dot{R}$ is always positive, but approaches zero as R approaches infinity.

Equation (20.20) can be rewritten so that we can integrate it and find an explicit function for $R(t)$. We take the square root of both sides of the equation, giving

$$R^{1/2} \, dR = \left(\frac{8\pi \, G\rho_0}{3} \right)^{1/2} dt$$

We integrate the left side from zero to R and the right side from zero to t to give

$$\frac{2}{3} R^{3/2} = \left(\frac{8\pi \, G\rho_0}{3} \right)^{1/2} t$$

(20.21)

This means that R is proportional to $t^{2/3}$. The universe always expands, but the rate of expansion becomes smaller and smaller.

We now look at the case $k > 0$. Since the first term on the right-hand side of equation (20.19) becomes smaller when R increases, a point will be reached eventually, at some finite R, where $\dot{R} = 0$. The expansion stops at some maximum scale factor $R_{\max}$. We can find $R_{\max}$ as the value of R, which makes $\dot{R} = 0$ in equation (20.19). That is,

$$0 = \frac{8\pi}{3} \frac{G\rho_0}{R_{\max}} - k$$

Solving for $R_{\max}$:

$$R_{\max} = \frac{8\pi}{3} \frac{G\rho_0}{k}$$

(20.22)

After $R_{\max}$ is reached, the universe starts to collapse. We say that the universe is *closed*.

We next look at the case $k < 0$. If k is negative then $-k$ is positive, and the right-hand side of equation (20.19) is always positive. As R gets very large, the first term on the right-hand side approaches zero, and $\dot{R}^2$ approaches $-k$. (Remember, $-k$ is a positive number.) This means that $\dot{R}$ approaches $(-k)^{1/2}$. The expansion continues forever, and we say that the universe is *open*.

We can think of an analogous situation of throwing a ball up in the air. If the total energy is negative, the ball will return to Earth. If the total energy is positive, the ball will escape, and its speed will remain positive. If the total energy is zero, the ball will reach infinity, but its speed will approach zero. We can think of k as being related to the energy of the spherical region of the universe that we are following.

In Fig. 20.4(b) we can see how the presence of gravity in an expanding universe affects the relationship between the Hubble time ($1/H_0$) and the true age of the universe. The presence of gravity means that the expansion has been slowing, so the universe was expanding faster in the past. That means it took less time to reach its current size than we would estimate from the current expansion rate. From Fig. 20.4(b) we see that H_0 is proportional to the current slope of $R(t)$.

Having established that the universe is expanding, we would now like to ask whether that expansion will continue forever. In other words, is the universe open or closed? We would like to have some quantity that we can measure to tell us. If we look back to our analogy of the ball thrown up in the air, if we know the position and velocity of the ball at some time, we also need to know its acceleration to know if the ball has sufficient energy to escape. Since the ball is slowing down, we want to know the deceleration of the ball.

For the universe, we define a *deceleration parameter*, whose value will tell us whether the universe is open or closed. We would like to define this parameter so that it is dimensionless (just as the scale factor is dimensionless), and is independent of the time t_0 that we choose for our reference epoch. The latter requirement says that the parameter should depend on quantities such as $\dot{R}/R$ and $\ddot{R}/R$. With these ideas in mind, we define the deceleration parameter as

$$q \equiv -\left(\frac{\ddot{R}}{R}\right)\left(\frac{R}{\dot{R}}\right)^2 \tag{20.23}$$

$$= -\frac{R\ddot{R}}{\dot{R}^2}$$

Since the expansion is slowing down, $\ddot{R}$ is negative, and q is positive. If we use equations (20.11), (20.16) and (20.17) to eliminate $\ddot{R}$ and $\dot{R}$, this becomes (see Problem 20.7)

$$q = \frac{4\pi}{3}\frac{G\rho}{H^2} \tag{20.24}$$

We can look at ranges of q for the three cases of k:

$k = 0$ We combine equations (20.11) and (20.20) to give

$$R^2H^2 = \frac{8\pi}{3}\frac{G\rho_0}{R}$$

or

$$H^2 = \frac{8\pi}{3}\frac{G\rho_0}{R^3}$$

$$= \frac{8\pi}{3}G\rho \tag{20.25}$$

where we have used equation (20.16) in the last step. Substituting equation (20.25) into the equation for q (equation 20.24), we have

$$q = 1/2$$

$k > 0$ In this case, q can be arbitrarily large. It even approaches infinity as R approaches R_{max} (since $\dot{R} = 0$ at R_{max}). This means that any value of q in the range

$$q > 1/2$$

will produce a closed universe.

$k < 0$ We have already seen that q must be greater than zero, so the range of q given by

$$0 < q < 1/2$$

will produce an open universe.

The deceleration will depend on the density of matter in the universe. We can define a *critical density*, ρ_{crit}, such that the universe is closed if $\rho > \rho_{crit}$ and open if $\rho < \rho_{crit}$. If $\rho = \rho_{crit}$, we will have $k = 0$, and the universe is on the boundary between open and closed. This last point allows us to find ρ_{crit}, since it is the density for $k = 0$, or $q = 1/2$. If we set $q = 1/2$ in equation (20.24) and solve for the density, ρ, we have

$$\rho_{crit} = \frac{3}{8\pi}\frac{H^2}{G} \tag{20.26}$$

It is convenient to define a *density parameter* Ω, which is the ratio of the true density to the critical density. That is

$$\Omega = \rho/\rho_{crit} \tag{20.27}$$

We can easily show (see Problem 20.8) that $\Omega = 2q$. If $H = 70$ km/s/Mpc, then ρ_{crit} is

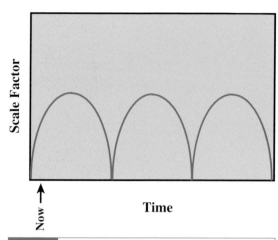

Fig 20.5. Oscillating universe. In this picture t_0 is the current time. The universe is the single cycle that contains t_0. If the universe is oscillating, then, after all the material comes back together, the expansion starts again.

$\sim 1 \times 10^{-29}$ g/cm^3. In the final section of this chapter, we will discuss observations that can determine the actual value of Ω.

There is a final point to consider if the universe turns out to be closed. After the expansion stops, a collapse will start. Eventually, all of the matter will come together into a dense, hot state for the first time since the big bang. Some people have taken to calling this event the *big crunch*. It is natural to ask what will happen after the big crunch. It has been suggested that the universe might reach a high density and then bounce back, starting a new expansion phase. If this can happen, then it might happen forever into the future, and might have happened for all of the past, as indicated in Fig. 20.5. Such a universe is called an *oscillating universe*.

If our universe turns out to be closed, can we tell if it is oscillating? Some theoreticians have argued that the big crunch/big bang in an oscillating universe strips everything down to elementary particles, and therefore destroys all information on what has come before. Others have argued that there are certain thermodynamic properties of the universe that might tell us if it is oscillating. Others have taken a wait-and-see attitude, pointing out that we will need a quantum theory of gravity to understand the densest state that is reached.

20.4 | Cosmology and general relativity

20.4.1 Geometry of the universe

When Einstein developed the general theory of relativity, he realized that it should provide a correct description of the universe as a whole. Einstein was immediately confronted with a result equivalent to equation (20.17), which says that if the density is not zero, the universe must be expanding or collapsing. This was before Hubble's work, and most believed in a static (steady-state) universe.

To get around this problem, Einstein introduced a constant, called the *cosmological constant*, Λ, into general relativity. It had no measurable effect on small scales, but altered results on cosmological scales. For example, in equation (20.17) the effect of the cosmological constant would be to replace the density ρ by $\rho - \Lambda/8\pi G$. This makes it possible to have a non-zero density, but a zero value for $\dot{R}$. Einstein withdrew the cosmological constant when he heard of Hubble's work, declaring the cosmological constant to be his biggest mistake. However, theoreticians have tended to keep it in the theory, and then formally set it to zero, or consider models with a non-zero Λ. The best determinations of various cosmological parameters, discussed below and in the next chapter, suggest that Λ may have a non-zero value.

Following Einstein's work, a number of people worked out cosmological theories, using different simplifying assumptions. The models are generally named after the people who developed them. The *de Sitter* models are characterized by $k = 0$ and a non-zero (positive) cosmological constant; the *Friedmann* models have a zero cosmological constant and also zero pressure (a good approximation at low density); *Lemaitre* models have non-zero density and a cosmological constant. As a result of his work, Lemaitre noted that (independent of the value of Λ) there must have been a phase in its early history when the universe was very hot and dense. This phase is called the *big bang*.

Many of the general relativistic results are similar to those we obtained in the previous section. This is because both depend on the fact that, in a spherically symmetric mass distribution, matter outside a sphere has no effect on the evolution of matter inside the sphere. One modification is the replacement of ρ by $\rho - \Lambda/8\pi G$ if you want a non-zero cosmological constant.

However, the general relativistic approach gives us a deeper insight by providing a geometric interpretation of the results. For example, the space-time interval, in spherical coordinates, becomes

$$(ds)^2 = (c\,dt)^2$$

$$-R^2\,(ct)\left[\frac{(dr)^2}{1 - kr^2} + r^2[(d\theta)^2 + \sin^2\theta(d\varphi)^2]\right]$$

$$(20.28)$$

(This is sometimes called the Robertson–Walker metric.) In this equation, $R(t)$ has the same meaning as before, and r, θ and φ are the usual spherical coordinates of objects such as galaxies. In cosmology, it is important to use what is known as a *comoving coordinate system*. This system expands with the universe. The form of this metric ensures that the models are homogeneous and isotropic (the cosmological principle). (Note that in books on general relativity, authors sometimes use different symbols: the scale factor is written as $a(t)$, and k becomes $1/R^2$, where R is the "radius of curvature" of the universe. These books also often take a system of units in which $c = 1$, so it does not appear explicitly in equations.)

In general relativity, whether the universe is open, closed or on the boundary tells us something about the geometry of space-time. That information is contained in the constant, k, which has the value zero for the boundary, $+1$ for a closed universe, and -1 for an open universe. However, k now tells us something about the curvature of space-time (Fig. 20.6). If $k = 0$, space-time is flat (Euclidean). The sums of the angles of

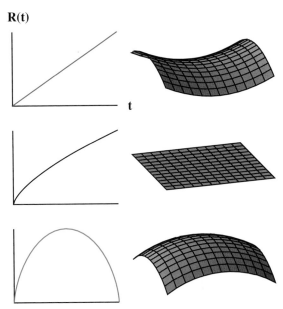

Fig 20.6. Schematic representations showing the relationship between cosmological model and the geometry of space-time. The top graph shows an open model, with negative curvature (like a saddle). The middle graph shows a universe that will expand forever, but is on the boundary, and the geometry is flat The bottom graph shows a closed model, and the geometry has a positive curvature. [© Edward L. Wright, used with permission]

triangles are always 180°. In this case space is infinite. If $k = +1$, space-time has a positive curvature, like the surface of a sphere. The sums of the angles of triangles are always greater than 180°. Space must be finite, just as the surface of a sphere is finite. Finally, if $k = -1$, we say that space-time has a negative curvature. The sums of the angles of triangles are always less than 180°. In this case, space is infinite. The relationship between geometry and the type of cosmological model is summarized in Table 20.1.

We can get a feel for the geometry of the universe by considering a two-dimensional analogy.

Table 20.1. | Parameters of various cosmological models.

Type	k	q	Ω	Curvature	Extent
Boundary	0	1/2	1	flat	infinite
Closed	+1	>1/2	>1	positive	finite
Open	−1	0 < q < 1/2	0 < q < 1	negative	infinite

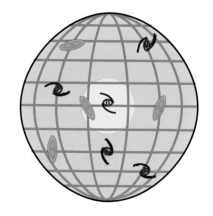

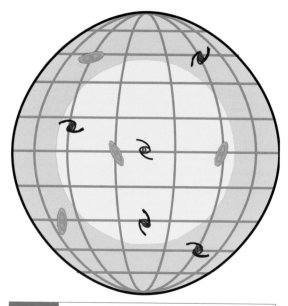

Fig 20.7. The universe as an expanding sphere. As the sphere expands, the coordinate system expands, and the radius of curvature changes. The dark circle with the lighter interior represents the part of the universe that we can see, with us at the center of the circle. A photon could have traveled from the circle to the center in the age of the universe. As the universe expands and ages, our horizon expands also. Since our horizon expands at the speed of light, new objects are always coming over the horizon.

We consider the universe as being confined to the surface of an expanding sphere, as shown in Fig. 20.7. One concept that we can now visualize is that we have a horizon due to the finite age of the universe t (as noted in our discussion of Olbers's paradox). We can only see light emitted

toward us within a distance equal to ct. This horizon is growing. We have seen that over small distances even the surface of a sphere appears flat. The curvature becomes apparent as you can survey larger areas. This means that, as our horizon grows, the curvature might become more apparent.

As the universe expands, we would like to keep track of the separation between any two comoving points. We start by considering nearby points at r and $r + dr$. The proper distance between those points is

$$\frac{R(t)\,dr}{(1-kr^2)^{1/2}}$$

If the points are far apart (like a distant galaxy to here), then we have to integrate that expression.

We can also use our analogy to see that it is meaningless to talk about the radius of the universe. In three dimensions, our sphere has a radius, but in two dimensions we can only talk about the surface. This is one reason why the scale factor $R(t)$ is a better way of keeping track of the expansion. Even though we cannot talk about the radius of the universe in a meaningful way, we can talk about the curvature of our surface. So we can talk about the radius of curvature of the universe. The larger the radius of curvature, the closer the geometry is to being flat.

Our expanding sphere analogy also tells us that it is not very meaningful to talk about the center of the universe. The sphere has a center, but it is not in the universe, which is the surface only. There is nothing special about any of the points on the surface of the sphere. If we go back in time to very small times, our sphere will be very small, and at $t = 0$ all the points are together, at the center. So the proper way to talk about the center is as a space-time event far in our past.

20.4.2 Cosmological redshift

We can also see that the redshift (Hubble's law) fits in as a natural consequence of the expansion (Fig. 20.8). As the universe expands, the wavelengths of all photons expand by the same

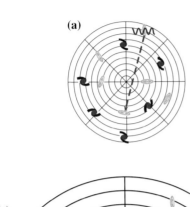

(a)

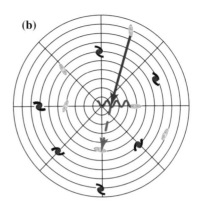

(b)

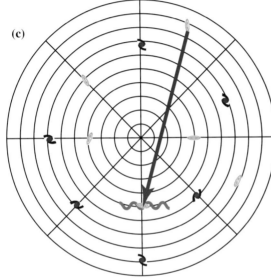

(c)

proportion that cosmic distances expand. That is, they expand in proportion to the scale factor. If radiation is emitted at wavelength λ_1 at epoch t_1, and detected at wavelength λ_2 at epoch t_2, then

$$\frac{\lambda_2}{\lambda_1} = \frac{R(t_2)}{R(t_1)} \qquad (20.29a)$$

If we let t_1 be some arbitrary time, and t_2 be the reference epoch t_0 (for which $R = 1$), this becomes

$$\frac{\lambda_0}{\lambda} = \frac{1}{R(t)}$$

We define the *redshift*, z, to be

$$z = \frac{\Delta\lambda}{\lambda}$$

where $\Delta\lambda = \lambda_0 - \lambda$, so $\lambda_0/\lambda = 1 + z$. This gives

$$1 + z = \frac{1}{R(t)} \qquad (20.29b)$$

Remember, since the radiation is emitted before the reference epoch, $R(t) < 1$, so $z > 0$.

We can derive an approximate expression for the redshift for radiation emitted some time Δt in the recent past, where $\Delta t \ll t_0$. Using a Taylor series, we have (see Problem 20.11)

$$\frac{1}{R(t_0 - \Delta t)} \cong \frac{1}{R(t_0)} + \Delta t \dot{R}(t_0) \qquad (20.30)$$

$$\cong 1 + \Delta t \dot{R}(t_0)$$

Combining this with equation (20.29b) gives

$$1 + z \cong 1 + \Delta t \dot{R}(t_0)$$

Setting $t = t_0$ in equation (20.12) gives us, $\dot{R}(t_0) = H_0$, so

$$z \cong H_0 \, \Delta t \tag{20.31}$$

If a photon takes Δt to reach us, it must have been emitted from a distance $d = c \, \Delta t$. Using this to eliminate Δt in equation (20.31) gives

$$cz \cong H_0 \, d \tag{20.32}$$

which is Hubble's law. (Remember, this approximation is for small Δt.)

It is important to note that equation (20.29) tells how to interpret the redshifts of distant galaxies. It is tempting to say that these galaxies are moving relative to us and their radiation is therefore Doppler-shifted. However, in computing a relative velocity for a Doppler shift, we take the difference between the velocities of the two galaxies. These two velocities must be with respect to a co-moving coordinate system. Therefore, apart from its peculiar motion, each galaxy's velocity is zero with respect to this coordinate system. Therefore, strictly speaking, there is no Doppler shift due to the expansion of the universe. The redshift arises as a result of the increase in wavelengths of all photons moving through an expanding universe. We therefore call it the *cosmological redshift*. Any additional motions with respect to the co-moving coordinates would produce a Doppler shift (red or blue) in addition to the cosmological redshift.

As a consequence of this, we should not directly interpret the redshift of a galaxy as giving a particular distance. The amount of redshift just tells us the amount by which the scale factor has changed between the time the photon was emitted and the time it was detected. For this reason, we often talk about the redshift of a particular galaxy, and don't bother to convert it to a distance. For example, we simply say that 3C273 is at $z = 0.15$. To convert a redshift to a distance we need a particular model for how $R(t)$ has evolved.

20.4.3 Models of the universe

In general relativity, the solutions for $R(t)$ are different from the Newtonian case. The equation in general relativity, analogous to equation (20.19), is called Einstein's equation. There are two parts relevant to our discussion of cosmology:

$$\left(\frac{\dot{R}}{R}\right)^2 = -\frac{kc^2}{R^2} + \frac{8\pi G\rho + \Lambda c^2}{3} \tag{20.33a}$$

and

$$\frac{\ddot{R}}{R} = -\frac{4\pi G}{3}\left[\rho + \frac{3P}{c^2}\right] + \frac{\Lambda c^2}{3} \tag{20.33b}$$

where P is the gas pressure, usually taken to be zero, except when the matter is hot and dense. Remember, the density at any time is ρ_0/R^3. So the mathematical effects of a non-zero density and a non-zero Λ are different, since the density term will have an extra factor of the variable R. We will look at separate cases below.

Since $H = (1/R)(dR/dt)$, we can write Einstein's equation as

$$3\left[H^2 + \frac{kc^2}{R^2}\right] = 8\pi G\rho + \Lambda c^2 \tag{20.34}$$

From these, the deceleration parameter becomes

$$q_0 = -\frac{1}{H_0^2}\left[\frac{\Lambda c^2}{3} - 4\pi G\left(\frac{\rho_0}{3} + \frac{P_0}{3c^2}\right)\right] \tag{20.35}$$

Note that this reduces to the classical case (equation 20.24) when $\Lambda = P = 0$.

The integration of these equations to give $R(t)$ is generally quite difficult. To simplify the situation, it is useful to look at limiting cases, namely zero cosmological constant and zero density. Zero cosmological constant would also be an approximate description of a universe with small cosmological constant, where the matter term dominates. Likewise, zero density would also approximately describe a case with low density where the cosmological constant dominates. In each case, we must also look separately at zero, positive and negative curvature. Below, we give results for various limiting cases without deriving them. You can verify that they are solutions by plugging them into the appropriate equations (see Problem 20.18). The results for some of these models are shown in Fig. 20.9.

Models with zero Λ are called Friedmann models. For a flat universe ($k = 0$), $R(t)$ in these models is given by

$$R(t) = (\text{const.})\left(\frac{t}{t_1}\right)^{2/3} \tag{20.36a}$$

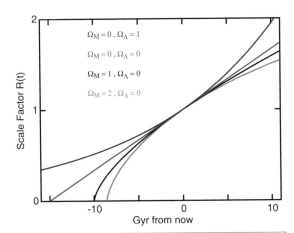

Fig 20.9. Scale factor vs. time for various cosmological models. Models are all chosen to have $R = 1$, now, and a Hubble constant $= 65$ km/s/Mpc (so they all have the same slope now). In terms of the density parameter (defined in equation 20.45), the models are (from top to bottom): $(\Omega_M = 0, \Omega_\Lambda = 1)$, $(\Omega_M = 0, \Omega_\Lambda = 0)$ $(\Omega_M = 1, \Omega_\Lambda = 0)$ $(\Omega_M = 2, \Omega_\Lambda = 0)$. [© Edward L. Wright, used with permission]

where t_1 is any convenient constant by which to scale the result. If we choose it to be t_0, the current age of the universe, taking the constant in front to be equal to unity makes $R = 1$ now. So, the flat model which describes our universe would be

$$R(t) = \left(\frac{t}{t_0}\right)^{2/3} \qquad (20.36b)$$

This result is the same as for the flat Newtonian case (equation 20.21).

For positive curvature ($k = +1$) and negative curvature ($k = -1$), the results are mathematically different from the Newtonian case, and are also complicated to express. They do have similar characteristics to their Newtonian counterparts. Namely, the positive curvature case produces a universe that expands, reaches some maximum R, and then contracts. The negative curvature produces an expansion that lasts forever.

Models with non-zero cosmological constant are called Lemaitre models. As we have said, for the sake of simplicity, we consider the empty Lemaitre models. One such model is also shown in Fig. 20.9 these models are useful for any universe dominated by the cosmological constant. As, we said above, these universes will behave roughly

like those with a mass density, $\rho_{EFF} = -\Lambda/8\pi G$, constant in space and time. A positive value of Λ behaves like a negative effective mass density (repulsion), and a negative value of Λ behaves like a positive effective mass density (attraction). So, for positive Λ, we would expect the expansion to accelerate, and for negative Λ we would expect the expansion to stop and reverse.

For zero curvature ($k = 0$), the result for $R(t)$ is

$$R(t) = (\text{const.}) \exp\left(\sqrt{\frac{\Lambda}{3}} t\right) \qquad (20.37a)$$

Note that for positive Λ, this corresponds to an exponential growth in the expansion rate. The flat model with zero density and a non-zero Λ is sometimes called the deSitter model.

For positive curvature ($k = +1$), there are solutions only for $\Lambda > 0$, $R(t) \geq \left(\frac{3}{\Lambda}\right)^{1/2}$, in which case

$$R(t) = \sqrt{\frac{3}{\Lambda}} \cosh\left(\sqrt{\frac{\Lambda}{3}} t\right), \qquad (20.37b)$$

where t is zero when $R(t)$ has its minimum value.

For negative curvature ($k = -1$)

$$R(t) = \begin{cases} \sqrt{\dfrac{3}{\Lambda}} \sin h\left(\sqrt{\dfrac{\Lambda}{3}} t\right) & \text{for } \Lambda > 0 \\[2ex] \sqrt{\dfrac{3}{-\Lambda}} \sin\left(\sqrt{\dfrac{-\Lambda}{3}} t\right) & \text{for } \Lambda < 0 \end{cases} \qquad (20.37c)$$

There are different ways in which we could define the distance, since we are dealing with objects whose separation changes between the time a photon is emitted at one galaxy and received in another. A convenient definition of distance in this case is that which we would associate with a distance modulus, $m - M$. This would tell us how to convert apparent brightnesses (or magnitudes) into absolute brightnesses (or magnitudes).

As light travels from a distant source, the observed brightness decreases as the photons from that source spread out on the surface of a sphere. Let the radius of that sphere be a. If the geometry of space-time is flat, the surface area of that sphere is $4\pi a^2$. So, the observed brightness falls as $1/a^2$, just as for light from nearby stars (Chapter 2). If the geometry is not flat, then falloff can be greater or less than $1/a^2$. So the relationship between observed brightness and distance depends on the geometry of the universe. If $\Lambda = 0$, and $q_0 > 0$, and if z is not too large, then the relationship between distance modulus and redshift is given by

$$m - M = 25 + 5 \log_{10}(cz/H_0) + 1.086z\,(1 - q_0) \quad (20.38)$$

where q_0 is the current value of the deceleration parameter. In this expression, cz/H_0 is in megaparsecs, accounting for the factor of 25 in front (see Problems 20.14 and 20.15).

The geometry of the universe also determines how the apparent angular size varies with distance. For flat geometry, and an object that is not too distant by cosmological standards, the angle subtended by an object of length L at distance D is L/D (in radians) as long as $L \ll D$. For an object at a cosmologically significant distance, the angular size is

$$\theta = \frac{L(1 + z^2)}{D_{\mathrm{L}}}$$

$$= \frac{L}{H_0 c}\,\frac{q_0^2(1 + z)^2}{\{q_0 z + (q_0 - 1)[(1 + 2q_0 z)^{1/2} - 1]\}} \quad (20.39)$$

where D_{L} is the luminosity distance.

It is useful to have an estimate of the relationship between the Hubble time, H_0^{-1}, and the true age of the universe, t_0. Obviously, this depends on which model we use to describe the universe.

For the Friedmann models ($\Lambda = 0$) the results are as follows.

For a flat universe ($k = 0$),

$$t_0 = (2/3)\,H_0^{-1} \quad (20.40a)$$

For positive curvature ($k = +1$),

$$0 < t_0 < (2/3)\,H_0^{-1} \quad (20.40a)$$

For negative curvature ($k = -1$)

$$(2/3)\,H_0^{-1} < t_0 < H_0^{-1} \quad (20.40c)$$

It is also useful to have an estimate of the relationship between the age of the universe when a photon was emitted and the redshift. In one model it is given by

$$t\,(\mathrm{Gyr}) = 10.5\,(H_0/65\ \mathrm{km/s/Mpc})\,(1 + z)^{-3/2} \quad (20.41)$$

20.5 | Is the universe open or closed?

In this section we look at evidence that might allow us to decide whether the universe is open or closed. It is impressive that we can even ask such a question and hope to achieve an answer.

The basic question is whether the actual density is less than or greater than the critical density. We could start by adding up the density of all the matter we can see, to find out if it gives $\rho > \rho_{\mathrm{crit}}$. However, we already know that there is a problem with dark matter, so, if we only include the visible matter, we will be missing a significant amount. Of course, if the visible matter is sufficient to close the universe, then we don't have to worry about the dark matter. It turns out that if the visible matter is insufficient to close the universe, then we have to account for the dark matter. It turns out that the density of visible matter is about 1% of the critical density.

If the universe is to be closed, the dark matter must do it. From Table 20.2, we see that the amount of dark matter required to close the universe is greater than the dark matter in clusters of galaxies. There also appears to be a trend towards more dark matter on larger scales. Therefore, we would not be surprised if there is enough dark matter to close the universe. However, in our attempt to see if the universe is open or closed, we can only include dark matter that we know is present by its gravitational effects on visible matter. We can therefore include the dark matter in galaxies, and clusters of galaxies, since we can detect its gravitational

Table 20.2. | Mass-to-light ratios on different scales.

Scale	M/L (solar units)
Milky Way to Sun	3
Spiral galaxy disk	10
Elliptical galaxy	30
Halo of giant elliptical	40
Rich cluster of galaxies	200
To close the universe	1200

effects. This still leaves us a factor of five short of closing the universe.

We have said that the best way to measure the mass of any object is to measure its gravitational effect on something. If we want to determine the mass of the Earth, we measure the acceleration of gravity near the surface. Therefore, instead of trying to find all of the matter needed to close the universe, we can look for its gravitational effects. We can try to measure the actual slowing down of the expansion of the universe to see if $-\ddot{R}$ is large enough to stop the expansion. When we do this, we are determining the current value of the deceleration parameter from its original definition (equation 20.23). Using the fact that $\dot{R}(t_0) = H_0$ (equation 20.12), it becomes

$$q_0 = -\frac{\ddot{R}(t_0)}{H_0^2} \qquad (20.42)$$

We don't actually try to measure $\ddot{R}(t_0)$. What we try to measure is the current rate of change of the Hubble constant $\dot{H}(t_0)$. We would therefore like to express $\ddot{R}(t_0)$ in terms of H_0 and $\dot{H}(t_0)$. We start with equation (20.11):

$$\dot{R}(t) = H(t)\,R(t)$$

Differentiating both sides with respect to t gives

$$\ddot{R}(t) = H(t)\,\dot{R}(t) + \dot{H}(t)\,R(t)$$

Setting $t = t_0$, and remembering that $\dot{R}\,(t_0) = H_0$, this becomes

$$\ddot{R}(t_0) = H_0^2 + \dot{H}(t_0) \qquad (20.43)$$

Substituting this into equation (20.42), we have

$$q_0 = -\left[\frac{\dot{H}(t_0)}{H_0^2} + 1\right] \qquad (20.44)$$

Equation (20.44) tells us that if we can measure the rate of change of the Hubble parameter, we can determine q_0. Unfortunately, measuring $\dot{H}(t_0)$ is not easy. This should not be a surprise since measuring H_0 is not easy. In principle, we can measure $\dot{H}(t_0)$ by taking advantage of the fact that we see more distant objects as they were a long time ago. If we can determine H for objects that are five billion light years away, then we are really determining the value of H five billion

years ago. If we include near and distant objects in a plot of Hubble's law, we should be able to see deviations from a straight line as we look farther back in time.

The difficulty comes in the methods used for measuring distances to distant objects. In our discussion of the extragalactic distance scale, we saw that for the most distant galaxies we can see, we cannot look at individual stars, such as Cepheids, within a galaxy. Instead, we must look at the total luminosity of a galaxy. We already know that the luminosities of galaxies change as they evolve. Galactic cannibalism provides us with the most spectacular example of this, but even normal galaxies change in luminosity with time. Therefore, if we calibrate the distance scale using the luminosities of nearby galaxies, we cannot apply this to more distant galaxies, precisely because we are seeing them as they were in the past. Before we can interpret observations of distant objects, we must apply theoretical evolutionary corrections. These corrections can be so large that they can make an apparently closed universe appear open or an open universe appear closed.

When we discussed the extragalactic distance scale, we mentioned that supernovae might provide very useful standard candles. In this case, the more reliable standards are those that arise in close binary systems (type I). Nearby examples tell us how to relate the light curve to the peak luminosity. So if we can compare the observed brightness with the known luminosity, we can calculate the distance. HST has been particularly useful for studying these objects, and can detect faint ones at cosmologically significant distances. These results suggest that the expansion of the universe may not be slowing very much. In fact, a few points are consistent with an acceleration in the expansion. This would be consistent with a non-zero cosmological constant. However, before these preliminary results are accepted, many more such objects need to be observed, and we need to examine whether these particular standard candles are not standard as thought.

An alternative approach is to measure the curvature of space-time by surveying the universe on a large scale. One way of carrying this out is with *radio source counts*. We divide the universe into shells, such as for our discussion of Olbers's paradox earlier in this chapter. We then count the number of radio sources in each shell. We use radio sources because we can see them far away. (With large, sensitive optical telescopes, optical counts are now being used also.) If the geometry of space-time is flat (Euclidean) the number of sources per shell will go up as r^2. If the geometry of space-time is curved, that curvature will become more apparent as we survey larger regions. Therefore, as we look farther away, we would expect to see deviations from the r^2 dependence. What is actually varying is the relationship between r and surface area. Of course, as we look far enough to see such deviations, we are also looking far back in time, and we are seeing radio sources as they were. Again, evolutionary corrections are necessary. The results so far are consistent with a flat universe.

There are also more indirect methods that have proved fruitful. These involve an understanding of the formation of elements in the big bang, and will be discussed in the next chapter. The results of these so far support a universe which is open. In addition, they only give information on the density of material that can participate in nuclear reactions, and may not include the dark matter.

One of the interesting aspects of this whole problem is that we should be so close to the boundary. Of all the possible values for the density of the universe, ranging over many orders of magnitude, we seem to be tantalizingly close to the critical density. Cosmologists have wondered whether this is accidental, or whether it is telling us something significant about the universe. They have noted that if Ω is not exactly unity, then it evolves away from unity as the universe becomes older. This means that for the actual density to be pretty close to the critical density

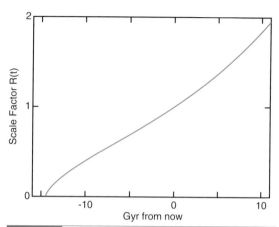

Fig 20.10. Scale factor vs. time for the cosmological model which best fits current data. The model has a Hubble constant of 65 km/s/Mpc and $\Omega_M = 0.3, \Omega_\Lambda = 0.7$. [© Edward L. Wright, used with permission]

now, it had to be *very* close to the critical density in the past.

If there is a non-zero cosmological constant, then it is possible to have a flat universe. Remember, we defined the density parameter for matter in equation (20.27) as

$$\Omega_M = \rho_M/\rho_{crit} \tag{20.45a}$$

We saw that if there is a non-zero Λ, then we can define an effective density due to that Λ, as $\Lambda/8\pi G$, so we can define a density parameter associated with Λ as

$$\Omega_\Lambda = \Lambda/8\pi G \, \rho_{crit} \tag{20.45b}$$

Then the total density parameter for the universe would be

$$\Omega_{TOT} = \Omega_M + \Omega_\Lambda \tag{20.45c}$$

Fig. 20.10 shows R vs. t for what is currently the best estimate of the model universe with $\Omega_{TOT} = 1$ ($\Omega_M = 0.3, \Omega_\Lambda = 0.7$). We will discuss the future of the universe in the next chapter when we look at the big bang.

Chapter summary

Cosmology is the study of the universe at the largest scale. It asks about the large-scale structure of the universe and how it has evolved. When we are talking about cosmological scales, the smallest building blocks are the galaxies.

One of the fascinating things about cosmology is that we can do normal astronomical observations to answer cosmological questions. The consideration of the simple question, "Why is the sky dark at night?" leads us to the profound conclusion that the universe must have a finite size, a finite age, or both.

In explaining the large-scale structure of the universe, we start with the simplest assumptions, that the universe is homogeneous and isotropic. This means that (on cosmological scales), the average properties are the same from place to place, and the appearance is the same in all directions. A class of theories that also include the assumption that the universe doesn't change with time, so-called steady-state theories, are no longer supported by observational evidence. Cosmological models in which the universe is becoming less dense as it expands have an era in the early universe when it was hot and dense. This era is called the big bang, and models that include it are called big-bang cosmological models.

In keeping track of the expansion of the universe, it is useful to deal with the scale factor, which tells us how much distances between galaxies have changed from one time to another.

As the universe expands, the wavelengths of all the photons in the universe increase by the same amount as the scale factor increases. This is called the cosmological redshift. It is the redshift that produces Hubble's law. In treating the structure of the universe, we must be sure to account for the effects of general relativity. The geometry of the universe may behave differently from that of a normal flat surface.

Each particle in the universe feels the gravitational attraction of all the other particles in the universe. Whether the universe will expand forever, or the expansion will eventually reverse, depends on the total density of material in the universe. This has turned out to be very hard to measure, though we have been able to determine that there is not nearly enough luminous matter to close the universe. So, if the universe is closed, then the dark matter must be responsible. The evolution of the universe may also be affected by a non-zero cosmological constant.

Questions

*20.1. How large a scale do we have to look at before the cosmological principle can be applied? How does this scale compare with the distance over which light could have reached us in the age of the universe?

20.2. Restate the argument in our discussion of Olbers's paradox using galaxies instead of stars as the sources of light.

20.3. In our discussion of Olbers's paradox, does it matter whether we talk about the appearance of the daytime or night time sky?

*20.4. Suppose that we were trying to invoke interstellar dust as a way out of Olbers's paradox by saying that it is the scattering by the dust that blocks out the distant light, not absorption. The dust will therefore not heat. Why doesn't this argument help?

20.5. How does the universe having a finite size or age save us from Olbers's paradox?

20.6. Does our motion towards the great attractor violate the part of the cosmological principle that the universe should appear isotropic?

20.7. What is the observational evidence that the universe is expanding?

20.8. Does Hubble's law rule out the steady-state models?

*20.9. What observations can we do to verify that the universe is isotropic?

20.10. If we lived in a contracting universe, would we still observe a cosmological redshift?

*20.11. How can we measure the curvature of the universe without getting outside it?

*20.12. What is the universe expanding into?

20.13. What are the advantages of using the scale factor $R(t)$ to keep track of the expansion of the universe?

20.14. Where is the center of the universe?

20.15. What are the various interpretations of the quantity k, discussed in this chapter?

20.16. If the universe is closed, can we distinguish a "one-time" universe from an oscillating universe?

20.17. Of the methods described for deciding whether the universe is open or closed, which ones rely on measuring the gravitational effects of all of the matter in the universe?

20.18. If the universe is expanding, how is it possible for that expansion to reverse?

20.19. Why do we say that if the universe is closed, then the dark matter must do it?

Problems

For all problems, unless otherwise stated, use $H_0 =$ 70 km/s/Mpc.

20.1. Suppose we detect radiation that was emitted by some galaxy far away. In the time the radiation traveled to reach us, its wavelength doubled. What happened to the scale factor of the universe in that time?

20.2. How much brighter would the sky be if it were uniformly filled with Suns, rather than the one we have? (Hint: think of the solid angle covered by the Sun relative to the whole sky.)

20.3. Show that, if the universe were infinite in age and extent the cosmological redshift is not sufficient to get us out of Olbers's paradox.

20.4. Estimate the rate of continuous creation required to keep the density constant (at 10^{-29} g/cm^3). Express your answer in protons/yr/km^3.

20.5. Show that the density of the universe is proportional to $1/R^3(t)$.

20.6. For the case $k < 0$, find an expression for $R(t)$ valid for large R. What are the limits on R for your expression to be valid?

20.7. Show that equation (20.24) follows from equations (20.11), (20.16) and (20.17).

20.8. Show that the density parameter Ω is twice the deceleration parameter q.

20.9. Rewrite equations (20.17) and (20.19) in terms of the density parameter, substituting for the critical density from equation (20.26).

20.10. If the current density of the universe is 1×10^{-29} g/cm^3, what value would be needed for the cosmological constant Λ in order for the universe to be static?

20.11. Show that equation (20.30) can be obtained by the appropriate use of a Taylor series.

20.12. What values of $\dot{H}(t_0)$ would be needed to make q_0 equal to (a) 0 to 1/2, (b) 1/2, (c) 1?

20.13. Derive an expression for the critical density (equation 20.26) without introducing the deceleration parameter q_0.

20.14. Show that if the distances are given by Hubble's law, then the distance modulus is given by equation (20.38) without the last term on the right.

20.15. Compare the distances obtained using Hubble's law and equation (20.38) with $q_0 = 1/2$ for objects with $z =$ (a) 0.1, (b) 1.0, (c) 3.5, (d) 5.0.

20.16. Estimate the distance modulus, $m - M$, for objects with $z =$ (a) 0.1, (b) 1.0, (c) 3.5, (d) 5.0.

20.17. Estimate the age of the universe at the time when radiation was emitted from objects with $z =$ (a) 0.1, (b) 1.0, (c) 3.5, (4) 5.0.

20.18. Suppose we observe an object that is 10 Mpc away. At what wavelength is the Hα line observed if (a) the object has no other motion, (b) the object has an additional motion away from us at 1000 km/s, (c) the object has an additional motion towards us at 1000 km/s.

Computer problems

20.1. For the case $k = 0$, find an expression for $\dot{R}$ vs. t, and plot a graph of your result.

20.2. For $k = 0$, what is the difference between the current age of the universe and the current value of the Hubble parameter?

20.3. For $k = 0$, how different is the Hubble parameter for objects with $z = 0.1, 1, 3, 5$ and 10^3?

20.4. For $k = 0$, plot a graph of the distance modulus $(m - M)$ vs. z.

20.5. Using equation (20.41), plot a graph of t in gigayears vs. z, for $H_0 = 50, 75$ and 100 km/s/Mpc.

Chapter 21

The big bang

In the preceding chapter, we noted that Lemaitre first pointed out that if the universe is expanding, then there must have been an era in the past when it was much denser than it is now. This hot, dense early era was named the *big bang* by Fred Hoyle, a steady-state cosmologist, in an attempt to ridicule the theory. The theory survived the ridicule, the name remained, and we now refer to all cosmological models with an evolving universe as 'big-bang cosmologies'. In this chapter, we will see what we can learn about conditions in the big bang, and what the relationship is between those conditions and the current state of the universe.

21.1 | The cosmic background radiation

Following the idea that the universe was very hot and dense, *George Gamow* suggested, in 1946, that when the universe was less than about 200 seconds old, the temperature was greater than one billion kelvin, hot enough for nuclear reactions to take place rapidly. In 1948, *Ralph Alpher, Hans Bethe* and Gamow showed (in a paper often referred to as the alpha/beta/gamma (for the names of the authors) paper) that these nuclear reactions might be able to explain the current abundance of helium in the universe. (We will discuss the synthesis of the elements in the next section.) In a more thorough analysis of the problem, Alpher and *Robert Herman*, in a classic paper published in 1948, found that the early universe should have been filled with radiation, and that the remnant of that radiation should still be detectable as a low intensity background of microwaves.

21.1.1 Origin of the cosmic background radiation

To help visualize the evolution of the early universe, we again rely on an analogy with an expanding sphere, as shown in Fig. 21.1. Remember, in this analogy, the universe is the surface of the expanding sphere. All particles and radiation must move along the surface. If you have a balloon, you can follow along with this analogy.

When the universe was young enough to have its temperature higher than 3000 K, the atoms were all ionized. The universe was a plasma of nuclei and electrons. The free electrons are particularly efficient at scattering radiation. They provided a continuum opacity for any radiation present. This means that radiation would not travel very far before getting scattered; the universe was opaque. The radiation therefore stayed in equilibrium with the matter. The spectrum of the radiation was that of a blackbody at the temperature of the matter. As the universe expanded, the density decreased, and the temperature decreased. As the matter cooled, the radiation also cooled. Then the point was reached at which the temperature dropped below 3000 K. (Various estimates place this at a time some 100 000 years after the expansion started.) At the lower temperature, the electrons and nuclei (mostly protons, or helium) combined to make atoms. This is called the *era of recombination*. The neutral atoms are very inefficient at absorbing radiation, except at a few narrow ranges of wavelengths corresponding to

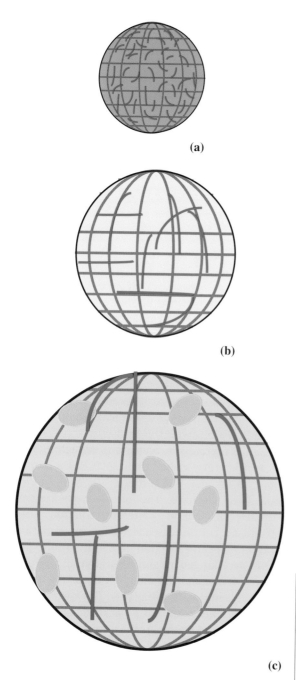

(a)

(b)

(c)

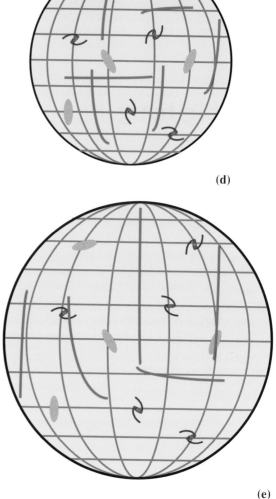

(d)

(e)

Fig 21.1. Diagram showing radiation and matter in an expanding universe. (Remember, this is a two-dimensional analogy.) (a) Before decoupling. The matter is dense and hot and the matter and radiation are in equilibrium. (b) At decoupling the universe is transparent, and the radiation now moves around without being absorbed. Photons are moving in all directions. (c) Protogalaxies are starting to form. Photons are moving in all directions and are redshifted as the universe expands. (d) Galaxies have formed, and photons are still moving in all directions. The redshift becomes larger as the universe expands. (e) Today, an inhabitant of any galaxy would see the redshifted photons coming at them from any direction.

spectral lines. For all practical purposes, the universe became transparent to the radiation. Since the radiation and matter no longer interacted significantly, we say that they were *decoupled*.

If we look at Fig. 21.1, we see that the last photons emitted by the plasma just before decoupling should still be running around the universe. A relatively small fraction of those photons have bumped into galaxies, and have been absorbed by the material in the galaxies. Anyone in one of those galaxies, looking around, should see radiation coming at them from all directions. The radiation does undergo one change. As the universe expands, all of the radiation is redshifted, a result of the cosmological redshift. This is illustrated in Figs. 21.1(d) and (e).

We can calculate what this redshifted blackbody radiation will look like. To do this, we note that the photons will be preserved; their frequencies will just change in a known way. For a blackbody at a temperature T, the energy density in photons with frequencies between ν and $\nu + d\nu$ is given by the Planck function,

$$U(\nu, T)\, d\nu = \left(\frac{8\pi h\nu^3}{c^3}\right)\left(\frac{1}{e^{h\nu/kT} - 1}\right) d\nu \qquad (21.1)$$

To find the number of photons per unit volume in frequencies between ν and $\nu + d\nu$, we take the energy density and divide by the energy per photon, $h\nu$. This gives

$$n(\nu, T)\, d\nu = \left(\frac{8\pi \nu^2}{c^3}\right)\left(\frac{1}{e^{h\nu/kT} - 1}\right) d\nu \qquad (21.2)$$

We assume that the radiation is emitted at some epoch t, with scale factor R, and we detect it at the reference epoch t_0. The observed wavelength λ_0 is related to the emitted wavelength λ by equation (20.29), as

$$\lambda_0/\lambda = 1/R$$

Since $\lambda = c/\nu$, the observed frequency ν_0 is related to the emitted frequency ν by

$$\nu/\nu_0 = 1/R$$

This means that

$$\nu_0 = R\nu \qquad (21.3)$$

In addition, photons emitted in the frequency range $d\nu$ will be observed in the frequency range $d\nu_0$, given by

$$d\nu_0 = R\, d\nu \qquad (21.4)$$

The photons emitted between ν and $\nu + d\nu$ are now observed between ν_0 and $\nu_0 + d\nu_0$. Also, all volumes increase by a factor of $1/R^3$. Combining all of these, we now have the number of observed photons per unit volume with frequencies between ν_0 and $\nu_0 + d\nu_0$ as

$$n(\nu_0, T)d\nu_0 = R^3\left(\frac{8\pi}{c^3}\right)\left(\frac{\nu_0}{R}\right)^2\left(\frac{1}{e^{h\nu_0/kRT} - 1}\right)\left(\frac{d\nu_0}{R}\right)$$

$$= \left(\frac{8\pi \nu_0^2}{c^3}\right)\left(\frac{1}{e^{h\nu_0/RkT} - 1}\right) d\nu_0$$

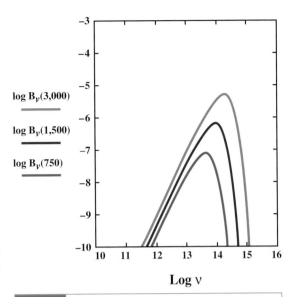

Fig 21.2. Spectrum of the radiation in the universe as it expands. We plot log intensity vs. log frequency, for three temperatures. 3000 K was the temperature when radiation and matter decoupled. 1500 K is when the scale factor increased by a factor of 2 after decoupling. 750 K is another factor of 2 increase in the scale factor.

Multiplying by $h\nu_0$ gives us back the energy density spectrum for radiation detected between ν_0 and $\nu_0 + d\nu_0$:

$$U(\nu_0, T)d\nu_0 = \left(\frac{8\pi h\nu_0^3}{c^3}\right)\left(\frac{1}{e^{h\nu_0/kRT} - 1}\right)d\nu_0 \qquad (21.5)$$

This looks just like a blackbody spectrum at temperature $T_0 = RT$. Therefore, the radiation will still have a blackbody spectrum, but will appear cooler by a factor of R. In Fig. 21.2, we see a few examples of how the spectrum evolves at different temperatures.

The redshift of the background radiation has an interesting consequence on the evolution of the universe. This is illustrated in Fig. 21.3. The energy density of the matter in the universe is $\rho_{mat}c^2$, where ρ_{mat} is the density of matter. We have already seen in Chapter 20 that the density is proportional to $1/R^3$. The number density of photons in the universe is also proportional to $1/R^3$. However, the redshift means that the energy per photon is proportional to $1/R$, so the energy density of radiation in the universe is proportional to $1/R^4$. This means that the energy density of radiation (integrated over

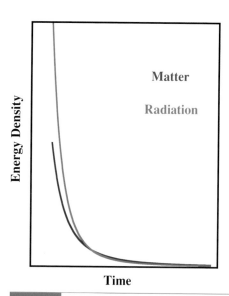

Fig 21.3. Density of radiation and matter as the universe ages. On the horizontal axis we plot the age of the universe. On the vertical axis we plot the energy density of matter in blue, and radiation in red.

21.1.2 Observations of the cosmic background radiation

Alpher and Herman (in 1948) gave an equation to calculate the temperature of the background radiation, and found a value of about 5 K. The actual value that you predict depends on knowing certain parameters, such as the Hubble parameter. These parameters were, obviously, not as well determined in 1948 as they are now. If one puts in the current values of the various parameters, the predicted value of the temperature of the background radiation drops slightly, to 2.7 K. Even with the slightly higher value, it seemed that this radiation would be very hard to detect. Since most of the radiation would be expected in the radio part of the spectrum, Alpher and Herman talked to radio astronomers, but the radio astronomers saw many difficulties. First, the signals would be very weak. Also, radio observations are easiest when you can compare the radiation in one direction with that in another direction. Since the background radiation would be the same everywhere, this basic technique could not be used. Also, at the time, there was a general feeling that the steady-state theory is correct, so there was not much motivation to carry out this difficult experiment.

Two physicists at the Bell Telephone Laboratories, *Arno Penzias* and *Robert Wilson* (Fig. 21.4), accidentally detected this radiation and reported their observations in 1965. Penzias and Wilson were unaware of the work of Alpher and Herman. They were using a very accurate radio telescope for both communications and radio astronomy. To have accurately calibrated results, they had to understand all sources of noise (interference) in their system. They found an unaccounted−for source of noise at a very low level. The noise seemed to be coming either from their system, or from everywhere in the sky. After carefully analyzing their system (including disassembling and reassembling certain parts, and even cleaning out bird droppings), they were confident that the noise was coming from everywhere in the sky. They only had a measurement at one wavelength, so they could not confirm the shape of the spectrum. However, they found that the intensity corresponds to a blackbody at a temperature of about 3 K.

At the same time, a group at Princeton, unaware of the work of Alpher and Herman, rederived their

the whole spectrum) drops more quickly than the energy density of matter. In the early universe, the energy density of radiation was greater than that of matter. We say that the universe was *radiation-dominated* at that time. Now the opposite is true; we live in a *matter-dominated* universe. The time when the radiation drops below the matter energy density is called the *crossover time*.

To understand the distribution of this radiation on the sky, we go back to our expanding surface analogy (Fig. 21.1). At the instant before the electrons and protons recombine, making the universe transparent, matter at all points is emitting photons in all directions. Since the matter becomes transparent, the photons continue running around in all directions on the surface of our expanding sphere. We see a steady stream of photons, not a brief flash as might be expected if we were looking at a localized explosion. The fact that the radiation is moving in all directions means that we see it coming from all directions. The radiation should appear isotropic. Also, cosmic background photons reaching us today are coming from one light day farther away than those that reached us yesterday.

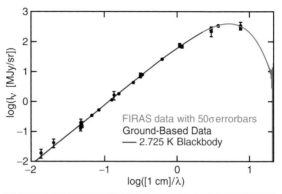

Fig 21.5. Spectrum of the cosmic background radiation as taken from ground-based observations. Most are at centimeter wavelengths where the Earth's atmosphere is very transparent. Measurements are the black data points with error bars. The smooth blue curve through those points is the best fit to the data. The red curve represents the COBE satellite (see below). The horizontal axis is essentially frequency plotted as $\log(1/\lambda)$, where λ is in centimeters. [© Edward L. Wright, used with permission]

Fig 21.4. Arno A. Penzias and Robert W. Wilson, who discovered the cosmic background radiation in 1967, in front of the telescope they used, at Bell Laboratories in New Jersey. [Reprinted with permission from Lucent Technologies' Bell Labs]

prediction, and set out to make the measurement. Someone familiar with both the Bell Laboratories and Princeton work put those two groups together. It was clear that Penzias and Wilson had found the cosmic background radiation. That confirmation came when the Princeton group carried out their observations at a different wavelength, and found that the radiation is present and that the spectrum is consistent with that of a blackbody. For their painstaking work in detecting this important signal, Penzias and Wilson were awarded the Nobel Prize for Physics in 1977.

The cosmic background radiation is of such significance in cosmology that its discovery led to extensive efforts to measure its properties as accurately as possible. An immediate consequence of its discovery was the death, for all practical purposes, of steady-state theories of the universe.

However, until it was shown unambiguously that the spectrum is that of a blackbody, there were still some ways for steady-staters to produce something like the background radiation. As Fig. 21.5 shows, the discovery and subsequent confirmations of the background radiation were on the long wavelength side of the peak of a 3 K blackbody spectrum (which is at about 1 mm). The most convincing observations would be to show that the spectrum does, indeed, turn over at wavelengths shorter than 1 mm.

However, even at wavelengths shorter than 1 cm, the Earth's atmosphere becomes sufficiently opaque that ground-based radio observations of sufficient precision are virtually impossible. More recently, observations from space have helped clarify the situation. Before we discuss those, however, an interesting experiment performed in the late 1960s (and still being improved upon) is worth some mention. This experiment involves optical observations of interstellar cyanogen (CN), which was first discovered in 1939.

The basic idea behind the CN experiment is shown in Fig. 21.6. The optical absorption lines are observed when the CN makes an electronic transition from its ground state to the first excited state. In Chapter 14 we saw that, for any electronic state,

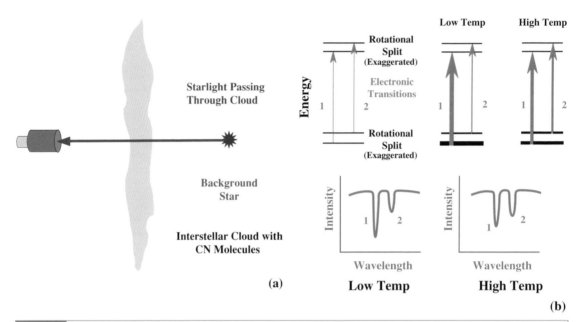

Fig 21.6. Using interstellar CN to study the cosmic background radiation. (a) We see what observations are done. The cloud with the CN is in between the Earth and a bright star. We take a spectrum of the star and see absorption lines from the CN. (b) We see how the strength of the absorption lines changes as we increase the temperature of the background radiation. In the upper left we see the energy levels. Electrons jumping from one level to another give us the big change in energy. Much smaller changes occur when the rate of rotation of the molecule changes. (These rotational energy shifts are exaggerated in the diagram.) Transitions 1 and 2 both correspond to the same jump by the electron, but they occur with the molecule in slightly different rotational states. In the upper right, we see the populations of the levels at low and high temperatures. The thicker the black line, the more populated the level. The red arrows indicate how strong the absorption is. The thicker the line, the stronger the absorption. In the lower panels, the resulting low and high temperature spectra are shown.

a molecule can be in many different rotational states. Normally, interstellar molecules are studied by radio observations of rotational transitions within the same electronic state. The effect of the rotational energy levels is to split the optical electronic transition into several lines. For example, there are lines corresponding to a transition from a given rotational state of the ground electronic state to the same rotational state in the excited electronic state. The wavelengths of the various transitions are slightly different. In practice, two transitions are observed, one from the ground rotational state and the other from the first excited rotational state.

The relative strength of the two optical lines is equal to the ratio of the populations in the two rotational states (in the ground electronic state), n_2/n_1. This is given by the Boltzmann equation as

$$\frac{n_2}{n_1} = \frac{g_2}{g_1} e^{-E_{21}/kT} \tag{21.6}$$

where g_1 and g_2 are the statistical weights of the two rotational levels, and E_{21} is the energy difference between the two rotational levels. The energy difference between the two levels in CN corresponds to the wavelength of 2.64 mm. If the CN is in a cloud of low density, there will be very few collisions with hydrogen to get molecules from the ground state to the next rotational state. (Collisions with electrons may also be important.) If there are no collisions, the CN can only get to the higher rotational state by absorbing radiation. If the CN is far from any star, the only radiation available is from the cosmic background. Under these circumstances, the populations will adjust themselves so the temperature in equation (21.6) is the temperature of the cosmic background radiation at 2.64 mm. (Of course, if the cosmic background radiation has a blackbody spectrum, its temperature at 2.64 mm should be the same as at any other wavelength.)

So, the CN sits in space, sampling the cosmic background radiation at a wavelength of 2.64 mm. This is a wavelength at which we could not directly measure the background temperature from the ground. The CN then modifies the light passing through the cloud. The relative intensity of these optical signals contains the information that the CN has collected on the cosmic background radiation. The optical signals easily penetrate the Earth's atmosphere. All we have to do is detect them and decode them. To decode them, we take the relative intensities as giving us n_2/n_1. We then solve equation (21.6) for T. As a farther refinement, we can check to see if there is any excitation of the CN due to collisions by electrons or atoms or molecules in the cloud. If there were, then the T that we measure would be greater than the background temperature. But, if this were happening, there would be a weak emission line at 2.64 mm. By looking for such a line and not finding it, we can rule out significant collisional excitation of the CN.

The most careful application of these techniques gives a temperature of the cosmic background radiation at 2.64 mm of 2.80 K, with an uncertainty of about 1%. This agrees well with the best direct longer wavelength measurements, indicating that, as we approach the peak, the radiation keeps its blackbody nature. It is possible to study the next rotational transition up at 1.32 mm, giving the background temperature even closer to the peak. The lower level of the second rotational transition does not have much population, so this is a very difficult experiment. However, the results are consistent with the blackbody nature continuing to the peak. There is another use to the CN technique. If we can observe CN in distant objects, we will be seeing it as it was in the past, and so the cosmic background temperature will be higher, and the CN excitation temperature will be higher. This gives us a check on our theories of how the temperature of the universe varies with z.

There is an interesting aside to this story. When interstellar CN was discovered in the late 1930s, the people who observed it also observed both optical spectral lines. They noted that the relative amounts of absorption corresponded to a temperature of about 3 K, but did not attach any significance to

Fig 21.7. Diagram showing COBE satellite. [NASA Goddard Space Flight Center and the COBE Science Working Group]

this. This shows us that, though many important discoveries in astronomy are made unexpectedly, simple luck is not enough. Confronted with the unexpected, the observer still must be able to recognize that something important is happening, and then be able to follow up the results.

The cosmic background radiation was considered so important that NASA decided to devote a whole satellite to its study. Thus, the *Cosmic Background Explorer Satellite*, or *COBE*, was launched in 1989. A diagram of the satellite is shown in Fig. 21.7. It carried instruments for making direct measurements of the radiation at wavelengths that were blocked by the atmosphere. The leader of the COBE team was *John Mather* of NASA's Goddard Spaceflight Center.

The COBE result for the spectrum is shown in Fig. 21.8. Measurements at many wavelengths are shown. Notice that the error bars are quite small. For comparison, we see the differences between observations and a 2.725 K blackbody. The agreement is spectacular. First, it is clear that the spectrum is truly that of a blackbody. The temperature of the background radiation has now been determined to better than 1%. There can be little doubt that this is the radiation predicted in 1948 by Alpher and Herman.

21.1.3 Isotropy of the cosmic background radiation

Earlier in this chapter, we said that the cosmic background radiation should appear the same no matter which way you look. That is, the radiation

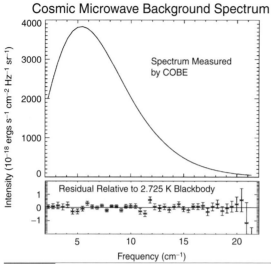

Fig 21.8. Diagram showing COBE spectrum of the background radiation with best fit to 2.725 K. The errors are so small that they do not show up on the data. The upper part of the diagram has the best fit spectrum, and the lower part has the error bars on an expanded scale. [NASA Goddard Space Flight Center and the COBE Science Working Group]

Fig 21.9. Origin of the dipole anisotropy of the cosmic background radiation. We see an observer in the center, moving to the right. This produces a Doppler shift to the blue on the right, to the red on the left.

should appear isotropic. The first observations found the radiation to be isotropic, but did not test this to a high degree of accuracy. More sensitive observations have looked for small deviations from perfect isotropy, called *anisotropies*. In describing any anisotropies in the background radiation, there are two relevant quantities: (1) how strong is the deviation, and (2) over what angular scale on the sky is the deviation seen?

One local source of anisotropy is the combined motion of the Earth, Sun, galaxy, local group and local supercluster with respect to the general Hubble flow, as indicated schematically in Fig. 21.9. This motion produces a Doppler shift, which varies with direction. If our net motion is with a speed v, in a particular direction, then the radial velocity away from the direction we are heading is

$$v_r = v \cos \theta \qquad (21.7)$$

In the direction in which we are heading ($\theta = 0°$), we get a maximum blueshift, and the radiation appears slightly hotter. In the opposite direction ($\theta = 180°$), we get a maximum redshift, and the radiation appears slightly cooler. Since this

anisotropy is characterized by a hot pole and a cold pole, with smooth variations in between, we call it the *dipole anisotropy*. We should point out that such an anisotropy doesn't violate special relativity by providing a preferred reference frame for the universe. We just happen to be measuring our velocity with respect to the matter that last scattered the background radiation. In fact, if we accurately determined Hubble's constant separately in different directions, we should measure the same anisotropy. (In fact, we do, in that this shows up in redshift surveys, primarily as our motion towards the great attractor, discussed in Chapter 18.)

As Fig. 21.10 shows, this dipole anisotropy has been observed by COBE. When we correct for the motions of the Earth and Sun within the galaxy, we find a Doppler shift corresponding to a motion of 600 km/s towards the great attractor.

There have also been searches for anisotropies on smaller angular scales. The goal of these studies is to learn more about the structure of the universe when the temperature was about 3000 K. After all, the background radiation retains an almost perfect record of that era. In looking for these anisotropies, we recognize that the universe

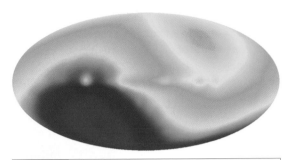

Fig 21.10. COBE measurements of the dipole anisotropy in the cosmic background radiation. This plot shows the whole sky on a single plot, with the galactic plane going horizontally across the middle. Bright regions are slightly hotter than the average temperature; dark regions are slightly cooler than the average temperature. [NASA Goddard Space Flight Center and the COBE Science Working Group]

Searches for these fluctuations have been carried out almost since the background radiation was discovered. However, the fact that they are so weak has made them quite elusive. The situation was resolved by COBE. In addition to measuring the spectrum of the radiation, COBE was also able to measure the distribution of radiation on the sky. The result of that map is shown in Fig. 21.11. In it, red regions are cooler than average, and blue regions are hotter than average. You can see that there is a mottled appearance, with structure all over the map. However, the signals are so weak, and the analysis is so difficult, that each bright spot you see does not exactly correspond to one bright area in the early universe. Each spot still has some important experimental uncertainties. However, the experimenters can use a statistical analysis of all of the bright spots, and tell us about their average properties. When this result was announced, cosmologists and the general public were quite excited. These fluctuations are so small that they are even a challenge to a satellite designed for the purpose. So, while the spectrum of the background radiation (like that in Fig. 21.8) was measured with great accuracy very early in the mission, the world had to wait another year to hear about the fluctuations. Of course, COBE can only sample on relatively large angular scales.

is not perfectly homogeneous. We are here, and our galaxy, cluster and supercluster are here. We can also see many other galaxies and clusters. For these objects to exist, there must have been very small seeds out of which they grew. That is, there must have been small concentrations of material around which more material could be attracted gravitationally. These small concentrations would also have been slightly hotter than their surroundings. This means that the radiation from these regions should appear slightly stronger than that from the surroundings. Theoreticians have suggested that these fluctuations in brightness will be very small, only about 10^{-5} of the strength of the radiation itself.

In order to study the anisotropies on a smaller scale, a balloon mission was launched over Antarctica in 1998. The project was known as Balloon Observations of Millimetric Extragalactic

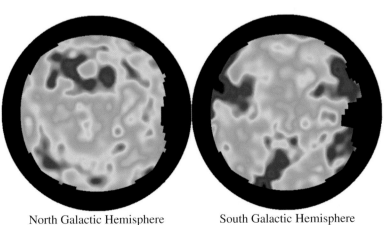

North Galactic Hemisphere   South Galactic Hemisphere

$-100\,\mu$K ▆▆▆▆▆ $+100\,\mu$K

Fig 21.11. COBE measurements of the small-scale anisotropies in the cosmic background radiation. The plot shows the whole sky viewed as two hemispheres. In this false color image, red areas are slightly cooler than the average and blue areas are slightly hotter. Notice that the largest fluctuations are 100 μK, out of a total 2.7 K signal. [NASA Goddard Space Flight Center and the COBE Science Working Group]

Radiation and Geophysics, BOOMERANG. The results are shown in Fig. 21.12. In Fig. 21.12(a), we show the map of roughly 3% of the sky that was observed. In Fig. 21.12(b), we look at how theoretical simulations show how the fluctuations should look for different geometries of the universe (closed, flat, open). A more detailed analysis of this type provides very strong evidence that the universe is flat. In Fig. 21.12(c), we look at the relative amplitude of fluctuations on different angular scales. Extrapolating these fluctuations back to the time of decoupling, this tells us about the seeds of various scales of structure that we see, as described in Chapter 18. In Fig. 21.12(d), we see how the BOOMERANG data fit into various other constraints on various cosmological parameters. Remember, flat universes are ones for which $\Omega_M + \Omega_\Lambda = 1$.

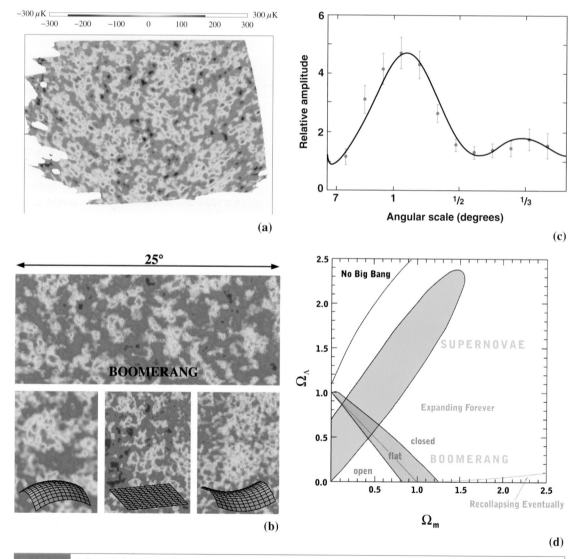

(a)

(b)

(c)

(d)

Fig 21.12. Balloon (BOOMERANG) measurements of smaller-scale anisotropies. (a) This image is almost 1800 square degrees, with the false color the same as in Fig. 21.11. (b) Models of fluctuations for different geometries of the universe. (c) Relative intensities of the fluctuations on various scales. (d) We see how the BOOMERANG data fit into various other constraints on various cosmological parameters. [The BOOMERANG Collaboration]

There is one remaining interesting problem with the isotropy of the background radiation. It is actually *too isotropic*. For the background to be isotropic, the early universe, apart from the fluctuations that would become galaxies, must have been quite uniform. We even incorporate this uniformity into our cosmological models. However, as illustrated in Fig. 21.13, conditions can only be identical at different locations if

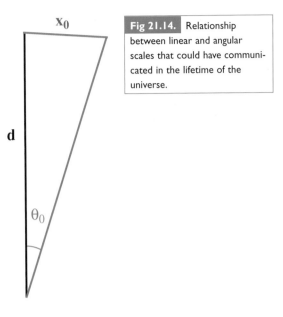

Fig 21.14. Relationship between linear and angular scales that could have communicated in the lifetime of the universe.

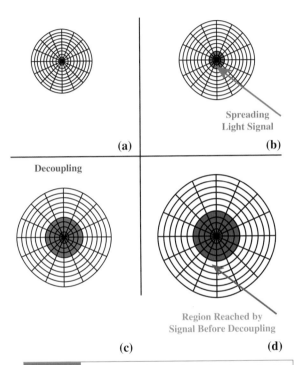

(a)   **(b)**

Decoupling

Spreading Light Signal

Region Reached by Signal Before Decoupling

(c)   **(d)**

Fig 21.13. Causality problems in the early universe. In this sequence, we look at the section of the universe that can be affected by light signals starting at some point just after the expansion started, traveling at the speed of light until decoupling. (a) The light signals start at the center of the grid. (b) As the universe expands (indicated by the larger grid), the light signals spread out in all directions. The green shaded area shows the region through which these signals have passed. (c) The expansion continues and the light signals spread farther. Eventually decoupling is reached. The green shaded area shows that region containing material that could have absorbed the radiation before decoupling. This is the only part of the universe that could have been affected before decoupling by the conditions at the center of the grid at the beginning. (d) After decoupling, no more material can be affected by the spreading light signals. The region that has been affected (shaded in purple) now simply expands with the universe. This is indicated by the fact that the shaded area in (d) covers the same number of grid sections as in (c).

they have some way of communicating with each other. Something must have regularized the structure in the early universe. However, such communication can travel no faster than the speed of light. Two objects separated by a distance greater than that which light can travel in the age of the universe cannot affect, or cause events to happen at, each other. This is called the *causality problem*.

We can use Fig. 21.14 to show the scales on which this is a problem. We must relate the current angular separation between two points with their linear separation at the time of decoupling. We now let the two points be separated by an angle θ_0 (where $\theta_0 \ll 1$ rad). If the points are a distance d from us, their current separation is

$$x_0 = \theta_0 d$$

If we are just seeing the light from these points, it must have been traveling for t_0, the current age of the universe. (We ignore the small difference between the current age of the universe and the time since decoupling, since decoupling occurred very early in the history of the universe.) This means that

$$d = ct_0$$

so

$$x_0 = \theta_0 ct_0$$

If we take our reference epoch to be now, so that $R(t_0) = 1$, and we let R be the scale factor at the time of decoupling, then the separation between the two points at the time of decoupling is R times their current separation, or

$$x = \theta_0 c t_0 R$$

The time for light to travel the distance x is x/c, or

$$\Delta t(x) = \theta_0 t_0 R$$

For these two points to communicate, $\Delta t(x)$ must be less than or equal to t, the age at decoupling. The farthest the two points can be is when $\Delta t(x) = t$, so

$$t = \theta_0 t_0 R$$

We now solve for θ_0, the maximum current angular separation between the two points that could have been causally connected before decoupling:

$$\theta_0 = t/R t_0 \qquad (21.8)$$

Example 21.1 Causality in the early universe
If the age of the universe at decoupling was 10^5 yr, and the current age is 1.5×10^{10} yr, find the maximum current angular separation between two points that can be causally connected.

SOLUTION
To use equation (21.8), we have to find R, the scale factor at the time of decoupling. We know that the temperature of the background radiation at any time t, $T(t)$, is proportional to $1/R(t)$. The background temperature is 3 K now and was 3000 K at decoupling, so R at decoupling must have been 10^{-3}. This means that

$$\theta_0 = \frac{(10^5 \text{ yr})}{(10^{-3})(1.5 \times 10^{10} \text{ yr})}$$

$$= 7 \times 10^{-3} \text{ rad}$$

This is approximately $0.4°$.

The above example tells us that points that appear more than half a degree apart on the sky cannot be causally connected. The background radiation should not appear smooth on scales larger than half a degree. However, it clearly does appear smooth on large scales. We will come back to this question in Section 21.3.

There is another source of small-scale anisotropy in the background radiation that is not actually cosmological. It has to do with the interaction of the radiation with clusters of galaxies as it passes

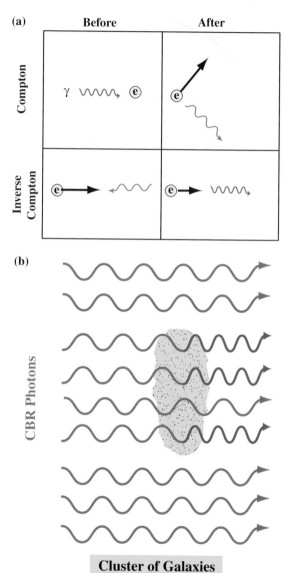

Fig 21.15. (a) Compton and inverse Compton scattering. In Compton scattering, a gamma-ray strikes a low energy electron and becomes a lower frequency photon, with the excess energy going to the electron. In inverse Compton scattering a low energy photon scatters off a high energy electron, with the photon gaining energy and the electron losing it. (b) The Sunyaev–Zeldovich effect. As low energy cosmic background photons strike the hot gas within a cluster of galaxies, inverse Compton scattering takes some of the photons from the low energy side of the 3 K blackbody peak and transfers them to the high energy side.

through. The galaxies in a cluster don't present a very large target, so most of the interaction is with the hot, low density intracluster gas. This gas is hot enough to be almost completely ionized. The background radiation interacts with the electrons in the gas.

In Fig. 21.15(a) we look at what happens when photons scatter off electrons, a process called *Compton scattering*. (*A. H. Compton* worked out the theory of the process and studied it in the laboratory as a demonstration of the photon nature of light.) Generally, in Compton scattering, the photon initially has more energy than the electron, so some energy is transferred from the photon to the electron. Since the photon loses energy, the wavelength increases. However, it is also possible for the electron to start out with more energy than the photon. In this case, called *inverse Compton scattering*, energy is transferred from the electron to the photon. The wavelength of the photon decreases.

The electrons in the intracluster gas have high energies, and the photons for the background radiation have very low energies. When they interact, we have inverse Compton scattering. The photons get a large boost in energy, going in as radio wavelength photons and coming out at shorter wavelengths. As shown in Fig. 21.15(b), this means that if we look in the directions of clusters of galaxies at radio wavelengths, some of the background radiation will have been removed. The background radiation should appear slightly weaker in the directions of clusters of galaxies, since some of the radio photons have been shifted to higher energy. This is called the *Sunyaev–Zeldovich effect*, after the two Russian astrophysicists who proposed it. This effect is very hard to detect, but it has now been detected.

21.2 | Big-bang nucleosynthesis

In the first three minutes of its existence, the universe was hot enough for nuclear reactions to take place. Fig. 21.16 shows how the abundances of various light elements changed during this time. Protons and neutrons combined to form deuterium (^{2}H), and two isotopes of helium (^{3}He and ^{4}He). The deuterium is very reactive, and most of it was used up as quickly as it could be

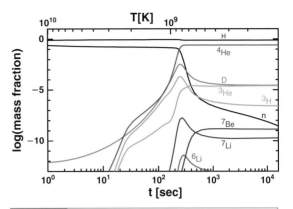

Fig 21.16. The formation of light elements in the early universe. On the horizontal axis, we show time. On the vertical axis we show the abundance of that isotope as time goes on. Each isotope is represented by a line of a different color, as marked. [© Edward L. Wright, used with permission]

formed. The ^{4}He is very stable, and, in those three minutes, 25% of the mass of the universe was tied up in ^{4}He. The ^{3}He makes up only 10^{-4} to 10^{-5} of the total mass. There is no stable nucleus with $A = 5$. This served as an effective barrier to the copious formation of still heavier elements.

We now look in a little more detail at what determined the abundance of ^{4}He. To do this we must look at the conditions that existed when the temperature was 10^9 K. This was probably at $t = 200$ to 300 s. We need to know the density then, and the relative number of neutrons and protons. Of course this depends on what happened before this time, when temperatures were even higher. We will see later in this chapter how we describe the physics in the first fraction of a second, but for He production, we only note that for $T > 10^{10}$ K (before 1 s) there was a thermal equilibrium between radiation and matter. At these temperatures, the energy density of radiation is much greater than that of matter (as discussed in the previous section). So the evolution of the scale factor, $R(t)$, is described by general relativity, where the energy density of radiation is the dominant energy. The evolution of R also depends on the number of different types of light particles, N_L, such as neutrinos. These light particles are important, because they move at or near the speed of light, so their motions dominate the equation of state for this relativistic gas. In a modification of the original Alpher and Herman paper, a few years later Alpher, Follin and Herman found that the best fit for this number is three. (Alpher and Herman treated the neutron abundance as a free parameter in their model, whereas Alpher, Follin and Herman calculated it by following the nuclear physics through the first seconds.) We will see below that these correspond to three types of neutrinos. At 10^{10} K, that equilibrium breaks down as the neutrons are frozen in at about 0.1 the number of protons. So we then need to follow nuclear reactions through 10^9 K.

The reactions that affected neutron abundance are:

$$n \rightleftarrows p + e^- + \bar{\nu}$$

$$e^+ + n \rightleftarrows p + \bar{\nu}$$

$$\nu + n \rightleftarrows p + e^-$$

The first reaction, with the right arrow is just the decay of a free neutron, which has a half-life of 10.6 min. The rates of all of these reactions can be expressed in terms of that half-life.

Because deuterium is so reactive, most of the neutrons that go into making deuterium will eventually end up in ^{4}He. The important reactions are:

$$^2H + {}^2H \rightleftarrows {}^3H + p$$

$$^2H + {}^2H \rightleftarrows {}^3He + n$$

$$^3H + {}^2H \rightleftarrows {}^4He + n$$

$$^3He + {}^2H \rightleftarrows {}^4He + p$$

If that is the case, then the mass fraction in ^{4}He is

$$Y = \frac{2n_n}{n_p + n_n}$$

$$= 2\left(\frac{n_n}{n_p}\right)\left(1 + \frac{n_n}{n_p}\right)^{-1}$$

The final abundance should also depend on n_B, the number density of baryons (particles such as neutrons and protons that participate in nuclear reactions) and n_γ, the number density of photons.

Even though each of these densities decreases as the universe expands, they each go as $1/R^3$, so their ratio doesn't change. (Remember the num-

ber of photons is not affected by the redshift, only their energy.) The ratio is about 10^{-10}. We will talk more below about the significance of this ratio.

In terms of these various factors, it is convenient to write these dependencies as

$$Y = (0.230) + (0.0011)\ln\left(\frac{10^{10} n_B}{n_\gamma}\right)$$

$$+ (0.013)(N_L - 3)$$

$$+ (0.014)(t_{1/2} - 10.6 \text{ min})$$

The final mixture of elements produced in the big bang depends on the density of material at $t = 1$ s, when the reactions started. We can relate the density at one second, $\rho(1 \text{ s})$, to the density now, ρ_0, by knowing $R(1 \text{ s})$. Fig. 21.17 shows the results of theoretical calculations of nucleosynthesis for

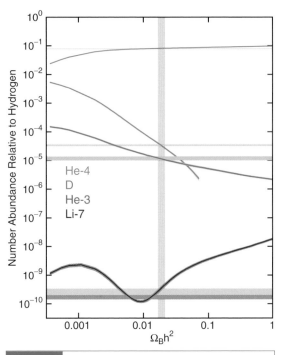

Fig 21.17. How the abundances of certain light elements depend on the density of the universe. On the horizontal axis, we plot the density of baryons (particles which can undergo nuclear reactions), expressed as the baryon density relative to the closure density. On the vertical axis, we plot the abundances of the various isotopes. The shaded area represents the best estimate of density from the abundance data. [© Edward L. Wright, used with permission]

models with different densities. These densities are expressed as the current density that would correspond to that early density. The stability of ^{4}He makes its abundance relatively insensitive to the density. However, the deuterium abundance drops off sharply with increasing density. This is because the deuterium is so reactive that a higher density provides it with more opportunities to react and be destroyed. There is a smaller density dependence in the ^{3}He.

The heavier elements (especially ^{7}Li) become more important at higher densities. This is because the high densities provide more reactions capable of building up the heavier elements. For densities greater than ρ_{crit}, C, N and O become somewhat abundant, but are still many orders of magnitude below their current observed abundances. This means that their current abundances of these heavier elements must be explained by production in stars.

The density dependence of the relative abundances of various nuclei provides us with a way of determining the density of the universe. We can measure the relative abundances of certain nuclei, such as D/H. We then use a diagram, like Fig. 21.17, to find the current density to which that ratio corresponds. In measuring the current abundances of any nucleus, it is important to account for any modification that has taken place since the big bang. Most important, the effect of nuclear processing of D in stars must be taken into account. The abundances that we see now are the result of big-bang and stellar nucleosynthesis. When we want to talk about the results of big-bang nucleosynthesis alone, we speak of *primordial abundances*.

It is important to note that this type of study only gives us the density of nuclear matter (protons and neutrons) in the universe. That is because it is only the nuclear matter that could have participated in nuclear reactions. For example, if the dark matter is in the form of neutrinos (which only interact via the weak interaction), then it may be that the nuclear matter is insufficient to close the universe, but the dark matter is still sufficient.

Most of the studies relating abundances to the density of the universe have involved deuterium. The net effect of stellar processing is to destroy deuterium. This means that the current D/H is less than the primordial value. On Earth, the D/H abundance ratio is about 10^{-4}. This means that for every deuterium nucleus there are 10 000 hydrogen nuclei. From Fig. 21.17, we see that this corresponds to a current density of 3×10^{-31} g/cm^3, which is much less than ρ_{crit}. Of course, the abundances in the Solar System may not be typical of the rest of the galaxy.

The next step is to study interstellar D, in whatever forms it is found. Direct observations of atomic D can be done in the Lyman α line (and other Lyman series lines) in the ultraviolet. These lines are observed in absorption against stars. Since interstellar extinction is so high in the ultraviolet, we cannot study clouds that are very far away. The results in our general area of the galaxy are similar to the Solar System value. Optical lines in the lines of sight to certain quasars give a value of about one D nucleus for every 40 000 H nuclei. This gives us an idea of values outside our galaxy.

Another possibility is the observation of the equivalent of the 21 cm line from atomic D. For deuterium, this line is shifted from 21 cm to 90 cm. Unfortunately, at this wavelength synchrotron radiation from the galaxy, whose intensity goes up approximately as λ^2, provides strong interference.

A final possibility is the radio observation of molecules containing deuterium, such as DCN. When we measure the abundances of these molecules, we find that the D abundance seems quite high. However, we now know that this is a result of chemical reactions (especially ion–molecule reactions) involving D and H proceeding at different rates for certain molecules. Thus, the abundances of the molecules don't reflect the true D/H abundance ratio. The observation of molecules with D substituted for H tells us more about interstellar chemistry than it does about cosmology.

The general conclusion of all of these D/H experiments is that the density of nuclear matter is probably about 5% of the critical density. If this result is correct, it doesn't mean that the universe is open. It just means that nuclear matter is not sufficient to close it. There is an additional note concerning ^{3}He. It also has an abundance that depends on density, though not as strongly as the D abundance does. If an atom of ^{3}He has one electron removed, a single electron remains. This ion behaves somewhat like a hydrogen atom. It has a transition analogous to the 21 cm line, at a wavelength of 9 cm. This line has been detected in galactic HII regions, but more detailed analyses are needed before these observations can yield cosmologically significant information.

In this section, we have discussed only what happened between one second and three minutes. Earlier than one second the universe was so hot that the internal structure of the neutrons and protons is important. Before we can understand what happened when the universe was less than one second old, we must look at some important features of elementary particle physics.

21.3 | Fundamental particles and forces

A major goal of physics over the centuries has been the search for the most fundamental building blocks of matter. We have progressed from earth, air, fire and water, through the atoms of Democritus, to Mendeleev's realization that the elements showed a regularity that was eventually explained by saying that the elements are not fundamental, but are made up of even smaller structures – nuclei and electrons. We have seen that the nuclei are made up of protons and neutrons.

21.3.1 Fundamental particles

Since the 1950s, physicists studying elementary particles have been able to use accelerators, like that shown in Fig. 21.18, to bring particles together at high energy. (Before the advent of accelerators, naturally produced cosmic rays were used.) When the energy of the collisions is just right, particles can be created out of the excess energy. As accelerators with more energy became available, increasing numbers of fundamental particles were found. Physicists began to suspect that these particles were no more fundamental than the 92 elements are.

To follow these results, we divide the various particles into groups, according to their ability to interact via the strong nuclear force. This is shown in Table 21.1. Particles that do not interact by the strong force, but still obey the Pauli exclusion principle, are called *leptons*. The most familiar lepton is the electron. The strongly interacting particles are called *hadrons*. We further divide the hadrons into the more massive *baryons*, and the less massive *mesons*. The most familiar baryons

Fig 21.18. Aerial view of Fermilab, the large particle accelerator outside Chicago. The ring has a diameter of over 6 km. [Fermilab photo]

Table 21.1. Classification of elementary particles.

Hadrons		Leptons
Mesons	Baryons	
π (pion)	n (neutron)	e, ν_e (electron, neutrino)
(+ others)	p (proton)	μ, ν_μ (muon, neutrino)
	(+ others)	τ, ν_τ (tau, neutrino)

Table 21.2. Quark properties.

Name	Symbol	Charge	Mass (GeV)
Up	u	$+(2/3)e$	0.35
Down	d	$-(1/3)e$	0.35
Strange	s	$-(1/3)e$	0.55
Charm	c	$+(2/3)e$	1.8
Top	t	$+(2/3)e$	4.5
Bottom	b	$-(1/3)e$	40.0

are the proton and neutron, and the most familiar meson is the pion.

The leptons appear to be the simplest group. In addition to the electron, the μ (mu) and τ (tau) particles are heavier versions of the electron. Their masses are 207 and 3660 times that of the electron, respectively. Neutrinos are also leptons. We think that there are three types of neutrinos, one to go with each of the other leptons (e, μ, τ). That is, in reactions for which we see an electron, we will see an electron neutrino, and so on. All of the evidence to date indicates that the leptons are truly fundamental. They seem to have no internal structure. We still don't know if the six leptons are all that there are. There is some evidence to suggest that there are no others, but we have been surprised before.

The hadrons do appear to have internal structure. This can be seen in experiments that have sufficiently high energy to probe the charge distribution within a proton. We now think that the hadrons are composed of particles called *quarks*. In the original theory of quarks, there were only three; now six have been found. That makes six quarks and six leptons, a balance which theorists seem to like.

The properties of the quarks are given in Table 21.2. Notice that they have fractional charges, coming in units of (plus and minus) one-third and two-thirds of the fundamental charge, *e*. However, the quarks can only combine in ways that produce integral net charges. Each quark has its own antiquark. All of the properties of a given antiquark (except mass) are the negative of those for the corresponding quark.

In the quark theory, any baryon is a combination of three quarks. For example, a proton is uud

and a neutron is udd. A meson is combination of a quark and an antiquark (not necessarily as the same type as the quark.) For example, a positive pion is ud, where the bar represents an antiparticle. The negative pion is ūd, and the neutral pion is uū. All known hadrons (and there are over 100) can be constructed by these simple rules. Note that, though there are six quarks, the only ones we need for everyday life (that is to make the proton and neutron) are u and d. Similarly, the only leptons that we need for everyday life are the electron and the electron neutrino.

The quark theory, originally proposed by *Murray Gell-Mann* (who received the 1969 Nobel Physics Prize for this work) was immediately quite successful, but there were a few important problems left. One was why only the particular combinations mentioned are allowed. The other is that, despite considerable effort, no one has been able to detect a free quark.

21.3.2 Fundamental forces

Along with the quest for fundamental particles, physicists are also trying to understand the forces with which the particles interact. The concepts of the forces and particles are intimately tied together. Without forces, particles would have no meaning, since we would have no way of detecting the particles. Our current thinking is that there are four fundamental forces, summarized in Table 21.3. They are arranged in order of strength, and their strength is given relative to the strong nuclear force. Short range forces are felt primarily on the scales of nuclei. Long range forces have a $1/r^2$ falloff in the force, which allows them to be felt over a large distance.

It also appears that particles are necessary as carriers of the forces, as summarized in Table 21.3.

Table 21.3. Forces and particles.

Force	Relative strength	Range	Carrier
Strong Nuclear	1	short	pion
Electromagnetic	10^{-2}	long	photon
Weak Nuclear	10^{-13}	short	W^+, W^-, Z
Gravity	10^{-40}	long	graviton (?)

For example, *quantum electrodynamics (QED)* is the theory that describes the electromagnetic force as being carried by photons. These photons may be real, or may only exist briefly on energy that can be borrowed because of the uncertainty principle (discussed in Chapter 8). These photons that live on borrowed energy are called *virtual photons*. The fact that the photon is massless leads to the electromagnetic force being long range. QED has been tested in many ways to very high accuracy, and is a very successful theory. The theory was developed in the 1940s by a number of physicists. One of the leaders in the field was *Richard Feynman* who came up with a way of visualizing the theory that also allowed him to carry out what had previously been very difficult calculations, which could be tested experimentally. Feynman shared the 1965 Nobel Prize in Physics for this work.

It is speculated that the gravitational interaction is carried by a massless particle, called the *graviton*. It is presumed to be massless, because gravity has the same long range behavior as electricity. However, no gravitons have ever been detected, and this theoretical framework is still being developed. We will see that the absence of quantum mechanical theory of gravity provides a limitation on how far back we can go in probing the big bang.

The strong nuclear force is carried by the pions. Of course, we now know that the pions themselves are not fundamental. They are each made of a quark and an antiquark. Since the pion has a mass, the strong nuclear force has a short range. In fact, the mass of the pion can be inferred from the range of the force (see Problem 21.9). The more massive the carrier, the greater the energy that must be temporarily 'borrowed' to produce the virtual particle. This means that the virtual particle lives a shorter length of time, and can travel a shorter distance in that lifetime.

The weak nuclear force, also a short range force, is carried by three particles. Two are the positive and negative W (for weak), and the third is a neutral particle, called the Z. These particles are much more massive than the pion, and the weak force is also a short range force. The W and Z particles have recently been detected with masses 80 and 90 times that of the proton.

21.3.3 The role of symmetries

In any study of particles and forces, symmetries play a very important role. Symmetries are important in many areas of physics. When we say that a system has a particular symmetry, we mean that the system looks the same after a certain transformation. For example, spherical symmetry means that the system looks the same, even if we rotate it through any angle, about any axis through one particular point (the center of the sphere). Recognizing symmetries can greatly simplify the solving of a problem. If a problem has a certain symmetry, then the solution must have the same symmetry. For example, in Fig. 21.19, we show a spherically symmetric charge distribution. If we want to find the electrical forces around it, those forces must have the same symmetric appearance.

Symmetries have an even deeper importance in physics. Whenever there is a symmetry, there is some quantity that is constant throughout the problem. This means that there is a conservation law. For example, the fact that the laws of physics cannot be changed by rotating our coordinate system leads to the conservation of angular momentum. The fact that the laws of physics are not changed by the translation of the origin of a coordinate system leads to the conservation of linear momentum. The fact that the laws of physics don't change with time leads to the conservation of energy.

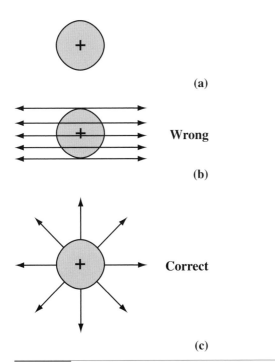

(a)

Wrong

(b)

Correct

(c)

Fig 21.19. Symmetry in an electricity problem. Suppose we wish to calculate the electric field due to the spherical charge distribution in (a). The resulting field must have the same symmetry as the charge distribution. The result in (b) is clearly wrong because, if we rotate the page, the charge distribution still looks the same, but the field direction changes. The type of distribution in (c) has the proper symmetry. A consideration of symmetry allows us to eliminate unreasonable answers.

We can understand the various forces by understanding what symmetries they have, or, equivalently, what conservation laws they obey. In general, a process will take place as long as it doesn't violate some conservation law. If a reaction that you think should take place does not, it means that there is some conservation law that you might not be aware of, and that the reaction would violate that conservation law. For example, before the quark theory had been proposed, there was a group of particles that should have decayed by the strong nuclear force, but did not. Because of this strange behavior, these particles were called 'strange' particles. It was proposed that there must be some property of these strange particles which had to be conserved. The particular decays would then have violated this conservation law. When the quark theory was proposed, the strange quark(s) was included to incorporate this property.

One interesting property of the weak interaction is that it doesn't obey all the conservation laws that the other forces do. Before this was realized, it was thought that conservation laws were absolute. The first symmetry found to be broken was that concerning *parity*, which has to do with the behavior of a system under a mirror reflection. It was realized by *T. D. Lee* and *C. N. Yang* in 1957 that it is possible to set up a beta decay experiment (beta decay taking place via the weak interaction) and the mirror image experiment, and achieve different results. This experiment was carried out by *C. S. Wu* a year later, and Yang and Lee shared the 1959 Nobel Prize in Physics for their prediction. Two other symmetries violated by the weak interaction are *charge conjugation* (the interchange of particles and antiparticles) and *time reversal*.

Sometimes, we find situations which are inherently symmetric, but somehow lead to an asymmetric result. They are called *spontaneous symmetry breaking*. For example, suppose we toss a coin in the air. As the coin is spinning, it has an equal probability of being heads or tails. As long as the coin stays in the air, the situation is symmetric between heads and tails. However, once the coin falls, the symmetry is broken. It is either heads or tails. Another example is a ferromagnet, shown in Fig. 21.20. If the magnet is heated above its critical temperature, it has spherical symmetry. There is no preferred direction. When we cool the material, it becomes a permanent magnet. Until it cools, all directions are equally probable. Once it cools, one direction is selected and the whole magnet cools, pointing in that direction. This is an example of a situation that is symmetric as long as it is hot enough, but cooling the system breaks the symmetry. Some circumstances in nature are symmetric as long as there is enough energy.

21.3.4 Color

We have already mentioned two problems with the quark theory. One is that there was no explanation for why the only allowed combinations are three quarks or one quark and one antiquark. The other is that we have not been able to detect

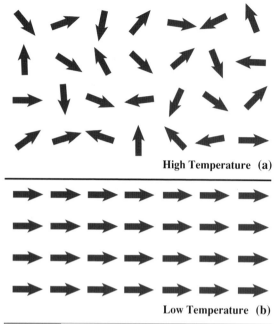

High Temperature (a)

Low Temperature (b)

Fig 21.20. Spontaneous symmetry breaking in a ferromagnet. (a) High temperature, above the critical temperature. Each arrow shows one iron atom's magnetic dipole orientation. Each is randomly oriented, so there is no preferred direction. (b) Low temperature. Now all the dipoles line up, so there is one preferred direction.

Each flavor of quark comes in each of the three colors. The six antiquarks come in corresponding anticolors.

The rule for combining quarks is that we can have only *colorless* combinations, like those shown in Fig. 21.21. One way of having a colorless combination is to have a quark of one color and an

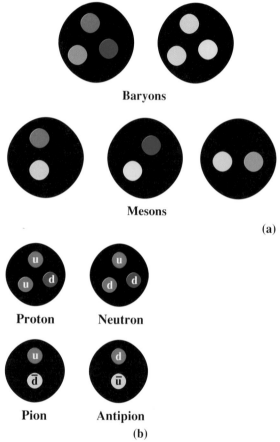

any free quarks. There was another problem with the original theory. Quarks have the same spin as electrons. Therefore, they should also obey the Pauli exclusion principle. However, some particles are observed which are clearly combinations of identical quarks, *uuu* for example, all in the ground state. This appears to be a violation of the exclusion principle.

The solution was to postulate an additional quark property that could be different for each of the three quarks in a baryon. This property is called *color*. This is just a convenient name, and has nothing to do with real color that we see. The color property is more like electric charge, except that it has three possible values (plus three antivalues) instead of one. (Electric charge is thought of as having one value and one antivalue.) Quark colors can be red (R), green (G) and blue (B). These colors relate to the ways in which quarks combine, and not to the properties of the particles they make up. To distinguish them from colors, we call the six quark types (u, d, s, c, t, b) *flavors*.

Fig 21.21. (a) Allowed quark combinations. These are the allowed color (and anticolor) combinations. Remember that each quark comes in six possible flavors (and each antiquark in six antiflavors) that give the particles their basic properties. In the top row, we see the combinations of three colors (or three anticolors) that give the baryons (and antibaryons). In the bottom row we see combinations of a color and anticolor that give the mesons. (b) Quark combinations for familiar particles. The proton is uud and the neutron is udd. In each case there is one quark of each color. (It doesn't matter whether the u or d is any particular color.) The pion is u$\bar{d}$ (with the u being some color and the $\bar{d}$ being the corresponding anticolor). The antipion is $\bar{u}$d (with the d being some color and the $\bar{u}$ being the corresponding anticolor).

antiquark with the corresponding anticolor. For example, a pion is $u\bar{d}$, where the possible color combinations are $R\bar{R}$, $G\bar{G}$, or $B\bar{B}$. This explains the mesons. The other possibility is to have one quark of each color. So for a proton (uud), the possible colors for three quarks are RBG, RGB, BGR, BRG, GRB, or GBR. We can see from this that a free quark would not be colorless, and is therefore not allowed by the theory. So the introduction of color has solved the three major problems mentioned above.

The properties involving color are more than a set of *ad hoc* rules. They have actually been derived from a mathematical theory, derived only on the assumptions of the types of symmetry that the theory should have. This theory is called *quantum chromodynamics*, or *QCD*. It is an analog to QED. In this theory, the strong force between hadrons is no longer a fundamental force. The real force is called the *color force*, acting between the quarks. Just as electric charge is a measure of the ability of particles to exert and feel the electromagnetic force, color measures the ability of particles to exert or feel the color force.

There are other analogies between the electric force and the color force. The requirement that particles be colorless is analogous to the statement that matter should be electrically neutral. For the most part, matter is neutral unless we provide a lot of energy to ionize atoms. As we will see below, the color force is so strong that the analogous process is not possible for quarks. There is another similarity. In the 19th century physicists thought that the force between neutral molecules, called the van der Waals force, was a fundamental force. After the development of the atomic theory in the early 20th century, it became clear that this was nothing more than the residual force between electrons and protons within the molecules. In the same way particle physicists now realize that the strong force between two particles is the residual of the color force among quarks making up the particles.

We have seen that quantum mechanical theories of forces require carrier particles. QCD is no different. In fact, the mathematical theory predicts the existence of a group of eight particles carrying the force. These particles are called *gluons*. There is a major difference between QED and

QCD. While the photons that carry the electromagnetic force have no electric charge themselves, the gluons that carry the color force have a color charge. This means that the gluons can interact with each other and with quarks. They can also change the colors of the quarks they interact with. This makes the theory very difficult mathematically. The detailed calculations that have characterized the success of QED have not yet been possible for QCD.

Another interesting feature comes out of QCD. The force between two quarks does not fall off with distance between the quarks. This means that the complete separation of two quarks requires an infinite amount of energy. Even if we had the energy available to drive two quarks far apart, we could not isolate a quark. This is illustrated in Fig. 21.22. As you pulled the two quarks apart, you would put enough energy into the system that you would simply create quark–antiquark pairs out of the energy. The new quark and antiquark would bind with the two quarks you were trying to pull

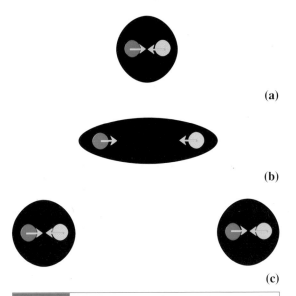

Fig 21.22. Quark confinement. The figure shows a quark and antiquark in a meson, as we try to separate them. In each part, the yellow arrows show the force between the quark and antiquark (which doesn't weaken as they get farther apart). (a) We just have the normal separation. (b) We are pulling them apart, doing work against the attractive force. (c) We have done so much work that we have created a new quark–antiquark pair, and now have two mesons instead of two free quarks.

apart, creating new combinations, but no free quarks. This leads to *quark confinement*, and tells us that we should not see any free quarks. Pulling quarks apart is like cutting a piece of string. Each piece will always have two ends. You can never get a piece with one end.

21.3.5 The unification of forces

Following his general theory of relativity, Einstein spent the latter part of his life attempting to 'unify' the forces of gravity and electromagnetism. By 'unify' we mean that we would like to explain all of the forces as really being manifestations of one larger force. The search is not new. After all, Maxwell unified the previously distinct forces of electricity and magnetism. Einstein's search for unification was not successful, and some thought that he was wasting his time because the problem had no solution. However, during the 1980s and 90s we have seen amazing progress towards this goal. This progress is based on mathematical theories, which are, in turn, based on the expectation that nature will obey certain symmetries. The progress in the unification of the forces is shown in Fig. 21.23.

The first of the recent successes was the unification of the electromagnetic and weak forces into one force, called the *electroweak force*. (For this work, *Sheldon Glashow*, *Abdus Salaam* and *Steven Weinberg* shared the Nobel Prize for Physics in 1979.) A major test of that theory was the prediction of the masses of the W and Z particles, carriers of the weak force, before their discovery.

In the electroweak force, the photon and the W (or Z) are essentially the same. You might wonder how two particles can be the same if one is very massive and the other is massless. The answer to this question tells us what we really mean by unification. In this case, the electromagnetic and weak forces appear to be the same when we deal with particles whose energies are much greater than the difference between the mass of the W and the mass of the photon (zero). At these high energies, the W mass arises from a small spontaneous symmetry breaking. However, we don't really notice the difference until the energy is so low that the mass of the W can no longer be ignored. This is another case of our maxim that nature is symmetric as long as there is enough energy.

The masses of the W and Z are about 80 GeV (being expressed as equivalent energy, where the proton mass is about 1 GeV). These masses are barely accessible in today's best particle accelerators. Therefore, we cannot even approach the range where particles will have energies much greater than this in order to see a world in which the electromagnetic and weak forces are the same. However, there must have been a brief instant in the early universe when the temperature was high enough for this symmetry to be present. This is one of the important connections between particle physics and cosmology, and we will explore it further in the next section.

The next step has been to unify the color (strong) force with the electroweak force, using the same mathematical tools. Theories that do this are called *grand unified theories*, or *GUTs*. These are still in the developmental stage. The carriers of this force are designated the X and Y particles. There masses are estimated to be about 10^{14} GeV, or 10^{12} times the mass of the W. This means that the differences between the color force and the electroweak force disappear when the energies are much greater than 10^{14} GeV. These energies are clearly beyond the range of accelerators that we could contemplate on the Earth. However, these conditions existed, ever so briefly, in the big bang.

One prediction of these GUTs is that the proton is not stable. Our normal understanding of the apparent stability of the proton is that it is the lowest mass baryon. The electroweak and

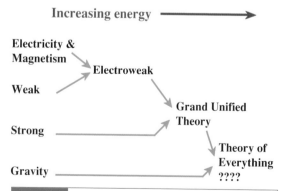

Fig 21.23. Unification of the forces. As we go from left to right in the figure, we see how, at higher energies, the forces become more unified.

color forces obey a conservation of baryons. Since a decay must always be to a lower mass particle, if a proton decayed the result would not be a baryon. Hence the stability of the proton. However, GUTs say that, at high enough energies, there is no difference between the electroweak and color forces. This means that the differences between hadrons and leptons should be lost. According to these theories, the proton would decay into a positron and a pion, with a half-life of about 10^{31} years. This doesn't mean that half of the protons wait this long, and then decay all at once. There should be a steady trickle of decays. The probability of any proton decaying in a given year is small, but there are many protons on the Earth, so we could actually hope to catch some decaying. So far, no such decays have been detected, indicating that the proton probably lives about ten times as long as the theory predicts. This means that there will have to be some modification of the theory.

Another prediction of the GUTs is the existence of *magnetic monopoles*. These are particles that would serve as the source of magnetic fields that are not due to the motions of charges. The magnetic field around a magnetic monopole would look like the electric field around a proton. The simplest GUT theory predicts that there should have been a large number of magnetic monopoles made in the big bang, and they should be as numerous as baryons. However, none have been detected. Again, some modification of the theory is required.

Even though there are still problems with GUTs, some theoreticians are forging ahead with the final step, the inclusion of gravity in the unified forces. These theories have been called *super GUTs*, or *supergravity*, or *theories of everything (TOE)*. Needless to say, they are still in the very speculative stages.

21.4 | Merging of physics of the big and small

21.4.1 Back to the earliest times

One of the most fascinating developments in the 1990s has been the interaction between two frontier areas of physics and astronomy – elementary particle physics and cosmology. In the preceding section, we saw that the big bang provides a unique opportunity for testing the predictions of various unified theories. It is as if an experiment was done for us, some 15 billion years ago, and the data are around in coded form. We only have to decode the results. For example, the current abundance of helium in the universe tells us that the three neutrino types we have so far are likely to be the only ones existing.

Some recent developments in particle physics may be able to help solve some of the outstanding problems in cosmology. We have already discussed most of these in one form or another. The problems that we would like to address include the following.

(1) Is the universe open or closed? If it is closed, what form does the dark matter take?
(2) Why is the universe so close to the boundary of being open or closed?
(3) Where did the magnetic monopoles go?
(4) How can the background radiation be so smooth on large scales?
(5) What caused the initial density concentrations that led to galaxy formation?
(6) Is there an excess of matter over antimatter? If so, how did it arise?
(7) For every baryon in the universe, there are approximately 10^{10} photons. Why is this?

To see how particle physics can help us answer these questions, we go back even farther than the epochs we have talked about already in this chapter. This is outlined in a time line, shown in Fig. 21.24. As we go through this scenario, notice the extremely short time scales.

We cannot say anything about the first 10^{-43} seconds. This time is called the *Planck time*. To describe phenomena on this time scale we need a quantum theory of gravity. It is even possible that on this short time scale, the continuous fabric of space-time breaks down.

From 10^{-43} s onward, the temperature was so high that GUTs are needed to describe the applicable physics. Since the electroweak and color forces are the same at these high energies, the differences between quarks and leptons disappear. (This is one reason why we think the number of quark types should equal the number of lepton types – currently thought to be six each.)

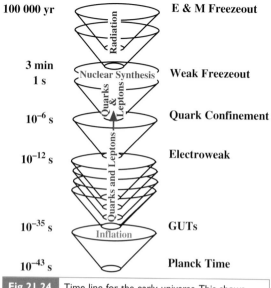

100 000 yr — Radiation — E & M Freezeout

3 min
1 s — Nuclear Synthesis — Weak Freezeout

Quarks & Leptons

10^{-6} s — Quark Confinement

Quarks and Leptons

10^{-12} s — Electroweak

Quarks and Leptons

10^{-35} s — Inflation — GUTs

10^{-43} s — Planck Time

Fig 21.24. Time line for the early universe. This shows events from the Planck time through the beginning of galaxy formation. Each magnification is a factor of about one million in time.

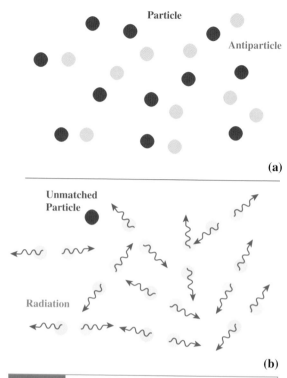

Particle

Antiparticle

(a)

Unmatched Particle

Radiation

(b)

Fig 21.25. Matter–antimatter imbalance and the excess of photons in the universe. In this figure there is one extra particle of matter for every ten matter–antimatter pairs. In the real universe it is one extra for ten billion! (a) The temperature is high enough that the particles and antiparticles are all present. (b) The temperature is low enough so that the particles and antiparticles annihilate and don't re-separate. For each annihilation, we get two photons. The lighter colored circles show the previous locations of particles and antiparticles.

During this period baryon number was not conserved. The breaking of symmetries involving mirror reflection, and matter–antimatter interchange resulted in a slight excess of matter over antimatter. This excess is one particle for every 10^{10} particle–antiparticle pairs produced. This tells us that our universe now has some matter, which is not balanced by an exact amount of antimatter. This answers question (6) listed above.

This slight excess of matter over antimatter also explains the observation that there are 10^{10} photons for every baryon in the universe. This is illustrated in Fig. 21.25. As the universe continued to cool, particles and antiparticles found each other and annihilated. For each pair that annihilated, two photons were produced to carry away the energy. It was like a cosmic game of musical chairs. For every 10^{10} particles that found a partner, there was one particle for which there was no antiparticle. That particle is still around, as well as the 10^{10} photons left from that number of annihilations. This means that for every baryon left today, there must be 10^{10} photons. This solves problem (7). This slight imbalance is very important. If there were perfect balance between matter and antimatter, the universe would now be all radiation, and we wouldn't be here.

At the end of this period, after only 10^{-35} seconds, the universe had 'cooled' to the point where the average energy was no greater than the mass of the X and Y particles. This means that the symmetry between the electroweak and color forces was broken. From that point on, baryon number was conserved.

By 10^{-12} seconds, the universe had cooled to a point where the electromagnetic and weak forces separate. This is because the average energy is comparable to the mass of the W. (Remember, we saw in the preceding section that this symmetry is broken when average energy is no longer much greater than the W–photon mass difference.) At this point, the weak force was as strong as the electromagnetic force. From that point on, the weak force became much weaker than the electromagnetic force. However, it was still hot enough for the quarks to move as a fluid. By 10^{-6} seconds, the quarks were confined in hadrons.

At 1 s, the temperature was low enough for the weak force to have weakened to the point where the neutrinos are rarely absorbed by matter. From that point on, the neutrinos and matter decoupled (just as radiation and matter decoupled after 100 000 years). These neutrinos should now be everywhere in the universe, just like the cosmic background radiation. After the neutrinos decoupled, there were still some electron–positron annihilations that added photons to the radiation. Thus the temperature of the neutrinos should be less than the 3 K of the radiation. In fact, it is estimated that the neutrinos are like a gas at 2 K. These low energy neutrinos are extremely hard to detect. However, there are so many of them that if it turns out that neutrinos have a very small rest mass, even as low as 20 eV (only about 1/40 000 of the mass of the electron), the total mass of neutrinos could exceed the mass of nuclear matter by a factor of ten. That could be enough to close the universe. It may also be that clumps of neutrinos formed the centers of attraction for the formation of clusters of galaxies.

We have now reached the regime that has already been discussed earlier in this chapter. Nuclear reactions continued for the first three minutes. By 100 000 years, the matter had cooled enough to become neutral, and the radiation decoupled, remaining to be detected as the cosmic background radiation.

21.4.2 Inflation

Particle physics has one more surprise for us, relating to the nature of the vacuum during the GUT era. Classically, we think of the vacuum as simply being 'nothing'. In quantum mechanics, it is not so simple. The vacuum is merely the lowest energy state. It is still possible for interesting things to happen. This is illustrated in Fig. 21.26. On the smallest scales, we have the ongoing creation and destruction of particle–antiparticle pairs.

It has been proposed that, during the GUT era, the nature of the vacuum changed. When this happened the universe underwent a phase change, like solid ice changing to liquid water, or

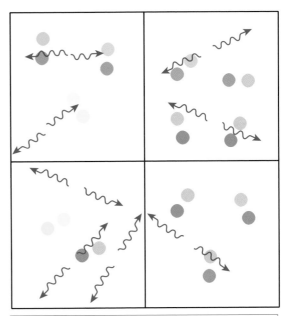

Fig 21.26. What might happen in a quantum mechanical vacuum. Each frame is at a slightly different time. We see particles/antiparticles and photons being created for brief instants.

liquid water changing to water vapor. According to the theory, the rearrangement resulted in an extremely rapid expansion of the universe. In this expansion, the scale factor had a rapid exponential growth, as if there were a large cosmological constant. R changed by more than 10^{40} in 10^{-32} s, as indicated in Fig. 21.27. This rapid growth was called *inflation*. Once the inflation ended, the normal expansion resumed.

Though it is still a speculative theory, inflation might explain some of our remaining cosmological puzzles. It solves the monopole problem (3) by saying that all of the monopoles were produced before inflation, and were carried far apart by inflation. Therefore, they are separated by large reaches of space, and we are not likely to run into many.

To see how it solves the flatness problem (2), let's go back to our expanding sphere analogy. Remember, we can only see things that are closer to us than the distance light could have traveled in the age of the universe. If there was a large inflation, our horizon doesn't include a very large piece of the universe. Since we can only see a small fraction of the surface, it appears flat.

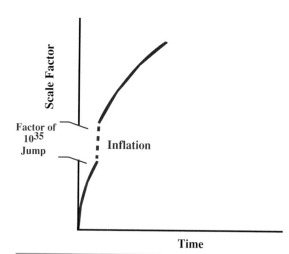

Fig 21.27. Change of the scale factor in an inflationary universe. Time is plotted on the horizontal axis and scale factor on the vertical axis. Notice that we have to put a break in the scale factor axis to show the whole change.

This is analogous to the Earth; when we only see a small piece of its surface, it appears flat too.

It solves the causality problem (4) by saying that everything within our horizon was so close together before inflation that it could all be causally connected.

It may even solve the galaxy formation problem (5) by saying that galaxies (or the matter to form them) condensed around irregularities in the distribution of phases as the inflation ended.

It must be remembered that these ideas are quite speculative. However, even if the specifics are wrong, they may have opened many new possibilities in our understanding of the universe. At this point theoreticians working on cosmological models now include some form of inflation as part of their standard model.

21.4.3 Galaxy formation

As mentioned in Section 21.1, we believe that galaxies grew out of small fluctuations in the density that were present before radiation and matter decoupled. (This is why anisotropies in the cosmic background radiation may tell us about the sizes of these fluctuations.) As we have said in Chapter 18, there are two basic galaxy formation scenarios: (1) the galaxies may have formed first and then gathered into clusters; (2) the initial condensations may have been larger, forming protoclusters first, with the galaxies then forming out of density fluctuations within the forming cluster. In this latter scenario, calculations show that the original large-scale structures would be flattened, and these are referred to as pancakes.

In deciding between these scenarios, it is important to understand the nature of the dark matter in the universe. Even if there is not enough dark matter to close the universe, there must be at least enough to account for the viral masses of clusters. Even this amount would place the density of dark matter as being significantly greater than the density of luminous matter. This means that the dynamics of the universe and galaxy formation is dominated by dark matter.

Models of galaxy formation must also include the effects of a non-zero cosmological constant, if there is one. That will affect the evolution of the universe on the largest scales, especially affecting the formation of the largest structures, superclusters. This gets around the problem, with cold dark matter explaining the largest structures. It now appears that we can simulate structures that we see on all scales with cold dark matter, and $\Omega_{MAT} = 0.3$, $\Omega_\Lambda = 0.7$. An example of one such stimulation is shown in Fig. 21.28.

21.4.4 Estimates of values of cosmological parameters

Here we summarize our current estimates of the values of various cosmological parameters needed to describe our big-bang universe. It should be understood that, while a consistent picture seems to be emerging from a variety of approaches, there is still a lot of work to be done, both observationally and theoretically.

Various approaches to measuring H_0 are giving consistent values near 70 km/s/Mpc. This corresponds to a Hubble time of 14 Gyr. The actual age of the universe for the currently accepted model parameters is also 14 Gyr.

The universe appears to be flat, as predicted by theories of inflation. The best observational evidence for this comes from the small-scale fluctuations in the cosmic microwave background radiation, as shown in Fig. 21.12. Remember, this radiation has been running through the universe back to $z = 1000$. For any given size fluctuations,

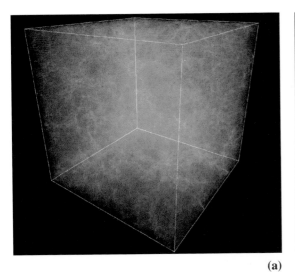

(a)

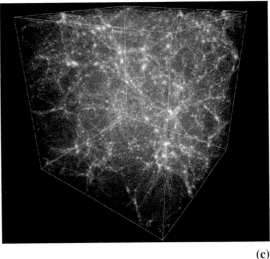

(c)

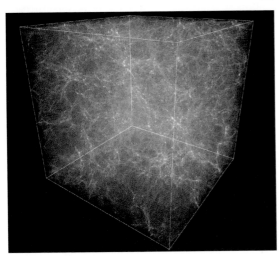

(b)

Fig 21.28. Simulation of large-scale structure formation and evolution in a cold dark matter universe with density parameter $\Omega_{MAT} = 0.3$, cosmological constant $\Omega_\Lambda = 0.7$, and Hubble constant $H = 65$ km/s/Mpc, inside a box of 154 Mpc in size. The calculations were performed on a Cray SVI supercomputer, with 2.7 million particles. (a) $z = 5$; (b) $z = 2$; (c) $z = 0$. [This simulation was produced at the Texas Advanced Computing Center by the University of Texas Galaxy Formation Group (Paul R. Shapiro, Hugo Martel and Marcelo Alvarez)]

curvature in space-time (positive or negative) would cause a distortion of how we view those fluctuations. The fluctuations in Fig. 21.12 are those that would be expected with no distortion, so they point to a flat universe.

If the universe is flat, this tells us that the total density parameter (defined in Chapter 20), $\Omega_{TOT} = 1$. This doesn't necessarily mean that there is enough matter for this. Ω_{TOT} only tells about the combined effects of matter (Ω_{MAT}) and a cosmological constant (Ω_Λ). So we know that $\Omega_M + \Omega_\Lambda = 1$. A number of other data suggest that Ω_M is approximately 0.3 and Ω_Λ is approximately 0.7. Remember, we saw in the previous

chapter that the effect of the cosmological constant is $\rho - \Lambda/8\pi G$, so it acts in the opposite sense of matter. While matter wants to bring galaxies together, the cosmological constant makes them want to fly apart. Observations (still preliminary) using supernovae to measure the deceleration parameter suggest that the expansion is not slowing down very quickly, and may be increasing, consistent with a significant cosmological constant.

If Ω_M is approximately 0.3, this is the combined effects of light and dark matter. From analysis of the results of big-bang nucleosynthesis, baryonic matter, neutrons and protons, is about 5% of the closure density. Only one-fifth of the baryons (or 1% of the closure density) is in the form of stars, galaxies, etc. It has been suggested that the rest is in the form of intergalactic ionized H. We still don't know the nature of the dark matter that makes up more than 25% of the closure density, but it appears that simulations with cold dark matter, and a non-zero cosmological constant, suggest that cold dark matter can explain most of the structures we see in the universe.

Chapter summary

When we work backwards to the conditions in the early universe, we find that there must have been an era in which the material was very hot and dense. This hot dense era is called the big bang. The debris of the big bang is all around us.

Just as the early universe was filled with hot matter, it was also filled with hot radiation. The radiation and matter stayed in contact as long as the radiation could scatter off the charged particles. When the temperature of the universe fell below 3000 K, the electrons and protons combined to form neutral atoms, and the universe became transparent. The radiation and matter decoupled. The radiation is still everywhere in the universe traveling in all directions. It is called the cosmic background radiation. As the universe expanded, the cosmological redshift increased the wavelengths of all of the photons, and the effect on the background radiation was to produce ever cooler blackbodies. The current temperature is about 2.7 K.

Following its accidental discovery, the background radiation has been studied extensively. Studies of the spectrum were made difficult by the Earth's atmosphere. Radio observations from the ground, some balloon and rocket observations, and some indirect measurements involving interstellar CN, were all employed to study the radiation until the Cosmic Background Explorer Satellite was launched. COBE provided definitive evidence that the spectrum is indeed that of a blackbody. These studies also showed that the radiation is very isotropic. There is a dipole anisotropy, due to the Earth's motion. COBE revealed very low level fluctuations in intensity all over the sky. It is believed that these fluctuations are a snapshot of the small density enhancements in the early universe that gave rise to the clusters of galaxies that we see today. Apart from these fluctuations, the radiation is actually too smooth. The degree of smoothness that we see would suggest that points that were farther than 100 000 light years apart would have had to communicate with each other in the first 100 000 years. This is called the causality problem.

One of the great successes of big-bang cosmologies is that they can explain the abundances of the light elements. For approximately the first three minutes the universe was hot enough for nuclear reactions to take place. Those reactions produced essentially all the helium and deuterium that we see in the universe today (though deuterium is destroyed and helium is produced in stars). The abundance of deuterium is particularly sensitive to the density of the material in the nuclear reactions. The denser the material, the less deuterium there is. The density of material in the first three minutes can be related to the current density, and can tell us if nuclear reacting material (protons and neutrons) can close the universe. The best estimate is that this material is only about 5% of what is needed to reach the critical density.

When the universe was less than one second old, the temperature was so high that matter was stripped down into its fundamental components, the elementary particles. The leptons, particles which do not feel the strong nuclear force, appear to be fundamental in that they have no apparent internal structure. There are six of these, with the familiar ones being the electron and electron neutrino. The hadrons, or strongly interacting particles, appear to be made up of various combinations of quarks. There are six known types (flavors) of quarks. The baryons (neutron, proton, etc.) are made up of three quarks; the mesons (pion, etc.) are made up of quark–antiquark pairs. Each quark can come in three different colors, which provide for the binding of the quarks.

In describing the elementary particles it is important to identify the fundamental forces. We think that each force is carried by a particle. (These particles can be real or virtual, meaning that they violate conservation of energy but for times too short to measure.) The long range electromagnetic force is carried by massless photons, as described by quantum electrodynamics (QED). The weak nuclear force is carried by the massive W and Z particles, and the strong nuclear force is carried by the pion. We now understand that this strong force is the residual of the color force among the quarks. That force is carried by gluons, as described by quantum chromodynamics (QCD). The mass of the carriers explains the short range of the forces. Attempts to unify the forces have been partially successful. The electromagnetic and weak forces appear to be manifestations of the electroweak force. Theories that combine the electroweak and QCD are called grand unified theories (GUTs). In each case, the unification means that the forces are equivalent when ener-

gies are much higher than the masses of the carriers of the forces. To the best of our knowledge, this has only occurred in the first fraction of a second of the big bang.

Applying particle physics to our study of the early universe has helped solve some cosmological problems. The excess of matter over antimatter, about one part in ten billion, arose from a slight asymmetry in certain weak interactions. This also explains why there are approximately ten billion photons for every baryon in the universe. The GUTs also suggest that, in the first fraction of a second, the universe went through a rapid expansion, or inflation, in which the scale factor increased by about 10^{35} almost instantaneously. This might explain why the universe appears to be flat and so isotropic. It might even explain how galaxy formation started.

Questions

21.1. When the temperature in the early universe fell to 3000 K, the matter recombined. Why is this an important event?

21.2. How is the existence of the cosmic background radiation related to the origin of the light elements?

*21.3. When the universe grows to twice its current size, what will the temperature of the background radiation be?

*21.4. When the cosmic background radiation first decoupled from the matter, in what part of the spectrum (radio, infrared, visible, ultraviolet, etc.) did the radiation peak?

21.5. Why is the cosmic background radiation hard to observe?

21.6. What are the similarities between CN molecules and space probes in studying the cosmic background radiation?

21.7. What is it learned from studying the spectrum of the cosmic background radiation rather than simply measuring its intensity?

21.8. Why was COBE able to measure properties of the cosmic background radiation that could not be measured in other ways?

21.9. What is the significance of the dipole anisotropy in the cosmic background radiation?

21.10. Does the dipole isotropy in the cosmic background radiation violate the part of the cosmological principle that states that the universe should be isotropic?

21.11. Why is it important to find small-scale anisotropies in the cosmic background radiation?

21.12. What do we mean when we say the universe is too isotropic?

21.13. What conditions were prevalent in the first three minutes that allowed nuclear reactions to take place?

21.14. Why is the abundance of deuterium particularly sensitive to the density during the time of nucleosynthesis?

21.15. How do we know that the C, N and O around us were not produced in the big bang?

21.16. Why is it so important to determine the relative abundance of deuterium in the universe?

21.17. If the deuterium measurements tell us that there aren't enough protons and neutrons in the universe to close it, how can it be that the universe might still be closed?

21.18. What makes us think that the hadrons are not fundamental particles?

21.19. What makes us think that the leptons might be fundamental particles?

21.20. What do we mean when we say that we only need the up and down quarks for everyday life?

21.21. Why are we not likely to see a free quark?

21.22. What are the successes of the quark theory?

21.23. What are the differences between quark color and flavor?

21.24. If the quarks have fractional charges, how do the particles that we see have integral charges?

21.25. What do we mean when we say we are trying to unify the forces?

21.26. How is the mass of the carrier of a force related to: (a) the range of the force, (b) the temperature at which that force might be unified with others?

21.27. How are we limited by the absence of a quantum mechanical theory of gravity?

*21.28. As the universe cooled (still in the first fraction of a second), previously unified forces became distinct. How can this be described as spontaneous symmetry breaking?

21.29. Why is it important that certain symmetries were not perfect in the early universe?

21.30. What is the difference between quantum electrodynamics (QED) and quantum chromodynamics (QCD)?

21.31. Work out the different combinations of u and d quarks that can make allowed particles. Give their electronic charge. (Note: the order doesn't matter; i.e. uud is the same as udu.)

21.32. What cosmological problems can be explained by inflation?

21.33. Why are there approximately 10^{10} photons in the universe for every baryon?

21.34. Why is there an excess of matter over antimatter in the universe?

21.35. If neutrinos had masses only about 1/40 000 that of the electron, how could they provide enough mass to close the universe?

Problems

For all problems, unless otherwise stated, use $H_0 = 70$ *km/s/Mpc.*

21.1. (a) What is the present energy density in the cosmic background radiation, and (b) what was it when it was emitted at a temperature of 3000 K?

21.2. (a) Suppose the scale factors at times t_1 and t_2 are R_1 and R_2, and the background radiation temperatures are T_1 and T_2. What is the relationship between T_1 and T_2 in terms of R_1 and R_2? (b) If we know that the background temperature now is 3 K, and we let $R = 1$, write an expression for the background temperature at any time when the scale factor is $R(t)$.

21.3. Assume that the energy density, integrated over all wavelengths, in the background radiation is given by the Stefan–Boltzmann law. Assume that we know the cosmological redshift, and show that these lead to the conclusion that the temperature must scale as $1/R$.

21.4. Use the Wien displacement law to find the peak wavelength of a blackbody for a temperature of (a) 3 K, (b) 3000 K.

*21.5. Compare the energy density in the cosmic microwave background with that in diffuse starlight. Assume that the diffuse starlight has a brightness temperature of 10 000 K and a filling factor of 10^{-14}.

21.6. In the CN experiment to study the cosmic background radiation, what must the population ratio be for the temperature to be 2.7 K? (Take $g_2/g_1 = 3$.)

21.7. For the dipole anisotropy, the fractional increase in temperature, $\Delta T/T$, is equal to $-v/c$, where v is our speed. Justify this relation using Wien's displacement law and the Doppler shift relation.

21.8. Use the result of the previous problem to predict $\Delta T/T$ resulting from the Earth's orbital motion.

21.9. (a) Assume that a force is carried by a virtual particle of mass m. Assume that this particle can exist for a time $h/2\pi mc^2$, and that it travels close to c. What is the approximate range of the force? (b) If the range of the strong force is 10^{-13} cm, what is the mass of the particle carrying the force? How does it compare with the mass of the pion?

21.10. If the electromagnetic force is carried by virtual photons, which can live for a time h/E, explain how the first can be felt at any range. (Hint: Think about photons of different wavelength and energy.)

21.11. What is the temperature at which the average kinetic energy is equal to the mass of the W?

21.12. (a) What is the temperature at which the average kinetic energy is equal to the mass of the X? (b) What is the mass of the X in grams?

21.13. If the average lifetime of the proton is 10^{31} yr, how much water would you have to watch to detect 100 proton decays in one year?

Computer problems

21.1. Make a graph comparing blackbody curves for $T = 2.75$ and 3.00 K.

21.2. Make a graph of the wavelength at which the cosmic background radiation spectrum peaks vs. z, for z ranging from 1 to 1000.

21.3. In the CN experiment to study the cosmic background radiation, draw a graph of the population ratio, n_2/n_1, as a function of temperature for an interesting range of temperature. (Take $g_2/g_1 = 3$.)

Part VI

The Solar System

In studying the Solar System, we find an important exception to our concept of astronomical objects being so remote that we cannot hope to visit them in the foreseeable future. People have already visited our nearest neighbor in the Solar System, the Moon, and brought back pieces to study in normal Earth-bound laboratories. Unmanned probes have landed on Venus and Mars and have visited all the other planets. Clearly, the opportunity for even limited close-up viewing has had a major impact on our understanding of the Solar System.

However, the study of the Solar System is not simply devoted to sending probes when we feel like it. The spacecraft have followed literally centuries of study by more traditional astronomical methods. By the time the first probe was launched to any planet, astronomers had already developed a picture of what they expected to find. Many of these pictures did not survive the planetary encounters, but they did provide a framework for asking questions, and for deciding what instruments were important to place on the various probes.

We have also had the advantage of having the Earth as an example of a planet to study. It has been possible to develop ideas about planetary surfaces, interiors, atmospheres and magnetospheres by studying the Earth. For that reason, we have devoted one whole chapter of this Part to the Earth, viewed not as our home base, but as just one planet. In studying the Earth, we will generate ideas which we will extend to studying other planets.

We will study the other planets in two groupings of similar planets, the inner and outer planets. Within each grouping, we don't study all aspects of a given planet before going on to the next planet. Instead, we study a given aspect, e.g. atmospheres, of all the planets for the group. This allows us to extend common ideas to similar objects, looking for similarities and differences.

We will also see how much of the physics we have used in other astronomical problems – orbits, energy transport, hydrostatic equilibrium, tidal effects, and using spectroscopy to study remote objects, to mention a few examples – fit very naturally into our study of the Solar System. Therefore, rather than trying to give a complete list of all the facts revealed by various probes, we emphasize the underlying physics.

Chapter 22

Overview of the Solar System

The Earth belongs to a group of nine planets, orbiting the Sun, called the *Solar System*. Each object follows its own orbit about the Sun. All of the planets orbit in the same direction. As large as the Earth seems to us, it is small compared to the distances between objects in the Solar System. This is true of the other planets, even those much larger than the Earth. For all practical purposes, the Solar System is vast emptiness, with a few small island oases.

If we could look at a side view of the Solar System, we would notice that the orbits are not very tilted with respect to that of the Earth. So, in a side view from the outside, the Solar System would look like a very thin disk. We call the plane of the Earth's orbit the *ecliptic*. The motion of the Earth around the Sun causes the Sun to appear to move against the background of fixed stars. That path is just the projection of the ecliptic onto the sky. The Earth's rotation axis is tilted (by 23.5°) so that the ecliptic does not line up with the Earth's equator.

We begin a brief tour of the Solar System by looking at the planets. A photograph of each planet is shown in Fig. 22.1. It is convenient to divide the planets into two groups, the *inner planets* and the *outer planets*. The inner four planets are Mercury, Venus, Earth and Mars. The giant outer planets are Jupiter, Saturn, Uranus and Neptune. The outer planets are much more massive than the inner planets. They also have very different compositions. The outermost planet is Pluto (though it does spend part of its orbit closer to the Sun than Neptune). Pluto is small, like the inner planets.

Most of the planets have moons orbiting them. Of course, the most familiar is our own Moon. Mercury and Venus are the only planets without any known moons. Mars has two small moons. Jupiter has four large moons and numerous smaller moons. Saturn has one large moon, and, like Jupiter, a collection of smaller moons. Uranus has three modest sized moons, and several smaller ones. Neptune has one large and a number of small moons, and, finally, Pluto has one moon that is relatively large compared to the size of the planet. All of these moons add to the diversity of objects that we can study in the Solar System.

In addition to planets and moons, there is a collection of smaller sized objects. Found mostly between the orbits of Mars and Jupiter is a collection of rocky bodies, called *asteroids*. They are mostly between the orbits of Mars and Jupiter. From time to time we see objects appear faintly in our sky that then brighten, and develop a tail, as shown in Fig. 22.2. These are *comets*. Occasionally the Earth runs into small debris left in its orbit. The material falls through the atmosphere and is heated. We see the glowing trail as a *meteor*. Occasionally the Earth suffers large impacts from these.

One of the goals in studying the Solar System is to find clues to its origin, and to put together a picture of that origin. We will defer that discussion until Chapter 27, after we have discussed all the material in the Solar System. For now, we note that we expect the Solar System to form as a biproduct of the formation of the Sun. The Solar System should have formed out of the disk that was part of the late stages of the formation of the

Fig 22.1. The planets. [NASA]

Sun. This explains why the orbits of the planets are almost in the same plane and why the orbital motions are in the same direction, preserving the angular momentum from the molecular cloud

and disk, as discussed in Chapter 15. Differences among the planets should be explainable by differences in temperature, density and composition as one goes farther out in the protosolar nebula.

22.1 | Motions of the planets

When we look at the night sky, it is clear that most of the objects maintain their relative positions. These are the stars. However, apart from the Sun and Moon, a small number of objects move against the background of fixed stars. These are the planets. The study of the motions of the planets has occupied astronomers for centuries. These motions do not appear simple. The planets occasionally seem to double back along their paths, as shown in Fig. 22.3. This doubling back is known as *retrograde motion*. Historically, any explanation of the motions of the planets had to include an explanation of this retrograde motion.

The earliest models of our planetary system placed the Earth at the center. This idea was supported by *Aristotle* in approximately 350 BC. His view was that the planets, the Sun and the Moon move in circular orbits about the Earth. Even though there is now ample evidence against this picture, one can see how placing the Earth at the center was a naturally simplifying assumption. The picture was modified by *Claudius Ptolemy*, in Alexandria, Egypt, around 140 AD. In order to explain retrograde motion, he added additional

Fig 22.2. Photograph of Halley's comet. [NOAO/AURA/NSF]

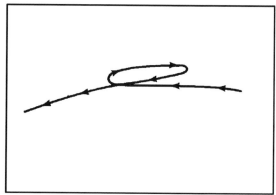

Fig 22.3. The apparent motion of Mars against the fixed background of stars. The loop when it doubles back is called retrograde motion.

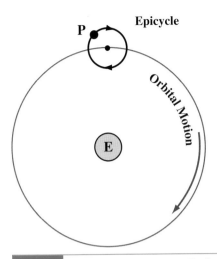

Epicycle

Fig 22.4. Epicycles. In this picture, the Earth is at the center. The planet, P, doesn't simply orbit the Earth. It goes around in a circle, which in turn orbits the Earth. If the planet's motion along the epicycle is faster than the epicycle's motion around the Earth, then the planet can appear to go backward for parts of each orbit. More layers of epicycles can be added to this picture.

circles, called *epicycles*. As shown in Fig. 22.4, each planet was supposed to move around its epicycle as the center of the epicycle orbits the Earth. To obtain a closer fit to the observed motions, higher order epicycles were added.

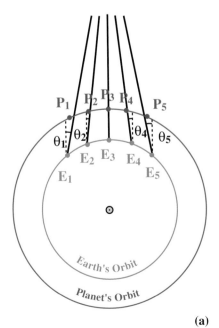

(a)

An opposing picture was supported by the 16th century Polish astronomer, *Nicholas Copernicus*. In the Copernican system, the Sun is at the center of the planetary system. This picture is therefore called the *heliocentric* model. Copernicus showed that the retrograde motion is an artifact, caused by the motion of the Earth. This is illustrated in Fig. 22.5. The Copernican system had the planets in circular orbits, not ellipses. Therefore, detailed predictions of planetary positions had small errors. To correct those errors, epicycles had to be added to the Copernican model, taking away from the simplicity of the picture.

When *Galileo Galilei* turned his newly invented telescope to the planets, he found that Venus does not appear as a perfect disk. It goes through a series of phases, similar to those of the Moon. The size of the disk also changes as the phase

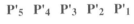

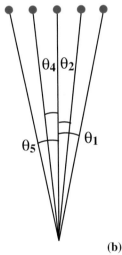

(b)

Fig 22.5. Retrograde motion in the heliocentric system. (a) The Sun is at the center. We consider the Earth at five positions E_1 through E_5 with the planet at P_1 through P_5 at the same times. We use the line of sight from the Sun through E_3 and P_3 as a reference direction. The dashed lines are all parallel to that direction, and the angles θ_1 through θ_5 keep track of the differences between the line of sight from Earth to the planet and the reference direction. We see that since the Earth is moving faster than the planet, the line of sight goes from being ahead of the dashed line to being behind the dashed line. (b) The view from Earth. The apparent position of the planet on the sky is indicated by P'_1 through P'_5. During this part of their orbits the planets appears to move backward on the sky.

changes. These observations can be explained easily in the heliocentric model, because Venus would not always be at the same distance from Earth. The phases result from the fact that we see differing amounts of the illuminated surface. There was no similar explanation in the Earth-centered system. Though Galileo was persecuted for holding that the heliocentric picture is the true one, his work had great influence on future scientific thought. Work switched from trying to find what was at the center of the planetary system to trying to understand how the planets, the Earth included, move around the Sun.

Extensive observations were carried out by *Tycho Brahe*, at Uraniborg, Denmark, late in the 16th century. Brahe moved to Prague in 1597, and died four years later. His results were taken over by an assistant, *Johannes Kepler*. Based on his analysis, Kepler published two laws of planetary motion in 1609, and a third law nine years later. Together, these are known as *Kepler's laws*. It is important to remember that these laws were based on observations, not on any particular theory.

Before discussing Kepler's laws, we look briefly at how we survey the Solar System, measuring the periods and sizes of orbits. We take advantage of certain geometric arrangements. These are shown in Fig. 22.6. We first look at planets that are closer to the Sun than the Earth. When the planet is between the Earth and the Sun, we say that it is at *inferior conjunction*, and it appears too close to the Sun in the sky to observe. As the planet moves in its orbit, the angle between it and the Sun (as seen from Earth) becomes larger. The planet appears farther and farther from the Sun. Eventually, since its orbit is smaller than the Earth's, it reaches a maximum apparent separation from the Sun. This is called the *greatest elongation*. At that point, the Earth, the Sun and the planet make a right triangle, with the planet at the right angle. After that the planet appears to get closer to the Sun, and when it is on the far side it is at *superior conjunction*. The pattern then repeats on the other side of the line from the Earth to the Sun. When the planet is on one side of the Sun it will appear east of the Sun in the sky, and when it is on the other side it appears west of the Sun. When it is west of the Sun, it rises and sets before the Sun, and it is therefore most easily visible in

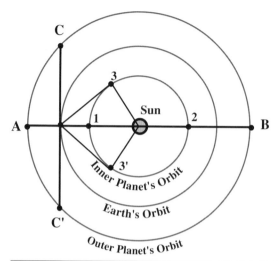

Fig 22.6. Configurations of the Earth and the inner and outer planets. Positions of the inner planets are indicated by numbers: (1) inferior conjunction, (2) superior conjunction, (3 and 3′) greatest elongation. Positions of the outer planets are indicated by letters: (A) opposition, (B) conjunction, (C) quadrature.

the morning. When it is east of the Sun, it rises and sets after the Sun, and is most easily visible in the evening.

We then look at planets that are farther from the Sun than the Earth. Let's start by looking at the planet when it is farthest from Earth, on the far side of the Sun. We say that the planet is simply at *conjunction*. At that point, it would be too close to the Sun in the sky to see. As it moves farther from that position it appears farther from the Sun on the sky. When it reaches a point where the Earth, Sun and planet make a right triangle, with the Earth at the right angle, we say that it is at *quadrature*. Notice that there is no limit on how far on the sky it can appear to get from the Sun. Eventually, it reaches the point where it is on the opposite side of the sky from the Sun. We call this point the *opposition*, and it is also the closest approach of the planet to Earth. When the planet is at opposition, it is up at night (since it is opposite to the Sun in the sky). Therefore, when a planet is favorably placed for observing, it is also closest to Earth and can be studied in the greatest detail.

When we talk about the orbital period of a planet, we mean the period with respect to a fixed reference frame, such as that provided by the stars. This period is called the *sidereal period* of the planet. However, we most easily measure the time it takes for the planet, Earth and Sun to come back to a particular configuration. This is called the *synodic period*. For example, the synodic period might be the time from one opposition to the next. How do we determine the sidereal period from the synodic period?

Suppose we have two planets, with planet 1 being closer to the Sun than planet 2. (For simplicity, we assume circular orbits.) The angular speed ω_1 of planet 1 is therefore greater than that of planet 2, ω_2. The relative angular speed is given by

$$\omega_{rel} = \omega_1 - \omega_2$$

Since $\omega = 2\pi/P$, where P is the period of the planet, the period of the relative motion of the two planets, P_{rel}, is related to P_1 and P_2 by

$$(1/P_{rel}) = (1/P_1) - (1/P_2) \tag{22.1}$$

Now we let one of the planets be the Earth, and express the periods in years. First we look at the Earth plus an inner planet. This means that P_1 is the period of the planet and P_2 is 1 yr. Equation (22.1) then becomes

$$(1/P_{rel}) = (1/P_1) - 1 \quad \text{(inner planet)}$$

Similarly for the Earth and an outer planet, equation (22.1) becomes

$$(1/P_{rel}) = 1 - (1/P) \quad \text{(outer planet)}$$

In each case P_{rel} is the synodic period and P is the sidereal period.

We now look at how the sizes of various planetary orbits are determined. The technique is different for planets closer to the Sun than the Earth and farther from the Sun than the Earth. Fig. 22.7 shows the situation for a planet closer to the Sun. When the planet is at its greatest elongation, it appears farthest from the Sun. The planet is then at the vertex of a right triangle, as shown in the figure. Since we can measure the angle E between the Sun and the planet, we can use the right triangle to write

$$\sin E = r/1 \text{ AU}$$

where r is the distance from the planet to the Sun. This equation can be solved for r to give us the

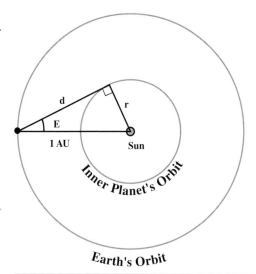

Fig 22.7. Diagram for finding the distance to an inner planet.

distance to the planet, measured in astronomical units.

Methods like this gives us distances in terms of the astronomical unit. Even if we don't know how large the AU is, we can still have all of the distances on the same scale, so we can talk about the relative separations of the planets. The current best measurement of the AU comes from situations like Fig. 22.7. We can now bounce radar signals off planets, such as Venus. By measuring the round-trip time for the radar signal (which travels at the speed of light), we know very precisely how far the planet is from the Earth. The right triangle in Fig. 22.7 gives us

$$\cos E = d/1 \text{ AU}$$

Since E is measured and d is known from the radar measurements, the value of the astronomical unit can be found. This distance is approximately 150 million kilometers (93 million miles). The exact value is accurate to within a few centimeters.

It is more complicated to find the distance to an outer planet. There are two different methods. The easier one was derived by Copernicus, but is not good for tracing out the full orbit. It just gives the distance of the planet from the Sun at one point in its orbit. Kepler's method of tracing the whole orbit is shown in Fig. 22.8. We make two observations of the planet, one sidereal period of

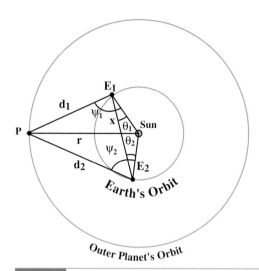

Fig 22.8. Diagram for finding distance to an outer planet.

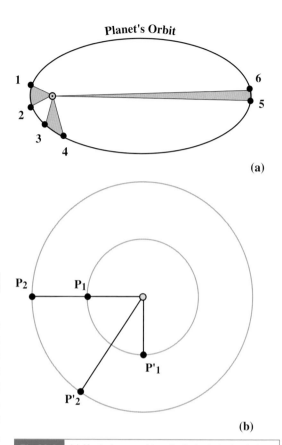

(a)

(b)

Fig 22.9. (a) Kepler's second law. Each shaded triangle has the same area. (b) Kepler's third law. In the time the inner planet moves from P_1 to P'_1 the outer planet moves from P_2 to P'_2

the outer planet apart. The Earth is at E_1 and E_2, respectively, when these are made. The angles ψ_1 and ψ_2 are directly determined. The angles θ_1 and θ_2 are known, as well as the distance x. (If the Earth's orbit were circular, then $\theta_1 = \theta_2$.) We then know $\psi_1 - \theta_1$ and $\psi_2 - \theta_2$, and can find d_1 and d_2, and, finally, r. The advantage of this method is that each point in the planet's orbit can be traced, with the observations overlapping in time.

We have already encountered Kepler's laws (Fig. 22.9) when we discussed the orbits of binary stars. After all, orbiting planets and orbiting stars must obey the same laws of physics.

The first law has to do with what types of paths the planetary orbits can be:

(1) *The planets move in elliptical orbits with the Sun at one focus.*

The second law, a consequence of angular momentum conservation, states:

(2) *A line from the Sun to a planet sweeps out equal areas in equal times.*

The Earth's orbit is not very eccentric, but we can still notice the effects of the Earth moving faster at some times than at others. By coincidence, the Earth is closest to the Sun just after the beginning of the year. Thus, it is slightly closer during the (northern) winter than during the (northern) summer. The result is that the number of days from the beginning of fall to the beginning of

spring is less than the number of days from the beginning of spring to the beginning of fall. This shows up in our calendar in two ways. First, the beginning of spring is usually on the 20th of March, while the beginning of fall is usually on the 22nd of September. Second, the shortest month, February, is in the middle of the winter.

The third law states:

(3) *The square of the period of an orbit (measured in years) is equal to the cube of the semi-major axis of the orbit (measured in AU).*

This follows from the inverse square law for gravity. (Actually, Newton deduced the inverse square law by seeing what gravitational law was necessary to give Kepler's third law. See Problem 22.5.)

There is nothing in Kepler's laws which tells us how far each planet should be from the Sun.

However, for over 200 years it has been known that there is a simple relationship giving, in AU, the semi-major axes of the actual planetary orbits. To generate this relationship, we take the series (0, 3, 6, 12, 24, . . .) in which each term is the double of the previous term. If we add four to each number, and then divide by ten, we obtain the approximate distances of the planets, out to Uranus, with an extra term giving the location of the asteroid belt. Though this is known as *Bode's law*, it is not really a law, in the sense that there is no physical basis for it. It may just be a mathematical curiosity.

22.2 | The motion of the Moon

Since the Moon orbits the Earth every 27.3 days (sidereal period), it is always changing its position with respect to the fixed stars, which serve as a backdrop. It is also changing its appearance, going through a full cycle of *phases* in the course of one month. The motion of the Earth around the Sun causes the phases to cycle in 29.5 days. (This number actually varies by up to 13 hours since the Earth doesn't move at a constant rate about the Sun.) The Moon rotates with the same period, so we always see the same face.

The Moon does not give off any light of its own. It shines by reflected sunlight. Therefore, the appearance of the Moon depends on the relative positions of the Earth, Sun and Moon. This is shown in Fig. 22.10. At any given time half of the Moon is illuminated. The changes in appearance are because different amounts of the illuminated side face the Earth. Let us follow it through one cycle. We start when the Moon is between the Earth and the Sun. Only the dark side of the Moon faces the Earth, and we see nothing. This is called the *new moon*. As the Moon moves over a little, a small piece of the illuminated side faces us and we see a crescent. Since the crescent is growing, we call it the *waxing crescent*. This appears to the east of the Sun (with the crescent side towards the Sun). This means that it is visible in the western sky at sunset. One-quarter of the way through the cycle, half of the visible side faces us, and we call it a *first quarter*. By that time, the Moon will have moved a quarter of the way across the sky,

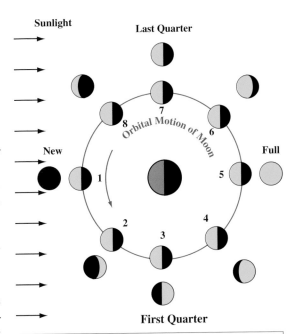

Fig 22.10. Lunar phases. The numbered images on the circle show the actual position and illumination of the Moon. The outer images show that appearance of the Moon, as seen from Earth. (1) New moon; (2) waxing crescent; (3) first quarter; (4) gibbous; (5) full moon; (6) gibbous; (7) last quarter ; (8) waning crescent.

and will appear high up at sunset. For the next quarter of a cycle, the visible part grows, and is called a *waxing gibbous*. Halfway through the cycle, we see the full illuminated side, and we call it a *full moon*. The full moon rises roughly as the Sun sets.

The second half of the cycle goes back through similar phases. For the third quarter of the cycle, the illuminated side becomes smaller, and we call it *waning gibbous*. Three-quarters of the way through the cycle we again see half of the illuminated face, and it is called the *last quarter*. For the last quarter, the half the we see is on the opposite side from that at the first quarter. The last quarter moon will be high in the sky at sunrise. Through the last quarter of the cycle, we see a *waning crescent*, getting smaller, and getting closer to the Sun. Finally, we return to the new moon.

The Moon's axis of rotation is inclined by 1.5° with respect to the plane of its orbit. The inclination contributes to an effect, known as *libration*, which allows us to see more than 50%

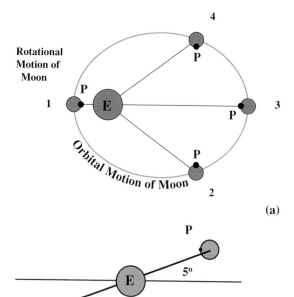

Rotational Motion of Moon

Orbital Motion of Moon

(a)

5°

(b)

2°

(c)

Fig 22.11. Librations, which allow us to see more than 50% of the Moon's surface. (a) The effect of the eccentricity of the Moon's orbit. The Moon rotates on its axis at a constant rate, but orbits the Earth at a variable rate, so at various times the rotation is ahead of or behind the orbital motion, allowing us to see an extra 6° around. Point P keeps track of the steady rotation of the Moon. We can see that the line from the Earth to the Moon passes through P at positions I and 3, but not at 2 and 4. (b) The effect of the inclination of the Moon's orbit, relative to the ecliptic, by 5°, allowing us to see over and under the poles. (c) The effect of the finite size of the Earth, allowing us to look from different directions, providing a 2° effect.

of the surface of the Moon. Another contribution to the libration comes from the fact that the Moon is so close to the Earth that observers on opposite sides of the Earth see the Moon to be rotated through approximately 2°. These effects, as shown in Fig. 22.11, allow us to see 59% of the Moon's surface.

Example 22.1 Forces on the Moon
Calculate the relative strength of the forces exerted on the Moon by the Earth and by the Sun.

SOLUTION
The force on each body is proportional to its mass, and inversely proportional to its distance from the Moon. Therefore

$$\frac{F_{SM}}{F_{EM}} = \frac{GM_\odot/r_{SM}^2}{GM_E/r_{EM}^2}$$

$$= \left(\frac{M_\odot}{M_E}\right)\left(\frac{r_{EM}}{r_{SM}}\right)^2$$

$$= \left(\frac{2 \times 10^{33}\,\mathrm{g}}{6 \times 10^{27}\,\mathrm{g}}\right)\left(\frac{3.85 \times 10^5\,\mathrm{km}}{1.50 \times 10^8\,\mathrm{km}}\right)^2$$

$$= 2.2$$

This means that the Sun exerts twice as great a force on the Moon as the Earth does. It is therefore not really proper to talk about the Moon orbiting the Earth. The Moon actually orbits the Sun, with the Earth causing the curvature of the Moon's orbit to change. This is shown in Fig. 22.12. Notice that the Moon's path is always concave toward the Sun. This is because the net force on the Moon is always inward, even when it is between the Earth and the Sun.

When the Moon passes between the Earth and the Sun, it is possible for it to block some of the sunlight. This is called an *eclipse of the Sun*, or a *solar eclipse*. If the Moon completely blocks the Sun,

Moon's Path

Earth-Moon Center of Mass

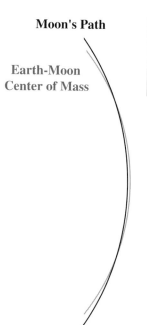

Fig 22.12. Orbit of the Moon, relative to the Sun. The Moon's orbit must always be concave toward the Sun.

we call it a *total eclipse*. In a total eclipse, the darkest part of the Moon's shadow, the *umbra* strikes the Earth. Otherwise it is a *partial eclipse*, and the lighter part of the shadow, or *penumbra*, passes over the Earth. When the Earth is between the Moon and the Sun, it can block the sunlight reaching the Moon. This is called an *eclipse of the Moon* or a *lunar eclipse*. There are also total and partial lunar eclipses, depending on whether the Moon passes through the Earth's umbra or penumbra. The relative placements of the Earth, Sun and Moon for both of these are shown in Fig. 22.13.

You might think that there would be a lunar eclipse at each full moon and a solar eclipse at each new moon. If the Moon's orbit were in the same plane as the Earth's orbit, this would be the case. However, the Moon's orbit is tilted by about 5° relative to the plane of the Earth's orbit. This is also shown in Fig. 22.13. Therefore, during most new moons, the Moon's shadow passes above or

below the Earth, and during most full moons the Moon passes above or below the Earth's shadow. We can only have an eclipse when the new or full moon occurs while the Moon is crossing the plane of the Earth's orbit, the ecliptic. The plane of the Earth's orbit intersects that of the Moon's orbit in a straight line, called the *line of nodes*. We can only have an eclipse when the Moon is close to a node. Also, we can only have an eclipse when the Moon is full or new.

This favorable arrangement occurs two times a year. These are called *eclipse seasons*. Each season is approximately 38 days long. Each new or full moon during the eclipse seasons results in at least a partial eclipse. The eclipse season is sufficiently long that any total eclipse of one object (Moon or Sun) must mean a partial eclipse of the other, either two weeks before or after. The direction of the plane of the Moon's orbit shifts around, going through a full cycle every 18.6 years. The amount of tilt doesn't change, but the direction of the tilt does. This results in an eclipse year that is actually 346.6 days long. Therefore, there may be up to seven (lunar plus solar) eclipses in a calendar year, of which two to five will be solar.

In a total lunar eclipse, the whole Moon passes through the darkest part of the Earth's shadow, the umbra. In some eclipses, the Moon only passes through the lighter (outer) part of the Earth's shadow, the penumbra. Penumbral eclipses are hardly noticeable. If the Moon is partly in the umbra, we see a partial eclipse. The Moon is never completely dark, even for a total lunar eclipse. Some sunlight is refracted (bent) by the Earth's atmosphere and illuminates the Moon. Since the atmosphere filters out the blue light better than the red, the Moon appears red. Particles in the Earth's atmosphere sometimes block sunlight more than at other times, and different eclipses have different amounts of light reaching the Moon. Eclipses just after major volcanic eruptions are particularly dark.

Lunar eclipses can be seen from any point on the nighttime side of the Earth. They are currently of limited scientific value. They used to provide astronomers with information on the thermal properties of the lunar soil. Astronomers could use radio and infrared observations to see how fast that soil cooled when the sunlight was

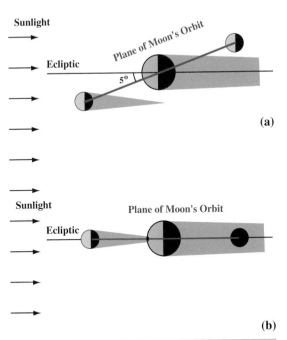

Fig 22.13. Eclipse geometry. Since the Moon's orbit is inclined by 5°, whether or not there is an eclipse depends on whether the Earth is in a part of its orbit where the new and full moon are in the ecliptic. (a) The Moon is out of the ecliptic at the new and full moon and no eclipse takes place. (b) The Moon is in the ecliptic at new and full moon and eclipses can take place. The times of year when eclipses are possible are called eclipse seasons.

removed. However, we now have samples of lunar material in the laboratory. As the Earth's shadow passes across the Moon, its shape serves as a reminder of the Earth's roundness.

Solar eclipses are of continuing scientific value, as was discussed in Chapter 6, when we talked about the Sun. As viewed from the Earth, the Sun and Moon cover almost exactly the same angle on the sky. This is just a coincidence. Since the Moon's orbit is elliptical, it is sometimes closer and sometimes farther from Earth. This means that usually the Moon covers a slightly larger region than the Sun, but occasionally the Moon is directly in line, but doesn't cover the Sun completely. A ring (or annulus) of the Sun shines around the edge of the Moon, so these are called *annular eclipses*.

When a total solar eclipse does occur, it is quite spectacular. First the Moon begins to cover the Sun slowly, creating a partial eclipse that engulfs more and more of the Sun. During any stage of a partial eclipse, there is still enough sunlight present to damage your eyes. You should not look directly at the partial stages of a solar eclipse. The easiest way to follow the progress is to use a small telescope to project an image of the Sun on a screen. Special solar filters can also be used, but *sunglasses, or exposed film, do not filter out the harmful radiation*. During totality it is safe to look at the Sun. With the bright disk of the Sun blocked, you can see the outer atmosphere, or corona, of the Sun. During totality, the sky becomes almost as dark as at night, and you can briefly see stars.

Since the Moon barely covers the Sun, at any instant, a total eclipse can only be seen over a small section of the Earth. As the Moon's shadow moves across the Earth, it traces out a thin band in which a total eclipse can be seen. Along the band, totality progresses from one end to the other. Since solar eclipses are only visible in limited areas, we generally have to plan to travel to see them. Therefore, it is important to know when and where you can see one.

22.3 | Studying the Solar System

What are the things that astronomers would like to know about the Solar System? Probably the most fundamental question is: *how was the Solar System formed, and how did it reach its current state?* Related to that are the questions of: *how did the Earth form and how did it reach the state where it can support life?* To answer these larger questions we must ask: *what are the constituents of the Solar System and what are their properties?* By properties we might be referring to mass, size, composition, structure. These are the questions that we will be addressing in this part of the book. To study the Solar System, we use traditional astronomical techniques, such as those discussed in Chapter 4. In addition, we have the advantage that we can send spacecraft to many parts of the Solar System.

In studying other planets, comparisons with the Earth are also interesting. For example, we think that any theory of atmospheric structure that can explain the Earth's atmosphere should, with the input of the right parameters, be able to explain that of Mars or Jupiter.

In studying the Solar System, theoretical tools now include sophisticated computer modeling of various processes. The processes include accretion of interstellar material to form the various planets, or the evolution of an atmosphere. These models are a great aid in interpreting complex observations.

We have undergone a great revolution in our observations of the Solar System. Ground-based observations, yielding images and spectra, have been quite useful. However, the opportunity afforded by space probes to take a close-up look has greatly added to our knowledge. Flybys have given us detailed pictures, and have allowed us to study spectral bands that do not penetrate the Earth's atmosphere. They can also directly sample the environment of the object being studied. Landings can provide even more information, allowing us to study conditions at a site. In the case of the Moon, we have had the additional luxury of bringing samples back for study in our laboratories. This has brought the Solar System into the realm of the geologist.

The Solar System actually provides us with a unique opportunity to visit the astronomical objects that we are studying. Throughout the chapters on the Solar System, we will constantly be comparing Earth-based and space-probe images of the members of the Solar System. The contrast is spectacular. You should remember, however, that this is not something that we can do for any of the other objects discussed in this book.

Also, as we discuss various members of the Solar System in detail, we will see that, as spectacular as the close-ups are, they complement a very active ground-based observing program. This is due, in part, to the fact that the space missions are expensive and of short duration. So, they give us detailed information, but for only a short period of time. Observations from the Earth can be carried out as long as we are willing to observe.

Because it is the closest, the Moon has been the subject of the most active space exploration. Both the United States and Soviet Union launched a variety of unmanned probes to the Moon in the late 1950s and 1960s. These started with crude craft that (intentionally) crashed on the surface, relaying brief series of images before impact. The US Ranger series was an example of this. With more sophisticated equipment, as well as more powerful booster rockets to lift the more sophisticated equipment, the probes were able to land, and send back data from the surface. The US Surveyor series carried out those missions.

These were all a prelude for astronauts exploring the Moon. Between 1969 and 1971, there were six manned lunar landings. Each carried out extensive studies of the surface and also returned samples to Earth for detailed laboratory analyses. The Soviet Union also had an unmanned spacecraft land on the Moon, and then return to Earth with rock and soil samples. All of these analyses are still continuing.

The rest of the Solar System is so far away that it has only been visited by unmanned probes, which signal their data back to Earth. The Voyager spacecraft had missions in which they flew close to planets, often a few planets in succession.

22.4 | Traveling through the Solar System

We have already seen that the distances between the planets are quite large, and that the Solar System is mostly empty space. However, by the standards of the distances to the nearest stars, the planets are quite close. They are close enough that the distances can be traveled by interplanetary probes. The shortest trips, to the Moon, take a few days. The longest trips, to the outer planets, take several years. By comparison, at the speeds of these probes, a trip to the nearest star would take about 20 000 years.

In this section, we look at the mechanics relating to space probes traveling to other planets. It is important to remember that during most of the flight of a planetary probe, it is unpowered. This means that it is simply in an orbit about the Sun. When we talk about orbiting space probes, the mass of the probe is much less than that of the Sun or any planet. We therefore do not have to consider the recoil of the Sun or any planet, making the problem a little easier.

Example 22.2 Orbit and escape from the Earth
What is the speed for a circular orbit just above the Earth's atmosphere? How does this compare with the escape velocity?

SOLUTION
For an object of mass m, in a circular orbit of radius r, around a planet of mass M, the gravitational force must provide the acceleration for circular motion, so

$$\frac{mv_{orb}^2}{r} = \frac{GMm}{r^2}$$

where v_{orb} is the orbital speed. Solving for v_{orb} gives

$$v_{orb} = \left(\frac{GM}{r}\right)^{1/2}$$

$$= \left(\frac{(6.67 \times 10^{-8}\ dyn\ cm^2/s^2)(6 \times 10^{27}\ g)}{(6.4 \times 10^8\ cm)}\right)^{1/2}$$

$$= 7.9 \times 10^5\ cm/s$$

$$= 7.9\ km/s$$

$$= 28\,000\ km/hr$$

The escape speed is that with which we must launch an object from the surface, such that it can move infinitely far away with zero kinetic energy. This means that we define potential energies to be zero when objects are infinitely far apart. The total energy of an object with the escape speed is zero. If v_{esc} is the escape speed, then

$$\left(\frac{1}{2}\right)mv_{esc}^2 - \frac{GMm}{r} = 0$$

Solving for v_{esc} gives

$$v_{\text{esc}} = \left(\frac{2GM}{r} \right)^{1/2}$$

Note that comparing v_{esc} and v_{orb} gives

$$v_{\text{esc}} = \sqrt{2}\, v_{\text{orb}}$$

Therefore, the escape speed for the Earth is 11 km/s.

In considering travel to other planets, we use elliptical orbits. After all, an object whose orbit touches both the Earth and Mars cannot be a circular orbit about the Sun. (We should note that parabolic or hyperbolic orbits can be used for one-time flybys, such as Pioneer and Voyager.) When we studied binary stars, we found that the velocity of a particle in an elliptical orbit of semi-major axis a, when the object is a distance r, from a mass M at the focus, is

$$v^2 = GM \left[\frac{2}{r} - \frac{1}{a} \right] \tag{22.2}$$

We can use this to find the total energy of the orbit,

$$E = \left(\frac{1}{2} \right) mv^2 - \frac{GMm}{r}$$

$$= \left(\frac{1}{2} \right) GMm \left[\frac{2}{r} - \frac{1}{a} \right] - \frac{GMm}{r} \tag{22.3}$$

$$= \frac{GMm}{2a}$$

Note that, if you are in an orbit of a given semi-major axis a, and fire a rocket such that your energy increases, you must go to a higher orbit, making your energy less negative. In the higher orbit, your speed will actually be less than it was in the lower orbit. Thus, firing the engine to "accelerate" the rocket has the interesting effect of reducing its orbital speed.

Another consequence of equation (22.3) is that the energy of an orbit depends only on the semi-major axis, and not on the eccentricity. If we launch a rocket, once its fuel is used up its energy is determined by its location and its speed. How, then, can we determine the eccentricity of the resulting orbit? The direction of motion when the fuel is used up determines the eccentricity. It

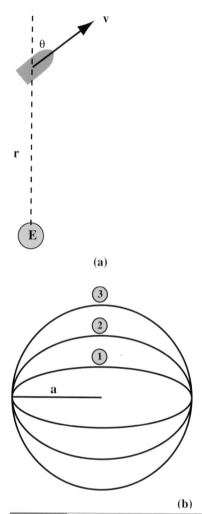

Fig 22.14. Angular momentum and orbital eccentricity. (a) If a rocket stops its powered flight at a distance r from the Earth while moving with a velocity $\mathbf{v}$, its energy is determined (as the sum of kinetic plus potential energies at that point), independent of the direction of motion. This means that the semi-major axis of the ellipse is determined. However, the direction of motion determines the eccentricity. (b) Ellipses with the same semi-major axis, but with different eccentricities.

also determines the total angular momentum, L. Referring to Fig. 22.14, the angular momentum is given by

$$L = mvr \sin\theta$$

The least eccentric orbit has the highest angular momentum for a given semi-major axis. It

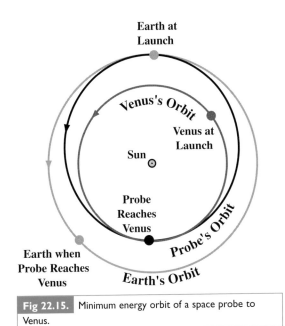

Earth at
Launch

Venus's Orbit

Venus at
Launch

Sun

Probe
Reaches
Venus

Probe's Orbit

Earth when
Probe Reaches
Venus

Earth's Orbit

Fig 22.15. Minimum energy orbit of a space probe to Venus.

should be noted that $\theta = 90°$ doesn't give a circular orbit unless the speed is the right speed for a circular orbit at that radius. However, $\theta = 90°$ means that the point where the fuel is spent will be the aphelion or perihelion of the orbit.

The trick in launching a planetary probe is to place it in an orbit that intersects the planet's orbit when the planet is at the point of intersection. In choosing the direction of launch, we can take advantage of the Earth's motion. There are many different orbits that can be chosen. However, the minimum energy orbit is the one that has the Earth at aphelion and the planet at perihelion (if the planet is closer to the Sun) or one that has the Earth at perihelion and the planet at aphelion (if the planet is farther from the Sun). The minimum energy orbit of probe to Venus is shown in Fig. 22.15.

Example 22.3 Minimum energy orbit to Venus
Find the semi-major axis and eccentricity of a minimum energy orbit from the Earth to Venus. Find the necessary launch speed and the time for the trip.

SOLUTION
In this case the Earth is at aphelion and Venus is at perihelion. The major axis of the orbit is the sum

of the distance of the Earth from the Sun and the distance of Venus from the Sun. Therefore, the semi-major axis is

$$a = (1/2)(R_E + R_V)$$

$$= 0.86 \text{ AU}$$

The eccentricity is defined by the separation between the foci, divided by $2a$. The Sun is at one focus, R_V from perihelion. Therefore, the symmetry of the orbit requires that other focus to be R_V from the aphelion. The distance between the foci is therefore

$$d = 2a - 2R_V$$

$$= R_E + R_V - 2R_V$$

$$= R_E - R_V$$

The eccentricity is therefore

$$e = \frac{R_E - R_V}{2a}$$

$$= \frac{R_E - R_V}{R_E + R_V}$$

$$= 0.16$$

At launch, $r = R_E$ so the launch speed is given by

$$v^2 = GM_\odot \left(\frac{2}{R_E} - \frac{2}{R_E + R_V} \right)$$

This gives $v = 27.3$ km/s. Note that the Earth's orbital speed is 30 km/s, so the probe must be going 2.7 km/s slower than the Earth. This means that we must launch the probe in the opposite direction from the Earth's motion.

We can find the length of the trip by using Kepler's third law. It tells us that the period of the orbit is given by

$$P^2 = a^3$$

For $a = 0.86$ AU, we find $P = 0.80$ yr. Therefore the trip to Venus will take half this time, or 0.40 yr (approximately five months).

Once we know how long the trip will take, we must launch when Venus is at the place in its orbit to take it to the rendezvous place in the travel time of the probe. The short period of time when a launch is possible is called a *launch*

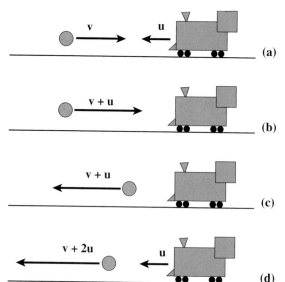

Fig 22.16. Analogy for gravity assist. (a) A low mass ball with speed v approaches a high mass train with speed u for a head-on collision. (b) The same situation viewed from the train. The train is at rest, and the ball has a speed v + u. (c) Still looking from the train, the elastic collision simply reverses the direction of the ball, so it is now moving away from the train at a speed v + u. (d) We revert to the view from the ground. If the ball is moving ahead of the train at a speed v + u, then the speed of the ball, relative to the ground, must be this speed, plus the speed of the train u, giving a speed v + 2u.

is much less than the mass of the train, and that the collision is completely elastic. The train is moving with speed u and the ball is moving with speed v. As viewed from the train, the ball is coming towards the train at a speed of u + v. In the elastic collision, the train is so massive that it doesn't recoil, so the ball simply reverses its velocity relative to the train. It is now moving away from the train at a speed u + v. As viewed from the ground, its speed is now v + 2u. In the collision, the ball has taken a little energy from the train (very small relative to the total energy of the train), and its speed is now greater than its initial speed by twice the speed of the train.

With a space probe, We cannot, of course, cause a space probe to bounce directly off Jupiter.

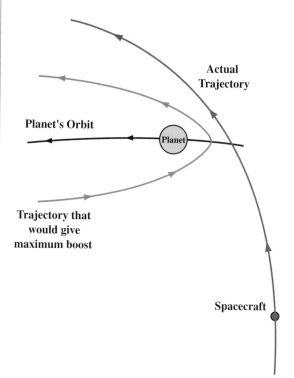

Fig 22.17. Trajectories for gravity assist. The space probe doesn't have to collide with the planet; it only has to make a close gravitational encounter. The maximum boost would come from a trajectory that comes the closest to a head-on collision. However, since the spacecraft is coming from well inside the planet's orbit, this is not practical. The actual trajectory still gives almost half of the maximum boost in speed.

window. The size of the window depends on how far we are willing to deviate from the minimum energy orbit. In particular, orbits with higher energy might be used since they have shorter travel times. Since the launch requires the correct relative positions of the Earth and Venus, the launch window opens once per synodic period of Venus. This explains the spacing of launches for a given planet.

There is a trick that can be used to minimize the energy needed to visit a planet beyond Jupiter. In this, we take advantage of an elastic gravitational encounter between the space probe and a massive planet. This is known as a *gravity assist*.

To understand how this works, let's look at an analogous situation, a ball bounced off the front of an incoming train. This is illustrated in Fig. 22.16. We assume that the mass of the ball

However, we can arrange the trajectory to have the same effect (Fig. 22.17). We cannot generally take advantage of a head-on collision, but the speed of the probe can be increased by an amount of the order of the speed of Jupiter. The technique has been used for the Voyager 1 and 2 missions, as shown in Fig. 22.18.

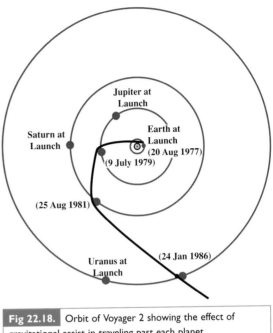

Jupiter at Launch

Saturn at Launch

Earth at Launch (20 Aug 1977)

(9 July 1979)

(25 Aug 1981)

Uranus at Launch

(24 Jan 1986)

Fig 22.18. Orbit of Voyager 2 showing the effect of gravitational assist in traveling past each planet.

Chapter summary

Most of the objects in the Solar System orbit the Sun close to the plane of the Earth's orbit, the ecliptic. There are nine planets, which fall into two distinct groupings. The four inner planets are small, with the Earth being the largest. The next four planets are larger, and have very different compositions than the inner planets. The outermost planet, Pluto, is small, like the inner planets. All but two of the planets have moons orbiting them. There is also a collection of smaller material in the Solar System. There are the asteroids, found mostly between Jupiter and Mars. There are also comets, which are usually faint, but occasionally brighten and develop a tail. The comets leave debris behind, which we see as meteors.

Since the Earth is orbiting the Sun, the motions of the planets, as viewed from Earth, appear complicated. However, they can be explained most easily in terms of a Sun-centered system. The orbits of the planets are ellipses, rather than simply being circles. The motions of the planets are summarized in Kepler's laws: (1) the planets move in elliptical orbits with the Sun at one focus; (2) a line from the Sun to a planet sweeps out equal areas in equal times;

(3) the square of the period of the orbit is simply related to the cube of the semi-major axis.

The Moon changes its position in the sky and its appearance as it orbits the Earth. The phases result from the fact that when we look at the Moon we are seeing it by reflected sunlight. Since the Moon's orbit is not in the plane of the ecliptic, there is not a solar eclipse and a lunar eclipse each month. However, approximately twice per year arrangements favorable for eclipses take place. In a solar eclipse, the Moon's shadow moves across a small band on the Earth. For observers in that band, the Moon covers the face of the Sun. In a lunar eclipse, the Moon passes through the Earth's shadow, and all observers on the night side of the Earth can see the eclipse. Some sunlight passing through the Earth's atmosphere gives the Moon a faint red glow, even during totality.

The Solar System is unique in that the objects are so close that we can visit them with space probes. (All other astronomical objects are too far away for this.) We have learned a lot from astronauts visiting the Moon, as well as from automated spacecraft. Robot spacecraft have visited most of the other planets and have sent back images as well as other data.

Questions

22.1. List the planets in order of increasing mass. Where do the most massive moons of each planet fit into this? Where do the largest moons of Jupiter, Saturn, Uranus and Neptune fit into this? See Appendix D for properties of the planets and satellites.

22.2. If you were going to construct a scale model of the Solar System, and the Earth was 1 m from the Sun in your model, how far are the other planets from the Sun in that model?

22.3. Why is the distinction between inner and outer planets important beyond the simple question of where the planets are located?

22.4. How many moons in the Solar System are more massive than ours? See Appendix D for properties of the planets and satellites.

22.5. Explain, in your own words, how retrograde motion is explained in a Sun-centered system.

22.6. In the philosophy of science, we are always told to use the simplest model that explains all of the available data. Compare the heliocentric and Earth-centered systems in terms of their simplicity.

22.7. Why was Galileo's use of the telescope important in supporting the heliocentric view of the Solar System?

22.8. Why is there a limit on the maximum angle that we can see between Venus and the Sun in the sky?

*22.9. Mercury's location in the sky makes it one of the most difficult planets to observe. Why is this?

22.10. The closest approach between Mars and Earth occurs when they are at opposition. However, that distance is not always the same from one opposition to the next. Why is that?

22.11. When Mars is at opposition (and close to the Earth for good observing), would you expect it to be up at night or during the day?

*22.12. Suppose the Earth and Mars are at opposition. Estimate how long it will be before they are at opposition again. Remember, both are moving. You might want to draw a diagram, and keep track of the motions of both planets in fixed time intervals.

22.13. Why is the frequency of launch windows for spacecraft to a planet dependent on the synodic period of that planet?

22.14. Suppose that you lived on Mars. (a) Describe how the Earth would move in your sky. (b) Would the Earth appear to go through phases, like Venus?

22.15. At what point in its orbit does a planet move (a) fastest, (b) slowest?

22.16. Why is the location of Jupiter the place where you would expect to find the most massive planet?

22.17. If we want to study Jupiter up close, we can send a spacecraft. Why is this not practical for studying the nearby stars?

Problems

22.1. Use the tabulated sidereal periods of the planets to find their synodic periods.

22.2. What are the angles of greatest elongation for Mercury and Venus?

22.3. Fill in the trigonometric steps in finding the distance to a planet using Kepler's method.

22.4. How does the angular momentum per unit mass vary with distance from the Sun?

22.5. Show how the inverse square law can be derived from Kepler's third law for circular orbits.

22.6. Show that viewers on opposite sides of the Earth see the Moon rotated through 2°.

22.7. What is the escape velocity from our point in the Solar System?

22.8. For a minimum energy orbit to Mars, find the semi-major axis, the eccentricity, and the time of flight. Draw a diagram showing the orbit, and the positions of the Earth and Mars at launch and landing.

22.9. Find the speed and distance from the center of the Earth for a satellite in synchronous orbit around the Earth. (A synchronous orbit is one that takes 24 hours.)

22.10. Express equation (22.2) in a form that gives v in terms of the escape velocity.

22.11. How does the kinetic energy of an object with the escape speed at a distance r compare with that of an object in a circular orbit of radius r.

22.12. If a ball of mass m bounces elastically off the front of a train of mass M, where $M \gg m$, with the ball initially moving with a speed v and the train initially moving at a speed u, what fraction of the kinetic energy of the train is lost in the collision?

22.13. Find the angular momentum of the Sun, and compare it with the angular momentum, orbital plus rotational, of all the planets.

22.14. Estimate the material that can be cleared out of the Solar System in 1 yr by a solar wind with a mass loss of 1 $M_\odot/10^6$ yr and a speed of 200 km/s. Assume that the material to be cleared out is at the position of the Earth.

22.15. Estimate the rate at which material can be swept out from the position of the Earth by radiation pressure from the Sun.

Computer problems

22.1. Assume that the Earth and Mars are both in circular orbits, and consider the part of their orbits where the Earth is overtaking Mars and passing it. Draw a graph of the apparent position of Mars in our sky as a function of time, from the time when the Earth is 15° behind Mars to the time when it is 15° ahead.

22.2. Given the sidereal periods of the planets, make a table showing their synodic periods.

22.3. Show that the planets obey Kepler's third law.

22.4. Show that the moons of Jupiter and those of Saturn obey Kepler's third law.

22.5. Find the masses of planets from satellite orbit data.

22.6. Find escape velocities for all the planets and for the largest moon of each planet.

22.7. For minimum energy orbits to Mercury, Mars, Saturn and Neptune find the semi-major axis, the eccentricity, and the time of flight. Draw a diagram showing the orbit, and the positions of the Earth and the planet at launch and landing.

Chapter 23

The Earth and the Moon

We start our discussion of the planets with the one with which we are most familiar, the Earth. In understanding the processes that are important on the Earth, both now and in the past, we are setting a framework for our understanding of the other planets. Therefore, in this chapter we will develop many of the ideas that should apply to all planets, both in terms of what properties are important and how we measure them.

23.1 | History of the Earth

23.1.1 Early history

The main steps in the history of the Earth are shown in Fig. 23.2. Somehow, the Earth accreted from the material in the original solar nebula. We will discuss more about the solar nebula in Chapter 27. Enough material collected together so that its own gravitational pull was able to keep most of the material from escaping. As particles fell towards a central core, they were moving closer together, so their gravitational potential energy decreased. This means that their kinetic energy increased. This kinetic energy was then available to heat the forming planet. In addition, heat was provided by the radioactive decay of potassium, thorium and uranium. Such decays led to heating, because the energetic particles – alpha, beta, gamma – were absorbed by the surrounding rock. The relatively massive alpha particles were particularly effective in this heating.

The heating resulted in a liquid, or molten, interior. Since materials are free to move in a liquid, heavier elements, such as iron and nickel,

Fig 23.1. In this photo of the Earth from space we can really appreciate that it is just another planet, one that has a moon. This image was taken on 16 Dec 1992, by the Galilelo spacecraft from a distance of 6.2 million km. [NASA]

sank into the center, while lighter elements, such as aluminum, silicon, sodium and potassium, floated to the surface. This process is called *differentiation*. The iron and nickel form the current core of the Earth. So much heat was trapped by the Earth that its core is still molten. The thorium and uranium were squeezed out of the core, and carried along in crystals to the surface. These radioactive materials provide heating close to the surface.

The molten core is responsible for the Earth's magnetic field. The rapid rotation of the Earth

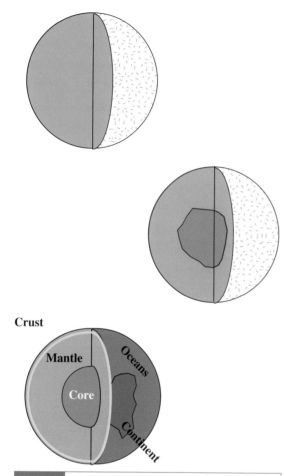

Crust

Mantle
Oceans
Core
Continent

Fig 23.2. Diagram showing steps in the formation and development of the Earth.

Fig 23.3. Cooled volcanic rock is an important component of the Earth's surface. [USGS]

potassium), and *sedimentary rocks*, that have been deposited gradually. Both types of rocks can be altered by high pressures, and are then called *metamorphic*. The lower part of the crust is composed of *gabbro* and *basalts*. These are dark rocks containing silicates of sodium, magnesium and iron. (Silicates are compounds in which silicon and oxygen are added to the other atoms in various combinations.) Gabbros have coarser crystals than do basalts. Larger crystals result from slower cooling.

An important clue to the Earth's interior comes from volcanic activity. This activity results from the heating of the crust, producing the molten material. Volcanic activity is very important in mountain building, as shown in Fig. 23.4. It also provides a means of transporting certain materials from the interior to the surface. In particular, we think that this is the source of carbon dioxide (CO_2), methane (CH_4) and water (H_2O), as well as sulfur-containing gases in our atmosphere. Visitors to active volcanoes on Earth, such as Kilauea on Hawaii, shown in Fig. 23.4, often note a strong sulfur smell.

As the water vapor was ejected into the atmosphere, the temperature was low enough for it to condense. Other gases dissolved in the water, and combined with calcium and magnesium leached from surface rocks. This had the important effect of removing most of the carbon dioxide from the Earth's atmosphere.

23.1.2 Radioactive dating

Much of what we know about the age of the Earth's crust comes from *radioactive dating*. In this

leads to a dynamo process. In this process, a small magnetic field plus convection creates electrical currents flowing through the fluid. These currents produce a larger magnetic field, and so on. This process takes energy from the Earth's rotation. The Earth's magnetic field is not fixed. The magnetic pole wanders irregularly. In addition, geological records indicate that the direction of the magnetic field has reversed every few hundred thousand years. The actual reversals take tens of thousands of years, and there is a period when the field is very weak. The causes of the reversals are not well understood.

The upper part of the Earth's crust is composed of several different types of rocks. There are *igneous rocks*, such as granite, which are formed in volcanos (and are enriched in silicon, aluminum and

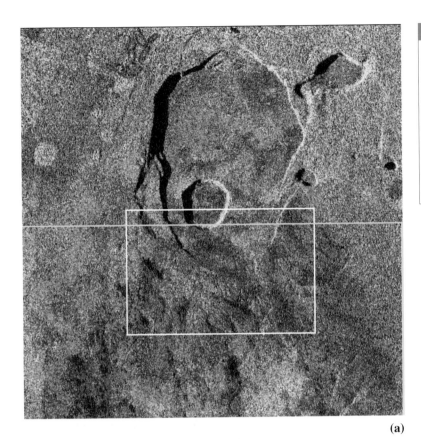

(a)

Fig 23.4. Images of an active volcano, Kilauea, in Hawaii. This is an image from space, to give you an idea of the quality images we can make from orbit around other planets. (a) An overhead view from radar studies on the shuttle Endeavor. This is an interferometric image. (b) A three-dimensional perspective (in false color) reconstructed from the radar images. [NASA]

(b)

technique, we are studying the products of radioactive decay. In radioactive dating we take advantage of the fact that if we start with some number N_0 of nuclei of a radioactive isotope, the number left after a certain time t is given by

$$N(t) = N_0\, e^{-t/\tau_e} \tag{23.1a}$$

The quantity τ_e is the time for the number in the sample to fall to $1/e$ of its original value. We can also write this expression in terms of the *half-life*,

$\tau_{1/2}$, which is the time for the number in the sample to fall to half of its original value,

$$N(t) = N_0\left(\frac{1}{2}\right)^{t/\tau_{1/2}} \tag{23.1b}$$

Comparing these two expression (see Problem 23.2), tells us that

$$\tau_{1/2} = (0.693)\tau_e$$

The behavior of $N(t)$ vs. t is shown in Fig. 23.5.

Half-lives of many nuclei are measured in the laboratory (see Problem 23.4). Therefore, if we know N_0 and $N(t)$, we can solve for t, the time that has elapsed since the sample had N_0 nuclei of the particular isotope. This is the essence of radioactive dating. It is important to choose an isotope whose half-life is comparable to the time period you are trying to measure. If the time period is much longer than the half-life, there will be too few left to measure. If the time period is much less than the half-life, very few decays will have

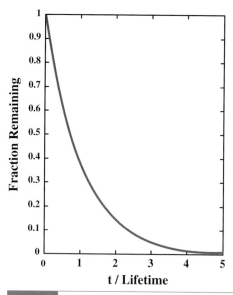

Fig 23.5. Radioactive decay. The vertical axis is the number of remaining nuclei N, divided by the original number N_0.

taken place. We believe that the Earth's crust is billions of years old, so the alpha decay of ^{238}U (uranium-238), with a half-life of 4.6 billion years, is ideal for studying its age.

A practical problem is that we often don't know how much of a given radioactive material the Earth started out with. This means that we must employ indirect methods, involving knowledge of the end products of the decays. For example, the alpha decay of ^{238}U is followed by a series of alpha decays, eventually leading to the stable isotope ^{206}Pb (lead-206). If we can assume that all of the ^{206}Pb on the Earth came from such decays, we could use the amount of ^{206}Pb as an indicator of the original amount of ^{238}U.

However, the Earth may have been formed with some ^{206}Pb, so we have to use even more involved techniques to correct for that effect. These techniques often involve measuring the relative abundances of certain isotopes, such as ^{204}Pb and ^{206}Pb. Such measurements require a means of separating the isotopes. This separation is much easier for gases than for solids. For this reason, we actually use a different age tracer. For example, ^{40}K (potassium-40) beta decays with a half-life of 1.3 billion years. The decay product is ^{40}Ar (argon-40). Since argon is a noble gas, it

doesn't stay bound to the rocks. We can collect the argon and study the relative abundances of various isotopes.

In considering the history of the Earth, we must remember that the atmosphere has eroded the surface, altering its characteristics. (As we will see below, the motions of the continents also alter the surface.) The oldest rocks that we see are dated at 3.7 billion years. (The oldest fossil cells are dated at 3.4 billion years.) We think that the surface underwent significant alteration 2.2 to 2.8 billion years ago. This was a period of volcanic mountain building with the Earth having a very thin crust. The crust was probably broken into smaller platelets.

The age of the Earth, even as an approximation, is an important reference. It gives us an idea of what the appropriate time scales are for understanding the Solar System. For example, if the Earth formed as a by-product of the formation of the Sun, then the Sun must be at least as old as the Earth. This tells us that in understanding the Sun and other stars, we must be able to explain how they can shine for billions of years. On the other end of the time scale, we can see that, as far back as ancient history seems to us, some 5000 years, it is but a blink on the time scales of the Earth's history.

At that time, the atmosphere was mostly water, carbon dioxide, carbon monoxide and nitrogen. It is thought that any ammonia and methane could not have lasted very long. There was little oxygen, since oxygen is a product of plant life. Since there was no oxygen, there was no ozone (O_3) layer to shield the Earth from solar ultraviolet radiation. We think that this ultraviolet radiation must have stimulated the chemical reactions to make the simplest organic compounds. It has been shown in the laboratory that such reactions are greatly enhanced in the presence of ultraviolet radiation. We will talk more about the early evolution of the Earth's atmosphere in Chapter 27.

23.1.3 Plate tectonics

The layer below the thin crust is kept heated by radioactive decay. The amount of heat is not sufficient to melt the material completely, but it keeps it from being completely solid. It has the

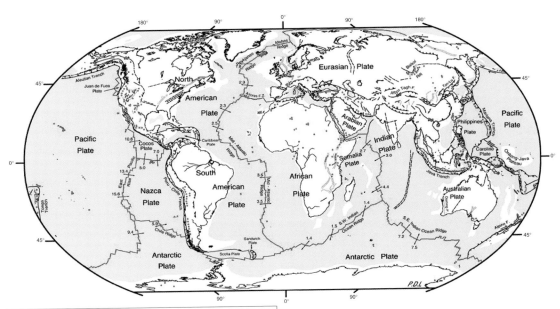

Fig 23.6. The main tectonic plates are outlined by areas of geological activity. Earthquakes and volcanos are marked by red dots. Arrows indicate the direction of motion of the plates. [NASA]

consistency of plastic. This means that if you press on it, the material doesn't deform instantly, as a liquid would. However, under a steady pressure, it will flow slowly. The solid layer above the plastic region is called the *lithosphere*. The Earth's lithosphere is broken into *plates*. The name is meant to suggest that they are much larger in extent along the surface than they are thick. The plates float on top of the plastic layer.

As they float, the plates move slowly. Since they carry the continents with them as they move, we refer to this motion as *continental drift*. The general term for any process involved in the movement or deformation of planetary surfaces is *tectonics,* so continental drift is also called *plate tectonics*. Fig. 23.6 shows the Earth's main tectonic plates. Their motion is being driven by material being forced up from below into some narrow gaps between the plates. One such region, shown in Fig. 23.7, is called the *mid-Atlantic ridge*. Throughout the ridge, fresh material is appearing on the sea floor, as the plates move away from the ridge.

The regions where the plates meet are characterized by a high level of geological activity.

When one plate is forced under another, the resulting upward pressure can build great mountain ranges, such as the Himalayas, as shown in Fig. 23.8. These plate boundaries also show a high frequency of volcanoes and earthquakes. The volcanoes result from material being pushed upward. The earthquakes result from the fact

Fig 23.7. A computer generated image of what the mid-Atlantic ridge would look like if there were no water in the ocean. [USGS]

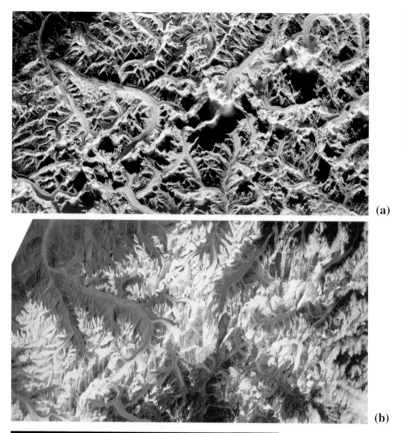

(a)

(c)

(b)

Fig 23.8. (a) Space radar image of Mt Everest, in the Himalayan mountain range. (b) Optical image. Both (a) and (b) were taken from the shuttle Endeavor (10 Oct 1997). Images show an area 70 × 38 km across. (c) Part of the San Andreas fault line in California. [NASA]

which this slippage takes place is called a *fault line*. One famous fault line – the San Andreas fault in California – is shown in Fig. 23.8(c).

23.2 | Temperature of a planet

The temperature of the Earth is determined by a balance between the energy absorbed from the Sun and the energy given off by the planet. For planets like the Earth, heat from the inside does not have much effect on the surface temperature. The planetary temperature for which these balance is called the *equilibrium temperature* of the planet. The actual energy transport might be complicated by the presence of an atmosphere, but we will first calculate the equilibrium temperature, ignoring atmospheric effects. In the absence of an atmosphere, the calculation is essentially the same as for that of an interstellar grain (see Chapter 14).

that slippage of the plates past each other is not smooth. For long periods the plates might not move as the pressure increases. Eventually, the pressure becomes too great, and there is a sudden movement along the boundary. The line along

We start by calculating the energy received per second. The energy per second given off by the Sun is its luminosity, $L_\odot$. (We could leave this in terms of the luminosity, or rewrite the luminosity as $4\pi R_\odot^2 \sigma T_\odot^4$.) At the distance d of the planet from the Sun, the luminosity is spread out over a surface area of $4\pi d^2$. This means that the luminosity per surface area is $L_\odot/4\pi d^2$. As seen from the Sun, the projected area of the planet is πR_P^2. The planet therefore intercepts a power equal to this area multiplied by the power per surface area. Finally, not all of the sunlight is absorbed by the planet. A fraction a, the albedo, is reflected. The amount absorbed is equal to $(1 - a)$ multiplied by the amount that actually strikes. In this calculation, we are assuming that the albedo is the same at all wavelengths. Therefore the power absorbed by the planet is

$$P_{abs} = \frac{L_\odot(1 - a)R_P^2}{4d^2} \tag{23.2}$$

We now look at the power radiated. We assume that the planet rotates fast enough that there is no great difference between day and night temperatures so we can treat the temperature of the planet as being the same everywhere. (This is a good approximation for the Earth but not for the Moon.) The power radiated per unit surface area is $e\sigma T_P^4$, where e is the *emissivity*. The emissivity can range from zero to one, and is one for a perfect blackbody. Multiplying this by the planet's surface area $4\pi R_P^2$, we obtain the total power radiated:

$$P_{rad} = 4\pi R_P^2 e\sigma T_P^4 \tag{23.3}$$

If we equate the power absorbed with the power radiated, we solve for the equilibrium temperature of the planet, giving

$$T_P = \left[\frac{L_\odot(1 - a)}{16\pi d^2\sigma e}\right]^{1/4} \tag{23.4}$$

This calculation doesn't account for the fact that the albedo and emissivity vary with wavelength. We must integrate the energy received over the spectral energy distribution of the Sun, and integrate the energy radiated over the spectral energy distribution of the planet. In each case, we incorporate a frequency dependent

albedo and emissivity inside the integral. For example, remembering the luminosity of the Sun,

$$L_\odot = 4\pi R_\odot^2 \int_0^\infty B_\nu(T_\odot)\, d\nu$$

where B_ν is the Planck function. The power absorbed by the Earth is then

$$P_{abs} = \left[\frac{\pi R_\odot^2 R_P^2}{d^2}\right]\int_0^\infty (1 - a_\nu)B_\nu(T_\odot)\, d\nu \tag{23.5}$$

Similarly, the power radiated is

$$P_{rad} = 4\pi R_P^2 \int_0^\infty e_\nu B_\nu(T_P)\, d\nu \tag{23.6}$$

Equating these gives

$$4\int_0^\infty e_\nu B_\nu(T_P)\, d\nu = \left[\frac{R_\odot^2}{d^2}\right]\int_0^\infty (1 - a_\nu)B_\nu(T_\odot)\, d\nu \tag{23.7}$$

This equation cannot be solved directly for T_P even if we know a_ν and e_ν. But since we know $T_\odot$, we can evaluate the right-hand side of the equation. We then try different values of T_P on the left-hand side, probably using a computer to evaluate the integral, until we find a value of T_P which makes the left-hand side equal to the right-hand side. To be even more rigorous, we should also account for the fact that the temperature is not constant across the surface of a planet.

We can take advantage of the fact that most of the Sun's energy is in the visible and most of the energy given off by the Earth is in the infrared. We can assume that there is a constant albedo, a_V, in the visible, and a constant emissivity, e_{IR}, in the infrared. Since a_V and e_{IR} are constant over the region of each integral where B_ν is significant, we can factor them out of the integral. The result is similar to equation (23.4):

$$T_P = \left[\frac{L_\odot}{16\pi d^2\sigma}\frac{(1 - a_V)}{e_{IR}}\right]^{1/4} \tag{23.8}$$

On the Earth, the albedos are different for the oceans and for land. They are also different for cloud cover. When we take these into account, the equilibrium temperature is 246 K. However, this is still not the temperature we measure at the ground. We have not yet considered the

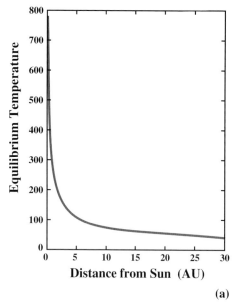

(a)

(b)

Fig 23.9. (a) Diagram showing graph of the equilibrium temperature vs. distance from the Sun. (b) Temperature variations across the surface of the Earth, as shown in near infrared images from space. This image is at a wavelength of 1 μm from the Galileo spacecraft at a distance of 2.1 million km. Lighter areas are giving off more infrared emission. [(b) NASA]

important effects of radiative transfer in the atmosphere.

We can also do this calculation for planets at any distance from the Sun, and we obtain the results shown in Fig. 23.9(a). Here we present a graph, in which we show the equilibrium temperature at different distances, and note the locations of the planets. Notice how this temperature decreases with increasing distance.

23.3 | The atmosphere

The Earth's atmosphere provides us with a multitude of phenomena whose complexity may make them seem beyond understanding. However, we can apply many of the basic physical ideas that we have already discussed for stars, such as hydrostatic equilibrium and energy transport, to form a reasonable understanding of those phenomena. With the aid of supercomputers, these ideas have been applied to the atmosphere in some detail. Also, the concepts that we develop in studying the Earth's atmosphere will be directly applicable to studying the atmospheres of other planets. In fact, one of the checks on computer models for the Earth's atmosphere is to see if they can predict the properties of the other atmospheres in the Solar System. In this section we look at some of the concepts common to all planetary atmospheres, and see how they apply to the Earth's atmosphere.

Though the Earth is a sphere, and the atmosphere is a spherical shell around it, the atmosphere is very thin (a few hundred kilometers) compared with the radius of the Earth (6380 km). This means that, if we stand on the ground, the effects of the curvature in the atmosphere are very small. This is shown in Fig. 23.10(a). This tells us that we can treat the atmosphere like a thin layer, and only worry about how things change as we go to higher altitudes. In studying the Earth's atmosphere, we want to understand how the pressure changes with altitude (the pressure distribution), how the temperature changes with altitude (temperature distribution), what the composition is, and how it changes with altitude, and how energy is transferred through the atmosphere.

In studying the atmosphere, it is convenient to divide it into layers, according to what conditions are prevalent. These layers are shown in part (b) of Fig. 23.10. Each of the layers has a different

(a)

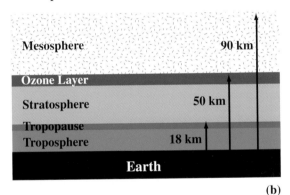

(b)

Fig 23.10. (a) Photograph showing the Earth and its atmosphere from space; note how thin the atmosphere is. (b) Diagram showing the layers in the atmosphere. [(a) NASA]

temperature distribution, as we will discuss in Section 23.3.2. The bottom layer, to which we are mostly confined, is called the *troposphere*. It is only 14 km thick. At the top of the troposphere is the thin *tropopause*. Above that is the *stratosphere*, in which some high altitude aircraft fly. At the top of the stratosphere is the ozone layer (which will be discussed below).

To relate the basic variables that describe a gas – temperature, density and pressure – we

must have an equation of state. We can treat planetary atmospheres as ideal gases, so the equation of state is simply

$$P = (\rho/m)kT \tag{23.9}$$

In this expression, m is the average mass per particle. For the Earth's atmosphere, this is approximately 29 times the mass of the proton, reflecting the fact that the atmosphere is mostly N_2 (molecular weight 28) and O_2 (molecular weight 32). For a surface temperature of 300 K, the density at the surface is 1.1×10^{-3} g/cm^3.

The pressure at the bottom of the atmosphere, near the surface of the Earth, is called *one atmosphere* or *one bar*. It is quite large, approximately 10^5 N for every square meter (10^6 dyn/cm^2 or 15 lb/in^2). Remember the weight of a typical person is about 750 N, so it is like having over 100 people stand on every square meter. We don't normally see the effects of this pressure because in most situations it tends to cancel. For example, for a wall, the air on opposite sides is pushing with equal and opposite forces, resulting in a net force on the wall of zero. We see the effects of the pressure if we remove it from one side, by using an air pump, for example.

23.3.1 Pressure distribution

The vertical distribution of pressure and density is governed by the condition of hydrostatic equilibrium, just as in stars. The weight of each layer is supported by the pressure difference between the bottom and the top of that layer. For stars, we treated the layers as spherical shells. We could do that for the Earth, also. However, the Earth's atmosphere is so thin that we can treat it as a plane parallel layer (see Problem 23.8).

The equation of hydrostatic equilibrium then becomes

$$\frac{dP}{dz} = -\rho g \tag{23.10}$$

We now have two equations, the equation of state and the equation of hydrostatic equilibrium. However, we have three unknowns, T, P and ρ. When we look at the temperature distribution, discussed later in this section, we find that, especially in the lower atmosphere, the temperature doesn't deviate by large amounts from its value at the ground, T_0. We therefore make the approximation that the temperature is constant. Knowing T, we can use equation (23.9) to substitute for the density in equation (23.10), so that the only remaining variable is pressure. This gives

$$\frac{dP}{dz} = -\left(\frac{mg}{kT_0}\right)P$$

To integrate this, we want all of the P dependence on one side and all of the z dependence on the other side. This gives

$$\frac{dP}{P} = -\left(\frac{mg}{kT_0}\right)dz$$

We now integrate this, with the limits on pressure being P_0, the surface pressure, and P, the pressure at the altitude of interest, and the limits on altitude being zero and z. That is,

$$\int_{P_0}^{P}\frac{dP'}{P'} = -\left(\frac{mg}{kT_0}\right)\int_0^z dz' \qquad (23.11)$$

Integrating and substituting the limits gives

$$\ln\left(\frac{P}{P_0}\right) = -\left(\frac{mg}{kT_0}\right)z$$

where we have used the fact that $\ln(P/P_0) = \ln(P) - \ln(P_0)$. If we raise e to the left-hand side, it should equal e raised to the right-hand side. Remembering that $e^{\ln x} = x$, this gives

$$\frac{P}{P_0} = e^{-(mg/kT_0)z}$$

The quantity kT_0/mg has dimensions of length, and is the distance over which the pressure falls to $1/e$ of its original value. We call this quantity the *scale height*, H, where

$$H = \left(\frac{kT_0}{mg}\right) \qquad (23.12)$$

In terms of the scale height, the pressure distribution becomes

$$P = P_0 e^{-z/H} \qquad (23.13)$$

The variation of pressure with altitude is shown in Fig. 23.11.

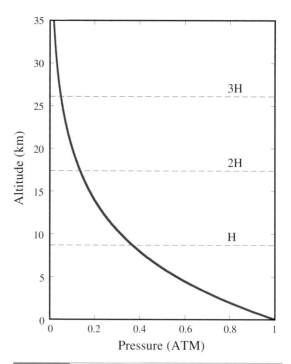

Fig 23.11. Pressure vs. altitude in the Earth's atmosphere. In plotting quantities about the atmosphere, we usually plot altitude on the vertical axis, even though it is the independent variable. The scale height is indicated by H.

Example 23.1 Scale heights
Compute the scale height for the Earth's atmosphere as well as that for an atmosphere of pure oxygen (O_2) and pure hydrogen (H_2).

SOLUTION
For a gas of molecular mass Am_p (where m_p is the proton mass), the scale height is

$$H = \frac{(300\ \text{K})(1.38 \times 10^{-16}\ \text{erg/K})}{(1.67 \times 10^{-24}\ \text{g})(980\ \text{cm/s})A}$$

$$= 2.53 \times 10^7\ \text{cm}/A$$

$$= 2.53 \times 10^2\ \text{km}/A$$

For $A = 29$, $H = 8.7$ km. For oxygen, $A = 32$, so $H = 7.9$ km; for hydrogen, $A = 2$, so $H = 125$ km.

The drop off shown in Fig. 23.11 explains several things. If we go up to an altitude of 2 km, the atmospheric pressure is already down to 80% of what it is at sea level. This explains why it is difficult to breathe, even at these "modest" mountain altitudes. As an example, Denver is at an altitude of 1.6 km, meaning that its pressure is only 83% that at sea level. It is also interesting to compare pressure changes with altitude with those associated with weather changes on the Earth. Severe storms are usually associated with regions of low pressure. However, even the most severe storms only have a drop in pressure to about 90% of normal pressure. More typically, the weather changes at sea level produce pressure changes of only a few percent. So these changes are much smaller than the changes that one encounters by climbing mountains. This is why we often use pressure meters as altimeters, that is, devices that tell us how high above sea level we are. Another important constituent in the Earth's atmosphere is water vapor. Equation (23.13) is not a good description of its distribution. This is because the atmospheric temperature is close to that at which water condenses. The water vapor may have a normal distribution at low altitudes. However, it may be almost totally absent at the cooler, higher altitudes.

23.3.2 Temperature distribution

The temperature distribution with altitude is shown in Fig. 23.12. Notice that it is more complicated than the pressure distribution. The complexity in the temperature distribution reflects the variety of mechanisms by which energy

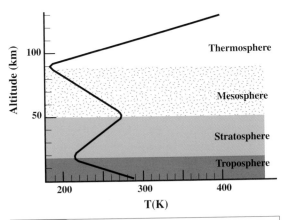

Fig 23.12. Temperature vs. altitude in the Earth's atmosphere. Layers are divided according to important energy balance mechanisms, so temperature behavior differs from layer to layer.

enters the atmosphere. The ultimate source of energy for the atmosphere is the Sun. However, it is not the direct source of heat over most of the atmosphere.

Most of the Sun's visible energy passes through directly to the ground, and is not absorbed by the atmosphere. This is true even if there are clouds. The clouds tend to scatter (rather than absorb) the visible radiation. This scattered light can either be directed back into space, and have its energy lost, or it can bounce around in the clouds, and eventually reach the ground. This explains why it is still light on a cloudy day, but not as light as on a clear day. The visible solar radiation reaches the ground, and some is reflected back (mostly from the oceans), and the rest is absorbed.

The heated ground then gives off radiation characteristic of its temperature, so the radiation from the ground is mostly in the infrared. Infrared radiation from the ground is then trapped in the lower atmosphere. Thus, the ground is the immediate source of energy for the lower atmosphere, explaining why the temperature in the lower atmosphere decreases as one moves farther from the ground. As the ground heats the air just above it, that air expands and rises. This convection is another means of energy transport from the ground to the lower atmosphere. (The above description is very simplified. In certain situations, "temperature inversions" are present in which warmer air is on top of cooler air, and there is little convection. With little convection, pollution can build up.)

The Sun's radiation that is absorbed goes into heating the ground. This process is shown in Fig. 23.13. If there were no atmosphere, the ground would heat up to the equilibrium temperature that we discussed above. However, the calculation of that equilibrium temperature was done on the assumption that all of the energy radiated by the Earth escaped into space. Since the Earth is at a temperature in the range 250–300 K, it is much cooler than the Sun, and its blackbody spectrum peaks at longer wavelengths. So, while the Sun gives off most of its energy in the visible part of the spectrum, the Earth gives off most of its energy in the infrared part. Many of the molecules in the lower atmosphere, especially

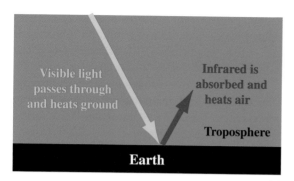

Fig 23.13. (a) Diagram showing the greenhouse effect. (b) Global (meaning averaged over all measuring stations) tropospheric deviations from monthly average temperatures, from 1979 to 2000. [(b) NASA]

(a)

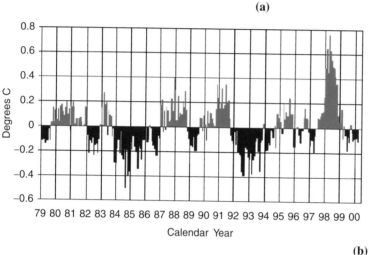

(b)

the water vapor, carbon monoxide and carbon dioxide, are very efficient at absorbing that infrared radiation. Therefore, instead of being radiated into space, some of the energy given off by the Earth is trapped in the lower atmosphere. This results in the surface of the Earth being hotter than if there were no atmosphere.

This effect, in which the visible light from the Sun heats the ground, and the infrared radiation from the ground heats the air above the ground, is called the *greenhouse effect*. The name results because this effect is similar to the way that greenhouses work. In a greenhouse, the window replaces the Earth's atmosphere. The window lets the visible radiation through, heating the ground. The ground then gives off infrared radiation, which is trapped by the windows, and the air inside the greenhouse is hotter than if there

were no glass. (In real greenhouses, the blocking of the wind is also important.) This effect also works in your house. Sunlight can pass through the windows, heating the interior, which produces infrared radiation, which is trapped by the windows. This explains how you can have useful solar heating, even in the winter.

On the Earth, the greenhouse effect is modest. It raises the temperature by about 25 K. We will see in the next chapter that the presence of large amounts of carbon dioxide on Venus has produced an extreme greenhouse effect on that planet. This leads us to worry that a similar thing could happen on Earth, if we build up the concentrations of gases that trap the infrared radiation near the ground. That is why atmospheric scientists are concerned over the by-products of human activity, from fires, to automobile and factory exhausts,

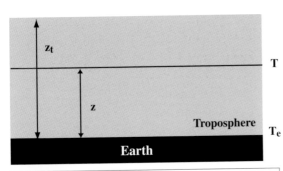

Fig 23.14. Radiative transport in the troposphere. We are considering material a height z above the ground, and the top of the troposphere is at a height z_t.

producing gases that will enhance the greenhouse effect. If there is an increase in these gases, then it is possible that the Earth's temperature will go into steady increase. This phenomenon is called *global warming*. The problem with measuring the effects of global warming is that the increase in any one year is small, and fluctuations due to variations in weather and climate cycles are much larger. However, atmospheric scientists are searching for steady trends (Fig. 23.13b).

The lower part of the atmosphere, heated by radiation from the ground, is called the *troposphere*. It is that part to which our daily life is confined. We now look at radiative energy transport from the ground into the troposphere, as outlined in Fig. 23.14. The ground is at a temperature T_e. We assume that the troposphere has a height z_t, and we are interested in energy reaching and leaving some intermediate height z.

The power per unit area reaching z from the ground is σT_e^4. The power per unit area radiated at z is σT^4, where T is the temperature at z. However, some of the radiation emitted at z does not escape. It is absorbed at a higher altitude, to be re-emitted. The energy density in the layer between z and z_t, due to the energy emitted at z is $4\sigma T^4/c$. Therefore, the energy per unit area in this layer is the energy per unit volume, multiplied by the thickness, $z_t - z$, so

energy/area $= (4\sigma T^4/c)(z_t - z)$

If all of this energy left z in a time t, the power per unit area is the energy per unit area, divided by t,

power/area $= (4\sigma T^4/ct)(z_t - z)$

Each time a photon is absorbed and re-emitted, its direction is changed in a random way. We say the photon does a *random walk*. If the walk is not random, all of the steps are in the same direction. If there are N steps of length L, the distance from the original point is NL. However, in a random walk, a lot of time is spent backtracking. It can be shown (see Problem 23.10) that for a random walk in one dimension (steps back and forth along a line), the distance from the origin after N steps of length L is $N^{1/2}L$, and for a three-dimensional walk the distance is $(N/3)^{1/2}L$. (In each case, the total distance traveled by the photons is NL. In the random walk, some of this is lost in the doubling back.)

Using this, the number of steps required for a photon to do a random walk of length $z_t - z$ is

$$N = 3[(z_t - z)/L]^2$$

where L is the photon mean free path. The length of each step is L, so the time for this number of steps is

$$t = NL/c$$
$$= (3/Lc)(z_t - z)^2$$

Using this, the power per area leaving z becomes

$$\frac{power}{area} = \frac{4\sigma T^4 (z_t - z)}{c(3L/c)(z_t - z)^2}$$
$$= \frac{4}{3}\sigma T^4 \frac{L}{(z_t - z)}$$

However, the path length $z_t - z$, divided by the photon mean free path L, is the optical depth τ (as discussed in Chapter 6), so

$$\frac{power}{area} = \frac{4}{3}\frac{\sigma T^4}{\tau}$$

Equating this to the power received from the ground σT_e^4, and solving for T, we have

$$T^4 = \left(\frac{3}{4}\right) T_e^4 \tau$$

A more accurate treatment of the radiative transfer, originally worked out by *Sir Arthur Eddington*, for the atmospheres of stars, modifies our result to

$$T^4 = \left(\frac{3}{4}\right) T_e^4 \left(\tau + \frac{2}{3}\right) \tag{23.14}$$

At the top of the troposphere, $\tau = 0$ (all photons escape upward). This means that $T = 0.84T_e$. If we use $T_e = 246$, then $T = 207$ K. The infrared optical depth from the bottom to the top of the troposphere is about 2, giving $T = 1.2T_e$, or about 293 K.

Convection also plays an important role in the energy transport in the lower atmosphere. As the air near the ground is heated, it expands. The buoyant force on the air will make it rise until it reaches air of its own density. The more rapidly the temperature falls with altitude, the more rapidly the pressure falls with altitude. A faster pressure fall-off means a larger pressure difference between the top and bottom of any parcel of air. The larger pressure difference provides a larger buoyant force, and the air rises more quickly, making convection more important. Therefore, convection becomes more important when the fall-off in temperature with altitude is larger.

As the air rises, it encounters lower pressure air and expands. The gas does work in the expansion. It takes no energy from its surroundings, so it cools. A process in which no energy is exchanged is called *adiabatic*. The convection process itself will modify the temperature gradient, dT/dz, to the value appropriate for an adiabatic process. If the temperature gradient by radiation is less than the adiabatic gradient, the temperature gradient is not enough to drive convection. However, if the temperature gradient is greater than the adiabatic gradient, convection will set in. Remembering that dT/dz is negative, we can say that the condition for significant convection is that

$$\left| \frac{dT}{dz} \right|_{rad} > \left| \frac{dT}{dz} \right|_{ad}$$

Example 23.2 Adiabatic temperature gradient
Find the value of dT/dz when energy flow is dominated by convection near the ground.

SOLUTION
We don't have to consider the whole atmosphere to do this. We only have to look at a small volume, or parcel, of air, and see how it behaves. Suppose we have a parcel of air with volume V_0 rising from just above the ground, where the temperature is T_0 and the pressure is P_0. As the parcel of air rises, its tem-

perature, pressure and volume are T, P and V. For an adiabatic process, these quantities are related by

$$PV^\gamma = P_0 V_0^\gamma$$

where γ is the ratio of the specific heats at constant pressure and constant volume. For a monatomic gas, $\gamma = 5/3$. For a diatomic gas, $\gamma = 7/5$. Since the Earth's atmosphere is mostly N_2 and O_2, we use 7/5. (If there is a lot of water vapor present, γ will be different.)
We can eliminate the volume using the ideal gas law

$$V = NkT/P$$

where N is the number of molecules in the parcel of air. This gives

$$P^{1-\gamma} T^\gamma = P_0^{1-\gamma} T_0^\gamma$$

For convenience, we will call this constant quantity C. Solving for T gives

$$T^\gamma = C P^{\gamma-1}$$

To obtain dT/dz, we differentiate both sides with respect to z, giving

$$\gamma T^{\gamma-1} \left(\frac{dT}{dz} \right) = C(\gamma - 1)P^{\gamma-2} \left(\frac{dP}{dz} \right)$$

$$= C(\gamma - 1)P^{\gamma-2} (-\rho g)$$

where we have used the hydrostatic law $dP/dz = -\rho g$. Solving for dT/dz, and substituting for C, we have

$$\frac{dT}{dz} = -\left(\frac{T_0 \rho_0}{P_0} \right) g \left(\frac{\gamma - 1}{\gamma} \right)$$

From the ideal gas law, $T\rho/P = m/k$, so

$$\frac{dT}{dz} = -\left(\frac{mg}{k} \right) \left(\frac{\gamma - 1}{\gamma} \right)$$

$$= 9.8 \text{ K/km}$$

As the warmer air rises from the ground, it must be replaced by cooler air from above. This must be replaced by cooler air from above. This leads to the general flow pattern shown in Fig. 23.15. This circulating flow is called a *convection cell*. We see this same pattern in a pot of boiling water, or in the Sun's atmosphere, causing the granular appearance.

The situation becomes more complicated when there is water vapor in the air. As the air rises and cools, it may pass through a temperature

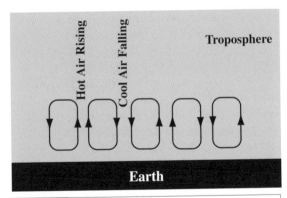

Fig 23.15. Convection cells. This motion carries hot air up and cold air down.

at which the water condenses into liquid droplets. When this happens, clouds form. Usually this happens at a particular altitude where the temperature is just right, so the cloud bottoms form a reasonably flat layer, as in Fig. 23.16(a). If the convection is weak, thin cloud layers will form. However, if the convection is strong, it can continue the cloud building upward, forming cumulus clouds. The convection can even be

strengthened by the energy released when water goes from a vapor to a liquid (the heat of vaporization). In the cases of particularly strong convection, thunderheads, like those in Fig. 23.16(b)

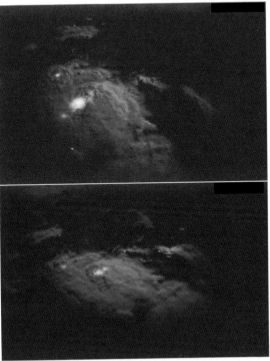

(c)

(a)

(d)

(b)

Fig 23.16. Weather phenomena resulting from convection. (a) Normal cloud layers result from convection. The cloud layers are the tops of the convection cells. In this photo, a thunderstorm is forming at a higher level in the background, with a smaller low level cloud in the foreground. (b) Lightning under thunderclouds. (c) Flashes above thunderstorms. (d) Very strong thunderstorms can lead to tornadoes. [(a), (b), (d) NOAA; (c) NASA]

are formed. Thunderheads tend to form on summer afternoons, when the solar heating creates a very large temperature gradient, driving very strong convection.

One place where solar energy is directly absorbed is the *ozone layer*. The ozone in the upper atmosphere is very efficient in absorbing the ultraviolet radiation from the Sun. This has two important effects: (1) it shields the lower atmosphere (especially us) from the harmful effects of that ultraviolet radiation, and (2) it provides a source of direct heating for the upper atmosphere. The direct heating makes the ozone layer hotter than the parts of the atmosphere immediately above or below it.

The shielding of the ultraviolet radiation is important, because if it is not blocked in the ozone layer, a significant amount will reach the ground. This excess solar ultraviolet radiation can be harmful to many living things, including people. There are concerns about the by-products of technology causing changes in the ozone layer, and possibly threatening the shielding effect. At first, concerns were focused on things that we do in the stratosphere, since that is just below the ozone layer, and chemicals that destroy the ozone could get from the stratosphere to the ozone layer quite easily. Normal commercial aircraft stay in the troposphere, but many high performance and supersonic aircraft (such as the Concorde) do fly in the stratosphere. More recently, theoretical and observational studies have shown that harmful chemicals can eventually make their way from the ground to the ozone layer. Recently, evidence has been found for a variable and large *ozone hole* over the Antarctic. This is indicated in Fig. 23.17.

23.3.3 Retention of an atmosphere

Even if a planet is formed with an atmosphere, that atmosphere will not necessarily be retained. Any molecules moving faster than the escape speed can escape from the top of the atmosphere. If escaping molecules are not replaced from below, by gases escaping from the planet's interior, the atmosphere will eventually be lost.

At any temperature the molecules with lower molecular mass will be moving faster, so these are the most likely to escape. We can calculate the value of A for which the molecular speed $(3kT/m)^{1/2}$ is equal to the escape speed for the planet $(2GM/R)^{1/2}$. Equating these, writing m as Am_p, and solving for A gives

$$A = \frac{3kTR}{2GMm_p}$$

For the Earth, $A = 0.06$. This tells us that there is no molecule for which the average speed is sufficient for escape. However, we have seen that because there is a distribution of speeds, many molecules move faster than the average.

The distribution of speeds, called a *Maxwell–Boltzmann distribution* is shown for hydrogen and oxygen in Fig. 23.18. The distribution is such that the number of molecules with speeds between v and $v + dv$ is

$$N(v)dv = (\text{constant})v^2\, e^{-mv^2/2kT}\, dv \qquad (23.15)$$

Note that we encountered this expression when talking about the fusion barrier in stars.

Note that if we go to higher and higher speeds, the probability becomes smaller and smaller but doesn't reach zero. If we go out to the escape speed, we find there are still some molecules moving at or faster than that speed. This means that a small fraction of the molecules can escape, even though the average molecule is securely bound to the Earth. The fraction of hydrogen molecules capable of escaping is greater than the fraction of oxygen molecules capable of escaping. This means that significantly more hydrogen will escape than oxygen.

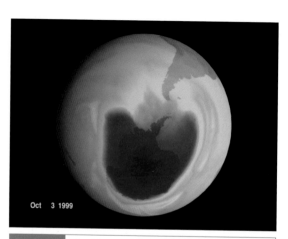

Oct 3 1999

Fig 23.17. This false color image shows the location of the ozone hole over the Antarctic. [NASA]

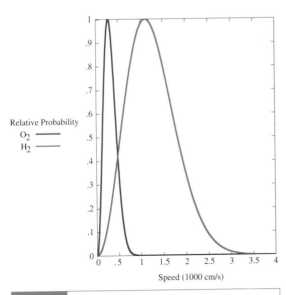

Fig 23.18. Maxwell–Boltzmann velocity distribution for hydrogen and oxygen molecules at $T = 300$ K. For each curve, the vertical axis is a relative probability of finding a molecule at a given speed. In each case, notice the large number of molecules with speeds greater than the average.

At first you might think that once the relatively small number of molecules moving faster than the escape velocity leave, the process is finished. However, the remaining molecules collide with each other, and re-establish the Maxwell–Boltzmann distribution. Since the escaping molecules took away more than the average energy per molecule, if there is no replacement of the energy, the new distribution will be at a lower temperature. The gas will be cooled by the escape of the faster molecules (just as a liquid is cooled by the evaporation of the faster molecules.) However, if there is a source of energy, such as sunlight or heat from the ground, the equilibrium can be established at the same temperature as before, and the same fraction of molecules will be moving faster than the escape speed. It is by this process that, over time, the atmosphere can escape. For the Earth, this escape has been much more rapid for the hydrogen than for the heavier molecules, such as oxygen. The Earth's atmosphere is indeed quite deficient in hydrogen. That is, the fraction of hydrogen in our atmosphere is much less than the fraction of hydrogen in the Solar System as a whole.

This escape can only take place from the highest levels of the atmosphere. A molecule lower down in the atmosphere might start out going faster than the escape speed, but it will collide with other molecules, losing energy before it can escape. The layer of the atmosphere from which the molecules can escape is called the *exosphere*. The thickness of the exosphere is taken to be equal to the average distance between collisions in the gas at high altitudes.

23.3.4 General circulation

Just as local air flows are in response to temperature differences which cause pressure differences, global air flows are subject to the same processes. The general tendency is for air to be heated at the equator, rise, and then flow toward the poles, where it cools, falls, and returns to the equator.

This simple pattern is disturbed by the effects of the Earth's rotation. Since the rotating Earth is an accelerating reference frame, we observe pseudo-forces. One of these is the familiar centrifugal force. It doesn't play an important role; however, the *coriolis force* does.

The origin of the pseudo-force is shown in Fig. 23.19. We are looking down from above the

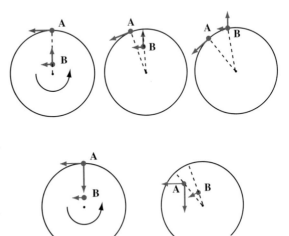

Fig 23.19. The coriolis pseudo-force. In the upper series, we look at an object thrown from a high latitude to the equator. A is the point aimed at on the equator, and B is the point where the object is launched. However, the horizontal motion at B is less than that at A, so the object reaches the equator behind A. We look at an object thrown from A to B, in the lower series. The object has a greater horizontal motion than does B, and therefore gets ahead of B.

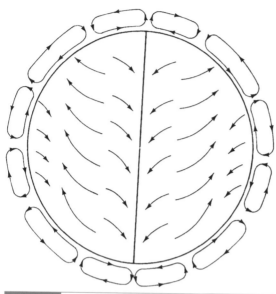

Fig 23.20. Circulation patterns on the Earth. The closed loops represent the major cells within the atmosphere, and the open arrows represent the general wind directions at the surface.

north pole. We are standing at point B, which is far from the equator. We want to aim at a target on the equator, at point A. We point directly at A and shoot. The bullet travels towards the equator with the speed that it left the gun. Its velocity has a slight horizontal component to the left due to the Earth's rotation. Object A has an even larger motion to the left, since it is at the equator, and is moving faster than we are. The result is that by the time the bullet reaches the equator, A has moved to the left of its path, and the shot misses, going behind the target. A similar thing happens if someone at A shoots at B. The bullet leaves A with a large horizontal speed, and misses B by getting ahead of it. To the observer on Earth, it is almost as if there is an additional force present.

In general, air flows from high to low pressure areas in the atmosphere (Fig. 23.20). Air traveling from the poles to the equator will reach the equator behind the point at which they were initially launched. Air traveling from the equator to the poles will arrive ahead of the point at which it was aimed. This produces a circulating pattern. Air flowing from a high to a low, going toward the equator, will lag behind the low. Air flowing in the other direction will get ahead. This results in a counterclockwise circulation around the

lows and clockwise circulation around the highs in the northern hemisphere. The opposite situation prevails in the southern hemisphere. This pattern is very evident in the circulation around hurricanes, which have very low pressure centers. A typical hurricane is shown in Fig. 23.21.

Recently, major improvements have been made in our understanding of the large-scale atmospheric circulation. Computers are used to model the Earth's atmosphere in considerable detail. The equations governing the fluid flow and energy balance are solved, taking such details as terrain into account. For such a model to be successful in predicting the movement of weather systems over periods of a few days, or even possibly weeks, it is important to have detailed information on conditions all over the Earth. Data are continuously collected from a variety of ground stations, from ships and weather buoys at sea, and from balloons sent up from various places on a regular schedule. Satellite observations are also important. The observations over the ocean are particularly important, since much of the energy that passes into the atmosphere is stored in the water. As the models are refined and the data gathering is improved, the predictions become better and better. We have now even reached the point where we can use the same approach to try to understand the global atmospheric properties of other planets, as we will see in the next few chapters.

(a)

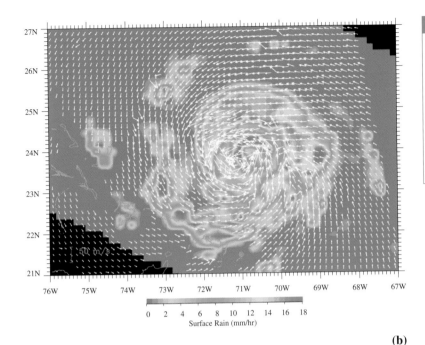

27N
26N
25N
24N
23N
22N
21N

76W 75W 74W 73W 72W 71W 70W 69W 68W 67W

0 2 4 6 8 10 12 14 16 18
Surface Rain (mm/hr)

(b)

23.4 | The magnetosphere

One of the early surprises of our unmanned space program was the discovery of belts of charged particles high above the Earth' surface. The particles are a source of low frequency synchrotron radiation. The radiation overwhelmed detectors on early spacecraft, making it appear as if no particles were present. However, *James van Allen* correctly interpreted the strange result as indicating the presence of large numbers of charged particles. We call these belts of charged particles the *van Allen radiation belts* (Fig. 23.22)

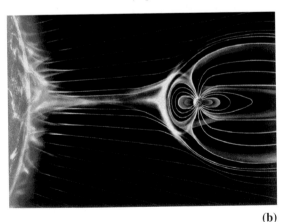

(b)

Rotation axis   **Magnetic axis**

(a)

Fig 23.22. Van Allen radiation belts and the Earth's magnetosphere. (a) The radiation belts are groups of trapped particles, concentrating into two bands, each with the shape of a doughnut that has been hollowed along the inner rim. (b) The Earth's magnetic field deflects the charged particles of the solar wind. The protected region is the magnetosphere, and the boundary is called the magnetopause. Just outside the magnetopause is a shock wave that looks like the bow wave when a ship plows through the water [NASA].

These particles are trapped by the Earth's magnetic field, and stay in spiral paths around the field lines. The region where there are large number of charged particles trapped by the Earth's magnetic field is called the *magnetosphere*. The region dominated by charged particles is also called the *ionosphere*. When we discussed solar activity (Chapter 6), we said that charged particles will follow helical paths around magnetic field lines. This is because the force on the particles is perpendicular to both the field lines and the velocity of the particle. This means that there can be no force along the field lines. The component of the velocity along the field, the drift speed, stays fixed, as the particles execute circular motion perpendicular to the field lines.

The situation is different if the particles are moving from a region of a weaker magnetic field to one of a stronger magnetic field, as illustrated in Fig. 23.23. The stronger field is represented by the field lines becoming closer together. We divide the magnetic field into two components,

B_{PAR}, parallel to the z-axis, along which the particles are drifting, and B_{PERP}, perpendicular to the z-axis. If the field were constant, B_{PERP} would be zero.

We look at the force on a proton moving in the indicated spiral path. (The argument works just as well for negatively charged electrons. See Problem 23.14.) Note that the sense of the spiral is determined by the charge of the particle. B_{PAR} doesn't change the z-component of the proton's speed. However, $v \times B_{PERP}$ points to the left. This is opposite to the direction of drift of the particle. The motion along the field lines is slowed, and eventually reversed. It is as if the particle struck a mirror. In fact, this phenomenon is called a *magnetic mirror*. The charged particles therefore spiral back and forth, trapped in the Earth's magnetic field.

The particles come from interplanetary space, mostly from the solar wind. In fact the Earth's magnetic field shields us from most of the solar wind by trapping the particles. Because of the dipole nature of the Earth's magnetic field, these charged particles get closer to the surface near the Earth's magnetic poles. In these regions, there is more of the normal atmosphere for the charged particles to strike. The fast moving charged particles lose some energy to the air as they pass through, and the air glows. We see this glowing air as an *aurora* (plural aurorae). An

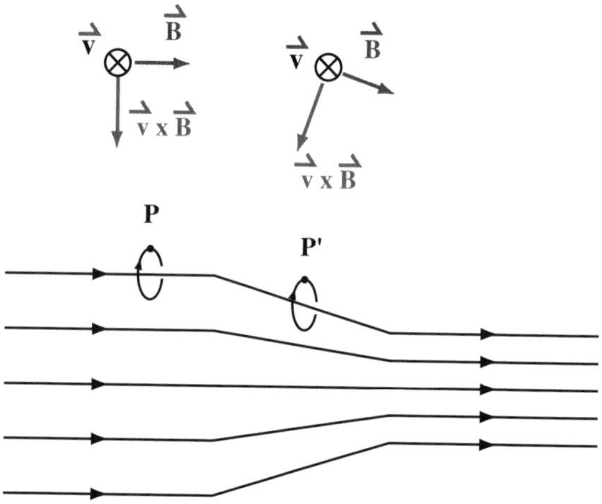

Fig 23.23. Magnetic mirror. Particles spiral around magnetic field lines. For simplicity, we look at positively charged particles, though the same argument can be carried out for negative particles. In the region where the magnetic field lines are parallel, on the left, the magnetic force on particle P is downward. This just keeps the particle orbiting. However, when the field lines become closer together, meaning that the field is increasing, the magnetic force on the particle at P' has a slightly rearward component. This is the force that slows and reverses the component of the motion parallel to the field lines.

Fig 23.24. When the solar wind particles penetrate the Earth's atmosphere, they cause the glowing aurorae. These are seen from above in an ultraviolet image. [NASA]

example is shown in Fig. 23.24. Since the magnetic mirrors are near the magnetic poles, the aurorae are most prominent close to the north and south poles. In addition, aurorae are strongest when there are a lot of charged particles arriving from the Sun.

23.5 | Tides

A number of phenomena on Earth depend on the fact that the gravitational force exerted by the Moon (or the Sun) on the Earth is slightly different at different parts of the Earth. As we have already discussed (Chapter 12), any effect which depends on the *difference* between the gravitational forces on opposite sides on an object is called a *tidal effect*.

If we have an object of mass m, a distance r from an object of mass M, the force on the object of mass m is

$$F = \frac{GMm}{r^2}$$

If we move the object, the force changes. The rate of change of the force is

$$\frac{dF}{dr} = -\frac{2GMm}{r^3}$$

The change in force, ΔF, in going from r to $r + \Delta r$ is

$$\Delta F = \left(\frac{dF}{dr}\right)\Delta r$$

$$= -\frac{2GMm}{r^3}\Delta r$$

The change in acceleration, Δa, is $\Delta F/m$, or

$$\Delta a = -\frac{2GM}{r^3}\Delta r \tag{23.16}$$

Note that the tidal effects fall as $1/r^3$, faster than the $1/r^2$ fall-off of the gravitational force itself.

Example 23.3 Tidal effects on the Earth
Compare the strength of the tidal effects exerted on the Earth by the Sun and the Moon.

SOLUTION
We can express the result as a ratio, using equation (23.16):

$$\frac{\Delta a_{SE}}{\Delta a_{ME}} = \left(\frac{M_\odot}{M_M}\right)\left(\frac{r_{ME}}{r_{SE}}\right)^3$$

$$= \left(\frac{2 \times 10^{33}\,\text{g}}{7.3 \times 10^{25}\,\text{g}}\right)\left(\frac{385 \times 10^3\,\text{km}}{150 \times 10^6\,\text{km}}\right)^3$$

$$= 0.46$$

Even though the Sun exerts a greater gravitational force on the Earth than the Moon, the closeness of the Moon makes its tidal effects greater.

The tidal effects of the Sun and Moon are responsible for the ocean tides on Earth (Fig. 23.25). We first look at how the Moon affects the Earth. We look at three points: (1) the point closest to the Moon, (2) the center of the Earth, and (3) the point farthest from the Moon. We see that $a_1 > a_2 > a_3$, as viewed from the rest frame of the Moon. However, as viewed from the Earth, all accelerations must be relative to a_2. In this frame the acceleration of (1) is $a_1 - a_2$ towards the Moon; that of (2) is zero; that of (3) is $a_3 - a_2$, and is therefore directed away from the Moon. This means that there will be a high tide on the side nearest the Moon and also on the side farthest from the Moon. We can think of the tide on the near side as the water on that side being pulled away from the Earth, and we can think of the tide on the far side as the Earth being pulled away from the water.

The Sun produces a similar effect, but only half as great in size. When the Moon and Sun pull along the same line, the difference between high and low tides is the greatest. When they pull at right angles to each other, the difference between high and low tides is the least. The tides on the oceans, where the water flows without obstruction, are well approximated by Fig. 23.25. However, in narrow or blocked waterways that run into the ocean, the flow of the water is quite complicated, and the time of high tide may be delayed by a few hours behind the time of the ocean high tide.

The height of the water tide is actually not found to be as large as one would calculate. This is because the Earth is not solid. It is therefore also distorted by these tidal forces. This constant

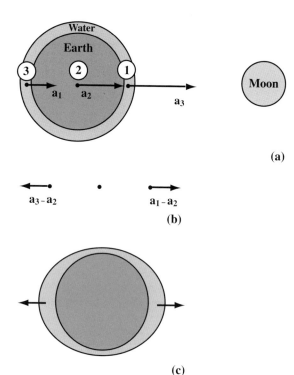

and is illustrated in Fig. 23.26. To see how this comes about, we just consider the effect of the Sun on the Earth. The Earth is not a perfect sphere, but has a larger diameter across the equator than between the poles. This oblate, or flattened, appearance is due to the Earth's rotation. We can think of the non-rigid Earth as deforming under the effects of the centrifugal force. Since the Earth is not spherical, the Sun can exert a torque on the Earth.

To see the direction of this torque, we idealize the Earth as a sphere with an extra band around the equator. The side of the band closest to the

Fig 23.25. Tides on Earth. (a) Different accelerations caused by the gravitational attraction of the Moon. (b) We express these accelerations relative to that of the center of the Earth. The result is that the water is pulled away from the Earth on the side facing the Moon (c), and the Earth is pulled away from the water on the far side.

reshaping of the Earth helps to heat its interior, and also dissipates the Moon's orbital energy. The Earth doesn't respond instantaneously to these tidal effects. Its rotation causes the bulge in the side near the Moon to get ahead of the Earth–Moon line. This means that the force can act to increase the Moon's orbital speed. This makes it move farther from the Earth. The corresponding force on the Earth causes its rotation to slow, keeping the total angular momentum of the Earth–Moon system fixed. Similarly, the Earth has exerted forces that distort the Moon. These forces, acting on the distorted Moon, act to slow its rotation, producing a rotation period equal to the Moon's orbital period.

The tidal forces that the Sun and Moon exert on the distorted Earth produce a torque that results in a continuous change in the direction of the Earth's rotation. This change is called *precession*,

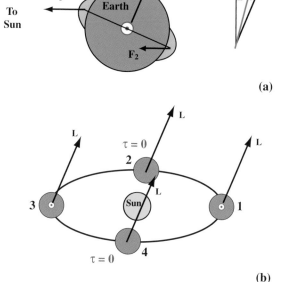

Fig 23.26. Precession caused by the Sun. (a) We treat the Earth as a sphere with an extra bulge around the equator, and look at the torque exerted by the Sun on the bulge. We see that F_1 is greater in magnitude than F_2, so there is a net torque on the Earth, whose direction, given by the right-hand rule, is out of the page. The torque would rotate the Earth, but the Earth already has some angular momentum, so the change in angular momentum must be added to the existing angular momentum, as shown on the right. The result is a new angular momentum whose magnitude is the same as before, but whose direction is different. (b) We show the torques at different points in the Earth's orbit. The torque can sometimes be zero, but when it is not zero it always points in the same direction.

Sun feels a slightly larger force than the side of the band farther from the Sun. When the Earth in is position 1 (first day of winter in the northern hemisphere), the greater force on the near side of the band produces a torque that points out of the page. Three months later, at position 2, the force on the near and far sides of the band both pass through the center of the Earth, so the torque is zero. Three months later, at position 3, the torque is again outward. Three months later, at position 4, it is again zero. The torque is sometimes zero, but is never in the opposite direction. Therefore, averaged over the year, there is a non-zero torque. The torque goes through two full cycles in its value over the course of the year. The effect of the Moon is essentially the same, with the torque going through two full cycles in its value each lunar orbit.

We now see the effects of this average torque $\langle \tau \rangle$. If the Earth's angular momentum is originally L, then after a short time dt the new torque L$'$ is given by

$$L' = L + dL$$

where

$$dL = \langle \tau \rangle \, dt$$

Note that dL is much smaller in magnitude than L, and is perpendicular to L, so it can only change the direction of L, not the magnitude. To change the direction of L the direction of the Earth's axis must change. The axis sweeps out a circle in the sky, returning to its current position in 26 000 years. Since the torque on the Earth is not constant in time, the precession doesn't occur smoothly. The small variations in the rate of precession are called *nutation*.

Precession has important observational consequences. Since the Earth's axis is changing orientation, the Earth's equatorial plane is doing the same. Therefore, the point at which the celestial equator and the ecliptic cross, the vernal equinox, changes its position in the sky. It has recently moved into the constellation Aquarius (the "Age of Aquarius"). Remember, for this precession the 23.5° tilt doesn't change, just the direction in which the axis points. Since the vernal equinox is the starting point of the astronomical coordinate system of right ascension and declination, the positions of all stars appear to change. It may seem like the change should be small, since it takes 26 000 years to go through a full cycle, but this amounts to 50 arc sec, or almost one-quarter minute of time each year.

23.6 | The Moon

The Moon provides us with the most spectacular example of a body that has been taken from the realm of remote sensing by traditional astronomical techniques to that of up-close study, including the luxury of bring samples back for studies in the laboratory. Various Earth- and space-based photographs are shown in Fig. 23.27.

Even from the Earth, we can see a variety of lunar features. The *highlands* are lighter colored. They contain mountains and valleys as well as long canyons, or *rilles*. The *maria*, once thought to be oceans because of their smooth, dark appearance, are more level. We can also see many different types of *craters*. Some have bright rays of ejected material. In some areas, we see several layers of catering, with younger craters appearing to cross the walls of older craters. One way of determining the relative age of surface features on the Moon is to look at the relative numbers of craters.

As we have said, lunar exploration has taken the Moon to the realm of the geologist. (In fact, the last crewed lunar landing included a geologist.) There were six manned lunar landings from 1969 through 1972, Apollos 11 through 17, with the exception of Apollo 13, which aborted its mission after an explosion en route. A variety of regions were visited, as shown in Fig. 23.28. These missions returned 382 kg of lunar rock. In the later flights, a rover vehicle allowed studies of extended regions around the landing sites. In addition to bringing back lunar samples, the astronauts left a variety of monitoring equipment on the Moon. This equipment included seismometers to detect moonquakes and meteor impacts and X-ray and radioactivity detectors. There were also unmanned Russian Luna flights (16, 20 and 24), which, from 1970 through 1976, returned 310 kg of rock from different sites. Some views are shown in Fig. 23.29.

Fig 23.27. Views of the Moon. (a) The near side, from the Galileo spacecraft on its way to Jupiter. (b) Apollo 12 photograph of the highland area crater Ptolemaeus (right center). The terraced crater with central peaks, in the center of the photo, is Herschel. (c) An Apollo 11 view of the crater Daedalus, on the lunar far side. This crater has a diameter of about 80 km. (d) Lunar Orbiter 5 image of a side view of the impact crater Copernicus. This crater is 93 km wide, and is located within the Mare Imbrium basin. [NASA]

23.6.1 The lunar surface

The lunar soil is in a layer ranging from 1 to 20 m in depth. It is called *regolith*, a combination of powder and broken rocky rubble. It is the result of meteoritic bombardment. We are talking about a large number of small meteorites that would have burned up in the Earth's atmosphere, so we see no corresponding material on the Earth. Larger meteor impacts have spread material around and mixed the material in a given region. The soil contains no water. It does have a high proportion of *refractory*, or high melting point, materials, such as calcium, aluminum and titanium.

The rocks are all basalts, meaning that they resulted from the cooling of lava. (Remember, the Earth has igneous rocks, but also has sedimentary rocks, such as limestone and shale.) The rocks contain some silicates (compounds containing silicon and oxygen, such as olivine) as well as

Fig 23.28. The locations of the various lunar landings (red = Luna; green = Apollo; yellow = Surveyor) are shown on this photograph of the near side of the Moon. [NASA]

oxides of iron and titanium. It also contains three new materials – named tranquilityite, armalcolite (after the three Apollo 11 astronauts, Armstrong, Aldren and Collins) and pyrroxferroite.

There are some differences between the maria and the highlands. The maria are basalts, like the lavas from the Hawaiian volcanoes. The highland rocks contain anorthosites, gabbro and norrite, which were probably cooled slowly deep within the Moon. There are also variations in element composition. The maria rocks have more titanium, magnesium and iron, while the highland rocks are rich in calcium and aluminum. The highland rocks are older. The maria rocks have ages in the range 3.1 to 3.8 billion years. These youngest rocks on the Moon are as old as the oldest rocks on the Earth. The highlands have rocks with ages ranging from 3.9 to 4.48 billion years. This latter number occurs in several places. There are also differences between the near and far side of the Moon, which is shown in Fig. 23.30. The far side has more craters and fewer maria. It also has a higher average altitude above the Moon's average radius.

Based on the properties of the rocks, and their distribution, it has been possible to deduce a scenario for the evolution of the lunar surface. Some of the parts of this scenario had been speculated on before the lunar explorations, but the dating of the lunar rocks gives exact time since certain important events. We think that the Moon formed 4.5 billion years ago. (Some small green rock fragments in the Apollo 17 samples are that old.) The surface underwent extensive melting and chemical separation for the next 200 million years. This melting resulted from the heat generated in extensive meteoritic bombardment, as well as some radioactivity. This bombardment created the large basins, hundreds of kilometers across. The bombardment eased approximately 4 billion years ago. Residual store heat and some radioactivity resulted in melting down to 200 km below the surface.

From 3.1 to 3.8 billion years ago (the ages of the rocks in the maria), lava rose and filled the basins, creating the maria. Since then the Moon has been quiet, except for the occasional meteor impact. The footprints left on the Moon will remain sharp for millions of years.

(a)

(b)

(c)

(d)

(e)

(f)

(g)

(h)

(i)

(j)

23.6.2 The lunar interior

Much of what we know about the lunar interior is deduced from lunar seismology. There are still small quakes, with about one-billionth the energy of a typical quake on Earth. Some of the quakes are the results of meteor impacts. By watching the propagation of seismic disturbances around and through the Moon, we can draw some conclusions about the interior. We can also learn about the interior from the amount of heat flowing through the surface. This heat flow rate is about one-third of the rate on the Earth.

The basic lunar structure is depicted in Fig. 23.31. The core has a radius of some 700 km. At this point, we are unsure about its composition

Fig 23.29. Photographs from Apollo missions. (a) The rising Earth viewed from Apollo 8. (b) The second person to step on the Moon, Edwin E. Aldren, Jr, steps off the Apollo 11 Lunar Module ladder as he prepares to walk on the surface (20 July 1969). This photo was taken by the first person on the Moon, Neil A. Armstrong. (c) This footprint left by the Apollo 11 crew will last for a long time as there is no water or air for erosion. (d) Wide angle view of the Apollo 17 landing site, Taurus-Littrow region (10 Dec 1972), including the Lunar Rover vehicle. (e) Orange soil found by geologist/ astronaut Harrison Schmitt (12 Dec 1972) during one of the Apollo 17 surface excursions. (f) Harrison Schmitt near a large boulder (13 Dec 1972). (g) A lunar soil sample from Apollo 12, back in the laboratory on Earth. This was from a core sample that collected dirt as far as 225 m below the surface. (h) Thin section of rock brought back by Apollo 12. Under polarized light, the lavender minerals are pyrexene; the black mineral is ilmenite; the white and brown, feldspar; and the remainder, olivine. (i) Photomicrograph of spheres and fragments in the "orange" soil brought back with Apollo 17. The magnified image shows particles in the 150–250 μm range. (j) Another Apollo 17 sample, a 32 g breccia. [NASA]

Fig 23.30. Photograph of the far side of the Moon, taken from the Galileo spacecraft on its way to Jupiter. [NASA]

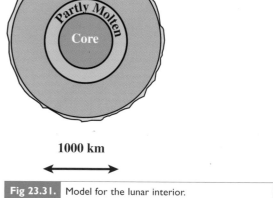

Fig 23.31. Model for the lunar interior.

or detailed structure. The absence of a magnetic field suggests that it is probably not molten. (However, some rocks show evidence for a magnetic field in the past.) Above the core is a partly molten zone 400 km thick. The deep quakes are thought to originate in this zone. Above this is a mantle, and above the mantle is a thin crust. The crust is variable in thickness, ranging from 0 to 65 km thick.

Anomalies in the orbits of early lunar spacecraft indicated that the mass of the Moon is not distributed uniformly in a sphere. Mass concentrations, or *mascons*, have been detected beneath the circular maria, such as Imbrium, Serenitatis, Crisium and Nectaris. We think that these resulted from volcanic lava filling the basins. The fact that these concentrations have survived the 3 billion years since the basin filling suggests that the lunar interior is quite rigid. If it were not, the mascons would have sunk a long time ago.

23.6.3 Lunar origin

Four basic theories of the origin of the Moon have been proposed.

Fission. In this theory, the Moon broke away from the Earth. This theory was particularly popular when it was realized that the Moon is the right size to have left behind the Pacific Ocean basin when it broke away. However, we now know that the plate tectonics on the Earth mean that the

Pacific Ocean has been in its current arrangement for a relatively short time, and that its current appearance is unrelated to its appearance some 4 billion years ago. The fission theory is supported by the fact that, when we allow for the effects of erosion on Earth, the surfaces of the Earth and Moon have some similarities. However, there are also some major composition differences. Also, the Moon's orbit is not in the plane of the Earth's orbit.

Capture. In this theory, the Moon passed close to the Earth, and was captured into its current orbit. One problem with this theory is that capture would more likely result in a much more eccentric orbit. It is also less likely that the orbit would lie so close to the plane of the ecliptic. Theoreticians have not been able to work out a reasonable dynamical scenario.

Condensation. In this theory, the Moon formed as a by-product of the formation of the Earth. Perhaps it gathered after the Earth was well under way in formation. Possibly, the Earth simply formed as a double planet. One problem with this idea is that we would expect the Moon to have an iron core, as does the Earth, and it does not.

Impact trigger. It is possible that a large object struck the Earth, and drove off enough material from the mantle and crust to form the Moon. This would explain why the Moon and Earth have similar crusts. This is currently the best accepted model.

Chapter summary

The Earth is the planet that we can study in the most detail. In this chapter, we looked at the properties of the Earth and the Moon. We also looked at the physical processes responsible for their current properties.

So much heat was trapped when the Earth formed that it still has a molten core of iron and nickel. Radioactive materials provide heating near the surface, keeping the mantle soft, like plastic. The radioactive materials also provide us with a way of dating various rocks. The oldest rocks on the Earth are 3.7 billion years old. The plastic nature of the material below the surface allows the continental plates to drift slowly. These plate tectonics are responsible for most of the geological activity that we see – volcanoes, earthquakes, and mountain building.

The temperature of the Earth (or any planet) is determined by the balance between the radiation absorbed from the Sun and that given off by the Earth. The atmosphere plays an important role in modifying that balance. The atmosphere is a very thin layer (relative to the radius of the Earth itself). The pressure is determined by hydrostatic equilibrium, and drops very quickly with altitude, falling to half of its sea level value at an altitude of about 6 km. The temperature distribution depends on the sources of heating for different parts of the atmosphere. In the ozone layer, solar ultraviolet radiation is absorbed, and directly heats that part of the atmosphere. Near the ground, the source of heating is the ground, either by infrared radiation from the ground, or convection of the air heated just above the ground. The trapping of infrared radiation near the ground produces the greenhouse effect, and raises the temperature near the ground by about 25 K. The temperature is high enough in the upper atmosphere for most of the hydrogen to have escaped. The general flow in the atmosphere is strongly influenced by the Earth's rotation.

The molten iron–nickel core provides the Earth with a magnetic field, similar to that of a dipole magnet. The field is strongest at the north and south magnetic poles (which don't coincide with the rotation poles). The magnetic field has the important role of trapping charged particles that hit the Earth from outer space. These charged particles give off a lot of long wavelength radio emission. They are also responsible for the glowing aurorae.

The Moon produces tidal effects on the Earth. These depend on the fact that the Moon's gravitational force is weaker on the side of the Earth away from the Moon than on the side closer to the Moon. This produces the water tides. The Sun has a similar effect, but it is not as large as the Moon's. The tidal effects also cause dissipation of some of the Moon's orbital energy, and have resulted in the Moon keeping the same face towards Earth. The Sun and Moon also cause the direction of the Earth's rotation axis to precess, like a top, with a period of 26 000 years.

The lunar surface is particularly interesting since it has been preserved over most of the history of the Moon. This is because the Moon has no atmospheric erosion. The lunar rocks show some differences between the maria and the highlands. An important difference is that the highland rocks are somewhat older (with the oldest being 4.48 billion years). The lunar interior is different from that of the Earth. In particular, the Moon does not have a molten core. The question of the lunar origin is still up in the air.

Questions

23.1. Why do we think that the Earth is more than a billion years old?

23.2. What does the fact that the Earth is differentiated tell us about its history?

23.3. Do you think that radioactive uranium would be useful for dating the bones of a cave person?

23.4. If the Earth and Moon formed together, why are the oldest rocks on the Moon older than the oldest rocks on Earth?

23.5. What does continental drift tell us about the Earth's interior?

23.6. What is the relation between continental drift and areas of geological activity (like earthquakes).

23.7. Suppose we have a space probe that starts at the Earth and goes outward. If there is no temperature control, describe what happens to the surface temperature of the spacecraft as it moves farther from the Sun.

23.8. If sunlight passes through the atmosphere, why does radiation from the ground become trapped?

23.9. How much of a difference does the Earth's atmosphere make in its temperature?

23.10. What keeps a balloon afloat? Draw a diagram showing the forces on a balloon.

23.11. If one atmosphere is such a large pressure, why doesn't the air outside a building drive the walls inward?

*23.12. The water in the oceans is also in hydrostatic equilibrium. This means that the water pressure increases as you go farther below the surface. A cubic meter of water weighs much more than a cubic meter of air. What does this tell you about how fast the pressure changes with depth in the water, relative to the changes in air?

23.13. What factors determine the temperature at a given altitude?

23.14. Why does the temperature decrease as we go up from the ground?

23.15. Why is the temperature in the ozone layer higher than it is just above or below that layer?

23.16. What is the importance of the ozone layer?

23.17. How does the greenhouse effect work?

23.18. Why is there concern about the possibility of global warming?

23.19. What is the difference between energy transport by convection and energy transport by radiation?

23.20. If the average speed of molecules in an atmosphere is less than the escape speed, how can the atmosphere escape?

*23.21. When water evaporates from your skin, you feel cool. Why is this? (Hint: Think about which molecules evaporate, and which are left behind.)

23.22. Why has more hydrogen escaped from our atmosphere than oxygen?

23.23. Why does the temperature of an atmosphere affect how much of it escapes?

23.24. Draw a diagram showing how the coriolis force acts in the southern hemisphere.

23.25. How do we know that the Earth has a magnetic field?

23.26. If you have a toy magnet that will stick to a refrigerator, do you think that the toy magnet or the Earth's magnetic field exerts a greater force on the refrigerator?

23.27. Describe the motion of a charged particle traveling in a uniform magnetic field.

23.28. What do we mean by a *magnetic mirror*?

23.29. Why are aurorae most prominent near the poles?

23.30. In general, what do we mean by a *tidal effect*?

23.31. Explain, in your own words, how there can be a high tide on the side of the Earth near the Moon, and on the opposite side of the Earth, at the same time.

23.32. What effect does precession have on the amount of tilt of the Earth's axis?

23.33. What effect does precession have on the apparent positions of stars in the sky?

23.34. What is unique about our ability to study the Moon?

23.35. How does the composition of the lunar soil differ from that of the Earth?

23.36. Why do we think that the Moon is more likely than the Earth to have rocks left over from the young Solar System?

23.37. What are the differences between the near and far sides of the Moon? Why do you think these differences occur?

23.38. How do we know about the lunar interior?

23.39. Some astronomers have suggested putting an observatory on the far side of the Moon. Why might this be a good location?

Problems

23.1. Show that if the number of nuclei in a sample is given by equation (23.1b), then, following any arbitrary starting time, the half-life is the time for the number to be reduced to half its value at that arbitrary time.

23.2. If the number of nuclei in a sample is given by equation (23.1a), show that the rate of radioactive decay (number of decays per second) is proportional to $N(t)$.

23.3. Show that $\tau_{1/2} = 0.693\,\tau_e$.

23.4. (a) How would you measure the half-life of an isotope whose half-life is long compared to your lifetime? (b) How would you determine the number of nuclei of that isotope in a particular sample (assuming you know the half-life)?

23.5. How would the equilibrium temperature of a planet be modified if the planet always kept the same side towards the Sun?

23.6. Express the equilibrium temperature of a planet as a function of that of the Earth and the distance of the planet, in astronomical units. Use your equation to construct a table of the equilibrium temperatures for the nine planets.

23.7. If we are interested in measuring distances through the atmosphere to within 1%, down to what angle from the zenith can we approximate the Earth's atmosphere as a plane parallel layer?

23.8. Assuming the pressure in the Earth's atmosphere varies according to equation (23.13),

find an expression for the column density of air $N(z)$ as a function of z. The column density is the number of molecules in a cylinder of unit area from the height z to the top of the atmosphere.

23.9. At what altitude is the pressure (1/2) atmosphere?

23.10. If after N steps of length L in a random walk in one dimension, the distance from the origin is $N^{1/2}L$, show that this distance is $(N/2)^{1/2}L$ and $(N/3)^{1/2}L$ for two- and three-dimensional walks, respectively.

23.11. Find the pressure distribution in an atmosphere where the temperature distribution is

$$T(z) = T_0 - bz$$

23.12. Using the fact that precession goes through a full cycle in 26 000 yr, calculate the average torque on the Earth. (Hint: It is necessary to calculate the angular momentum of the Earth.)

23.13. When we applied hydrostatic equilibrium to the Earth's atmosphere, we found that the pressure falls off exponentially. However, when we apply it to the ocean, the pressure varies only linearly with depth. How can you account for the difference?

23.14. Show that positively and negatively charged particles are reflected the same way by a magnetic mirror. (Hint: Remember that oppositely charged particles spiral magnetic field lines in opposite directions.)

Computer problems

23.1. Suppose we have two radioactive nuclei, one with a half-life of 1000 yr and the other with a half-life of 2000 yr. Suppose we start with equal amounts of both nuclei. Make a graph showing the abundance of each element from the start point to until 10 000 yr after. Also make a graph of the ratio of the abundances (shorter/longer half-life).

23.2. Calculate the equilibrium temperatures for all nine planets.

23.3. Make a graph showing the variation of partial pressures with altitude for O_2, CO and CO_2, assuming a temperature of 300 K.

23.4. Make a graph showing the variation of partial pressure with altitude for O_2, for temperatures of 275 K, 300 K and 325 K.

23.5. Calculate the normalization for the Maxwell–Boltzmann distribution by integral = N.

23.6. Reproduce Fig. 23.18 for temperatures of 275 K and 350 K.

Chapter 24

The inner planets

The Solar System naturally divides into two groups of planets, separated by the asteroid belt. The four inner planets have many things in common with the Earth, whereas the next four planets present worlds of an entirely different type. (Pluto is an additional enigma.) In this chapter, we look at Mercury, Venus and Mars, comparing their properties with each other, and with the Earth.

24.1 | Basic features

24.1.1 Mercury
Mercury is the closest planet to the Sun, and is not much larger than our Moon. There is an interesting story concerning its rotation period. Since Mercury is so close to the Sun, we never have a really good view of it, and surface features are hard to recognize. By noting the positions of large surface features, it appeared that the rotation period was 88 days, the same as the planet's orbital period. This would have meant that Mercury always keeps the same face towards the Sun (just as the Moon keeps the same face towards the Earth). Since Mercury is so close to the Sun, it seemed plausible that some tidal effect could keep its rotation period synchronized with its orbital period.

However, the situation was corrected following radar observations. Radio waves were bounced off Mercury and then detected back on Earth. The planet's rotation causes a spread in the Doppler shifts of the reflected waves. From the amount of spread, we can tell how fast the planet is rotating. Astronomers were surprised to find that Mercury's rotation period is 59 days, two-thirds of the 88 day orbital period. This relationship means that when Mercury is favorably placed for observations, it often has the same side towards Sun, making it seem that it always had the same side towards the Sun. This simple ratio is probably no accident. It may result from the varying tidal effects, as Mercury has a rather eccentric orbit.

Our only close-ups of Mercury have come from the Mariner 10 spacecraft, which made three flybys of Mercury. The orbit of Mariner 10 was arranged to bring it back close to Mercury periodically. The closest flyby was within 300 km, and allowed a very detailed study of surface features. Mariner 10 photos of the whole planet are shown in Fig. 24.1.

24.1.2 Venus
Venus has been referred to as our "sister planet". Its size is very close to that of Earth. It should also encounter somewhat similar solar heating conditions. It receives twice as much solar energy as does the Earth. For some time it was thought that Venus might be a good candidate for finding life. At the very least, it is a good candidate for testing theories which explain various aspects of the Earth. Study of the surface is hindered by thick clouds.

The planet's rotation period is 243 days. The sense of the rotation is opposite to that of the orbital motion. This is called *retrograde rotation*. If the planets simply condensed out of a rotating nebula, conservation of angular momentum tells

(a)

(b)

More recently the Pioneer Venus Orbiter produced radar maps of most of the surface. Earth- and space-based images of Venus are shown in Fig. 24.2.

24.1.3 Mars

Mars has long held our fascination. Except during dust storms, there is no thick cloud cover, so we have had a comparatively good view of the surface. Particularly intriguing to Earth-based observers were color changes with season that suggested some vegetation. The axis of Mars is tilted by 25°, so we would expect the seasons to be like those on Earth. The white polar caps also change size with the season, raising the possibility that they hold water ice. In addition, the rotation period is very close to that of the Earth.

Mars has also been the subject of extensive exploration. The Mariner spacecraft flew by, sending back photos of a barren, crater-marked surface. Earth- and space-based images of Mars are

us that all of the planets should be rotating on their axes in the same sense as they are orbiting about the Sun. It has been suggested that the drag of a heavy atmosphere and tidal effects of the Sun could be responsible for the unusual rotation of Venus.

Exploration of Venus by spacecraft has been quite extensive. Especially notable are the Soviet Venera landers, which also sampled the atmosphere on the way down to the surface. In addition, the Pioneer Venus probes (in 1978) provided a wealth of information. One spacecraft went into orbit and used radar to map the surface features. The other sent probes through the atmosphere.

(a)

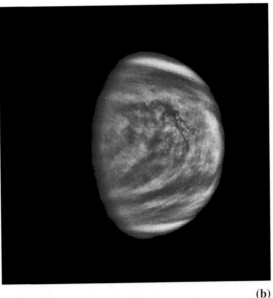

(b)

Fig 24.2. (a) HST image of Venus. (b) Galileo image of Venus from a distance of 1 million km. This has been enhanced to show small-scale cloud structure. [(a) STScI/NASA; (b) NASA]

shown in Fig. 24.3. Two Viking spacecraft left orbiters around Mars and also sent landers to the surface. These landers sampled the soil, and even carried out a preliminary search for life. Some chemical effects were found that mimic some simple aspects of certain living things, but no evidence for life was found.

24.1.4 Radar mapping of planets

Since we obtain so much information from radar mapping of planetary surfaces, it is useful to go over the basic ideas behind it. A raised surface feature may show up as a stronger than average radar echo. However, the problem is to determine where on the surface the feature is. There are two effects that help us locate the feature. One is the time delay for the signal returning to Earth, and the other is the Doppler shift. These are illustrated in Fig. 24.4.

We first look at the time delay. Since the surface is round, different parts of the surface are different distances from our radio telescope. These different distances mean that the light (or radio wave) travel times will be different for waves bouncing off different parts of the surface. We can express time delays relative to that of a wave bouncing off the closest point. According to the figure, the extra distance that the signal has to travel is $2x$, so the time delay is

$$\Delta t = 2x/c$$

We can see that

$$x = R - y$$
$$= R - R\cos\theta$$
$$= R(1 - \cos\theta)$$

This means that the time delay is

$$\Delta t = \left(\frac{2R}{c}\right)(1 - \cos\theta) \tag{24.1}$$

If we measure the time delay, and we know the planet's radius, R, then we can solve equation (24.1) for θ. This does not give a point uniquely. There is a whole ring of points that all have the

Fig 24.3. (a) HST image of Mars at closest approach to Earth. (b) Space-based image of Mars. [(a) STScI/NASA; (b) NASA]

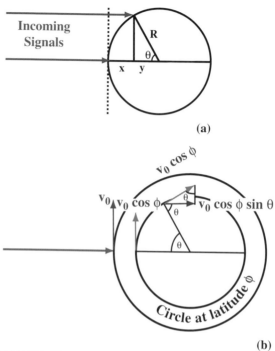

Fig 24.4. Radar mapping of a planet: (a) time delay; (b) Doppler shift.

same θ. As viewed from the radio telescope, lines of constant time delay appear as concentric rings about the closest point.

We now look at the Doppler shift. If a point on the equator moves with a speed v_0, then a point at latitude φ moves with speed (see Problem 24.2)

$$v(\varphi) = v_0 \cos \varphi$$

Now we view this point from the above the pole, assuming the line from the point to the pole makes an angle θ with the line from the pole to the telescope. The Doppler shift depends on the radial velocity, $v_r(\theta, \varphi)$, which is given by

$$v_r(\theta, \varphi) = v(\varphi) \sin \theta$$

$$= v_0 \cos \varphi \sin \theta \qquad (24.2)$$

As seen from the telescope, lines of constant Doppler shift form concentric rings about the point on the equator that is just appearing from the back side, and the point on the equator that is just about to disappear.

By combining time delay and Doppler shift data, we limit the source of the echo to two possible points. These are the two points where the time delay circle intersects the Doppler shift circle. The remaining ambiguity in the location of the feature can be removed by observing at a different time where the feature's location and Doppler shift have changed. More recently, orbiting spacecraft have also been used for higher resolution radar mapping on Venus and Mars.

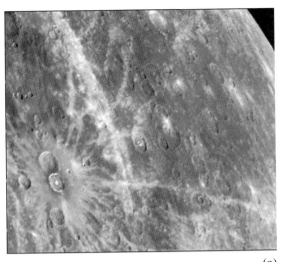

(c)

Fig 24.5. Surface features of Mercury. (a) Mariner 10 image of rayed crater. (b) Mariner 10 image of large crater. (c) Mariner 10 image of scarp. [NASA]

(a)

(b)

24.2 | Surfaces

The Mariner 10 spacecraft has mapped 50% of Mercury's surface. Some images are shown in Fig. 24.5. The third and closest encounter shows features 50 m across. An obvious feature is extensive cratering. The surface is reminiscent of the Moon, but there are also significant differences. The craters on Mercury are flatter than those on the Moon. The larger surface gravity on Mercury means that there is more slumping in the sides of the craters. The surface may also be more plastic than that of the Moon. There is evidence for some erosion, most likely by micrometeorites.

There are a number of unusual features. For example, there are long fractures, called *scarps*. These indicate some large-scale compression of the surface. There are also a series of irregular features called *weird terrain*. These were probably caused by the shocks from impacts by large objects.

Infrared observations (also from Mariner 10) indicate that the surface has a layer of fine dust, which is several centimeters thick. These observations also indicate surface temperatures ranging from about 700 K on the day side to 100 K on the night side. This large temperature difference tells us that there is very little heat flow, via soil or atmosphere, from the warm side to the cold side. The extreme temperature differences from day to night can introduce stresses that weaken the surface features.

The total cloud cover on Venus (Fig. 24.2b) means that we must rely on radar mapping to see the large-scale surface features. Maps have now been made of 93% of the surface, and are shown in Fig. 24.6. Original maps were made from ground-based telescopes, particularly the large dish at Aricebo (Fig. 4.27). The more recent data are from orbiting radar.

Venus is relatively flat. Approximately 60% of the surface lies within 500 m of the average radius, and only 5% of the surface deviates by more than 2 km from this radius. However, the total range of elevation, about 13 km, is comparable to that on

Fig 24.6. Radar maps of Venus from the Magellan spacecraft. (a) A full hemisphere. (b) Mosaic of large impact craters. This image has been processed to resemble a cloud-free optical image.

(a)

(b)

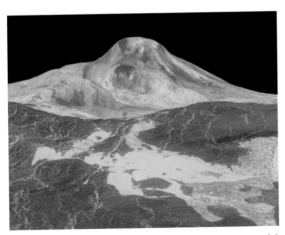

(c)

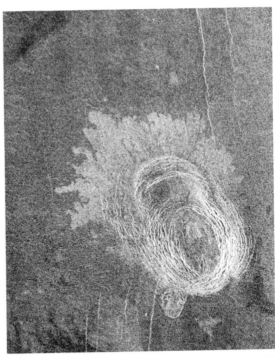

(e)

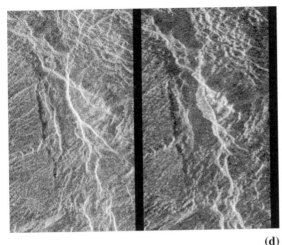

(d)

Fig 24.6. (*Continued*) (c) Three-dimensional perspective of Maat Mons. (d) Aphrodite Terra. The images are before (left) and after (right) a landslide. (e) Sag caldera Sachs Patera. [NASA]

Earth. About 20% of the planet's surface is covered by lowland plains, and about 10% by true highlands. The remaining 70% is described as rolling uplands. These uplands have a variety of features, some of which may be large impact craters. There is evidence for volcanoes, some of which may still be active.

The maps show two large "continents". By using the term "continent" we do not mean to imply that there is evidence for plate tectonics on Venus. One continent is called *Terra Ishta*. It is in the northern hemisphere. It is approximately the size of the United States, and several kilometers above the mean radius of the planet. It has a high mountain, *Maxwell Montes*, which is 12 km above the mean radius (as compared with 9 km for Mt Everest on Earth). The western part is a plateau some 2500 km across. The other continent, *Aphrodite Terra*, is twice as large as Terra Ishta, and

has a rougher terrain. It has a large highland region with deep (3 km) depressions, hundreds of kilometers wide and 1000 km long.

Photographs from the surface show angular rocks. This was surprising, since it was expected by some that the thick atmosphere would lead to considerable erosion, smoothing the rocks. However, a very low wind speed has been found on the surface, so the erosion is not as great as originally thought. The soil is basalt, providing additional evidence of volcanic activity. The surface temperature is roughly constant at 750 K.

Even before Viking, many of the early myths about Mars had been dispelled. Mariner 4 showed a cratered surface. Mariners 6 and 7 showed some signs of erosion. Mariner 9 arrived during a planet wide dust storm. We now know that it is the dust storms that we have been interpreting as seasonal changes in vegetation. When the dust cleared, the Mariner 9 cameras showed a variety of features, including volcanoes, canyons, craters, terrace areas, and channels. Many of these are shown in Fig. 24.7. Surface views from landers are shown in Fig. 24.8.

The largest volcano, *Olympus Mons*, rises 25 km and is some 600 km across at the base. This means that it has the very shallow slopes typical of shield volcanoes on Earth. An example of such volcanoes is the Hawaiian chain. However,

Olympus Mons is much higher than Mauna Kea, an impressive statistic, in view of the fact that Earth is larger than Mars. We think that this tells us that there are no plate tectonics on Mars. On Earth, the moving plates keep the sites of volcanic eruptions moving. New lava goes into

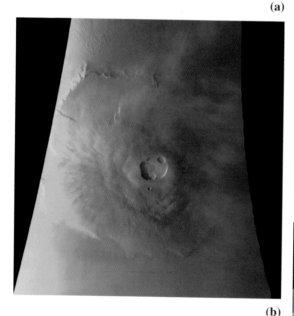

(a)

(b)

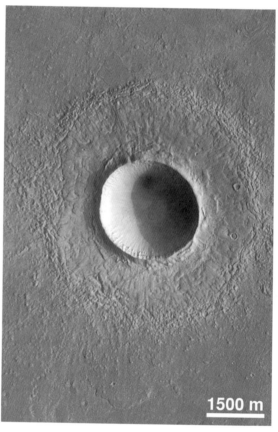

1500 m

(c)

Fig 24.7. Aerial views of Mars. (a) Valles Marinaris, from the Mars Global Surveyor. On the right is a close-up of the area outlined on the left. (b) Overhead view of Olympus Mons, in true color, from the Mars Global Surveyor. (c) Martian crater on northern Elysium Planitia, from the Mars Global Surveyor. This is twice the diameter of the Arizona meteor crater. (d) Three-dimensional view of northern polar cap, from the Mars Global Surveyor.

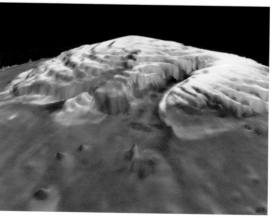

(d)

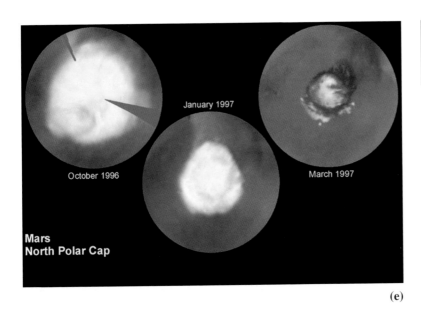

October 1996

January 1997

March 1997

Mars
North Polar Cap

(e)

making new mountains, not making old mountains larger. Therefore, we see a chain in Hawaii, but only one large mountain on Mars.

The channels are interesting, because they may have held water in the past. If they did, it must have been long ago, since the channels are cratered. This means that water has not eroded the channels since the heavy cratering in the Solar System, possibly 4 billion years ago.

Mars shows a difference between its northern and southern hemispheres. The southern hemisphere has more craters and is 1 to 3 km above the mean radius of the planet. It also seems to have an old part, with many craters, and a less old part. The northern hemisphere has volcanic plains around large volcanoes.

An interesting large feature is the *Tharsis ridge*, which is a large bulge. It is the largest area of

(a)

(b)

Fig 24.8. Surface views of Mars. (a) Enhanced color image of Viking 1 landing site. (b) Ice at Viking 2 landing site, Utopian Planitia.

Fig 24.8. (*Continued*) (c) Trench excavated by Viking 1 surface sampler. (d) Twin peaks at Pathfinder landing site. (e) Flatlands as seen by Pathfinder rover. (f) Sojourner rover tracks in compressible soil. [NASA]

(c)

(d)

(e)

(f)

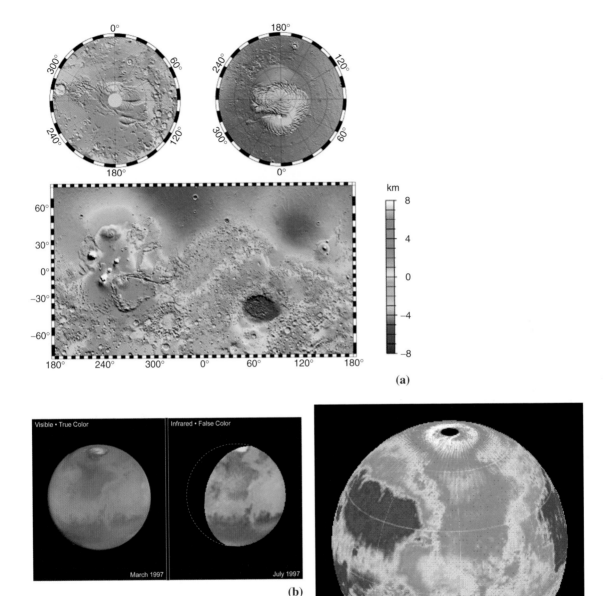

Fig 24.9. (a) Maps of Mars global topography as measured by Mars Global Surveyor orbiter. (b) Hubble Space Telescope visible and infrared images. (c) Global temperature map of Mars, as measured by the thermal emission spectrometer on the Global Surveyor orbiter. [(a), (c) NASA; (b) STScI/NASA]

volcanic activity, and has many young craters. It also has a number of rills and fractures. A large equatorial canyon, *Valles Marineris*, extends away from Tharsis. There is evidence for wall collapse and channel formation. The whole ridge may be uplifted crust.

The seasonal changes in the polar caps have also been of considerable interest. It has long been speculated that water is stored there in the winter and released in the summer. The release is by sublimation, a phase change directly from the solid to gas phase. However, the caps never sublime totally, with some part remaining through the summer.

It now appears that the part that never sublimes is composed of water. The growth of the caps in the winter is due to the freezing of CO_2. (On Earth, we call frozen CO_2 dry ice; it is used to keep things cold because it has a lower freezing point than water.)

24.3 | Interiors

24.3.1 Basic considerations

Since we cannot directly observe the interior of a planet, we must come up with indirect methods for determining the interior structure. We briefly go over some types of evidence that we can use for studying planetary interiors.

The average density of a planet can give us information. For example, consider the simple structure shown in Fig. 24.10. The planet has a core with density ρ_C and radius R_C, and a mantle with density ρ_M and radius R_M, the radius of the planet. In this case the mass of the planet is

$$M_P = \left(\frac{4\pi}{3}\right)[R_C^3 \rho_C + (R_P^3 - R_C^3)\rho_M]$$

The average density of the planet is its mass, divided by its volume,

$$\langle \rho \rangle = \frac{M_P}{\left(\frac{4\pi}{3}\right)R_P^3}$$

$$= \rho_C \left(\frac{R_C}{R_P}\right)^3 + \rho_M\left[1 - \left(\frac{R_C}{R_P}\right)^3\right]$$

(24.3)

If we know the material that is likely to make up the mantle and the core, we can estimate ρ_M and ρ_C. The average density is easily determined, so we can find (R_C/R_P).

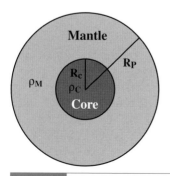

Fig 24.10. Planet with core and mantle.

Additional information can come from the rotational inertia, I, since it depends on how the mass is distributed. We can measure the rotational inertia by seeing the effects that perturbing torques have on the planet. The rotational inertia for the case shown in Fig. 24.10 is

$$I = \left(\frac{2}{5}\right)\left(\frac{4\pi}{3}\right)[R_C^3 \rho_C R_C^2 + R_P^3 \rho_M R_P^2 - R_C^3 \rho_M R_C^2]$$

$$= \left(\frac{8\pi}{15}\right)[\rho_C R_C^5 + \rho_M R_P^5 - \rho_M R_C^5]$$

(24.4)

$$= \left(\frac{8\pi}{15}\right)R_P^5\left\{\rho_C\left(\frac{R_C}{R_P}\right)^5 + \rho_M\left[1 - \left(\frac{R_C}{R_P}\right)^5\right]\right\}$$

For planets on which we can place seismometers, we can learn about interior activity. We can also learn about how seismic disturbances move through the interior. For example, one Apollo experiment involved allowing a spent Lunar Module to crash into the Moon, going on to measure the seismic disturbances with seismometers that had been left behind. Seismic waves travel at different speeds through different materials. They will also be reflected from boundaries between different materials. Analyzing these waves can tell us about the composition and size of various interior sections.

Additional information comes from the magnetic field, which would require a molten iron core, and some rotation to stir the core. Heat flow measurements can tell us how close radioactive material is to the surface. This tells us whether the mantle is well mixed or differentiated.

In Fig. 24.11 we show the interior structures of the four inner planets. In order to make the

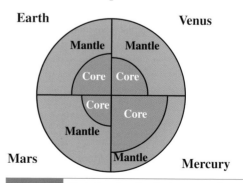

Fig 24.11. Model interiors for the four inner planets. In each case, the sizes are relative to the planet's radius, so the total radius of each planet is the same in this figure.

comparison more meaningful, we have scaled the results to the size of the planet. In each case we can see what fraction of the interior is occupied by the core and what fraction by the mantle.

We can also carry out theoretical modeling of planetary interiors, just as we do for stellar interiors. Just as in stars, planetary interiors must be in hydrostatic equilibrium, meaning that

$$\frac{dP}{dr} = -\frac{\rho G M(r)}{r^2}$$

where $M(r)$ is the mass interior to r. If the density can be approximated as being constant, then

$$M(r) = \left(\frac{4\pi}{3}\right)\rho r^3 \qquad (24.5)$$

so the equation of hydrostatic equilibrium takes the form

$$\frac{dP}{dr} = -\left(\frac{4\pi}{3}\right)G\rho^2 r \qquad (24.6)$$

We can use equation (24.6) to estimate the central pressure of a planet (just as we did for the Sun), by taking the pressure at the surface to be zero and integrating:

$$\int_0^{P(r)} dP' = -\left(\frac{4\pi}{3}\right)G\rho^2 \int_{R_P}^{r} r'\, dr'$$

to get

$$P(r) = \left(\frac{2\pi}{3}\right)G\rho^2 (R_P^2 - r^2)$$

In comparing this result with more sophisticated models of the Earth, we find that it underestimates the central pressure by a factor of about 2. (Among other things, the assumption of constant density is an oversimplification.)

24.3.2 Results

Even though Mercury's surface looks like that of the Moon, Mercury's density is higher. We estimate that an iron core makes up about 70% of Mercury's mass. This is a greater percentage than for either the Earth or the Moon. This is because Mercury formed under higher temperature conditions (being closer to the Sun) than did the Earth or Moon. The radius of the core is about 1800 km, leaving a 700 km mantle. The core radius is 72% of that of the planet.

A surprising result of the Mariner 10 studies was the discovery of a weak magnetic field around Mercury. This field is about 1% the strength of the Earth's field. It is surprising since Mercury's low mass would suggest that the core is not hot enough to be molten. Also, the rotation is so slow that the core is not stirred up very much. There is also radioactivity coming from near the surface, suggesting a differentiated interior. This differentiation would have also required a period of melting of the interior.

The surface plains may have resulted from volcanic flooding, suggesting an active past. Mercury also has scarps, unlike the Moon. These scarps may have resulted from a contraction of the surface, something that would have also required a molten history for the planet.

The interior structure of Venus is believed to be very similar to that of the Earth. There are some composition differences. We think that the core of Venus formed later than that of the Earth. The lithosphere is also about twice the thickness of that of the Earth. Venus also has no measurable magnetic field. This may also be an effect of the planet's slower rotation.

The density of Mars is 3 g/cm^3, much lower than that of the Earth. This suggests that the core cannot be very large. It is about 1200 km in radius, meaning that it is only about 40% of the planetary radius. The core is probably a combination of iron and iron sulfide. If it is all iron, it is probably even smaller than currently estimated. We have already said that the existence of large volcanoes indicates that there are no plate tectonics, also arguing for a cooler interior than the Earth.

24.4 | Atmospheres

The atmospheric compositions of Venus, Earth and Mars are shown in Fig. 24.12.

Not much of an atmosphere was expected on Mercury, and only a small amount of gas was found. The surface pressure is 10^{-15} atmospheres. The gas was detected by ultraviolet spectroscopy.

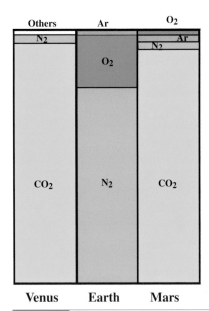

Fig 24.12. Diagram showing the atmospheric compositions of Venus, Earth and Mars.

This small amount of gas is 98% helium. This light atom should have escaped long ago. This suggests that it is being replaced continuously. Two possible sources are the solar wind and certain types of radioactive decay. Most of the remaining 2% of the gas is hydrogen. This may also come from the solar wind. In addition, there are small traces of oxygen, carbon, argon, nitrogen and xenon.

The atmosphere of Venus is very interesting, especially in comparison with that of the Earth. It was first observed in 1761 by the Russian astronomer Lomonosov, who saw the backlighted atmosphere when Venus passed between the Earth and across the Sun.

The surface pressure is 90 atmospheres, much higher than that of the Earth. The atmosphere is 96% carbon dioxide (CO_2) and 3.5% nitrogen (N_2). The total amount of nitrogen on Venus is comparable to that in the Earth's atmosphere. However, on Earth, nitrogen is the primary constituent. Venus also has very small quantities of water, sulfur dioxide, argon, carbon monoxide, neon, hydrogen chloride and hydrogen fluoride.

The large amount of CO_2 produces a very strong greenhouse effect on Venus. The 750 K is some 400 K higher than the temperature would be without an atmosphere. It is intriguing that two planets could start out so close in conditions and end up so different. The crucial difference between Venus and Earth seems to be the extra sunlight, making Venus initially somewhat warmer than the Earth. On Earth, water condensed, whereas on Venus it remained as a gas, and escaped. On Earth, the water kept the CO_2 bound up in the rocks, in the form of various carbonates. With the water on Venus this couldn't happen, and the CO_2 stayed in the atmosphere. (The amount of CO_2 in the rocks on Earth is comparable to the amount of CO_2 in the atmosphere of Venus.)

Once Venus had more CO_2 in its atmosphere than the Earth did, the greenhouse effect heated the lower atmosphere. This heating released more CO_2 into the atmosphere, increasing the greenhouse effect. This situation is called a *runaway greenhouse effect*. A small difference in initial conditions ends up with a large difference in final conditions.

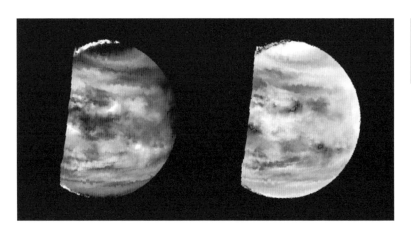

Fig 24.13. Images of low level clouds on Venus as seen in the near infrared by the Galileo orbiter. [NASA]

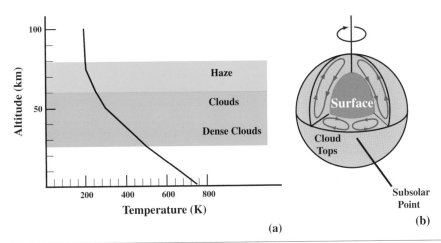

(a)

(b)

Fig 24.14. (a) Temperature distribution in the atmosphere of Venus. The region below the dense clouds is clear. (b) Atmospheric circulation on Venus. There is only one cell from the equator to the pole. There is a break in the east–west flow, with the air rising at the point receiving the maximum solar energy.

In Fig. 24.14(a) we show the temperature at different altitudes. Above the clouds (about 50 km above the surface), the greenhouse effect is no longer very strong, and the temperature drops to about 300 K. The clouds are made up of sulfuric acid. The sulfuric acid droplets are about 2 μm. There is a lot of sulfur in the atmosphere because there is no water to remove it. The haze above these clouds enhances the greenhouse effect. There are three cloud layers between 48 and 80 km above the surface. The presence of some water in the lower altitudes has washed the sulfuric acid away, and the lower atmosphere is clear. Many were surprised by the clarity of the Venera pictures from the surface. There is also some lightning below the clouds.

There is a general westward circulation of the winds, as shown in Fig. 24.14(b). The wind speed is a modest 1 m/s near the ground, growing to a substantial 100 m/s near the cloud tops. There is a general pole to equator circulation, but it is not broken into cells as it is on Earth. This circulation minimizes the temperature difference between the equator and the poles to a few kelvin.

The surface pressure on Mars is 0.007 atmosphere. This is sufficient for the atmosphere to be of some importance. The atmosphere is 95% CO_2, leading to a small greenhouse effect. The temperatures are raised by about 5 K above the value they would have without an atmosphere. The atmosphere is 2.7% nitrogen and 1.6% argon. There are traces of oxygen, carbon monoxide, water vapor, neon, krypton, xenon and ozone. There are blue/white clouds, made up of carbon dioxide and water. Most of the water, however, is tied up permanently in the northern polar cap. We think that Mars had a more plentiful atmosphere in the past, possibly with a pressure equal to that on Earth. At that time, there was probably more water. The earlier atmosphere probably came from an era of volcanic activity.

Strong winds sweep over the planet, lifting large amounts of dust into the air. There are a

4/30/1999 00:10:10 UTC

Fig 24.15. Clouds on Mars. A mid-summer storm in the northern hemisphere. [NASA]

number of small dust storms and a few global dust storms over the Martian year. Some atmospheric scientists have pointed to dust storms on Mars as providing a test of theories that say that the aftermath of a nuclear war on Earth would produce enough dust in the atmosphere to cool our planet (the so-called "nuclear winter").

The density in the Martian atmosphere is so low that the winds are not efficient at energy transport. Therefore, large temperature variations can exist across the planet. For example, in the winter the poles are as cold as 150 K, which is cold enough to freeze CO_2. At its maximum extent, the CO_2 cap extends almost halfway to the equator. The similarity between the rotation period of Mars and Earth produces similarities in the circulation patterns. However, since there are no oceans on Mars, and since the atmosphere is so thin, the surface responds more quickly to heating changes. Therefore, the hottest point on Mars is always the point closest to the Sun, the *subsolar point*. This results in a circulation pattern in which one large cell spans the equator. There is also a large day/night temperature difference, a few tens of kelvin.

The Viking landers also carried out a search for microscopic life on Mars. Three different experiments were carried out to try to detect changes in small samples of material as a result of metabolism of microscopic life. Evidence for some unusual chemical reactions was found, but there is no evidence for living organisms.

Fig 24.16. View of Phobos from the Mars Global Surveyor orbiter. The largest crater, Stickney, is 10 km in diameter. [NASA]

24.5 | Moons

No moons have been found for Mercury and Venus. Mars has two moons that are much smaller than the Earth's. One is shown in Fig. 24.16. Phobos is 22 km across, and orbits Mars in 7 hr 40 min. Demos is 15 km across and orbits in 30 hr. Since one moon orbits faster than Mars's rotation period and the other is slower, from the surface they appear to move across the Martian sky in opposite directions. Since the orbital period of Demos is close to Mars's rotation period, an observer on Mars would see that it takes 137 hr to make a complete cycle of the sky. It takes almost three days to go from rising to setting, and then another three days to reappear.

Each moon has an irregular shape. They appear dark, like certain asteroids. This has led to conjecture that these two satellites were asteroids that were captured by Mars.

Chapter summary

In this chapter we looked at the properties of the inner planets of the Solar System.

We saw what could be learned from ground-based observations. We saw how some of the observations could lead to misinterpretations, such as the rotation period of Mercury. We saw how radar mapping is a useful tool, both from the Earth and from spacecraft.

We discussed the wealth of information that has been obtained by flybys, orbiters and landers. Venus and Mars have been visited extensively. In the case of Venus, the spacecraft have allowed us to probe the dense cloud cover that hides the surface.

We have seen certain similarities among the inner planets. For example, there is extensive cratering on Mercury, Mars and our Moon. However, there are also many differences. For example, the Earth is the only planet to have an interior structure that produces plate tectonics. We see very thin atmospheres on Mars and Mercury, while Venus has a much thicker atmosphere than the Earth. Of course there is a large runaway greenhouse effect on Venus, probably initiated by the fact that, with a slightly higher temperature than the Earth, the CO_2 was not bound into the rocks.

Questions

24.1. How do the sizes of the four inner planets compare? Where does the Moon fit in?

24.2. Of the inner planets, which have surfaces that are most affected by the atmosphere?

24.3. Why is it not surprising that the surface of Mercury has similarities with that of the Moon?

24.4. Once we have a spacecraft orbiting Venus, why is it better to make a radar map of the surface than to take a series of photographs?

*24.5. Why do you think that radar observations of a planet become more difficult as the planet is farther from Earth?

24.6. The best Mariner 10 maps of Mercury have a resolution of 50 m. If a spacecraft made a map of the Earth with the same resolution, what types of structures could be discerned?

24.7. If Mercury has no atmosphere, what is the mechanism for erosion of the surface?

24.8. What is the difference between continents on Venus and on Earth?

24.9. Venus is often called the Earth's sister planet. What features do the Earth and Venus have that are similar? Which features are very different?

24.10. Compare the largest mountains on Venus, Earth and Mars.

24.11. If the atmosphere of Mars is so thin, why is there so much erosion of the surface?

24.12. What causes the seasonal variations in the appearance of Mars?

24.13. What is the evidence that there was once water on Mars?

24.14. What is the evidence that there are no plate tectonics on Mars?

24.15. What makes the polar caps on Mars change their appearance from summer to winter?

24.16. How do we learn about the interior of a planet?

24.17. Compare the significant features of the interiors of the four inner planets.

24.18. Why was the discovery of a magnetic field on Mercury surprising?

24.19. Describe the runaway greenhouse effect that took place on Venus.

24.20. The Earth and Venus are similar in many ways, yet their atmospheres evolved so differently. Why is that?

24.21. How might an increase of carbon dioxide in the Earth's atmosphere produce a greater greenhouse effect?

24.22. Contrast the clouds on Venus with those on Earth.

Problems

24.1. Compare the tidal effects of the Sun on Venus with the tidal effects of the Sun and Moon on the Earth.

24.2. For radar mapping of a planet, what time and frequency resolution are needed in measur-ing the time delays and Doppler shifts to provide a resolution of 1 km/s on the surface?

24.3. How much would a 1 GHz radar signal be spread in frequency by the rotation of Mercury?

24.4. Approximate the central pressures of the four inner planets and compare them.

24.5. How far under water to we have to go on Earth to obtain a pressure of 90 atmospheres?

24.6. Using the mass, radius and core radius of the Earth (given in Chapter 23), find the ratio of the core material to the density of the mantle material.

24.7. Estimate the adiabatic temperature gradients near the surfaces of Venus and Mars.

24.8. To what altitude do we have to go in the Venus atmosphere to reach a pressure of 1 atmosphere?

24.9. Show that if a point on the equator of a planet moves with speed v_0 due to rotation, a point at latitude φ moves with speed $v_0 \cos \varphi$.

Computer problems

24.1. Make a graph of pressure vs. altitude in the Venus atmosphere. Assume that the surface pressure is 90 atmospheres, the main constituent is CO_2, and that the temperature is a constant 600 K.

24.2. Make a graph of pressure vs. altitude in the Mars atmosphere. Assume that the surface pressure is 0.007 atmospheres, the main constituent is CO_2, and that the temperature is a constant 200 K.

Chapter 25

The outer planets

In the outer planets, we find a considerable contrast with the four inner planets. We therefore study them as a group, comparing surfaces, interiors and atmospheres. The relative sizes of the outer planets (and Earth) are shown in Fig. 25.1.

25.1 | Basic features

Jupiter, shown in Fig. 25.2, is by far the most massive planet in the Solar System. It is 318 times as massive as the Earth, and is a respectable 0.1% as massive as the Sun. (The rest of the planets together only have 129 Earth masses.) Jupiter's density is much lower than that of the inner planets, 1.3 g/cm^3 vs. 5.4, 5.3, 5.5 and 3.9 for Mercury, Venus, Earth and Mars, respectively. Its density is only slightly greater than that of liquid water. This suggests that the composition of Jupiter is basically different from that of the inner planets. This is due, in part, to the larger gravity, 2.54 g at the cloud tops. The larger gravity means that the lighter gases have been retained.

The atmosphere is 85% hydrogen and 15% helium, with a variety of trace constituents. This composition is much closer to that of the Sun than it is to the inner planets. The motions in the atmosphere are affected by the planet's rapid rotation. The period is 9.92 hr at the equator. The rotation period is greater at the poles. The rapid rotation produces a large coriolis force. This is manifested in the appearance of bands and spots, such as the *Great Red Spot*, which is 14 000 by 30 000 km, and has persisted for centuries.

The energy output of Jupiter is interesting. It seems to radiate over 50% more energy than it receives from the Sun. In addition it is also a strong source of non-thermal (synchrotron) radio emission.

Space exploration has greatly improved our view of Jupiter. Pioneers 10 and 11 and Voyagers 1 and 2 have provided us with spectacular images as well as a variety of other observations. These close-up observations have also added to the already known extensive moon system. They also revealed a ring around the planet.

Saturn's mass is 30% that of Jupiter, or 95.2 times that of the Earth. It has the lowest density of the planets, 70% the density of liquid water. With this low density, its acceleration of gravity at the cloud tops is only slightly greater than that on Earth, 1.07g. Its composition is similar to that of Jupiter.

Like Jupiter, it rotates rapidly, in 10.7 hr at the equator, and slower at the poles. Through a telescope, we can see bands in the cloud structure. However the bands do not show as much contrast as those of Jupiter.

Of course, Saturn (Fig. 25.3) is best known for its prominent rings. From Earth, we see the rings as three main structures, and can deduce that they are very thin. Like Jupiter, Saturn has an extensive moon system, with 18 moons having been identified to date. Pioneers 10 and 11 as well as Voyagers 1 and 2 have revealed a great complexity in the structures of the rings in addition to surprising views of the moons.

Uranus (Fig. 25.4) is the closest planet that has not been known since ancient times. It was

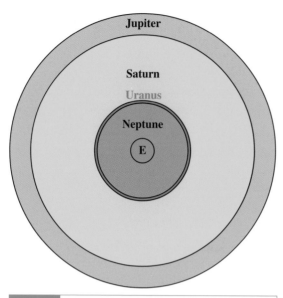

Fig 25.1. Relative sizes of Jupiter, Saturn, Uranus, Neptune and Earth.

Notice the very faint contrast between the different bands.

The discovery of Neptune (Fig. 25.5) makes an interesting story. After following the orbit of Uranus, astronomers found that it did not move exactly in its predicted path. The perturbation

(a)

(b)

Fig 25.2. (a) HST image of Jupiter. (b) Voyager 1 image of Jupiter taken from a distance of 54 million km. [(a) STScI/NASA; (b) NASA]

discovered in 1781 by William Herschel. Its mass is considerably less than that of Jupiter and Saturn, but at 14.6 Earth masses, it is definitely not in the class of the inner planets. Its density is 1.2 g/cm^3, close to that of Jupiter and Saturn. The difference in sizes reflects a difference in the composition of Uranus and Jupiter. Its surface gravity is slightly less than that of the Earth, at 0.87 g.

From ground-based telescopes, we can tell that Uranus has a high albedo. That is, it reflects a lot of sunlight. This suggests a coating of clouds. It is difficult to identify features for the purpose of measuring the rotation period. Values from 12 to 24 hr have been proposed, but the most recently accepted value is 16 hr. The rotation axis is tipped so much that it is almost in the plane of its orbit. The rotation is retrograde. It is in the opposite direction from the planet's orbital motion. Infrared observations of the clouds suggest a surface temperature of 58 K, a very cold place.

Uranus has five main moons and a number of smaller ones. In addition, a ring system has been discovered accidentally during an ocultation of a star by Uranus. We had our first close-ups from space when Voyager 2 flew by in January 1986. Some images from Voyager 2 are shown in Fig. 25.4.

(a)

(b)

Fig 25.3. (a) HST image of Saturn. (b) Saturn image from Voyager 1, at a distance of 31 million km. [(a) STScl/NASA; (b) NASA]

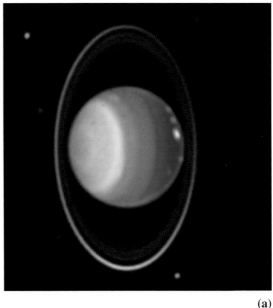

(a)

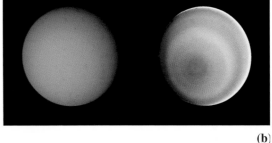

(b)

Fig 25.4. (a) HST image of Uranus. (b) Uranus taken from Voyager 2, from a distance of 9.1 million km. The left image shows Uranus as it would appear to the eye. The picture on the right is enhanced to show the cloud cover. [(a) STScl/NASA; (b) NASA]

of a planet beyond Uranus was suspected. Calculations of the possible location of the planet were carried out independently in 1845 by *John C. Adams*, in England, and *Urbain Leverier*, in France. Adams presented his calculations to the Astronomer Royal, who was not impressed, and did not carry out the easy observations that would have been necessary to test the idea. Leverier had no better luck in France. Finally, based on Leverier's calculations, *Johannes Galles*, in Germany, carried out the observations, and found Neptune in 1846. We credit both Adams and Leverier with the successful prediction. As an interesting aside, recent readings of Galileo's notes indicate that he may actually have observed Neptune, noting its changing position relative to

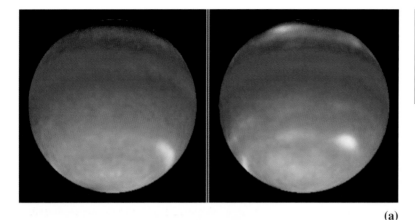

Fig 25.5. (a) HST images of Neptune (showing opposite hemispheres). (b) Voyager 2 image of Neptune, from a distance of 6.1 million km. [(a) STScI/NASA; (b) NASA]

(a)

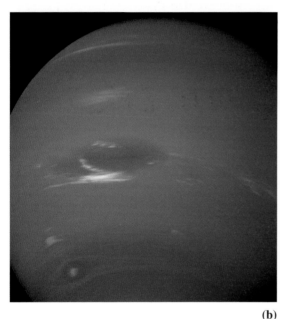

(b)

25.2 | Atmospheres

Jupiter's atmosphere contains hydrogen and helium in the same proportion as in the Sun's atmosphere. This suggests that Jupiter has its original atmosphere. With its large mass, it has been able to hold even the lightest atoms. A number of minor constituents have been identified as well. NH_3 (ammonia) and CH_4 (methane) are the most prominent. In addition, C_2H_6 (ethane), C_2H_2 (acetylene), H_2O, PH_3, HCN (hydrogen cyanide) and CO (carbon monoxide) have been identified.

The temperature distribution is shown in Fig. 25.6. The temperature is 125 K at the cloud tops. As you go down from there, the temperature increases by about 2 K for every kilometer that you drop. Above the cloud tops, the pressure increases as you go up. The emitted radiation is approximately the same at all latitudes. This is true despite the fact that the solar heating is greatest at the equator. This may mean that winds are effective at distributing heat from the equator to the poles. Such large winds would require large temperature differences to drive them. We do not see these differences in the upper atmosphere. This means that the transport must take place in the lower atmosphere. Another possibility is that Jupiter has an internal heat source that supplies more heat to the poles than to the equator.

Example 25.1 Energy from Jupiter
Compare the energy given off by Jupiter with the energy it receives from the Sun. Assume that Jupiter radiates like a 125 K blackbody, the temperature of the cloud tops.

the stars, but did not have enough observations to identify it as a new planet.

Neptune's mass is 25.2 times that of the Earth, similar to that of Uranus. From occultations, we can tell that its radius is 3.88 Earth radii. From these numbers, its density turns out to be 1.6 g/cm^3, slightly greater than that of Uranus. The acceleration of gravity on its surface is slightly greater than on Earth at 1.14g. Its rotation period is also hard to determine, with published values ranging from 17 to 26 hr. The currently accepted value is 25.8 hr. We can deduce the presence of an atmosphere by the rate at which starlight dims during occultations. Neptune has two larger and six smaller moons.

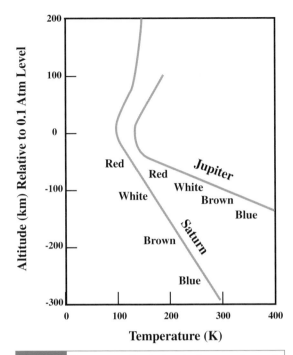

Fig 25.6. Temperature distributions in the atmospheres of Jupiter and Saturn. Since we don't know where the surfaces are, we use as our reference point the place where the pressure is 0.1 atmosphere. The colors written along each curve indicate the dominant color material at that temperature.

SOLUTION

Jupiter's luminosity is given by its surface area $4\pi R_J^2$, multiplied by the total power per unit area σT_J^4, or

$$L_J = (4\pi R_J^2)(\sigma T_J^4)$$

The power received from the Sun is the solar luminosity $(4\pi R_\odot^2)(\sigma T_\odot^4)$, divided by the area of a sphere at the distance of Jupiter from the Sun, $4\pi d^2$, where d is the distance of Jupiter from the Sun, and multiplied by the projected area of Jupiter πR_J^2, or

$$P_{rec} = \frac{(4\pi R_\odot^2)(\sigma T_\odot^4)(\pi R_J^2)}{(4\pi d^2)}$$

Taking the ratio of these gives

$$\frac{L_J}{P_{rec}} = 4\left(\frac{T_J}{T_\odot}\right)^4\left(\frac{d}{R_\odot}\right)^2$$

$$= 4\left(\frac{125}{5500}\right)^4\left(\frac{(5.2)(1.5\times10^8\text{ km})}{7\times10^5\text{ km}}\right)^2$$

$$= 1.3$$

The situation is even worse than this, because Jupiter doesn't absorb all of the sunlight that strikes it. This indicates that Jupiter gives off more radiation than it receives from the Sun. Jupiter must have an internal energy source.

The fact that the temperature fall-off with altitude below the clouds is close to the adiabatic rate suggests that convection is an important form of energy transport. (Remember, we saw that, on Earth, when there is a rapid drop in temperature with altitude, then convection can be very strong.) Because of the temperature distribution, we think that there are three major cloud layers, resulting from the fact that different constituents condense at different temperatures and pressures. We think that the highest cloud layer is ammonia ice, the middle layer is ammonium hydrosulfide (NH_4SH), in the form of crystals, and the lowest layer is water, in a mixture of liquid and ice.

The varying conditions mean that the chemistry is different at different altitudes. We think that the different cloud colors reflect different compositions, a result of the varying chemistry. The different colors come from different temperature ranges, and therefore from different levels. For example, the blues correspond to the warmest regions, and are therefore closest to the surface. They are probably only seen through holes in the higher layers. Brown, white and red come from progressively lower temperatures, meaning they come from progressively higher levels. We have still not identified all of the compounds responsible for the various colorations. Other factors besides temperature also affect the chemistry. For example, some regions have more lightning than others, and the energy from the lightning can help certain chemical reactions go.

The east–west winds are quite substantial. They flow at about 100 m/s near the equator, and at about 25 m/s at the higher latitudes. The winds flow in alternating east–west and west–east bands. These alternating wind patterns correspond to the alternating color bands. On Jupiter, there are five or six pairs of alternating bands in each hemisphere. For comparison, on Earth, there are only the westward (in the northern hemisphere) trade winds at low latitudes and the eastward jet stream at high latitudes.

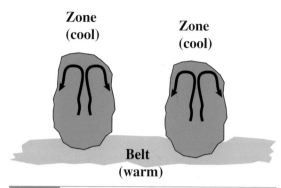

Zone (cool)   **Zone (cool)**

Belt (warm)

Fig 25.7. Convection patterns that produce the belts and zones.

The bright colored bands are called *zones*. They appear to have gas rising, as shown in Fig. 25.7. The dark colored bands are called *belts*. They appear

to have gas falling. The tops of the belts are about 20 km lower than the tops of the zones. The belts are brighter in the infrared, indicating a higher temperature. The stability of the large-scale cloud patterns has not been understood. The locations of the bands are constant, even though their colors occasionally change. Close-up pictures have shown small circular regions, called *eddies*, which are shown in Fig. 25.8. We generally associate eddies with the dissipation of energy. We would therefore expect the patterns to wash out. The fact that the east–west bands are stable indicates that their flow must continue deeper into the atmosphere than the eddies.

The most famous cloud feature on Jupiter is the Great Red Spot, shown in Fig. 25.8. It has been observed for more than 300 years. It covers more

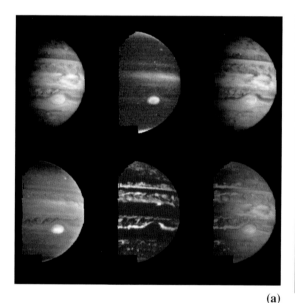

(a)

(b)

Fig 25.8. Images of Jupiter's atmosphere. (a) Multilevel clouds from near IR images from Galileo orbiter. These images are at different IR wavelengths, and the differences are because of the different opacities at each wavelength. The top left and right images are at 1.6 and 2.7 μm, respectively, and show relatively clear views deep into the atmosphere. The middle image is at 2.2 μm and shows high altitude clouds and haze. The lower left and center images are at 3 and 5 μm, showing deeper clouds. The false color image at the lower right is a composite showing the clouds at different layers. (b) True color mosaic of the belt–zone boundary. (c) The Great Red Spot, from Galileo orbiter images at four wavelengths: upper left is violet (415 nm); upper right is IR (757 nm); lower left is IR (732 nm); lower right is IR (886 nm) at a wavelength of strong methane absorption.

(c)

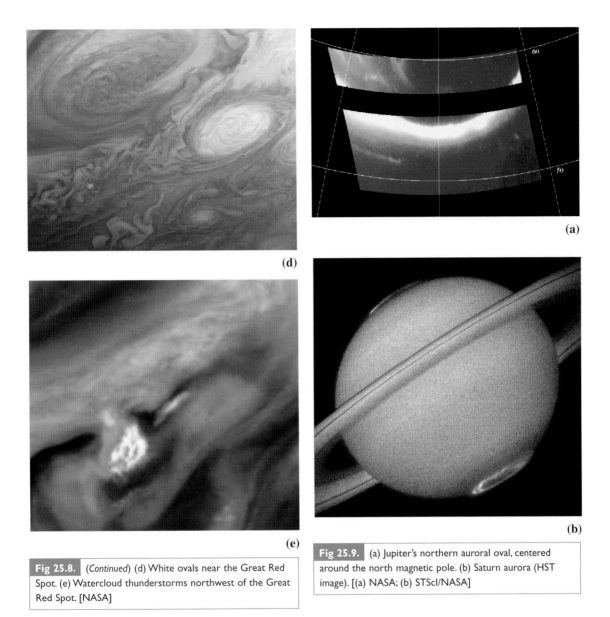

(d)

(a)

(e)

(b)

Fig 25.8. (*Continued*) (d) White ovals near the Great Red Spot. (e) Watercloud thunderstorms northwest of the Great Red Spot. [NASA]

Fig 25.9. (a) Jupiter's northern auroral oval, centered around the north magnetic pole. (b) Saturn aurora (HST image). [(a) NASA; (b) STScI/NASA]

than 10° in latitude. It is also surrounded by white ovals, which have flows that would dissipate energy very quickly. In trying to explain its stability, there are two questions that must be answered. (1) How can it be maintained for so long as a stable fluid flow? (2) What is the energy source to replace the energy lost in the eddy flow around the spot? The most successful model has been to say that the spot is analogous to a hurricane on Earth. Large vertical convection currents allow it to draw energy from the latent heat of condensing materials below. On Earth, hurricanes draw energy from that released when water vapor is condensed. This is why hurricanes intensify when they pass over large bodies of water. There are different models for the details of how the red spot works, but it does seem that, with the conditions in Jupiter's atmosphere, such a storm should be stable for hundreds of years!

Jupiter's has a strong magnetic field, whose effects are felt far out into space (Fig. 25.9a). At the top of the cloud layer the magnetic field has ten times the strength of the Earth's field. As far as 7 million km in front of Jupiter, the solar wind is affected by this magnetic field, and is deflected

(a)

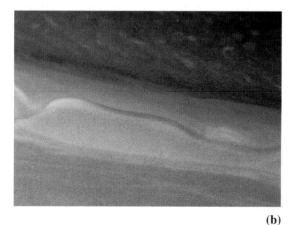

(b)

Fig 25.10. Images of Saturn's atmosphere. (a) Voyager 1 image of Saturn's red spot. (b) Ribbonlike structure (Voyager 1 image). [NASA]

to flow around the planet. The region with no solar wind stretches as much as 700 million km beyond Jupiter, and even Saturn passes through it! (By comparison, the Earth creates a region with no solar wind that is only 1% as long.) As with the Earth's magnetic field, Jupiter traps charged particles, creating an active ionosphere. This is the source of strong radio emission.

It should not be surprising that we find many similarities between the atmosphere of Saturn (Fig. 25.10) and that of Jupiter. However, there are certain differences. These differences arise from the fact that Saturn is farther from the Sun than is Jupiter. Saturn's lower gravity and lower rotational speeds are also important.

The temperature is 95 K at the cloud tops. Saturn gives off approximately twice as much energy as it receives from the Sun. The temperature rises as one goes deeper into the atmosphere (as shown in Fig. 25.6). The rate of temperature change with altitude is about half of that on Jupiter.

The winds on Saturn are much greater than on Jupiter. They are about 450 m/s at the equator and about 100 m/s at higher latitudes. There is also an alternating pattern (as on Jupiter), with the speeds alternating between 0 m/s and about 100 m/s. Saturn does not have as many bands as Jupiter does. The lower temperature means that there are chemical differences. This is evidenced by the fact that the bands don't show as much contrast as those of Jupiter. Voyager showed the equivalent of the Great Red Spot on Saturn. It had not been seen from the Earth because of the lower color contrast.

The atmospheres of Uranus (Fig. 25.11) and Neptune (Fig. 25.12) are hydrogen rich, like those of Jupiter and Saturn. However, Uranus and Neptune contain a higher proportion of heavy materials. This is because those lower mass planets probably retained less of their original hydrogen than did Jupiter and Saturn.

The temperatures of Uranus and Neptune are almost the same, about 57 K. One would expect Neptune to be cooler than Uranus, but it appears to give off more heat than it receives from the Sun. It is thought that this may be related to the way in which the atmosphere traps the heat.

Both atmospheres contain methane. This molecule has been identified from its infrared spectrum. Both planets have a greenish color, and it is believed that this color comes from the methane. There is a difference in the cloud content of the two atmospheres. Uranus has an atmosphere that is cold and clear to great depths. There appear to be no clouds or haze in the lower atmosphere. There may be some higher up. Neptune has a variable haze. We think that the haze is composed of aerosol particles or methane ice crystals. A comparative study of these two atmospheres, especially

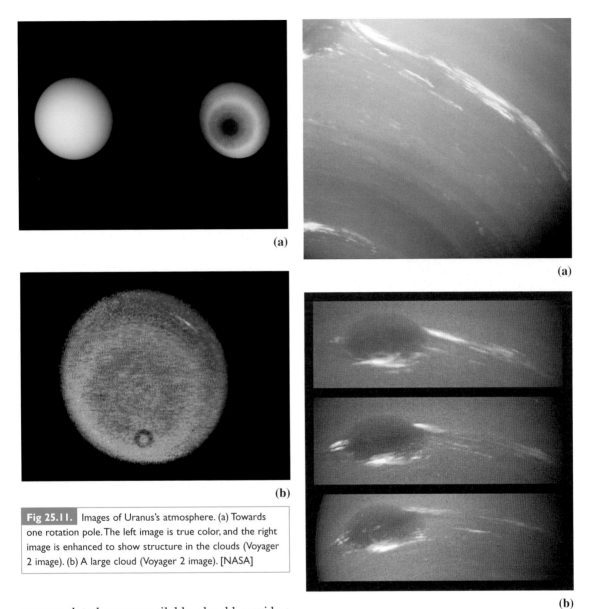

Fig 25.11. Images of Uranus's atmosphere. (a) Towards one rotation pole. The left image is true color, and the right image is enhanced to show structure in the clouds (Voyager 2 image). (b) A large cloud (Voyager 2 image). [NASA]

Fig 25.12. Images of Neptune's atmosphere. (a) True color image of clouds (Voyager 2 image). (b) Changes in great dark spot (Voyager 2 image). [NASA]

as more data become available, should provide a good test of our ability to model planetary atmospheres with enough detail to predict or explain the differences.

Voyager 2 provided a good view of Uranus from a distance of five planetary radii. There is very little structure in the atmosphere. However, there is a polar haze, probably composed of methane and acetylene. There is also an extended hydrogen (atomic and molecular) corona. Ultraviolet emission from this corona had previously been detected from Earth-orbiting satellites. The rotation period of the atmosphere appears to be just under 17 hr.

The wind speeds seem to be in the range of tens of kilometers per hour.

Voyager 2 also revealed an interesting temperature distribution. The equator is cooler than either of the poles. In addition, the pole facing the Sun is cooler than the shadowed pole. This is obviously an interesting problem in energy transport.

25.3 | Interiors

We cannot directly sample the interiors of these outer planets. However, we can use physical laws to construct computer models of the interior. We can then see which models produce results that agree with observations. In Fig. 25.13 we show the internal structures of Jupiter, Saturn, Uranus and Neptune. To allow comparison, they are all scaled relative to the size of the planet. In that way, we can compare what fraction of the interior is made up of the various sections.

For Jupiter, the outermost layer is a hydrogen–helium envelope. With that, extending from a radius 10 000 to 54 000 km, is a liquid region. This liquid is hydrogen. The pressure in this region rises to 40 million times the atmospheric pressure on the Earth's surface. Under these conditions, the hydrogen forms into a metal. This metallic hydrogen contains 73% of the planet's mass. Within the liquid region is the core. The core may be made of rock and ice materials, though it has been suggested that the hydrogen

continues right to the center. The pressure is 80 million bar; the temperature is about 25 000 K. The core contains only 4% of Jupiter's mass.

The transition from the gaseous region to the liquid region is probably a gradual one. The transition from the normal liquid hydrogen to the metallic hydrogen probably takes place over a small change of radius. We think that the interior has excess energy stored from the time of the collapse of the planet. The energy is so large that it has not all escaped yet. This is probably the source of the excess energy that Jupiter gives off.

The general structure of Saturn is probably very similar to that of Jupiter, as shown in Fig. 25.13. There are some differences, however. Saturn has a larger core, containing about 26% of the mass. The central pressure is 50 million bar, and the central temperature is about 20 000 K.

There is a smaller metallic hydrogen zone. The range of radii is from 16 000 to 28 000 km. The temperature in this zone ranges from 9000 to 12 000 K. This zone contains about 17% of Saturn's mass.

Our understanding of Saturn's excess heat is not as good as that for Jupiter's. Some other explanation is needed. It has been suggested that some of the energy comes from helium condensing and sinking through the less dense material towards the core.

Uranus and Neptune have higher densities than Jupiter and Saturn. This suggests a different composition. The cores are rock. The rock is mostly silicon and iron. Over the core is a mantle. This mantle probably contains liquid water, ammonia and methane. Over the mantle is a crust of hydrogen and helium. It may be in the form of high density gas. The central pressure is about the same for both planets, about 20 million bar. The central temperature is about 7000 K.

25.4 | Rings

Jupiter, Saturn, Uranus and Neptune have ring systems. Saturn's, shown in Fig. 25.14, has been known since Galileo, while those of Jupiter, Uranus and Neptune are recent discoveries. We first review the basic properties of the rings, and then consider the effects that are important in shaping them.

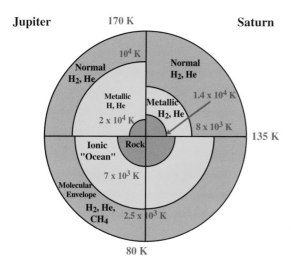

Jupiter   170 K   **Saturn**

Uranus   **Neptune**

Fig 25.13. Interiors of Jupiter, Saturn, Uranus and Neptune. These are from model calculations, and are expressed relative to each planet's radius. The numbers at each boundary are estimated temperatures.

(a)

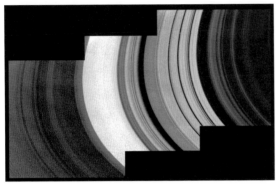

(b)

25.4.1 Basic properties

Saturn's rings (Fig. 25.14) shine by reflected sunlight. Therefore, the brighter areas are those with more material to reflect the light. Therefore, the brighter areas are those of higher optical depth, meaning a greater amount of reflected sunlight.

Three main rings are apparent. The A ring is the farthest from the planet. It has a width of 20 000 km, ranging from 2.02 to 2.27 times the planetary radius. Its thickness is less than 200 m. Approximately half of the light striking this ring is reflected back. The B ring is in the middle and is the brightest. It extends from 1.52 to 1.95 planetary radii. The A and B rings are separated by a gap, called *Cassini's division*. The innermost ring is the C ring. It is the darkest. It extends from 1.23 to 1.52 planetary radii.

Additional rings have been found from the ground and from spacecraft. The D ring is a faint ring inside the C ring (1.11 to 1.23 radii). The E ring is a very faint ring, outside the A ring (3 to 8 radii). The F ring is just beyond the A ring (2.37 radii), and was discovered by Pioneer 11. Finally, the G ring is a faint ring 2.8 radii from the planet.

The rings are composed of individual particles, rather than being solid structures. We can tell this from the spectra of the rings. The Doppler shifts vary across the rings, as shown in Fig. 25.15. These

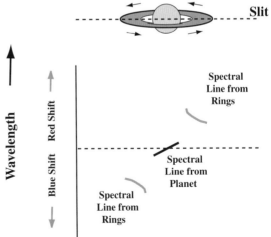

(c)

Fig 25.14. Saturn's rings. (a) Voyager 1 image of the unlit side, so the B ring appears black. (b) Voyager 1 mosaic. Cassini's division appears to the right of center as five bright rings with a substantial dark gap. (c) Spokes in the B ring (Voyager 1 image). [NASA]

Fig 25.15. Simulated spectra of Saturn's rings, showing the change in Doppler shift as one goes farther out in the ring. In the upper diagram the placement of a slit is indicated. Below, the schematic spectrum shows the Doppler shift of a particular line as one moves across the slit.

Fig 25.16. Jupiter's ring. Galileo orbiter image of main rings and halo. The top image is exposed to show the halo. [NASA]

Fig 25.17. Uranus's rings, detailed structure (Voyager 2 image). [NASA]

variations are those to be expected for individual particles orbiting Saturn, with orbital speeds being given by Kepler's third law. The particle sizes range from a few centimeters to 10 meters. The total mass of the rings is poorly known, but is estimated to be in the range 10^{17} to 10^{19} kg. This is only about one-millionth of the Moon's mass. Infrared observations suggest that the particles are ices of water and other molecules.

The Voyager flybys produce spectacular pictures of the rings. We find that there is an extensive pattern of detailed structure, as shown in Fig. 25.14. Each ring is divided into smaller rings.

The first evidence for rings around Jupiter, shown in Fig. 25.16, came from Pioneer 11 in 1974. The discovery was confirmed by Voyager 2. The rings are different from those of Saturn. They fall within a narrow range of radii, 1.72 to 1.81 planetary radii. Their width is 6000 km. Their thickness is less than 30 km. The rings are very faint. The rings appear smooth, except for a 600 km wide enhancement at 1.79 planetary radii.

The particles are much smaller than in Saturn's ring, being a few micrometers across. They have low albedos (reflectivities), suggesting that they are silicates, rather than ice. Such small particles are not expected to stay in the ring for very long, so it seems that they must be replaced continuously. It is possible that some ring material is being thrown off the closest large moon, Io, and from two embedded smaller moons.

The March 1974 discovery of a ring system around Uranus (Fig. 25.17) was a great surprise. An occultation of a star by Uranus was being observed. Some observations were being carried out from

the air, in the Kuiper Airborne Observatory. Normally, this observatory is used for infrared observations. However, in the case of the occultation, it was used for optical observations, with the airplane providing a means of getting the telescope to a favorable viewing point. Shortly before Uranus was supposed to block the starlight, a number of brief dips in the starlight were noticed. These dips were repeated after the occultation. They were the result of the rings passing in front of the star.

This observation, and many follow-up observations, have revealed a system of nine narrow rings. They extend from 1.60 to 1.95 planetary radii. Each ring is about 10 km wide, except for one which is about 100 km wide. The ring particles orbit with a range of eccentricities. There are also some changes in width. For example, one ring varies in distance from the planet by about 800 km, and in width from 20 to 100 km. The gaps between the rings are quite clear. The rings themselves have a low albedo, making them difficult to see.

Voyager 2 provided close-up information on the rings. New rings were discovered as well. Radio observations of one ring showed that the particles in it are probably larger than 30 cm across. Backlighted views show the regions away from the

Fig 25.18. Neptune's rings (Voyager 2 image). [NASA]

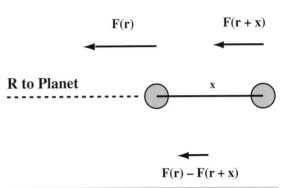

Fig 25.19. Roche limit. We compare the attractive forces between the two spheres with the difference between the forces that the planet exerts on each sphere.

known rings to be filled with very thin rings. The particles in these rings may be only a micrometer across. Voyager 2 also provided images of Neptune's rings; one is shown in Fig. 25.18.

25.4.2 Ring dynamics

These ring systems pose a number of interesting questions One obvious question is that of their origin. Other questions center around their structure. Why are they so thin? Why are there multiple rings? Why is the appearance so different from planet to planet? All of these questions relate to the dynamics of the particles in the rings. We must understand the forces that perturb the motions of the ring particles, and the response of those particles to those forces.

When we study rings, our first dynamical consideration is the role of tidal effects. The result is to cause different parts of an object to accelerate differently. If some particles are very close to a massive object, the tidal effects due to the massive object can prevent the particles from staying together under the influence of their mutual gravitational attraction. We define the *Roche limit* of a planet to be the minimum distance that an object can be from that planet and still be held together by *gravitational forces only*. We are within the Roche limit of the Earth, but we are not torn apart. That is because we are held together by electrical forces, not gravitational forces.

We can get a feeling for the Roche limit by looking at the simple situation in Fig. 25.19. We have an object represented by two small particles of equal mass m, a distance x apart. One particle is a distance R from a planet of mass M and the other object is a distance $R + x$ from the planet. We assume that $x \ll R$. The attractive force between the two particles is

$$F_{att} = G\frac{m^2}{x^2}$$

The tidal force F_{tid} is the difference between the forces exerted by the planet on the two objects. We can write this as

$$F_{tid} = \left(\frac{dF}{dR}\right)\Delta R$$

Since the force on the particle of mass m a distance R from the planet is

$$F = G\frac{mM}{R^2}$$

we have

$$\frac{dF}{dR} = -2G\frac{mM}{R^3}$$

This makes the tidal force

$$F_{tid} = -2G\frac{mMx}{R^3}$$

where we have taken $\Delta R = x$.

We find the limiting value of R, the Roche limit, by equating the magnitudes of the tidal force and the attractive force between the particles. This gives

$$2G\frac{mMx}{R^3} = G\frac{m^2}{x^2}$$

Simplifying, we have

$$2\frac{M}{R^3} = \frac{m}{x^3}$$

We further note that the mass of the two particles is $2m$, and the volume they occupy is approximately x^3, so m/x^3 is approximately the density ρ. Making this substitution and solving for R gives the Roche limit as

$$R_{\text{Roche}} = \left(\frac{4M}{\rho}\right)^{1/3} \qquad (25.1)$$

In calculating the Roche limit for a planet, we have to enter the value for the density of the material that is trying to hold itself together. Obviously, the greater the density, the closer it can venture to the planet without disruption. As a conservative limit, we often take the density of the planet itself. That is, we find the Roche limit for an object with the same density as the planet. In general, an object just forming around a planet will have a lower density than that of the planet, so it must be farther away than objects as dense as the planet.

We find that all of the ring systems lie within the Roche limits for their respective planets. This leads to the idea that rings are made up of particles that would have formed moons. However, the particles were too close to the planet for the gravitational attraction to allow the moons to form. The arrangements of various rings relative to the Roche limits are shown in Fig. 25.20.

This begins to give us some picture of how rings may have evolved. The scenario is illustrated in Fig. 25.21. As each planet started to form, the material around it formed into a disk. For the material far enough from the forming planet, the collection into a moon was possible. However, for the material inside the Roche limit, a moon could not form. Therefore, the material continued to orbit in a disk. Not all the particles were originally in the thin disk. However, those that weren't had to pass through the disk twice per orbit. During these passages, collisions with particles in the disk changed their orbits. Eventually, the orbits were changed sufficiently that the stragglers joined the disk. The flattening was not

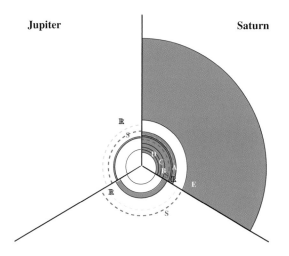

Fig 25.20. Rings and Roche limits for Jupiter, Saturn and Uranus. Again, each diagram is given in units of that planet's radius. The shaded areas are actual rings. (For Saturn, the rings are indicated by their letter designations.) The dashed line labeled R is the Roche limit for material with a density of 1 g/cm³. The dashed line labeled S is the radius for synchronous orbit about the planet. Certain electromagnetic effects change sign at this radius.

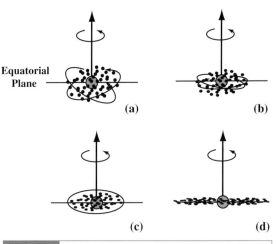

Fig 25.21. Ring evolution. (a) Particles start in a rotating cloud. Some sample orbits are shown. Particles in such orbits collide, and the collisions tend to reduce the motions perpendicular to the equatorial plane. (b) The cloud flattens as a result of these collisions. (c) Eventually most of the cloud is in a disk, with a few particles having large motions perpendicular to the equatorial plane. However, the particles pass through the plane twice per orbit, and collisions eventually bring them into the plane, producing the thin distribution shown in (d).

total, since collisions in three dimensions always leave some residual random motion in the direction perpendicular to the plane.

Within the plane of the disk, the ring could spread out over a range of radii. One mechanism for this spreading involved collisions between particles in which one was closer to the planet than the other. The closer one would overtake the farther one in their orbits. A gravitational encounter would then slow the inner one and speed up the outer one. This would cause the inner one to fall farther in and the outer one to move farther out, meaning a spreading of the ring. Another effect, Poynting Robertson drag, discussed in Chapter 27, causes smaller particles to move inward after losing energy under the impact of photons.

The final appearance of the rings around a planet depends on the position of the ring relative to the orbits of the planet's satellites. The gravitational effects of the satellites sculpt the rings by perturbing the orbits of the ring particles. This may have the effect of producing the detailed structure in Saturn's rings, or of confining the rings to some range of radii, as happened in the case of Uranus. Satellites embedded within the rings can serve as sources of particles in the ring, replacing particles removed by other effects. The moons can also remove particles from the ring.

The effects of multiple moons sculpting the rings are particularly important when the moons have *resonant orbits*. By this we mean that the ratios of the orbital periods are equal to the ratios of small integers, as illustrated in Fig. 25.22. This means that certain configurations of the satellites repeat with regularity, on a relatively short time scale. Thus, the perturbing effects of those repeating arrangements are strongly reinforced. (Similarly, if we push a swing in resonance with its motion, the effects of pushing are reinforced.) An example is found in the asteroid belt between Mars and Jupiter. There are gaps, called *Kirkwood gaps*, in which few asteroids are found. They correspond to orbits whose period would be related to Jupiter's by ratios of small integers. In Jupiter's moon system, the periods of the orbits of Io, Europa and Ganymede have the ratio 1:2:4. These are important in influencing

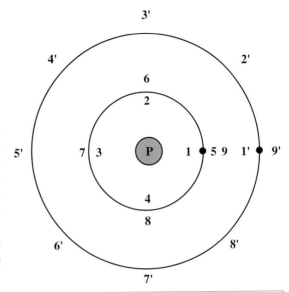

Fig 25.22. Resonant orbits. In this case, the inner satellite has half the period of the outer satellite. Positions of the inner satellite are marked 1 through 9, while the positions of the outer satellite are marked 1′ through 9′. The important point is that at every orbit of the outer satellite, the two are close together, so any perturbation can be amplified, much the way pushing a swing at its resonant frequency can amplify its motion.

the appearance of Jupiter's ring. (It also has important effects on Io, as we will see in the next section.)

Some other effects of satellites on rings are shown in Fig. 25.23. In Saturn's rings the Cassini division is at a radius where the orbital period would be half that of Mimas. It has been suggested that various resonances are important in producing the many gaps seen in the A ring as well as establishing the boundaries of the A and B rings. *Shepherding satellites* may confine rings, such as the F rings (shown in Fig. 25.24) but other rings are confined without any such satellites. It has also been suggested that satellites called *guardian satellites* produce a spiral pattern that clear the regions of resonant orbits (Fig. 25.23b). In addition, a satellite near a ring can push it away. This accounts for the narrowing of the ring of Uranus.

Another interesting effect results when two small moons have orbits very close to each other, as shown in Fig. 25.23(a). The inner moon overtakes

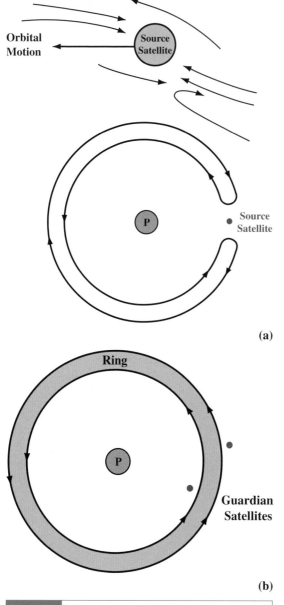

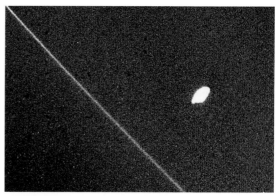

Fig 25.24. Rings and satellites. Saturn's F ring with shepherding satellite (Voyager 2 image). [NASA]

the more massive moon, the less massive moon executes a *horseshoe orbit*.

A moon within a ring has a similar effect on particles. As seen from the moon, particles approach from the outer leading side of the inner trailing side. Their paths are altered by tidal effects of the planet, since the rings are inside the Roche limit. This causes some of the particles to follow looped paths, similar to a horseshoe orbit. Since the tidal effects are directed towards the planet, particles have a hard time sticking to the moon on the side of the moon closest and farthest from the planet, but can stick to the leading and trailing edges.

25.5 | Moons

The planets that we have discussed in this chapter provide us with an interesting variety of moons. In this section, we discuss the most important of those moons, planet by planet.

The four largest moons of Jupiter were discovered by Galileo, and are therefore called *Galilean satellites*. They are, in order of distance from Jupiter: Io, Europa, Ganymede and Callisto (Fig. 25.25). The smallest, Europa, has a radius of 1561 km, and the largest, Ganymede, has a radius of 2631 km. The ones that are closest to Jupiter are denser. The densities of Io and Europa are comparable to those of Mars and the Moon. These moons are close enough to Jupiter to be shielded from the solar wind by Jupiter's magnetic field.

Fig 25.23. Satellites and rings. (a) Source satellites. The upper figure shows the motions of particles relative to the satellite. The lower figure shows the full orbit of the particles that are turned around near the satellites. (b) Guardian satellites. Satellites just inside or outside the rings confine the rings.

the outer moon. In a gravitational encounter, the inner moon is pulled out, and the outer moon is pulled in. The moons actually exchange orbits. The process repeats when they overtake again. As viewed from the rotating system orbiting with

Fig 25.25. Galileo orbiter images of the Galilean satellites: Ganymede, Callisto, Io and Europa. [NASA]

They are massive enough to affect each other's orbits. We have already mentioned the resonance among the inner three (with periods being in the ratio 1:2:4). This resonance keeps Io's orbit elliptical. It also produces large tidal disruptions of Io, which dissipate energy in Io's interior, keeping it hot.

The higher densities of Io and Europa suggest that they contain a significant amount of silicate rock. The lower densities of Ganymede and Callisto suggest that they contain much less rock. Jupiter condensed far enough from the Sun so that the temperature was low enough for water ice to form. We therefore expect the non-rock part to be water ice. The composition differences among the four satellites could have occurred because Jupiter itself was an important heat source as it collapsed. Thus, the nearer satellites were heated by Jupiter more than the farther ones. This would explain why the nearer ones do not have any water ice.

In Fig. 25.26, we compare the basic internal structure of these four moons. Of the four, Callisto's density is the least well known. Models suggest that it is 40% to 60% silicate (by mass). This silicate is concentrated in a core. Over the core is a 850 km thick section of liquid water, with a 250 km thick crust of water ice. Ganymede is 60% to 80% silicates, but has a basically similar structure. Its high albedo (0.4) also points to an ice surface. Europa is thought to have a silicate core, with a water layer and a thin (150 km thick) crust. Its high albedo (0.6) is also indicative of an ice surface. Io has a molten silicate interior and a frozen sulfur dioxide (SO_2) and sulfur surface, with active volcanoes. We now look at these moons in some more detail.

One obvious feature of Io, shown in Fig. 25.27, is its yellow color. We think that this results from the sulfur in the surface material. There are no impact features. The surface is volcanic. We have even obtained pictures of volcanic eruptions in progress. The volcanic flows might also contain sulfur. They may be basalts colored with sulfur. There are also many calderas, over 200 of which are more than 20 km across. A few large shield volcanic mountains (like Olympus Mons on Mars) are evident. There are also non-volcanic mountains. One consequence of this volcanic activity is that the surface is altered on short time scales.

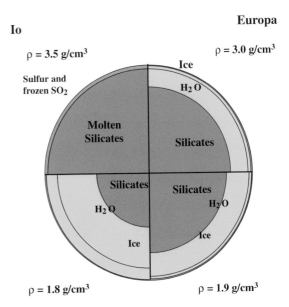

Fig 25.26. Structures of the Galilean satellites, from theoretical models. Again, each is plotted relative to the radius of the particular satellite.

(a)

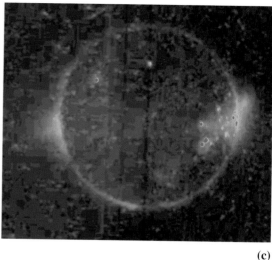

(c)

Fig 25.27. Galileo orbiter images of Io. (a) True color global image. (b) Ongoing volcanic eruption at Tvarshtar Catina. (c) This volcanic eruption stands out as Io is eclipsed by Jupiter, and is dark. [NASA]

(b)

most geologically active areas on Earth have heat flows as high as 1.7 W/m^2.

Io is also strongly affected by Jupiter's magnetic field. As Io orbits the planet, Jupiter's magnetic field sweeps by at 57 km/s. This has the

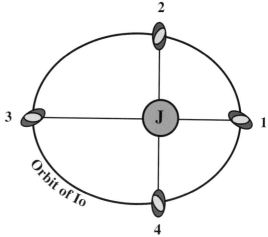

Fig 25.28. Orbital resonances and tidal heating of Io. An orbital resonance with Europa keeps Io's orbit elliptical. This means that, even though Io rotates at a constant rate, equal to its orbital period, it doesn't go around the orbit at a constant rate. If we think of Io as an ellipsoid, with an inner and outer section, the effect of the torques caused by Jupiter is for the inner layer to be out of phase with the outer layer. The two layers move against each other and generate heat.

The extensive volcanic activity is unexpected for such a small object. We would not expect such a small object to have a molten interior. However, we have already seen that the gravitational effects of the other moons, enhances by orbital resonances, distort Io. As shown in Fig. 25.28, this distortion changes its orientation as the moon moves around in its orbit. The effect is to cause internal friction, heating the interior. This results in a very large heat flow through the surface, about 2 W/m^2. For comparison, the average heat flow on Earth is 0.06 W/m^2, although some of the

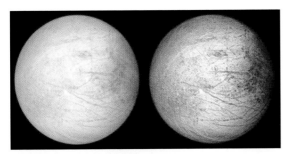

(a)

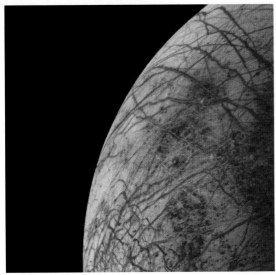

(b)

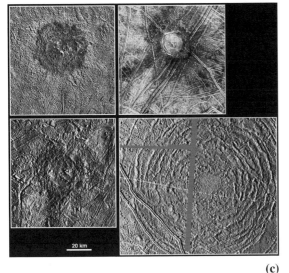

(c)

Fig 25.29. Images of Europa. (a) True (left) and enhanced (right) color views of the whole moon (Galileo orbiter image). (b) Voyager 2 closest approach. (c) Large impact structures (Galileo orbiter image). [NASA]

same effect as the moving magnetic field in an electric generator. It produces a large potential difference. In the case of Io, about 600 000 V (see Problem 25.9)! The ionized gas between Io and Jupiter is a good conductor, so currents flow parallel to the magnetic field. These currents are as high as one million amperes! This also results in bursts of radio emission, explaining why Jupiter is a strong radio source. The bursts are more frequent when Io is in certain positions.

Io also has an atmosphere. It is quite irregularly distributed, with more gas being over the warmer areas. This suggests that the atmosphere is being replaced by the volcanic activity. A major constituent is sulfur dioxide. Io also has a dense ionosphere, with a density of about 10^4 to 10^5 particles/cm^3. A large cloud has also been detected around Io. This cloud was first detected from Earth by observations of sodium, and is referred to as the *sodium cloud*. However, sodium was just one the easiest elements to observe, and other constituents are present. The cloud extends tens of thousands of kilometers along the orbit.

Our knowledge of Europa, shown in Fig. 25.29, is less detailed. The closest flybys have produced pictures with only 4 km resolution. However, we can tell that the surface is relatively flat. There is a complicated pattern of lines. There is also a lack of large craters, suggesting a young surface.

(a)

Fig 25.30. Galileo orbiter images of Ganymede. (a) Trailing hemisphere.

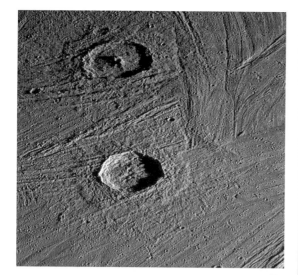

(b)

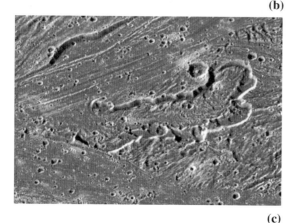

(c)

Fig 25.30. (*Continued*) (b) Fresh impact craters. The image covers an area 142 × 132 km. (c) Calderas. [NASA]

The surface is probably made of ice. We think that the lines are tension patterns in the ice. This tension may have resulted from an expansion of the surface, probably by about 5%. There are some dark patches; these are probably composed of silicates.

Ganymede, shown in Fig. 25.30, also has an ice surface. It is covered with irregular light and dark regions. The dark regions are heavily cratered, indicating that they are the older part of the surface. The craters in these regions are relatively flat. The lighter regions are grooved. The grooves appear to be alternating ridges and troughs. The pattern suggests tension, as on Europa.

Callisto, shown in Fig. 25.31, has a heavily cratered surface. These craters are also flat, typical of an icy surface. An unusual feature is the large ring structures. These are probably the result of violent impacts in the past.

The moon of Saturn that we know the best is its largest, Titan, shown in Fig. 25.32. It has the

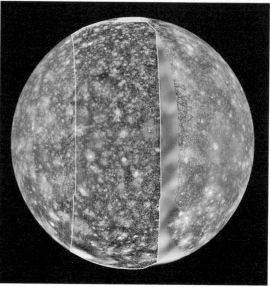

(a)

(b)

Fig 25.31. Images of Callisto. (a) Combined Voyager and Galileo mosaic. (b) Possibly an oblique impact (Galileo orbiter image). [NASA]

Fig 25.32. Titan's haze layer (Voyager 1 image). [NASA]

(a)

(b)

(c)

Fig 25.33. Saturn's moons. (a) Dione (Voyager 1 image). (b) Iapetus (Voyager 2 image). (c) Enceladus (Voyager 2 image).

most significant atmosphere of all the moons in the Solar System. Voyager 1 passed within 5000 km of Titan, giving us a good opportunity to study it. Some of the other moons are shown in Fig. 25.33. Titan's density is quite low, only twice that of liquid water, suggesting a mixture of rock and ice. Its composition and structure have been affected by heat given off by Saturn during its formation. The model of its interior, shown in Fig. 25.34, has 55% of the mass in a rock core. It is possible that internal heating has taken place. Tidal distortion should not be as important for Titan and Saturn as for Io and Jupiter. There may be some radioactivity, but it seems more likely that the energy has been stored from the time that the moon formed. Beyond the core, there are probably layers of different structures of ice.

Titan's atmosphere contains methane (CH_4). This was originally determined from the Earth. However, closer observations have shown that nitrogen (N_2) makes up 80 to 95% of the atmosphere. It has also been suggested that there is some argon. Ultraviolet radiation triggers a chemistry that produces traces of ethane, acetylene and

(d)

Fig 25.33. (*Continued*) (d) Tethys (Voyager 2 image). [NASA]

The surface pressure is 1.6 atmospheres. The surface temperature is 93 K. The highest temperature in the atmosphere is about 150 K, in the upper atmosphere. The lowest temperature is about 70 K, about 40 km above the ground. An interesting feature is that it contains the temperature (90.7 K) at which the three phases – gas, liquid and solid – of methane can coexist. We might therefore expect to find a frozen methane surface, with methane oceans and methane clouds in the atmosphere. Others have speculated on oceans of ethane rather than methane.

Voyager 2 provided our first good look at the five known moons of Uranus, and discovered a number of smaller moons. Close-ups of some of the moons are shown in Fig. 25.35. As with

ethylene. These are found in the lower atmosphere. Higher up, hydrogen cyanide (HCN) is formed. There is no oxygen, since it is tied up in the frozen water. The chemistry produces a smog, possibly with some seasonal variation.

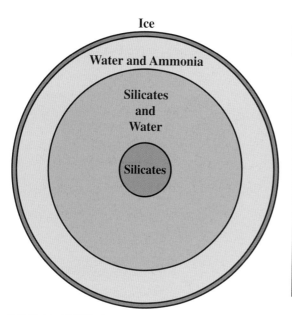

Fig 25.34. Internal structure of Titan, as deduced from theoretical models.

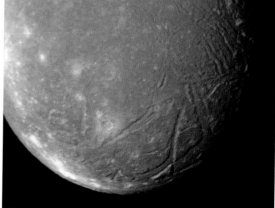

(a)

(b)

Fig 25.35. Voyager 2 images of the Moons of Uranus. (a) Ariel; (b) Titania;

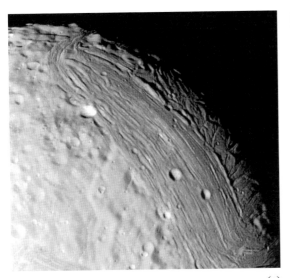

Fig 25.36. Global color mosaic of Triton (Voyager 2 image). [NASA]

(c)

(d)

Fig 25.35. (Continued) (c) Miranda, showing a variety of terrains; (d) Oberon. [NASA]

Jupiter and Saturn, the moons of Uranus show a variety of features. Oberon has large albedo variations over its surface. It is cratered, and a small mountain was photographed on the limb. Titania has craters, ridges, rills and scarps. Its cratering is sparse compared to that of other moons. Ariel has many bright reflected regions as well as deep scarps and valleys. There are many filled in features, indicating an active surface. Some of the most intriguing structures are found on Miranda. It has three distinct types of surface: an old surface, a scored region, and a region with very complex structures.

Neptune's largest moon, Triton, is shown in Fig. 25.36.

Chapter summary

In this chapter, we looked at the outer planets in the Solar System. They are much larger than the inner planets, and have structures that are very different from those of the inner planets.

Because of their large masses, these planets, especially Jupiter and Saturn, have been able to retain much of their initial supply of hydrogen, giving them compositions that are dominated by hydrogen. The minor constituents play important roles in the atmospheric structure, energy transfer and appearance.

Both Jupiter and Saturn give off more energy than they receive from the Sun. It is speculated that this comes from the gravitational potential energy liberated during the collapse of the planets.

The general circulation on Jupiter and Saturn is quite complicated, producing bands and varying cloud and wind flow. Large, hurricanelike systems, such as the Great Red Spot, are found. The atmospheres of Uranus and Neptune are not as dense or as active.

The interiors of Jupiter and Saturn are mostly liquid hydrogen, with the possibility of metallic hydrogen in the center. Uranus and Neptune are thought to have rocky cores.

We looked at the properties of the rings around these planets. We showed how tidal effects may be important in ring formation. We also looked at the dynamical effects that provide the intricate structure of the rings and gaps.

Finally, we looked at the major moons of these planets. These moons provide a diversity of surface and atmospheric phenomena.

Questions

25.1. What are the major differences between the four inner planets and the four outer planets (i.e. not including Pluto)?

25.2. What are the similarities and differences between Jupiter and Saturn?

25.3. What are the similarities and differences between Uranus and Neptune?

25.4. Why is it hard to measure the rotation period of Uranus?

*25.5. The outer planets are much more massive than the Earth, yet the accelerations of gravity on their surfaces are not much different than that on Earth. How can that be?

25.6. The atmosphere of Jupiter contains a high proportion of hydrogen. This is not true of the Earth's atmosphere. Why is this?

25.7. Why would you expect Jupiter's atmosphere to have a composition similar to that in the early Solar System?

25.8. What accounts for the different colors in the bands on Jupiter (or Saturn)?

25.9. What is the explantion for the Great Red Spot?

25.10. Compare the compositions of the atmospheres of Jupiter and Saturn.

25.11. Compare the compositions of the atmospheres of Uranus and Neptune.

25.12. What is the source of the excess energy that Jupiter gives off?

25.13. What are the differences in the ring systems of Jupiter, Saturn, Uranus and Neptune?

25.14. What is the relationship between planetary rings and the Roche limit?

25.15. Standing on the Earth, we are within its Roche limit. Why aren't we torn apart by tidal effects?

25.16. If Saturn's rings were known for centuries, why did it take so long to find the rings around Jupiter, Uranus and Neptune?

25.17. Contrast the ways in which the rings of Jupiter and Uranus were discovered.

25.18. What is the evidence that Saturn's rings are made of many small particles, rather than being solid?

25.19. How do moons affect the appearance of planetary rings? What is the evidence that this is happening?

25.20. What is the evidence that Io has a molten interior?

25.21. What is the cause of Io's molten interior?

25.22. Contrast the properties of the largest moon around each of the giant outer planets.

25.23. List the 15 most massive objects in the Solar System (not including the Sun), in order of decreasing mass.

25.24. List the moons of the Solar System in order of decreasing size. Where do the smallest planets fit in?

Problems

25.1. Find the gravitational potential energy of Jupiter. Assume that this amount of energy has been released over a period of 4 billion years. How does the average rate of energy release compare with that in the sunlight received by Jupiter?

25.2. Confirm Kepler's third law for the moons of Jupiter. Use the information to derive Jupiter's mass.

25.3. Calculate the rate of temperature fall-off for an adiabatic process near the Jupiter cloud tops. Compare it with the similar number for Saturn.

25.4. From the data given in the chapter, show that Saturn gives off approximately twice as much power as it receives from the Sun.

25.5. What is the ratio of solar energy per second per unit surface area reaching Uranus to that reaching Neptune?

25.6. Voyager 2 was five planetary radii from Uranus. At that time, what angle was subtended by the planet as viewed from the spacecraft?

25.7. Show that all the rings of Jupiter, Saturn and Uranus lie within the Roche limits for these planets.

25.8. In deriving the Roche limit, we ignored the fact that the particles are orbiting the main planet. This introduces pseudo-forces in the rest frame of the orbiting material. How do these pseudo-forces affect the Roche limit calculation?

25.9. Use Faraday's law to derive an expression for the potential difference across a planet by a magnetic field B sweeping across the planet's surface at a speed v. Take the planet radius to be R.

25.10. Compare the magnitude of the tidal effects that Jupiter exerts on Io with those that Saturn exerts on Titan.

25.11. Estimate the adiabatic temperature gradients at the altitude of minimum temperature for both Jupiter and Saturn.

Computer problem

25.1. (a) Show that the major moons of each outer planet obey Kepler's 3^{third} law. (b) Calculate the mass of each outer planet from the orbital data of the most massive moon.

Chapter 26

Minor bodies in the Solar System

There are a vast number of smaller objects in our Solar System, not as substantial as our Moon, but which provide important clues on the history of the Solar System. These are *asteroids*, *comets* and *meteoroids*. We have also included the ninth planet, Pluto, in this chapter. As we will see below, recent determinations of Pluto's mass make it by far the least massive planet, and it has more properties in common with the other less massive objects in the Solar System.

26.1 | Pluto

Pluto was discovered in 1930, following an extensive search, by *Clyde Tombaugh*. The search was initiated by *Percival Lowell* after it was thought that a planet beyond Neptune might be perturbing Neptune's orbit. Calculations narrowed the range of possible locations on the sky, and a search was carried out. As Fig. 26.1 shows, Pluto doesn't stand out very well against the background of stars. It is detectable as a planet only by its very slow motion with respect to the stars.

For Pluto to have a perturbing effect on other planets, its mass must be greater than that of the Earth. For this reason, since its discovery, Pluto's mass has been overestimated. We now know that its mass is much less than previously thought, and that it has no measurable effect on other planets. In a sense, Pluto's discovery was accidental. It was a result of an extensive search of a particular region in the sky. For this reason, other searches have been carried out for a "tenth planet", none with success.

Pluto's mass is now known reasonably accurately. This is because a moon was discovered orbiting Pluto in 1978. The moon is named Charon, and is shown in Fig. 26.2. By studying its orbital motion we can determine Pluto's mass. Actually, using the Hubble Space Telescope, it has been possible to look at the motions of both Pluto and Charon about their common center of mass. From this it has been deduced that Pluto's mass is 1.3×10^{25} g, and Charon's mass is about 1/12 of that. Pluto's mass is only about 1/500 of the Earth's mass, or one-fifth that of our Moon.

Pluto's size has been estimated from its failure to occult certain stars. (See Section 26.4 for how this technique is used for asteroids.) However, our best measurements now come from optical interferometry techniques. Using this size, and the measured mass, we find that Pluto has a very low density, about 0.5 to 1.0 g/cm^3. Charon's density is even lower, only 20% greater than that of liquid water. This suggests that its composition is similar to that of the moons of the giant planets. It has been suggested that Pluto's surface is frozen methane and that its atmosphere is also composed of methane.

Pluto's size, density and orbit raise questions about its status as a planet. Its orbit is the most eccentric of the planets. It even spends part of its orbit closer to the Sun than Neptune. It has been suggested that Pluto may actually be an escaped moon of Neptune. This would explain its small size, low density, and crossing of Neptune's orbit. However, when we trace back the orbits of

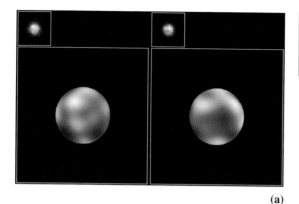

(a)

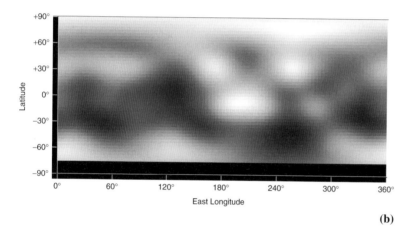

(b)

Fig 26.1. (a) HST images of Pluto, showing both hemispheres. The insert images are the unprocessed images. (b) Surface map of Pluto, based on the HST images. [STScI/NASA]

Neptune and Pluto, we find no time when they were actually close together. Thus, if Pluto did escape from Neptune, its orbit must have been perturbed since then. Thus, Pluto's origin is still a mystery.

26.2 | Comets

Every now and then, a spectacular comet, like those shown in Fig. 26.3, is visible for a few weeks,

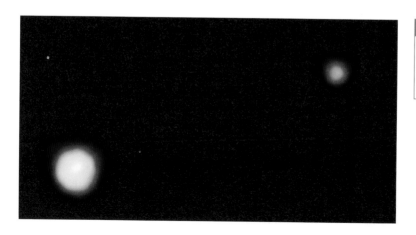

Fig 26.2. Pluto, with Charon. This is a near infrared image (2.2 μm). The separation between the objects is 0.9 arc sec. [STScI/NASA]

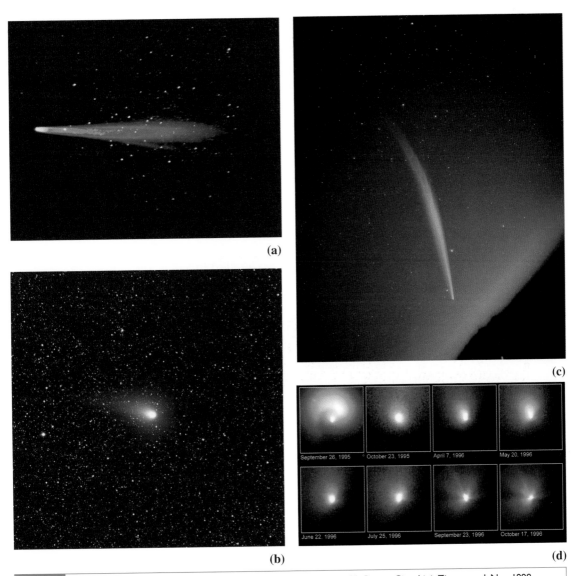

Fig 26.3. (a) Comet Halley in 1910. This is a computer-reconstructed image. (b) Comet Giacobini–Zinner, on 1 Nov, 1998. (c) Comet Ikeya–Seki, 1996. (d) HST time sequence showing the evolution of the core of Hale–Bopp (1995). [(a)–(c) NOAO/AURA/NSF; (d) STScI/NASA]

and we are reminded of this phenomenon. However, most comets are faint, and do not attract attention. They are still very important in our understanding of the history of the Solar System. Over 600 comets are known. Their masses are less than one-billionth of the mass of the Earth.

Figure 26.4 shows the basic structure of a comet. The smallest part is called the *nucleus*. It is only a few kilometers across. One way of determining sizes is by bouncing radio waves off the surface and measuring the strength of the

reflected signal. Since the early 1950s the conventional picture, advanced by *Fred L. Whipple*, is that the nucleus is a "dirty snowball". It contains dust, plus ices of water, carbon dioxide, ammonia and methane. When a comet is close to the Sun, material is ejected from the nucleus. This material acts like the exhaust of a rocket, and provides thrust for the comet. This may actually alter the orbit of the comet. It is one of the reasons why the exact prediction of comet orbits is difficult.

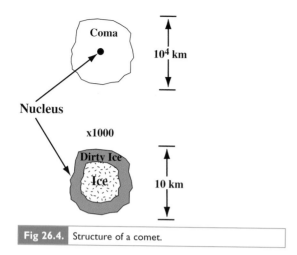

Fig 26.4. Structure of a comet.

Outside the nucleus is an extended region of gas and dust, called the *coma*. It is 10^5 to 10^6 km in extent, and shines by reflected sunlight. The material in the coma flows outward at about 0.5 km/s. The outflow of gas drags it away from the nucleus. The sunlight reflects off both gas and dust in the coma. The coma can also emit radiation from excited gas. Spacecraft observations have shown Lyman alpha emission in the ultraviolet. These indicate that the hydrogen cloud is up to a factor of ten larger than the coma itself. It is thought that this hydrogen comes from the breakup of water molecules and OH radicals by solar ultraviolet radiation. Spectra of the coma have indicated the presence of a number of simple molecules, such as NH_3, H_2O, OH and NH.

When a comet moves relatively close to the Sun, it may develop a large *tail*. This tail can be up to 1 AU in extent, but is of such low density that it doesn't contain an appreciable fraction of the mass of the comet. Only about 1/500 of the mass of the comet is in the tail.

There are actually two tails, as shown in Fig. 26.5. The *gas tail* is blown straight out by the interaction of the solar wind and the comet. The gas tail always points away from the Sun. Variations in the solar wind produce a varied appearance along the length of the tail. A number of molecular ions have been detected in the gas tail: CO^+, CO_2^+, CH^+, CN^+, N_2^+, OH^+ and H_2O^+. It also contains some more complex molecules, such as formaldehyde (H_2CO). Since it contains ions, it is called the *ion tail*. Emission from

CO^+ is responsible for the blue color of the tail. The gas tail can be up to 10^8 km long. (This is almost 1 AU.) The *dust tail* is material that is left behind in the orbit. We see it as a smooth curve, tracing out the comet's orbit. It is ejected by pressure from sunlight. When it is free of the comet, the dust continues in a Keplerian orbit, perturbed by radiation pressure. The dust tail can be up to 10^7 km long. Sometimes, a tail appears to be pointing toward the Sun. This is an illusion, caused by the appearance of the tail pointing away from the Sun, as viewed from the Earth in particular positions with respect to the comet and the Sun.

We can estimate the effect of radiation pressure on the dust grains. The source of the radiation pressure is the momentum carried by photons. For a photon of energy E, the momentum is

$$p = E/c \tag{26.1}$$

If this photon is absorbed by an object, then all of the momentum is transferred to the object. If the photon is reflected back off an object, then the momentum delivered to the object is twice this. That is, since the photon reverses direction the magnitude of its momentum change is $2p$. Suppose we want to calculate the pressure a distance r from the Sun. Imagine a spherical shell at this radius. The force F on the shell is just the momentum per second carried by the photons reaching the shell. That is

$$F = dp/dt$$

The pressure P is the force, divided by the area of the shell $4\pi r^2$, so

$$P = \frac{dp/dt}{4\pi r^2}$$

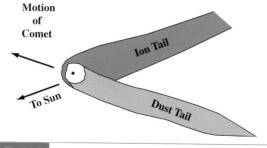

Fig 26.5. The two comet tails.

By equation (26.1),

$$dp/dt = \frac{dE/dt}{c}$$

But dE/dt is just the energy per second emitted by the Sun, or the solar luminosity $L_\odot$. Therefore

$$P = \frac{L_\odot}{4\pi r^2 c} \qquad (26.2)$$

At a distance of 1 AU from the Sun, the pressure is 5×10^{-5} dyn/cm^2, which is 5×10^{-11} atm.

Despite this small pressure, if a grain is small enough, the force on the grain due to radiation pressure can exceed the gravitational attraction by the Sun on the grain. Small grains are important, because the gravitational force depends on the mass of the grains, which is proportional to d^3 (where d is the grain size), and the radiation pressure force is pressure multiplied by area, so it goes as d^2. Therefore, the ratio of forces

$$\frac{F_{rad}}{F_{grav}} \propto \frac{d^2}{d^3}$$

$$\propto \frac{1}{d}$$

Since the ratio of the forces is proportional to $1/d$, the radiation pressure is more effective for smaller grains. Note that equation (26.2) tells us that the radiation pressure force is proportional to $1/r^2$ (where r is the distance from the Sun). This is the same dependence as the gravitational force on r. Therefore, the ratio of the forces is independent of the distance from the Sun. It only depends on grain size.

Example 26.1 Radiation pressure
For grains of a given density 1 g/cm^3, find the grain size for which gravity and radiation pressure are equal.

SOLUTION
The gravitational force on a grain of mass m is

$$F_{grav} = \frac{GmM_\odot}{r^2}$$

$$= \frac{G(4\pi/3)d^3\,\rho M_\odot}{r^2}$$

The radiation pressure is

$$F_{rad} = \frac{L_\odot}{4\pi r^2 c}(\pi d^2)$$

Equating these and solving for d gives

$$d = \frac{3L_\odot}{16\pi GM_\odot\,\rho c}$$

$$= \frac{3(4 \times 10^{33}\ \text{erg/s})}{16\pi(6.67 \times 10^{-8}\ \text{dyn cm}^2/\text{g}^2)(2 \times 10^{33}\text{g})(1\ \text{g/cm}^3)(3 \times 10^{10}\ \text{cm/s})}$$

$$= 6 \times 10^{-5}\ \text{cm}$$

For grains smaller than this, the radiation pressure will push out more strongly than gravity pulls in.

Our current picture of the origin of comets places them in a cloud, called the *Oort cloud*, some 50 000 AU from the Sun. The cloud is like a spherical shell around the Solar System. The existence of such a cloud is suggested by the fact that comets appear to be bound to the Solar System (no hyperbolic orbits), and that they come from all directions, orbiting with equal frequency in either direction relative to the orbital motions of the planets. The cloud may contain 10^{12} to 10^{13} comets, giving it a total mass in the range of one to ten Earth masses. From time to time, one of the objects in the cloud has its orbit severely perturbed, and starts to head for the inner Solar System. At this time, the comet is just the material that will be seen as a nucleus when the comet is in the inner Solar System. There is no coma or tail. As the comet moves closer to the Sun, it develops the coma, and then the tail, as discussed above. The coma begins to appear when the comet is about 3 AU from the Sun. At this point, the temperature is about 215 K, which is right for the sublimation of water ice to form water vapor.

The comet is brightest at perihelion. The tail has its longest physical extent then. However, the apparent length of the tail depends on the viewing angle from the Earth. Therefore, the tail may not appear to be longest at perihelion. Once the comet has passed through the inner Solar System, it continues outward in its orbit. Most of the orbits are ellipses, so the comets will return, unless the orbit is perturbed as the comet passes near a planet. Orbital eccentricities and periods

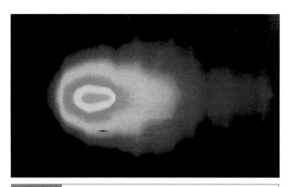

Fig 26.6. IRAS image of comet IRAS–Araki–Alcock. [NASA]

vary considerably. For example, Comet Encke has a period of 3.3 years and Comet Kohoutek has a period of 80 000 years. It is thought that most short-term comets have had their orbits severely perturbed by Jupiter.

The appearance of a particular comet near perihelion is often hard to predict in advance. For example, it was predicted that Comet Kohoutek (1974) would be very bright. This was based on the fact that it appeared to be bright when it was far away. The comet was not as bright as predicted. It is now speculated that the comet was making its first pass by the Sun, and therefore behaved differently than a comet making a return visit. On the other hand, Comet West (1975) put on a better show than expected. People who remember the spectacular view of Halley's comet in 1910 were disappointed by an unfavorable viewing angle in 1986. The closest approach of Halley's comet to Earth in 1986 was much farther than in 1910. In addition, the view in 1986 was marred for many observers by the spread of light pollution in the previous 76 years.

Some comets pass quite close to the Earth. For example, IRAS–Araki–Alcock (1983) passed within 4.6 million km (Fig. 26.6). This comet was of interest because of the simultaneous discovery by the IRAS satellite, in the infrared, and two more traditional ground-based optical observers.

Astronomers and the general public were treated to a rare astronomical phenomenon when a large comet, which had actually broken into fragments, struck Jupiter in 1994. The comet is called Shoemaker–Levy, for its co-discoverers. When its orbit was analyzed, it was found that it was on a collision course with Jupiter.

The Hubble Space Telescope provided very sharp images of the comet. One of these is shown in Fig. 26.7. The comet had broken into several fragments that were seen stretched out along its orbit. It was speculated that this break-up was caused by the tidal effects of Jupiter. That is, the side of the comet that was closer to Jupiter was pulled with a stronger gravitational force than the side that was farther away. This would first simply stretch the comet, but could eventually pull it apart into pieces of various sizes.

Astronomers and the rest of the news followers were fascinated with the possibility of witnessing such a catastrophic event as a comet striking another planet. Also, since the comet had broken into pieces, there would be a series of impacts over a period of several hours. Apart from the spectacular nature of such an event, it would allow astronomers to study the effects of this impact on the Jovian atmosphere. This would allow atmospheric specialists to test their theories about the structure and composition of the Jovian atmosphere. It would also allow them to study the general effects of such a catastrophic event, with an eye towards understanding how such events might affect other planets, including the Earth.

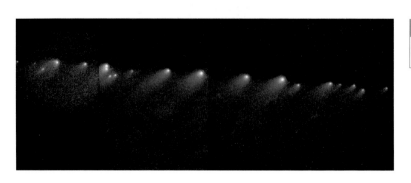

Fig 26.7. Photograph (from HST) of fragments of Comet Shoemaker–Levy. [STScI/NASA]

This encounter was also fascinating because it could be predicted, and the times and locations of impacts could be calculated. Normally, by their very nature, catastrophic events (from earthquakes to supernovae) happen with little or no warning. In this case, astronomers could prepare well in advance to watch the events unfold.

In the anticipation of the collisions, there was one cautionary aspect. The impact was going to occur on the side of Jupiter facing away from the Earth. So we could not watch the actual impacts from Earth (or even from a telescope in orbit around the Earth). We would have to wait for Jupiter's rotation to bring the impact around to the side facing the Earth. There was concern that if the impacts were not very strong, then their effects would quickly fade, and we would not see too much by the time the impact sites came into view. However, Jupiter rotates very quickly (taking only ten hours to make one rotation). It was calculated that the impacted sites would rotate into view roughly an hour after each impact. The other major uncertainty was about the masses of the fragments. The greater the masses, the more they would affect the atmosphere.

When the impacts occurred very few people were disappointed. At the high speed of the impact (remember, kinetic energy is $(1/2)mv^2$), the masses of most of the fragments were sufficient to cause major disruptions, as shown in Fig. 26.8. As each impact site came around into view, a plume of material could be seen that had been ejected from the atmosphere. Also, infrared observations confirmed that the impact sites were hotter than their surroundings.

In the aftermath of these spectacular events, many were wondering about how such an impact would affect the Earth, and how often such impacts on the Earth might occur. There is already growing speculation that the impact of an object from space caused the extinction of the dinosaurs, some 70 million years ago.

Comets are important in our understanding of the Solar System. We think that the Oort cloud is left over from the material that condensed to form the Solar System. The current idea is that the comets formed near Jupiter and Saturn, and were ejected out to the Oort cloud location. The cloud is far enough out to be unaffected by

heating in the Solar System itself. Therefore, the composition of comets should reflect the composition of the original solar nebula. That is one reason why there was considerable activity in launching spacecraft to fly near comets as they were close to the Earth, including Halley's comet in 1986.

Also far out in the Solar System is a group of objects called *Kuiper Belt objects*. These icy objects are found between the orbit of Neptune (30 AU) out to 50 AU. So far, more than 300 have been discovered, and it is estimated that there are at least 70 000 such objects larger than 100 km across. Unlike the Oort cloud, Kuiper Belt objects are more tightly confined to the ecliptic, forming a thick band. It is likely that these are left over

(a)

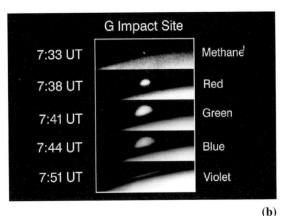

(b)

Fig 26.8. HST images of Jupiter during and after the Comet Shoemaker–Levy impacts. (a) Jupiter after the fragment G impact. (b) Images (at various wavelengths) of fragment G impact spot.

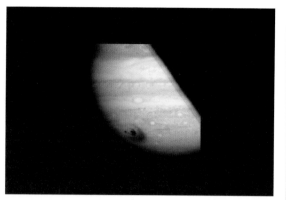

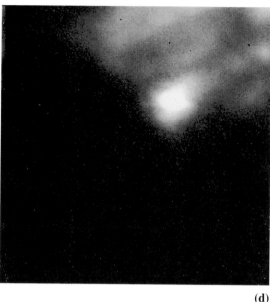

(c)

(d)

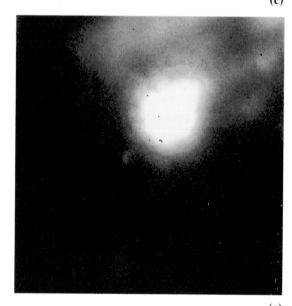

(e)

from the formation of the Solar System, providing information on the conditions under which the outer part of the Solar System was formed. We also think that the Kuiper Belt is the source of short period comets (as opposed to the long period comets that come from the Oort Cloud).

26.3 | Meteoroids

Meteoroids are small chunks of matter left in space. They are up to tens of meters in diameter. When the Earth encounters a meteoroid (Fig. 26.9), the meteoroid may fall through the Earth's atmosphere. It is then heated by the friction between the air and the meteoroid. It glows brightly as it streaks across the sky, as shown in Fig. 26.10. At this point, we refer to it as a *meteor*. Most meteors burn up as they pass through the atmosphere. However, some do reach the ground. The ones that reach the ground are called *meteorites*. The largest meteorites produce craters, including the large one shown in Fig. 26.11.

Some very small meteoroids, much less than 1 mm across, may settle into the upper atmosphere, to be collected by balloons. These are of interest because there are many more small meteoroids than large ones.

Most meteoroids that produce meteors are probably the debris of comet tails. They are therefore left behind in the orbit of the comet, as shown in Fig. 26.12. When the Earth crosses the comet's orbit, we see a large number of meteors – a *meteor shower*. These showers occur at the same time each year, since they represent the passage of the Earth through the orbit of the comet. This scenario explains why we see most meteors after midnight. After midnight, an observer is on the side of the Earth facing in the direction of the orbital motion of the Earth.

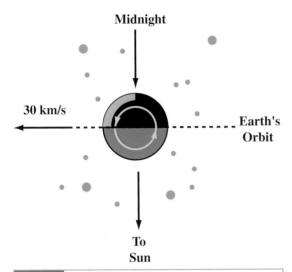

Fig 26.9. Earth moving through meteoroids. Note that the activity is greatest when the observer is on the leading edge of the Earth, which occurs after midnight. The maximum effect is actually at dawn, so the best time to see meteors is just before it begins to get light in the morning.

Since we think that comets are left over from the formation of the Solar System, meteorites give us a chance to examine that material directly. Two different compositions have been identified. One group, called *irons*, are mostly iron and nickel. The other type, called *stones*, have an appearance similar to ordinary rocks and are hard to find on the ground unless the fall has been witnessed. There is another type, *stony irons*, which is a combination of two types of material. Most meteorites are stones.

Most of the stones contain small rounded glasslike particles, called *chondrules*. Meteorites containing chondrules are called *chondrites*. The others are called *achondrites*. Chondrites with large amounts of carbon are called *carbonaceous chondrites*. It is believed that these are the oldest meteorites. An example of a famous carbonaceous chondrite is the *Murchison meteorite*, which fell in Australia in 1969. This meteorite contains amino acids of a type (left-handed vs. right-handed) not found on Earth. The largest carbonaceous chondrite is the Allende meteorite, which fell on Mexico in 1969. We think that the carbonaceous chondrites were never strongly heated after formation. They therefore preserve the original material out of which the Solar System formed. The Allende meteorite has centimeter-sized inclusions of minerals rich in calcium and aluminum.

It has been suggested that there is some relationship between asteroid types and meteorite

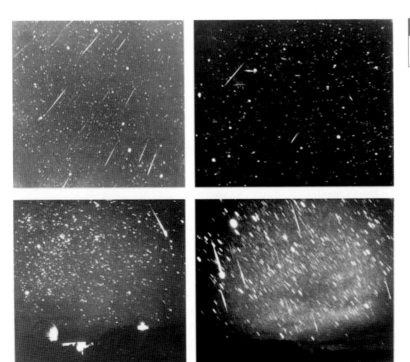

Fig 26.10. Views of the Leonid meteor shower. [NOAO/AURA/NSF]

Fig 26.11. Meteorite crater on Earth. This is the Barringer Crater, just east of Flagstaff, Arizona. The bowl has a 1 km diameter. Notice the elevated rim. It is the first terrestrial crater recognized as coming from a meteor impact. [USGS]

types. In this picture, the C type asteroids are like the carbonaceous chondrite meteorites. The S type asteroids are like the ordinary chondrites or the stones.

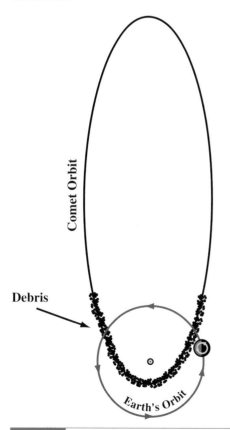

Fig 26.12. Meteoroids and comets. The meteoroids are debris left by the comet along its orbit. We see a meteor shower as the Earth passes through the debris.

Radioactive dating tells us that meteorites are 4.5 billion years old. This confirms our idea that they are part of the debris left over from the formation of the Solar System. Studies are now being done of the relative abundances of various elements to understand the composition of the material out of which the Solar System formed. So far, some unusual abundances have been found. For example, xenon-129 is formed by the beta decay of iodine-129, with a half-life of 17 million years. A lot of xenon-129 has been found, suggesting a large production of iodine-129 just prior to the formation of the Solar System. However, this is still quite speculative, and some claim that they can explain the abundances without invoking any unusual events.

26.4 | Asteroids

The distribution of asteroids is shown in Fig. 26.13. Most of the asteroids lie in a band between the orbits of Mars and Jupiter. This band is called the *asteroid belt*. Over 3000 asteroids have been cataloged to date, but there are many more. The ones that have been cataloged are the brightest, and presumably the largest. We would expect there to be many more small ones. The combined mass of the asteroids is less than that of the Moon.

The sizes of the asteroids are determined from stellar occultations, as shown in Fig. 26.14. The orbits of many asteroids are well known. When we talk about the orbit of an asteroid (or any other object), we are talking about the path of its center. When the center comes close to passing in front of a star, the asteroid can only occult the star if the asteroid is large enough. When an occultation is expected, astronomers from various parts of the Earth watch. If they all see the occultation, then the asteroid is larger than some size. If none sees the occultation, the asteroid is smaller than some size. If some see it and others don't, the size of the asteroid can be determined quite accurately.

Of the asteroids that have been studied in this way, only six are larger than 300 km, 200 are larger than 100 km, and there are many smaller than 1 km. If we know the size and how bright they appear, we can estimate the albedo of the

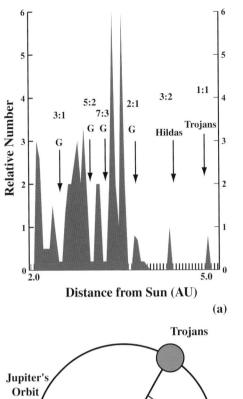

Fig 26.13. (a) Radial distribution of asteroids. The places marked G are gaps, with the orbital period relation to Jupiter given above. The two groups, the Hildas and Trojans, are also indicated. (b) Locations of the Trojans, at two Lagrangian points of the Jupiter–Sun system.

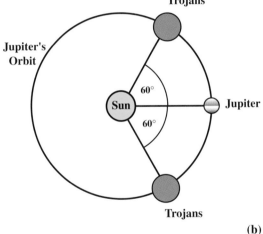

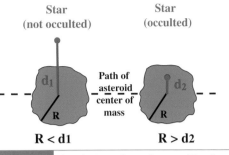

Fig 26.14. Occultations of stars by asteroids tell us about asteroid sizes.

where the gravitational forces of the Sun and Jupiter and the pseudo-force, the centrifugal force, sum to zero. It is thought that Phoebe, Saturn's outermost moon, which is only 200 km across and quite dark (low albedo), is a captured asteroid. It has also been suggested that the smaller moons of Jupiter, as well as the moons of Mars, are captured asteroids. Astronomers think that some of the larger asteroids would have a similar appearance to Phobos and Deimos.

Asteroids are classified according to their surface types. E types have very high albedos. They are rare and are found at the inner edge of the belt. S types have lower albedos. They are more abundant and are found from the inner part to the center of the belt. M types are also abundant, being found in the middle of the belt. They have moderate albedos. The most abundant are the C types, which are found in the outer parts of the belt. They are characterized by a low albedo. There is some correspondence between asteroid types and meteorite types.

The brightness of asteroids varies with time. Studies of their light curves show periodic variations with up to a factor of three change in brightness. This is interpreted as indicating either an irregular shape or an irregular surface coverage. In the latter case, the asteroid would have a dark and a light side. In some cases, elongated shapes have been detected.

The origin of asteroids is still not clear. For a long time, it had been speculated that the asteroid belt was the remainder of a planet that was destroyed. The total mass of the asteroids, only 2% of the mass of the Earth, would not provide a very

asteroid (see Problem 26.7). Albedos for asteroids fall in the range 3% to 50%.

Not all of the asteroids are in the asteroid belt. Some, the *Apollos*, cross the Earth's orbit. Another group, known as the *Trojans*, is found in Jupiter's orbit, one-sixth of an orbit ahead and behind the planet. This is a point, called a *Lagrangian point*,

massive planet. However, it does seem likely that the large tidal effects of Jupiter kept a planet from forming too close to Jupiter. The asteroids may therefore may have been debris that would have formed into a planet had Jupiter not prevented it.

Chapter summary

In this chapter we looked at Pluto, as well as asteroids, comets and meteoriods. These small objects give us important clues to the history of the Solar System.

Pluto was discovered in a search for a planet perturbing the orbit of Neptune. (We now know that those perturbations are not real.) As we have learned more about Pluto, especially through the discovery of a moon, whose orbital size and period give us Pluto's mass as about one-fifth that of our own Moon, we have come to revise our understanding of its role in the Solar System. It is also the planet with the most eccentric orbit, spending some of its orbit inside that of Neptune.

We think that comets may provide us with the best record of material out of which the Solar System formed. This is because they reside well beyond the orbit of Pluto, and only make brief visits to the inner Solar System. We saw how the appearance of a comet changes as it approaches the Sun. We saw the processes that are important to the development of the tail.

For the most part, meteoroids are the debris left behind by comets. When they fall to Earth, they provide us with information on the composition of material in the forming Solar System.

We saw that most asteroids are between Mars and Jupiter, but there are a few notable exceptions. Even within the asteroid belt, there are differences in composition as one moves farther from the Sun.

Questions

26.1. How does the mass of Pluto compare with that of the larger moons in the Solar System?

26.2. Why was the discovery of a moon around Pluto important in telling us about Pluto's mass?

26.3. Why are some astronomers questioning Pluto's status as a planet?

26.4. How do we measure the size of an asteroid?

26.5. If you see a bright asteroid, what might you conclude about it?

26.6. If comets have very little mass, how can they be seen so easily when they are near the Earth?

26.7. What does the term "dirty snowball" mean when applied to comets?

26.8. If you see a comet in the sky with two tails, how can you tell which is the gas tail and which is the dust tail?

26.9. Why do we think that comets may preserve a record of the early Solar System?

26.10. Why are comets brightest when they are close to the Sun?

26.11. Draw a diagram to show how a tail going away from the Sun could be viewed from Earth as pointing towards the Sun.

26.12. Why do we think that comets come from far out in the Solar System (i.e. the Oort cloud)?

26.13. What is the relationship between meteoroids and comets?

26.14. Why are meteor showers most active between midnight and dawn for any observer?

26.15. Explain why the presence of xenon-129 in large quantities in a meteoroid means that large quantities of iodine-129 were produced just before the formation of the meteoroid.

Problems

26.1. What is the force on the Earth due to the Sun's radiation pressure? How does that compare with the Sun's gravitational force on the Earth?

26.2. It has been suggested that radiation pressure from the Sun could be used to propel a large spacecraft toward the outer Solar System. How large a sail would you need to provide an acceleration of 1 m/s^2 for a 10^6 kg spacecraft?

26.3. Suppose an asteroid is a distance d from Earth. Its center of mass is going to pass within an angle θ of the star. How large does the asteroid have to be to occult the star?

26.4. How far can the center of mass of a 100 km radius asteroid pass from the direct line between the Earth and a star and still have the asteroid occult that star, if the asteroid is 3 AU from Earth?

26.5. Use the mass of a typical comet and the mass of the Oort cloud to estimate the number of comets in the Oort cloud.

26.6. (a) Estimate the kinetic energy of a 100 m diameter object, of density 5 g/cm^3, striking the Earth with a speed equal to the escape speed from the Earth. (b) Find one phenomenon on Earth that has a comparable energy associated with it. (c) Why is the escape speed a reasonable estimate? (Hint: Think of the speed an object would have if it fell from far away.)

26.7. For an asteroid of radius r and albedo a a distance d from the Earth, and a distance R from the Sun, find an expression for the amount of reflected sunlight reaching the Earth. You may treat the asteroid as a disk, oriented so the sunlight will be reflected toward the Earth.

Computer problem

26.1. Suppose a grain 10^{-3} cm in extent starts at rest near the Earth. It is pushed outward by the Sun's radiation pressure. How fast is it moving by the time it reaches Mars?

Chapter 27

The origin of life

In this chapter we look at the steps that led up to life on Earth, starting with the formation of the Solar System. We then look at the possibilities of finding life on other planets, both within the Solar System, and around other stars.

27.1 | Origin of the Solar System

One of our goals in studying the Solar System is understanding how it formed. As we studied the planets we saw that they provide many clues to the Solar System's history. In this section, we briefly outline some of the ideas that have been proposed. Any theory on the formation of the Solar System should be able to explain such things as the fact that the planets' orbits are approximately in the same plane, and the fact that the planets orbit in the same direction. In addition, it must be able to explain the distribution of angular momentum in the Solar System. Also, the different compositions and appearances of the planets must be explained.

Historically, two basic scenarios have been discussed. In one, the Solar System formed as a by-product of the Sun's formation. The material left over from the Sun's formation is the material out of which the planets formed. The idea was first discussed by *Rene Descartes* in 1644, and was elaborated upon by *Immanuel Kant*, and farther by *Pierre Simon de Laplace*, who was the first to take the effects of angular momentum into account. In the other scenario, originally proposed by *Georges Leclerc de Buffon*, the material to form the planets was ripped from the Sun by the effects of a passing object, possibly a comet.

Because of our present understanding of star formation, we now think that the Solar System is the remnant of the material that collapsed to form the Sun (Fig. 27.1). The original cloud might have been spherical. However, it must have been rotating, since we know that the Solar System has angular momentum. As we saw in Chapter 15, the result of the rotation is that collapse perpendicular to the axis of rotation is retarded, while that parallel to the axis of rotation continued. This means that the spherical cloud flattened to form a disk. It is the disk out of which the planets probably formed. Once the planets had formed, the debris not included in the planets was mostly cleared away by a very strong wind from the Sun. This would have been when the Sun was going through a T Tauri phase, and its wind would have been much stronger than it is today. The peak mass loss rate may have been 1 $M_\odot/10^6$ yr. The wind carried sufficient energy and momentum to sweep out the debris and stop the infall into the solar nebula.

Attempts have been made to calculate the minimum amount of material in the solar nebula. This is the amount of material in the planets, plus that which escaped during the formation. In understanding how the nebula produced planets, there is a problem involving angular momentum distribution. The Sun has only 2% of the angular momentum in the Solar System (see Problem 27.1), but it would be expected that most of the angular momentum is in the central condensation. To explain this, it has been proposed that the material to form the planets fell slowly into the cloud around the already forming Sun.

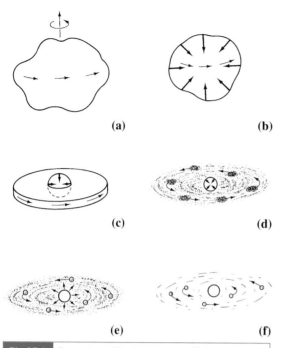

Fig 27.1. Formation of the Solar System. (a) A rotating interstellar cloud. (b) The cloud begins to contract. Since the angular momentum is conserved, the rotation becomes faster. (c) The rotation is fast enough to slow the collapse perpendicular to the axis of rotation, so a disk forms. The center is collapsing fastest, forming a denser concentration that will eventually become the Sun. (d) When the rotation prevents farther collapse of the disk, it breaks up into smaller clumps, so that some of the angular momentum is taken up by the orbital motion of the clumps. The clumps can then collapse. (e) Clumps of material gather together, forming planets, as the proto-Sun begins to radiate, and generate a large wind. (f) The wind clears debris from the Solar System.

In following the evolution of the solar nebula, we must keep track of three types of materials: gases, ices and rocks. Most of the mass was in the gas (as most of the mass of the interstellar medium is in gas). However, gas cannot be held to a growing planet by gravity, so it escapes from all but the largest objects. The ices are water (H_2O), carbon dioxide (CO_2) and nitrogen (N_2), along with some ammonia (NH_3) and methane (CH_4). These make up 1.4% of the mass of the Solar System. The rocks are iron oxides and silicates of magnesium, aluminum and calcium. Some of the iron was metallic and some of it was in iron sulfide (FeS). They can only be destroyed at high temperatures, in excess of 2000 K. They make up 0.44% of the mass (not including the Sun) in the Solar System. They are particularly prominent in the inner planets, while the ices are prominent in the outer planets. Comets provide us with the best clues on the initial composition of the rocks and ices.

The accretion of the nebula probably took place over 10 000 to 100 000 years. The first step in the process was for small grains to clump together. The grains collided, sometimes making larger ones, and sometimes breaking into smaller ones. The process produced many grains about 1 cm in size. These grains were large enough to settle through the gas in the plane of the nebula. This brought the clumps closer together, and allowed for even more collisions. Calculations indicate that the thin sheet of grains could then clump into objects with sizes of a few kilometers (essentially asteroid sized objects). About 1000 of these could then form a group held together by their own gravity. At that point, the groups were spinning too fast to collapse completely. Eventually these groups served as the cores for farther condensation of bodies orbiting at the same distance from the Sun.

Different parts of the Solar System then evolved differently because of the fall-off in solar radiation with distance from the Sun. The collapsing nebula had a higher temperature in the center (near the forming Sun) than at the edge. In equation (23.4) we found the equilibrium temperature of a planet as a function of the solar luminosity and its distance from the Sun. This expression works as well for dust particles. (Indeed, the calculation that led to equation (23.4) was essentially the same as that which we used to calculate the temperature of interstellar grains, in Chapter 14.) So, from equation (23.4), we see that the temperature falls off as the square root of the distance from the Sun. When the temperature was about 3000 K near the center, it was a few hundred kelvin in the regions of planetary formation. It also falls by a factor of about five between the orbits of Venus and Neptune. Therefore, different materials condensed at different distances from the center.

Another factor affecting the nature of forming planets was a fall-off in the density of material as one goes farther from the Sun. As we saw in Chapter 15, when even a uniform interstellar cloud collapses, it develops a higher density in the center than at the outside. In fact, ultimately the highest density center becomes the star. In the higher density regions near the center, the material is also moving faster, as a result of infall, converting gravitational potential energy into kinetic energy. The higher density and higher speeds near the center meant that collisions also played an important role in shaping the gas. As a result of the temperature and density variations, we can think of planetary formation as occurring in three zones: (1) the terrestrial planets, (2) the giant planets, and (3) comets.

Near the Sun, the temperature was too high for most of the gas (especially the H_2) to have survived the star formation process. So solid materials had to be involved. We think that the original building blocks for the terrestrial planets where chondrules (discussed in Chapters 23 and 26). These were heated to temperatures of 1500 to 1900 K, and then cooled. Chondrules that we can study in meteors suggest that, to give their particular structure, the heating and cooling took place very quickly, possibly over a few hours. This would mean that the early solar nebula was rocked with energetic events. These chondrules would have had a range of sizes, but typical sizes might have been a few millimeters. There were so many of these chondrules, and their relative speeds were so low (about 1 m/s), that they eventually began to stick together. We don't know what provided the attractive force. It has been speculated that the so-called van der Waals force could have been involved. This is a very weak electrical attraction between neutral objects when the charges can move so that the centers of positive and negative charge are in different places. This created small aggregates of chondrules, which could also grow by sweeping up dust. Eventually they grew to sizes of about 1 km. At that point, we call them *planitesimals*.

Even though there were many planitesimals, they were distributed over a large volume of space, so encounters between planitesimals were rare, maybe once per thousand years. Computer simulations show that these collisions eventually made larger objects, and after about 20 000 years

several Moon-sized objects should have appeared. After about ten million years, these objects collected to form most of the four terrestrial planets, though these planets probably continued to sweep up planitesimals for 100 million years. These collisions were constantly reforming the surfaces of the planets through violent events.

The outer edge of the inner zone is the asteroid belt. There is a large gap between Mars and Jupiter suggesting that there was room for another planet to form. It is the one gap in Bode's law (discussed in Chapter 22). We don't expect a gap, since we expect that the material in the solar nebula would have been falling off gradually in abundance, and we know there was enough material farther out to form the giant planets. The most likely explanation is that the early formation of the very massive Jupiter prevented the formation of a planet. This could have been either by Jupiter somehow preventing the formation of the more massive planitesimals, or by Jupiter somehow removing them after they had formed. We do know that Jupiter has been effective at removing objects from certain resonant orbits, e.g. the Kirkwood gaps discussed in Chapter 26.

In the second zone, material was far enough out for water ice to exist. Since O is more abundant than the elements that are important in dust grains (e.g. Si, Mg, Fe), particles of water ice (essentially snowflakes) would have been more abundant than dust particles in the second zone. It is thought that Jupiter and Saturn formed initially from planitesimals made up primarily of water ice. These planitesimals would have formed in a manner similar to those for the rocky planitesimals that formed the terrestrial planets. However, once Jupiter and Saturn had enough material to exert strong gravitational forces, then they would have collected all of the interstellar material (mostly gas and a little dust) that was near them. This resulted in two very massive planets. Also, the planets had compositions reflected in the interstellar medium 4.5 billion years ago. So the compositions of Jupiter and Saturn are essentially the same as that of the Sun, meaning that they have primarily hydrogen.

Uranus and Neptune formed in the outer parts of the second zone. The icy planitesimals would have filled a larger volume of space, meaning fewer collisions, and less chance for growth than the ones that started Jupiter and Saturn. There would therefore have been less gravity to hold interstellar gas in. Furthermore, the density of interstellar gas was lower the farther one got from the center of the solar nebula. So, Uranus and Neptune are just the result of the buildup of icy planitesimals, and are dominated by ices. Their compositions are therefore different from those of Jupiter and Saturn. (Fig. 25.13 illustrates some of the differences in composition between Jupiter and Saturn and Uranus and Neptune.)

Most of the satellite systems probably grew from a disk forming around the planet. This process repeated the formation of the rocky planets on a smaller scale. The satellites whose orbits are close to the ecliptic and are not too eccentric were probably made in this way. Satellites with very inclined or eccentric orbits may have been captured.

In the third zone, beyond Neptune, ice/rock planitesimals were formed. However, they fill such a large volume of space that gravitational encounters are very rare. This means that they cannot collect into a planet. The ones from Neptune's orbit out to 50 AU formed the Kuiper Belt. Those farther out formed the Oort cloud. As we discussed in Chapter 26, occasionally one of these objects has its orbit perturbed, and enters the inner Solar System as a visible comet.

The scenario that we have discussed probably left a large amount of debris around the planets. However, the Solar System is now relatively clean. Where did the leftover material go? We think that the Sun went through a stage when its wind was much stronger than it is now, much like the winds in T Tauri stars. The peak mass loss rate may have been 1 $M_\odot/10^6$ yr. The wind carried sufficient energy and momentum to sweep out the debris and stop the infall into the solar nebula.

Collisions among the particles also helped to clear the Solar System. Some of the debris crashed directly into planets, leaving the craters that we still see on Mercury, Mars and the Moon. Other collisions led to ejection of bodies from the Solar System. Finally, the momentum carried by the sunlight itself may have helped clear the

Solar System. The momentum carried by the light is the energy, divided by the speed of light. That is,

$$p = E/c$$

$$= h/\lambda$$

The radiation would have been able to sweep out small dust particles, and also carry away the orbital angular momentum of the dust. Finally, some of the collisions could also reduce the orbital energy of the dust, allowing it to fall into the Sun. This is known as the *Poynting–Robertson effect*. In these collisions, aberration of light, discussed in Chapter 7, makes it appear to the dust particle that the light from the Sun is coming from slightly ahead in the orbit. Thus, the particles lose energy, and eventually spiral into the Sun.

27.2 | Chemistry on the early Earth

It is very likely that the atmosphere, surface and oceans of the early Earth were very different from their present condition. We have very little direct evidence, and this has been an area of considerable disagreement.

As the Earth was forming, there was still a lot of debris in the Solar System, so the Earth was subjected to a much greater rate of bombardment than it is today. As the conditions became less violent, the Earth cooled enough for water to form on the surface. The continents were beginning to form. There was considerable volcanic activity, which introduced large quantities of hydrogen sulfide (H_2S) into the atmosphere.

There was no free oxygen, O_2. The oxygen in today's atmosphere is the result of plant life. (Remember, there is a symbiotic relationship between animal and plant life. Plants consume CO_2, which is given off by animals, and through photosynthesis convert sunlight into energy, and give off O_2 as a by-product. The O_2 is then used by the animals.) The absence of free oxygen also meant that there could be free iron (Fe) on the surface. Today any free Fe would react with O to make iron oxide (rust).

For a long time it was thought that the atmosphere contained a lot of hydrogen in various forms, especially molecular hydrogen (H_2), methane (CH_4), ammonia (NH_3) and water vapor (H_2O). More recently, it has been suggested that there was more carbon monoxide (CO) and carbon dioxide (CO_2) than methane, and more nitrogen (N_2) than ammonia. These more recent ideas also suggest that there was not very much molecular hydrogen. There is some disagreement over how much phosphorous was available. We will see below that P is an important component in some biotic molecules.

There is also some disagreement over what the temperature was on the early Earth. Estimates range from freezing to boiling. The Sun gave off 25% less energy 4 Gyr ago than it does now. However, there was no ozone layer to absorb solar ultraviolet, so a larger fraction would have penetrated to the surface.

There may have been frequent lightning in the early Earth's atmosphere. The effects of such lightning were simulated in a laboratory at the University of Chicago, in the early 1950s, by a graduate student, *Stanley L. Miller*, under the supervision of his advisor *Harold Urey*. (Urey had won the 1934 Nobel Prize in Chemistry for his discovery of deuterium.) The *Miller–Urey experiments* helped chemists understand how the first prebiotic molecules may have formed in the Earth's atmosphere. To simulate the effects of lightning, they carried out their experiment in a sealed glass tube with electrodes and produced repeated electrical discharges.

Miller and Urey started with a mixture of methane, ammonia and hydrogen. The effects of evaporation and condensation of the early oceans were simulated by recycling water through the system. They ran the experiment for a few days at a time and then analyzed what had been produced. After some runs they found simple organic molecules important as building blocks of life, including amino acids. At the time they did these experiments, it was thought that the early atmosphere contained large amounts of methane, ammonia and hydrogen. As we said above, there is now some disagreement over how abundant those molecules actually were. Alternative means have been investigated in which the simple amino acids formed in various clay deposits.

More recently, it has been suggested that molecules like those that came out of the Miller–Urey experiments could have been formed in the interstellar medium as part of the molecular cloud from which the Sun (and Solar System) formed. As we saw in Chapter 14, a rich collection of interstellar molecules has been found, including some simple prebiotic molecules. It had been thought that, even if these molecules were made in interstellar space, they would not survive the process of star formation, at least not in the inner Solar System. However, we know that some of this interstellar material has been preserved, as comets in the Oort cloud or the Kuiper belt. Some comets that passed close to the early Earth could have left some of this material behind, for it to sink into the atmosphere, and then eventually find its way to the surface.

There is some evidence that such materials might survive a meteoritic impact. Remember, meteors are the debris of comets, left in their orbits. The most famous example is the Murchison meteorite (mentioned in Chapter 26), which fell in 1969, in Murchison, Australia. It was found to contain a number of amino acids which are not likely to have been made on Earth.

27.3 | Origin of life on Earth

Whether as a result of conditions like those simulated by the Miller–Urey experiments, or as a result of deposition from comets, it is possible that the early atmosphere was enhanced in these prebiotic organic molecules. So the question is how we go from these simple organic molecules to the life that is around us now. Evidence suggests that there was a gap of almost 1 Gyr between the formation of the Earth and the appearance of the first multicell organisms.

At the molecular and cellular level, a fundamental tenet of life is the ability to replicate itself. That replication allows for the development of multicell organisms. It also allows for generations of organisms. Imperfect replications are called *mutations*. Mutations can have no effect, provide a change for the better (more able to survive), or provide a change for the worse (less able to survive). According to Darwin's theory of evolution, the increased survival rate of the beneficial mutations means that the better trait is passed on to future generations.

At the molecular level, for most life as we know it, the replication is carried out by a very large molecule, *deoxyribose nucleic acid (DNA)*. It is shaped like a ladder twisted into a double helix structure. This structure was first worked out in 1953 by *James Crick* (England) and *William Watson* (USA), and they won the Nobel Prize in Chemistry for their work. There are four possible parts, called *nucleotides*, for each position of the ladder. The nucleotides are made of subunits, called base, sugar and phosphate. In assembling a chain of nucleotides, sugars link to phosphates, so they alternate in the chain. The bases dangle off to the side. Each base can only have a particular base partner on the other side of its ladder rung. These partners are called *base pairs*. The sequence of bases determines the genetic information that is carried in the DNA, and governs the replication.

There is another nucleic acid, *ribonucleic acid (RNA)* . The current purpose of RNA is to facilitate the replication of the DNA by transferring information from the DNA to proteins. RNA consists of chains of up to a few thousand nucleotides. Each nucleotide has a phosphate, ribose (which is a five-carbon sugar), and one of four bases (adenine, guanine, cytosine and utracil). All four bases have a flat ringlike structure. By comparison, the sugar in DNA is deoxyribose (ribose without an oxygen), and one of the bases, utracil, is replaced by thymine (utracil to which a methyl group, CH_3, has been added). So you see there is a very close relationship between DNA and RNA. It is possible that historically RNA played a role in the development of DNA.

So, the question is, how do you go from a prebiotic soup to the complex DNA in less than 1 Gyr? One suggestion is that, once the simple amino acids formed, they would have had so many chances to react over half a billion years, that it was possible to form DNA molecules by random chance. One argument in favor of this is that since there was no life, there were no predators. Therefore, once formed, simple amino acids could stay for hundreds of millions of years without being destroyed. However, realistic calculations show that the likelihood of random formation is very small.

Most organic chemists working on this problem have decided that it is more likely that the DNA we have today is the end product of a series of well defined steps, building up more complex classes of molecules. While each of these steps may have taken some time, none was as improbable as the direct formation of DNA, and even the sequence of events is much more likely than the direct formation. Of course, there is no agreement on what those steps were. This is in part due to the fact that the initial conditions are not well known. Also, the current state is so far removed from the initial conditions that there are many equally plausible ways of getting here. We will just briefly note some of the more prevalent ideas.

It is generally agreed by those looking at the possible large steps that would lead from simple amino acids to DNA, that RNA is an important intermediary. In fact, one might be tempted to note that amino acids are the constituents of proteins, and proteins are the constituents of RNA, so the amino acids could have formed into proteins and the proteins formed into RNA. However, those studying how RNA works today have noted that RNA is a catalyst for the synthesis of proteins. This means that the RNA would have had to form first.

The formation of RNA without making proteins first is quite difficult (given the complexity of RNA). Chemists working on the problem have focused on the idea of finding enzymes that might serve as catalysts for this process. Remember, a catalyst is something that helps promote some reaction but is not changed in the process. Some catalytic reactions have been proposed which may have created RNA on a time scale of less than a year. Once the RNA formed, it could begin the process of replication. Furthermore, one by-product of such a process was the formation of certain proteins.

Once the chemicals are available, the development of life requires the formation of cells. Cells are the basis of all life we know now, and one of the questions that is still being addressed is when the cells first developed. In the first RNA that developed, replications that directly produced surviving molecules were favored. With the development of cells, a replication (and some variation) could be favored because it produced something which could help the cell survive. There are two different views on when cell walls began to appear. One is early in the process, and the other is late, about 3.8 Gyr ago. Different processes for the formation of RNA favor one or the other picture.

Cell walls are made of a lipid bilayer (double molecular level). It is about 5×10^{-7} cm thick. The molecules that make up these typically have two ends, one that attracts water and the other that attracts fat. Membranes grow by adding more material to a pre-existing membrane.

27.4 | Life in the rest of the Solar System?

If the development of life on Earth did not require a special set of circumstances, then we expect life to have started elsewhere in the galaxy. It is therefore of interest to search for life elsewhere, and the obvious starting place is our Solar System. Finding even primitive life elsewhere in the Solar System would indicate that the Earth is not just one lucky case, and would give us hope of finding it widespread in the galaxy. Also, finding certain types of life elsewhere in the Solar System would give us insights into how life actually formed on the Earth. When we talk about searches for life, we generally mean "life as we know it". That is, life based on carbon bearing (organic) molecules. (Science fiction writers have speculated on other forms of life, such as silicon based, but there is no current evidence to suggest that searching for such life forms would be fruitful.) Development of carbon based life generally requires water, so in choosing places to search for life, we would look at places with evidence for water, at least in the past. Such life would also require the presence of an atmosphere.

Let us think about how we might look for life elsewhere in the Solar System. Some used to suggest that "canals" on Mars were evidence for intelligent life, or that changing colors with seasons suggested vegetation. We now know that those canals don't really exist. They are artifacts of visually connecting unconnected features. The color changes are real, but are due to dust storms, which vary with season. So, we must look for life in much more subtle ways. The life we find may be microscopic, or it may be extinct, having left fossils. So we use remote sensing to select likely sites of current or former microscopic life, and then inspect those sites with various landing equipment. That landing equipment must have instruments capable of detecting the life or some by-product of its existence. For example, we may look for the results of respiration. Such searches may be limited in the sense that they are looking for some particular organism or by-product.

Lunar soil samples, returned to Earth by Apollo astronauts, have been extensively studied in the laboratory, with no evidence for extraterrestrial life. For the first few missions, astronauts were kept in a quarantine for an extended period of time, because of the fear that they might carry some form of previously unknown contagion. That practice was limited when the first few missions revealed no signs of microbial life. Not finding life on the Moon should not be surprising, as the Moon lacks both an atmosphere and water.

Mars is potentially an interesting place to look for life, either current or fossil. That is because we think that prior to 3.5 Gyr ago, when life was emerging on Earth, conditions on Mars were similar to those on Earth. There is evidence for abundant liquid water on Mars, in the form of rivers, lakes and possibly larger bodies, like oceans. We might ask how far the early, prebiotic, chemistry proceeded on Mars. Did such a chemistry develop so far as to lead to life – replicating molecules? If such early life started, how did it evolve? Is it still present or did it die off? If it is still present, we can look for it directly. If it died off, we can still look for fossil evidence.

The first attempts to answer these questions were made by the Viking landers. In designing experiments to look for chemical signs of life, e.g. respiration or photosynthesis, you have to make a decision about what chemicals you will look for. This requires making assumptions about the kind of life we are looking for. So, as a starting point, the Viking experiments were designed to look for microbial life with a chemistry similar to that on Earth. These experiments did not yield evidence for existing life "as we know it" at either site. A more extensive analysis of this data suggests that there is some evidence for organic chemical activity, but Martian life is not the only possible explanation. This shows some of the difficulties in designing and interpreting remote experiments to answer such subtle questions.

We are now early in the next phase of this search on Mars. From orbital mapping, we look for places that show evidence for an abundance of water in the past, as well as temperate weather conditions to promote chemical activity. (It has even been suggested that Mars had sources of heated water, like geothermal springs on Earth.) From these orbital missions landing sites have been chosen for current and future landers. These landers will use increasingly sophisticated experiments to explore the chemistry of the terrain surrounding the landing sites. Eventually, material from promising sites will be returned to Earth on unmanned spacecraft. These samples will be studied extensively in terrestrial laboratories.

There is already a small source of Mars surface material on Earth. These are rocks that were thrown off the surface of Mars by meteoritic impact, and then happened to strike the Earth, like other meteors that the Earth encounters. The hard part is to distinguish rocks from Mars from those that come from normal meteor showers. If we can study them in the laboratory, we find that their chemistry is generally like that at the Viking lander site. This strongly suggests that they are from Mars. One meteor "observatory" on Earth is in Antarctica, as it is easy to pick out rocks against the white snow/ice background. One object found in 1984 was not classified as Martian until 1993. It was studied extensively for the next two years, and the researchers found microscopic fossils, which they concluded may have come from Mars. However, other groups studying this meteor suggested that these fossils may have been contamination from Earth. This shows how difficult these experiments can be.

27.5 | Other planetary systems?

The notion that the Solar System formed as a by-product of the formation of the Sun has a number of interesting consequences. One is that, if planetary systems are a natural by-product of star formation, we should be able to find many other planetary systems in our galaxy. As you might suspect, looking for planets around a distant star is a formidable observational challenge for a number of reasons. Any radiation given off by the planets (either by reflected starlight or emitted far infrared and radio emission) would be very weak, especially at large distances. This is complicated by the fact that it is much weaker than the radiation from the star in that system. The linear separation between a planet and the star it orbits is not very large, so the angular separation is small, even for relatively nearby systems (a few parsec away). The masses of the planets are much less than the stars they orbit, so the recoil motion of the star is also very small.

To illustrate the problem, consider Example 27.1.

Example 27.1 Appearance of the Solar System

Assume we are observing the Solar System from a distance of 10 pc. (a) What is the angular separation between Jupiter and the Sun? (b) Estimate Jupiter's apparent magnitude. (c) What is the angular amplitude of the Sun's motion in response to the gravitational force exerted on it by Jupiter? (d) By how much will a spectral line from the Sun be Doppler-shifted due to Jupiter's orbital motion?

SOLUTION

(a) We can generalize equation (2.17) to tell us that if an object is a distance $D(pc)$ from us, and has a linear separation $R(AU)$, then its angular separation (in arc sec) is

$$\Delta\theta(\text{arc sec}) = R(AU)/D(pc)$$

So for Jupiter (5.2 AU from the Sun),

$$\Delta\theta(\text{arc sec}) = 5.2/10$$

$$= 0.5 \text{ arc sec}$$

(b) To estimate the brightness of Jupiter, we first calculate how much sunlight hits it. The fraction, f, of sunlight hitting Jupiter is the ratio of the solid angle of Jupiter as viewed from the Sun ($\pi R_j^2/d_j^2$) divided by the solid angle of a full sphere (4π), so

$$f = (R_j^2/4\ d_j^2) = 8 \times 10^{-9}$$

If the albedo of Jupiter is 0.5, then half of this is reflected, meaning that Jupiter is 4×10^{-9} as bright as the Sun. This makes its absolute magnitude $2.5\log_{10}(4 \times 10^{-9}) = 21$ mag fainter than the Sun. The absolute mag of the Sun is $+4.8$, so the absolute mag of Jupiter is approximately $+26$, and at a distance of 10 pc, the apparent magnitude equals the absolute magnitude, so the apparent magnitude is $+26$, barely at the limit of our sensitivity, and it must be seen next to the Sun, which is 21 mag brighter and only 0.5 arc sec away.

(c) The ratio of the radii of the orbit of Jupiter and that of the Sun about the Sun/Jupiter center of mass is just the ratio of the masses:

$$r_2/r_1 = m_1/m_2$$

$$= (1.9 \times 10^{30})/(2 \times 10^{33})$$

$$= 10^{-3}$$

So the radius of the Sun's orbit about the center of mass is

$$r_2 = (10^{-3})(5.2 \text{ AU})$$

$$= 5.2 \times 10^{-3} \text{ AU}$$

At a distance of 10 pc, the angular size is

$$\Delta\theta(\text{arc sec}) = 5.2 \times 10^{-3}/10$$

$$= 5.2 \times 10^{-4} \text{ arc sec}$$

(d) To find the orbital velocities, we use equation (5.23) and solve for the sum of the speeds:

$$(v_1 + v_2)^3 = \frac{2\pi G(m_1 + m_2)}{P}$$

$$= \frac{2\pi(6.67 \times 10^{-8} \text{ dyn cm}^2/\text{g}^2)(2.0 \times 10^{33} \text{ g})}{(11.9 \text{ yr})(3.16 \times 10^7 \text{ s / yr})}$$

$$= (2.2 \times 10^{18} \text{ cm/s})^3$$

Taking the cube root gives

$$v_1 + v_2 = 1.3 \times 10^6 \text{ cm/s} = 13 \text{ km/s}$$

The ratio of the speeds is equal to the ratio of the masses (equation 5.12),

$$v_2/v_1 = m_1/m_2$$

$$= (1.9 \times 10^{30})/(2 \times 10^{33})$$

$$= 10^{-3}$$

Solving for the speed of Jupiter, we have

$$v_1 = 1.3 \times 10^{-2} \text{ km/s} = 13 \text{ m/s}$$

So $v/c = 4.3 \times 10^{-8}$.

For comparison, remember that the Doppler shifts and broadening of typical interstellar lines were a few kilometers per second, so the recoil of the Sun due to Jupiter's motion is much less than that. The Doppler shift is independent of the distance. However, the farther away the object, the fainter the signal, so the harder it is to measure the Doppler shift.

This illustrates the difficulty of detecting planets around other stars. Shining by reflected sunlight, Jupiter would appear just at our detection threshold, even if it wasn't swamped by the direct light from the Sun which would be less than an arc second away. This does suggest, however, that if you were going to detect direct radiation from a planet, you might do better in the infrared where the blackbody radiation from the planet peaks. The Sun still gives off much more radiation, but the imbalance is less (see Problem 27.4).

Directly seeing the motion of the Sun about the center of mass would also be very difficult. The best hope is to look for the Doppler shift caused by that motion. As we discussed for binary stars (Chapter 5), that motion is best observed if we are in the plane of the orbit. We would observe a variation in the star's Doppler shift that looked like a sine wave with a period equal to the orbital period of the planet. Just as for binary stars, if the orbit is inclined, you still see periodic motion, but the range of Doppler shifts is reduced by the sine of the inclination angle. If more than one planet is present (e.g. Jupiter and Saturn) you would see a more complicated pattern that comes from adding two sine waves with different periods, amplitudes and phases (see Problem 27.5).

The technique which has proved most successful has been looking for the variations in the Doppler shifts of nearby stars. A group headed by *Geoffrey Marcy* and *R. Paul Butler* has studied a large number of potential systems. Other groups have also made independent measurements, giving more confidence, as the measurements are difficult. These groups have studied more than 1000 stars. This comprises a nearly complete sample of Sun-like stars within 30 pc of us. They have found evidence for a planet in more than 90 systems (so far). More recently, they have found a few systems with evidence for more than one planet.

Marcy and Butler's observations are done at Lick Observatory using a 3 m telescope, and on Mauna Kea, using the 10 m Keck-I telescope (discussed in Chapter 4). Because the variations in the Doppler shifts are so small, very accurate spectrometers had to be used. Before the starlight enters the spectrometer, it passes through a tube of iodine gas. The iodine produces a series of narrow absorption lines superimposed on the stellar spectrum, providing a very accurate calibration, and making it easier to detect variations in Doppler shifts. The spectra are also corrected for things like the motion of the Earth around the Sun, the motion of the observatory about the center of the Earth, and perturbations on the Earth's motion due to the other planets. Once a variation consistent with a planet is detected, its period is measured. The central stars of these systems are nearby and are well studied, so their properties are well known. If the mass of the central star is known, then the period and Kepler's third law can be used to determine the semi-major axis of the planet's orbit. By measuring the amplitude of the Doppler shift, we can tell how much the star is moving about the center of mass, so we can determine the mass of the planet, assuming that we are in the plane of its orbit. If we are not in the plane of its orbit then the derived mass is a lower limit.

Some sample data are shown in Fig. 27.2 The dots show the data points, with error bars to indicate the uncertainties in the measurements. The dashed lines show the best fit to the data. The masses are expressed as $M_{JUP}/\sin(i)$, since we don't know the inclination angle. So, these numbers are lower limits to the true mass. They are expressed as M_{JUP}, since that is a convenient reference.

The orbital periods range from 3 days to 15 yr. and the Doppler shift variations range from 11 to 1800 m/s. The orbital semi-major axes range from 0.038 to 6.0 AU. Orbits range from almost circular to eccentricities of 0.93. $M/\sin(i)$ ranges from 0.22 to 14.7 Jupiter masses. The distribution of masses is shown in Fig. 27.3.

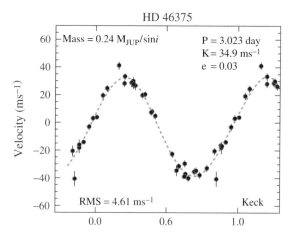

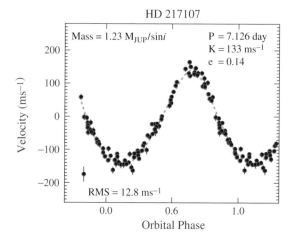

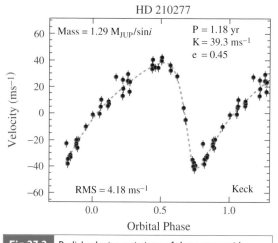

Fig 27.2. Radial velocity variations of three stars with evidence for planets. The horizontal axis is phase within the orbit, relative the listed period. Observations over many periods are combined in this way. [Geoffrey Marcy, University of California at Berkeley]

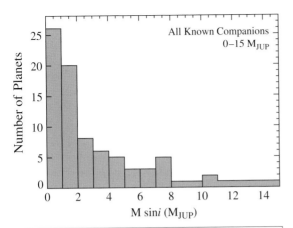

Fig 27.3. Mass distribution for extrasolar planets. [Geoffrey Marcy, University of California at Berkeley]

This small sample has already raised a number of questions. For example, $5M_{JUP}$ planets are found closer to their stars than Mercury is to the Sun. There are also very few very massive planets, even though these would have been very easy to detect. Also compared to our Solar System, a number of very eccentric orbits have been found.

There is also a way to look for systems that might be forming planets. We saw earlier in this chapter that we think that planetary systems form from the disks that are a by-product of the star formation process. In Chapter 15, we saw how the presence of these disks could be inferred from the existence of bipolar outflows. That is, the disks collimate the outflow. Remember, for the bipolar flows, the presence of the disks has only been inferred in most cases, though there are a few examples where we see objects that could be the collimating disks. Even those disks are hard to see because they are small and subtend small angles. The disks that will form planets around a solar mass star are even smaller. For example, a 1000 AU disk at the 500 pc distance of the Orion Nebula, the nearest extensive star forming region, would subtend an angle of 2 arc sec. Of course, this is large compared to the size of Jupiter, so it would be much easier to see than even a giant planet.

These disks are best seen in the infrared for a number of reasons. They are cooler than the protostars they surround, so they give off relatively more radiation in the infrared than in the visible. Also, since they are often deep inside molecular clouds, with tens of magnitudes of visual extinction, they are hard to see in the visual. Of course, since they are small, we must use infrared observations with very good angular resolution. HST has provided the opportunity to carry out these observations, and a few samples are shown in Fig. 15.26.

These disks are important because they allow us to study the stage between the collapse of a molecular cloud to form a star and the formation of a planetary system. Of course, in order to study these disks in detail, we would like to be able to do high resolution spectral line observations, so we can trace the velocity structure of the disks. This is something that will be most easily done on the Atacama Large Millimeter Array (ALMA, Fig. 4.32) when it is finished.

27.6 | Searches for extraterrestrial intelligence

A number of recent discoveries suggest that life may be more common in the galaxy than many had thought. It appears that organic molecules can form and survive in even the hostile environment of interstellar space. It also appears that planetary systems form as a natural by-product of star formation, so there might be a large number of potential hosts to life. It is natural to suggest that if life has had sufficient time to evolve on a planet, then it might lead to intelligent life at some point. Such intelligent life might, knowingly or accidentally, give off evidence of its existence. These thoughts have fueled the push for *searches for extraterrestrial intelligence (SETI)*.

The important issues in SETI are (1) what is the likelihood that detectable extraterrestrial civilizations exist (or how many exist), and (2) what is the best strategy for detecting them. We look briefly at both issues.

The development (and survival) of a detectable extraterrestrial civilization depends on a number of factors. The important factors were first put down by *Frank Drake* (Cornell) in the famous *Drake equation*, which gives the number of civilizations in our galaxy that would be able to contact each other as

$$N = R_* f_P n_e f_\ell f_i f_C L \qquad (27.1)$$

where R_* = the rate at which stars are forming, f_P = the fraction of stars that have planets, n_e = the number of planets per planetary system with conditions suitable for life to develop, f_ℓ = the fraction on which life actually exists, f_i = the fraction of life forms that develop intelligence, f_C = the fraction of intelligent species that choose to communicate and L = the average lifetime of the civilization after they reach a technological state.

We can estimate some of these quantities and make guesses at others. The low mass star formation rate in the galaxy is about 10/yr. Theoretical considerations involving angular momentum suggest that all stars should form either as multiple stars (binaries etc.) or have planets (or both). Since roughly half the stars are binaries, we might suggest that at least the other half should have planets. As a more conservative estimate, initial searches have revealed planets in roughly 10% of the systems searched, so we take this factor as 0.1. The number of planets per system with conditions suitable for life is hard to estimate. It can also include large moons, like Titan or Europa, so for our Solar System it might be five (Earth, Venus, Mars, Titan, Europa). Let's be optimistic and take this number to be one. The fraction on which life exists might be 0.1, and the fraction of those on which intelligent life exists, could be 0.01. We can guess that half of intelligent civilizations would choose to communicate. The final step is the lifetime of the civilization during which communication is possible. For us, communication has only been possible for the last 50 years (since the development of radar in WW II). It is very possible that a civilization does not live long after developing technology, either because it might kill itself off in a nuclear war (or accident) or might simply pollute the planet or use up its natural resources. If a civilization survived long enough to develop space travel, then they might be able to relocate when they made their home planet uninhabitable. Such a civilization might be more outward looking and more interested in communication. Such a civilization might survive 10^6 yr, but many others might only

survive 100 yr. So, let's try an average lifetime of 10^5. Putting all these together gives $N = 50$. Obviously, the uncertainties are large, but this discussion gives us a feel for what the considerations are.

If 50 civilizations are uniformly spread over the disk of the galaxy (Problem 27.6) then the mean separation between civilizations is 1 kpc, or 30 000 ly. This means that any radiation takes 30 000 yr to get from one civilization to the next. This number depends on $N^{1/2}$, so it does not depend very strongly on your assumptions that went into N. Spacecraft travel much slower than the speed of light. Even at 0.1 c (a pretty ambitious speed) it would take 300 000 yr to get from civilization to civilization. This is way too long to make it useful. This is why even astronomers who think that there may be abundant life in the galaxy do not believe that we are being visited by alien spacecraft.

The most obvious form of communication is electromagnetic radiation. Only moderate wavelength (1 to 30 cm) radio waves pass through interstellar space with little attenuation. This physical fact doesn't change with position in the galaxy, so we can presume that a civilization would use radio waves in that range for such communication. As we saw in Chapter 4, radio transmitters and receivers are relatively narrow band, typically covering ~100 MHz at a time (see Problem 27.7). We must therefore make some guess, then retune the receiver and repeat the observations at nearby frequencies, so we cover a reasonable frequency range. We would want to avoid strong interstellar lines (like the 21 cm line) because we could not distinguish a natural from an artificial signal. Some have suggested between 21 cm and 18 cm (the OH line), far enough away from either line so that you would not see it red- or blueshifted into your search window.

You must also decide where to look. This is because radio telescopes have beams that only look at a small fraction of the sky. One strategy of SETI projects has been to look in great detail in the directions of nearby stars that may have planetary systems and have environments like our Solar System. Less time per location could be expended in looking systematically over large parts of the sky. So this is one of the great technological problems of SETI. You must search over two spatial dimensions (on the sky) and frequency. Sensitive searches are being carried out at Aricebo, taking advantage of the large surface (now upgraded), and the ability to use a number of receivers simultaneously, so we can cover more frequency ranges. More continuous coverage is being done using smaller telescopes.

Another problem is how to recognize a signal from an extraterrestrial intelligent source. The recent development of fast relatively inexpensive computers has made it possible to search signals for regular patterns or variations that could not be natural. Remember, when pulsars were first discovered, some thought that they might be signals from extraterrestrial civilizations.

There is one other possibility that has been suggested – neutrinos. They travel at the speed of light. They pass through interstellar space virtually unattenuated. Detectors don't have to be tuned to a narrow energy (frequency) range to detect them. Their energy can be deduced after they are detected. The same is true for direction. Detectors for neutrinos from natural extraterrestrial sources (e.g. supernovae) are just large tanks of water (or part of the ocean). They don't have to be pointed. They will detect neutrinos from any direction. After it is detected, the direction can be inferred. Of course, as we have seen, generating and detecting neutrinos is very hard. On the other hand our neutrino technology is 100 years behind our radio technology. (Who would have predicted 100 years ago where our radio technology is today?) On the time scale of the evolution of civilizations 100 years is not very much.

Chapter summary

In this chapter we looked at the steps that led to the origin of life, starting with the formation of the Solar System.

We think that the Solar System formed as a by-product of the formation of the Sun, from the material left behind in a disk orbiting the proto-Sun. A decrease in temperature and density of the material led to different formation scenarios for the inner planets, the outer planets and the comets. The final stages of formation must have included a strong solar wind to clear out the debris that was not part of the planets.

Because of abundant water and a reasonable temperature range, we think that prebiotic organic molecules formed on the early Earth. There are still uncertainties about the mechanisms, including the relative importance of electrical activity (as analyzed in the Miller–Urey experiments) and seeding from early comets. It is also not clear how one gets from these prebiotic molecules to DNA, which is the basis for our replicating life. Despite the large amount of time available, it is not likely that a random set of reactions would have led to DNA. It is more likely that a series of well defined steps was involved.

We think that conditions may have been right for the formation of life in other parts of the Solar System. The presence of water would have been important in that formation. We think that early Mars had conditions much like the early Earth, including large amounts of water, so it is possible that life may have formed on early Mars in much the same way as it did on Earth. Searches for traces of that life are in their early stages. There has been some suggestion that a rock from Mars, found in Antarctica, may have signs of microbial life, though there is also a possibility of contamination from the Earth.

In looking for life beyond the Solar System, the first step is to find other planetary systems. We saw how difficult it is to detect planets orbiting even a nearby star. The angular separations are very small. The planets are much fainter than their host stars. They are also much less massive, so it is hard to measure the Doppler shift in the stars as they move in response to the orbital motion of the planets. We described some experiments which are just beginning to yield evidence for such Doppler shifts in a few nearby systems.

The final question we looked at is the possibility of detecting extraterrestrial intelligence. We first looked at the factors affecting the number of such civilizations that we might detect. The factors are very uncertain, but there may be of the order of tens of such systems in our galaxy. This means that they are probably too far apart to visit with spacecraft, but we can detect them by radio waves. Searching for such signals requires mapping the sky a little bit at a time, and also covering a reasonable frequency range for detection.

Questions

27.1. If a planet had formed in the asteroid belt, do you think it would have been more like Mars or Jupiter? Justify your answer.

27.2. It appears that some planets come in twinlike pairs, Earth–Venus, Jupiter–Saturn, Uranus–Neptune. Do you think this is an accident?

27.3. Looking at each term in the Drake equation (27.1), make an argument why it might be higher or lower than taken for the calculation in this chapter.

Problems

27.1. Calculate the angular momentum of the Sun, and compare it with that of the rest of the Solar System.

27.2. How many Moon-mass objects would it take to form each of the inner planets?

27.3. (a) How much would the Sun move (angular motion) in response to the orbital motion of a planet with ten Jupiter masses, orbiting 50 AU from the Sun, as viewed from 10 pc away? (b) What would the Doppler shift be?

27.4. What are the relative brightnesses of Jupiter and the Sun at the wavelength where Jupiter's blackbody spectrum peaks?

27.5. For an extrasolar planet, using dv (the variation in the Doppler shift), the period, P, and the mass of the star M, find the radius of the planet orbit, and the mass of the planet (a) if we are in the plane of the orbit, (b) if the orbit is tilted.

27.6. If there are N civilizations uniformly spread over the disk of our galaxy (with a radius of 15 kpc), what is the mean separation between civilizations (ignore the thickness of the plane)?

27.7. How many 100 MHz slices would it take to cover the whole spectrum from 1 cm to 30 cm?

Computer problems

27.1. Calculate the far IR luminosity of Jupiter and compare it with that of the Sun.

27.2. Make a graph of the Doppler shift for two planets (e.g. Jupiter and Saturn).

27.3. Use the data in Fig. 27.2 to derive masses for the three planets whose data are given.

Appendix A Glossary of symbols

a albedo
          semi-major axis of ellipse
$\mathbf{a}$ acceleration
A angle to convergent point of moving
          cluster
          area
          atomic mass number
          extinction

b bottom quark
          distance at which a trajectory would pass an
          object if there were no deflection
          semi-minor axis of ellipse
B blue filter
$B_\nu(T)$ Planck function at temperature T
$\mathbf{B}$ magnetic field
BC bolometric correction

c charm quark
          speed of light

d distance from Earth to astronomical object
          down quark
          telescope diameter
d deuteron
d_0 reference distance for absolute magnitudes
dM mass of thin shell

e charge of electron
          charge of proton
          eccentricity of ellipse
          emissivity
e^+ positron
e^- electron
E energy
$\mathbf{E}$ electric field

f energy flux
          focal length
          fraction of radiation absorbed
$\mathbf{F}$ force

g acceleration of gravity
g_i statistical weight of the ith state
G universal gravitation constant

h Planck's constant
H Hubble parameter
H_0 current value of Hubble parameter
H hydrogen
HI atomic hydrogen
HII ionized hydrogen
$H\alpha$ first Balmer line

H_2 molecular hydrogen

i inclination of orbit
          intensity
I moment of inertia

J angular momentum
          angular momentum quantum number

k Boltzmann constant
          constant, giving curvature of universe
K kinetic energy
KE kinetic energy

ℓ galactic longitude
          path length for absorption
L luminosity
          mean free path
$L_\odot$ solar luminosity
$\mathbf{L}$ angular momentum

m magnitude
          mass
          order of interference maximum
m_e electron mass
m_g gravitational mass
m_i inertial mass
M_J Jeans mass
m_n neutron mass
m_p proton mass
m_r reduced mass
M absolute magnitude
          mass
$M(r)$ mass interior to radius r
$M_\odot$ solar mass

n density
          index of refraction
          principal quantum number
n neutron
n_i level population (i can be any index)
N column density
          number of lines on a grating
          number of neutrons in a nucleus

p parallax angle
p proton
$\mathbf{p}$ momentum
P period of orbit
          power
          pressure
          probability
PE potential energy

q	deceleration parameter	z	zenith distance (angle away from zenith)
	electric charge		redshift $(\Delta\lambda/\lambda_0)$
q_0	current value of deceleration parameter	Z	number of protons in a nucleus (atomic
r	radius		number)
R	distance from center of galaxy	α	alpha particle, or helium nucleus
	radius	β	beta particle (electron or positron)
	rate		v/c
	ratio of total-to-selective extinction	γ	gamma-ray
	resolving power of prism or grating		photon
	Rydberg constant	γ	ratio of specific heats
	scale factor in cosmology		special relativistic factor $(1 - \beta)^{-1/2}$
R_J	Jeans length	δ	increment
R_S	Schwarzschild radius	Δ	increment
$R_\odot$	solar radius	Δs	space-time interval
		Δv	velocity spread
s	strange quark	$\Delta\lambda$	wavelength shift or spread
t	time	$\Delta\nu$	frequency shift or spread
	top quark	$\Delta\theta$	telescope angular resolution due to
t_{ff}	free-fall time		diffraction
t_{rel}	relaxation time	ϵ	rate of energy generation, per unit mass
T	temperature	Φ_B	magnetic flux
T_i	ionization temperature	κ	opacity
T_k	kinetic temperature	λ	wavelength
T_x	excitation temperature	λ_{max}	wavelength at which spectrum peaks
u	up quark	λ_0	rest wavelength
	speed	Λ	cosmological constant
U	potential energy	μ	proper motion
	ultraviolet filter	ν	frequency
		ν	neutrino
v	speed	$\bar{\nu}$	antineutrino
v_a	orbital speed at aphelion or apastron	ν_0	rest frequency
v_{esc}	escape velocity	ρ	density
v_p	orbital speed at perihelion or periastron	ρ_{crit}	density to close the universe
v_r	radial velocity	σ	cross section
v_{rms}	root-mean-square speed		Stefan–Boltzmann constant
v_T	tangential velocity	τ	optical depth
$\mathbf{v}$	velocity	τ_e	lifetime to fall to 1/e of initial value
V	visible filter	$\tau_{1/2}$	half-life
	volume	ω	angular speed
W	W particle	Ω	angular speed
			ratio, ρ/ρ_{crit}, for the universe
$\mathbf{x}$	position		
X_r	rth ionization state of element X		

Appendix B | Physical and astronomical constants

Physical constants

speed of light	$c = 2.99792456 \times 10^{10}$ cm/s
	$= 2.99792456 \times 10^{5}$ km/s
gravitation constant	$G = 6.6732 \times 10^{-8}$ dyne cm^2/g^2
Boltzmann constant	$k = 1.3806 \times 10^{-16}$ erg/K
Planck's constant	$h = 6.6262 \times 10^{-27}$ erg s
Stefan–Boltzmann constant	$\sigma = 5.6696 \times 10^{-5}$ erg/cm^2 K^4 s
Wien displacement constant	$\lambda_{max}T = 2.89789 \times 10^{-1}$ cm K
Rydberg constant	$R = 1.097373 \times 10^{5}$/cm
Avogadro's number	$N_A = 6.022169 \times 10^{23}$/mol
atomic mass unit	$u = 1.66053 \times 10^{-24}$ g
mass of proton	$m_p = 1.6726 \times 10^{-24}$ g
mass of neutron	$m_n = 1.6749 \times 10^{-24}$ g
mass of electron	$m_e = 9.1096 \times 10^{-28}$ g
mass of hydrogen atom	$m_H = 1.6735 \times 10^{-24}$ g
charge of proton	$e = 4.8033 \times 10^{-10}$ esu
Bohr radius	$a_0 = 5.29177 \times 10^{-9}$ cm

Astronomical constants

astronomical unit	1 AU $= 1.4959789 \times 10^{13}$ cm
	$= 1.4959789 \times 10^{8}$ km
parsec	1 pc $= 3.0856 \times 10^{18}$ cm
	$= 3.0856 \times 10^{13}$ km
	$= 3.2615$ ly
light year	1 ly $= 9.4605 \times 10^{17}$ cm
solar mass	$M_\odot = 1.9891 \times 10^{33}$ g
solar radius	$R_\odot = 6.9598 \times 10^{10}$ cm
	$= 6.9598 \times 10^{5}$ km
solar luminosity	$L_\odot = 3.83 \times 10^{33}$ erg/s
Earth mass	$M_E = 5.977 \times 10^{27}$ g
Earth radius (equatorial)	$R_E = 6.37817 \times 10^{8}$ cm
	$= 6.37817 \times 10^{3}$ km
Earth–Moon distance	$R_{EM} = 3.84403 \times 10^{10}$ cm
	$= 3.84403 \times 10^{5}$ km
Moon mass	$M_M = 7.35 \times 10^{25}$ g
Moon radius	$R_M = 1.738 \times 10^{8}$ cm
	$= 1.738 \times 10^{3}$ km
galactic center–Sun distance	$R_0 = 8.5$ kpc
orbital speed of the Sun about the galactic center	$v_0 = 220$ km/s

Appendix C | Units and conversions

Prefixes and symbols for powers of 10

Standard form	Symbol	Prefix
10^{-18}	a	atto
10^{-15}	f	femto
10^{-12}	p	pico
10^{-9}	n	nano
10^{-6}	μ	micro
10^{-3}	m	milli
10^{-2}	c	centi
10^{-1}	d	deci
10^{1}	da	deca
10^{2}	h	hecto
10^{3}	k	kilo
10^{6}	M	mega
10^{9}	G	giga
10^{12}	T	tera

Length

1 in = 2.54 cm (exact)
1 m = 1.094 yd
1 km = 0.6214 mi
(See Appendix B for astronomical length units.)

Energy

1 joule = 10^7 erg
1 erg = 6.242×10^{11} eV
$m_p c^2$ = 938.3 MeV
$m_e c^2$ = 0.511 MeV
1 jansky (Jy) = 10^{-26} W/m^2 Hz

Appendix D | Planet and satellite properties

Tables D.1 and D.2 were compiled from material appearing in NASA's 'Planetary fact sheet'.

Table D.1. Planet properties.

Planet	a (AU)	Orbital period (yr)	i (deg)	e	Rotation period	Axis tilt (deg)	Mass (Earth)	Radius (Earth)
Mercury	0.387	0.241	7.00	0.205	58.8^d	0	0.055	0.383
Venus	0.723	0.615	3.40	0.007	-244^d	-2	0.815	0.949
Earth	1.000	1.000	0.00	0.017	23.9^h	23.5	1.000	1.000
Mars	1.52	1.88	1.90	0.094	24.6^h	28.2	0.107	0.533
Jupiter	5.20	11.9	1.30	0.049	9.9^h	3.1	318	11.2
Saturn	9.58	29.4	2.50	0.057	10.7^h	26.7	95.2	9.45
Uranus	19.2	83.7	0.78	0.046	-17.2^h	-82.1	14.5	4.01
Neptune	30.1	163.7	1.78	0.011	16.1^h	28.3	17.1	3.88
Pluto	39.2	248	17.2	0.244	6.39^d	122	0.002	0.19

a = Semi-major axis of orbit; i = inclination of orbit; e = eccentricity of orbit. Rotation period is sidereal. 1 AU = 149 600 000 km. Mass and radius are given as a fraction of Earth's. Mass of Earth = 5.997×10^{27} g. Radius of Earth (equatorial) = 6378 km.

Table D.2. Satellite properties.

	a (10³ km)	Orbital period (days)	e	i (deg)	Mass (10²⁰ kg)	Radius (km)
Earth						
Moon	385	27.3	0.055	18.3–28.6 V[a]	730 000	1738
Mars						
Phobos	9.38	0.319	0.018	1.0	0.09	13
Deimos	23.5	1.26	0.002	2.0	0.02	7.5
Jupiter[b]						
Galilean satellites						
Io	422	1.77	0.000	0.04	893	1822
Europa	671	3.55	0.000	0.47	480	1561
Ganymede	1070	7.16	0.001 V	0.21	1482	2631
Callisto	1880	16.7	0.007	0.51	1076	2410
Lesser satellites						
Metis	128	0.298	0	0.02	0.01	20
Adrastea	129	0.295	0	0.03	0.002	10*
Amalthea	181	0.489	0.003	0.4	0.075	75*
Thebe	222	0.670	0	0	0.008	50*
Leda	11 200	239	0.16 V	26.7	0.000 06	5
Himalia	11 500	251	0.16 V	27.6	0.095	85
Lysithea	11 720	260	0.110 V	29.0	0.0008	12
Elara	11 790	260	0.22 V	24.8	0.008	40
Ananke	21 280	610 R[c]	0.24 V	149	0.0004	10
Carme	23 400	702 R	0.25 V	165	0.001	15
Pasiphae	23 620	708 R	0.41 V	15	0.003	18
Sinope	23 940	724 R	0.25 V	158	0.008	14
Saturn[a]						
Greater satellites						
Mimas	184	0.942	0.020	1.5	0.0005	200*
Enceladus	238	1.37	0.004	0.0	0.001	248*
Tethys	295	1.89	0.000	1.9	0.01	527*
Dione	377	2.74	0.002	0.0	0.02	560
Rhea	527	4.52	0.001	0.4	0.034	764
Titan	1222	15.9	0.029	0.3	1.8	2575
Hyperion	1481	21.2	0.104	0.4		142*
Iapetus	3561	79.3	0.028	14.7 V	0.026	718
Lesser satellites						
Pan	133	0.575		0.0	0.000 03	10
Atlas	138	0.602	0.002	0.3	0.0001	17*
Prometheus	139	0.613	0.004	0.0	0.0033	50*
Pandora	142	0.629	0.004	0.0	0.0020	45*
Epimetheus	151	0.694	0.009	0.3	0.0054	60*

Janus	152	0.695	0.007	0.1	0.0192	90*
Calypso	295	1.89		0.0	0.0004	12*
Telesto	295	1.89		0.0	0.0007	10*
Helene	377	2.74	0.005	0.2	0.0003	16*
Phoebe	13 000	550 R	0.163	175	0.072	110*

Uranus

Major satellites

Miranda	130	1.41	0.003	4.2	0.66	235
Ariel	191	2.52	0.003	0.3	13.5	580
Umbriel	266	4.14	0.005	0.3	11.7	585
Titania	436	8.71	0.002	0.1	35.2	789
Oberon	583	13.5	0.001	0.1	30.1	762

Lesser satellites

Cordelia	49	0.33				20
Ophelia	53	0.37				21
Bianca	59	0.44				27
Cresida	61	0.46				40
Desdimona	63	0.47				32
Juliet	64	0.49				47
Portia	66	0.51				68
Rosalind	70	0.56				36
Belinda	75	0.62				40
Puck	86	0.77				81
Caliban	7320	580 R	141			48
Stephano	8002	677 R	144			10
Sycorax	12 179	1283 R	159			95
Prospero	16 418	1993 R	152			15
Setebos	17 459	2202 R	158			15

Neptune

Naiad	48	0.29				29
Thalassa	50	0.31				40
Despina	52	0.33				74
Galatea	62	0.43				79
Larissa	74	0.55				95
Proteus	118	1.12				210
Triton	355	5.88	0.000	20	214	1353
Nereid	5513	360	0.75	28	0.2	170

Pluto

Charon	17	6.39	0.0		14 600	400

a = semi-major axis of orbit; i = inclination of orbit; e = eccentricity of orbit.

[a]For Jupiter and Saturn there are additional smaller satellites, which are not listed here.

[b]V = variable.

[c]R = retrograde.

*Indicates an average size for a non-spherical object.

Appendix E Properties of main sequence stars

Spectral type	M_v	$B-V$	T (K)	M_{BOL}	$M/M_\odot$	$R/R_\odot$	$L/L_\odot$
O5	−6	−0.45	35 000	−10.6	39.8	17.8	3.2×10^5
B0	−3.7	−0.31	21 000	−6.7	17.0	7.6	1.3×10^4
B5	−0.9	−0.17	13 500	−2.5	7.1	4.0	6.3×10^2
A0	+0.7	0.0	9 700	0.0	3.6	2.6	7.9×10^1
A5	+2.0	+0.16	8 100	+1.7	2.2	1.8	2.0×10^1
F0	+2.8	+0.30	7 200	+2.7	1.8	1.4	6.3
F5	+3.8	+0.45	6 500	+3.8	1.4	1.2	2.5
G0	+4.6	+0.57	6 000	+4.6	1.1	1.05	1.3
G5	+5.2	0.70	5 400	+5.1	0.9	0.93	7.9×10^{-1}
K0	+6.0	+0.54	4 700	+5.8	0.8	0.85	4.0×10^{-1}
K5	+7.4	+1.11	4 000	+6.8	0.7	0.74	1.6×10^{-1}
M0	+8.9	+1.39	3 300	+7.6	0.5	0.63	6.3×10^{-2}
M5	+12.0	+1.61	2 600	+9.8	0.2	0.32	7.9×10^{-3}

Appendix F | Astronomical coordinates and timekeeping

Coordinate systems

When we want to locate a star, or any other astronomical object, we only need to specify its direction. We don't need its distance. We therefore need only two coordinates, two angles, to locate an astronomical object. Sometimes, it is convenient to think (as the ancients did) of the stars as being painted on the inside of a sphere, the celestial sphere. Just as we can locate any place on the surface of Earth with two coordinates, latitude and longitude, we need two coordinates to locate an object on the celestial sphere.

We choose coordinate systems for convenience in a particular application. In general, to set up a coordinate system we first identify an equator and then choose coordinates that correspond to latitude and longitude.

A convenient system for any particular observer is the *horizon system*. The horizon becomes the equivalent of the equator in that system. The angle around the horizon, measured from north, through east, south and west, is the *azimuth*. The angle above the horizon is called *elevation*. Instead of elevation, we can use the zenith distance, which is the angle from the zenith (overhead) to the object. From their definitions, we can see that the sum of the zenith distance and the elevation is always 90°. The azimuth ranges from 0° to 360°, and the elevation from −90° to 90° (with negative elevations being below the horizon). The problem with this system is that, as the Earth rotates, the azimuths and elevations of the stars change in a complicated way. It would not be very useful to prepare a catalog of stars, just giving their azimuths and elevations.

One solution is to use a coordinate system based on the projection of the Earth's equator onto the celestial sphere, the *celestial equator*. Such a system is called an *equatorial coordinate system*. The angle above or below the celestial equator is called the *declination*, and it is designated by the symbol "δ". It ranges from −90° to 90°. As the Earth rotates, the declinations of objects don't change. Also, the declination doesn't change as observers move around the Earth. There are two ways to measure the other coordinate.

(1) We can measure the angle, going westward, from the observer's meridian. This is called the *hour angle*, H. We can measure it from 0° to 360°, but it is convenient to express it in units of time, from 0 to 24 hr. As the Earth rotates, the hour angle of each object changes, but in a simple way, increasing by one hour for each hour that the Earth rotates. In addition, at any instant, observers at different longitudes will measure different hour angles for the same object. (The hour angles differ by the difference in longitudes.)

(2) To achieve a coordinate system that is the same for all observers and doesn't change with time, we must fix that coordinate system with respect to a location on the celestial sphere. We choose as our reference point one of the two intersections of the celestial equator and the ecliptic (the Sun's path around the sky). These two points are called *equinoxes*. (When the Sun is at either equinox, all observers on Earth have a 12 hr day and a 12 hr night. This occurs on the first day of spring and the first day of fall.) We choose the *vernal equinox*, the point where the Sun is on the first day of spring, as our starting point for the coordinate, *right ascension*, designated by the symbol α. The right ascension is measured from 0 to 24 hr, increasing from west to east. That is, objects with higher right ascensions cross an observer's meridian later than objects with lower right ascensions.

One effect of the Earth's precession is to move the equinoxes by about 50 arc sec per year. Thus, the origin of our coordinate system is drifting. Therefore, in compiling a catalog of objects, the time, or *epoch*, at which the coordinates apply must be specified. The epoch is usually put in parentheses after the α and δ, for example, $\alpha(2000)$ and $\delta(2000)$. An observer can then calculate where the objects will be on the date they are to be observed (a relatively simple calculation). To keep things simple, we generally agree on standard epochs, and keep our catalogs on a common standard. Catalogs that are just coming out use the standard epoch 2000. By changing the standard epoch every 50 years, we don't have to change too often, and we are able to keep the catalog coordinates reasonably close to the actual coordinates. (In 50 years, the origin will have moved by less than one degree.)

Other coordinate systems are useful for studying particular sets of objects. For example, in studying the Solar System, ecliptic coordinates are useful. In this system, the *ecliptic latitude* β is measured above or below the ecliptic (from −90° to 90°), and the *ecliptic longitude* λ is measured around the ecliptic, starting at the vernal equinox and increasing eastward (from 0° to 360°).

A coordinate system that is useful for studying galactic structure is the *galactic coordinate system*. (We touched on this briefly in Chapter 16.) The *galactic latitude b* is measured above and below the galactic plane. The *galactic longitude ℓ* is measured from the galactic center, increasing in the same direction as the right ascension.

Once an object is located in one coordinate system, those coordinates can be transformed into any of the other systems. The equations for those transformations are beyond the scope of this brief summary, but the calculations can be done on a hand calculator, and certainly by a computer that would be involved in pointing a large telescope.

Timekeeping

It is natural to use the Earth's rotation as a basis for timekeeping. We can keep track of the Earth's rotation by noting the motion of the stars. For each rotation of the Earth the stars make a full circle in the sky. We can therefore measure time by choosing a star, or other point in the sky, and seeing the fraction of the daily circle that it has made.

We choose the reference point to be the vernal equinox. We measure a time, called *local sidereal time (LST)*, by the progress of the vernal equinox. When the vernal equinox is at an observer's meridian, the LST is zero for that observer. One hour later, the LST is 1 hr; in addition, the hour angle of the vernal equinox is 1 hr. This means that the LST *is simply the hour angle of the vernal equinox*. When the LST time is 1 hr, objects with a right ascension of 1 hr will be on the meridian. This means that the *LST is also equal to the right ascension of the object that happens to be on the meridian*. Observers at two different points on Earth will have different LSTs. The LSTs will differ by the longitude difference between the two observers.

Most of our civil timekeeping is referenced to the Sun. We could define a local solar time, based on the hour angle of the Sun. This is what a sundial measures, but this is not very useful for civil systems. For uniformity, civil systems utilize time zones. This means that the Sun can be as much as a half an hour ahead or behind your local time, even more for some very wide time zones.

As the Earth moves around the Sun, the Sun appears projected against a changing background of stars (progressing through the constellations of the zodiac). This means that the right ascension of the Sun increases by an average of $3^m56.56^s$ per day. The time for the Sun to

go from one passage of your meridian to the next is this much longer than the time for a star to go from one passage to the next. Therefore, a day by the Sun, a *solar day*, is longer, by this amount, than a day according to the stars, a *sidereal day*.

Another problem with solar time is that the right ascension of the Sun does not change smoothly. This is the result of two effects. (1) The Earth's orbit is elliptical, so the Earth moves faster when it is closer to the Sun, and slower when it is farther away. This variable speed is mirrored in the apparent motion of the Sun against the background of stars. (2) The Sun moves along the ecliptic, which makes a 23.5° angle with the celestial equator. Therefore, the right ascension and declination of the Sun are both changing. Even if the Sun were to move along the ecliptic at a constant rate, its right ascension would change at a variable rate. Because of these two effects, we define a fictitious object, called the *mean Sun*, which moves along the celestial equator at a uniform rate. Time kept by the mean Sun is called the *mean solar time*, and it is the time that would be kept by a clock. The relationship between the mean Sun and the real Sun is given by a quantity called the *equation of time*. This quantity is depicted graphically by the distorted figure "8" that appears in the empty areas of some globes. This is called an *analemma*.

To the extent that the solar time is used in astronomical timekeeping, it is usually *universal time (UT)*, which is the mean solar time at Greenwich, England. It is often useful to convert from UT to LST for any observer. This is done with the aid of a publication, such as the *Astronomical Almanac*, published by the US Naval Observatory. The *Almanac* gives, for each date, the LST at Greenwich at 0^h UT. To this, we add the UT of interest, multiplied by a factor to account for the difference between sidereal and solar times. This gives the LST at Greenwich at the UT of interest. We then subtract the longitude of the observer L. We can write this as

$$LST(L, UT) = LST(0, 0) - L + UT(1 + 2.738 \times 10^{-3})$$

Where LST(0,0) is the LST at Greenwich at 0^h UT.

All of this is complicated by the effects of precession, and the wobble of the Earth, known as *nutation*. While *true sidereal time* is the actual hour angle of the vernal equinox, our sidereal clocks really keep a *mean sidereal time*. There is also a problem in the definition of a year. A *sidereal year* is the time for the Sun to return to the same place with respect to the fixed background of stars. We could use this definition, but after a long time the precession will cause the seasons to occur in different months. For this reason, we use a *tropical year*, defined as the time it takes for the Sun to travel from vernal equinox to vernal equinox, remembering that the equinox moves while it is happening. A sidereal year has 365.2564 mean solar days, while a tropical year has 365.2422 mean solar days.

Our definition of the year also brings us back to a definition for a universal time, which really has a con-stant rate. We base ephemeris time on the rate of the mean Sun at the beginning of the year 1900. *Ephemeris time* is a certain fraction of the tropical year 1900 (ephemeris year), which contains 365.242 199 mean solar days, so the ephemeris second is defined as 1/31 556 925 974 of an ephemeris year.

Appendix G | Abundances of the elements

Atomic number	Symbol	Element	Atomic weight	Relative abundance by number[a]
1	H	hydrogen	1.008	1.00
2	He	helium	4.003	1.45×10^{-1}
3	Li	lithium	6.939	1.00×10^{-9}
4	Be	beryllium	9.013	2.51×10^{-10}
5	B	boron	10.812	6.31×10^{-10}
6	C	carbon	12.012	3.02×10^{-4}
7	N	nitrogen	14.007	9.12×10^{-5}
8	O	oxygen	16.000	6.76×10^{-4}
9	F	fluorine	18.999	2.51×10^{-7}
10	Ne	neon	20.184	2.75×10^{-4}
11	Na	sodium	22.991	1.66×10^{-6}
12	Mg	magnesium	24.313	2.88×10^{-5}
13	Al	aluminum	26.982	1.91×10^{-6}
14	Si	silicon	28.09	2.95×10^{-5}
15	P	phosphorus	30.975	3.39×10^{-7}
16	S	sulphur	32.066	1.66×10^{-5}
17	Cl	chlorine	35.454	2.51×10^{-7}
18	Ar	argon	39.949	4.17×10^{-6}
19	K	potassium	39.103	7.59×10^{-8}
20	Ca	calcium	40.08	1.66×10^{-6}
21	Sc	scandium	44.958	8.13×10^{-10}
22	Ti	titanium	47.90	6.61×10^{-8}
23	V	vanadium	50.944	6.03×10^{-9}
24	Cr	chromium	52.00	2.40×10^{-7}
25	Mn	manganese	54.940	1.26×10^{-7}
26	Fe	iron	55.849	7.94×10^{-6}
27	Co	cobalt	58.936	5.25×10^{-8}
28	Ni	nickel	58.71	8.51×10^{-7}
29	Cu	copper	63.55	4.47×10^{-8}
30	Zn	zinc	65.37	1.91×10^{-8}
31	Ga	gallium	69.72	2.82×10^{-10}
32	Ge	germanium	72.60	1.51×10^{-9}
33	As	arsenic	74.924	2.0×10^{-10}
34	Se	selenium	78.96	1.6×10^{-9}
35	Br	bromine	79.912	4.0×10^{-10}
36	Kr	krypton	83.80	1.6×10^{-9}
37	Rb	rubidium	85.48	2.24×10^{-10}
38	Sr	strontium	87.63	5.62×10^{-10}
39	Y	yttrium	88.908	2.51×10^{-10}
40	Zr	zirconium	91.22	2.5×10^{-10}
41	Nb	niobium	92.91	5.0×10^{-11}
42	Mo	molybdenum	95.95	8.32×10^{-11}
43	Tc	technetium	99.0	

44	Ru	ruthenium	101.07	3.31×10^{-11}
45	Rh	rhodium	102.91	6.03×10^{-12}
46	Pd	palladium	106.4	1.78×10^{-11}
47	Ag	silver	107.874	5.0×10^{-12}
48	Cd	cadmium	112.41	3.16×10^{-11}
49	In	indium	114.82	7.9×10^{-12}
50	Sn	tin	118.70	3.55×10^{-11}
51	Sb	antimony	121.78	4.0×10^{-11}
52	Te	tellurium	127.61	1.0×10^{-10}
53	I	iodine	126.909	2.5×10^{-11}
54	Xe	xenon	131.30	1.0×10^{-10}
55	Cs	caesium	132.91	1.3×10^{-11}
56	Ba	barium	137.35	1.29×10^{-10}
57	La	lanthanum	138.92	2.5×10^{-11}
58	Ce	cerium	140.13	4.0×10^{-11}
59	Pr	praseodymuim	140.913	6.3×10^{-12}
60	Nd	neodynium	144.25	3.2×10^{-11}
61	Pm	promethium	147.0	
62	Sm	samarium	150.36	1.0×10^{-11}
63	Eu	europium	151.96	5.0×10^{-12}
64	Gd	gadolinium	157.25	1.3×10^{-11}
65	Tb	terbium	158.930	2.5×10^{-12}
66	Dy	dysprosium	162.50	1.6×10^{-11}
67	Ho	holmium	164.937	3.2×10^{-12}
68	Er	erbium	167.27	7.9×10^{-12}
69	Tm	thulium	168.941	1.3×10^{-12}
70	Yb	ytterbium	173.04	1.3×10^{-11}
71	Lu	lutecium	174.98	2.0×10^{-12}
72	Hf	hafnium	178.50	4.0×10^{-12}
73	Ta	tantalum	180.955	2.0×10^{-12}
74	W	tungsten	183.86	1.3×10^{-11}
75	Re	rhenium	186.3	4.0×10^{-12}
76	Os	osmium	190.2	2.0×10^{-11}
77	Ir	iridium	192.2	1.6×10^{-11}
78	Pt	platinum	195.10	4.0×10^{-11}
79	Au	gold	196.977	5.0×10^{-12}
80	Hg	mercury	200.60	7.9×10^{-12}
81	Tel	thallium	204.38	3.2×10^{-12}
82	Pb	lead	207.20	4.0×10^{-11}
83	Bi	bismuth	208.988	5.0×10^{-12}
84	Po	polonium	210.0	
85	At	astatine	211.0	
86	Rn	radon	222.0	
87	Fr	francium	223.0	
88	Ra	radium	226.05	
89	Ac	actinium	227.0	
90	Th	thorium	232.047	2.0×10^{-12}
91	Pa	protactinium	231.0	
92	U	uranium	238.03	1.0×10^{-12}

(Continued)

Atomic number	Symbol	Element	Atomic weight	Relative abundance by number[a]
93	Np	neptunium	237.05	
94	Pu	plutonium	242.0	
95	Am	americium	242.0	
96	Cm	curium	245.0	
97	Bk	berkelium	248.0	
98	Cf	californium	252.0	
99	Es	einsteinium	253.0	
100	Fm	fermium	257.0	
101	Md	mendelevium	257.0	
102	No	nobelium	255.0	
103	Lr	lawrencium	256.0	
104	Rf	rutherfordium	261.0	
105	Ha	hahnium	262.0	

[a]Abundances are by number, relative to hydrogen. These represent the best determinations of solar or Solar System abundances. No entry means that the abundance is not well determined.

Index